Fundamentals of Classical and Statistical Thermodynamics

Fundamentals of Classical and Statistical Thermodynamics

Bimalendu Narayan Roy
The Polytechnic of Namibia, Windhoek, Namibia
(formerly at Brunel University, UK)

JOHN WILEY & SONS

Copyright © 2002 by John Wiley & Sons Ltd,
Baffins Lane, Chichester,
West Sussex PO19 1UD, England

National 01243 779777
International (+44) 1243 779777

e-mail (for orders and customer service enquiries): cs-books@wiley.co.uk
Visit our Home Page on http://www.wiley.co.uk
or
http://www.wiley.com

Other Wiley Editorial Offices

John Wiley & Sons, Inc., 605 Third Avenue,
New York, NY 10158-0012, USA

Wiley-VCH Verlag GmbH, Pappelallee 3,
D-69469 Weinheim, Germany

John Wiley & Sons (Australia) Ltd, 33 Park Road, Milton,
Queensland 4064, Australia

John Wiley & Sons (Asia) Pte Ltd, 2 Clementi Loop #02-01,
Jin Xing Distripark, Singapore 0512

John Wiley & Sons (Canada) Ltd, 22 Worcester Road,
Rexdale, Ontario M9W 1L1, Canada

Library of Congress Cataloging-in-Publication Data

Roy, Bimalendu Narayan.
   Classical and statistical thermodynamics / Bimalendu Narayan Roy.
     p. cm.
   Includes bibliographical references and index.
   ISBN 0-470-84316-0 (pbk.: acid-free paper) ISBN 0 470 84313 6 (cased)
    1. Thermodynamics. I. Title.
   QC311 .R66 2001
   536′.7–dc21

2001040144

British Library Cataloguing in Publication Data

A catalogue record for this book is available from the British Library

ISBN 0-470-84313 6 (Hardback)
ISBN 0-470-84316 0 (Paperback)

Typeset in 10/12pt Times by Thomson Press (India) Ltd, Chennai

This book is dedicated to the victims of racism, my son Niladri and to the unfading loving memory of my deceased brother Purnendu. Like my parents the late Dr. Sachindra Nath and Prativa Roy, Purnendu too had an engrossing influence on me in my childhood. He was not only a brother but also an untiring playmate and a constant companion in my early education. His demise at a very young age was an excruciating experience for me.

As an engineering student, Niladri's indomitable and insatiable curiosity and interest in thermodynamics has been a driving force behind me in writing this book. Indeed, he unwittingly but quite pedantically initiated some of the book's discussions.

Contents

Appendices

List of Symbols

a	van der Waals constant, or activity
A	Area, or Helmholtz free energy, or work content
b	van der Waals constant
c	Concentration of solutions, or velocity of light
$C_V(\text{el})$	Electronic heat capacity of a substance
C_V	Heat capacity of a substance at constant volume
C_P	Heat capacity of a substance at constant pressure
E	Efficiency of the Carnot refrigerator
$\bar{E}$	Average energy of a substance
E_{av}	Statistical average energy of a substance
E_F	The Fermi energy
E_k	The energy of a microstate (the total energy of the system)
f	Fugacity
$f(E)$	The Fermi distribution function
F	Force, or degree of freedom in the phase rule, or Faraday constant
$F(T_D/T)$	The Debye function
g	Acceleration due to gravity
$g(E)$	Density of state, or the number of single particle states
g_i	Degeneracy of ith state
G	Gibbs free energy, or thermodynamic potential, or gravitational constant
G^0	Standard free energy of a substance
h	Planck constant, or hour, or height
H	Enthalpy, or heat content, or heat of reaction, or magnetic field intensity
H^0	Standard enthalpy of a substance
I	Ionization potential, or moment of inertia of a rigid body
k	Boltzmann constant
K	Kelvin temperature, or equilibrium constant of a chemical reaction, or Henry's law constant, or thermal conductivity of a substance
l	Distance, or latent heat per gram of weight
L	Length, or latent heat per mole

m	Mass of a substance, or concentration of a solution in molality
m_e	Mass of an electron
m_0	Reduced mass of a substance
M	Molecular weight, or concentration of a solution in molarity
n	Number of moles
$n(E)$ or $n(\varepsilon)$	Energy distribution in the Fermi or boson gas
N	Number of molecules or atoms or particles, or concentration of a solution in normality
N_{ex}	Number of bosons in the excited state when energy $\varepsilon = 0$
N_0	Avogadro number
p	Partial pressure of a gas, or pressure
p	Momentum
P	Atmospheric pressure or pressure exerted by a gas, or number of phases in the phase rule
$\mathscr{P}$	Probability
Q	Thermodynamic heat, or reaction quotient of a chemical reaction
r	Compression ratio of the Diesel or Otto engine
R	Universal gas constant
s	Spin of an atom
S	Distance, or absolute entropy, or ionic strength
S^0	Standard entropy
S_0	Entropy at the absolute zero of temperature
t	Time or temperature other than absolute
T	Temperature
T_b	The Bose–Einstein temperature
T_B	The Boyle temperature
T_D	The Debye temperature
T_E	The Einstein temperature
T_F	The Fermi temperature
T_i	Inversion temperature
U	Internal energy of a substance
U_1 or U_2	Energy of system 1 or system 2 before they are placed in thermal contact
U_1' or U_2'	Instantaneous energy of system 1 or system 2
v	Specific volume of a substance, velocity
v_F	The Fermi velocity of electron
v_L	Longitudinal velocity of sound in a solid
v_m	The most probable speed of a molecule
v_T	Transverse velocity of sound in a solid
V	Volume of a substance
V_m	Molar volume of a substance
V_0	Standard volume of an ideal gas
w	Weight of a substance
W	Work done by or on a system, or thermodynamic probability

x	Mole fraction
Z	Compressibility factor, or partition function
α	Degree of dissociation, or coefficient of thermal expansion
β	Coefficient of volume expansion, or a constant equal to $1/kT$
γ	Heat capacity ratio (C_P/C_V), or coefficient of activity or fugacity
Δx	Change in x, where x is any number or quantity
ε	Energy eigenvalue, or static energy
ε_t	Translational energy of an atom or a molecule
ε_r	Rotational energy of an atom or a molecule
ε_v	Vibrational energy of an atom or a molecule
κ	Force constant
η	Viscosity, or efficiency of an engine
$\eta(\mathbf{v})$	Maxwell velocity distribution
η_{Diesel}	Efficiency of the Diesel engine
η_{Otto}	Efficiency of the Otto engine
θ	Reduced temperature, or simply temperature
μ	Chemical potential, or Joule–Thomson coefficient
μ^*	Reduced electrochemical potential, or the reduced Fermi level
ν	Frequency of an operator
υ	Vibrational quantum number
ρ	Density of a substance
Ω	Number of microscopic realization of a macroscopic state
Ω_T	Total number of accessible microstates of a system

Preface

Both classical thermodynamics and statistical physics have seen many exciting developments in recent years. I became interested in statistical physics at my postgraduate stage when I came into contact with the late Professor Satyendra Nath Bose of Calcutta University, India (the founder of the Bose–Einstein statistics and the discoverer of the *boson* particle). Although at that young age it was not at all easy for me to grasp the extraordinary and brilliant idea underlying the Bose–Einstein statistics, I became fascinated in the subject through numerous discussions with Professor Bose. He had an unusual ability to inspire his students both inside and outside the classroom. It is virtually impossible to describe in words how much I was influenced by his epoch-making discovery which still baffles many erudite scientists.

Actually, *Fundamentals of Classical and Statistical Thermodynamics* is a revised version of the book *Principles of Modern Thermodynamics*. I wrote the first draft of the original book in the very early years of my career but could not publish it until 1995. In 1990 I approached Adam and Hilger (now known as Institute of Physics Publishing, UK); the Commissioning Editor Mr Roger Cooper realized the potential of the book and wrote to me 'upon reading your brief outline, my first reaction is very positive'. He commissioned the book without delay but left IOP soon after and Mr Tony Wayte took over as the Commissioning Editor. He put an awful lot of effort into completing the necessary preliminaries leading to printing. Unfortunately, he too moved before the book went in press. Thereafter, the manuscript went through the gauntlet of many obstacles until it appeared in the published form in 1995. However, since IOP had no plan for a second edition of the book, it was out of print after only a few years despite its success and popularity. Hence, the idea of writing the present book evolved and John Wiley & Sons, Ltd agreed to publish it.

Certainly, statistical mechanics is mathematical in nature and often described by students as a difficult and somewhat esoteric subject. However, this notion has its origin in the way the subject is presented to students. Traditionally, statistical mechanics is considered as one component of a trilogy, the other members being thermodynamics and kinetic theory. Nowadays, there is a general tendency to group these subjects together; however, some proponents of statistical mechanics strongly feel that this subject should stand on its own merit and be presented independently of the other two components. A danger of this approach is that the students generally fail to see the interrelationships among thermodynamics, kinetic theory and statistical mechanics. In view of this, I have made a positive attempt to weave all the three elements of the trilogy into an integrated fabric, bearing in mind that all three are concerned with heat transfer as the state of a system changes in some manner determined by restrictions which are imposed on it.

Thermodynamics has been able to describe, with remarkable accuracy, the macroscopic behaviour of a huge variety of systems over the entire range of experimentally accessible temperatures, i.e. 10^{-4} to 10^6 K. In fact, it provides a truly universal theory of matter in the aggregate, yet the whole subject is based on only four laws. Even though these laws appear rather simple, their interpretations are profound; they provide important tools for studying the behaviour and stability of systems in equilibrium and non-equilibrium.

Thermodynamics is restricted to a consideration of the macroscopic properties predicted by the laws, completely independently of any assumptions regarding the atomic structure of the substance comprising the aggregated matter. In relation to thermodynamics, the main goal of statistical mechanics is to develop a set of mathematical tools which are capable of predicting the thermodynamic functions applicable to a given thermodynamic system, taking as the starting point the molecular structure and the intermolecular forces within the aggregated matter. Once the thermodynamic functions, appropriate for the system being considered, have been determined by statistical mechanics, then the thermodynamic laws and relationships can be used to obtain whatever information is desired about the system. Thus, the areas covered by thermodynamics and statistical mechanics are dependent on each other, mutually complementary, and in many ways overlap each other; there are no artificial boundaries between them.

Statistical mechanics does not concern itself with the behaviour of an individual molecule or atom, but only attempts to discover the average or most probable result of the combined behaviour of all the particles forming the substance. In a system, the number of macroscopic parameters (e.g. the volume, the temperature and the number of particles) is certainly insufficient to specify the detailed microscopic state of the system. The microscopic states are generally referred to as energy states (momentum and velocity are also considered). These energy states are strictly eigenstates or stationary states determined by the application of quantum mechanics to the system's microstates. There is a very large number of distinct stationary microstates, all of which are consistent with the particular macroscopic equilibrium thermodynamic state being investigated. The fundamental assumption of statistical mechanics is that the value of any physical quantity that describes an equilibrium macroscopic thermodynamic system is found by taking the average of the physical quantity over all accessible stationary microstates of the system which are consistent with the macroscopic system. Then, the task of statistical mechanics is to develop a procedure which enables us to calculate the macroscopic properties of a system being considered from the molecular properties of the substance comprising the system.

While the subject matter of this book is traditional, the method of presentation is new and designed to develop an understanding of the interrelationships among thermodynamics, kinetic theory and statistical mechanics beginning with well defined functions. The first half of the book develops thermodynamics as a subject in its own right, while the second half is devoted to the fundamentals of statistical mechanics, equilibrium statistical mechanics and applications.

From my teaching experience for over 25 years I have identified, *inter alia*, two problems: (a) there is no single textbook on thermodynamics or statistical mechanics which comprehensively covers the subject, and (b) no author has written solidly on all the important topics which are vital to the understanding of both classical thermodynamics and statistical mechanics. Most books on thermodynamics and statistical physics are either too hard for average students, too rigorous in treatment, or inadequate and vague to provide an in-depth knowledge of the subjects.

This book discusses the fundamentals of classical equilibrium thermodynamics, thermal physics, kinetic theory and statistical mechanics. It will be a good first book for students of physics, chemistry, materials science and engineering, and also a reference book for practising scientists, technologists, engineers and lecturers. The objective of combing all the above subjects in one book is to (i) show the interrelationships among these subjects, (ii) interpret the thermodynamic quantities and laws in terms of statistical mechanics, (iii) develop the Boltzmann distribution and the approach to thermodynamics on the basis of partition function, and (iv) apply the principles of statistical mechanics to thermodynamics.

The aim of the book is to present a difficult subject in a simple, coherent and easily understandable way for beginners. In many ways, thermodynamics, statistical mechanics and quantum mechanics are virtually inseparable. Yet, students learn these as independent subjects. I have attempted to show, in a simple way, their interrelationships and what these subjects stand for. In fact, statistical mechanics has developed out of the application of the theory of probability to the analysis of the motion of large assemblages of molecules, and has led to a better understanding of the foundation of classical thermodynamics. However, it is possible to begin the study of statistical mechanics the other way round, i.e. with the general principles of thermodynamics which are solidly grounded on experiment. I have adopted the latter approach. I think my approach, the method of presentation and, to a great degree, the material presented are distinctive enough to make this book an exposition of the fundamental concepts of classical thermodynamics and statistical mechanics. The text is tailored to the needs of today's students of physics, chemistry, applied science, engineering, etc., as well as practising technologists, engineers and scientists. Lecturers will find the book very useful. A large number of worked examples have been included to enable the reader to appreciate the application of the concepts and theories to real-life problems. This aspect of the book has received much more extended treatment than that given in any other book.

The book is intended to be a learning text. I tried to be thorough and meticulous. However, because of the limited time which students spend in formal education, it is not expected that they will appreciate all of it. I hope students will have a fair understanding of the theory and rationale of applications of thermodynamics and statistical mechanics in their formal use of the book, so that deeper knowledge and insight can be gained in self-instructional modes.

One of the main themes of the book is to show how the basic notions of atomic theory lead to a coherent conceptual framework capable of describing and predicting the properties of macroscopic systems. The book is suitable for junior undergraduate and postgraduate students. This book is not intended only for students with advanced standing or a group of esoteric readers. The distinguishing features of the book are that: (i) it contains much more information than can be usually found in any single book of its kind, (ii) it places equal emphasis on every aspect of the subjects, tries to show the physical significance of the mathematical formalisms and relates these to what happens in our physical world, (iii) it includes a very large number of numerical examples in the text and, on average, 20 to 25 different types of easy to very challenging exercises in each chapter, with complete solutions, and (iv) the important equations are boxed for easy reference. The exercises probe somewhat deeper into the intellectual content of the subjects, with the aim of bringing out the connections between different concepts. Also, the exercises are designed to extend and supplement the information in the text.

The book has covered a wide range of topics in some depth. However, it is written in such a way that any part of it can be skipped without losing continuity, and the lecturer can select the areas most relevant to the needs of his course. Two criteria have determined the sequence of the topics: first, to let ideas emerge naturally and spontaneously from the facts, and second, to treat more difficult material later in the book to allow the reader to acquire the sophistication needed to follow it.

The book has been written with those students in mind who, unencumbered by any prior familiarity with the subject, are encountering it for the first time from the vantage point of their previous knowledge of elementary atomic properties of matter. Each chapter is independent, yet there is a strong interrelationship among them. The reader who completes up to Chapter 9 will learn the fundamentals of classical thermodynamics and its applications. The prerequisites needed for these chapters are only elementary knowledge of atomic physics or chemistry. Similarly, the reader who studies only Chapters 12 to 14 will learn the fundamentals of statistical mechanics without having to learn thermodynamics. The requirements for these chapters are a mere knowledge of some atomic and quantum physics or chemistry in their most unsophisticated form. The mathematical tools necessary in both cases are simple derivatives and integrals, together with a familiarity with some common mathematical functions. The level of mathematics has been deliberately kept to a minimum. New derivations of many standard formulae are given in a lucid way to clarify their physical significance without sacrificing logical rigour.

Chapters 1 to 9 and 12 to 14 are intended for junior undergraduates while the remaining chapters will suit senior undergraduate and postgraduate students. Following the discussion of free energy, the criteria of equilibrium and stability have been introduced to lay the foundation for treating mixtures, phase equilibria and phase transitions. In Chapters 12 to 14, fundamentals of statistical mechanics have been discussed, and the interpretations of thermodynamics quantities and laws on the basis of statistical approach have been given.

Appropriate diagrams, data tables, footnotes and reference have been generously used throughout the book. An extensive glossary of all the symbols frequently used in the text, a list of useful physical and chemical constants and their numerical values, a table of SI units together with the conversion factors, and plethora of relevant mathematical formulae are included for ready reference.

If this book is useful and interesting to those who wish to learn the fundamentals of modern thermodynamics and statistical mechanics, I shall deem my effort successful.

Bimalendu Narayan Roy

Acknowledgements

This book is an outcome of many years of my lecturing to undergraduate and postgraduate students in the USA, the UK, Asia and Africa. Needless to say that in writing this book a free blending of my own knowledge of thermodynamics and statistical mechanics with that of many other authors has unavoidably occurred. I acknowledge, without any reservation, such indirect contributions and ingress of any text, however small, from another source without my knowledge. I shall be failing in my obligation if I do not extend my sincere thanks to several generations of my students whose canny criticisms have been a constant source of corrections and modifications in the early stages of the manuscript. Also, I take this opportunity to acknowledge the direct or indirect encouragement of many colleagues in different countries in writing this book. I am particularly grateful to Prof. Reed A. Howald of Montana State University, USA. Chapters 1 to 11 of this book have been adopted from the original book *Principles of Modern Thermodynamics*. Prof. Howald read these chapters at the pre-publication stage of the original book and made many useful comments which, indeed, helped to improve the scientific quality of these chapters in the present book.

The process of bringing this book in its present state has not been smooth; it has run the gauntlet of various discouraging and disappointing circumstances. However, through the publishing experience and interest of John Wiley & Sons, particularly Dr. Andrew Slade, Senior Publishing Editor, this book has now appeared in the published form. I thank Dr. Slade, his Editorial Assistant Rachael Ballard, Copy Editor Tim Rouse, Production Editor Robert Hambrook and this team. It was a pleasure to work with all of them.

I take this opportunity to extend my very sincere thanks to Dr. Tjama Tjivikua, Rector of the Polytechnic of Namibia, Namibia, for his ungrudging moral and material support in this work.

Finally, I shall be failing in my obligation if I do not express an especial acknowledgement to my wife Manju. She missed me for a long time while the book was being written but helped me in the preparation of the manuscript, diagrams, etc. The result of her patience, forbearance and sacrifice is this book.

1
Thermodynamic Laws, Symbols and Units

1.1 INTRODUCTION

Thermodynamics is a physical science concerned with the transfer of heat and the appearance or disappearance of work attending various conceivable chemical and physical processes. The processes which are subject to thermodynamic considerations include not only the natural phenomena that occur about us every day but also controlled chemical reactions, the performance of engines and even hypothetical processes such as chemical reactions that do not occur but can be imagined. Thermodynamics is exceedingly general in its applicability, and this makes it a powerful tool for solving many kinds of important problem. The thermodynamic methods do not make any assumptions as to the atoms and molecules. The only quantities and concepts which enter thermodynamics are the experimental properties of matter such as pressure, volume, temperature and composition. Such properties are the properties of matter in bulk rather than of individual isolated molecules and are, therefore, called *macroscopic properties* as opposed to *microscopic properties*. Thermodynamics provides a convenient and powerful method of relating, systematizing and discussing such properties.

Historically, the science of thermodynamics was developed to provide a better understanding of heat engines, with particular reference to the conversion of heat into useful work. To facilitate the development of thermodynamics, we find it convenient and sometimes necessary to define numerous other concepts that are derived from, and related to, those of heat and work. Among the concepts to be so defined are the energy and entropy functions, which are suggested by certain laws of thermodynamics; these laws also provide a postulatory basis for the logical development of the subject. Other functions, such as enthalpy and free energy, are also defined, mainly for convenience.

1.2 THE USEFULNESS OF THERMODYNAMICS

By applying the laws and principles of thermodynamics it is, for example, possible to predict whether or not a particular chemical process can take place under any given conditions. The amount of energy that must be put into the process and its maximum yield can also be determined. The effect of changes in these conditions upon the equilibrium state can also be predicted. Equations may be written correlating physical and chemical properties of substances. The measurement of the heat evolved during the combustion of all kinds of fuel has obvious practical significance for the fuel technologists.

Also, formulae and laws discovered experimentally can be derived theoretically, e.g. the laws of chemical equilibrium, or an equation for the variation in vapour pressure with temperature. There are several other applications and uses of this particular branch of basic science. We shall discuss these in some detail later.

1.3   LIMITATIONS OF THERMODYNAMICS

We have already stated that thermodynamics is exceedingly general in its applicability, and this makes it a powerful tool for solving many kinds of important problem, but at the same time this generality renders it incapable of answering many of the specific questions that arise in connection with those problems. Thermodynamics can often tell us, for example, that a process will occur but not how fast it will occur, and it can often provide a quantitative description of an overall change in state without giving any indication of the character of the process by which the change might take place. Furthermore, thermodynamics does not provide the deep insight into chemical and physical phenomena that is afforded by microscopic models and theories.

1.4   THE LAWS OF THERMODYNAMICS

Thermodynamics provides the most general and efficient methods for studying and understanding complex physical and chemical phenomena. Thermodynamics is the science of heat and temperature and, in particular, of the laws governing the conversion of heat into mechanical, electrical or other macroscopic forms of energy. An important characteristic of thermodynamics is that it permits the derivation of relationships between different laws of nature, even though the laws themselves are not a consequence of thermodynamics.

Thermodynamics is phenomenological, concerning macroscopic quantities such as pressure, temperature or volume. It is both the strength and the weakness of thermodynamics that the relationships based upon it are completely independent of any microscopic (i.e. on the atomic and molecular level) explanation of chemical and physical phenomena. The strength is that thermodynamic relationships are not affected by the changes to which microscopic explanations are applied. On the contrary, the conclusions of atomic and molecular theories must not contradict those of thermodynamics, so that thermodynamics can be used as a guide in the development of microscopic theories.

The weakness is that thermodynamics does not provide the deep insight into chemical and physical phenomena that is afforded by microscopic models and theories. Although thermodynamics is completely self-contained, it is nevertheless possible to find a microscopic interpretation of it in statistical mechanics, which provides considerable insight and is of great value for a full understanding of thermodynamics.

Classical thermodynamics is based on the four laws of thermodynamics. In thermodynamics we are concerned with the behaviour of vast quantities of particles in the substances that we study. The laws of thermodynamics are the laws of the generalized behaviour of the particles. These laws are as follows.

1. *The zeroth law* which deals with temperature and temperature scale; this law is seldom considered because similar consideration in terms of the second law is possible.

2. *The first law* deals with macroscopic properties, work, energy, enthalpy, etc.

3. *The second law* mainly deals with entropy, a property most fundamentally responsible for the behaviour of matter.

4. *The third law* deals with the determination of entropy.

All these laws will be discussed in detail later.

1.5 THERMODYNAMIC SYSTEM

The substance or substances involved in physical and/or chemical changes is defined as the system. All other objects which may act on the system are called the surroundings. The system may be complex or it may be very simple and may consist only of a homogeneous piece of matter, such as a certain volume of air. What is included in a system is a matter of convenience that depends on the question at issue. Boundaries of the system may be real or they may be imaginary. Physical boundaries may be part of the system and their properties may or may not be important; they may be movable or fixed. A system may be open, closed, thermally isolated or mechanically isolated.

It is important to specify the system. For example, it may be a liquid inside a vessel, bounded by the surface of the liquid and by the walls that may, but need not, be part of the system. This system is open because liquid may escape into the atmosphere as vapour or further substance may be added from the surroundings. In another situation a vessel may be provided with a tight cover, with the liquid and the gas phase inside considered to be the system. This system would be closed. If heat is evolved during a process, the walls may take up some of the heat and might be counted as part of the system, or the walls may be made of heat-insulating material and then the system is adiabatic.

1.6 STATES AND STATE FUNCTIONS

The state of a system can be defined completely by the four observable macroscopic properties of matter known as the variables of state, namely pressure, volume, temperature and composition. In fact, any two of these variables are sufficient to fix the state. The chosen independent variables are called the *state variables* or *variables of state* while the other properties which depend on them are called *state functions* or *functions of state*. State variables are not necessarily independent of one another. For example, in a specified amount of gas, volume and density are determined by the pressure and temperature. There are two important properties of state functions or variables of state.

1. Once we specify the state of a system by giving the values of all or a few of the state variables, e.g. P, V, T, and composition, then all other properties such as mass m, density ρ, viscosity η and refractive index μ, are fixed. The relationship between the dependent and independent variables of state is known as *the equation of state*.

2. When the state of a system is altered, the change in any state function depends only on the initial and final state of the system, and not on how the change is brought about. For example, when a gas is compressed from an initial pressure P_1 to a final pressure P_2 the change ΔP in pressure is given by $\Delta P = P_2 - P_1$. Only the initial and final values of the pressure determine ΔP. Any intermediate values which ΔP may have assumed in changing from P_1 to P_2 are immaterial.

1.7 EQUILIBRIUM STATES

There are some conditions of a thermodynamic system which cannot be described in terms of state variables. Let us consider a certain volume of a gas confined in a cylinder with a movable piston. When the piston is at rest (i.e. not moving within the cylinder), the state of the gas can be specified by giving the values of its pressure and temperature. If the gas is suddenly compressed, its state cannot be described in terms of one pressure and temperature. While the piston is moving, the gas in the immediate vicinity of the piston is compressed and heated, whereas the gas at the far end of the cylinder is not. There is then no such thing as the pressure or temperature of the gas as a whole. Conditions in which the state variables are changing with time and space (called non-equilibrium states) are not dealt with in thermodynamics. Thermodynamics deals only with equilibrium states in which the state variables have values that are uniform and constant throughout the whole system. Some of the criteria for an equilibrium state are as follows.

(a) The mechanical properties of the system must be uniform and constant, i.e. there must be no unbalanced forces acting on or within the system, since any unbalanced forces would cause the volume to change continuously and we would be unable to specify the state of the system.

(b) The chemical composition of a system at equilibrium must be uniform, and there must be no net chemical reactions taking place. Any net chemical change would inevitably change the density, temperature, etc., of the system and make it impossible to specify its state.

(c) The temperature of the system must be uniform and the same as the temperature of its surroundings. Because of the temperature difference, heat tends to flow until this difference disappears. In any system where heat is flowing, the macroscopic properties are not uniform and may change with time. Therefore, such a system cannot be in an equilibrium state.

1.8 TEMPERATURE AND THERMAL INTERACTIONS

Temperature and thermal interactions are characteristic of thermodynamics. Two bodies are said to have different *temperatures* if they change their properties when in contact with each other, even though interchange of matter, or mechanical, chemical or electrical interaction, etc., is prevented. For example, when a hot metal wire is immersed in cold water, the wire is found to shorten, and the water changes its density. When there are no further changes, there is equality of temperature, i.e. *thermal equilibrium* exists. Interactions that cause equalization of the temperature but are not of mechanical, chemical, electrical or similar nature are called *thermal interactions*.

In order to allow different workers to reproduce each other's experiments, we need some reliable standard temperature scale based on an easily measurable property of a readily available material. Experiments show that, at low densities, all gases at constant volume have the same dependence of pressure on temperature. Since all ideal gases respond identically to a given temperature change, it is convenient to define temperature in terms of the properties of ideal gases rather than of liquids. Temperature is then that quantity which depends linearly on the pressure of an ideal gas held at constant volume. This can be expressed as

$$P = P_0 + P_0 \alpha t \tag{1.1}$$

or

$$t = \frac{1}{\alpha} \frac{P - P_0}{P_0} \tag{1.2}$$

where t is the temperature, P_0 is the gas pressure at zero degree on the temperature scale and $1/\alpha$ is a constant which depends on the size of the degree. When the freezing point (when saturated with air of 0.1 MPa pressure) and the boiling point (phase transformation from liquid to vapour) of water are assigned the value 0° and 100°, respectively, $1/\alpha = 273.15$, and the temperature scale is called the *centigrade perfect gas scale*.

The ratio of two pressures P_1 and P_2, which correspond to the temperatures t_1 and t_2, is

$$\frac{P_1}{P_2} = \frac{P_0(1 + \alpha t_1)}{P_0(1 + \alpha t_2)} = \frac{1 + \alpha t_1}{1 + \alpha t_2} \tag{1.3}$$

Now, since $1/\alpha = 273.15$, we can define a new scale by the equation $T = t + 1/\alpha = t + 273.15$. Then we would have

$$\frac{P_1}{P_2} = \frac{T_1}{T_2} \tag{1.4}$$

The quantity T is called the absolute or Kelvin temperature and is written T K. This is directly proportional to the volume V of an ideal gas at constant pressure: $T = \text{constant} \times V$.

The degree Kelvin is defined by the statement that the *triple point* of pure water, at which ice, liquid water and vapour are in equilibrium, is at 273.16 K. Because the zero of the Kelvin scale is fixed, only one fixed point is required for this definition. (The ice point is 0.01 K below the triple point of water and therefore is at 273.15 K.)

The universal aspect of the absolute scale defined by $T = \text{constant} \times V$ is not only that it makes possible a particularly simple formulation of the ideal gas law but also that many other laws of nature assume a similarly simple form through its use. An important example is the second law of thermodynamics, as will be seen later.

It should be noted that the triple point of water, to which the present definition of the Kelvin scale refers, is an invariant fixed point that has no man-made aspects such as the freezing and boiling points of water have; the boiling point is related to atmospheric pressure, and the freezing point is related to the amount of air dissolved in water. Furthermore, on this Kelvin scale the difference between the freezing and boiling points of water is no longer exactly $100°$ by definition, even though the assignment of $273.16\,K$ to the triple point of water makes this difference close to $100°$. In other words, the temperature $t' = T - 273.15$ is not identical with the centigrade temperature, even though it is very close to it. To recognize this point, the scale t' is called the *Celsius scale*. In our discussions no distinction between degree centigrade and degree Celsius will be made, and the approximation $T = t + 273.15$ will be completely acceptable.

The only aspect of the absolute Kelvin scale that is not universal is that the size of the degree is related to the properties of water. Any other size of the degree would, of course, do just as well, and indeed the Rankine scale is just as absolute as the Kelvin scale. The ratio of the temperatures on any two absolute scales must, however, be constant.

When $T = 0$, the corresponding volume V of an ideal gas becomes zero. This is, of course, an extrapolation, and there is no substance known that behaves as an ideal gas near $0\,K$. This has no bearing on the concept of an absolute zero of the temperature scale, but temperature measurements near $0\,K$ would be difficult or impossible, if ideal-gas behaviour were the only basis of the absolute scale.

The reason for the importance of the Kelvin temperature in thermodynamics is the fact that it is independent of the detailed structure or properties of any material.

1.9   REVERSIBLE AND IRREVERSIBLE PATHS

These two concepts are of fundamental importance in thermodynamics. Changes of state may be brought about along reversible or irreversible paths. A reversible path is one that may be followed in either direction. At any point the direction may be reversed by a small change in a variable such as the temperature or pressure. To understand these two terms, let us consider the following example.

Let us imagine that there is a cylinder containing a gas and that the cylinder is equipped with a piston capable of moving backwards and forwards and exerting pressure on the gas. If the pressure exerted by the gas equals the external pressure, there is equilibrium. At the equilibrium state, by a slight decrease in the piston pressure the gas can be made to expand; also by a slight increase of the piston pressure the gas can be made to contract. Now, let us imagine that the gas is expanded by slowly decreasing the piston pressure so that at all times it is very slightly smaller than the pressure exerted by the gas on the piston. A P–V diagram of this process is as shown in Figure 1.1. P_{int} indicates the pressure exerted by the gas on the piston and P_{ext} indicates the pressure exerted by the piston on the gas as the gas expands from state 1 to state 2. The smaller the difference between P_{int} and P_{ext}, i.e. the difference between the pressures of the piston and the gas, the closer the lower curve (broken curve) lies to the upper curve (full curve) and the slower the rate at which the gas expands. In the idealized case where the two curves coincide, expansion takes an infinite time, because there is no pressure difference left to act on the piston. In fact, in this limiting case, expansion can be changed into compression by an infinitesimal increase in the outside pressure, and compression into expansion by an infinitesimal decrease in the outside pressure. Only under these circumstances is the expansion of the gas reversible. Hence, we can say that reversible

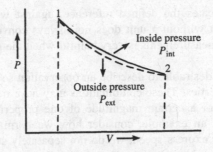

Figure 1.1 Schematic diagram of pressure versus volume

expansion of the gas is an idealized concept that can be carried out in theory only. Any actual expansion is irreversible.

A reversible process is generally an idealized concept that represents the limit of a sequence of irreversible processes, for which the parameter changes required to reverse the direction become smaller and smaller. At the limit it proceeds infinitely slowly through a sequence of equilibrium states.

Examples of typical irreversible processes are

1. the flow of heat from a hot to a cold body,

2. the expansion of a gas into a vacuum,

3. the temperature increase of a system by the performance of friction work, or

4. the dissolving of sugar in water.

Irreversibility does not imply that it is impossible to force a process in the reverse direction but only indicates that such a reversal cannot be achieved simply by changing parameters by infinitesimal amounts.

1.10   SOME USEFUL SYMBOLS

There is quite a large number of symbols used in thermodynamics and statistical mechanics for various terms. It is useful to familiarize oneself with some of these symbols at this stage. Different workers use different symbols for the same quantity. It does not really matter very much as far as the subject of thermodynamics and statistical mechanics is concerned, but for convenience it is advisable to use the internationally accepted symbols. A list of some of the important symbols which will be frequently used later is given at the beginning of this book.

1.11   UNITS

Whenever we make a measurement of some property associated with an object or a phenomenon, it is always important to express the measurement in the form of a number and a unit. The unit indicates the defined reference, against which the measurement is compared. A measurement without a unit does not always convey the intended meaning. The inclusion of units of measurements is very helpful when the measurements are used in calculations.

In chemistry, it is often desirable to describe an observation so that it includes combinations of fundamental properties. These combinations may have more than one unit and are often expressed in terms of a certain magnitude of one property compared with a fixed magnitude of another. As an example, consider how we normally express the speed of motion. When we want to express speed, we do not separately state the distance travelled and the time involved, instead we state the magnitude of the distance travelled in a specific unit of time, i.e. rather than say that a speed was 400 miles in 2 h we normally express such a speed as 200 miles per hour or 200 miles h^{-1}. We express the distance travelled per unit time. This expression of one observation per unit amount of another is merely a convenient way to state a relationship between the two properties. Expressions that relate properties to one another are very important in chemistry and physics.

It is very important to include the units in a measurement and any factor used in the calculations. By including the units we can make sure that the unwanted units cancel out and the proper unit remains. Units can be treated as algebraic quantities and therefore it is possible to cancel out, to multiply and to divide units. The treatment of units in calculations is very important but can be confusing. When the same unit appears in the numerator and denominator of a calculation, the unit may be cancelled out. When the same unit appears in two or more factors involved in multiplication operations, the units can be multiplied to form square units, cubic units, etc.

There are several accepted systems of expressing units, e.g. the fps system (foot–pound–second system or the British system), the cgs system (centimetre–gram–second system or the French system) and the mks system (metre–kilogram–second system). In chemistry it is still common to use cgs units for the measurement of physical quantities, although many chemists have now changed to the mks system. In the cgs system the units of length, mass and time are the centimetre, the gram and the mean solar second, respectively; the unit of force is 1 dyn, i.e. $1\,g \times 1\,cm \times 1\,s^{-2}$, and the unit of energy is 1 erg, i.e. $1\,g \times 1\,cm^2 \times 1\,s^{-2}$.

In many instances, physical chemists, in describing experimental results, use secondary units, e.g. atmospheres, litre-atmospheres and calories. To the extent that the units of force and energy are so small in the cgs system, secondary units are almost forced on one. In order to have a coherent system of units which can all be defined without duality, recommendations have been made that chemists, as well as physicists, adopt the mks system [Système Internationale des unités (SI units)]. The SI system has seven fundamental and six derived units. These are:

Fundamental
for length: the metre
for mass: the kilogram
for time: the mean solar second
for temperature: the kelvin
for amount of substance: the mole
for electric current: the ampere
for luminescence: the candela

Derived

for energy: the newton-metre or joule
for force: the newton
for pressure: the pascal
for power: the watt
for potential difference: the volt
for electricity: the coulomb

In chemistry, the generally accepted unit quantity of matter is the gram-atom or gram-mole per litre. In the mks system, the corresponding terms would be the kilogram-mole and kilogram-moles per cubic metre. The relationship between the cgs and mks units is as follows:

$$1 \text{ kg mol} \equiv 1000 \text{ g mol}$$
$$1 \text{ kg mol m}^{-3} \equiv 10^{-3} \text{ g mol } \ell^{-1}$$
$$\equiv 1 \text{ mmol } \ell^{-1}.$$

Many chemists feel that, although concentrations expressed in the mks system are quite convenient, the kilogram-mole is rather a large unit for many laboratory procedures.

1.12   CONVERSION FACTORS

Often, a measurement can be expressed in terms of more than one unit. Distances can be measured and expressed in units of inches, feet, metres or kilometres. Obviously, the size of a given distance would be the same, irrespective of which unit is used to express the measurement of the distance. Usually, the relationship between two different units in the measurement of the same property is defined or can be deduced. This is called a conversion factor. Any relationship between two units used to measure the same property can be expressed in this form. What good does it do to us to express such relationships in the form of conversion factors? As we shall see, these factors are very useful in chemistry and physics when we have to solve numerical problems. When a quantity is multiplied by a conversion factor, it is not fundamentally changed. However, we would not perform this multiplication without a reason. Conversion factors that express a relationship between units can be used to convert a measurement expressed in terms of one unit to a measurement in terms of another. By multiplying a measurement by a conversion factor, we would not fundamentally alter the measurement but would only change the measurement from one unit to another. Of course, converting a measurement from one unit to another changes the numerical part of the measurement as well as the unit.

Many problems in chemistry and physics can be solved by using conversion factors. Any quantity can be multiplied by any number of these factors without fundamentally changing it. The relationships expressing the amount of one property per unit amount of another are also conversion factors, since they represent an equivalence between the two properties. For example, when we state a speed, such as 40 m s^{-1}, this indicates that 40 m is the distance travelled in 1 s. Thus an expression such as this can be considered to be a property conversion factor and can be used to convert from one property to another.

A conversion factor is often expressed in terms of one quantity per unit amount of another. However, any conversion factor can be expressed in a form that is the reciprocal of the normal form. The reciprocal or inverse of a factor is formed by interchanging the numerator with its units and the denominator with its units. Very often the reciprocal of a factor is needed to solve a problem. For example, we can express the relationship between feet and inches as 12 in/1 ft or, in the reciprocal form which is found by inverting the above factor, as 1 ft/12 in. The form in which we express a factor depends on how we want to use it. That is, the first factor above can be used in the conversion of measurement from units of feet to units of inches, while the second factor can be used in the conversion of a measurement from units of inches to feet.

Whenever we solve a problem by the conversion factor method, we must decide what factors are needed and whether to use the factors in one form or in its reciprocal form. This decision depends on which units are to be cancelled and what unit we want in the final answer. When the problem is set up, the desired units should result. If not, the problem is probably set up incorrectly or the proper units are not used on all factors. The problem is set up by stating the quantity to be converted and then multiplying this quantity by the necessary factors. The choice of factors depends on the problem.

A list of some standard units and relevant conversion factors is given in Appendix 4.

1.13 SUMMARY OF DEFINITIONS

The following terms are commonly used in thermodynamics. It is, therefore, convenient to know their definitions.

When chemical and physical changes take place, a change in energy occurs. The study of these energy transformations is known as thermodynamics. When they originate from a chemical process the subject is specifically known as *chemical thermodynamics*.

System
The substance (or substances) involved in physical and/or chemical changes is known as the system. There are four types of system in thermodynamics.

1. *Open system*. In such a system, exchange of energy and matter occurs with its surroundings.

2. *Closed system*. In such a system, exchange of energy may occur but no transfer of matter occurs between the system and its surroundings.

3. *Thermally isolated system*. In such a system, no exchange of energy (in the form of heat) takes place.

4. *Mechanically isolated system*. In such cases, no work is done on the system or by the system.

Surroundings
This is defined as the regions outside the boundaries of the system which may act on the system.

Process

The actual change that occurs in a system and the manner of its occurrence is known as the process. A process may be *physical* or *chemical*. Magnetization of an iron bar is a physical process but rusting of iron is a chemical process.

Reversible process

In order that a process be thermodynamically reversible it must be carried out very very slowly (infinitesimally slowly) so that the system remains in temperature and pressure equilibrium with its surroundings. A system undergoing such a change can be completely restored to its initial state. This concept has nothing to do with reversible chemical reactions. It is, indeed, a hypothetical or an ideal process which cannot be achieved in practice.

Irreversible process

In such a process a property of the system differs by a finite amount from one instant to another and the system cannot return to its original state. Such processes are real or natural processes.

Spontaneous process

A spontaneous process is one which takes place under a given set of conditions without application of any force, e.g. spreading of solute through solvent from regions of high concentration to regions of low concentration. Such processes are also irreversible processes.

Isothermal process

When a reversible process occurs at a constant temperature (i.e. no change in temperature occurs), it is said to be an isothermal process. In such a process an exchange of heat between the system and its surroundings occurs to maintain the temperature constant.

Adiabatic process

An adiabatic process is thermally isolated so that no heat can enter or leave the system. In the case of a gas undergoing expansion or compression, the temperature and the volume adjust themselves to maintain equilibrium as the pressure is changed. The system does external work but, because it is thermally insulated, the necessary energy comes from the kinetic energy of the gas molecules. As the kinetic energy of the gas molecules decreases, there is a corresponding decrease in temperature. When a gas is compressed adiabatically, there is a rise in temperature. The essential characteristic is that no heat is absorbed in an adiabatic process.

Isentropic process

A reversible adiabatic process is called an isentropic process.

Isobaric process

Iso means 'same' and baros means 'weight' and hence an isobaric process is a constant-pressure process, i.e. a process carried out at a constant pressure.

Isochoric process
The Greek word 'chora' means place. So an isochoric process is one where the volume remains constant throughout the process, or in other words the process is carried out at constant volume.

Cyclic process
If the initial state is designated by 1 and the final state by 2, and if states 1 and 2 coincide, then the process is called a cyclic process. This process leads from a given state through a sequence of changes back to the original state.

Macroscopic properties
These are the properties of matter which are obvious to us and they are, naturally, the first features that we use to describe a physical situation. Examples are volume, pressure and temperature.

Thermal equilibrium
When two systems are in contact and there is no change in thermal behaviour, the systems are said to be in thermal equilibrium. For example, when a hot metal wire is immersed in cold water, the wire is found to shorten, and the water changes its density. These changes are caused by the exchange of heat. When there are no further changes, it is said that thermal equilibrium exits.

Thermal interactions
Interactions that cause equalization of the temperature but are not of a mechanical, chemical, electrical or similar nature are called thermal interactions. Such interactions may, of course, be the source of mechanical, electrical or other changes.

State
The state of a system is described by specifying the values of all relevant macroscopic variables, so that the system could be precisely duplicated from this information.

State variables or state functions
The macroscopic quantities that are used to specify the state of a thermodynamic system are called the state variables because their values depend only on the condition or the state of the thermodynamic system. Volume, temperature, pressure and density are state variables for a homogeneous system, but work and heat are not state variables.

Equation of state
The relationship between temperature, pressure and volume of a given amount of a substance is called the equation of state, which depends, of course, on the nature of the substance. Mathematically, an equation of state may be denoted by $V = f(T, P, n)$ and is read as 'V is a function of T, P and n', where V is the volume, T is the absolute temperature, P is the pressure and n is the number of moles. The ideal-gas equation $PV = nRT$ is an example of an (idealized) equation of state.

Extensive variables

Those variables that are proportional to the amount of matter are called extensive variables, e.g. volume V and heat capacity.

Intensive variables

Those variables that are independent of the amount of matter are called intensive variables, e.g. temperature T, pressure P and viscosity η.

Equilibrium state

This is a state of a system in which the state variables, such as temperature, pressure and volume, have values that are uniform and constant throughout the whole system.

Heat

This is a form of energy mainly due to temperature. Heat can be transferred solely because of a temperature difference between a system and its surroundings; it usually produces a rise in temperature when it enters a system. This does not necessarily mean that temperature changes can be brought about only by the transfer of heat. This implication is correct, however, if no work is involved in the process. Heat is denoted by Q or q and it is considered positive when it is added to the system or when it crosses the boundary from the surroundings into the system. The unit of heat is the joule.

Work

Work is defined as the product of a force multiplied by the distance through which it acts, i.e. $W = FS$ where W is the work, F the force and S the distance. Work can also be defined as a method by which energy is transferred from one mechanical system to another. It is denoted by W and its units are the joule, the erg and the litre-atmosphere. The joule is the unit in the mks system, the erg is the unit in the cgs system and the litre-atmosphere is commonly used for pressure volume work. 1 J is the amount of work performed by a force of 1 N along a path of 1 m (see Section 1.11). 1 J is thus also $1 \, kg \, m^2 s^{-2}$. 1 erg equals the amount of work performed by 1 dyn along a path of 1 cm. 1 erg is thus $1 \, g \, cm^2 s^{-2}$. Work is said to be positive when it is done by the surroundings on the system, and negative otherwise.

Expansion work

This is defined as the work performed when a system expands against an outside pressure. Such work does not impart kinetic energy to the movable parts of the boundary and is considered to be negative because it is performed on the surroundings.

Energy

This is defined as the capacity for doing work. Normally, it is denoted by U and its units are the erg, the joule and the calorie. There are several types of energy, namely potential energy, kinetic energy, internal energy, external energy, etc. Internal and external energies have importance in thermodynamics.

Internal energy

This is the energy possessed by all substances in varying amounts according to the motion and special arrangement of the particles making up the atoms and the molecules. Its absolute value cannot be measured for a given system but its increment ΔU is positive when the internal energy of a system increases. The increase ΔU in internal energy when a system changes from state A to state B does not depend on how the change is accomplished. It is simply the difference between the final and the initial energy: $\Delta U = U_B - U_A$.

External energy
This is the product of the pressure P and the volume V of a system. It can be regarded as the energy a substance possesses by virtue of the space it occupies.

Kinetic energy
This is the energy that a substance possesses because of its motion. Mathematically, it is one-half of the mass m of a body multiplied by the mean square of the velocity c of all the molecules. At constant temperature the kinetic energy is constant for a particular gas.

Heat absorbed
This is the term applied to the transition of internal energy from one substance to another when it is due to a temperature difference. It is denoted by the symbol Q.

Heat capacity
This is defined as the amount of heat required to raise the temperature of 1 mol of a system by 1°C. It is denoted by the symbol C. The heat capacity can vary with temperature and is best defined in the differential form $C = dQ/dT$, where dQ is the heat absorbed when the temperature is increased by dT degrees.

Molar heat capacity
If the system is 1 mol of a pure substance, the heat capacity is known as the molar heat capacity of that substance.

Specific heat capacity
If the system is 1 g of a substance, the heat capacity is called the specific heat capacity, so that the molar heat capacity is the specific heat capacity multiplied by the molecular weight. For substances not consisting of molecules such as NaCl, the heat capacity is known as the *formal heat capacity*, which is the specific heat capacity multiplied by the formula weight. This is also usually referred to as the *molar heat capacity*.

Heat capacity C_V at constant volume
If a gas is heated with its volume constant, then the measured heat capacity is called the heat capacity at constant volume and is denoted by the symbol C_V. At constant volume the heat absorbed increases the energy of the system when the temperature is raised from T_1 to T_2, i.e. $C_V(T_2 - T_1) = U_2 - U_1$. For a very small change dT in temperature, the heat capacity at constant volume is equal to the rate of change of internal energy with temperature, i.e. $C_V = dU/dT$.

Heat capacity C_P at constant pressure

If a gas is heated with the pressure fixed, the measured heat capacity is called the heat capacity at constant pressure and is denoted by the symbol C_P. When heat is supplied to a system at constant pressure, expansion occurs and therefore work is done against the applied pressure. Consequently, more heat is required to produce a 1°C rise in temperature at constant pressure than at constant volume. The extra heat needed goes into the work done in expansion, i.e. $C_P = C_V +$ work done in expansion. However, work done in expansion is given by $P(dV/dT)$, i.e. pressure multiplied by the increase in volume V for a rise in temperature of 1°C. Also, from above, $C_V = dU/dT$. Hence $C_P = dU/dT + P(dV/dT)$. We shall show later that $H = U + PV$, where H is the enthalpy, U is the internal energy, P is the pressure and V is the volume of the gas concerned. Then, differentiating with respect to T, we have $dH/dT = dU/dT + P(dV/dT)$ (since P is constant). Then, substituting in the equation $C_P = dU/dT + P(dV/dT)$, we have $C_P = dH/dT$. Thus, the heat capacity at constant pressure is equal to the rate of change of the heat content, or heat of reaction, with temperature.

Enthalpy

Enthalpy is a thermodynamic quantity defined by the equation $H = U + PV$, where H is the enthalpy. Its units are those of energy. We shall develop some of the important properties of enthalpy later.

Exothermic process

If heat is liberated in a chemical process, it is known as an exothermic process. An example is the combustion of benzene:

$$C_6H_6(\ell) + 7\tfrac{1}{2}O_2(g) = 3H_2O(\ell) + 6CO_2(g) + 3278.42\,kJ \text{ at } 18°C$$

This shows that the products of the reaction have a lower heat content than the reactants.

Endothermic process

If heat is absorbed in a chemical process, it is known as an endothermic process. An example of such a process is the conversion of liquid water to steam:

$$H_2O(\ell) = H_2O(g) - 44.17\,kJ \text{ at } 18°C$$

This reaction has taken place with absorption of heat. Water in the gaseous form has higher heat content than the liquid.

Heat of reaction

The heat of reaction is the difference in the heat contents of the products and the reactants at a constant pressure and at a definite temperature. This is denoted by ΔH and is measured in joules.

Heats of reaction and all similar data are quoted at 18 or 25°C in order that they may be readily compared and to facilitate calculations involving them.

Heat of formation

The heat of formation is the increase ΔH in the heat content when 1 mol of a substance is formed from its *elements*. The heat of formation of a substance is the same value as that of the heat of reaction but with the sign reversed. The heat of formation of any free element is always regarded as zero.

Heat of combustion

The heat of combustion is the change in the heat content resulting from the complete combustion of *1 mol* of a substance.

Heat of neutralization

The heat of neutralization of an acid or alkali is the heat evolved when *1 gram-equivalent* is neutralized at a particular dilution.

For all strong acids and bases, its value is approximately 57.36 kJ. This is because the process of neutralization merely results in the formation of water according to the equation $H^+ + OH^- = H_2O$.

Heat of solution

The heat of solution of a substance is the change in heat content when *1 mol* is dissolved in a large volume of water.

Heat of fusion

When a solid substance melts, the increase in its enthalpy is called the heat of fusion.

Heat of vaporization or sublimation

The increase in enthalpy that accompanies the vaporization or sublimation of a liquid or a solid is called the heat of vaporization or sublimation.

Entropy

This is the most important term in thermodynamics but at the same time its concept is rather difficult to understand. There is no straightforward definition for entropy but generally the entropy of a system is defined in terms of a differential equation for infinitesimal change dS in entropy. When a very small quantity dQ of heat is absorbed at the absolute temperature T, the increase in entropy is given by $dS = dQ/T$ for a reversible condition, by $dS > dQ/T$ for a spontaneous process and by $dS < dQ/T$ for a nonspontaneous process. The dimensions of entropy are energy and temperature and the values are usually quoted in joules per degree kelvin per mole of the substance concerned.

EXERCISES

1.1   Define macroscopic and microscopic properties of matter.

1.2   Discuss the limitations and usefulness of thermodynamics.

1.3   Why is it important to express a measurement in the form of a number and a unit? Discuss the various accepted systems of expressing units of measurements.

1.4 What are the base or fundamental units of length, time, mass, temperature, amount of substance, electric current and luminous intensity in the Système International des unités (SI units)?

1.5 What is a conversion factor? Discuss the usefulness and importance of conversion factors in solving numerical problems.

1.6 Determine the conversion factor between kilometres and miles, between kilometres per hour and miles per hour and between calories and joules.

1.7 The velocity of light in vacuum is $2.98 \times 10^8 \, ms^{-1}$. A light year is the distance that light travels in 1 year and an astronomical unit (AU) is the average distance from the Sun to the Earth $(1.50 \times 10^8 \, km)$. Calculate

(a) the number of metres in 1 light year

(b) the number of astronomical units in 1 light year.

1.8 State whether the following statements are true or false and, if false, correct the statements.

(a) Thermodynamics is dependent on the microscopic properties of matter such as work, and energy.

(b) Thermodynamics provides a quantitative description of the overall change in a chemical or physical process.

(c) Thermodynamics provides a deep insight into chemical and physical processes.

(d) Because of its general applicability, thermodynamics provides information about not only the yields but also the velocity of any chemical process.

(e) Thermodynamics provides a convenient and powerful method of relating, systematizing and discussing the macroscopic properties.

(f) The subject of classical thermodynamics is based on the four laws of thermodynamics.

(g) By applying the laws and principles of thermodynamics it is not possible to predict whether a particular chemical process can take place under any given conditions.

(h) Thermodynamics cannot tell us that a process will occur but can tell us how fast it will occur.

1.9 What are the units of the following quantities in the British, the cgs and the mks systems: length; mass; time; temperature; force; energy; pressure; heat; acceleration; density; volume?

1.10 Briefly discuss the importance and usefulness of units and conversion factors in thermodynamics.

1.11 The reaction $Zn + 2HCl = ZnCl_2 + H_2$ is carried out in a closed vessel kept in a thermostat bath. A study of the chemical process is made by thermodynamics.

(a) What is the system in the study?

(b) What is the surrounding?

(c) Is the system thermally isolated?

(d) Is the system mechanically isolated?

(e) Is the system in question an open system?

(f) Is our body an open system?

1.12   The following processes are said to be thermodynamic processes. Identify the type of each process, where P, V and T are the pressure, volume and temperature, respectively.

(a) A gas is expanded from the initial state P_1, V_1, T_1 to a final state P_2, V_2, T_2 such that there is no exchange of heat, i.e. $\Delta Q = 0$.

(b) A gas is expanded from a state P_1, V_1, T_1 to another state P_2, V_2, T_1.

(c) A gas is compressed from a state P_2, V_2, T_2 to another state P_2, V_1, T_1.

(d) A gas is expanded from a state P_2, V_1, T_1 to another state P_2, V_2, T_2.

(e) A gas is subjected to the process (c) above followed by the process (d) above.

1.13   The following variables are related to a homogeneous system: pressure; volume; temperature; heat; heat capacity; internal energy; density; number of moles; work.

(a) Classify these variables as intensive or extensive.

(b) Classify these as state or non-state variables.

(c) If the homogeneous system in question is an ideal gas, specify the state of this system by means of these variables.

1.14   (a) Define and distinguish between the following terms: specific heat at constant volume C_V; specific heat at constant pressure C_P; enthalpy H and heat Q.

(b) By definition $H = U + PV$. So, for a change in enthalpy this expression becomes $\Delta H = \Delta U + \Delta(PV) = \Delta U + W$. Is this proof acceptable? If the proof is not satisfactory, make the necessary amendments.

(c) The enthalpy change for the reaction

$$H_2(g) + \tfrac{1}{2}O_2(g) = H_2O(g)$$

is $-240\,kJ$ at 25°C.

(i) What is the measured enthalpy change called?

(ii) Classify the reaction as endothermic or exothermic.

(iii) State whether the reactants or the products have more energy.

1.15   Consider a system which expands against the outside pressure. Will there be any work done in this process? If so, what type of work is done by whom? Will this work involve any exchange of kinetic energy between the system and its surroundings?

1.16   Define and distinguish between the heat of reaction and the heat of formation.

1.17   A gas is compressed from a state P_2, V_2, T_2 to another state P_2, V_1, T_1. How would you carry out an expansion reversibly and irreversibly?

1.18   It is stated that the state of a system can be defined completely by the four observable macroscopic properties of matter known as the variables of state. What are these variables of state? Are these variables of state independent of each other?

1.19   When we specify the state of a system by giving the values of all or a few of the state variables, do we impose any restriction on the other properties of that system? Is it necessarily true that, when the state of a system is altered, the change in any state function depends only on the initial and final state of the system, and not on how the change is brought about?

1.20   Are there any conditions of a thermodynamic system which cannot be described in terms of state variables? Give an example of such condition. What are these conditions called?

1.21   A certain volume of gas is confined in a cylinder with a movable piston. If the gas is suddenly compressed, can its state be described in terms of one pressure and temperature? If not, explain why.

1.22   What are the criteria for an equilibrium state?

1.23   If two bodies change their thermal properties when in contact with each other, even though interchange of matter, or electrical, mechanical or chemical interaction, etc., is prevented, would you consider that these two bodies are at different temperatures? If so, when would you expect these two bodies to attain thermal equilibrium?

1.24   If the temperature t and pressure P_0 of a gas are expressed algebraically by the equation $P = P_0 + P_0 \alpha t$, how does the temperature depend on the pressure? If the freezing point and the boiling point of water are assigned the values $0°$ and $100°$ respectively, what would be the value of $1/\alpha$?

1.25   If P_1 and P_2 are the pressures of a gas corresponding to the temperatures t_1 and t_2, respectively, show that $P_1/P_2 = T_1/T_2$, where $T = t + 1/\alpha$.

1.26   Is there any distinction between degree centigrade and degree Celsius? Is there any aspect of the absolute Kelvin scale that is not universal? Why is the Kelvin temperature so important in thermodynamics?

1.27   'A reversible path is one that may be followed in either direction.' Discuss and illustrate.

1.28   How does a reversible process differ from an irreversible process? Is a reversible process an idealized concept that represents the limit of a sequence of irreversible processes for which the parameter changes required to reverse the direction become smaller and smaller? Does irreversibility imply that it is impossible to force a process in the reverse direction?

2

Some Further Preliminary Aspects of Thermodynamics

2.1 WORK

The simplest representation of work involves the operation of a force through a distance in such a way as to produce an increase in the potential or kinetic energy of some object. In our thermodynamic discussions we shall find it more convenient to think of a pressure operating through a change in volume rather than a force acting through a distance, even though the two are substantially equivalent. Work exists only at the time that it is performed and thus is a form of energy transfer. In thermodynamics, only macroscopic work is considered, and not work on an atomic level. Work is denoted by the symbol W and is considered positive when performed by the surroundings on the system and negative otherwise.

Let us consider a system consisting of a substance contained in a cylinder of cross-section A and fitted with a movable frictionless piston as shown in Figure 2.1. Let us imagine that at some instant the piston is at some distance, say l, from the closed end of the cylinder. If the piston is moved through a distance dl against the external pressure P_{ext}, then the work dW performed by the system would be $F\,dl$, where F is the force required to move the piston through the distance dl. However, force = pressure × area, and therefore $F = P_{ext}A$. Then $dW = -P_{ext}A\,dl$. The minus sign arises from the fact that for a positive dl (i.e. an expansion) the system performs work on the surroundings which by our convention represents negative work. Since $A\,dl$ is the volume increase dV of the system, then $dW = -P_{ext}\,dV$.

For a finite change in volume the external pressure P_{ext} need not remain constant during the change in volume. It is, however, supposed to be known for each point of the path along which the system expands. It may be given as a function of the temperature T and the volume V that the system is to assume: $P_{ext} = f(T, V)$. Thus, for a finite change in volume

$$W = -\int_1^2 P_{ext}(V, T)\,dV \tag{2.1}$$

The limits of the integral denote the initial and final states of the system.

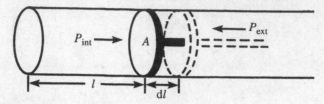

Figure 2.1   A cylinder containing a substance and fitted with a movable frictionless piston

In the important special case in which the outside pressure remains constant during a finite expansion of the system, P_{ext} may be brought outside the integral sign, so that

$$W = -P_{ext} \int_1^2 dV = -P_{ext}(V_2 - V_1) = -P_{ext}\Delta V \qquad (2.2)$$

[In our later discussions we shall simply write P for P_{ext} so that, unless specified, all P will indicate external pressure.] Then,

$$W = -P\Delta V \qquad (2.3)$$

This equation shows that, when $P = 0$, i.e. when the external pressure is zero, the work $W = 0$. *A system that expands into a vacuum therefore performs no work.*

Let us consider the reversible expansion of an ideal gas at constant temperature T, in which the external pressure P_{ext} is successively reduced so that it always balances the internal pressure P_{int}. By the ideal gas equation

$$P_{ext} = P_{int} = \frac{nRT}{V} \qquad (2.4)$$

we have

$$W = -\int_1^2 P_{ext}\, dV = -nRT \int_1^2 \frac{dV}{V} = -nRT \ln\frac{V_2}{V_1} \qquad (2.5)$$

Figure 2.2 represents this process. The shaded area represents the work performed by the system and is, therefore, equal to $-W$. As will be seen later, heat has to be supplied to the system to keep the temperature of the expanding gas constant.

When a system contracts, W is positive because the surroundings now perform work on the system. For a cyclic process involving pressure–volume work only, W is the area enclosed by the path in a P–V diagram (Figure 2.3). W is a positive area when the path corresponds to counter-clockwise rotation, and a negative area for clockwise rotation.

A given change in the state of a gas from V_1 and P_1 to V_2 and P_2 at constant T may, of course, be achieved in many ways, and the W values obtained may be different for any two of these ways. The gas may be expanded along a reversible path. This is indeed the path for which the integral $\int_1^2 P_{ext}\, dV$ for the change in state considered is a maximum. All along this path, $P_{ext} = P_{int}$ (except for an infinitesimal amount), and at no point could P_{ext} be increased (thus increasing the integral) without changing the expansion into a compression. The outside pressure may, however, be made smaller than P_{int} for any portion of the path. For this changed portion of the path the contribution to the integral would be decreased, and in addition the

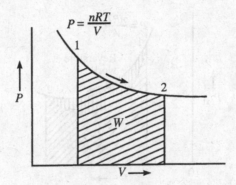

Figure 2.2   Schematic diagram of pressure versus volume for isothermal reversible expansion of an ideal gas

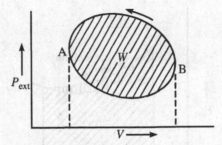

Figure 2.3   Schematic diagram of pressure versus volume for an ideal gas undergoing a cyclic process

expansion would not be reversible because it could not be changed into a compression by an infinitesimal increase of P_{ext}. It is apparent that the reversible path is the only path for which the integral has its maximum value. For any actual path, this integral, the work $-W$ performed by the system on the surroundings, is less than this maximum value. Therefore,

$$-W_{actual} \leqslant -W_{reversible} \qquad (2.6)$$

2.1.1   Expansion

In this case the external pressure (Figure 2.4, broken curve) is always below the internal pressure (full curve), reaching the full curve only in the limiting case of a reversible expansion. The lower shaded area thus represents the irreversible actual work, and the upper shaded area the reversible work. The area represents $-W$ because work is done on the surroundings.

2.1.2   Compression

In this case the external pressure is always above the internal pressure, except in the idealized limit of reversible compression. The lower shaded area in Figure 2.5 represents

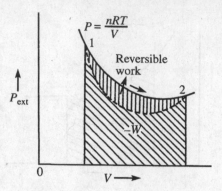

Figure 2.4   Schematic diagram of pressure versus volume for reversible expansion of an ideal gas

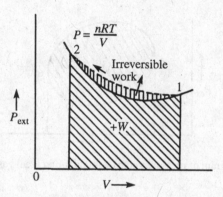

Figure 2.5   Schematic diagram of pressure versus volume for reversible compression of an ideal gas

the reversible work, and the upper shaded area the irreversible work. The areas now represent $+W$ because work is done on the system.

In any actual case, P_{ext} must be larger than P_{int} to overcome frictions and to make the piston move (Figure 2.5). The work W_{act} performed on the system is, therefore, larger than the work that would be necessary for compression along the idealized path of a continuous chain of equilibrium states. Therefore, in this case

$$W_{actual} \geqslant W_{reversible} \tag{2.7}$$

However, Equations 2.6 and 2.7 are the same when the signs are taken into consideration. It is true for both compression and expansion that for a given change in state the work performed by the system is a maximum for a reversible path:

$$W_{act} \leqslant - W_{rev} = (-W)_{max}. \tag{2.8}$$

A comprehensive discussion of compression, expansion and maximum work is given in section 4.7.

2.2   HEAT AND HEAT CAPACITY

According to our definition, heat is a form of energy mainly due to temperature. The meaning of the word 'heat' is somewhat confused by a tendency to think of heat as a real substance which is 'contained' or which 'flows'. This confusion is overcome if we realize that heat, like work, is not a material entity, but a method of energy transfer. The distinction between work and heat is that work is energy transferred by means of a mechanical link between a system and its surroundings, while heat is energy transferred solely because of a temperature difference between a system and its surroundings.

There are several means by which macroscopic work can be done on an assembly, causing its energy to change. However, it is possible to transfer energy to an assembly in ways which are not observable as macroscopic work. Let us consider the assembly in Figure 2.6. If we imagine the atoms in the walls surrounding the assembly as small vibrating masses, then some will be moving towards the boundary and some will be moving away. As a result there might be no observable motion. However, the interaction of these atoms with the molecules can cause a change in the energy of individual particles in the assembly and hence a total change in the energy of the whole assembly. The macroscopically observable work is zero. So, we must have some other way to account for the energy change. This leads us to conceive of heat as another general method for macroscopic energy transfer. Any energy change in an assembly that cannot be accounted for in terms of observable work is considered to be due to energy transfer as heat.

The sign convention for heat is the same as that for work. Heat that is going into a system from the surroundings is positive, whereas heat flowing out from a system into the surroundings is negative. Heat always flows from a body of higher temperature to a body of lower temperature. Any actual heat transfer requires a temperature difference, however small. Like all actual processes, heat flow is an irreversible process. The transfer of heat is reversible only at temperature equality when system and surroundings are at equilibrium. It is, like all reversible processes, an idealized process.

Heat and work are not stored within matter: they are 'done on' matter or 'done by' matter. A quantity of heat Q is positive when it is added to the system, in other words when it crosses the boundary from the surroundings into the system. The work W, on the contrary, is said to be positive when it appears in the surroundings. Energy is what is stored but heat and work are two ways of transferring energy across the boundaries of an assembly. Once energy has been imparted to an assembly, it is impossible to tell whether the energy was transferred as heat or as work. Hence the term *heat of a substance* is as meaningless as *work of a substance*.

A quantity of heat Q is a product of two terms, namely the difference (ΔT) in temperature (a potential factor in which, the greater the ΔT, the greater the possibility of thermal energy

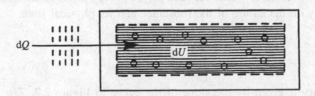

Figure 2.6   Energy transfer not observable as macroscopic work

transfer) and heat capacity C (which is a measure of the responsiveness of the system to the temperature difference ΔT). Thus,

$$Q = C\Delta T \tag{2.9}$$

or

$$C(J\,°C^{-1}) = \frac{Q(J)}{\Delta T(°C)} \tag{2.10}$$

This is a defining equation for a general heat capacity. It shows that the heat capacity of a system depends not only on the heat-absorbing ability but also, in a simple way, on the amount of substance present. We shall discuss this in detail later.

2.2.1   Heat depends on the path

Just as the work associated with a given change in state depends upon the means by which the change was brought about, so the heat attending a given change in state also depends on the path. Since the change in internal energy is independent of the process, the heat Q is different for different processes. Therefore, W is not a function of the state of the system. Hence, as we said before, it is meaningless to speak of the 'heat in a system' or 'the heat of a system'. Let us select some arbitrary reference state, such as the state at absolute zero. We say that the 'heat in the system' is zero in this state. Then the 'heat in the system' in any other state would be the heat flowing into the system when it is brought from absolute zero to this state. However, the heat is different for every different process by which the system is brought from the reference state and, therefore, no unique value can be assigned to the 'heat in a system'. Although it cannot be illustrated by simple graphical means, it is nevertheless true that the heat associated with a given change in state can, like work, have any value whatsoever from negative to positive infinity.

2.2.2   Heat capacity

As already defined in Chapter 1, heat required to raise the temperature of 1 mol of a system by 1°C is called its *heat capacity C*. If the system is 1 mol of a pure substance, the heat capacity is defined as *molar heat capacity* of that substance. The analogous quantity for 1 g of substance is called the *specific heat capacity*, so that the molar heat capacity is the specific heat capacity multiplied by the molecular weight.

Heat capacity, or in other words the ratio of the energy input to the temperature increase, depends strongly on the nature of the substance and its physical state:

$$C = \frac{dQ}{dT} \tag{2.11}$$

The heat capacity is given by the slope of the curve in Figure 2.7. The observation of Q versus T may be made either at constant pressure or at constant volume. Heat capacity at

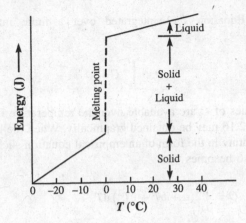

Figure 2.7   Schematic diagram of energy versus temperature

constant volume is denoted by C_V and that at constant pressure is denoted by C_P. For gases the difference between C_P and C_V is considerable, while it is much smaller for liquids or solids. Experimental information about the C_P of a substance may often be represented with considerable precision by an empirical equation. With some substances, such as $H_2O(g)$, the value of C_P is nearly a linear function of temperature and, therefore, may be represented by the equation

$$C_P = a + bT \tag{2.12}$$

One must be cautious in assigning physical meaning to the parameters of empirical equations such as Equation 2.12. It might appear that $C_{T=0} = a$, but this inference is not supported by the experimental evidence which extends only from 100°C to higher temperatures. An empirical equation is valid only for the range of experimental conditions for which its parameters have been evaluated.

In order to represent the data for other substances it may be necessary to add a third term to Equation 2.12, yielding

$$C_P = a + bT + cT^2 \tag{2.13}$$

or

$$C_P = a + bT + cT^{-2} \tag{2.14}$$

or

$$C_P = a + bT + cT^{-1}. \tag{2.15}$$

Again, the constants a, b and c in such an equation are chosen for best fit to the experimental values of C_P.

It is possible that one may reverse the operations described above to compute the heat required to increase the temperature of a substance over any interval within the range of validity of the data. Equation 2.11 integrated over a finite interval of temperature becomes

$$Q = \int_{T_1}^{T_2} C \, dT. \tag{2.16}$$

If the experimental values of C are available over the temperature interval of interest, the value of Q in Equation 2.16 may be obtained graphically. When the heat capacity is known as a function of temperature in the form of an empirical equation such as Equation 2.13, the integral in Equation 2.16 becomes

$$Q = \int_{T_1}^{T_2} (a + bT + cT^2) \, dT \tag{2.17}$$

$$= a(T_2 - T_1) + \tfrac{1}{2}b(T_2^2 - T_1^2) + \tfrac{1}{3}c(T_2^3 - T_1^3). \tag{2.18}$$

The molar heat capacity of a substance depends on its temperature. For example, it takes $75.22 \, \text{J mol}^{-1 \circ}\text{C}^{-1}$ to raise the temperature of $1 \, \text{mol}$ of water from 25 to 26°C, and $75.8 \, \text{J mol}^{-1 \circ}\text{C}^{-1}$ to raise it from 95 to 96°C. It is, therefore, more precise to define the molar heat capacity C_x by the differential relationship

$$dQ_x = nC_x \, dT \tag{2.19}$$

where dQ_x is the heat required to raise the temperature of n mol of substance from T to $T + dT$ along the path x.

The change in C_P and C_V with temperature, in a range of some ten or even hundred degrees Celsius, is usually small and neglected unless indicated otherwise. Thus, if n mol of a substance are to be heated (or cooled) isobarically from T_1 to T_2, the amount of heat involved is

$$Q \approx nC_P(T_2 - T_1) = nC_P \, \Delta T. \tag{2.20}$$

If Q is positive, the heat has to be supplied; if Q is negative, it is liberated.

2.3 ENTHALPY

The concept of enthalpy will be discussed in detail later; here we shall only try to understand its preliminary concept very briefly. The quantity enthalpy was introduced into thermodynamics by the American physicist and physical chemist J. Willard Gibbs, who was one of the founders of modern chemical thermodynamics.

Chemical reactions are usually performed in vessels open to the atmosphere, so that the pressure during a reaction remains constant. If gases are evolved or used up, or the volume of the system changes in any other way during the reaction, the resulting reversible pressure–volume work done on the surroundings is, by Equation 2.2,

$$-\int_1^2 P \, dV = -P(V_2 - V_1) = -P\Delta V. \tag{2.21}$$

For a reaction with volume increase or positive ΔV, the work $P\Delta V$ done on the surroundings is of no particular interest but must nevertheless be taken into account, because the energy for it has to come from somewhere. Conversely, for a reaction with volume loss, or negative ΔV, the surroundings perform the work $-P\Delta V$ on the system automatically. So, this implies that it would be handy to have a method to deal with this $P\Delta V$ work implicitly. Let the total work performed on the system be separated into two terms: the pressure–volume work $-P\Delta V$ and the remainder W', called the net work. $W = W' - P\Delta V$. What is thus needed is a formulation in which only the net work W' performed on the system appears explicitly:

$$W' = W + P\Delta V$$

This can be accomplished for isobaric processes by introducing a new function enthalpy H (the Greek word *enthalpein* means to warm), defined by

$$H = U + PV \qquad (2.22)$$

where U, P and V are the internal energy, pressure and volume, respectively. We shall show later that for constant-pressure processes the use of enthalpy H indeed takes pressure–volume work into account automatically.

This definition assures us that it is a function only of the state of a system, since U, P and V are state functions. The units of enthalpy must be those of U.

We know that $PV = nRT$, where n is the number of moles, R is the universal gas constant and T is the temperature. Then enthalpy can be expressed as

$$H = U + nRT. \qquad (2.23)$$

So, for 1 mol of an ideal gas (i.e. $n = 1$)

$$H = U + RT. \qquad (2.24)$$

Obviously, this is also a function of temperature only.

2.4 ENTROPY

Although the introduction of the concept of entropy at this stage will be rather premature, a familiarity with this very important term of thermodynamics will perhaps help in the understanding of the detailed treatment of the term later. The concept of entropy is, so to say, abstract and rather philosophical to some extent. In general, it is defined by saying that there is a quantity S, called entropy, which is a function of the state of a system. The general mathematical expression for the difference ΔS in entropy, between states 1 and 2 of a system is

$$\Delta S = \int_1^2 \frac{dQ_{reversible}}{T}. \qquad (2.25)$$

This equation says that, to compute ΔS, move the system from state 1 to state 2 by means of a reversible path. For each infinitesimal step of the reversible path, compute dQ/T and the sum of these quantities is ΔS. The expression is very simple for a process done at constant temperature:

$$\Delta S = \int_1^2 \frac{dQ_{rev}}{T} = \frac{1}{T} \int_1^2 dQ_{rev} = \frac{Q_{rev}}{T}. \tag{2.26}$$

It is very important to realize that the entropy change must always be computed by taking the system from state 1 to state 2 by a reversible path. However, since S is a function of state, the ΔS of the system is independent of the path taken and depends only on the initial and final states of the system. Although these two statements may seem contradictory, in fact they are not, because

$$\frac{dQ_{rev}}{T} \neq \frac{dQ_{irrev}}{T}. \tag{2.27}$$

The entropy change is also independent of the path but it is equal to $\int (dQ/T)$ only when the process is carried out reversibly. It is the quantity $\int (dQ/T)$ which depends on how the process is done, and not the ΔS of the system.

2.5   KINETIC THEORY OF GASES

In order to understand some thermodynamic treatments based on the kinetic theory of gases it would be appropriate for the reader to know the fundamental aspects of this theory here. The idea that the properties of gases such as diffusibility, expansibility and compressibility could be explained by considering the gas molecules to be in continuous motion occurred to several scientists as far back as the mid-eighteenth century (Bernoulli in 1738, Poule in 1851 and Kronig in 1856). In the second half of the nineteenth century, Boltzmann, Maxwell, Clausius and others developed this hypothesis into the detailed kinetic theory of gases. To develop a molecular theory of gases some assumptions had to be made. These are the properties ascribed to the molecules of an ideal gas. The assumptions are summarized below.

For *ideal gases* they are as follows.

1. A gas is composed of separate particles called molecules and it can be represented by a simple 'model'.

2. Gaseous particles, whether atoms or molecules, behave like point centres of mass, i.e. the volume actually occupied by the individual molecules of an ideal gas under ordinary conditions is insignificant compared with the total volume of the gas.

3. These point centres of mass are far apart and most of the time do not exert any force (attraction or repulsion) on one another, except when near the temperature at which they become liquids.

4. They are in continuous, completely random motion in a straight line.

For *real gases* the following hold.

(i) The volume occupied by the molecules under ordinary conditions may not be negligible compared with the total volume of the gas.

(ii) The forces (attractive or repulsive) exerted by the molecules on one another may not be negligible.

The true test of the kinetic molecular theory is its ability to describe the behaviour of a gas quantitatively. We shall see later that the ideal gas law equation $PV = nRT$ can be derived from the assumptions of the theory. This led chemists and physicists to believe that the assumptions of the theory indeed represent the true properties of gas molecules.

2.5.1   Derivation of the fundamental equation of the kinetic theory of gases

Let us consider N molecules of an ideal gas, each having a mass m, confined in a cubic box of side L cm, as shown in Figure 2.8. The velocity with which each molecule travels inside the box can be resolved along X, Y and Z directions to give $\dot{x}$, $\dot{y}$ and $\dot{z}$, respectively. Since each side of the box is of length L, the volume V of the box is $V = L^3$. There are innumerable molecules within this volume. To make it easy, let us consider only one of these molecules first and later we shall generalize the results. Let us consider the jth molecule and take its velocity component along the X axis as $\dot{x}_j$.

According to the assumption (4) above, this jth molecule moves along the X axis in a straight line until it collides with the wall perpendicular to the X axis, Upon collision with this wall, the direction of its motion will reverse along the X axis but the magnitude of its velocity component $\dot{x}_j$ will remain the same.

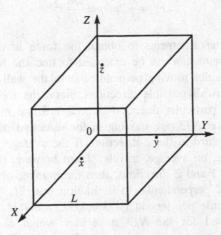

Figure 2.8   A cubic box of side L containing N molecules of an ideal gas

It is apparent from Figure 2.8 that the time required by the jth molecule to travel the length L is $L/\dot{x}_j$ s. Therefore, this molecule will collide $\dot{x}_j/L$ times per unit time with the wall perpendicular to the X axis. By Newton's second law of motion,

$$\text{force } F = \text{mass } m \times \text{acceleration } a. \tag{2.28}$$

Acceleration is defined as change in velocity per unit time, i.e. $a = \dot{x}_j/t$, where t is the time. Therefore, the force on the jth molecule is

$$F_j = m\frac{\dot{x}_j}{t} = \frac{m\dot{x}_j}{t} = \frac{\text{momentum}}{\text{time}}. \tag{2.29}$$

Initially, a molecule travelling in the positive X direction with velocity $\dot{x}_j$ has momentum $m\dot{x}_j$. After collision with the wall perpendicular to the X axis its momentum is $-m\dot{x}_j$, because the velocity is reversed in direction but unchanged in magnitude. This gives the change in momentum as

$$\Delta(m\dot{x}_j) = \text{final momentum} - \text{initial momentum}$$
$$= -m\dot{x}_j - m\dot{x}_j = -2m\dot{x}_j. \tag{2.30}$$

The change in momentum imparted to the wall perpendicular to the X axis is negative with respect to the change in momentum of the molecule, i.e. $2m\dot{x}_j$, because the momentum is conserved in any impact. Therefore, the force exerted on the walls perpendicular to the X axis is given by

$$\text{force} = \text{number of impacts per unit time}$$
$$\times \text{change in momentum per impact}$$

or

$$F_j = \frac{\dot{x}_j}{L}2m\dot{x}_j = \frac{2m}{L}\dot{x}_j^2. \tag{2.31}$$

We have developed the arguments to obtain the force in one direction due to one molecule only. These arguments can be extended to find the force on each wall of the box due to all the molecules moving perpendicular to the wall concerned. Although the molecules are moving in all possible directions, since the motion of the molecules is entirely random and no particular direction is preferred, we may assume that one-third of the total molecules, i.e. $N/3$, are moving in the direction of the X axis, one-third in the Y direction and one-third in the Z direction. If the average speed of the molecule is u cm s^{-1} and a molecule, on average, travels $2L$ cm between two consecutive collisions along any one of the X, Y and Z directions, then the number of times that it will collide per second with the wall perpendicular to its motion is $u/2L$. Then the average change in momentum per molecule per second is $2mu(u/2L) = mu^2/L$. Hence, the total change in momentum per second for the $N/3$ molecules which can collide with the wall perpendicular to their direction of motion is $(N/3)(mu^2/L)$. This represents the average force on the wall concerned, because force may be defined as change in momentum per unit time. Now

$$\text{pressure } P = \frac{\text{force}}{\text{area}}. \tag{2.32}$$

The area of any wall of the box is L^2. Therefore,

$$P = \frac{N}{3} \frac{mu^2}{L} \frac{1}{L^2} = \frac{N}{3} \frac{mu^2}{L^3}. \tag{2.33}$$

L^3 is the volume V of the box. Then,

$$P = \frac{N}{3} \frac{mu^2}{V}$$

or

$$PV = \tfrac{1}{3}Nmu^2. \tag{2.34}$$

Equation 2.34 is the fundamental equation of the kinetic theory of gases. It should be pointed out here that the equation is only as exact as the ideal gas laws themselves. However, it holds for real gases in very good approximation at ordinary pressures. Equation 2.34 may be written

$$PV = \frac{2N}{3} \frac{mu^2}{2} = \tfrac{2}{3}N \text{ KE} \tag{2.35}$$

where KE represents the average kinetic energy of the molecule, which is directly proportional to the temperature T, in kelvin, i.e. KE = constant $\times T$, or KE $= \alpha T$. In addition, the number N of molecules is proportional to the number n of moles of molecules. Then N = constant $\times n$, or $N = \beta n$. Substituting for KE and N in Equation 2.35 gives

$$PV = \tfrac{2}{3}\beta n\alpha T = n\tfrac{2}{3}\alpha\beta T. \tag{2.36}$$

Clearly, $\tfrac{2}{3}\alpha\beta$ is a constant and, if we write $\tfrac{2}{3}\alpha\beta = R$, Equation 2.36 becomes

$$PV = nRT. \tag{2.37}$$

If the quantity R is referred to as the *universal gas constant*, Equation 2.37 shows that from the assumptions of the kinetic theory of gases we can derive the ideal-gas equation. As this equation describes the behaviour of gases very well, the assumptions used in its derivation are valid.

2.5.2 Some useful derivations

Strictly speaking, the gas laws apply exactly only to those gases whose molecules do not attract each other and which occupy no appreciable part of the whole volume. However, there are no gases that have these properties. Therefore, we can speak of such hypothetical gases only as *ideal* or *perfect gases*. In a later section we discuss deviations from the ideal-gas laws. However, under ordinary conditions the deviations from the ideal-gas laws are so little that they may be neglected. We now intend to deduce the ideal-gas laws from Equation 2.37.

2.5.2.1 Boyle's law

This law states that, at constant temperature, the pressure of a given mass of an ideal gas is inversely proportional to its volume, i.e. $P \propto 1/V$ when T is constant. In other words, $PV =$ constant when T is constant.

At any given temperature, the average squared speed u^2 is constant; so also are N and m in Equation 2.34. This means that at constant temperature the right-hand side of Equation 2.34 is constant, i.e. $PV =$ constant, which is Boyle's law.

2.5.2.2 Charles's law

This law states that the volume of a given mass of an ideal gas is directly proportional to temperature at constant pressure, i.e. $V =$ constant $\times T$ when P is constant. This equation can be derived from Equation 2.34 as follows.

At constant pressure, $m \propto T$, and N and u^2 are constant. Then Equation 2.34 can be written

$$V = \frac{P}{3} N u^2 T = \text{constant} \times T. \tag{2.38}$$

2.5.2.3 Kinetic energy of a mole of an ideal gas

If N_0 is the Avogadro number, $N = nN_0$ where n is the number of moles and N is the number of molecules. Substituting this into Equation 2.34 gives

$$PV = \tfrac{1}{3}nN_0 mu^2 = \tfrac{2}{3}nN_0 \tfrac{1}{2}mu^2. \tag{2.39}$$

For an ideal gas, $PV = nRT$. Therefore,

$$\tfrac{2}{3}nN_0 \tfrac{1}{2}mu^2 = nRT$$

or

$$N_0 \tfrac{1}{2}mu^2 = \tfrac{3}{2}RT. \tag{2.40}$$

We know from Equation 2.35 that $\tfrac{1}{2}mu^2$ represents the average kinetic energy of a single molecule. Therefore, $N_0 \tfrac{1}{2}mu^2$ is the total kinetic energy of 1 mol of an ideal gas. Hence,

$$\text{KE of 1 mol of ideal gas} = \tfrac{3}{2}RT \tag{2.41}$$

Equation 2.41 shows that the kinetic energy of translation of the molecules of an ideal gas depends only on the absolute temperature, and that at a given temperature the lighter molecules have the greater speeds.

2.5.2.4   The Boltzmann constant

Equation 2.40 can be rewritten in the form

$$\tfrac{1}{2}mu^2 = \frac{3}{2}\frac{R}{N_0}T$$

or

$$\tfrac{1}{2}mu^2 = \tfrac{3}{2}kT \qquad (2.42)$$

where $k = R/N_0$ is called the Boltzmann constant.

Equation 2.42 tells us that the temperature is a measure of the average kinetic energy of a single molecule. It also indicates that the molecules of all ideal gases have the same kinetic energy at the same temperature.

2.5.2.5   Root mean square speed

We can rewrite Equation 2.42 in the form

$$u^2 = \frac{3kT}{m}$$

or

$$\sqrt{u^2} = \sqrt{\frac{3kT}{m}} \qquad (2.43)$$

The quantity $\sqrt{u^2}$ is called the root mean square speed of a molecule. Although this is not the same as the average speed u, the difference between the two is so small that for most purposes they can be equated. For an actual gas, u is $(8\pi/3)^{1/2}$ times the value of $\sqrt{u^2}$.

It is apparent from Equation 2.43 that the root mean square speed depends on the mass of the molecule.

2.5.2.6   Ratio of the root mean square speeds

Equation 2.42 shows that, if two ideal gases are at the same temperature, their molecules have the same kinetic energy. Then for two molecules 1 and 2 of two ideal gases we may write

$$\tfrac{1}{2}m_1u_1^2 = \tfrac{1}{2}m_2u_2^2 \qquad (2.44)$$

or

$$\frac{u_1^2}{u_2^2} = \frac{m_2}{m_1} \tag{2.45}$$

or

$$\frac{\sqrt{u_1^2}}{\sqrt{u_2^2}} \simeq \left(\frac{m_2}{m_1}\right)^{1/2} \tag{2.46}$$

This shows that the ratio of the root mean square speeds of two molecules of two ideal gases is equal to the square root of the inverse ratio of molecular masses.

2.5.2.7 Relationship between the density of an ideal gas and the root mean square speed

Equation 2.34 can be written as

$$P = \tfrac{1}{3} N \frac{m}{V} u^2 \tag{2.47}$$

The product Nm is the total mass of the ideal gas. Therefore, Nm/V represents the density ρ of the gas. Thus, Equation 2.47 becomes

$$P = \tfrac{1}{3} \rho u^2. \tag{2.48}$$

2.5.2.8 Avogadro's law

Rewriting Equation 2.34 for two ideal gases 1 and 2

$$P_1 V_1 = \tfrac{1}{3} N_1 m_1 u_1^2 \tag{2.49}$$

$$P_2 V_2 = \tfrac{1}{3} N_2 m_2 u_2^2. \tag{2.50}$$

If equal volumes of the two gases are considered at the same pressure, Equations 2.49 and 2.50 may be equated to give

$$\tfrac{1}{3} N_1 m_1 u_1^2 = \tfrac{1}{3} N_2 m_2 u_2^2. \tag{2.51}$$

Again, if these two gases are at the same temperature, their molecules have the same kinetic energy, i.e. $\tfrac{1}{2} m_1 u_1^2 = \tfrac{1}{2} m_2 u_2^2$. Therefore, from Equation 2.51, $N_1 = N_2$, which means equal volumes of two different ideal gases at the same pressure and temperature contain equal numbers of molecules. This is Avogadro's law.

We have seen earlier that from the assumptions of the kinetic theory of gases we can derive the ideal-gas equation $PV = nRT$. This, together with the fact that Avogadro's law can be derived from the kinetic theory of gases, indicates that the model characterized by the initial assumptions is realistic.

2.6 MODIFICATION OF KINETIC THEORY FOR REAL GASES

The kinetic theory of gases accounts for the properties of ideal gases but it requires some modification before it is applicable to real gases, because real gases do not in actual practice obey the gas laws. In Section 2.5 we assumed a model on certain assumptions to deduce the fundamental equation of the kinetic theory of gases. The assumptions (2) and (3) in this model require correction in order to apply the model to real gases. In Section 2.5 we stated that the following holds for real gases.

(i) The volume occupied by the molecules under ordinary conditions may not be negligible compared with the total volume of the gas.

(ii) The force exerted by the molecules on one another may not be negligible.

Let us now consider these assumptions in some detail.

(i) *Volume of molecules*: Although it has been assumed that the volume of the molecules is negligible compared with the total volume of the gas, it has been necessary to postulate a collision diameter to represent the proximity of closest approach of the molecules on collision. Therefore, it seems that, even though the molecules have no volume, they behave kinetically. It is possible to estimate the effective volume from collision diameter of a gas. Such estimated effective volumes are of the order of 3×10^{-23} mℓ per molecule. At ordinary temperature and pressure there are about 2.7×10^{19} molecules mℓ^{-1}. This gives the average volume of a molecule as 3.7×10^{-20} mℓ. So, under normal temperature and pressure, the volume of the molecules is less than 0.1% of the total volume of the gas.

It has been found that at a pressure of about 10 MPa the molecules would occupy a volume which is only about 12 times their own volume. The effect of this increase could not be ignored. The volume of a gas can be reduced by cooling and applying pressure until it transforms into a liquid and then solidifies. In the solid form, however, there is considerable resistance to any further compression. Therefore, it seems that there is a limiting volume which may be taken as approximately equal to that of the molecules. The effect of the finite size of the molecules will be to reduce the available free space for movement. Consequently, the free space will be less than the volume of the gas. As a result, the number of impacts on the walls of the container, and therefore the pressure on them, will be greater than that obtained by the use of the kinetic theory of gases.

(ii) *Force exerted by the molecules on one another*: Liquids possess the cohesive property. The fact that gases can be transformed into liquids implies that the molecules of a gas possess attractive force. The consequence of attractive forces among the molecules is to make the pressure less than that to be expected theoretically. The Joule–Thomson experiment (see Section 5.6.3) provides direct experimental proof of molecular attraction. The general observation of this experiment is that a gas cools itself on passing through a porous plug. In passing from a high to a low pressure the gas does not perform any external work. Therefore, the cooling observed must be due to the work done in overcoming the mutual attraction of the molecules. Hydrogen and helium are found to exhibit a reversed Joule–Thomson effect at ordinary temperatures. However, if previously cooled to low temperatures, they are found to behave normally. Even in hydrogen and helium, molecular attractions are present but, except at low temperatures, other factors tend to mask this effect.

2.7 THE VAN DER WAALS EQUATION

In 1873, J. D. van der Waals made the first successful attempt to modify the ideal-gas equation in order to correct for the volume and molecular attraction. When gases at ordinary pressures and temperatures are compressed, the volume is reduced by reducing the intermolecular space. At high pressures the molecules are brought so closely together that the actual volume which they occupy is a comparatively large fraction of the total volume, including the empty intermolecular space. If the pressure is further increased, only a fraction of the entire volume is compressed because the volume occupied by the molecules themselves is not compressible. Therefore, it is apparent that at very high pressures the entire volume is not inversely proportional to the pressure, as would follow from Boyle's law. However, a point to remember is that even at moderately high pressures the molecules do not occupy most of the volume. At comparatively low pressures, the molecules of a real gas do not exert any force on one another, simply because they are widely separated. Therefore, at relatively low pressures, real gases behave like ideal gases. On the contrary, at high pressures the molecules are brought closer together and consequently the attractive force between the molecules increases, which has the same effect as an increase in external pressure. It then follows that, if external pressure is applied to a volume of gas, particularly at low temperatures, the decrease in volume is slightly greater than would be obtained by pressure only. This effect is more pronounced at low temperatures for the simple reason that at such temperatures the molecules move more slowly and have less tendency to move apart on colliding with one another.

To derive the ideal-gas equation $PV = nRT$ from the kinetic theory of gases, a number of assumptions were made. Van der Waals modified the ideal-gas equation to take into account that two of these assumptions may not be valid. The modified assumptions which are applicable to real gases are as follows.

(i) The volume of the molecules may not be negligible in relation to the volume V occupied by the gas.

(ii) The attractive forces between the molecules may not be negligible.

Let us consider a molecule inside a gas. It is surrounded by other molecules which are equally distributed in all directions. Therefore, the surrounding molecules exert no resultant attractive force on this molecule. As the molecule under consideration approaches the walls of the container in which the gas is confined, the uniform average distribution of the surrounding molecules changes, in which the gas molecules exist only on one side of the molecule under consideration, so that there is an attractive force tending to pull the molecule inwards. Therefore, it seems that the moment that any molecule is just about to strike the walls of the container, and thereby contribute its share in the total gas pressure, the molecules behind it in the bulk of the gas exert a force tending to pull the molecule away from the wall. The measured pressure P is thus less than the ideal pressure given by the kinetic theory of gases. It is, therefore, necessary to add a correction term P' to the measured pressure P. Thus, the ideal pressure is $P + P'$. Van der Waals considered that the pressure defect P' is proportional to the product of firstly the number of molecules striking unit area of the wall per second at any instant and secondly the number per unit volume behind them. For a given volume of gas, both these numbers are proportional to the density ρ of the gas. Therefore, the total attractive force which is related to the correction term P' is proportional to ρ^2. If V is the volume occupied by 1 mol of gas, then $\rho \propto 1/V$. Thus, $P' \propto \rho^2 \propto 1/V^2$, i.e. $P' = a/V^2$, where a is a constant for the particular gas. The corrected pressure is then

$$P + P' = P + \frac{a}{V^2} \tag{2.52}$$

The term a/V^2 is a measure of the attractive force of the molecules in the bulk of the gas; it is called the *cohesion pressure*. It is of importance to the properties of liquids.

All molecules have a particular diameter or volume because repulsive forces occur where they approach very closely. Hence, they cannot be compressed indefinitely. In order to obtain the ideal volume of the space inside a container occupied by the molecules, a correction term b, known as the *co-volume*, must be subtracted from the measured volume V. This means that the corrected volume is $V - b$. The term b is a factor depending on the actual volume of the molecules. On a first consideration, it might be thought that b was itself equal to the volume of the molecules but this is not the case. According to van der Waals, the co-volume is equal to four times the actual volume of the molecules; other slightly different estimates have also been made.

The product of the ideal pressure and volume of 1 mol of a gas should equal RT. Therefore, on application of the pressure and volume corrections considered above, it follows that

$$\left(P + \frac{a}{V^2}\right)(V - b) = RT. \tag{2.53}$$

When V is large, both b and a/V^2 become negligible and the van der Waals equation reduces to the ideal-gas equation $PV = RT$ (for $n = 1$).

At low pressures, the correction a for intermolecular attraction is more important than the correction b for molecular volume. At high pressures and small volumes, the correction for the volume of the molecules becomes important, since the molecules are relatively incompressible and they form an appreciable part of the total volume. It is interesting to note that at some intermediate pressure the two corrections cancel one another and apparently the gas follows the ideal-gas law $PV = nRT$ over a small range of pressures.

We can say that the gas laws are applicable only to gases whose molecules do not attract one another and do not occupy an appreciable part of the whole volume. In real life there are no gases which have these properties; so we can speak of such hypothetical gases only as *ideal* or *perfect* gases. Under ordinary conditions the deviations from the gas laws are so little that they may be ignored.

A van der Waals gas can be liquefied and it possesses a 'critical point' which is given by the conditions

$$\left(\frac{\partial P}{\partial V}\right)_T = 0, \quad \left(\frac{\partial^2 P}{\partial V^2}\right)_T = 0$$

The critical pressure, volume and temperature are given by

$$P_c = \frac{a}{27b^2} \qquad V_c = 3b \qquad T_c = \frac{8a}{27bR}.$$

Also

$$\frac{P_c V_c}{RT_c} = \frac{3}{8}.$$

2.7.1   The van der Waals constants

It was shown by van der Waals that the values of the constants a and b in Equation 2.53 could be calculated from the coefficients of thermal expansion and of compression of the gas, and substitution of the values in Equation 2.53 gave better agreement with experiment than the simple gas equation. Values of a and b can, of course, be evaluated by substituting known values of P and V in the van der Waals equation and then solving the resultant simultaneous equations. These constants can also be calculated from the Joule–Thomson effect and from the critical constants of the gases.

Both a and b vary to some extent with temperature. The value of a increases with the ease of liquefaction of the gas, which is to be expected if it is a measure of intermolecular attraction. It is important to remember that the values of a and b depend on the units of volume and pressure. If the pressure is expressed in atmospheres and the volume in litres, then a is in square litres atmosphere per square mole and b is in litres per mole (actually $a = PV^2$ and $b = V$).

At extremely low pressures, V is very large. Under such conditions both a and b in the van der Waals equation may be neglected. The term a/V^2 is very small and b is only a tiny fraction of V at extremely low pressures. Then Equation 2.53 becomes $PV = RT$. This means that under these conditions the gas obeys the ideal-gas law, which may be considered as representing the limiting behaviour of gases at extremely low pressures. At slightly higher pressures, it is generally possible to ignore b relative to V. Then Equation 2.53 becomes

$$\left(P + \frac{a}{V^2}\right)V = RT \tag{2.54}$$

or

$$PV = RT - \frac{a}{V}. \tag{2.55}$$

Equation 2.55 shows that $PV < RT$ and decreases with increasing pressure because V decreases with increasing pressure and hence a/V becomes greater.

At moderately high pressures, V is relatively small. Under these conditions, it is not justifiable to ignore b; however, a/V^2 is small compared with the high value of P. Then Equation 2.53 becomes

$$P(V - b) = RT \tag{2.56}$$

or

$$PV = RT + Pb. \tag{2.57}$$

This equation shows that PV is now greater than RT and increases with increasing P.

Let us consider Figure 2.9 which represents the curves of PV against P for hydrogen and nitrogen at 273.15 K. The dip in the curve of PV against P for nitrogen is due to the a/V^2 term in the van der Waals equation, whereas the rise in this curve after passing through a minimum is due to the b term. Hydrogen, and for that matter helium as well, have exceptionally small values of a in comparison with b (for hydrogen, $a = 0.245\ \ell^2$ atm mol^{-1} and $b = 2.67 \times 10^{-2}\ \ell$ mol^{-1}; for helium, $a = 0.034\ \ell^2$ atm mol^{-1} and $b = 2.36 \times 10^{-2}\ \ell$ mol^{-1}). Because of this, the influence of b predominates, and PV increases as P is raised even for small values of b.

As the temperature is raised, a point is reached at which the minimum, as seen in Figure 2.9, occurs (theoretically it occurs at $P = 0$); no minimum is observed at still higher temperatures and the curve becomes linear. Any gas then behaves in the same manner as hydrogen or helium do at ordinary temperatures. The minimum so obtained in the PV versus P plot when $P = 0$ is called the *Boyle point* T_B. The value of T_B can be obtained from the van der Waals equation. Rewriting Equation 2.53, we obtain

$$P = \frac{RT}{V - b} - \frac{a}{V^2} \tag{2.58}$$

or

$$PV = RT \frac{V}{V - b} - \frac{a}{V}. \tag{2.59}$$

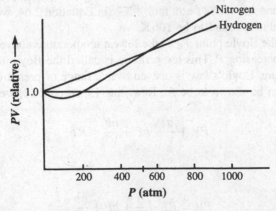

Figure 2.9 *PV* versus *P* diagram for hydrogen and nitrogen

Differentiating Equation 2.59 with respect to P and keeping temperature constant give

$$\frac{\partial}{\partial P}(PV)_T = \left(\frac{RT}{V-b} - \frac{RTV}{(V-b)^2} + \frac{a}{V^2} \right) \left(\frac{\partial V}{\partial P} \right)_T . \tag{2.60}$$

[Note: The right-hand side of Equation 2.59 has been partially differentiated with respect to V first, taking both R and T constant, and V has been partially differentiated with respect to P.]

In order to obtain a minimum in the curve of PV versus P we must have $\partial(PV)/\partial P = 0$. This condition implies that we may equate the expression in the first pair of large parentheses on the right-hand side of Equation 2.60 to zero, i.e.

$$\frac{RT}{V-b} - \frac{RTV}{(V-b)^2} + \frac{a}{V^2} = 0 \tag{2.61}$$

or

$$RT = \frac{a}{b} \left(\frac{V-b}{V} \right)^2 . \tag{2.62}$$

If the minimum is to occur when $P = 0$, then V will be extremely large, i.e. infinite, and Equation 2.62 becomes (replacing T by the Boyle point T_B)

$$RT_B = \frac{a}{b} \tag{2.63}$$

or

$$T_B = \frac{a}{Rb} . \tag{2.64}$$

At all temperatures greater than the Boyle point the value of PV will always increase as pressure increases. Since the values of a for hydrogen and helium are very small, it can be easily seen that, for these two gases, T_B will be relatively low. If we substitute the values of a and b for hydrogen and $R = 0.082\,\ell$ atm $mol^{-1} K^{-1}$ in Equation 2.64, we obtain $T_B = 112\,K$. The experimental value is found to be $106\,K$.

We have defined the Boyle point T_B as the lowest temperature above which PV increases continuously with increasing P. This temperature is called the Boyle temperature because, above this temperature, Boyle's law is obeyed over a range of pressure. Using the van der Waals equation it can be shown to be so. Rewriting Equation 2.53, we find that

$$PV + \frac{a}{V} - bP + \frac{ab}{V^2} = RT$$

or

$$PV = RT - \frac{a}{V} + bP + \frac{ab}{V^2} . \tag{2.65}$$

If the pressure is not too large, ab/V^2 may be neglected because both a and b are small. Also, since $PV = RT$, we may replace V by RT/P in a/V. Then, Equation 2.65 becomes

$$PV = RT - \frac{aP}{RT} + bP \qquad (2.66)$$

$$= RT + P\left(b - \frac{a}{RT}\right) \qquad (2.67)$$

If the temperature T in Equation 2.67 is the Boyle point, then we may replace T by T_B and, substituting the value of T_B from Equation 2.64 into Equation 2.67, we obtain

$$PV = RT_B. \qquad (2.68)$$

This equation shows that the ideal-gas laws are obeyed, if the pressure is not too high.

2.8 VIRIAL EQUATION OF STATE

Suppose that the pressure P and the volume V of n mol of a gas held at a constant temperature are measured over a wide range of values of the pressure, and the product Pv, where $v = V/n$, is plotted as a function of $1/v$. The relation between Pv and $1/v$ may be expressed by means of a power series or *virial expansion* of the form

$$Pv = A'\left(1 + \frac{B'}{v} + \frac{C'}{v^2} + \frac{D'}{v^3} + \cdots\right) \qquad (2.69)$$

where $A', B', C', \ldots$, are called virial coefficients. A' is the first virial coefficient, B' the second virial coefficient, etc., and these coefficients all depend on the temperature and the nature of the gas. This type of equation is useful for purposes of extrapolation. All the known forms of equation of state, i.e. the Clausius equation of state, the Berthelot equation of state, the Dieterici equation of state, etc., can be expressed in the virial form if desired, the virial coefficients being determined by the constants of the various equations. Of all the virial coefficients, the second virial coefficient B' is the most important; it depends on molecular interaction.

In the pressure range 0–40 atms, the relation between Pv and $1/v$ is almost linear, so that only the first two terms in the expansion are significant. In general, the greater the pressure range, the greater are the number of terms in the virial expansion.

There are other forms of the virial equation. One of these forms, which is commonly used, is

$$PV = RT + BP + CP^2 + \cdots \qquad (2.70)$$

where $B, C, \ldots$ are called the second, third, $\ldots$ virial coefficients. $B', C', \ldots$ in Equation 2.69 are related to $B, C, \ldots$ in Equation 2.70. It should be noted here that the notation $B', C', \ldots$, for the virial coefficients is not the customary notation. Generally, all second virial coefficients are denoted by B, third virial coefficients by C, and so on. It is important for the reader to know which form of equation is being used in order to interpret the meaning of the coefficients.

The virial equation may be derived from some assumptions. For example, if we assume T and P to be independent variables, then PV is a function of T and P. Mathematically, $PV = f(T, P)$ where f denotes some function. As $P \rightarrow 0$, $PV \rightarrow RT$ and, even at a pressure of several atmospheres, PV does not deviate vastly from RT. Therefore, it is possible to expand $f(T, P)$ in a Maclaurin series about $P = 0$ at constant temperature. Thus,

$$f(T, P) = f(T, 0) + \frac{\partial}{\partial P} f(T, P)_{P=0, T=T} P + \frac{1}{2!} \frac{\partial^2}{\partial P^2} f(T, P)_{P=0, T=T} P^2 + \cdots$$

where $f(T, 0) = RT$. If we take

$$\frac{\partial}{\partial P} f(T, P)_{p=0, T=T} \equiv B$$

$$\frac{1}{2!} \frac{\partial^2}{\partial P^2} f(T, P)_{P=0, T=T} \equiv C$$

and so on, then the Maclaurin series becomes

$$PV = f(T, P) = f(T, 0) + BP + CP^2 + \cdots$$

or

$$PV = RT + BP + CP^2 + \cdots.$$

This equation is identical with the virial Equation 2.70. If there are no attractive and repulsive intermolecular forces within a mass of gas, then it can be shown from statistical mechanics that the virial coefficients $B, C, \ldots$ identically vanish, and the virial equation reduces to $PV = RT$ (for 1 mol of gas), which is the ideal-gas law. Therefore, we can say that a gas which obeys $PV = RT$ at all pressures has no intermolecular forces acting between its molecules. However, it should be remembered that, in reality, null intermolecular forces never occur. Consequently some properties of real gases do not approach their ideal-gas values as pressure tends to zero.

EXERCISES 2

2.1 Show that a system that expands into a vacuum performs no work.

2.2 Is it true for both compression and expansion that for a given change in state the work performed by the system is a maximum for a reversible path?

2.3 Calculate the amount of work done when a substance expands by $1\,\text{cm}^3$ against the standard atmospheric pressure.

2.4 Calculate the difference between ΔH and ΔU when 1 mol of water is boiled at 100°C and 1 atm. The volume of liquid water may be neglected and the volume of 1 mol of water at 100°C may be taken as $0.03\,\text{m}^3$, assuming that steam behaves ideally.

2.5 When a block of metal of mass 1 kg at 400 K is dropped in 0.5 kg of water at 294 K, the temperature of the water rises to 300 K. If the heat capacity of water is $4200\,\text{JK}^{-1}\,\text{kg}^{-1}$, calculate the heat capacity of the metal.

2.6 Calculate the amount of work done by 1.0 kg of water when it is heated to steam at 100°C. Assume that the pressure remains constant at 0.1 MPa.

2.7 Does heat attending a given change in state depend on the path? Is it meaningful to use the term 'heat of a substance'?

2.8 Define the heat capacity of a substance at constant volume and constant pressure. Does the heat capacity of a substance depend on its nature and physical state? If the heat capacity of a substance is known as a function of temperature in the form of an empirical equation, is it possible to calculate the heat required to increase the temperature of that substance over any temperature interval within the range of validity of the data?

2.9 Define the term enthalpy. Is it a function only of the state of a system?

2.10 Show that for a finite change in volume of an ideal gas the amount of work done is given by the equation

$$W = - \int_{V_1}^{V_2} P \, dV$$

where V_1 and V_2 are the initial and final volumes. Hence show that a system which expands in a vacuum performs no work.

2.11 Consider an ideal gas enclosed in a cylinder equipped with a movable piston as follows.

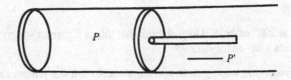

(a) Suggest a method to stop heat flow in or out of the system. What is such a system called?

(b) If the internal pressure P is equal to the external pressure P', what would be the condition of the system?

(c) Assume that the pressure P' on the piston is greater than the internal pressure P and that the piston moves to the left without heat flowing in or out of the system. Explain what happens to the motion of the gas molecules inside the system, the energies of these molecules and their average energies. Which observable variables of the system can be used to detect the overall change in average energy of the molecules?

2.12 State whether the following statements are correct.

(a) Heat content is an extensive property.

(b) The internal energy and the enthalpy are not functions of the temperature.

(c) A thermodynamic process is said to be isothermal, if there is no exchange of heat between the system and its surroundings.

(d) A thermodynamic process is said to be adiabatic, if the temperature remains constant.

(e) Along any reversible path, at any point the direction may be reversed without changing any variable.

(f) When a system expands, work is done on the system.

(g) For a cyclic process, work is positive when the path corresponds to clockwise direction.

(h) For an actual path, the work done by a system on its surroundings is more than that for a reversible path.

(i) The molar heat capacity of a substance is the specific heat capacity multiplied by the number of moles of that substance.

(j) The molar heat capacity of a substance depends on the temperature.

(k) For a cyclic process, the change in the internal energy is zero.

(l) For an isothermal process, the change in the internal energy is the maximum.

(m) For an adiabatic process, the change in the internal energy is equal to the amount of heat supplied.

2.13 'A substance is at 200° of heat.' Do you think that this is a correct statement? Is it possible to conceive of matter in a zero-energy state?

2.14 A substance is receiving energy as heat from a source but the temperature of the substance remains constant. Explain what is happening to the energy transferred as heat.

2.15 An isolated body at a temperature is allowed to interact with another isolated body at a higher temperature until they reach thermal equilibrium. Comment on the energy of the systems at the equilibrium condition.

2.16 Does heat depend on path? Show that, when the heat capacity of a body is known as a function of temperature in the form of an empirical equation, the quantity of heat Q is given by

$$Q = \int_{T_1}^{T_2} (a + bT + cT^2 + \cdots)dT.$$

2.17 Suppose that 1 mol of a perfect gas at 25°C and 0.5 MPa pressure is expanded irreversibly against the constant external pressure of 0.1 MPa until the internal pressure has been lowered to 0.1 MPa. The expansion is an isothermal expansion by addition of heat as required. Calculate the quantity of work done in the process.

2.18 The same gas considered in the above problem is now expanded reversibly from 0.5 to 0.1 MPa at 25°C (isothermal expansion). Calculate the amount of work done in this process.

2.19 State the postulates on which the kinetic theory of gases is based. What modifications are made in the postulates in dealing with real gases? How are these modifications represented in the van der Waals equation?

2.20 Using the concepts of the elementary kinetic theory of gases, derive the equation for the root mean square velocity of the molecules of an ideal gas in terms of its pressure and density.

2.21 (a) Using a simple treatment of the kinetic theory of gases and stating all the necessary assumptions, derive an equation for the pressure exerted by an ideal gas on the walls of its container.

(b) Use this equation to deduce Boyle's law and Charles's law.

2.22 Using the fundamental equation of the kinetic theory of gases derived in problem 2.21 (a), show that the kinetic energy of a mole of an ideal gas is given by $\frac{3}{2}RT$ where R is the universal gas constant and T the temperature. Hence show that, if two ideal gases are at the same temperature, their molecules have the same kinetic energy.

2.23 The fact that the ideal-gas laws and Avogadro's law can be derived from the kinetic theory of gases indicates that the model characterized by the initial assumptions is realistic. Discuss the model.

2.24 The kinetic theory of gases accounts for the properties of ideal gases. Can this theory account for the properties of real gases? If not, what modifications are required to make the theory applicable to real gases.

2.25 Consider a real gas at

(i) high pressure and small volume

(ii) high pressure and low temperature

(iii) low pressure and high temperature.

For which of these sets of conditions will this real gas behave most like an ideal gas? Can this gas be expected to deviate from the ideal behaviour for any of the above conditions?

2.26 The pressure exerted by the molecules of hydrogen on the walls of its container is $1.013 \times 10^5 \, \text{N m}^{-2}$. The density of hydrogen at standard pressure is $0.0899 \, \text{kg m}^{-3}$. Calculate the root mean square speed of hydrogen molecules.

2.27 Calculate the root mean square speed of nitrogen molecules at standard conditions.

2.28 Consider a given molecule in an ideal gas at constant temperature. Can this molecule double its speed under this condition? Explain your answer.

2.29 Express the van der Waals equation of state in the virial form.

2.30 Using the van der Waals equation, calculate the volume of a mole of a gas at 300 K and 10.1 MPa pressure. The values of a and b in the van der Waals equation are $0.137 \, \ell^2 \, \text{MPa mol}^{-2}$ and $0.032 \, \ell \, \text{mol}^{-1}$, respectively.

2.31 If the virial equation is given in the form

$$\frac{PV}{RT} = 1 + \frac{B'}{V} + \frac{C'}{V^2} + \cdots$$

find, for a van der Waals gas, B' in terms of the van der Waals constants.

3
Zeroth Law, Thermal Equilibrium and Thermometry

3.1 INTRODUCTION

Having gathered the basic ideas about the definitions of thermodynamic terms, quantities and units we shall now study the subject in some detail within the scope of this book. The first step to studying thermodynamics in detail is to discuss the laws of thermodynamics and the related topics. We have already said that there are four laws of thermodynamics. Let us discuss these four laws one by one.

3.2 ZEROTH LAW AND THERMAL EQUILIBRIUM

The zeroth law or the law of thermal equilibrium is an important principle of thermodynamics. The importance of this law to the temperature concept was not fully realized until after the other parts of thermodynamics had reached an advanced state of development. Hence the usual name the zeroth law.

The zeroth law is defined as follows: *Two systems which are both in thermal equilibrium with a third system are in thermal equilibrium with each other.*

The temperature concept can be made precise by the statements firstly that systems in thermal equilibrium with each other have the same temperature, and secondly that systems not in thermal equilibrium with each other have different temperatures. The zeroth law, therefore, gives us an operational definition of temperature which does not depend on the physiological sensation of 'hotness' or 'coldness'. The zeroth law is based on the experience that systems in thermal contact are not in complete equilibrium with one another until they have the same degree of hotness, i.e. the same temperature.

Let us now try to understand what the zeroth law actually means. To illustrate this law let us consider two samples of gas. It should be clearly understood here that the argument does not depend, in the least, on whether gases (real or ideal), liquids or solids are chosen. Let the first sample, which we shall call sample 1, be confined in a volume V_1, and the second sample be confined in volume V_2; the pressures at the beginning are P_1 and P_2, respectively. At the beginning, the two systems are separated from each other and are in complete internal

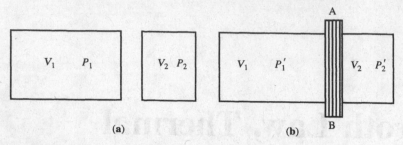

Figure 3.1. (*a*) Systems separated; (*b*) systems in thermal contact

equilibrium. The volume of each container is fixed. This is represented schematically in Figure 3.1.

Let us imagine that the two systems are brought in contact through a wall AB as shown in Figure 3.1(b). Now two possibilities may arise:

1. when in contact through the wall the systems may not influence each other at all, or

2. they may influence each other.

In case (1), the wall would be defined as an insulating or adiabatic wall. It should be noted that in this situation the pressures of the two systems are not affected by placing them in contact. In case (2), i.e. when the systems do influence each other when they are in contact, we shall observe that the pressures will change with time, finally reaching two new values P'_1 and P'_2; then no change in pressure readings will occur with time. In this case the wall is a thermally conducting wall and the systems are in thermal contact. After the properties of the two systems in thermal contact settle down to values which no longer change with time, we can say that the two systems are in *thermal equilibrium.*

Now let us consider three systems A, B and C arranged as in Figure 3.2(a). In Figure 3.2(a), systems A and B are in thermal contact, and systems B and C are in thermal contact. When this composite system is allowed sufficient time, the systems will come to thermal equilibrium. Then A is in thermal equilibrium with B, and B is in thermal equilibrium with C. Now, if we separate A and C from contact with B and place them in thermal contact with each other as shown in Figure 3.2(b), we shall observe that no changes in the properties of A

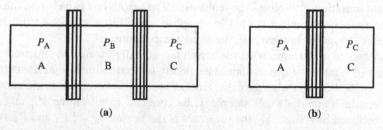

Figure 3.2 (*a*) Three systems A, B and C which are in thermal contact; (*b*) two systems A and C of A, B and C in thermal contact

and C occur with time. Therefore, A and C are in thermal equilibrium with each other. This is stated in the zeroth law of thermodynamics.

3.3   TEMPERATURE AND THERMOMETRY

We have referred to the temperature of a system in the preceding sections without giving a precise definition of this quantity or specifying how it is measured. The concept of temperature, like that of force, originated in man's sensory perceptions, but man's temperature sense, like his force sense, is unreliable and restricted in range; out of the primitive concepts of relative hotness and coldness an objective science of thermometry has developed.

We have already defined equality of temperature within a single isolated system, i.e. we postulate that all parts of an isolated system eventually come to and remain at the same temperature or, in other words, the temperature is the same at all points of a system which is in thermodynamic equilibrium. Now suppose that we have two isolated systems, both in thermodynamic equilibrium. If we bring the two systems in contact, can we say, without actually measuring the temperatures of the two systems, that they are in thermal equilibrium? The answer is 'No'. Man's sensory perception is not sufficiently accurate to measure the temperatures of these two systems in contact. Hence, it would be extremely desirable to have some means of knowing whether the two assemblies would be in thermal equilibrium, if they were brought into contact.

Temperature is considered as a property of matter which allows us to make this determination, and we say that two systems in thermal equilibrium have the same temperature. Two systems not in thermal equilibrium have different temperatures, and energy transfer as heat may take place from one system to the other. Therefore, we add to the concept of temperature the idea that energy transfer as heat always takes place from a system at a higher temperature to a system at a lower temperature when two systems are in thermal contact. This sets the direction of increasing temperature. Our experience further tells us that this energy transfer will tend to equalize the temperatures. We now require a scale for measuring temperature, and it will have to be in terms of observable quantities, such as volume, pressure and electrical resistance.

Let us choose a system, the thermometer, having some property x which is conveniently measurable and varies reasonably rapidly with temperature. This thermometer is allowed to come to thermal equilibrium with a system whose temperature is reproducible, e.g. melting ice. The value of x is measured, and following the degree Celsius system this value is marked on the thermometer as zero. The thermometer is next allowed to come to thermal equilibrium with another system having a reproducible temperature, e.g. water boiling under 0.1 MPa pressure. The property x is measured, and again following the degree Celsius system we mark the new position as 100 on the thermometer. Between 0 and 100 we place 99 evenly spaced marks. If we like, the space above 100 and below 0 may also be divided into intervals of the same size as those between 0 and 100. The thermometer is ready now.

To measure the temperature of any body, the thermometer is allowed to come to thermal equilibrium with the body. The temperature of the body will be indicated on the thermometer. The property chosen as the thermometric property must continually increase or decrease in value as the temperature rises or falls in the range of application of the thermometer. The thermometric property should not have a maximum, minimum or stationary value in the temperature range in which the thermometer is to be used. This procedure can be reduced to a formula by which the temperature can be calculated from the measured value of the thermometric property x. Let x_0 be the value at the ice point, and let x_{100} be the value at the steam point. These points are separated by 100°C. Then,

$$\frac{dx}{dt} = \frac{x_{100} - x_0}{100 - 0} = \frac{x_{100} - x_0}{100}. \tag{3.1}$$

The right-hand side of Equation 3.1 is a constant. Multiplying both sides by dt and integrating, we have

$$x = \frac{x_{100} - x_0}{100}t + C \tag{3.2}$$

where C is a constant of integration. Now, at $t = 0$, $x = x_0$. Using these values, Equation 3.2 becomes $x_0 = C$. Hence substituting this value of C in Equation 3.2, we have

$$x = \frac{x_{100} - x_0}{100}t + x_0$$

or

$$t = \frac{x - x_0}{x_{100} - x_0}100. \tag{3.3}$$

This is the thermometric equation. From the measured value of the thermometric property x the temperature on this particular scale can be calculated.

3.4   OBJECTION TO THE ABOVE PROCEDURE

An objection may be raised to the above procedure on the grounds that it appears necessary that the thermometric property be a linear function of the temperature. This objection is not valid since we have no way of knowing whether a property is linear with temperature until we have chosen some method of measuring the temperature. The method of operation, in fact, by its very nature automatically makes the thermometric property a linear function of the temperature measured on that particular scale. This reveals a real difficulty associated with thermometry. A different scale of temperature would be obtained for each different property which is chosen as the thermometric property. Even with one substance, different scales of temperature will be obtained depending on which property is chosen as the thermometric property.

This situation can be rectified by searching for a class of substances, all of which have some property which behaves in much the same way with temperature. Gases could be classed as such substances. For a given change in temperature, the relative change in pressure at constant volume (or the relative change in volume under constant pressure) has nearly the same value for all real gases. So, we may use an ideal gas in the thermometer and define an *ideal-gas scale of temperature*. However, it is to be remembered that, in spite of its utility, the ideal-gas scale of temperature does not solve the difficulty. The ideal-gas scale is a generalization but the scale depends upon the properties of a hypothetical substance.

3.5 ABSOLUTE OR THERMODYNAMIC TEMPERATURE SCALE

The above-mentioned predicament may be eliminated using the second law of thermodynamics. By using this law it is possible to establish a temperature scale which is independent of the particular properties of any substance, real or hypothetical. This scale is the *absolute* or the *thermodynamic temperature scale*, also called the *Kelvin scale* after the inventor Lord Kelvin. By choosing 100° between the ice point and the steam point and with the usual definition of 1 mol of substance, the Kelvin scale and the ideal-gas scale become numerically identical.

Having overcome the fundamental difficulties, we use all sorts of thermometer. If the temperatures of two bodies A and B are measured with different thermometers, the thermometers must agree that the temperature $T_A > T_B$ or $T_A = T_B$ or $T_A < T_B$. The different thermometers need not agree in the numerical value of either T_A or T_B. If it is necessary, the reading of each thermometer can be translated into the temperature in kelvins, in which case the numerical values will agree.

3.6 A COMPARISON OF TEMPERATURE SCALES

The zero of the thermodynamic temperature scale (to which the Kelvin scale is equivalent) always occurs at the same condition, regardless of what value we choose for the triple point (of water) temperature.

The Celsius temperature scale employs a degree of the same magnitude as that of the Kelvin scale but its zero point is shifted to 273.15 K. This makes 0 K correspond to −273.15°C. Thus, the Celsius temperature of the triple point of water is 0.01°C. Therefore, if t denotes the Celsius temperature

$$t = T - 273.15 \text{ K} \tag{3.4}$$

where T is in kelvin. The Celsius temperature t_s at which steam condenses at 0.1 MPa pressure is

$$t_s = T_s - 273.15 \text{ K}. \tag{3.5}$$

There are two other scales in common use in engineering. The Rankine temperature T_R (measured in degrees Rankine) is proportional to the Kelvin temperature according to the relation

$$T_R = \tfrac{9}{5}T. \tag{3.6}$$

A degree of the same size as that of the degree Rankine is used in the Fahrenheit scale t_F but with the zero point shifted according to the relation

$$t_F = T_R - 459.67°\text{R}. \tag{3.7}$$

Substituting Equations 3.4 and 3.6 into 3.7, we obtain

$$t_F = \tfrac{9}{5}t + 32°\text{F} \tag{3.8}$$

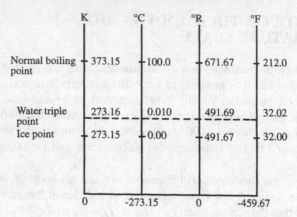

Figure 3.3 Comparison of the Kelvin, Celsius, Rankine and Fahrenheit scales of temperature

It follows from Equation 3.8 that the Fahrenheit temperature of the ice point ($t = 0°C$) is 32°F and that of the steam point ($t = 100°C$) is 212°F. The 100°C or 100 K between the ice point and the steam point correspond to 180°F or 180°R. It is evident that the degree Fahrenheit (which is the same size as the degree Rankine) is only 5/9 of the degree Celsius, and the temperatures on the two scales are related by the equation

$$t°C = \tfrac{5}{9}(t°F - 32°) \tag{3.9}$$

The Kelvin, Celsius, Fahrenheit and Rankine temperature scales are compared in Figure 3.3.

Example 3.1

Find the temperature of the ice and steam points on the Rankine or Fahrenheit absolute scale, in which the difference $T_s - T_i$ between the steam and ice points is 180°.

Solution

We know that $0°F \equiv 459.67°R$. The ice point on the Fahrenheit scale starts at 32°F. Therefore, 32°F will correspond to $459.67 + 32°R = 491.67°R$.

Again, since the difference between steam and ice points is 180°, so the steam point will correspond to $491.67 + 180 = 671.67°R$.

3.7 THE INTERNATIONAL PRACTICAL TEMPERATURE SCALE

The International Committee of Weights and Measures defined a temperature scale which is called the *International Practical Temperature Scale*. This scale is based on a number of fixed and reproducible equilibrium temperatures (fixed points) to which numerical values are

Table 3.1   International Practical Temperature Scale

Fixed point, temperature of equilibrium ($P = 0.1$ MPa)	Temperature (°C)
Between liquid oxygen and its vapour (oxygen point)	−182.97
Between ice and air-saturated water (ice point) (fundamental fixed point)	0.00
Between liquid water and its vapour (steam point) (fundamental fixed point)	100.00
Between liquid sulphur and its vapour (sulphur point)	444.60
Between solid and liquid silver (silver point)	960.80
Between solid and liquid gold (gold point)	1063.00

assigned, and on specified formulae for computing temperatures from the indications of specified instruments. The scale conforms as closely as possible to the centigrade thermodynamic scale. It is not intended to use as a substitute for the centigrade thermodynamic scale, which is accepted as fundamental. Its practical importance lies in its reproducibility rather than its accuracy. A list of the primary and fundamental fixed points is given in Table 3.1.

3.8   TYPES OF THERMOMETER

When we talk about thermometers, our initial thought is of a mercury-in-glass thermometer. This is a good device for measuring temperature but unfortunately it is of almost no use for measuring high temperatures or low temperatures. Alcohol-in-glass thermometers are good devices for measuring low temperatures but above 78°C they are useless. So scientists all over the world were looking for thermometers that would be suitable for measuring both high and low temperatures, e.g. gas thermometers (constant-volume and constant-pressure gas thermometers), electrical resistance thermometers and thermocouples. The lowest ideal-gas temperature that can be measured with a gas thermometer is about 0.5 K, provided that low-pressure He^3 is used. At moderately high temperatures, gas thermometers tend to be of no use.

Among the electrical resistance thermometers, the platinum resistance thermometer is most commonly and widely used. This thermometer may be used for very accurate work within the range from −253 to 1200°C. The most widely used temperature-measuring device is the thermocouple. The range of a thermocouple depends on the materials of which it is composed. A platinum–rhodium thermocouple has a range from 0 to 1600°C. The advantage of a thermocouple is that it achieves thermal equilibrium quite rapidly with the system whose temperature is to be measured, because its mass is small. It follows temperature changes easily but it is not so accurate as a platinum resistance thermometer.

EXERCISES 3

3.1 The normal body temperature is 98.6°F. What is the normal body temperature on the Celsius scale?

3.2 If the temperature of dry ice is −79°C, what would be its temperature on the Fahrenheit scale?

3.3 If the normal boiling point of hydrogen is $-252.8°C$, calculate the normal boiling point of hydrogen in degrees Fahrenheit and degrees Rankine.

3.4 At what temperature would the readings of Fahrenheit and Celsius thermometers be the same?

3.5 The triple point of water is $0.01°C$.

(a) What is the value of $0°C$ on the Kelvin scale?

(b) If another temperature unit is $5/9$ of the kelvin, how many of these other units are equivalent to $100\,K$?

(c) Is the size of the kelvin greater or smaller than or equal to the degree Celsius?

3.6 Why is the zeroth law called the zeroth law instead of naming it the first law of thermodynamics? What is the significance of this law?

3.7 Briefly discuss how the concept of thermal equilibrium led to the concept of thermometry. Do you think that this development is restricted in any way?

3.8 The ice point and the normal boiling point of water are $273.15\,K$ and $373.15\,K$, respectively. What are the corresponding values of these two points in degrees Celsius, degrees Rankine and degrees Fahrenheit?

4

The First Law
of Thermodynamics

4.1 INTRODUCTION

Having gathered the basic ideas about the definitions of thermodynamic terms, quantities and units, we now proceed to study the subject in some detail. This is achieved within the limited scope of this book by first discussing the laws of thermodynamics followed by some of the important applications and topics related to the laws.

4.2 HISTORY OF THE FIRST LAW OF THERMODYNAMICS

The first law of thermodynamics is an extension of the principle of the conservation of mechanical energy. Such as extension became reasonable after it was shown that expenditure of work could cause the production of heat. The first quantitative experiments on conversion of work to heat were carried out by Benjamin Thompson. He suggested that heat might arise from the mechanical energy expended and was able to estimate the heat that would be produced by a horse working for an hour. In modern units his value for this *mechanical equivalent of heat* would be $0.183\,\mathrm{cal\,J^{-1}}$. In 1799, Humphrey Davy provided further support for Thompson's theory, showing that even in the absence of air latent heat could be provided by mechanical work.

By about 1840 the law of conservation of energy was accepted in purely mechanical systems, the interconversion of heat and work was well established, and it was understood that heat was simply a form of motion of the smallest particles composing a substance.

Julius Robert Mayer established in 1842 the following: 'From application of established theorems on the warmth and volume relationships of gases, one finds that the fall of a weight from a height of about 365 m corresponds to the warming of an equal weight of water from 0 to 1°C'. This figure related mechanical units of energy to thermal units. The conversion factor is called *the mechanical equivalent of heat J*. Hence, $w = Jq$. In modern units, J is usually given as joules per calorie. To lift a mass of $1\,\mathrm{g}$ to a height of $365\,\mathrm{m}$ requires $365 \times 10^2 \times 981\,\mathrm{erg}$ of work, or $3.58\,\mathrm{J}$. To raise the temperature of $1\,\mathrm{g}$ of water from 0 to 1°C requires $1.0087\,\mathrm{cal}$. The value of J calculated by Mayer is therefore $3.56\,\mathrm{J\,cal^{-1}}$.

The accepted modern value is $4.184\,\mathrm{J\,cal^{-1}}$. Mayer was able to state the principle of the conservation of energy, the first law of thermodynamics, in general terms, and to give one rather rough numerical example of its application.

The exact evaluation of J and the proof that it is a constant independent of the method of measurement was accomplished by Joule. Although Mayer was the father of the philosophy of the first law of thermodynamics, the precise experiments carried out by Joule firmly established the law on an experimental or inductive foundation.

4.3   DEFINITION AND FORMULATION OF THE FIRST LAW OF THERMODYNAMICS

The philosophical argument of Mayer and the experimental work of Joule led to a definite acceptance of the conservation of energy, the first law of thermodynamics. There are several definitions of the first law. A few of the accepted definitions of this law are given below.

1. The total amount of energy of an isolated system remains constant but may change from one form to another.

2. When an amount of energy of one form disappears, an equivalent amount of energy of other forms appears.

3. Energy cannot be created or destroyed.

4. 'Perpetual motion' cannot be realized.

It is worthwhile to add a few words about perpetual motion. A perpetual-motion machine is one which runs for ever. The promise of such a machine is irresistible. Perpetual motion proposals fall into two types. A perpetual-motion machine of the first type actually produces energy. Some of this energy is lost through frictional losses, while the rest is accessible as free energy. On the other hand, a perpetual-motion machine of the second type does not create or destroy energy. Such a machine runs for ever through complete elimination of friction. However, any attempt to withdraw any amount of energy from it will slow it down. Perpetual-motion machines of the first and the second type contravene the first and the second laws of thermodynamics, respectively.

Advocates of perpetual motion of the first type generally agree with scientists about the laws of force and torque that operate within a perpetual motion machine of the first type. Newton's laws are easily appreciated when it is accepted that unhindered motion is a natural state of a body, that it changes its velocity only when it is forced to do so by a force acting on it and that slowing down is not attributable to the natural state but is the consequence of frictional forces. The same people tend to disagree that Newton's laws also imply conservation of energy. If there is no force acting on the body, there will be no change in velocity nor any change in energy.

An interesting point about perpetual-motion machines is that such machines are invariably cyclic. By cyclic we mean that after sufficient operation the machine goes back to its initial state. Because a non-cyclic machine is almost impossible to work, cyclicity is definitely a matter of immense convenience. Energy balance over a full cycle is necessary. An energy balance can be achieved by introducing an acceleration mechanism over one part of the cycle and a fine deceleration process in another part. A more refined machine can be achieved which interchanges energy between its various forms, such as heat, latent heat of evaporation, motion and electromagnetism.

Perpetual-motion machines of the second type are equally interesting. Such machines in one sense exist in real life. It follows from Newton's laws that an isolated system in motion, on which no force or torque is acting, exhibits precisely perpetual motion of the second kind. An example of perpetual motion of the second type is the orbiting of electrons around the atomic nucleus. The Earth's motion around the Sun is another example. These examples show that perpetual motion of the second type is common on atomic and celestial scales; however, such a motion is not common in everyday life. The best known example of perpetual motion in everyday life is superconductivity, in which a current circulates ceaselessly in a wire loop without a battery. A second example is a superfluid having no viscosity. Perpetual motion in the thermodynamic variables, such as pressure, temperature and volume, is impossible.

It is worthwhile to note that the earliest known perpetual-motion proposal is found in an Indian Sanskrit manuscript dating as far back as the first half of the fifth century. This proposal refers to a wheel having sealed holes which are drilled in radially from the circumference and partly filled with mercury. The wheel itself is free to rotate about a horizontal axis. Once the wheel is set in rotational motion, it is expected to retain its rotation. Presumably, the inventor believed that the extra moment, generated by the presence of mercury in the holes on the descending side of the wheel moving under centrifugal force to the end at the circumference, imparts sufficient momentum to maintain the rotation of the whole thing. This idea formed the basis of many other proposals later put forward in Europe. In these proposed prototypes, weights attached to the wheel circumference disposed themselves farther from the horizontal axis on the descending than the ascending side of the wheel.

Any scepticism about perpetual motion is justifiable. For more than a century it remained a field where theory and experiment had success to back up each other in unratifying the phenomenon. The irony of perpetual motion is that human beings operate according to Newton's laws at an unconscious level almost all the time. The saga of the perpetual-motion machine exemplifies human effort and a sluggish advancement to enlightenment, with theory and experiment thriving side by side.

We can use the principle of conservation of energy to define a function U called the *internal energy*. Let us imagine that a closed system undergoes a process by which it passes from state A to state B as shown in Figure 4.1. If the only interaction with its surroundings is in the form of transfer of heat Q to the system, or performance of work W on the system, the change in U will be

$$\Delta U = U_B - U_A = Q + W. \tag{4.1}$$

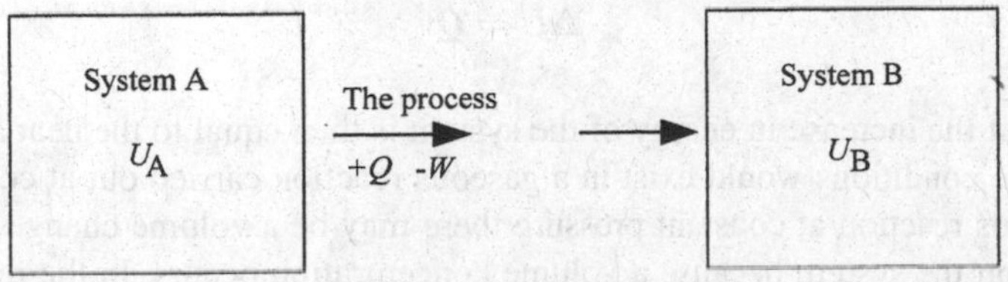

Figure 4.1 shows System A with U_A, the process with $+Q$ $-W$, leading to System B with U_B.

Figure 4.1   A closed system undergoes a process to pass from state A to state B

Note: In Equation 4.1 we have defined W as the work done on the system and Q is added to the system. If we had defined W as work done by the system, Equation 4.1 would become $\Delta U = Q - W$. For an isolated system, we have by definition $W = Q = 0$ and therefore $\Delta U = 0$.

Now, the first law of thermodynamics states that this energy difference ΔU depends only on the initial and final states, and not on the path followed between them. Both Q and W have many possible values, depending on exactly how the system passes from A to B, but $Q + W = \Delta U$ is invariable and independent of the path. If this were not true, it would be possible, by passing from A to B along one path and then returning from B to A along another, to obtain a net change in the energy of the closed system in contradiction to the principle of conservation of energy. Therefore, the change (ΔU) in internal energy must be equal to the energy absorbed in the process from the surroundings in the form of heat Q minus the energy lost to the surroundings in the form of external work (W) *done by the system*. **Remember that the work done by the system is negative**. The work done may be due to expansion or electrical work. We can, therefore, say that Equation 4.1 is a mathematical expression of the first law. For a differential change, Equation 4.1 becomes

$$dU = dQ + dW. \tag{4.2}$$

The first law has often been stated in terms of the universal human experience, i.e. it is impossible to construct a perpetual-motion machine, namely a machine that produces useful work by a cyclic process with no change in its surroundings. To see how this experience is embodied in the first law, let us consider a cyclic process from state A to state B and back to A again. If perpetual motion were ever possible, it would sometimes be possible to obtain a net energy increase $\Delta U > 0$ by such a cycle. That this is impossible can be ascertained from Equation 4.1 which indicates that, for any such cycle, $\Delta U = (U_B - U_A) + (U_A - U_B) = 0$. A more general way of expressing this fact is to say that for any cyclic process the integral of dU vanishes:

$$\oint dU = 0. \tag{4.3}$$

4.4 REACTIONS AT CONSTANT VOLUME AND CONSTANT PRESSURE

When the system does no work or no work has been done on it, the work term is zero, i.e. $W = 0$. Then, from Equation 4.1

$$\Delta U = Q. \tag{4.4}$$

This shows that the increase in energy of the system is then equal to the heat absorbed in the reaction. These conditions would exist in a gaseous reaction carried out at constant volume.

For a gaseous reaction at constant pressure there may be a volume change. In such cases, work is done on the system because a volume concentration occurs. In the reverse reaction, if any (e.g. $2CO + O_2 \rightleftharpoons 2CO_2$) expansion occurs, work would be done by the system. If no other kind of work is done, then

$$\Delta U = Q - P\Delta V \tag{4.5}$$

where P is the pressure (which is constant) and ΔV is the change in volume of the system during the process.

If the volume increase is due to the formation of 1 mol of a perfect gas, then from the relation $PV = nRT$ (here n is the number of moles equal to 1), we have $P\Delta V = RT$. Then Equation 4.5 becomes

$$\Delta U = Q - RT. \tag{4.6}$$

Example 4.1

When 1 g-atom of pure iron is dissolved in dilute HCl at 18°C, the heat liberated is 87.03 kJ. Calculate the energy change ΔU of the system.

Solution

Since heat is liberated, the products of the reaction possess less internal energy than the reactants. This indicates that the amount of heat evolved is negative. The basic equation to be used should be Equation 4.6. For the reaction $Fe + 2H^+ = H_2(g) + Fe^{2+}$, we know that 1 mol of gas is produced which acts as an ideal gas, and the amount left dissolved in the solution is negligible. Here $RT = 2.415$ kJ. It is given that the heat liberated $Q = -87.03$ kJ. Substituting these values in Equation 4.6, we obtain $\Delta U = -89.44$ kJ.

The energy function is determined to the extent of an arbitrary additive constant; it has been defined only in terms of the difference in energy between one state and another. Sometimes, as a matter of convenience, we may adopt a conventional standard state for a system and set its energy in this state equal to zero, e.g. we might choose the state of the system at 0 K and 0.1 MPa pressure as our standard. Then the energy U in any other state would be the change in energy in going from the standard state to the state in question.

4.5   TO SHOW THAT ENERGY IS CONSERVED IN ANY CHANGE IN STATE

Let us consider two systems A and B in contact with each other, the composite system $A + B$ being isolated from the rest of the universe. In the composite system we can choose A as our system and B as its surroundings. If A undergoes a change in its state, a quantity of heat Q will flow from B and a quantity of work W will flow to B. According to Equation 4.1, for system A we have

$$U_{A(final)} = U_{A(initial)} + Q + W. \tag{4.7}$$

If we had chosen B as the system and A as the surroundings, then a quantity of heat $-Q$ would have flowed from A and a quantity of work $-W$ would have flowed to A. Then, for the system B

$$U_{B(final)} = U_{B(initial)} + (-Q) + (-W). \tag{4.8}$$

Adding Equations 4.7 and 4.8, we have

$$(U_A + U_B)_{final} = (U_A + U_B)_{initial}$$

or

$$(U_A)_{final} - (U_A)_{initial} = -[(U_B)_{final} - (U_B)_{initial}]$$

or

$$\Delta U_A = -\Delta U_B. \tag{4.9}$$

The total energy of the two systems is the same after the change in state as it was before. Alternatively, the increase ΔU_A in energy of A is just balanced by the decrease $-\Delta U_B$ in energy of B. The energy of the system plus the energy of the surroundings are conserved in the change in state. If we imagine that B is so large as to encompass the entire universe excluding only system A, then A and B together would compose the Universe. Thus, we have the well known statement of the *first law of thermodynamics due to Clausius*: *The energy of the Universe is constant.*

4.6   CONCEPT OF ENERGY

Having discussed the first law of thermodynamics in terms of energy, it has now become necessary to have a broader concept of energy. We have already defined energy as the capacity for doing work, but this definition is of little value since work itself is defined in terms of energy. Chemical thermodynamics deals with the internal energy U, energy possessed by the system by virtue of the mass and motion of the molecules, intermolecular forces, chemical composition, etc. Any energy that the system possesses because of other considerations is ignored.

The second feature of energy considerations is concerned with the change rather than its absolute value. The mathematical expression for this concept is $\Delta U \equiv U_2 - U_1$. This indicates that the finite change ΔU in total internal energy is the difference between the energy in the final state and that in the initial state. The most significant aspect of this kind of relation is that the energy change depends only on the initial and final states and is independent of the path linking these states. Figure 4.2 represents this fundamentally important idea. The absolute internal energy of the system (with unspecified origin) is plotted for the initial and final states. The three processes shown begin and end in the same energy states regardless of the intermediate stages through which the system passes. The three processes start at the energy level U_1 and end at the energy level U_2. Therefore, any change ΔU in internal energy must be equal to $U_2 - U_1$. Let us consider each process separately. For the process (*a*), obviously the total change in the internal energy is $\Delta U = U_2 - U_1$. In the process (*b*), the system starts at U_1, goes to U_3, then to U_4 and ends at U_2. Therefore, the total change in the

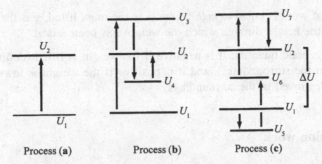

Process (a)   Process (b)   Process (c)

Figure 4.2   A diagram showing that the energy change depends only on the initial and final states

internal energy is $\Delta U = (U_3 - U_1) + (U_4 - U_3) + (U_2 - U_4)$, i.e. $\Delta U = U_2 - U_1$. Finally, in the process (c), the system goes to U_5, then to U_6 to U_7 and ends at U_2. In this case, the total change in the internal energy is $\Delta U = (U_5 - U_1) + (U_6 - U_5) + (U_7 - U_6) + (U_2 - U_7)$, i.e. $\Delta U = U_2 - U_1$. These calculations show that the change in the internal energy of any system depends only on the initial and the final state and is independent of the path followed to arrive at the final state. There are many forms that the energy can assume but it is convenient to classify all energy entering or leaving the system as either heat or work.

4.6.1   Some properties of energy

(i) We have already stated that, for a specified change in state, the change ΔU in the internal energy of a system depends only on the initial and final states of the system and not on the path connecting those states. Although both Q and W depend on the path, $Q + W = \Delta U$ is independent of the path.

(ii) The energy is an extensive state property of the system. Under the same conditions of temperature and pressure, 10 mol of the substance composing the system has ten times the energy of 1 mol of the substance. The energy per mole is an intensive state property of the system.

(iii) Energy is conserved in all transformations. A perpetual-motion machine is a machine which by its action creates energy by some transformation of a system. The first law of thermodynamics asserts that it is impossible to construct such a machine.

4.7   WORK

The concept of work is of fundamental importance in thermodynamics. The definition of work is given in Section 1.13; several features should be noted in the definition of work.

1. Work appears only at the boundary of a system.

2. Work appears only during a change in state.

3. Work is manifested by an effect on the surroundings.

4. The quantity of work is equal to mgh where m is the mass lifted, g is the acceleration of gravity and h is the height through which the weight has been raised.

5. Work is an algebraic quantity; it is negative if the weight is lifted (defined as work has been produced in the surroundings), and it is positive if the weight is lowered (defined as work has been destroyed in the surroundings).

4.7.1 Expansion work

If a system increases its volume against an opposing pressure, a work effect is produced in the surroundings. This work is defined as expansion work. Let us consider a quantity of gas contained in a cylinder fitted with a weightless frictionless movable piston D. The temperature of the whole system is maintained constant and the space above the piston is evacuated so that no air pressure is pushing down the piston. The positions of the piston at the initial and final stages are shown in Figure 4.3(a) and 4.3(b), respectively. A small mass M is placed on the piston. The initial state of the system is described by T, P_1 and V_1, and the final state by T, P_2, V_2, where T is the temperature, P is the pressure and V the volume enclosed by the piston and the cylinder. The difference between the initial and the final position of the piston is h. Work is produced during the transformation since the mass M in the surroundings has been raised through a vertical distance h against the force Mg of gravity. The quantity of work W produced is given by

$$W = Mgh. \tag{4.10}$$

If the area of the piston is A, then the downward pressure acting on the piston is $Mg/A = P_0$, the pressure which opposes the motion of the piston. Thus, $Mg = AP_0$. Putting this in Equation 4.10, we have

$$W = AP_0h. \tag{4.11}$$

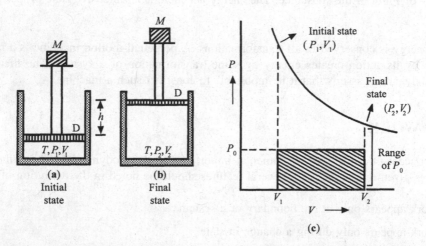

Figure 4.3   Schematic diagram of work produced in a single-stage expansion

The product Ah is simply the additional volume enclosed by the boundary in the change in state. Therefore, $Ah = V_2 - V_1 = \Delta V$. Then, Equation 4.11 becomes

$$W = P_0(V_2 - V_1). \tag{4.12}$$

This work W is represented graphically by the shaded area in the $P-V$ diagram (Figure 4.3(c)). The curve in this figure is the isotherm of the gas. The mass M can have any arbitrary value from zero to some definite upper limit. It is evident that P_0 can have any value within the range $0 < P_0 < P_2$. Therefore, the quantity of work produced may have any value between zero and some definite upper limit.

We conclude that *work is a function of the path*. We should remember that P_0 is arbitrary and not related to the pressure of the system. The sign of W is determined by the sign of ΔV, since $P_0 = Mg/A$ is always positive. In expansion, ΔV is positive; so $W = -P\Delta V$ is negative, i.e. the work is done by the system on the surroundings; the weight rises. In compression, ΔV is negative; so $W = -P\Delta V$ is positive, i.e. the work is done by the surroundings on the system; the weight falls. Equation 4.12 is correct only if P_0 is constant throughout the change.

So far we have discussed expansion work done in one single step only. It is possible to perform expansion work in several steps. Let us imagine that a large weight was first placed on the piston during the first part of the expansion from V_1 to some intermediate volume V'. In such a two-stage expansion we still apply Equation 4.12 to each stage of the expansion, using, of course, different values of P_0 in each stage. Then the total work produced is the sum of the work produced in each step, i.e.

$$W = W_1 + W_2 = P_0'(V' - V_1) + P_0''(V_2 - V') \tag{4.13}$$

where W_1 is the work done in the first stage, W_2 is the work done in the second stage, P_0' is the opposing pressure in the first stage and P_0'' is the opposing pressure in the second stage.

The quantity of work W produced in the two-stage expansion is represented by the shaded areas in Figure 4.4 for the special case $P_0'' = P_2$.

Comparing Figures 4.3(c) and 4.4 we find that for the same change in state the two-stage expansion produces more work than the single-stage expansion could possibly produce. If the heats had been measured, we would have found different quantities of

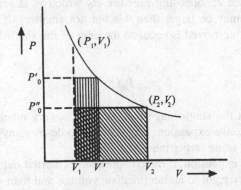

Figure 4.4 Schematic diagram of work produced in a two-stage expansion

heat associated with the two paths. We conclude that *in a multistage expansion the work produced is the sum of all the small amounts of work produced in each state*. If P_0 is constant as the volume increases by an infinitesimal amount dV, then the small quantity of work dW is given by

$$dW = -P_0 \, dV. \tag{4.14}$$

The total work done in the expansion from V_1 to V_2 is

$$W = -\int_1^2 dW = -\int_{V_1}^{V_2} P_0 \, dV. \tag{4.15}$$

Equation 4.15 is the general expression for expansion work for any system. Once P_0 is known as a function of the volume of the system concerned, the integral is evaluated by the usual methods.

4.7.2 Compression work

We have seen that, in expansion, work is produced. On the other hand, in compression, work is destroyed. The amount of work destroyed in compression is computed using the same equation that is used to compute the work produced in expansion. In compression, the final volume is less than the initial volume as contrary to expansion. So, in every stage, ΔV is negative; therefore, the total work destroyed is positive. The sign is automatically considered by the integration process, if the volume of the final state is the upper limit and the volume of the initial state is the lower limit in the integral Equation 4.15. To compress a gas, a larger weight is required on the piston than was lifted in the expansion. Thus, more work is destroyed in the compression of a gas than is produced in the corresponding expansion.

To consider the amount of work destroyed in compression, let us consider the same system as before. However, now the initial state is the expanded state T, P_2, V_2, while the final state is the compressed state T, P_1, V_1, the temperature T being the same in both the states. Figure 4.5 illustrates the system.

If the gas is to be compressed to the final volume V_1 in one stage, we must use a mass M large enough to produce an opposing pressure P_0 which is at least as great as the final pressure P_1 (the mass may be larger than this but not smaller). If we choose M such that $P_0 = P_1$, then the work destroyed is equal to the area of the shaded rectangle in Figure 4.5 with a negative sign, i.e.

$$W = -P_0(V_1 - V_2). \tag{4.16}$$

The work destroyed in the single-stage compression is very much greater than the work produced in the single-stage expansion. It is possible to destroy any greater amount of work in this compression by using larger masses.

Just as in the case of expansion, if the compression is carried out in two stages (compressing first with a lighter weight to an intermediate volume and then with a heavier weight to the final volume) less work is destroyed. The total amount of work destroyed would be equal

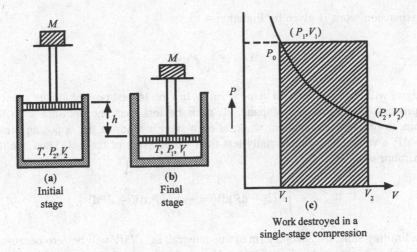

Figure 4.5   Schematic diagram of work destroyed in a single-stage compression

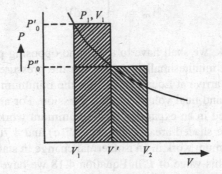

Figure 4.6   Schematic diagram of work destroyed in a two-stage compression

to the sum of the individual amounts of work done in each stage. The total work destroyed is the area of the shaded rectangles in Figure 4.6.

4.7.3   Maximum and minimum quantities of work

We have already seen that in a two-stage expansion more work can be produced than in a single-stage expansion. It appears that, if an expansion is done in many stages using a large mass at the beginning and reducing it as the expansion proceeds, even more work can be produced. This is true but there is a limitation to this procedure. The weights to be used must not be so large as to compress the system instead of allowing it to expand. By carrying out the expansion in a progressively larger number of stages, the work produced can be increased up to a definite maximum value. (This is true only if the temperature is constant along the path of the change in sate. If the temperature changes along the path, there is no upper limit to the work produced.)

Similarly, in a multistage compression, less work is destroyed than that destroyed in a single-stage compression.

The expansion work is given by Equation 4.15:

$$W = -\int_{V_1}^{V_2} P_0 \, dV.$$

The integral will have a maximum value when P_0 has the largest possible value at each stage of the process. If the gas is to expand, P_0 must be less than the pressure P of the gas. Therefore, to obtain the maximum work, at each stage we adjust the opposing pressure to $P_0 = P - dP$, a value just infinitesimally less than the pressure of the gas. Then, we have for the maximum work W_m:

$$W_m = -\int_{V_1}^{V_2} (P - dP)dV = -\int_{V_1}^{V_2} (P \, dV - dPdV). \qquad (4.17)$$

In the limiting state, the second term of this integral, i.e. $dPdV$ will be zero because it is an infinitesimal of higher order than the first. Hence, for the maximum work in expansion

$$W_m = -\int_{V_1}^{V_2} P \, dV. \qquad (4.18)$$

For the minimum work, we shall have to adjust the opposing pressure, at each stage, to $P_0 = P + dP$, a value just infinitesimally greater than the pressure of the gas. By the same argument we shall again arrive at Equation 4.18 for the minimum work for compression, if V_1 and V_2 are the initial and final volumes in compression. For an ideal gas, the maximum quantity of work produced in an expansion or the minimum work destroyed in a compression is represented by the shaded areas in Figures 4.7(a) and 4.7(b).

The maximum or minimum work in an isothermal change in state can be easily evaluated, since $P = nRT/V$. Using this value of P in Equation 4.18 we have

$$W_{max} \text{ or } W_{min} = -\int_{V_1}^{V_2} \frac{nRT}{V} dV = -nRT \int_{V_1}^{V_2} \frac{dV}{V} = -nRT \ln\left(\frac{V_2}{V_1}\right). \qquad (4.19)$$

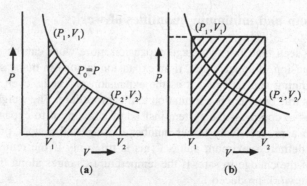

Figure 4.7 Schematic diagram of (a) the maximum work produced in an expansion and (b) the minimum work destroyed in a compression for an ideal gas

In the expansion process, $V_2 > V_1$; so the logarithm of V_2/V_1 is positive. In the compression process, $V_2 < V_1$; so the logarithm of V_2/V_1 is negative. Thus, the sign of W is taken care of by itself.

The application of Boyle's law to a perfect gas at a constant temperature shows that $P_1V_1 = P_2V_2$ or $V_2/V_1 = P_1/P_2$. Substituting this relation into Equation 4.19, we have

$$W = -nRT \ln \left(\frac{P_1}{P_2} \right). \tag{4.20}$$

We can now calculate the maximum work of expansion of a perfect gas from knowledge of either the initial and final volumes (Equation 4.19), or the initial and final pressures (Equation 4.20).

In the case of a perfect gas, expanding into a vacuum would not result in either absorption or liberation of heat, i.e. $Q = 0$. Also, since no work is done, $W = 0$. Hence, from the first law, $\Delta U = Q + W = 0 + 0 = 0$, i.e. $\Delta U = 0$. Thus, *when a perfect gas expands at constant temperature there is no change in the internal energy U which is therefore independent of its volume*. This is true, no matter where the expansion takes place. Since ΔU is zero for an isothermal expansion or compression, the heat absorbed is equal to the negative work done by the perfect gas:

$$Q = -W. \tag{4.21}$$

Combining Equations 4.19–4.21, we have

$$Q = -W = -nRT \ln \left(\frac{V_2}{V_1} \right) = -nRT \ln \left(\frac{P_1}{P_2} \right) \tag{4.22}$$

Q is the heat absorbed in the isothermal reversible expansion of n mol of a perfect gas from state 1 to state 2.

Example 4.2

What is the maximum work in joules per mole for the expansion of 1 mol of a perfect gas from 10^{-3} to 5×10^{-3} m^3 at 25°C?

Solution

The basic equation to be used for this problem is

$$W = -RT \int_{V_1}^{V_2} \frac{dV}{V} = -RT \ln \left(\frac{V_2}{V_1} \right)$$

n being 1. Now, $R = 8.31 \, \text{JK}^{-1} \text{mol}^{-1}$, $T = 25°\text{C} + 273 = 298 \, \text{K}$, $V_1 = 1 \times 10^{-3} \, \text{m}^3$ and $V_2 = 5 \times 10^{-3} \, \text{m}^3$. Therefore,

$$W = -8.31 \times 298 \times 2.303 \log \left(\frac{5}{1} \right) = -3.986 \times 10^3 \, \text{J mol}^{-1}$$

4.7.4   Reversible and irreversible transformation

Let us consider the same system as before, i.e. a quantity of gas is confined in a cylinder at a constant temperature T. Let us first consider expansion of the gas from the state T, P_1, V_1 to the state T, P_2, V_2 and then compress the expanded gas to the original state. Thus, the gas is subjected to a *cyclic transformation*. Let us perform this cycle by two different processes and calculate the net work effect W_c for each process.

Process I is a single-stage expansion with $P_0 = P_2$ followed by a single-stage compression with $P_0 = P_1$. The work produced in the expansion is, from Equation 4.15,

$$W_{exp} = -P_2(V_2 - V_1) \qquad (4.23)$$

and the work produced in the compression is

$$W_{comp} = -P_1(V_1 - V_2). \qquad (4.24)$$

The net work effect in the cycle is the sum of these two, i.e.

$$W_c = W_{exp} + W_{comp} = -(P_2 - P_1)(V_2 - V_1). \qquad (4.25)$$

Since $P_2 - P_1$ is negative and $V_2 - V_1$ is positive, the overall sign of W_c is negative. The negative sign implies that net work has been destroyed in the cycle.

Process II is a multistage expansion with $P_0 = P$ followed by a multistage compression with $P_0 = P$. The work produced in expansion is, from Equation 4.18,

$$W_{exp} = -\int_{V_1}^{V_2} P \, dV$$

and the work produced in compression is

$$W_{comp} = -\int_{V_2}^{V_1} P \, dV.$$

Then, the net work effect in the cycle is

$$W_c = W_{exp} + W_{comp} = -\int_{V_1}^{V_2} P \, dV - \int_{V_2}^{V_1} P \, dV$$

$$= -\int_{V_1}^{V_2} P \, dV + \int_{V_1}^{V_2} P \, dV = 0. \qquad (4.26)$$

We can easily conclude that in the first process the system has been restored to its initial stage but the surroundings have not been restored. On the other hand, in the second process the system is restored to its initial state, and so also are the surroundings, since no net work is produced.

Now, if a system undergoes a change in state through a specified sequence of intermediate states and is then restored to its original state by traversing the same sequence of states in reverse order and if the surroundings are also brought to their original state, the transformation in either direction is reversible; the corresponding process is a *reversible process*. If the surroundings are not restored to their original state after the cycle, *the transformation and the process are irreversible*. With these definitions in mind, the first process is an irreversible process, whereas the second process is a reversible process.

It is to be noted that in the reversible process the internal equilibrium of the gas is disturbed only infinitesimally and in the limit not at all. Therefore, at any stage in a reversible transformation, the system does not depart from equilibrium by more than an infinitesimal amount. Reversible processes are not real but ideal processes. Real processes are always irreversible. The fundamental characteristic of every irreversible (and therefore every real isothermal cyclic) transformation is that, if any system is kept at a constant temperature and subjected to a cyclic transformation by irreversible processes (real processes), a net amount of work is destroyed in the surroundings. We shall see later that this is in fact a statement of the second law of thermodynamics. The general conclusion reached is valid regardless of how the system is constituted. By appropriate modification of the argument, the general conclusions reached could be shown to be correct for any kind of work, e.g. electrical work or work done against a magnetic field. To calculate the quantities of these other kinds of work, we would not, of course, use the integral of pressure over volume, but the integral of the appropriate force over the corresponding displacement.

EXERCISES 4

4.1 Show analytically that energy is conserved in any change in state.

4.2 Suggest a process which can be carried out such that $\Delta U = Q = W = 0$.

4.3 Analytically prove that the internal energy change of a system depends only on the initial and final states of the system and is independent of the path linking these states.

4.4 Can you establish that the thermodynamic work is a function of the path? Show that, for the same change in state, the work produced in a multistage irreversible expansion is more than that produced in a single-stage expansion.

4.5 Prove that, in an irreversible process for the same change in state, the work destroyed in a single-stage compression is very much greater than the work produced in a single-stage expansion.

4.6 Establish that the work destroyed in a multistage compression is less than that destroyed in a single-stage compression.

4.7 The volume of 1 mol of an ideal monatomic gas, initially at 0.202 MPa and 298.15 K, is doubled by

(a) reversible isothermal expansion and

(b) irreversible isothermal expansion

by suddenly dropping the external pressure to 0.101 MPa and allowing the gas to expand against this constant external pressure. Calculate ΔU, W and Q for each process. Sketch both the paths on a $P-V$ diagram and show the work involved in both cases. (*Hint*: consider the volume of an ideal gas at standard temperature and pressure).

4.8 Show that for an isothermal reversible expansion of an ideal gas the amount of work done by the gas is given by

$$W = -\int_{V_1}^{V_2} P \, dV = -nRT \ln \left(\frac{V_2}{V_1} \right) = -nRT \ln \left(\frac{P_1}{P_2} \right).$$

4.9 Consider 2.0 mol of an ideal gas having a volume $V_1 = 3.50\,\text{m}^3$ at a temperature

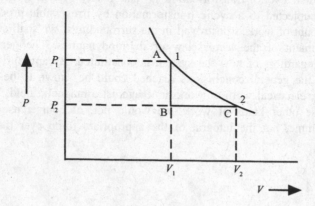

$T_1 = 300\,\text{K}$. This gas is allowed to expand to a new volume $V_2 = 7.0\,\text{m}^3$ at a temperature $T_2 = T_1$. The process is carried out (a) isothermally and (b) along the path ABC as shown in the figure. In the process (b) the pressure is allowed to drop at constant volume and the volume increases at constant pressure. For each process, calculate the work done by the gas, the quantity of heat added to the gas and the consequent change in the internal energy of the gas.

4.10 1.0 kg of water is boiled to steam at 100°C and at a constant pressure of 0.101 MPa. Calculate the work done and the change in internal energy in this process. Assume that the heat required to boil 1.0 kg of water is 2260 kJ.

4.11 A gas obeys the equation of state $P(V-b) = RT$, where b is a constant. For this gas sketch

(a) a plot of V versus T at constant P (isobar)

(b) a plot of P versus V at constant T (isotherm)

(c) a plot of P versus T at constant V (isochore).

5

Some Consequences
of the First Law

5.1 HEAT CAPACITY

We have already defined heat capacity as the amount of heat required to raise the temperature of 1 mol of a system by 1°C. It is denoted by C. Equation 2.10 is a general definition of heat capacity and it implies that the heat capacity of a system depends not only on the heat-absorbing ability but also, in a simple way, on the amount of substance present. It is, therefore, necessary to define heat capacity in terms of material in the system. Two notations are used as follows.

1. *Specific heat capacity*. Heat absorbed by 1 g of material that undergoes a rise in temperature of 1°C.

2. *Molar heat capacity*. Heat absorbed by 1 g mol of material that undergoes a rise in temperature of 1°C.

This shows that molar heat capacity is the specific heat capacity multiplied by the gram-molecular weight. A further distinction is to be made between heat capacities depending on whether the volume or pressure remains constant during the heating process. The heat effect (Q) in a process for a change in temperature from T_1 to T_2 can be calculated from the relation

$$Q = C\Delta T = C(T_2 - T_1) \tag{5.1}$$

if it is assumed that the heat capacity C, which generally varies with temperature, remains essentially constant between T_1 and T_2. This situation is analogous to the calculation of PV work in a constant-pressure process, and Q can be graphically represented by the area under the curve as shown in Figure 5.1(a).

Integration must be used if C varies significantly with temperature. This is illustrated in Figure 5.1(b). If we use the same arguments as before, the heat absorbed is the area under the curve:

$$Q = \int_{T_1}^{T_2} (C \text{ as a function of } T)\, dT \tag{5.2}$$

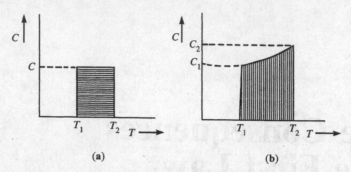

Figure 5.1   Schematic diagram of the variation in heat capacity of an ideal gas: (a) C is invariant with temperature; (b) C is a function of temperature

Note: For many substances, C can be expressed by an empirical power series of the form $C = a + bT + cT^{-2}$. In such a case

$$Q = \int_{T_1}^{T_2} (a + bT + cT^{-2})\mathrm{d}T = a(T_2 - T_1) + \frac{b}{2}(T_2^2 - T_1^2) - c\left(\frac{1}{T_2} - \frac{1}{T_1}\right)$$

for 1 mol of material.

If the heat capacity cannot be expressed conveniently as a function of temperature, Q can be evaluated by mechanical area determination in a graph of heat capacity versus temperature over the range of interest. If a suitable function is available, the integration can be performed numerically by the use of integration rules.

To obtain a quantity which is characteristic of a substance and which does not depend on how much of that substance one considers, we define the specific heat capacity c of a substance as the ratio of the true heat capacity C of a system composed of that substance to the mass m of the system, i.e.

$$c = \frac{C}{m}. \tag{5.3}$$

The true molar specific heat capacity would then be defined as

$$c = \frac{C}{n} \tag{5.4}$$

where n is the number of moles.

5.2   SPECIFIC HEAT CAPACITY

So far we have talked about C as though it were a well-defined quantity but, of course, the heat required to pass from one specified state of system to another will vary with the path taken. The heat capacity can then become a well defined quantity only by virtue of a specification of path. In fact, we find useful two species of heat capacities, corresponding to two simply specified paths, namely that at constant volume and that at constant pressure.

(i) At *constant volume*, over a temperature range in which the heat capacity is constant, Equation 5.1 becomes

$$Q_V = C_V \Delta T \tag{5.5}$$

but $Q_V = \Delta U$ by definition. Hence,

$$\Delta U = C_V \Delta T. \tag{5.6}$$

If, over the temperature range concerned, C_V is not constant, then from Equation 5.2

$$\Delta U = Q_V = \int_{T_1}^{T_2} (C_V \text{ as a function of } T) \, dT. \tag{5.7}$$

For n mol,

$$\Delta U = n \int_{T_1}^{T_2} C_V \, dT = nC_V(T_2 - T_1) = nC_V \Delta T \tag{5.8}$$

if C_V is constant.

(ii) At *constant pressure*, we have, similarly to Equation 5.5,

$$Q_P = C_P \Delta T \tag{5.9}$$

but Q_P is, by definition, the heat content or enthalpy (which will be discussed later). Then,

$$\Delta H = C_P \Delta T. \tag{5.10}$$

If C_P varies over the temperature range concerned, then, substituting as before, we have

$$\Delta H = Q_P = \int_{T_1}^{T_2} (C_P \text{ as a function of } T) \, dT. \tag{5.11}$$

For n mol,

$$\Delta H = n \int_{T_1}^{T_2} C_P \, dt = nC_P(T_2 - T_1) = nC_P \Delta T \tag{5.12}$$

if C_P is constant.

5.3   RELATIONSHIP BETWEEN C_P AND C_V

When an ideal gas is heated at constant volume, no work is done, and all the heat invested goes into increasing the temperature of the gas. If an ideal gas is heated at constant pressure, although the same amount of heat is still required to raise the temperature of the gas, an additional investment of heat is required for the $P\Delta V$ work done by the heated gas as, at constant pressure, it expands against the atmosphere. Then, if C_V is the heat required to increase the temperature of 1 mol of an ideal gas by 1°C at constant volume, and C_P is the corresponding quantity at constant pressure, we have

$$C_P = C_V + P\Delta V. \tag{5.13}$$

However, from the ideal-gas law we have $PV = nRT$, where the symbols have their usual meanings. Then, for 1 mol of gas, i.e. for $n = 1$, $PV = RT$. This can also be written as, for small changes in volume V and temperature T, $P\Delta V = R\Delta T$. By definition of 'heat capacity', the quantity of material is 1 mol and the temperature increase is 1°C. Hence,

$$P\Delta V = R\Delta T = R(1) = R. \tag{5.14}$$

Combining Equations 5.13 and 5.14 gives

$$C_P = C_V + R. \tag{5.15}$$

This equation has been abundantly confirmed for real gases at reasonably low pressures. The ratio of the 'heat capacities' is given by

$$\gamma = \frac{C_P}{C_V} = \frac{C_V + R}{C_V}. \tag{5.16}$$

This equation gives an interesting criterion for distinguishing monatomic and polyatomic gases. From the kinetic theory of gases we know that

$$PV = \tfrac{1}{3}Nmu^2 \qquad \text{(see Equation 2.34).} \tag{5.17}$$

Here P is the pressure resulting from the bombardment of the walls of the container by the particles of the gas, N is the number of particles, m is the mass of each particle, u^2 is mean square velocity and V is the volume of the gas. Now, if we consider just 1 mol of an ideal gas, then N becomes the Avogadro number. Then, rewriting Equation 5.17 as

$$\tfrac{3}{2}PV = N\tfrac{1}{2}mu^2 \tag{5.18}$$

we note that the right-hand side of Equation 5.18 represents the kinetic energy of all the particles present. However, from the ideal-gas law we know that $PV = nRT$. Hence, for 1 mol of gas ($n = 1$)

$$PV = RT. \tag{5.19}$$

Combining Equations 5.18 and 5.19, we have

$$\begin{aligned}
\tfrac{3}{2}RT = N\tfrac{1}{2}mu^2 \\
= \text{kinetic energy of translational motion of molecules of} \\
\text{1 mol of any ideal gas at any temperature } T.
\end{aligned} \tag{5.20}$$

Let us consider 1 mol of an ideal monatomic gas heated at constant volume. Because of the heating, we are adding energy to the gas and this added energy will increase the kinetic energy of the gas particles. Then, for this gas, C_V will simply be the increment in the kinetic energy of 1 mol as a result of 1°C rise in temperature. Thus,

$$\text{kinetic energy} = \tfrac{3}{2}R\Delta T = \tfrac{3}{2}R = C_V. \tag{5.21}$$

Putting $C_V = \tfrac{3}{2}R$ in Equation 5.15, we have

$$C_P = C_V + R = \tfrac{3}{2}R + R = \tfrac{5}{2}R \tag{5.22}$$

Again, putting $C_V = \tfrac{3}{2}R$ in Equation 5.16,

$$\gamma = \frac{C_P}{C_V} = \frac{C_V + R}{C_V} = \frac{\tfrac{3}{2}R + R}{\tfrac{3}{2}R} = 1.67$$

or

$$\gamma = 1.67. \tag{5.23}$$

5.3.1 Principle of equipartition of energy

The measured heat capacities for diatomic and triatomic gases increase with increased number of atoms per molecule. A simple explanation for this is the fact that the internal energy includes translational kinetic energy, as well as other forms of energy. The molecules can have translational, rotational and vibrational kinetic energy. We have said that in the case of a monatomic gas the energy input goes into increasing the energy of translational motion. On the other hand, in the case of polyatomic gases, part of the energy input is used to increase the energies of rotational and vibrational motions. Because of this diversion, the total heat input required to produce the standard increase in kinetic energy corresponding to 1°C increase in temperature is greater for a polyatomic gas than for a monatomic gas.

This concept will be better understood, if the idea of 'degree of freedom' is introduced, by which we mean the number of independent ways that molecules can possess energy. Let us consider a monatomic gas. It can have only three degrees of freedom, as an atom can have velocity along the x, y and z axes. These are considered to be three independent motions because any change in any one of these velocity components would not affect the other components. Let us now consider a diatomic gas as shown in Figure 5.2. In Figure 5.2(a) the atoms can rotate vertically, whereas in Figure 5.2(b) they can rotate horizontally, i.e. the two atoms can rotate about two different axes. However, rotation about another axis passing through the two atoms would give rise to very little energy because the moment of inertia is insignificant. Therefore, like a monatomic molecule, a diatomic molecule too has the same three degrees of freedom associated with translational kinetic energy, but additionally it has two more degrees of freedom associated with rotational kinetic energy, i.e. a total of five degrees of freedom. It can be seen from Table 5.1 that C_V for diatomic gases is about

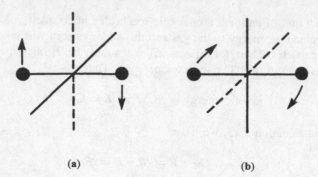

(a)   (b)

Figure 5.2 Rotation of atoms of a diatomic gas: (*a*) vertical rotation; (*b*) horizontal rotation

Table 5.1 Heat capacity ratio of some important gases at 0.1 MPa pressure

Gas	Specific heat $(kJ\,kg^{-1}\,K^{-1})$		Molar heat capacity $(J\,mol^{-1}\,K^{-1})$		$C_P - C_V$ $(J\,mol^{-1}\,K^{-1})$	γ
	C_V	C_P	C_V	C_P		
Monatomic						
He	3.138	4.812	12.468	20.794	8.326	1.67
Ne	0.619	1.029	12.468	20.794	8.326	1.67
A						1.67
Hg						1.67
Na						1.67
Diatomic						
H_2		0.014				1.40
N_2	0.740	1.038	20.753	29.079	8.326	1.40
O_2	0.648	0.912	21.046	29.413	8.368	1.40
Triatomic						
CO_2	0.640	0.833	28.451	36.945	8.494	1.30
H_2O	1.464	2.017	25.941	34.309	8.368	1.32
Polyatomic						
C_2H_6	1.435	1.724	43.095	51.672	8.577	1.20

five-thirds of that for a monatomic gas. This is exactly in the same ratio as their degrees of freedom. This led to an important idea, the *principle of equipartition of energy.*

This principle was theoretically derived by Maxwell using statistical mechanics. It states that energy is equally shared by the active degrees of freedom and, in particular, each active degree of freedom of a molecule has an average energy which equals $\frac{1}{2}kT$. Hence, the average energy of a monatomic gas molecule would be $\frac{3}{2}kT$ and that of a diatomic gas molecule $\frac{5}{2}kT$. Thus, the internal energy of a diatomic gas would be $U = \frac{5}{2}nRT$, where n is the number of moles, R is the universal gas constant and T is the temperature. More complex molecules have more degrees of freedom. It is now clear that, the greater the degrees of freedom, i.e. 'rotational and vibrational motions' active in absorbing energy, the greater will be C_V and the smaller the ratio $\gamma = C_P/C_V = 1 + R/C_V$. Since all but the monatomic gases have such extra degrees of freedom fully operative at and above room temperature, $\gamma = 1.67$ becomes a highly distinctive mark of a monatomic gas.

It is found that, at very low temperatures, C_V for H_2 has a value of $\frac{3}{2}R$. For all other diatomic gases, C_V has a value of $\frac{5}{2}R$, indicating that they have only five degrees of freedom. On the other hand, at very high temperatures, C_V is about $\frac{7}{2}R$ as if there are seven degrees of freedom. These observations make the situation complicated. However, the explanation for this is that, at low temperatures, the molecules possess primarily translational kinetic energy without any rotational energy. Thus, only three degrees of freedom prevail. On the other hand, at high temperatures, the five degrees of freedom, namely three translational and two rotational, together with two additional degrees of freedom are active. The two additional degrees of freedom originate from the vibration of the two atoms as if they were connected by a spring; one degree of freedom is contributed by the vibrational motion and the other by the potential energy of vibrational motion. The potential energy of vibrational motion of a spring is given by $\frac{1}{2}kx^2$, where k is the spring constant and x is the length. At room temperature, these two degrees of freedom are not active. Using the quantum theory, Einstein was able to explain why fewer degrees of freedom are operative at lower temperatures.

It is worthwhile to mention that the principle of equipartition of energy can be applied to solids. At high temperatures, the molar heat capacities of all solids are close to $3R$. Although some degrees of freedom are not operative at low temperatures, at high temperatures each atom apparently has six degrees of freedom. In a crystalline solid, each atom can vibrate about its equilibrium position as if it were connected to each of its neighbours by a spring. Therefore, it can have three degrees of freedom due to kinetic energy and another three due to potential energy of vibration in each of the three x, y and z directions.

Heat capacity ratio of some important gases at 0.1 MPa pressures are given in Table 5.1.

Note: The difference between C_P and C_V for liquids and solids is rather small, and, except for where high accuracy is required, it is sufficient to take $C_P = C_V$. The reason for this is that the thermal expansion coefficients of liquids and solids are very small, so that the volume change on increasing the temperature by 1°C is very small; correspondingly the work produced by the expansion is small and little energy is required for the small increase in the spacing of the molecules. Almost all the heat withdrawn from the surroundings goes into increasing the energy of the chaotic motion and so is reflected in a temperature increase which is nearly as large as that in a constant-volume process.

5.4 HEAT CONDUCTION

When two parts of a substance are maintained at different temperatures and the temperature of each small volume element of the intervening substance is measured, experiment shows a continuous distribution of temperature. The transport of energy between neighbouring volume elements by virtue of the temperature difference between them is known as *heat conduction*.

Let us consider a piece of material in the form of a slab of thickness d and of area A. One face is maintained at temperature T and the other at $T + dT$. The heat Q that flows perpendicular to the faces for a time t is measured. The experiment is repeated with other slabs of the same material but with different values of d and A. The results of such experiments show that, for a given value of dT, Q is proportional to the time and the area. Also, for a given time and area, Q is proportional to the ratio dT/d provided that both dT and d are small. This result may be mathematically represented by

$$\frac{Q}{t} \propto A \frac{dT}{d}. \tag{5.24}$$

This is only approximately true when dT and d are finite but is rigorously true in the limit as dT and d approach zero. If we generalize this result for an infinitesimal slab of thickness dx, across which there is a temperature difference dT and introduce a constant of proportionality K, the fundamental law of heat conduction becomes

$$\frac{dQ}{dt} = -KA \frac{dT}{dx}. \tag{5.25}$$

The derivative d$T/$dx is called the *temperature gradient*. K is called the *thermal conductivity*. The minus sign is introduced so that the positive direction of the flow of heat coincides with the positive direction of x. For heat to flow in the positive direction of x, this must be the direction in which T decreases.

A substance with a large thermal conductivity is known as a *thermal conductor*, and one with a small value of K as a *thermal insulator*. The numerical value of K depends on a number of factors, one of which is the temperature. If the temperature difference between parts of a substance is small, K can be considered to be almost constant throughout the substance. There are three simple cases that can be handled in an elementary way. In all cases, we shall assume that K is constant throughout the conducting substance.

(a) *Linear flow of heat perpendicular to the faces of a slab.* If the temperature difference $T_1 - T_2$ and the thickness d are small, then

$$Q = KA \frac{T_1 - T_2}{d} \tag{5.26}$$

(b) *Radial flow of heat between two coaxial cylinders.* Let us imagine that the conducting material lies between an inner cylinder of radius r_1 and an outer cylinder of radius r_2, both of length l. If the inner cylinder is maintained at constant temperature T_1 and the outer at T_2, there will be a steady radial flow of heat at the constant rate Q. Let us consider the flow of this

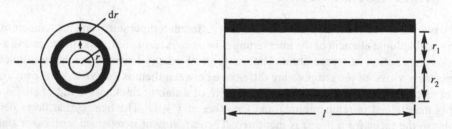

Figure 5.3   Radial flow of heat between two coaxial cylinders

amount of heat across a cylindrical shell of material bounded by the cylinders at r and $r + dr$ (Figure 5.3). Let T be the temperature at r and $T + dT$ at $r + dr$. The area of the shell is

$$A = 2\pi r l \tag{5.27}$$

and the temperature gradient is dT/dr. Hence,

$$Q = -K2\pi r l \frac{dT}{dr} \tag{5.28}$$

and

$$dT = -\frac{Q}{2\pi l K} \frac{dr}{r}. \tag{5.29}$$

Integrating Equation 5.29 between r_1 and r_2, we have

$$T_1 - T_2 = \frac{Q}{2\pi l K} \ln\left(\frac{r_2}{r_1}\right). \tag{5.30}$$

(c) *Radial flow of heat between two concentric spheres.* If the conducting material lies between an inner sphere of radius r_1 held at constant temperature T_1 and an outer sphere of radius r_2 held at constant temperature T_2, there will be a steady radial flow of heat at a constant rate Q. Considering this flow across the spherical shell bounded by the spheres at r and $r + dr$, we have

$$Q = -KA\frac{dT}{dr} = -K4\pi r^2 \frac{dT}{dr} \tag{5.31}$$

and

$$dT = \frac{-Q}{4\pi K} \frac{dr}{r^2}. \tag{5.32}$$

Integrating Equation 5.32 between r_1 and r_2, we have

$$T_1 - T_2 = \frac{Q}{4\pi K}\left(\frac{1}{r_1} - \frac{1}{r_2}\right). \tag{5.33}$$

If $T_1 - T_2$ is small, K is almost equal to the thermal conductivity at the mean temperature. When the substance to be investigated is a non-metal, the same general method is used. Since the thermal conductivity K is equal to $(Q/A)(dx/dT)$, it has units

$$\frac{J\,s^{-1}}{m^2}\frac{m}{°C} = \frac{W}{m\,°C} = W\,m^{-1}\,°C^{-1}.$$

Lengths are usually expressed centimetres. Further discussion about thermal conductivity is not within the scope of this book.

Example 5.1

Consider two thin concentric spherical shells of radius 5 cm and 15 cm, respectively. Their annular cavities are filled with charcoal. When energy is supplied at a steady rate of 10.8 W to a heater at the centre, a temperature difference of 50°C is set up between the spheres. Find the thermal conductivity of charcoal.

Solution

Given that $T_1 - T_2 = 50°C$, $Q = 10.8$ W, $r_1 = 5$ cm and $r_2 = 15$ cm and if K is the thermal conductivity of charcoal, then substituting these values in Equation 5.33, i.e.

$$T_1 - T_2 = \frac{Q}{4\pi K} \left(\frac{1}{r_1} - \frac{1}{r_2} \right)$$

gives

$$50°C = \frac{10.8\,\text{W}}{4 \times 3.14\,\text{K}} \left(\frac{1}{5} - \frac{1}{15} \right)$$

or

$$K = \frac{10.8 \times 2}{4 \times 3.14 \times 50 \times 15}\,\text{W cm}^{-1}\,°\text{C}^{-1}$$
$$= 2.3 \times 10^{-3}\,\text{W cm}^{-1}\,°\text{C}^{-1}$$

5.5 REVERSIBLE AND IRREVERSIBLE PROCESSES

Thermodynamic processes are often classified as either *isothermal* or *adiabatic*. In isothermal processes we allow whatever transfer of heat is required to keep the temperature of the system constant. In an adiabatic process no heat enters or leaves the system. These two main classes of processes are again divided into four:

1. reversible isothermal process

2. irreversible isothermal process

3. irreversible adiabatic process

4. reversible adiabatic process

We shall now discuss these processes systematically.

5.5.1   Reversible isothermal process

Let us consider a fixed mass of an ideal gas enclosed in a cylinder fitted with a piston. To constitute a reversible process, let us imagine that the internal pressure P of the gas is balanced by the external pressure on the piston at every step; in other words, the difference between the internal and the external pressures in infinitesimally small. Let us allow the gas to expand very very slowly, keeping the temperature constant all the time by either withdrawing heat from or supplying heat to the system. Let P_1, V_1, T_1 be the initial state and P_2, V_2, T_1 be the final state. This situation is schematically represented in Figure 5.4.

Now for perfect-gas behaviour, $(\partial U/\partial V)_T = 0$. Hence, for expansion at constant temperature, $\Delta U = 0$. From the first law we therefore have $Q = -W$. Also, the work done by the gas during a small change $\mathrm{d}V$ in volume is $\mathrm{d}W = -P\,\mathrm{d}V$. In a reversible change, all the work is used in pushing back the piston, and no work is used to increase the kinetic energy of the gas or in frictional losses. Therefore, for the complete change in volume from V_1 to V_2, the total work is given by

$$W = -Q = -\int_{V_1}^{V_2} P\,\mathrm{d}V.$$

For an ideal gas, $PV = nRT$. Then the above integral becomes

$$W = -Q = -nRT\int_{V_1}^{V_2}\frac{\mathrm{d}V}{V} = -nRT\,\ln\left(\frac{V_2}{V_1}\right).$$

Since by Boyle's law, at constant temperature, $V_2/V_1 = P_1/P_2$, we have

$$W = -Q = -nRT\,\ln\left(\frac{V_2}{V_1}\right) = -nRT\,\ln\left(\frac{P_1}{P_2}\right).$$

This equation can also be written as

$$P_2 = P_1\exp\left(\frac{W}{nRT}\right). \tag{5.34}$$

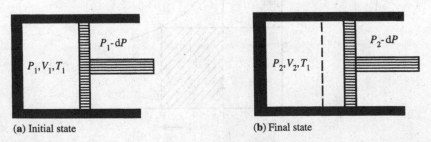

(a) Initial state

(b) Final state

Figure 5.4   Schematic diagram of reversible isothermal expansion of an ideal gas

5.5.2   Irreversible isothermal process

We know that the change in any state function is independent of the path taken between the states. From this we conclude that ΔH and ΔU for irreversible expansion are the same as for the corresponding reversible expansion. Hence, as before, $\Delta U = 0$ and $\Delta H = 0$ for the irreversible isothermal expansion of an ideal gas.

Since work and heat depend on the path taken between the states, it is not possible to calculate these quantities without knowing how the process is done. Let us imagine that the irreversible expansion is carried out by suddenly dropping the external pressure P_{ext} from its initial value P'_{ext} to any value P''_{ext}. Since this decrease is sudden, there will be no noticeable volume change for this decrease in external pressure. So, we can argue that, when the external pressure reaches the new value P''_{ext}, the expansion will proceed against this new external pressure P''_{ext} held constant. Schematically we can represent this path as shown in Figure 5.5.

As before, the work done during this isothermal irreversible expansion is given by

$$W = -\int_{V_1}^{V_2} P_{ext}\, dV = -P''_{ext} \int_{V_1}^{V_2} dV = -P''_{ext}(V_2 - V_1). \tag{5.35}$$

Since $\Delta U = 0$, the heat associated with the irreversible isothermal process is

$$-Q = W = -P_{ext}(V_2 - V_1). \tag{5.36}$$

From Figure 5.5 it is clear that the work done by the system in the irreversible isothermal expansion is less than that in the corresponding reversible isothermal expansion that occurs between the same two states. This can be shown mathematically as follows.

Let us consider a system that can do mechanical work against its surroundings, e.g. a gas enclosed in a cylinder. Let us denote the gas by A and the surroundings by B, having

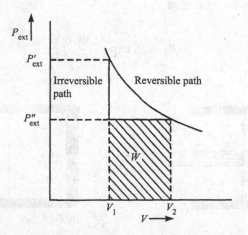

Figure 5.5   Schematic diagram of isothermal irreversible expansion of an ideal gas

pressures P_A and P_B, respectively. Suppose that the gas is allowed to expand at constant temperature and pressure against the external pressure in two different ways.

Firstly, the expansion is done *irreversibly* with $P_A > P_B$. Let $P_A = P_B + \Delta P$. The work done by the system is

$$W_{irrev} = -P_B \Delta V = -(P_A - \Delta P)\Delta V$$

or

$$W_{irrev} = -P_A \Delta V + \Delta P \Delta V.$$

Secondly, the same expansion is carried out *reversibly* with $P_A = P_B + dP$, where dP is an infinitesimal amount of pressure. The work done by the system is

$$W_{rev} = -P_B \Delta V = -(P_A - dP)\Delta V$$

or

$$W_{rev} = -P_A \Delta V + dP \Delta V.$$

Since dP is an infinitesimal quantity, we safely neglect the quantity $dP\Delta V$, so that

$$W_{rev} = -P_A \Delta V.$$

This shows clearly that W_{rev} is greater than the corresponding work W_{irrev} done in the irreversible expansion by the finite quantity $\Delta V \Delta P$. Therefore, the work done on the surroundings is a maximum for the reversible process; work lost as heat is infinitesimal. For the irreversible process, the work done on the surroundings is less than the maximum; the difference appears as heat. Conversely, if the work is done on the system, this work is again a maximum for the reversible change. So, we can generalize the statement that *the numerical value of the work done on the surroundings, whether positive or negative, is always a maximum in reversible process*. This is, in fact, true for any kind of work and for any path. Since only irreversible processes are observable (the reversible processes are the hypothetical processes) we conclude that *all observable processes produce less work than the maximum work, and result in dissipation of work as heat*.

5.5.3   Adiabatic irreversible change

With no heat entering or leaving the system, for which $Q = 0$ by definition, any work performed by the system must be done at the expense of its internal energy. The Equation 4.1 reduces to

$$\Delta U = W. \tag{5.37}$$

From the previous discussions we know how to express W as an integral of P and dV, but it is not very obvious how to express ΔU for an expansion, in which pressure, volume and

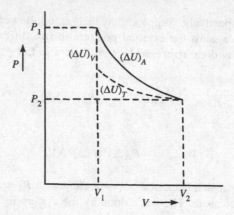

Figure 5.6   Schematic diagram of adiabatic irreversible expansion of an ideal gas

temperature are not constant. However, we can determine ΔU for the overall change by summing (over a series of steps that lead to the final state) ΔU-values readily evaluated for steps under any imposed conditions. Let us imagine the adiabatic expansion as shown in Figure 5.6.

In the first stage the ideal gas is cooled at constant volume to its final temperature T_2; in the second stage, with the reintroduction of all the heat hypothetically removed in the first stage, the gas expands isothermally to its final volume V_2. For the overall adiabatic change we then have

$$(\Delta U)_A = (\Delta U)_V + (\Delta U)_T \qquad (5.38)$$

where the subscript A represents adiabatic change, but we know that the isothermal expansion of an ideal gas produces no change in internal energy. So, we can write $(\Delta U)_T = 0$. Also, from Equation 5.8 we know that

$$(\Delta U)_V = n \int_{T_1}^{T_2} C_V \, dT.$$

Then Equation 5.38 becomes

$$(\Delta U)_A = n \int_{T_1}^{T_2} C_V \, dT + 0 = n \int_{T_1}^{T_2} C_V \, dT. \qquad (5.39)$$

Substituting Equation 5.39 into our general Equation 5.37 for adiabatic change,

$$\Delta U = n \int_{T_1}^{T_2} C_V \, dT = - \int_{V_1}^{V_2} P \, dV = W. \qquad (5.40)$$

Example 5.2

5 mol of an ideal monatomic gas, with the specific heat at constant volume being 20.92 J mol^{-1} K^{-1}, expands irreversibly but adiabatically from an initial pressure of 2.02 MPa against a constant external pressure of 0.101 MPa until the temperature drops from the initial value of 27°C to a final value of 7°C. How much work has been done in the process? What is the final volume?

Solution

From the problem we have $n = 5$ mol, $C_V = 20.92$ J mol^{-1} K^{-1}, $T_1 = 27 + 273 = 300$ K, gas pressure $P = 2.02$ MPa, $T_2 = 7 + 273 = 280$ K and $P_0 = 0.101$ MPa. Substituting these values into Equation 5.40, i.e. into

$$\Delta U = nC_V(T_2 - T_1)$$

we have

$$\Delta U = (5\,\text{mol})(20.92\,\text{J mol}^{-1}\,\text{K}^{-1})\,[(280 - 300)\,\text{K}] = -2.09\,\text{kJ}.$$

Converting this value of ΔU from joules to litre megapascals, we have $\Delta U = -20.631$ MPa. This must equal the work done in the adiabatic expansion. Then,

$$W = -P_0(V_2 - V_1) = 0.101(V_2 - V_1) = -20.63\,\ell\,\text{MPa}.$$

Now, using the perfect-gas law, we have

$$PV_1 = nRT_1$$

where V_1 is the initial volume of the gas and the temperature $T_1 = 300$ K. Therefore,

$$V_1 = \frac{nRT_1}{P} = \frac{(5\,\text{mol})(0.082 \times 0.101\,\ell\,\text{MPa mol}^{-1}\,\text{K}^{-1})(300\,\text{K})}{2.02\,\text{MPa}}$$

$$= \frac{5 \times 8.2 \times 3}{20}\,\ell = 6.15\,\ell.$$

Now, knowing V_1, we can easily find the value of V_2 from $(0.101\,\text{MPa})\,(V_2 - V_1) = -20.63\,\ell\,\text{MPa}$. Thus, $V_2 - 6.15 = 20.63\,\ell$ or $V_2 = 26.78\,\ell$. Hence, the amount of work done is 20.63 ℓ MPa and the final volume is 26.78 ℓ.

5.5.4 Adiabatic reversible change

For adiabatic change we have, from Equation 5.40 for ideal gases,

$$nC_V\,dT = -P\,dV = \frac{-nRT}{V}\,dV \tag{5.41}$$

since $PV = nRT$ from the ideal gas law. Rearranging Equation 5.41 gives

$$\frac{dT}{T} = -\frac{R}{C_V} \frac{dV}{V}.$$

Assuming that C_V and R are constants, and integrating over 1 and 2, we have

$$\ln\left(\frac{T_2}{T_1}\right) = \frac{-R}{C_V} \ln\left(\frac{V_2}{V_1}\right) = \frac{R}{C_V} \ln\left(\frac{V_1}{V_2}\right) = \ln\left[\left(\frac{V_1}{V_2}\right)^{R/C_V}\right]$$

or

$$\frac{T_2}{T_1} = \left(\frac{V_1}{V_2}\right)^{R/C_V}$$

or

$$T(V)^{R/C_V} = \text{constant}. \tag{5.42}$$

Equations 5.41 and 5.42 are applicable only in the case of ideal gases. For *van der Waals gases*, the counterpart of Equation 5.41 would be

$$nC_V \, dT = -\left(P + \frac{a}{V^2}\right) dV = -\frac{nRT}{V-b} \, dV \tag{5.43}$$

where a and b are van der Waals constants. Rearranging Equation 5.43 gives

$$\frac{dT}{T} + \frac{R}{C_V} \frac{dV}{V-b} = 0.$$

Integrating, we find that

$$\ln T + \frac{R}{C_V} \ln(V-b) = \ln(\text{constant})$$

or

$$\ln T + \ln[(V-b)^{R/C_V}] = \ln(\text{constant})$$

or

$$T(V-b)^{R/C_V} = \text{constant}. \tag{5.44}$$

Since the gas necessarily obeys its equation of state in any reversible process, the relations between T and P and between P and V can be found from the foregoing equations by eliminating V or T between them and the appropriate equation of state. Thus, for an ideal gas,

$$PV = nRT$$

or

$$V = \frac{nRT}{P}$$

or

$$dV = \frac{nR}{P} \, dT - \frac{nRT}{P^2} \, dP \text{ (by partial differentiation).}$$

Substituting this value of dV in Equation 5.41, we have

$$nC_V \, dT = -P \, dV = -nR \, dT + nRT \, \frac{dP}{P}$$

or

$$(C_V + R) \frac{dT}{T} = R \, \frac{dP}{P}.$$

Integrating and rearranging, we obtain

$$T^{(C_V+R)} P^{-R} = \text{constant}$$

or

$$TP^{-R/(C_V+R)} = \text{constant.} \tag{5.45}$$

Again, since for an ideal gas

$$\frac{T_1}{T_2} = \frac{P_1 V_1}{P_2 V_2}$$

Equation 5.45 may be written as

$$\frac{T_1}{T_2} = \left(\frac{P_2}{P_1} \right)^{-R/(C_V+R)}$$

or

$$\frac{P_1 V_1}{P_2 V_2} = \left(\frac{P_2}{P_1} \right)^{-R/(C_V+R)}$$

or

$$PV^{(C_V+R)/C_V} = \text{constant.} \tag{5.46}$$

We have already proved in Equation 5.15 that $C_P - C_V = R$ and in Equation 5.16 that $\gamma = C_P/C_V$. Hence, substituting these relations into Equations 5.42, 5.45 and 5.46 and rearranging, we have

$$TV^{\gamma-1} = \text{constant} \tag{5.47}$$

$$TP^{(1-\gamma)/\gamma} = \text{constant} \tag{5.48}$$

$$PV^{\gamma} = \text{constant.} \tag{5.49}$$

For a van der Waals gas, the equation of state is

$$\left(P + \frac{a}{V^2}\right)(V - b) = nRT$$

or

$$V - b = \frac{nRT}{P + a/V^2}$$

or

$$dV = \frac{nR\,dT}{P + a/V^2} - \frac{nRT\,d(P + a/V^2)}{(P + a/V^2)^2}.$$

Substituting this value of dV into Equation 5.43, i.e.

$$nC_V\,dT = -\left(P + \frac{a}{V^2}\right)dV$$

or

$$nC_V\,dt = -nR\,dT + \frac{nRT}{P + a/V^2}\,d\left(P + \frac{a}{V^2}\right)$$

or

$$(C_V + R)\frac{dT}{T} = R\frac{d(P + a/V^2)}{P + a/V^2}.$$

Integrating and rearranging give

$$T\left(P + \frac{a}{V^2}\right)^{-R/(C_V + R)} = \text{constant.} \tag{5.50}$$

Again, differentiating both sides of the van der Waals equation and rearranging, we have

$$dT = \frac{1}{nR}\left[\left(P + \frac{a}{V^2}\right)dV + (V - b)\,d\left(P + \frac{a}{V^2}\right)\right].$$

Substituting this value of dT into Equation 5.43 and rearranging, we find that

$$\frac{-R}{C_V}\left(P + \frac{a}{V^2}\right) dV = \left(P + \frac{a}{V^2}\right) dV + (V - b)\, d\left(P + \frac{a}{V^2}\right)$$

or

$$\left(\frac{R + C_V}{C_V}\right) \frac{dV}{V - b} = -\frac{d(P + a/V^2)}{P + a/V^2}.$$

Integrating and rearranging, we obtain

$$(P + a/V^2)(V - b)^{(C_V+R)/C_V} = \text{constant}. \tag{5.51}$$

As in the case of ideal gases, using the relations $C_P - C_V = R$ and $C_P/C_V = \gamma$, Equations 5.44, 5.50 and 5.51 can be written as

$$T(V - b)^{\gamma-1} = \text{constant} \tag{5.52}$$

$$T\left(P + \frac{a}{V^2}\right)^{(1-\gamma)/\gamma} = \text{constant} \tag{5.53}$$

$$\left(P + \frac{a}{V^2}\right)(V - b)^\gamma = \text{constant}. \tag{5.54}$$

The work done in an adiabatic expansion of an ideal gas is given by

$$dW = -\int_{V_1}^{V_2} P\, dV.$$

Replacing P from Equation 5.49, this becomes

$$dW = -\text{constant} \int_{V_1}^{V_2} V^{-\gamma}\, dV$$

or

$$W = -\frac{1}{1 - \gamma}\left[CV^{(1-\gamma)}\right]_{V_1}^{V_2} \tag{5.55}$$

where C is a constant. From Equation 5.49 it is obvious that $P_1V_1^\gamma = P_2V_2^\gamma = C$. Hence, in putting the upper limit in Equation 5.55, we take $C = P_2V_2^\gamma$ while for the lower limit we take $C = P_1V_1^\gamma$. Then we have

$$W = -\frac{1}{1 - \gamma}(P_2V_2 - P_1V_1). \tag{5.56}$$

The minus sign simply implies that the work is done by the system.

We know that in an adiabatic process the work is done wholly at the expense of the internal energy of the system, since no heat flows into or out of a system. Hence, for an ideal gas

$$W = \Delta U = U_1 - U_2 \tag{5.57}$$

but $U_1 - U_2 = C_V(T_1 - T_2)$. Then Equation 5.57 becomes

$$W = C_V(T_1 - T_2) \tag{5.58}$$

if C_V remains constant over the temperature range. In such a case, the corresponding work done by a van der Waals gas is given by

$$W = C_V(T_1 - T_2) - a\left(\frac{1}{V_1} - \frac{1}{V_2}\right). \tag{5.59}$$

Example 5.3

0.5 mol of a monatomic gas fills a $1\,\ell$ container to a pressure of 2.02 MPa. It is allowed to expand reversibly and adiabatically until a pressure of 1.01 MPa is reached. Calculate the final volume and temperature. What would be the work done in this expansion, given that, for a monatomic gas, γ is 1.66?

Solution

From the problem we have initial gas pressure $P_1 = 2.02\,\text{MPa}$, initial gas volume $V_1 = 1\,\ell$, final gas volume $V_2 = ?$, final pressure $P_2 = 1.01\,\text{MPa}$ and $\gamma = 1.66$. Substituting these values into Equation 5.49, i.e. $P_1 V_1^\gamma = P_2 V_2^\gamma$, we obtain

$$(2.02\,\text{MPa})(1.0\,\ell)^{1.66} = (1.01\,\text{MPa})V_2^{1.66}$$

or

$$V_2 = 1.52\,\ell$$

Then the final temperature T_2 will be given by

$$T_2 = \frac{P_2 V_2}{nR} = \frac{(1.01\,\text{MPa})(1.52\,\ell)}{(0.5\,\text{mol})(0.082 \times 0.101\,\ell\,\text{MPa}\,\text{mol}^{-1}\text{K}^{-1})} = 370.73\,\text{K}.$$

Note: Owing to the decrease in temperature of a gas in an adiabatic expansion the pressure falls off more sharply than in the corresponding isothermal expansion. As a result, an isothermal expansion yields a greater final volume and a greater work output.

We may now summarize the thermodynamic changes in isothermal and adiabatic expansions of ideal gases. This is given in Table 5.2.

Table 5.2 Summary of the thermodynamic changes in isothermal and adiabatic expansion of ideal gases

Expansion	Reversible	Free (irreversible)	Actual
Isothermal	$T_1 = T_2$ $W = -nRT \ln\left(\dfrac{V_2}{V_1}\right)$ $\Delta U = 0$ $\Delta H = 0$ $Q = nRT \ln\left(\dfrac{V_2}{V_1}\right)$	$T_1 = T_2$ $W = 0$ $\Delta U = 0$ $\Delta H = 0$ $Q = 0$	$T_1 = T_2$ $0 < W < -nRT \ln\left(\dfrac{V_2}{V_1}\right)$ $\Delta U = 0$ $\Delta H = 0$ $0 < Q < nRT \ln\left(\dfrac{V_2}{V_1}\right)$ $Q = -W$
Adiabatic	$T_2 < T_1$ $W = C_V(T_2 - T_1) > 0$ $\Delta U = C_V(T_2 - T_1) < 0$ $\Delta H = C_P(T_2 - T_1) < 0$ $Q = 0$	$T_2 = T_1$ $W = C_V(T_2 - T_1) = 0$ $\Delta U = 0$ $\Delta H = 0$ $Q = 0$	$T_2 < T_1$ $W = C_V(T_2 - T_1) > 0$ $\Delta U = C_V(T_2 - T_1) < 0$ $\Delta H = C_P(T_2 - T_1) < 0$ $Q = 0$

5.6 ENTHALPY

We have seen that no mechanical work is done during a process carried out at constant volume, since V is constant and $dV = 0$; therefore $W = -P\,dV = 0$. Then from the first law it follows that the increase in energy equals the heat absorbed at constant volume, i.e.

$$\Delta U = Q_V. \tag{5.60}$$

If the pressure is held constant and no other work is done except $P\,\Delta V$ work, we have

$$\Delta U = U_2 - U_1 = Q + W = Q_P - P(V_2 - V_1)$$

or

$$(U_2 + PV_2) - (U_1 + PV_1) = Q_P \tag{5.61}$$

where Q_P is the heat absorbed at constant pressure. However, according to the definition of enthalpy H, $U + PV = H$. Hence, Equation 5.61 becomes

$$H_2 - H_1 = \Delta H = Q_P \tag{5.62}$$

This shows that the heat absorbed in any reversible isobaric process is equal to the difference between the enthalpies of the system in the end states of the process.

Enthalpy, like energy U or temperature T, is a function of state of the system alone and is independent of the path through which that state is reached. This follows from the equation $H = U + PV$, because U, P and V are all state functions. The specific enthalpy, per mole or per unit mass, is given by

$$h = u + Pv \tag{5.63}$$

where u is the specific internal energy and v is the specific volume. We have

$$H = U + PV$$

or

$$dH = dU + P\,dV + V\,dP.$$

Since $dW = -P\,dV$, we have, from the first law, $dU = dQ + dW = dQ - P\,dV$ and therefore

$$dU + P\,dV = dQ.$$

Hence,

$$dH = dQ + V\,dP. \tag{5.64}$$

In a process at constant pressure,

$$(dQ)_P = C_P(dT)_P$$

and $dP = 0$. Therefore, Equation 5.64 becomes

$$(dH)_P = C_P(dT)_P. \tag{5.65}$$

For an infinitesimal change,

$$\left(\frac{\partial H}{\partial T}\right)_P = C_P. \tag{5.66}$$

For a finite process at constant pressure,

$$(H_2 - H_1)_P = (\Delta H)_P = (U_2 - U_1)_P + P(V_2 - V_1). \tag{5.67}$$

Enthalpy plays the same role in isobaric processes as the internal energy does in isometric processes, i.e.

$$
\begin{aligned}
C_P &= \left(\frac{\partial H}{\partial T}\right)_P & Q_P &= H_2 - H_1 \\
C_V &= \left(\frac{\partial U}{\partial T}\right)_P & Q_V &= U_2 - U_1
\end{aligned}
\tag{5.68}
$$

Again from the definition $H = U + PV$, for a finite change we may write

$$H_2 - H_1 = \Delta H = \Delta U + \Delta(PV). \tag{5.69}$$

If we are dealing with ideal gases, then $PV = nRT$, or $\Delta(PV) \simeq \Delta(nRT)$. Hence, replacing $\Delta(PV)$ by $\Delta(nRT)$ in Equation 5.69, we have

$$\Delta H = \Delta U + \Delta(nRT). \tag{5.70}$$

5.6.1   Enthalpy change for van der Waals gases

Treatment for calculating work, heat, energy and enthalpy of van der Waals gases is exactly the same as that for ideal gases but we make use of the van der Waals equation of state

$$\left(P + \frac{a}{V^2}\right)(V - b) = nRT$$

and therefore P and V in the case of ideal gases are replaced by $P + a/V^2$ and $V - b$, respectively, in the case of van der Waals gases. Hence, work done by 1 mol of a van der Waals gas can be obtained by replacing P with $[RT/(V-b) - a/V^2]$ in the integration

$$W = -\int_{V_1}^{V_2} P\, dV$$

and then integrating in the normal way. Therefore,

$$W = -\int_{V_1}^{V_2} \left(\frac{RT}{V - b} - \frac{a}{V^2}\right)\, dV$$

or

$$W = -\left[RT\, \ln\left(\frac{V_2 - b}{V_1 - b}\right) + \frac{a}{V_2} - \frac{a}{V_1}\right]. \tag{5.71}$$

Without using the second law of thermodynamics it is difficult to express in a simple form the change in energy in an isothermal expansion. To avoid this difficulty we shall here make use of a deduction (see Equation 6.29), i.e.

$$P + \left(\frac{\partial U}{\partial V}\right)_T = T\left(\frac{\partial P}{\partial T}\right)_V.$$

For van der Waals gases, this equation becomes

$$\left(\frac{\partial U}{\partial V}\right)_T = \frac{a}{V^2}. \tag{5.72}$$

Therefore,

$$\Delta U = \int_{V_1}^{V_2} \frac{a}{V^2}\, dV = -\frac{a}{V_2} + \frac{a}{V_1}. \tag{5.73}$$

Then, from the first law and with the help of Equations 5.71 and 5.73,

$$Q = \Delta U - W = RT\, \ln\left(\frac{V_2 - b}{V_1 - b}\right). \tag{5.74}$$

To calculate the change ΔH in enthalpy, we integrate each of the terms in $dH = dU + d(PV)$. The equation of state of a van der Waals gas may be written as (for $n = 1$)

$$P = \frac{RT}{V - b} - \frac{a}{V^2}$$

or

$$PV = \frac{RTV}{V - b} - \frac{a}{V}.$$

Therefore,

$$
\begin{aligned}
\Delta(PV) &= RT\left(\frac{V_2}{V_2 - b} - \frac{V_1}{V_1 - b}\right) - \frac{a}{V_2} + \frac{a}{V_1} \\
&= bRT\frac{V_1 - V_2}{(V_2 - b)(V_1 - b)} - \frac{a}{V_2} + \frac{a}{V_1} \\
&= bRT\left(\frac{1}{V_2 - b} - \frac{1}{V_1 - b}\right) - \frac{a}{V_2} + \frac{a}{V_1}.
\end{aligned}
\tag{5.75}
$$

Now, combining Equations 5.69, 5.73 and 5.75 we have

$$\Delta H = bRT\left(\frac{1}{V_2 - b} - \frac{1}{V_1 - b}\right) - \frac{2a}{V_2} + \frac{2a}{V_1}. \tag{5.76}$$

It is interesting to see how the energy of a gas depends on its volume, or how the enthalpy of a gas depends on pressure. For this purpose, Joule and Gay–Lussac carried out a series of experiments and later Joule and Thomson carried out another set of experiments. These two experiments are now known as the Joule experiment and the Joule–Thomson experiment.

5.6.2 The Joule experiment

The experimental procedure is to allow the system to come to thermal equilibrium, to open a valve, to re-establish thermal equilibrium and then to observe the final temperature. The experiment is carried out as follows.

Two containers, one containing a gas at a specified pressure and temperature and the other evacuated, joined by a tube provided with a stopcock are immersed in a calorimeter bath (Figure 5.7). The stopcock is opened, the gas is allowed to expand freely from one container into the other and then the net temperature change attending the process is measured after reaching thermal equilibrium. It is necessary to stir the liquid in the calorimeter bath in order to attain uniformity in temperature for the purpose of avoiding localized heating and cooling.

It is observed that, when ordinary gases at moderate pressures are subjected to the Joule experiment, the net temperature change is very very small so that we can assume that the temperature change is actually zero. Then we can say the net quantity of heat Q entering or leaving the system during the experiment is zero, i.e. $Q = 0$. From the experimental set-up it

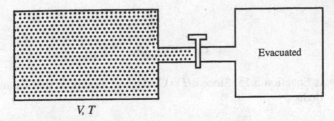

Figure 5.7 Two containers, one containing a gas at a specified pressure and temperature and the other evacuated, connected by a tube fitted with a stopcock

is apparent that no weights have been lifted nor has any piston swept out volume changes. So, we can say that no work is done during the process, i.e. $W = 0$. In a free expansion, no work is done by the gas on its surroundings. However, it is true that, while the expansion is taking place, the gas remaining in the left container is doing work on the gas that has already flowed into the container on the right but this work is done by one part of the gas on the other part; this is not work that is done by the system as a whole on its surroundings. As both Q and W are equal to zero, we conclude that any change ΔU in internal energy is also zero. *This implies that the energy of a gas at a given temperature is independent of its volume.*

In the above discussion we have assumed that there is no heat effect. Very accurate experiments of this type show that actually there is a small heat effect involved. This heat effect (per mole of gas) decreases as the initial pressure is reduced. This implies that, in the limiting state when the pressure is zero, the energy of the gas will no longer depend on its volume. *Therefore, we conclude that the energy of an ideal gas is a function of temperature only.* (It should be noted that, even in the limiting state of zero pressure, a gas should behave ideally.)

If the energy U is a function of temperature only, i.e. $U = f(T)$, then C_V is also a function of T only. Hence, from Equation 5.68 we can write

$$dU = C_V \, dT = f'(T) \, dT. \tag{5.77}$$

If C_V is not constant over the temperature change involved, then C_V may be replaced by an empirical relation in terms of temperature. If C_V is constant over the temperature change, we obtain, on integrating Equation 5.77,

$$U = C_V T + \text{constant} \tag{5.78}$$

For 1 mol of an ideal gas, $H = U + PV = U + RT$. As the right-hand side of this equation is necessarily a function of temperature, we conclude that enthalpy is also a function of temperature. So, we can write

$$dH = dU + d(RT) = C_P \, dT$$
$$= dU + R \, dT = C_P \, dT$$
$$= C_V \, dT + R \, dT = C_P \, dT \text{ (using Equation 5.77)}$$

or

$$C_P = C_V + R.$$

This is the same as Equation 5.15. Since $dH = C_P\, dT$, and assuming that C_V and therefore C_P are constant, we deduce

$$H = C_P\, dT + \text{constant.} \tag{5.79}$$

This is identical with Equation 5.78.

5.6.3   The porous plug or the Joule–Thomson experiment

As the Joule experiment was not capable of measuring precisely the extremely small changes in temperature in a free expansion, Joule and Thomson (later Lord Kelvin) devised another better experiment during 1850–60. They devised their experiment in such a way that the temperature change due to expansion of a gas would not be masked by the comparatively large heat capacity of the surroundings. The results of this experiment provide information about intermolecular forces. This information can be used to reduce gas thermometer temperatures to the Kelvin scale, and also the temperature drop produced in the experiment can be used in the liquefaction of gases such as hydrogen and helium. The schematic representation of the experimental set-up is shown in Figure 5.8.

A cylindrical tube, insulated to prevent any transfer of heat to the surroundings, is fitted with two pistons and a porous plug which is capable of allowing gas to flow slowly through it. The tube A is initially filled with a certain amount of a gas at temperature T_1, volume V_1 and pressure P_1; tube B is empty. The gas is then allowed to flow slowly through the plug in such a way that its pressure in A is kept constant at P_1 by the movement of the piston towards the plug. At the same time the piston in B is adjusted in such a way that the low pressure $P_2\,(< P_1)$ is kept constant. Let the final volume in B, after all the gas has streamed through the porous plug, be V_2 and its temperature T_2. The significant datum obtained in this experiment is the change in temperature due to flow of the gas through the porous plug. This can be obtained by measuring temperatures T_1 and T_2.

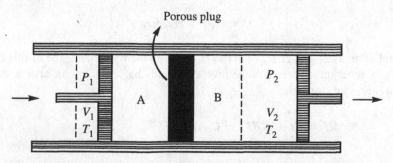

Figure 5.8   Schematic arrangement of the porous plug experiment

Since the whole system is insulated in such a way that there is no heat transfer to the surroundings, we have $Q = 0$. Therefore, the change in internal energy is equal to the work done by the system. The total work done is

$$W = P_1 V_1 - P_2 V_2$$

Then,

$$\Delta U = U_2 - U_1 = P_1 V_1 - P_2 V_2$$

or

$$(U_2 + P_2 V_2) - (U_1 + P_1 V_1) = 0$$

or

$$H_1 - H_2 = 0 \ (\text{or } H_2 - H_1 = 0)$$

i.e.

$$\Delta H = 0.$$

This shows that the Joule–Thomson experiment is carried out under constant-enthalpy conditions. When the gas involved is perfect, H is a function of T only, and therefore $\Delta H = 0$ implies that $\Delta T = 0$, or $T_2 = T_1$. There is thus no temperature change for a perfect gas. For an imperfect gas it generally depends on whether T_2 is greater or smaller than T_1. If T_2 is smaller than T_1, the gas will be cooled and, if T_2 is greater than T_1, the gas will be heated. The crucial temperature is called the *Joule–Thomson inversion temperature*. Above this temperature there will be heating; below this temperature there will be cooling upon Joule–Thomson expansion.

The Joule–Thomson effect provides a sensitive method of studying gas imperfection by measuring the temperature changes produced by a pressure drop. It has great technical value, as the important methods of gas liquefaction are based on it. For most gases the Joule–Thomson inversion temperature lies above the room temperature, so that no pre-cooling is necessary before they are allowed to expand through a plug. Helium and hydrogen are exceptions; these are heated when Joule–Thomson expansion occurs at room temperature.

The *Joule–Thomson coefficient* is defined as the change in temperature per unit change in pressure when the enthalpy is constant. In terms of partial derivatives

$$\mu = \left(\frac{\partial T}{\partial P} \right)_H . \tag{5.80}$$

From this equation it is evident that, if the gas cools in the process of streaming through the plug, the Joule–Thomson coefficient μ is positive, because the pressure always decreases in the experiment. Conversely, a negative Joule–Thomson coefficient implies an increase in temperature.

If we consider H as a function of temperature and pressure, the total differential of H is given by

$$dH = \left(\frac{\partial H}{\partial T}\right)_P dT + \left(\frac{\partial H}{\partial P}\right)_T dP.$$

However, $(\partial H/\partial T)_P = C_P$. Therefore,

$$dH = C_P \, dT + \left(\frac{\partial H}{\partial P}\right)_T dP.$$

As in the Joule–Thomson experiment there is no change in enthalpy, i.e. $dH = 0$, we have

$$0 = C_P \, (dT)_H + \left(\frac{\partial H}{\partial P}\right)_T (dP)_H$$

or

$$0 = C_P \left(\frac{\partial T}{\partial P}\right)_H + \left(\frac{\partial H}{\partial P}\right)_T$$

or

$$\left(\frac{\partial H}{\partial P}\right)_T = -C_P \left(\frac{\partial T}{\partial P}\right)_H = -C_P \mu. \tag{5.81}$$

In the Joule–Thomson experiment, $dH = 0$. So, it follows from

$$dH = \left(\frac{\partial H}{\partial P}\right)_T dP + \left(\frac{\partial H}{\partial T}\right)_P dT$$

that

$$\left(\frac{\partial T}{\partial P}\right)_H = -\left(\frac{\partial H}{\partial P}\right)_T \Big/ \left(\frac{\partial H}{\partial T}\right)_P. \tag{5.82}$$

It can be shown that $(\partial H/\partial P)_T = V - T(\partial V/\partial T)_P$ and we already know $(\partial H/\partial T)_P = C_P$. Substituting these in Equation 5.82 gives

$$\left(\frac{\partial T}{\partial P}\right)_H = \frac{T(\partial V/\partial T)_P - V}{C_P}. \tag{5.83}$$

If C_P is assumed to be constant over a small temperature range, then Equation 5.83 can be written as

$$\Delta T = \frac{T(\partial V/\partial T)_P - V}{C_P} \Delta P. \tag{5.84}$$

This is the equation for the differential Joule–Thomson effect, ΔT being the increase in temperature for a change ΔP in pressure in the proximity of the temperature T. In their experiment, Joule and Thomson found that the decrease in temperature was proportional to the difference in pressure on the two sides of the porous plug. This is in agreement with Equation 5.84.

From our discussion we now conclude that the Joule–Thomson coefficient should be zero for an ideal gas. Most gases at room temperature have a positive Joule–Thomson coefficient. As the temperature increases, μ decreases to zero and then changes sign. This change in sign occurs at the Joule–Thomson inversion temperature, i.e. at this temperature $\mu = 0$; below this temperature, μ is negative. Below certain limiting pressures, there are two inversion temperatures characteristic of each gas and dependent on pressure. Between the upper and lower inversion temperatures, μ is negative; below the lower and above the upper inversion temperatures, μ is positive; at the inversion temperatures, μ is zero.

5.7 THE JOULE–THOMSON COEFFICIENT FOR VAN DER WAALS GASES

We have discussed how the energy of a perfect gas depends on its volume, and also how the enthalpy of such gases depends on pressure. Let us now consider van der Waals gases. For 1 mol of gas, the van der Waals equation of state becomes

$$\left(P + \frac{a}{V^2}\right)(V - b) = RT$$

or

$$P = \frac{RT}{V - b} - \frac{a}{V^2}. \tag{5.85}$$

Expanding $(V-b)^{-1}$ and neglecting the higher-order terms, Equation 5.85 can be written

$$P = \frac{RT}{V} + \frac{bRT}{V^2} - \frac{a}{V^2}. \tag{5.86}$$

We have already deduced that, if C_V is constant for a temperature change, then

$$\left(\frac{\partial U}{\partial T}\right)_V = C_V. \tag{5.87}$$

It can be shown that (see later)

$$\left(\frac{\partial U}{\partial V}\right)_T = -P + T\left(\frac{\partial P}{\partial T}\right)_V. \tag{5.88}$$

Differentiating Equation 5.86 at constant volume and neglecting higher-order terms, we have

$$\left(\frac{\partial P}{\partial T}\right)_V = \frac{R}{V} + \frac{Rb}{V^2}. \tag{5.89}$$

Combining Equations 5.86, 5.88 and 5.89 gives

$$\left(\frac{\partial U}{\partial V}\right)_T = \frac{a}{V^2}. \tag{5.90}$$

It is interesting to note that, for an ideal gas, $(\partial U/\partial V)_T = 0$. We can write Equation 5.90 as

$$(dU)_T = \frac{a}{V^2}(dV)_T. \tag{5.91}$$

Since a is very small, we can assume that the contributions to the internal energy given by Equations 5.87 and 5.91 are additive even for a finite change in volume and temperature. Hence,

$$U = \int C_V \, dT + \int \frac{a}{V^2} \, dV = U_0 + C_V T - \frac{a}{V} \tag{5.92}$$

where U_0 is a constant of integration.

It is interesting to note that, for a van der Waals gas, U is not a function of temperature only, whereas for an ideal gas, U is a function only of temperature. (We shall prove this later.) U_0 is the specific internal energy at temperature T_0. Again, the van der Waals equation for 1 mol of gas may be written as

$$\left(P + \frac{a}{V^2}\right)(V - b) = RT$$

or

$$PV = RT - \frac{a}{V} + bP + \frac{ab}{V^2}.$$

Neglecting the second-order term ab/V^2, we have

$$PV = RT - \frac{a}{V} + bP. \tag{5.93}$$

From the definition of enthalpy we have $H = U + PV$. Now, substituting the values of U and PV from Equations 5.92 and 5.93, we obtain

$$H = \left(U_0 + C_V T - \frac{a}{V} \right) + \left(RT - \frac{a}{V} + bP \right).$$

If we assume, very approximately, that $PV = RT$, then $V = RT/P$, and the above equation of enthalpy may be expressed as a function of T and P as

$$H = U_0 + (C_V + R)T + P\left(b - \frac{2a}{RT} \right). \tag{5.94}$$

Let us now consider the temperature change ∂T when 1 mol of a van der Waals gas is subjected to a Joule–Thomson expansion for a pressure change ∂P. Referring to Figure 5.8, if we assume that the temperature and pressure in chamber A are now $P_1 = P + \partial P$ and $T_1 = T + \partial T$, respectively, while in chamber B we have $P_2 = P$ and $T_2 = T$, we shall still have a Joule–Thomson expansion at constant enthalpy. So, we may write

$$H(P + dP, \ T + dT) = H(P, T). \tag{5.95}$$

The left-hand side of this equation can be expanded by Taylor expansion as

$$H(P + dP, \ T + dT) \simeq H(P, T) + \left(\frac{\partial H}{\partial P} \right)_T \partial P + \left(\frac{\partial H}{\partial T} \right)_P \partial T.$$

Combining this equation with Equation 5.95 gives

$$\left(\frac{\partial H}{\partial P} \right)_T \partial P + \left(\frac{\partial H}{\partial T} \right)_P \partial T \simeq 0$$

or

$$\frac{\partial T}{\partial P} \simeq -\left(\frac{\partial H}{\partial P} \right)_T \Big/ \left(\frac{\partial H}{\partial T} \right)_P. \tag{5.96}$$

Note: It is assumed that $\partial P/P$ and $\partial T/T$ are small, very much smaller than 1.

However, $(\partial H/\partial T)_P = C_P$. Differentiating both sides of Equation 5.94 with respect to P, we obtain

$$\left(\frac{\partial H}{\partial P} \right)_T = b - \frac{2a}{RT}.$$

Hence, we have

$$\frac{\partial T}{\partial P} = -\frac{b - 2a/RT}{C_P}.$$ (5.97)

Therefore, for a Joule–Thomson expansion of a van der Waals gas,

$$b - \frac{2a}{RT} = \left(\frac{\partial H}{\partial P}\right)_T = -C_P \left(\frac{\partial T}{\partial P}\right)_H.$$ (5.98)

Since, at the inversion temperature T_i,

$$\mu = \frac{\partial T}{\partial P} = -\frac{b - 2a/RT_i}{C_P} = 0$$

or

$$T_i = \frac{2a}{Rb}.$$ (5.99)

At this temperature the Joule–Thomson effect is zero; no change in temperature occurs on expansion. Below T_i there is cooling and above it heating for Joule–Thomson expansion.

Since ∂P is always negative for an expansion, from Equation 5.97 it is obvious that ∂T is positive or negative depending on whether $2a/RT$ is smaller or greater than b. Hence, we can say that, for $2a/RT < b$, there is Joule–Thomson heating and, for $2a/RT > b$, there is cooling on isenthalpic expansion.

Evidently the Joule–Thomson effect depends on both a and b, even though it may depend only on a as it is the cohesion force a/V^2 against which work is to be done.

To find the difference between C_P and C_V, we partially differentiate both sides of Equation 5.94 with respect to T. Then,

$$\left(\frac{\partial H}{\partial T}\right)_P = (C_V + R) + \frac{2aP}{RT^2}.$$

However, we have already proved that $(\partial H/\partial T)_P = C_P$. Therefore,

$$C_P = C_V + R + \frac{2aP}{RT^2}$$

or

$$C_P - C_V = R + \frac{2aP}{RT^2}.$$ (5.100)

This shows that in the case of a van der Waals gas the difference between C_P and C_V is greater than that in the case of an ideal gas.

EXERCISES 5

5.1 An ideal monatomic gas is allowed to expand slowly until its pressure decreases to exactly half its original value. Calculate the extent of volume change if: (a) the expansion is carried out adiabatically, and (b) the process is isothermal. For this gas, $\gamma = \frac{5}{3}$.

5.2 A model of an oxygen (O_2) gas molecule is shown below.

(a) What type of motion can we expect this molecule to have?

(b) What would be the internal energy U of this molecule in terms of its kinetic and potential energies?

(c) If this molecule behaves ideally, calculate the values of C_P and γ, assuming that $C_V = \frac{5}{2}R$.

(d) Calculate the increase in internal energy and enthalpy of 1 mol of O_2 when its temperature is raised from 27 to 227°C. What assumptions, if any, do you make in your calculations?

(e) Calculate the amount of heat required to raise the temperature of 1 mol of O_2 from 27 to 227°C (i) at constant pressure and (ii) at constant volume.

The heat capacity of O_2 at constant pressure is given by $C_P = (25.6 + 1.4 \times 10^{-3}T)\,\text{J}\,\text{K}^{-1}\,\text{mol}^{-1}$.

5.3 (a) Show that, at constant volume over a temperature range in which the heat capacity is constant, the change in internal energy of 1 mol of an ideal gas is given by $\Delta U = C_V \Delta T$.

(b) Show that, at constant pressure over a temperature range in which the heat capacity is constant, the change in enthalpy of 1 mol of an ideal gas is given by $\Delta H = C_P \Delta T$.

5.4 Deduce the mathematical relations $C_P - C_V = R$ and $C_P/C_V = \gamma$ where R is the universal gas constant and γ is a constant. How would you interpret γ? Show that for a monatomic ideal gas the value of γ is 1.67.

5.5 (a) It is observed that monatomic gases have higher γ-values than those of polyatomic gases. Explain why there is such a difference.

(b) 'Except for work of great accuracy it is sufficient to take $C_P = C_V$ for liquids and solids.' How would you justify this?

5.6 Show that the radial flow of heat between two coaxial cylinders of radii r_1 and r_2 and of length l is given by $Q = 2\pi l K(T_1 - T_2)/\ln(r_2/r_1)$ where T_1 and T_2 are the temperatures of the cylinders, and K is the thermal conductivity.

5.7 Deduce a general equation for radial flow of heat between an inner sphere of radius r_1 held at constant temperature T_1 and an outer sphere of radius r_2 held at constant temperature T_2.

5.8 A gas obeys the equation of state given by

$$\left(P + \frac{a}{V^2}\right)(V - b) = nRT$$

where P, V and T are the pressure, volume and temperature, respectively, R is the universal gas constant, n is the number of moles of the gas, and a and b are the van der Waals constants. Calculate W, ΔU, Q and ΔH for 1 mol of this gas for an isothermal reversible expansion from a volume V_1 to another volume V_2.

5.9 A gas obeys the equation of state $PV = nRT + nB(T)P$, where $B(T)$ is a constant which depends on temperature. Find W, ΔU, Q and ΔH.

5.10 The second virial coefficient B of an inert gas varies approximately linearly with temperature and has values $-21.0\,\text{m}\ell\,\text{mol}^{-1}$ at 0°C and $-11.0\,\text{m}\ell\,\text{mol}^{-1}$ at 50°C. Calculate ΔU, Q, W and ΔH for a reversible isothermal expansion of 1 mol of this gas, from 1.01 to 0.101 MPa at 25°C (a) assuming that the gas behaves ideally, and (b) using the equation of state $P[V - B(25°C)] = RT$.

5.11 Show that for a reversible isothermal process the work done by an ideal gas may be expressed as $P_2 = P_1 \exp(W/nRT)$ where P_1 and P_2 are the initial and final pressures, T is the temperature and n is the number of moles of the gas.

5.12 A certain gas obeys the equation of state $PV = RT + bP$.

(a) Derive expressions for heat and work in a reversible isothermal expansion of the gas.

(b) Derive expressions for heat and work in a reversible adiabatic expansion of the gas.

(c) Calculate ΔU, Q and W when 1 mol of the gas expands reversibly at 300 K from 0.220 to 20.02 ℓ (i) isothermally and (ii) adiabatically. Assume that $b = 0.020\,\ell\,\text{mol}^{-1}$, $C_V = \frac{3}{2}R$ and $\gamma = \frac{5}{3}$.

5.13 (a) 1 mol of an ideal gas is allowed to expand against a piston which supports 0.041 MPa pressure. The initial pressure is 1.013 MPa and the final pressure is 0.041 MPa, the temperature being kept constant at 0°C. Calculate W, Q, ΔU and ΔH.

(b) If the expansion is carried out in such a way that the confining pressure at all times is less than the gas pressure, find the values of W, Q, ΔH and ΔU.

(c) Plot P versus V using the system variables given above and identify by shading the areas corresponding to the work obtained above.

5.14 (a) 2 moles of an ideal monatomic gas are allowed to expand reversibly and adiabatically from 10 ℓ and 0°C to a final volume of 100 ℓ. Calculate (i) the pressure and the temperature of the gas after the expansion and (ii) Q, W, ΔU and ΔH for the expansion.

(b) The same amount of this gas is allowed to expand by suddenly releasing the pressure to 0.101 MPa and then allowing it to expand against this constant pressure. Calculate the temperature and the pressure of the gas after the expansion and W, Q, ΔU and ΔH for the expansion.

5.15 The rate of heat flow through a slab of cork of 0.152 m diameter and thickness 0.0016 m is 2.877 kJ min^{-1}. If the surface temperatures of the slab are 100 and 20°C, calculate the value of the thermal conductivity of the cork.

5.16 A wall 10 cm thick made of wood is to be insulated with a layer of asbestos. Calculate the thickness of asbestos that will be required to reduce the heat loss to 20% of the loss through the wall alone. Assume that in both cases the inside surface is at 30°C and the outside at 0°C. The thermal conductivities of wood and asbestos are $1.464 \times 10^{-3} \, J \, s^{-1} \, cm^{-1} \, {}^\circ C^{-1}$ and $3.90 \times 10^{-4} \, J \, s^{-1} \, cm^{-1} \, {}^\circ C^{-1}$, respectively.

5.17 The thermal conductivity of the material of a wall having a uniform thickness d is a linear function of temperature and is given by $K = a + bT$, where a and b are constants. Derive an equation for the rate of heat flow through the wall when the surface temperatures are constant at T_1 and T_2.

5.18 A sphere with a diameter of 60 cm is covered with some insulating material so that the external diameter of the composite sphere is 75 cm. Calculate the rate of heat leakage through the insulator when the temperatures of the inner and outer surfaces of the insulation are 300°C and 100°C, respectively, given that the thermal conductivity of the insulation is $1.22 \times 10^{-3} \, J \, s^{-1} \, cm^{-1} \, {}^\circ C^{-1}$.

5.19 Determine the rate of heat transfer per centimetre length between two cylindrical bodies, the radii of the inside and outside surfaces being 5 cm and 30 cm, respectively. The inner and outer surface temperatures are 200°C and 150°C, respectively. Assume that both the bodies are made of the same material having a thermal conductivity $1.5 \times 10^{-3} \, J \, s^{-1} \, cm^{-1} \, {}^\circ C^{-1}$.

5.20 Show that for an adiabatic reversible change in the case of ideal gases the following equations hold.

(i) $T(P)^{-R/(R+C_V)} = \text{constant}$

(ii) $P(V)^{(R+C_V)/C_V} = \text{constant}$

(iii) $T(V)^{R/C_V} = \text{constant}$

Hence, show that they can be written as follows.

(a) $T(P)^{(1-\gamma)/\gamma} = \text{constant}$

(b) $PV^{\gamma} = \text{constant}$

(c) $T(V)^{(\gamma-1)} = \text{constant}$

5.21 Show that for an adiabatic reversible change in the case of van der Waals gases the following equations apply.

(i) $T(V - b)^{R/C_V} = \text{constant}$

(ii) $T\left(P + \dfrac{a}{V^2}\right)^{-R/(C_V+R)} = \text{constant}$

(iii) $\left(P + \dfrac{a}{V^2}\right)(V - b)^{(C_V+R)/C_V} = \text{constant}$

Hence, show that they can be transformed to the following.

(a) $T(V - b)^{\gamma-1} = $ constant

(b) $T\left(P + \dfrac{a}{V^2}\right)^{(1-\gamma)/\gamma} = $ constant

(c) $\left(P + \dfrac{a}{V^2}\right)(V - b)^{\gamma} = $ constant

5.22 Show that the heat absorbed in any reversible isobaric process is equal to the difference between the enthalpies of the system in the end state of the process.

5.23 Describe a suitable experiment to establish that the energy of an ideal gas at a given temperature is independent of its volume but is dependent on temperature.

5.24 Discuss the significance of the Joule–Thomson coefficient. How would you distinguish an ideal gas from a non-ideal gas just by measuring the Joule–Thomson coefficient?

5.25 Show that for an ideal gas the Joule–Thomson coefficient is given by $(\partial H/\partial P)_T = -\mu C_P$.

5.26 Define the Joule–Thomson inversion temperature. Discuss the behaviour of the Joule–Thomson coefficient with reference to the inversion temperature.

5.27 Show that for a van der Waals gas the Joule–Thomson coefficient is given by $b - 2a/RT = -\mu C_P$, where a and b are the van der Waals constants. Hence, show that in the case of such a gas the inversion temperature T_i is given by $T_i = 2a/Rb$.

5.28 Show that the difference between C_P and C_V in the case of a van der Waals gas is given by $C_P - C_V = R + 2aP/RT^2$. How does this difference between C_P and C_V compare with that in the case of ideal gases?

5.29 Show that the work done in an adiabatic expansion of an ideal gas is given by

$$W = -\frac{1}{1 - \gamma}(P_2 V_2 - P_1 V_1) = C_V(T_1 - T_2)$$

where T_1, P_1, V_1 and T_2, P_2, V_2 are the temperature, pressure and volume at the initial and final states, respectively.

5.30 Prove that the change in enthalpy of a van der Waals gas in a reversible isobaric process is given by

$$\Delta H = bRT\left(\frac{1}{V_2 - b} - \frac{1}{V_1 - b}\right) - \frac{2a}{V_2} + \frac{2a}{V_1}$$

5.31 1 mol of an ideal monatomic gas is carried through the cycle shown in the figure below. It consists of steps A, B and C involving states 1, 2 and 3. Assume that the steps are all reversible, given that $C_P = 20.92\,\text{J}\,°\text{C}^{-1}\text{mol}^{-1}$ and $C_V = 12.55\,\text{J}\,°\text{C}^{-1}\text{mol}^{-1}$.

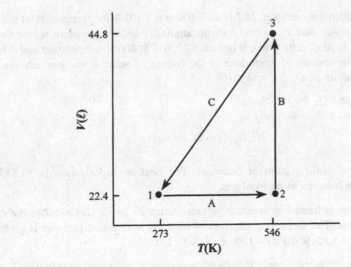

Fill in the following tables.

State	P (MPa)	V (l)	T (K)
1			
2			
3			

Step	Name of process	Work (J)	Q (J)	ΔU (J)
A				
B				
C				
Cycle				

5.32 18.02 g of liquid water is enclosed under a frictionless weightless piston at 100°C and 0.101 MPa pressure. The pressure above the piston is decreased slightly below 0.101 MPa and the water is allowed to vaporize isothermally until it is all vaporized. For the process to happen, a quantity of 40.710 kJ heat is required. The specific volume (i.e. volume per unit weight) of water and steam at 100°C and 0.101 MPa are 1.043 ml and 1677 ml, respectively.

(a) Calculate the work associated with the vaporization, and ΔU and ΔH for the process. If the volume of liquid water is neglected, what will be the amount of work done?

(b) Calculate Q, ΔU and ΔH for the process, if the piston is removed and the water is allowed to vaporize freely and isothermally into an evacuated space of such volume that the pressure reaches 0.101 MPa when all the water is vaporized.

5.33 1 mol of hydrogen gas occupies 24.79 ℓ at 300 K and 0.101 MPa pressure. Heat is added to this system in such a way that P(system) $= P$(surroundings) and the pressure remains constant during a change into another state where it occupies 33.10 ℓ at 400 K temperature and 0.101 MPa pressure. Calculate the amount of work done in the change of state. If the gas behaves ideally, what would be the values of Q, ΔU and ΔH?

5.34 Calculate Q_P and Q_V for the change

$$2C_6H_6(\ell) = 2C_6H_6(g)$$

at 80.1°C (the normal boiling point of benzene). The heat of vaporization is 30.8 kJ mol^{-1}. Assume that benzene behaves as an ideal gas.

5.35 14 g of nitrogen is heated at constant pressure from 25 to 75°C. Calculate the enthalpy change for this process. The molar heat capacity of nitrogen at constant pressure is given by the equation $C_P = 26.96 + 5.85 \times 10^{-3}T - 3.38 \times 10^{-7}T^2$ J °C^{-1}.

5.36 86.9 kJ heat is liberated when 1 g-atom of pure iron is dissolved in dilute HCl at 18°C. Calculate the change of energy of the system assuming that the liberated hydrogen behaves as an ideal gas.

5.37 In the vaporization of 10 g of cyclohexane (C_6H_{12}) at an external pressure of 0.053 MPa and at 334 K, 3.573 kJ were required. The vapour pressure of cyclohexane at this temperature is 0.053 MPa. Calculate the change in internal energy and enthalpy in the vaporization process.

5.38 The internal energy for hydrogen gas at 298.15 K and 0.1 MPa (101.3 kN m^{-2} = 1 atm) can be written as

$$U(\text{total}) = U(\text{translational}) + U(\text{rotational}) + U(\text{vibrational})$$
$$+ U(\text{electrons}) + U(\text{nuclear})$$

Which of the terms in the above sum are difficult to determine? What does this illustrate? Calculate the thermodynamic value of the internal energy of hydrogen gas under the given conditions.

5.39 Prove that for an ideal gas having constant heat capacities the gradient of a P–V graph for a reversible adiabatic process is negative. Also show that this slope has a larger absolute value than the slope of a P–V graph for an isotherm at the same values of pressure and volume.

5.40 Which of the substances Fe, $H_2O(\ell)$, Br$^-$, He, Cl(g), an oxygen atom and HCl have $U = H = 0$ at 298.15 K and 0.1 MPa?

6

Some Applications of the First Law

6.1 ENERGY EQUATION OF A SYSTEM

The energy equation of a system is the relation between the internal energy U and the state variables. This equation cannot be derived analytically from the equation of state of the system. In order to proceed systematically, we shall consider each pair of state variables (pressure, volume and temperature) in turn as independent and derive the relevant energy equation.

6.2 TEMPERATURE AND VOLUME INDEPENDENCE

Let us consider the internal energy of a system as a function of temperature T and volume V, i.e. $U = f(T, V)$. Then,

$$dU = \left(\frac{\partial U}{\partial T}\right)_V dT + \left(\frac{\partial U}{\partial V}\right)_T dV. \tag{6.1}$$

Let us consider those systems for which $dW = -P\,dV$. Substituting this and the value of dU from Equation 6.1 into Equation 4.2, we have

$$dQ = \left(\frac{\partial U}{\partial T}\right)_V dT + \left[P + \left(\frac{\partial U}{\partial V}\right)_T\right] dV. \tag{6.2}$$

This equation is a general equation and can be applied to any substance and to any reversible process.

(i) *For a process at constant volume*

$$dV = 0 \qquad (dQ)_V = C_V (dT)_V$$

Then, from Equation 6.2

$$C_V(dT)_V = \left(\frac{\partial U}{\partial T}\right)_V (dT)_V. \tag{6.3}$$

Therefore,

$$C_V = \left(\frac{\partial U}{\partial T}\right)_V. \tag{6.4}$$

This shows that measurement of C_V gives the rate of change in internal energy with temperature at constant volume, and $(\partial U/\partial T)_V$ may be replaced by C_V in any equation in which it occurs, even if the equation refers to a process in which the volume is not constant. Hence, with the help of Equations 6.2 and 6.4 for any reversible process, the first law can be written

$$dQ = C_V\, dT + \left[P + \left(\frac{\partial U}{\partial V}\right)_T\right] dV \tag{6.5}$$

(ii) *For a process at constant pressure*

$$(dQ)_P = C_P(dT)_P. \tag{6.6}$$

Then, Equation 6.5 becomes

$$C_P(dT)_P = C_V(dT)_P + \left[P + \left(\frac{\partial U}{\partial V}\right)_T\right](dV)_P. \tag{6.7}$$

Dividing both sides by $(dT)_P$; we have

$$C_P = C_V + \left[P + \left(\frac{\partial U}{\partial V}\right)_T\right]\left(\frac{\partial V}{\partial T}\right)_P. \tag{6.8}$$

(iii) *For a process at constant temperature*, $dT = 0$ and then Equation 6.5 becomes

$$(dQ)_T = \left[P + \left(\frac{\partial U}{\partial V}\right)_T (dV)_T\right] = P(dV)_T + \left(\frac{\partial U}{\partial V}\right)_T (dV)_T. \tag{6.9}$$

This equation implies that the heat supplied to a system in an isothermal reversible process equals the sum of the work done by the system and the increase in its internal energy.

Note: We can say that in an isothermal process a system behaves as if it had an infinite heat capacity, since any amount of heat can flow into or out of it without producing a change in temperature.

(iv) *For a reversible adiabatic process*, the system is thermally insulated and $dQ = 0$. Then, Equation 6.5 becomes

$$C_V \, dT = -\left[P + \left(\frac{\partial U}{\partial V}\right)_T\right] dV. \tag{6.10}$$

Dividing both sides by dV, we have

$$C_V \frac{\partial T}{\partial V} = -\left[P + \left(\frac{\partial U}{\partial V}\right)_T\right]. \tag{6.11}$$

6.3   TEMPERATURE AND PRESSURE INDEPENDENCE

Let us consider the internal energy of a system as a function of temperature and pressure, i.e. $U = f(T, P)$. In this case we have, for internal energy and volume,

$$dU = \left(\frac{\partial U}{\partial T}\right)_P dT + \left(\frac{\partial U}{\partial P}\right)_T dP \tag{6.12}$$

$$dV = \left(\frac{\partial V}{\partial T}\right)_P dT + \left(\frac{\partial V}{\partial P}\right)_T dP. \tag{6.13}$$

With the help of Equations 6.12 and 6.13 the first law can be written

$$dQ = \left[\left(\frac{\partial U}{\partial T}\right)_P + P\left(\frac{\partial V}{\partial T}\right)_P\right] dT + \left[\left(\frac{\partial U}{\partial P}\right)_T + P\left(\frac{\partial V}{\partial P}\right)_T\right] dP. \tag{6.14}$$

As before, we shall consider four cases.

(i) *For a process at constant pressure*

$$dP = 0 \qquad (dQ)_P = C_P(dT)_P$$

(from Equation 6.6). Putting these in Equation 6.14 gives

$$C_P = \left[\left(\frac{\partial U}{\partial T}\right)_P + P\left(\frac{\partial V}{\partial T}\right)_P\right] \tag{6.15}$$

Note: Although $C_V = (\partial U/\partial T)_V$, it is not true that $C_P = (\partial U/\partial T)_P$.

Putting Equation 6.15 into Equation 6.14, we have

$$dQ = C_P \, dT + \left[\left(\frac{\partial U}{\partial P}\right)_T + \left(\frac{\partial V}{\partial P}\right)_T\right] dP. \tag{6.16}$$

This equation is analogous to Equation 6.5.

(ii) *For a process at constant volume*

$$(dQ)_V = C_V(dT)_V \tag{6.17}$$

Putting Equations 6.15 and 6.17 in Equation 6.14, we have

$$C_V(dT)_V = C_P(dT)_V + \left[\left(\frac{\partial U}{\partial P}\right)_T + P\left(\frac{\partial V}{\partial P}\right)_T\right](dP)_V$$

or

$$C_V = C_P + \left(\frac{\partial U}{\partial P}\right)_T + P\left(\frac{\partial V}{\partial P}\right)_T\left(\frac{dP}{dT}\right)_V. \tag{6.18}$$

This equation is analogous to Equation 6.8.

(iii) *For a process at constant temperature*, $dT = 0$; then from Equation 6.14,

$$(dQ)_T = \left[\left(\frac{\partial U}{\partial P}\right)_T + P\left(\frac{\partial V}{\partial P}\right)_T\right](dP)_T \tag{6.19}$$

This equation implies that the heat supplied equals the sum of the increase in internal energy and the work done.

(iv) *For an adiabatic process.* $Q = 0$; then from Equation 6.16

$$C_P\, dT = -\left[\left(\frac{\partial U}{\partial P}\right)_T + P\left(\frac{\partial V}{\partial P}\right)_T\right]dP \tag{6.20}$$

$$C_P\frac{\partial T}{\partial P} = -\left[\left(\frac{\partial U}{\partial P}\right)_T + P\left(\frac{\partial V}{\partial P}\right)_T\right]. \tag{6.21}$$

Note: It is worthwhile mentioning that in the case of Equations 6.10, 6.11, 6.20 and 6.21, the entropy (to be discussed later) of the system remains constant throughout the process.

6.4 PRESSURE AND VOLUME INDEPENDENCE

As before, in such cases we shall consider the internal energy of a system as a function of pressure and volume, i.e. $U = f(P, V)$. Then,

$$dU = \left(\frac{\partial U}{\partial P}\right)_V dP + \left(\frac{\partial U}{\partial V}\right)_P dV. \tag{6.22}$$

Putting Equation 6.22 in Equation 4.2, we have

$$dQ = \left(\frac{\partial U}{\partial P}\right)_V dP + \left[P + \left(\frac{\partial U}{\partial V}\right)_P\right]dV. \tag{6.23}$$

This equation is again perfectly general and can be applied to any substance whatsoever and to any reversible process. As before, we shall consider four cases.

(i) *For a process at constant volume*

$$dV = 0 \qquad (dQ)_V = C_V(dT)_V.$$

Substituting these into Equation 6.23, we have

$$C_V(dT)_V = \left(\frac{\partial U}{\partial P}\right)_V dP.$$

Hence, Equation 6.23 becomes

$$dQ = C_V(dT)_V + \left[P + \left(\frac{\partial U}{\partial V}\right)_P\right]dV. \tag{6.24}$$

This equation is analogous to Equation 6.5.

(ii) *For a process at constant pressure*

$$dP = 0 \qquad Q_P = C_P(dT)_P.$$

Substituting these into Equation 6.23, we have

$$C_P(dT)_P = \left[P + \left(\frac{\partial U}{\partial V}\right)_P\right](dV)_P.$$

Hence, Equation 6.23 becomes

$$dQ = C_P(dT)_P + \left(\frac{\partial U}{\partial P}\right)_V dP. \tag{6.25}$$

(iii) *For a process at constant temperature*, $dT = 0$, and substituting this in Equation 6.24, we have

$$(dQ)_T = \left[P + \left(\frac{\partial U}{\partial V}\right)_P\right](dV)_T$$

or

$$(dQ)_T = P(dV)_T + \left(\frac{\partial U}{\partial V}\right)_P (dV)_T. \tag{6.26}$$

This equation is analogous to Equation 6.9.

(iv) *Finally, for an adiabatic process*, $Q = 0$ and, substituting this in Equation 6.24, we have

$$C_V(dT)_V = -\left[P + \left(\frac{\partial U}{\partial V}\right)_P\right]dV$$

or

$$C_V \left(\frac{\partial T}{\partial V} \right)_V = -\left[P + \left(\frac{\partial U}{\partial V} \right)_P \right].$$ (6.27)

This is analogous to Equation 6.11.

We now have expressed all six of the partial derivatives of U in terms of the experimentally measurable properties of substance, i.e. C_P, C_V and the state variables. We have also obtained three equations for the heat flowing into a system in an isothermal process, and three forms of differential equation of an adiabatic process. All the above equations are entirely general and not restricted to any particular substance, such as an ideal gas.

Example 6.1

Using the first law and related definitions, show that

(i)
$$C_V = \frac{-(\partial U/\partial V)_T}{(\partial T/\partial V)_U}$$

(ii)
$$\left(\frac{\partial U}{\partial T} \right)_P = C_P - P \left(\frac{\partial V}{\partial T} \right)_P$$

(iii)
$$C_P - C_V = \left[P + \left(\frac{\partial U}{\partial V} \right)_T \right] \left(\frac{\partial V}{\partial T} \right)_P.$$

Solution

(i) By definition, $C_V = (\partial U/\partial T)_V$. In addition, if U is a function of T and V,

$$dU = \left(\frac{\partial U}{\partial T} \right)_V dT + \left(\frac{\partial U}{\partial V} \right)_T dV.$$

If U is a constant, $dU = 0$ and then

$$\left(\frac{\partial U}{\partial T} \right)_V dT = -\left(\frac{\partial U}{\partial V} \right)_T dV$$

or

$$C_V \, dT = -\left(\frac{\partial U}{\partial V} \right)_T dV$$

or

$$C_V = \frac{-(\partial U/\partial V)_T}{(\partial T/\partial V)_U}$$

since

$$\left(\frac{\partial T}{\partial V}\right)_U = \frac{1}{(\partial V/\partial T)_U}.$$

(ii) By definition, $C_P = (\partial H/\partial T)_P$ and $H = U + PV$. So,

$$C_P = \frac{\partial}{\partial T}(U + PV)_P = \left(\frac{\partial U}{\partial T}\right)_P + P\left(\frac{\partial V}{\partial T}\right)_P$$

or

$$\left(\frac{\partial U}{\partial T}\right)_P = C_P - P\left(\frac{\partial V}{\partial T}\right)_P.$$

(iii) If U is a function of T and V, we have

$$dU = \left(\frac{\partial U}{\partial T}\right)_V dT + \left(\frac{\partial U}{\partial V}\right)_T dV.$$

On the assumption that P is a constant and dividing both sides of the above equation by dT,

$$\left(\frac{\partial U}{\partial T}\right)_P = \left(\frac{\partial U}{\partial T}\right)_V + \left(\frac{\partial U}{\partial V}\right)_T \left(\frac{\partial V}{\partial T}\right)_P.$$

Now, by definition, $H = U + PV$, $C_P = (\partial H/\partial T)_P$ and $C_V = (\partial U/\partial T)_V$. Then

$$C_P - C_V = \frac{\partial}{\partial T}(U + PV)_P - \left(\frac{\partial U}{\partial T}\right)_V$$

$$= \left(\frac{\partial U}{\partial T}\right)_P + P\left(\frac{\partial V}{\partial T}\right)_P - \left(\frac{\partial U}{\partial T}\right)_V.$$

Substituting for $(\partial U/\partial T)_V$ from above, we obtain

$$C_P - C_V = \left[P + \left(\frac{\partial U}{\partial V}\right)_T\right]\left(\frac{\partial V}{\partial T}\right)_P.$$

6.5   INTERNAL ENERGY OF GASES

We have seen that the equations derived in Section 6.2 all contain the expression $P + (\partial U/\partial V)_T$. Therefore, for any numerical calculations of the Equations 6.5, 6.7, 6.8 and 6.9 we have to compute the quantity $(\partial U/\partial V)_T$. If C_P, C_V and the coefficient β of expansion could be determined experimentally with precision, $(\partial U/\partial V)_T$ (i.e. the variation in internal energy with volume at constant temperature) could be computed from the relation

$$\left(\frac{\partial U}{\partial V}\right)_T = \frac{C_P - C_V}{\beta V} - P \tag{6.28}$$

where β is the true coefficient of volume expansion and is given by

$$\beta = \frac{1}{V}\frac{dV}{dT}.$$

It is difficult to measure C_V precisely; therefore, this method of computing $(\partial U/\partial V)_T$ is not practical.

We have already seen that Joule's experiment is a direct method, although not highly precise, for measuring the variation in internal energy with volume at constant temperature. There again, the problem is that this experiment shows that $(\partial U/\partial V)_T$ is very small for real gases, and we may arbitrarily take it as zero for ideal gases.

To avoid these difficulties we shall assume a result that can be derived from the second law of thermodynamics. We shall show later that, for any substance,

$$P + \left(\frac{\partial U}{\partial V}\right)_T = T\left(\frac{\partial P}{\partial T}\right)_V. \tag{6.29}$$

Using this relation and the relevant equation of state, $(\partial U/\partial V)_T$ can be computed for any substance. For example, for 1 mol of an ideal gas

$$PV = RT$$

or

$$P = \frac{RT}{V}$$

or

$$\left(\frac{\partial P}{\partial T}\right)_V = \frac{R}{V}$$

or

$$T\left(\frac{\partial P}{\partial T}\right)_V = \frac{RT}{V}.$$

Then, from Equation 6.29

$$\left(\frac{\partial U}{\partial V}\right)_T = T\left(\frac{\partial P}{\partial T}\right)_V - P = \frac{RT}{V} - P = P - P = 0. \tag{6.30}$$

This shows that *at constant temperature the internal energy of an ideal gas is independent of volume or density.*

Again,

$$\left(\frac{\partial U}{\partial P}\right)_T = \left(\frac{\partial U}{\partial V}\right)_T \left(\frac{\partial V}{\partial P}\right)_T = 0 \tag{6.31}$$

(from Equation 6.30). This shows that *at constant temperature the internal energy of an ideal gas is independent of pressure as well.* We now conclude that the internal energy of an ideal gas is a function of its temperature only, and that at a given temperature it is the same whether the gas occupies a large or small volume, or whether it is subjected to a large or small pressure.

According to the hypothesis there is no attraction or repulsion among the molecules of an ideal gas and, therefore, there is no potential energy acting between them. So, their total energy (identified as the internal energy) is the sum of their kinetic energies only. However, as we already know that kinetic energy is a function of temperature only, the internal energy is also a function of temperature only. Hence, for an ideal gas, Equation 6.4 becomes

$$C_V = \frac{dU}{dT} \tag{6.32}$$

or

$$dU = C_V \, dT.$$

On integrating,

$$U = U_0 + \int_{T_0}^{T} C_V \, dT \tag{6.33}$$

where U_0 is a constant of integration defined as the internal energy at some arbitrary reference temperature T_0 (not necessarily equal to absolute zero). As U_0 is unknown, there is no justification for taking $U_0 = 0$ at absolute zero.

If C_V is constant, the right-hand side of Equation 6.33 can be easily computed by bringing C_V outside the integral sign and then integrating in the normal way. If C_V is not constant over the temperature range concerned, it must be replaced by an empirical relation in terms of T before the integral is computed. When C_V is constant, we have

$$U = U_0 + C_V(T - T_0). \tag{6.34}$$

6.5.1   Internal energy of van der Waals gases

We may write the equation of state of 1 mol of a van der Waals gas as

$$P = \frac{RT}{V-b} - \frac{a}{V^2}.$$

Differentiating partially with respect to T, keeping V constant, gives

$$\left(\frac{\partial P}{\partial T}\right)_V = \frac{R}{V-b}$$

or

$$T\left(\frac{\partial P}{\partial T}\right)_V = \frac{RT}{V-b}.$$

Then, from Equation 6.29,

$$P + \left(\frac{\partial U}{\partial V}\right)_T = T\left(\frac{\partial P}{\partial T}\right)_V = \frac{RT}{V-b} = P + \frac{a}{V^2}.$$

Therefore,

$$\left(\frac{\partial U}{\partial V}\right)_T = \frac{a}{V^2}. \qquad (6.35)$$

Combining Equations 6.1, 6.4 and 6.35 gives

$$dU = C_V \, dT + \frac{a}{V^2} \, dV.$$

Integrating, we obtain

$$U = U_0 + \int_{T_0}^{T} C_V \, dT + \left(\frac{a}{V_0} - \frac{a}{V}\right) \qquad (6.36)$$

where U_0 is an integration constant defined as the specific internal energy at an arbitrary reference temperature T_0 and volume V_0.

This shows that *the internal energy of a van der Waals gas depends on its volume and temperature*. Since the van der Waals constant b is a measure of molecular diameter only, it does not affect the energy, and hence it does not appear in the energy equation.

As before, if C_V is constant over the temperature range concerned, Equation 6.36 can be computed by bringing C_V outside the integral sign, and we have

$$U = U_0 + C_V(T - T_0) + \frac{a}{V_0} - \frac{a}{V}. \qquad (6.37)$$

On the other hand, if C_V is not constant over the temperature range, it must be replaced by an empirical relation in terms of T before the integral is computed.

As a matter of interest, Equations 6.34 and 6.37 can be represented graphically by taking the internal energy U along the vertical axis, and T and V along mutually perpendicular horizontal axes.

Example 6.2

The equation of state of a van der Waals gas is given by

$$\left(P+\frac{a}{V^2}\right)(V-b)=RT.$$

Calculate W, ΔU, Q and ΔH of this gas for an isothermal reversible expansion from an initial volume V_1 to a final volume V_2.

Solution

From the given equation of state,

$$P=\frac{RT}{V-b}-\frac{a}{V^2}$$

Therefore,

$$W=-\int_{V_1}^{V_2}P\,dV=-\int_{V_1}^{V_2}\left(\frac{RT}{V-b}-\frac{a}{V^2}\right)dV$$

or

$$W=-RT\ln\left(\frac{V_2-b}{V_1-b}\right)+\left(\frac{a}{V_1}-\frac{a}{V_2}\right).$$

From Equation 6.35,

$$\left(\frac{\partial U}{\partial V}\right)_T=\frac{a}{V^2}.$$

Integrating both sides gives

$$\Delta U=\int_{V_1}^{V_2}\frac{a}{V^2}dV=a\left(\frac{1}{V_1}-\frac{1}{V_2}\right).$$

Hence, from the first law $\Delta U=Q+W$, we have

$$Q=\Delta U-W=a\left(\frac{1}{V_1}-\frac{1}{V_2}\right)+RT\ln\left(\frac{V_2-b}{V_1-b}\right)-\left(\frac{a}{V_1}-\frac{a}{V_2}\right)$$

or

$$Q = RT \ln\left(\frac{V_2 - b}{V_1 - b}\right).$$

Again, by definition, $H = U + PV$. Therefore, $\Delta H = \Delta U + \Delta(PV)$. The given equation of state can be rearranged as

$$P = \frac{RT}{V - b} - \frac{a}{V^2}.$$

Now, multiplying both sides by V, we obtain

$$PV = RT \frac{V}{V - b} - \frac{a}{V}$$

Then,

$$\Delta(PV) = RT\left(\frac{V_2}{V_2 - b} - \frac{V_1}{V_1 - b}\right) - \left(\frac{a}{V_2} - \frac{a}{V_1}\right)$$

$$= RT \frac{V_2(V_1 - b) - V_1(V_2 - b)}{(V_2 - b)(V_1 - b)} - \left(\frac{a}{V_2} - \frac{a}{V_1}\right)$$

$$= \frac{-RTb(V_2 - V_1)}{(V_2 - b)(V_1 - b)} - \left(\frac{a}{V_2} - \frac{a}{V_1}\right)$$

$$= \frac{-RTb[(V_2 - b) - (V_1 - b)]}{(V_2 - b)(V_1 - b)} - \left(\frac{a}{V_2} - \frac{a}{V_1}\right)$$

$$= RTb\left(\frac{1}{V_2 - b} - \frac{1}{V_1 - b}\right) - \left(\frac{a}{V_2} - \frac{a}{V_1}\right).$$

Therefore, $\Delta H = \Delta U + \Delta(PV)$ becomes

$$\Delta H = RTb\left(\frac{1}{V_2 - b} - \frac{1}{V_1 - b}\right) + 2a\left(\frac{1}{V_1} - \frac{1}{V_2}\right).$$

6.6 THE CARNOT CYCLE

In 1824 a French engineer, N. L. Sadi Carnot (1796–1832), investigated the principles governing the transformation of thermal energy (i.e. heat) into mechanical energy (i.e. work). In practice, it is impossible to convert heat completely into work. This shows that heat is different from other forms of energy. Whereas all other forms of energy are completely interconvertible, and all forms of energy may be completely transformed into heat, heat itself is not fully convertible to any other form of energy. The efficiency of a machine is given by the fraction of the heat absorbed by the machine which it can convert into work, or any other form of energy. For example, if a machine absorbs heat Q_2 at the higher temperature T_2 and gives out heat Q_1 at the lower temperature T_1, then according to the first law

$Q_2 - Q_1 = W$, i.e. the difference between Q_2 and Q_1 has been converted into work W. Then the efficiency η of the machine is given by

$$\eta = \frac{Q_2 - Q_1}{Q_2} = \frac{W}{Q_2}. \tag{6.38}$$

We have already seen that the amount of work done in a reversible change is the maximum. Therefore, it follows that a machine operating reversibly will produce the maximum efficiency. This problem is simplified by the theorem given by Carnot. His theorem states: 'All periodic machines operating reversibly between the same temperatures of source and sink have the same efficiency.' This implies that, if the machine operates reversibly, its efficiency is independent of the nature of the working substance or substances used or its mode of operation. This theorem may be proved in the following way.

Let us consider two machines 1 and 2 which operate reversibly between the same temperatures but have different efficiencies. Let us assume that, in each cycle, machine 1 can convert a portion W of the heat Q_2 absorbed at the higher temperature T_2 and give out the remainder $Q_2 - W = Q_1$ at the lower temperature T_1. In the case of machine 2, in each cycle it converts a portion W' ($W' < W$) of the heat Q_2 absorbed at T_2 into work and gives up an amount $Q_2 - W' = Q_1'$ (where $Q_1' > Q_1$) at T_1. We now combine these two machines such that machine 1 works in the direct manner and machine 2 operates in the reverse direction. This coupling of the machines is permissible because we have assumed that both are reversible machines. We can compare the quantities of heat and work in each complete cycle of the machines (Table 6.1). On completion of a cycle the two machines will be back to their original state, and the net result is

$$\text{work done} = W + (-W') = W - W'$$

$$\text{heat absorbed at } T_1 = Q_1' - Q_1 = (Q_2 - W') - (Q_2 - W) = W - W'.$$

This shows that the heat absorbed at T_1 is equal to the work done, i.e. the coupled machines, on completion of their respective cycles, can convert heat absorbed at the same temperature completely into work, but this is not possible. When we discuss the second law of thermodynamics at a later stage, we shall see that a direct consequence of this law is that the two reversible machines must have the same efficiency.

Because all reversible machines must have the same efficiency, it is not necessary to consider two machines. Carnot's cycle lends itself to simple theoretical treatment. It consists of four reversible steps; therefore it is a reversible cycle. In this machine we consider a working substance which is confined in a cylinder fitted with a frictionless and weightless piston. The working substance may be a solid, liquid or gas (ideal or non-ideal), and it may even change from one phase to another during the cycle. It is

Table 6.1 Quantities of heat and work for each machine

Machine	Heat absorbed at T_2	Heat given up at T_1	Work done
1	Q_2	Q_1	W
2	Q_2	Q_1'	$-W'$

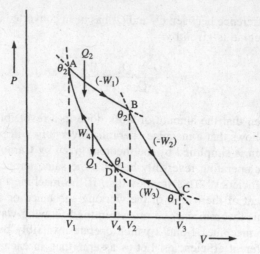

Figure 6.1   Schematic diagram of the Carnot cycle

assumed that two very large heat reservoirs are available at temperatures θ_2 and θ_1, and also that it is possible to surround the cylinder by an insulator, thus allowing adiabatic processes to be carried out. In the cyclic process the working substance takes in heat reversibly from a reservoir at a constant temperature θ_2, expands first isothermally and reversibly and then adiabatically and reversibly to a lower temperature θ_1 (partly converted into work), gives up heat reversibly to a reservoir at the same temperature θ_1 and undergoes compressions, first isothermally and then adiabatically and reversibly to its original state, so that the entire process constitutes a complete cycle. We have written the temperatures as θ_1 and θ_2 to indicate that they are empirical temperatures, measured on any convenient scale. The diagram of a Carnot cycle in the P–V plane is an area bounded by two isothermals and two adiabatics. If the working substance is a gas, the cycle has the general appearance shown in Figure 6.1. Since the mass of the system is fixed, the state can be described by any two of the three state variables P, T and V. In this figure, stage A to B and stage B to C are expansion stages where the gas is allowed to expand isothermally and adiabatically. Stage C to D and stage D to A are compression stages where the gas is compressed isothermally and adiabatically to its original state. The Carnot cycle is a hypothetical process, in which an ideal machine absorbs heat from its surroundings and converts it into work. The steps in the working of the engine for one complete cycle are as follows.

1. The gas is placed in contact with the heat reservoir at temperature θ_2 and allowed to expand reversibly and isothermally at θ_2 from V_1 to V_2; the path is represented by AB. It absorbs heat Q_2 from the hot reservoir and does work $-W_1$ on its surroundings. (The negative sign indicates only that the work was done by the system.)

2. The gas, insulated from any heat reservoir, expands reversibly and adiabatically $(Q=0)$ from V_2 to V_3, i.e. from B to C, and does work $-W_2$ as its temperature falls from θ_2 to θ_1.

3. The gas is placed in contact with the heat reservoir at θ_1 and compressed reversibly and isothermally at θ_1 from V_3 to V_4, i.e. from C to D. In this process, work W_3 is done upon it, and it gives up heat $-Q_1$ to the heat reservoir.

4. The gas, insulated from any heat reservoir, is compressed reversibly and adiabatically ($Q=0$), from V_4 to the initial volume V_1, i.e. from D to A. In this process, work W_4 is done on it, and its temperature rises from θ_1 to θ_2.

As a concrete example of a Carnot cycle, let us consider the Carnot engine shown in Figure 6.2. Let the working substance be an ideal gas enclosed in a cylinder tightly fitted with frictionless piston. The cylinder walls and the piston are perfect heat insulators and the base is a heat conductor. There are two heat reservoirs at temperatures θ_2 and θ_1, and there are a heat-insulating stand and a work reservoir. Let us start the cycle with the gas at temperature θ_2 corresponding to the point A in Figure 6.1.

Step 1: The cylinder is placed on the reservoir at temperature θ_2 and the gas is allowed to expand isothermally and reversibly from volume V_1 to volume V_2. In this process, some work is done by the substance and delivered to the work reservoir (not shown in Figure 6.2). A quantity of heat Q_2 is absorbed from the reservoir. (The amount of expansion is arbitrary.) Since the gas is ideal and the process is an isothermal process, $\Delta U=0$, and by the first law the energy changes of the surroundings during this step can be expressed as

$$Q_2 = -W_1 = -\int_{V_1}^{V_2} P\,dV = -nR\theta_2 \ln\left(\frac{V_2}{V_1}\right). \tag{6.39}$$

Step 2: The cylinder is next placed on the insulating stand so that it is completely thermally insulated. Now the gas is allowed to expand adiabatically and reversibly from volume V_2 to volume V_3 until its temperature drops to θ_1. In this stage, more work is done by the gas but no heat is absorbed. Since the gas is ideal and the process is adiabatic, $Q=0$. So, by the first law, $\Delta U=W$. Therefore,

$$W_2 = \Delta U = nC_V(\theta_1 - \theta_2). \tag{6.40}$$

Step 3: The cylinder is then placed on the reservoir at temperature θ_1 and the gas is compressed isothermally from volume V_3 to V_4. Work is now done on the gas, and a quantity of heat Q_1 flows from the gas to the cold reservoir. Since the gas is ideal and the temperature

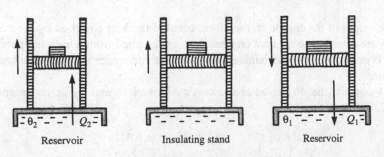

Figure 6.2   Schematic diagram of the Carnot engine

is constant, $\Delta U = 0$. Therefore, by the first law

$$Q_1 = -W_3 = -\int_{V_3}^{V_4} P\, dV = -nR\theta_1 \ln\left(\frac{V_4}{V_3}\right). \tag{6.41}$$

Step 4: Now the cylinder is again placed on the insulating stand so that it is completely thermally insulated. The gas is compressed adiabatically from volume V_4 to the original volume V_1. In this adiabatic compression the temperature rises to θ_2, the original temperature. Work is done on the gas and no heat is absorbed or given out. So $Q = 0$. Therefore,

$$\Delta U = W_4 = nC_V(\theta_2 - \theta_1). \tag{6.42}$$

It can be easily seen from Figure 6.1 that the work done by the system in the two expansion stages, from A to B and B to C, is greater than the work done on the system in the two compression stages, from C to D and D to A. Let W be the net work done. The system absorbs heat Q_2 from the reservoir at temperature θ_2. The net heat absorbed by the system is, therefore, $Q = Q_2 - Q_1$. In addition, since the system is carried out through a cyclic process, there is no change in its internal energy, i.e. for the whole cycle $\Delta U = 0$.

The total work produced in the cycle is the sum of the individual work in each cycle. So, by adding Equations 6.39–6.42

$$W = W_1 + W_2 + W_3 + W_4 = nR\left[\theta_2 \ln\left(\frac{V_1}{V_2}\right) - \theta_1 \ln\left(\frac{V_4}{V_3}\right)\right]$$

$$+ nC_V[(\theta_1 - \theta_2) + (\theta_2 - \theta_1)] = nR\left[\theta_2 \ln\left(\frac{V_1}{V_2}\right) - \theta_1 \ln\left(\frac{V_4}{V_3}\right)\right]. \tag{6.43}$$

Hence, from the first law,

$$\Delta U = Q + W = (Q_2 - Q_1) + (W_1 + W_2 + W_3 + W_4) = 0$$

or

$$Q = -W$$

i.e.

$$Q_2 - Q_1 = -nR\left[\theta_2 \ln\left(\frac{V_1}{V_2}\right) - \theta_1 \ln\left(\frac{V_4}{V_3}\right)\right]. \tag{6.44}$$

The work output of the engine is, therefore, equal to the heat absorbed by the system. These processes are common to all heat engines; heat is absorbed from a high-temperature source, part of it is converted into mechanical work, and the remainder is rejected as heat at a lower temperature.

Since V_1 and V_4 lie on the adiabatic curve AD, and V_2 and V_3 lie on another adiabatic curve BC, it follows that

$$\left(\frac{V_4}{V_1}\right)^{\gamma-1} = \frac{\theta_1}{\theta_4} \qquad \left(\frac{V_3}{V_2}\right)^{\gamma-1} = \frac{\theta_1}{\theta_4}.$$

Therefore,

$$\frac{V_4}{V_1} = \frac{V_3}{V_2} \qquad \frac{V_2}{V_1} = \frac{V_3}{V_4}.$$

Substituting in Equation 6.43 gives

$$W = nR\left[Q_2 \ln\left(\frac{V_1}{V_2}\right) - \theta_1 \ln\left(\frac{V_1}{V_2}\right)\right] = nR(\theta_2 - \theta_1) \ln\left(\frac{V_1}{V_2}\right). \qquad (6.45)$$

Dividing Equation 6.45 by Equation 6.39, we get

$$\frac{W}{Q_2} = \frac{\theta_2 - \theta_1}{\theta_2}. \qquad (6.46)$$

Replacing W by $Q_2 - Q_1$ we get

$$\frac{Q_2 - Q_1}{Q_2} = \frac{\theta_2 - \theta_1}{\theta_2}. \qquad (6.47)$$

It is worth noting that in the isothermal steps the maximum work is obtained on expansion and the minimum work is done in compression of the gas. In the adiabatic steps, $\Delta U = W$, and the work terms are determined only by the initial and final states. For the whole cycle we may summarize the changes as shown in Table 6.2.

The efficiency of a reversible Carnot engine operating in reversible cycles is given by the ratio of the work output to the heat input and is denoted by η. Hence,

$$\eta = \frac{\text{work output}}{\text{heat input}} = \frac{Q_2 - Q_1}{Q_2} = 1 - \frac{Q_1}{Q_2} = \frac{\theta_2 - \theta_1}{\theta_2} = 1 - \frac{\theta_1}{\theta_2}. \qquad (6.48)$$

It should be noted that no heat engine is 100% efficient. If the rejected heat were included as part of its output, the efficiency of every engine would be 100%. The above definition of efficiency applies to every type of heat engine; it is not restricted to the Carnot engine only. Since every step in this cycle is carried out reversibly, the maximum possible work is obtained for the particular working substance and temperature considered.

It can be shown that no other engine working between the same two temperatures can convert thermal energy to mechanical energy with a greater efficiency than does the Carnot engine. Other reversible engines, in fact, will have the same efficiency as the Carnot engine.

Table 6.2 Summary of the thermodynamic quantities in various steps of the Carnot cycle

Change	Heat absorbed by the gas	Work done by the gas
Step 1 A→B, isothermal	$Q_2 = -W_1$	$W_1 = -nR\theta_2 \ln\left(\frac{V_2}{V_1}\right)$
Step 2 B→C, adiabatic	0	$W_2 = \Delta U = nC_V(\theta_1 - \theta_2)$
Step 3 C→D, isothermal	$Q_1 = -W_3$	$W_3 = -nR\theta_1 \ln\left(\frac{V_4}{V_3}\right)$
Step 4 D→A, adiabatic	0	$W_4 = \Delta U = nC_V(\theta_2 - \theta_1)$

The Carnot cycle efficiency can, therefore, be used to make an estimate of the maximum conversion of heat to work that can be expected for a real engine.

If the Carnot cycle is operated in a counter-clockwise direction, the directions of all the arrows in Figures 6.1 and 6.2 are reversed but there will be no change in the magnitudes of Q_2, Q_1 and W. Heat Q_1 is removed from the low-temperature source, work W is given up by the work reservoir, and heat Q_2 is delivered to the high-temperature reservoir.

Example 6.3

1 mol of a perfect gas is subjected to all the steps of a reversible Carnot cycle. Calculate T, P, U and H at the end of each step. Assume that C_P and C_V are constants.

Solution

Step 1: Since this is an isothermal expansion, the temperature at the end of this step is the same as the initial temperature. Therefore, $T_1 = T_2$. In addition, from the equation of state of an ideal gas, namely $PV = nRT$, we have $P_2V_2 = RT_2 = RT_1$. Then $P_2 = RT_1/V_2$. Since for an isothermal change there is no change in internal energy, $U_1 = U_2$, and consequently $H_1 = H_2$.

Step 2: This is an adiabatic expansion (refer to Figure 6.1). For an adiabatic change, $TV^{\gamma-1} = \text{constant}$. Hence,

$$T_2 V_2^{\gamma-1} = T_3 V_3^{\gamma-1}$$

or

$$T_3 = T_2 \left(\frac{V_2}{V_3}\right)^{\gamma-1} = T_1 \left(\frac{V_2}{V_3}\right)^{R/C_V}$$

as $\gamma = C_P/C_V$ and $C_P = C_V + R$. Again, for an adiabatic change, $PV^\gamma = \text{constant}$. Then,

$$P_3 V_3^\gamma = P_2 V_2^\gamma = \frac{RT_1}{V_2} V_2^\gamma$$

or

$$P_3 = \frac{RT_1 V_2^{\gamma-1}}{V_3^\gamma} = \frac{RT_1 V_2^{R/C_V}}{V_3^{C_P/C_V}}$$

$$U_3 - U_1 = C_V(T_3 - T_1) = C_V\left[T_1\left(\frac{V_2}{V_3}\right)^{R/C_V} - T_1\right]$$

or

$$U_3 = U_1 + C_V T_1\left[\left(\frac{V_2}{V_3}\right)^{R/C_V} - 1\right]$$

$$H_3 - H_1 = C_P(T_3 - T_1) = C_P\left[T_1\left(\frac{V_2}{V_3}\right)^{R/C_V} - T_1\right]$$

or

$$H_3 = H_1 + C_P T_1 \left[\left(\frac{V_2}{V_3} \right)^{R/C_V} - 1 \right].$$

Step 3: This is an isothermal compression; therefore there is no change in temperature, internal energy and enthalpy. Hence,

$$T_4 = T_3 = T_1 \left(\frac{V_2}{V_3} \right)^{R/C_V}$$

$$P_4 V_4 = RT_4 = RT_1 \left(\frac{V_2}{V_3} \right)^{R/C_V}$$

or

$$P_4 = \frac{RT_1}{V_4} \left(\frac{V_2}{V_3} \right)^{R/C_V} \qquad U_4 = U_3 = U_1 + C_V T_1 \left[\left(\frac{V_2}{V_3} \right)^{R/C_V} - 1 \right]$$

and

$$H_4 = H_3 = H_1 + C_P T_1 \left[\left(\frac{V_2}{V_3} \right)^{R/C_V} - 1 \right].$$

Step 4: This is an adiabatic compression. At the end of this step the system returns to the original state. Therefore,

$$T_1 = T_1, \; U_1 = U_1, \; H_1 = H_1$$

and

$$P_1 = RT_1/V_1.$$

6.7 THE CARNOT REFRIGERATOR

Considering what we have said in the last paragraph of the previous section, we now have a Carnot refrigerator rather than a Carnot engine. In this case, heat is pumped out of a system at a low temperature (e.g. the interior of a household refrigerator), mechanical work is done (by the motor) and the heat equals the sum of the mechanical work; the heat removed from the low-temperature reservoir is liberated at a high temperature (and is absorbed by the air in the room). Table 6.3 shows the sign of the quantities work and heat in the two modes of operation.

In the case of a Carnot refrigerator the efficiency is commonly termed the 'coefficient of performance' and denoted by E. This is given by the ratio of heat extracted from the low-temperature reservoir to the work destroyed. Therefore,

$$E = \left| \frac{Q_1}{W} \right| = \frac{Q_1}{Q_1 - Q_2} = \frac{\theta_1}{\theta_2 - \theta_1}. \tag{6.49}$$

Table 6.3 Signs of work and heat

Cycle	Q_2	Q_1	W
Forward (the Carnot cycle)	+	−	+
Backward (the Carnot refrigerator)	−	+	−

It appears from Equation 6.49 that, as the lower temperature θ_1 decreases, the coefficient of performance drops off very rapidly. The coefficient of performance of a refrigerator, unlike the efficiency of a heat engine, can be much larger than 100%.

6.8 THERMODYNAMIC EFFICIENCY

We have already proved that *all engines operating in a reversible and cyclic manner between the same two temperatures will possess the same thermodynamic efficiency, whatever the working substance*. Equations 6.46 and 6.47 give the efficiency of a reversible engine. The efficiency of a reversible heat engine or any reversible machine is thus determined by the temperature of the heat introduced and the temperature of the heat discharged. The efficiency will be higher, the greater the difference between θ_1 and θ_2. A point to remember is that there is no change in entropy in the operation of such an engine.

From Equation 6.47 it appears that, in order to have an engine of 100% efficiency, either the temperature θ_2 must be infinite or the temperature θ_1 must be very very small. The first requirement is obviously impractical, although engines are run at as high a temperature as possible to increase the efficiency. The second requirement, as we shall see later when the third law of thermodynamics is dealt with, is another impossibility. From Equation 6.48 we have

$$\eta = \frac{Q_2 - Q_1}{Q_2} = \frac{\theta_2 - \theta_1}{\theta_2}$$

or

$$1 - \frac{Q_1}{Q_2} = 1 - \frac{\theta_1}{\theta_2}$$

or

$$\frac{Q_2}{\theta_2} - \frac{Q_1}{\theta_1} = 0. \qquad (6.50)$$

This shows that the absolute magnitudes of the quantities of heat absorbed and rejected are proportional to the temperatures of the heat reservoirs in either cycle. We may generalize Equation 6.50 as

$$\sum_{\text{cycle}} \frac{Q}{\theta} = 0. \qquad (6.51)$$

If the ideal gas in the Carnot cycle is considered as being carried through a series of infinitesimally small steps throughout the cycle, then Equation 6.51 may be written

$$\int \frac{dQ}{\theta} = 0 \qquad (6.52)$$

In view of the fact that the whole process has been carried out reversibly we may recollect that the left-hand side of Equation 6.52 is, according to the definition, nothing but entropy. Hence, we have

$$\int dS = 0. \qquad (6.53)$$

Example 6.4

A refrigerator operates in a cyclic and reversible manner between 0 and 27°C. How much work must be done in freezing 1 ℓ of water? Also, calculate the coefficient of performance, given that the heat of fusion of water is 334.72 J g^{-1}.

Solution

T_2 is the higher temperature equal to $27 + 273 = 300$ K. T_1 is the lower temperature equal to $0 + 273 = 273$ K. Q_2 is the heat given out at temperature T_2. Q_1 is the heat absorbed at temperature T_1. However,

$$Q_1 = \text{mass of water} \times \text{heat of fusion of water}$$
$$= 1000 \times 334.72 = 334.72 \text{ kJ.}$$

We shall show later that the absolute magnitudes of the quantities of heat absorbed and rejected are proportional to the temperatures of the heat reservoirs in either cycle, i.e. $Q_2/Q_1 = T_2/T_1$. Hence, $Q_2 = Q_1(T_2/T_1)$. Therefore,

$$Q_2 = \frac{334.72 \times 300}{273} = 367.82 \text{ kJ.}$$

The work done

$$W = Q_2 - Q_1 = 33.10 \text{ kJ.}$$

The coefficient of performance

$$E = \frac{Q_1}{Q_1 - Q_2} = \frac{334.72}{33.10} = 1011\%.$$

6.9 THERMODYNAMIC SCALE OF TEMPERATURE

In our foregoing discussion no mention was made of the actual temperatures that were qualitatively represented by θ_2 and θ_1. We have shown in Section 6.8 that the efficiency of the cycle must depend only on these qualitative temperatures and not upon the particular substance carried through the cycle. The fact that the ratio Q_1/Q_2 depends only on the temperatures θ_1 and θ_2 and not on the substance used suggests a logical definition of absolute temperature not requiring the concept of an ideal gas. Consequently, if we simply agree to define the ratios of absolute temperatures as the absolute values of the corresponding heat quantities in a reversible Carnot cycle, we arrive at a scale that is fully compatible with the ideal-gas scale without assuming the existence of ideal gases. Hence, from Equation 6.50 we define the thermodynamic temperature as

$$\frac{\theta_2}{\theta_1} = \left| \frac{Q_2}{Q_1} \right|. \tag{6.54}$$

Here also we may choose any size for a unit difference on the scale but conventionally the unit is chosen in terms of the freezing and boiling points of water. So, if we agree that 100 units represent the difference between the temperatures of melting ice and boiling water at 0.1 MPa pressure, our definition of absolute temperature becomes complete. The thermodynamic scale thus eliminates reference to the properties of an ideal gas.

If we now denote the absolute temperatures by T_1 and T_2, then with the help of Equation 6.54 we may write

$$\frac{T_2}{T_1} = \left| \frac{Q_2}{Q_1} \right|. \tag{6.55}$$

Clearly, we now have

$$\frac{\theta_2}{\theta_1} = \frac{T_2}{T_1}. \tag{6.56}$$

This implies that θ is proportional to T, or $\theta = CT$, where C is a constant of proportionality. If we take θ_2 and T_2 as the boiling point of water, and θ_1 and T_1 as the melting point of ice on the thermodynamic and the absolute temperature scales, respectively, then we have

$$\theta_2 - \theta_1 = 100 = T_2 - T_1. \tag{6.57}$$

From Equations 6.56 and 6.57 we conclude that

$$\theta = T. \tag{6.58}$$

Hence, the thermodynamic temperature scale is identical with the ideal-gas scale but the concept of an ideal gas is not necessary, although it will continue to remain convenient in our theoretical discussion.

6.10   THE DIESEL ENGINE

Many cyclic processes do not operate on a Carnot cycle. Two common examples of such processes are the Diesel and the Otto cycle of gasoline engines. However, any reversible cyclic process is subjected to the same thermodynamic analysis as a Carnot engine. We first discuss the Diesel cycle with an ideal gas having a constant heat capacity as the working substance. Figure 6.3 schematically shows the steps involved. It consists of a constant-pressure and a constant-volume step, and two adiabatic steps. In the constant-pressure and constant-volume steps heat is transferred.

Step 1→2: This is given by the path AB. For this adiabatic change we have

$$Q_{12} = 0 \qquad -W_{12} = \Delta U_{12} = C_V(T_2 - T_1).$$

Step 2→3: This is the path BC which is a constant-pressure process. For this process

$$W_{23} = P_2(V_3 - V_2) = P_3(V_3 - V_2)$$
$$\Delta U_{23} = C_V(T_3 - T_2)$$
$$Q_{23} = \Delta U_{23} + W_{23} = C_V(T_3 - T_2) + P_2(V_3 - V_2).$$

Step 3→4: This is again an adiabatic process given by the path CD, for which

$$Q_{34} = 0 \qquad -W_{34} = \Delta U_{34} = C_V(T_4 - T_3).$$

Step 4→1: This process is a constant-volume process given by the path DA, and for this

$$W_{41} = 0 \qquad Q_{41} = \Delta U_{41} = C_V(T_1 - T_4).$$

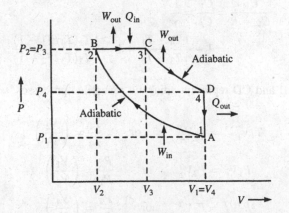

Figure 6.3   Schematic diagram of the steps of the Diesel engine

For a complete cycle, the total changes in the heat, work and internal energy are

$$W_{cycle} = W_{12} + W_{23} + W_{34} + W_{41} = -C_V(T_2 - T_1) + P_2(V_3 - V_2)$$
$$- C_v(T_4 - T_3)$$
$$Q_{cycle} = Q_{12} + Q_{23} + Q_{34} + Q_{41} = C_V(T_3 - T_2) + P_2(V_3 - V_2)$$
$$+ C_V(T_1 - T_4)$$
$$\Delta U_{cycle} = \Delta U_{12} + \Delta U_{23} + \Delta U_{34} + \Delta U_{41} = C_V(T_2 - T_1) + C_V(T_3 - T_2)$$
$$+ C_V(T_4 - T_3) + C_V(T_1 - T_4) = 0.$$

The thermal efficiency of the entire cycle is

$$\eta = \frac{\text{total work done in the cycle}}{\text{total heat input}} = \frac{W_{cycle}}{Q_{in}} = \frac{W_{12} + W_{23} + W_{34}}{Q_{23}}$$
$$= \frac{-C_V(T_2 - T_1) + P_2(V_3 - V_2) - C_V(T_4 - T_3)}{C_V(T_3 - T_2) + P_2(V_3 - V_2)}.$$

Since the working substance is an ideal gas, we have $P_2V_2 = RT_2$ and $P_3V_3 = RT_3$. In addition, from Figure 6.3, $P_2 = P_3$. Substituting these values into the above equation, we get

$$\eta = \frac{-C_V(T_2 - T_1) + R(T_3 - T_2) - C_V(T_4 - T_3)}{C_V(T_3 - T_2) + R(T_3 - T_2)}$$
$$= \frac{C_V(T_3 - T_2) + R(T_3 - T_2) - C_V(T_4 - T_1)}{(C_V + R)(T_3 - T_2)}$$
$$= \frac{(C_V + R)(T_3 - T_2) - C_V(T_4 - T_1)}{(C_V + R)(T_3 - T_2)}$$
$$= 1 - \frac{C_V(T_4 - T_1)}{(C_V + R)(T_3 - T_2)}.$$

Then, because

$$\frac{C_V}{C_V + R} = \frac{C_V}{C_P} = \frac{1}{\gamma}$$
$$\eta = 1 - \frac{1}{\gamma}\frac{T_4 - T_1}{T_3 - T_2} = 1 - \frac{1}{\gamma}\frac{T_1(T_4/T_1 - 1)}{T_2(T_3/T_2 - 1)}. \tag{6.59}$$

Since the paths AB and CD represent adiabatic reversible processes, we have

$$T_1V_1^{\gamma-1} = T_2V_2^{\gamma-1} \quad \text{or} \quad \frac{T_1}{T_2} = \left(\frac{V_2}{V_1}\right)^{\gamma-1}$$

$$P_1V_1^{\gamma} = P_2V_2^{\gamma} \quad \text{or} \quad \frac{P_2}{P_1} = \left(\frac{V_1}{V_2}\right)^{\gamma}$$

$$P_3V_3^{\gamma} = P_4V_4^{\gamma} \quad \text{or} \quad \frac{P_3}{P_4} = \left(\frac{V_4}{V_1}\right)^{\gamma}.$$

Hence,

$$\frac{P_4}{P_3} = \left(\frac{V_3}{V_4}\right)^{\gamma} = \left(\frac{V_2 V_3}{V_2 V_4}\right)^{\gamma} = \left(\frac{V_2 V_3}{V_1 V_2}\right)^{\gamma} \tag{6.60}$$

(because $V_1 = V_4$). In addition, $P_1 V_1 = RT_1$ and $P_4 V_4 = RT_4$. Then,

$$\frac{T_4}{T_1} = \frac{P_4 V_4}{P_1 V_1} = \frac{P_4}{P_1} \tag{6.61}$$

(because $V_1 = V_4$) or

$$\frac{T_4}{T_1} = \frac{P_4}{P_1} \frac{P_2 P_3}{P_2 P_3} = \frac{P_2 P_4}{P_1 P_3} \tag{6.62}$$

(because $P_2 = P_3$). Substituting the values of P_2/P_1 and P_4/P_3 into Equation 6.62, we obtain

$$\frac{T_4}{T_1} = \left(\frac{V_3}{V_2}\right)^{\gamma}. \tag{6.63}$$

Again, $P_2 V_2 = RT_2$ and $P_3 V_3 = RT_3$. This gives

$$\frac{T_3}{T_2} = \frac{P_3 V_3}{P_2 V_2} = \frac{V_3}{V_2} \tag{6.64}$$

(because $P_2 = P_3$). Substituting the values of T_1/T_2, T_4/T_1 and T_3/T_2 into Equation 6.59, we find that

$$\eta = 1 - \frac{1}{\gamma}\left(\frac{V_2}{V_1}\right)^{\gamma-1} \frac{(V_3/V_2)^{\gamma} - 1}{(V_3/V_2) - 1} \tag{6.65}$$

The ratio $V_1/V_2 = r$ is called the *compression ratio* and $V_3/V_2 = r_c$ is called the *cut-off ratio*. Equation 6.65 thus becomes

$$\eta = 1 - \frac{1}{\gamma}\left(\frac{1}{r}\right)^{\gamma-1} \frac{r_c^{\gamma} - 1}{r_c - 1}. \tag{6.66}$$

6.11 THE OTTO CYCLE

Let us now consider the Otto cycle operating reversibly with an ideal gas of constant heat capacity as the working substance. The Otto cycle consists of two constant-volume steps and two adiabatic steps. During the constant-volume steps, heat is transferred. The cycle is schematically represented in Figure 6.4.

Step 1→2: This is an adiabatic change given by the path AB. For this change,

$$Q_{12} = 0 \qquad - W_{12} = \Delta U_{12} = C_V(T_2 - T_1).$$

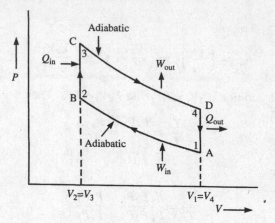

Figure 6.4   Schematic diagram of the steps of the Otto cycle

Step 2→3: This is constant-volume process represented by the path BC, for which we have

$$W_{23} = 0 \qquad Q_{23} = \Delta U_{23} = C_V(T_3 - T_2).$$

Step 3→4: This is the second adiabatic change given by the path CD. For this change,

$$Q_{34} = 0 \qquad - W_{34} = \Delta U_{34} = C_V(T_4 - T_3).$$

Step 4→1: This is again a constant volume process represented by the path DA and for this

$$W_{41} = 0 \qquad Q_{41} = \Delta U_{41} = C_V(T_1 - T_4).$$

The total quantities of heat, work and internal energy for the entire cycle are given by

$$W_{\text{cycle}} = W_{12} + W_{23} + W_{34} + W_{41} = -C_V(T_2 - T_1) - C_V(T_4 - T_3)$$
$$Q_{\text{cycle}} = Q_{12} + Q_{23} + Q_{34} + Q_{41} = C_V(T_3 - T_2) + C_V(T_1 - T_4)$$
$$\Delta U_{\text{cycle}} = \Delta U_{12} + \Delta U_{23} + \Delta U_{34} + \Delta U_{41}$$
$$= C_V(T_2 - T_1) + C_V(T_3 - T_2) + C_V(T_4 - T_3) + C_V(T_1 - T_4) = 0.$$

The thermal efficiency of the cycle is given by

$$\eta = \frac{W_{\text{cycle}}}{Q_{\text{in}}} = \frac{W_{12} + W_{34}}{Q_{23}} = \frac{-C_V(T_2 - T_1) - C_V(T_4 - T_3)}{C_V(T_3 - T_2)}$$
$$= 1 - \frac{T_4 - T_1}{T_3 - T_2}. \tag{6.67}$$

Since the paths AB and CD are reversible adiabatic changes, we may write

$$T_1 V_1^{\gamma-1} = T_2 V_2^{\gamma-1} \quad \text{or} \quad \frac{T_1}{T_2} = \left(\frac{V_2}{V_1}\right)^{\gamma-1}$$

and

$$T_3 V_3^{\gamma-1} = T_4 V_4^{\gamma-1} \quad \text{or} \quad \frac{T_4}{T_2} = \left(\frac{V_3}{V_4}\right)^{\gamma-1}.$$

However, from Figure 6.4, $V_1 = V_4$ and $V_2 = V_3$. Therefore,

$$\frac{T_1}{T_2} = \left(\frac{V_2}{V_1}\right)^{\gamma-1} \qquad \frac{T_4}{T_3} = \left(\frac{V_2}{V_1}\right)^{\gamma-1}$$

Then,

$$\frac{T_1}{T_2} = \frac{T_4}{T_3}$$

or

$$T_4 = \frac{T_1 T_3}{T_2}. \tag{6.68}$$

Substituting the value of T_4 from Equation 6.68 into Equation 6.67, we obtain

$$\eta = 1 - \frac{T_1 T_3/T_2 - T_1}{T_3 - T_2} = 1 - \frac{T_1 T_3 - T_1 T_2}{T_2(T_3 - T_2)} = 1 - \frac{T_1}{T_2} = 1 - \left(\frac{V_2}{V_1}\right)^{\gamma-1}. \tag{6.69}$$

As in the case of the Diesel engine, the ratio V_1/V_2 is defined as the *compression ratio r*. Hence, Equation 6.69 becomes

$$\eta = 1 - \left(\frac{1}{r}\right)^{(\gamma-1)}. \tag{6.70}$$

This equation implies that the value of η increases, i.e. the efficiency of the Otto cycle increases as the compression ratio increases. When $r \to \infty$, $\eta \to 1$.

We can now compare the ideal thermal efficiencies of the Otto and Diesel cycles. Comparison of Equations 6.66 and 6.70 shows that, for a fixed value of r, η_{Diesel} and η_{Otto} differ only in the term that multiplies $(1/r)^{\gamma-1}$. In the case of η_{Otto} this term is 1, whereas for η_{Diesel} this is $(r_c^{\gamma} - 1)/\gamma(r_c - 1)$. Since both γ and r_c are greater than unity, this term is always greater than unity. From this, we can conclude that, if an ideal gas is used as the working substance and if it is compared at the same value of r_c, a reversible Diesel cycle is less efficient than a reversible Otto cycle. However, there are practical problems, such as pre-ignition, for real engines operating on the Otto cycle. These problems can reduce the compression ratios for such engines, whereas Diesel engines can achieve greater compression ratios and thereby greater efficiencies.

Example 6.5

An engine operating on the Otto cycle has an ideal gas as the working substance. The pressure and the temperature at the point A (see Figure 6.4) are 0.1 MPa and 300 K, respectively, and those at the point B are 1.5 MPa and 2500 K, respectively. Sufficient quantity of heat is introduced in step B→C. Calculate the quantities of heat and work for all the steps of the cycle and the total work in the entire cycle. Then calculate the efficiency of the engine, given that for the working substance $C_V = 21.0 \, \mathrm{J \, mol^{-1} \, K^{-1}}$ and $C_P = 30.0 \, \mathrm{J \, mol^{-1} \, K^{-1}}$.

Solution

We have $T_1 = 300 \, \mathrm{K}$ and $T_3 = 2500 \, \mathrm{K}$; we need T_2 and T_4. Since step A→B is adiabatic and reversible

$$T_1 P_1^{(1-\gamma)/\gamma} = T_2 P_2^{(1-\gamma)/\gamma}$$

or

$$T_2 = T_1 \left(\frac{P_1}{P_2} \right)^{(1-\gamma)/\gamma} = T_1 \left(\frac{P_2}{P_1} \right)^{(\gamma-1)/\gamma}$$

However, $\gamma = C_P/C_V = 30.0/21.0 = 1.42$. So $(1-\gamma)/\gamma = 0.296$. Then,

$$T_2 = 300 \left(\frac{1.5}{0.1} \right)^{0.296} = 668.7 \, \mathrm{K}$$

From Equation 6.68,

$$\frac{T_4}{T_1} = \frac{T_3}{T_2} \quad \text{or} \quad T_4 = \frac{T_1 T_3}{T_2}.$$

Hence,

$$T_4 = \frac{300 \times 2500}{668.7} = 1121.6 \, \mathrm{K}.$$

For *Step A→B*

$$Q_{AB} = 0 \qquad -W_{AB} = \Delta U_{AB} = C_V (T_2 - T_1) = 7742.7 \, \mathrm{J \, mol^{-1}}.$$

For *Step B→C*

$$W_{BC} = 0 \qquad Q_{BC} = \Delta U_{BC} = C_V (T_3 - T_2) = 38\,457.3 \, \mathrm{J \, mol^{-1}}$$

For *Step C→D*

$$Q_{CD} = 0 \qquad -W_{CD} = \Delta U_{CD} = C_V(T_4 - T_3) = -28\,946.4\,\mathrm{J\,mol^{-1}}.$$

For *Step D→A*

$$W_{DA} = 0 \qquad Q_{DA} = \Delta U_{DA} = C_V(T_1 - T_4) = -17\,253.6\,\mathrm{J\,mol^{-1}}.$$

The thermal efficiency is given by

$$\eta = \frac{W_{cycle}}{Q_{in}} = \frac{W_{AB} + W_{CD}}{Q_{BC}} = \frac{-7742.7 + 28\,946.4}{38\,457.3} = 0.55.$$

The total work done in the entire cycle is

$$W_{cycle} = W_{AB} + W_{BC} + W_{CD} + W_{DA} = W_{AB} + W_{CD}$$
$$= -7742.7 + 28\,946.4\,\mathrm{J\,mol^{-1}} = 21\,203.7\,\mathrm{J\,mol^{-1}}.$$

It seems that the engine converts only 55% of the heat supplied into work and the remaining 45% is wasted as heat into the surroundings. This lost heat is $Q_{DA} = -17\,253.6\,\mathrm{J\ mol^{-1}}$.

EXERCISES 6

6.1 Show that $(\partial U/\partial T)_V$ may be replaced by C_V in any equation in which it occurs, even if the equation refers to a process in which the volume is not constant.

6.2 Show that for any reversible process the first law can be written as $dQ = C_V\,dT + [P + (\partial U/\partial V)_T]\,dV$ when the internal energy of the system is expressed as a function of temperature and volume.

6.3 Prove that, when the internal energy of a system is expressed as a function of temperature and pressure only, then, for a constant-pressure process,

$$dQ = C_P\,dT + \left[\left(\frac{\partial U}{\partial P}\right)_T + P\left(\frac{\partial V}{\partial P}\right)_T\right]dP.$$

6.4 Prove that, when the internal energy of a system is expressed as a function of temperature and pressure only, then, for a constant-temperature process,

$$dQ = \left[\left(\frac{\partial U}{\partial P}\right)_T + P\left(\frac{\partial V}{\partial P}\right)_T\right]dP.$$

6.5 Prove that, when the internal energy of a system is expressed as a function of pressure and volume, then, for a process at constant temperature,

$$(dQ)_T = P(dV)_T + \left(\frac{\partial U}{\partial V}\right)_P (dV)_T.$$

6.6 Show that, at constant temperature, the internal energy of an ideal gas is independent of volume or density.

6.7 Show that, at constant temperature, the internal energy of an ideal gas is independent of pressure.

6.8 For a perfect gas, the pressure, temperature and volume are related by $PV = RT$. What are the values of $(\partial P/\partial V)_T$, $(\partial P/\partial T)_V$ and $(\partial V/\partial T)_P$?

6.9 Show that the internal energy of a van der Waals gas depends on its volume and temperature.

6.10 The coefficient β of thermal expansion is defined by the equation

$$\beta = \frac{1}{V}\left(\frac{\partial V}{\partial T}\right)_P.$$

Obtain an expression for β for a gas which obeys the van der Waals equation.

6.11 A working substance takes in heat reversibly from a heat reservoir at a constant temperature θ_2, expands at first isothermally and reversibly and then adiabatically and reversibly to a lower temperature θ_1. It is then compressed at first isothermally and reversibly and thereafter adiabatically and reversibly to its original state, so that the entire process constitutes a cycle. Deduce expressions for the total work done and the total heat absorbed by the system. Hence show that the efficiency is given by $\eta = (\theta_2 - \theta_1)/\theta_2$.

6.12 Calculate the efficiency of a steam engine operating with a condenser at 30°C and a boiler at 200°C.

6.13 We have showed that the efficiency of a Carnot engine would be higher, the greater the difference between the higher and lower temperatures. Which would you consider as the more convenient way to increase the efficiency: to increase the temperature θ_2 keeping θ_1 fixed, or to decrease the temperature θ_1 keeping θ_2 fixed?

6.14 A refrigerator operating at 30°C is employed to maintain a cold storage tank at −10°C. What is the minimum amount of work required to withdraw 4.184 kJ from the tank?

6.15 A Carnot engine absorbs heat from a reservoir at a temperature of 100°C and rejects heat to a reservoir at a temperature of 0°C. Calculate the amount of work done, the heat rejected and the efficiency when the engine absorbs 1000 J from the high-temperature reservoir.

6.16 A refrigerator having a coefficient of performance one-half that of a Carnot refrigerator is operated between reservoirs at temperatures of 200 and 400 K. It absorbs 600 J from the low-temperature reservoir. Calculate the amount of heat rejected to the high-temperature reservoir.

6.17 Show that the absolute magnitudes of the quantities of heat absorbed and rejected are proportional to the temperatures of the heat reservoirs in the case of both forward and reversed Carnot cycles.

6.18 Show that the thermodynamic temperature scale is identical with the ideal-gas temperature scale.

6.19 A Carnot refrigerator is operated between two reservoirs at 0 and 100°C. If 1 kJ is absorbed from the low-temperature reservoir, calculate the amount of heat rejected to the high-temperature reservoir. Then calculate the coefficient of performance.

6.20 A freezer is kept in an ambient temperature of 300 K. In order to maintain the temperature of the freezer box at 240 K, heat is to be removed from it at $1250\,\mathrm{J\,s^{-1}}$. Calculate the maximum coefficient of performance of this freezer.

6.21 A reversible heat engine operates between a heat reservoir at 680 K and another heat reservoir at 273 K. Calculate the amount of work done by the engine when 250 kJ of heat is transferred to the heat engine from the reservoir at 680 K.

6.22 1 mol of an ideal monatomic gas undergoes a reversible Carnot cycle with V_1 and V_2 as $20\,\ell$ and $40\,\ell$, respectively, $\theta_1 = 27°C$ and $\theta_3 = -73°C$. Calculate the changes in the pressure, volume, temperature, internal energy and enthalpy at each step of the cycle.

6.23 Sketch the following diagrams for the reversible Carnot cycle of an ideal gas with constant C_P and C_V:

(a) pressure versus volume

(b) temperature versus pressure

(c) internal energy versus pressure

(d) enthalpy versus pressure

(e) volume versus temperature

(f) volume versus internal energy

(g) internal energy versus enthalpy

(h) internal energy versus temperature

(i) enthalpy versus temperature.

6.24 Calculate the maximum possible efficiency of a steam engine which operates between 600 and 300°C. Can you comment on the exhaust temperature of this engine?

6.25 A heat engine operating between 1000 and 25°C produces a quantity of work which is entirely used to run a refrigerating machine operating between 0 and 25°C. Calculate the ratio of the heat absorbed by the engine to that absorbed by the refrigerating machine. Assume ideal operation in both cases.

6.26 Derive the following identities:

(a)
$$C_V = -\left(\frac{\partial U}{\partial V}\right)_T \left(\frac{\partial V}{\partial T}\right)_U$$

(b)
$$\left(\frac{\partial U}{\partial V}\right)_P = C_P \left(\frac{\partial T}{\partial V}\right)_P - P.$$

Hence show that, for 1 mol of an ideal gas,

$$\left(\frac{\partial U}{\partial V}\right)_P = \frac{PC_V}{R}.$$

(c)
$$\left(\frac{\partial U}{\partial T}\right)_P = C_P - P\left(\frac{\partial V}{\partial T}\right)_P$$

6.27 Show that for an ideal gas,

$$\left(\frac{\partial U}{\partial V}\right)_T = T\left(\frac{\partial P}{\partial T}\right)_V - P.$$

Hence show that, for 1 mol of a van der Waals gas,

$$\left(\frac{\partial U}{\partial V}\right)_T = \frac{a}{V^2}.$$

What is the physical significance of this result in the van der Waals equation?

7

The Second Law
of Thermodynamics
and the Concept of Entropy

7.1 DEFINITION OF THE SECOND LAW
OF THERMODYNAMICS

We have seen that the first law of thermodynamics enables us to calculate the energy changes of a chemical reaction, but it does not enable us to predict whether a reaction can or cannot occur. The second law of thermodynamics does enable us to predict whether a reaction can or cannot occur. From experience we know that heat flows spontaneously from a hot body to a colder body until a uniform temperature is reached, but it is not possible to reverse the process. If a machine were constructed to achieve this reverse process it would not be 100% efficient. Alternatively, we can say that no engine has ever been developed that converts the heat extracted from one reservoir into work without rejecting some heat to a reservoir at a low temperature. This concept is the basis of the second law of thermodynamics. This law has been stated in several ways.

(i) *The principle of Thomson (Lord Kelvin)* states: 'It is impossible by a cyclic process to take heat from a reservoir and to convert it into work without simultaneously transferring heat from a hot to a cold reservoir.' This statement of the second law is related to equilibrium, i.e. work can be obtained from a system only when the system is not already at equilibrium. If a system is at equilibrium, no spontaneous process occurs and no work is produced. Evidently, Kelvin's principle indicates that the spontaneous process is the heat flow from a higher to a lower temperature, and that only from such a spontaneous process can work be obtained.

(ii) *The principle of Clausius* states: 'It is impossible to devise an engine which, working in a cycle, shall produce no effect other than the transfer of heat from a colder to a hotter body.' A good example of this principle is the operation of a refrigerator.

(iii) *The principle of Planck* states: 'It is impossible to construct an engine which, working in a complete cycle, will produce no effect other than raising of a weight and the cooling of a heat reservoir.'

(iv) *The Kelvin–Planck principle* may be obtained by combining the principles of Kelvin and of Planck into one equivalent statement as the Kelvin–Planck statement of the second law. It states: 'No process is possible whose sole result is the absorption of heat from a reservoir and the conversion of this heat into work.'

In these statements the phrase 'working in a cycle' specifies that the working substance returns exactly to its initial state; therefore, the process can be carried out repeatedly.

The second law is not a deduction from the first law but a separate law of nature, referring to an aspect of nature different from that contemplated by the first law. The first law denies the possibility of creating or destroying energy, whereas the second law denies the possibility of utilizing energy in a particular way. The continuous operation of a machine that creates its own energy and thus violates the first law is called *perpetual motion of the first kind*. A cyclic device which would continuously abstract heat from a single reservoir and convert that heat completely to mechanical work is called a *perpetual-motion machine of the second kind*. Such a machine would not violate the first law (the principle of conservation of energy), since it would not create energy, but economically it would be just as valuable as if it did so. Hence, the second law is sometimes stated as follows: '*A perpetual motion machine of the second kind is impossible.*'

7.2 THE CONCEPT OF ENTROPY

Just as the first law of thermodynamics suggested a definition of a function of the state of the system called '*energy*', similarly the second law suggests the definition of another function. Let us consider the process illustrated in Figure 7.1. Such a process will not happen unless we apply some external force to the system. Heat will not flow 'uphill', even though the amount of thermal energy transferred could be calculated quite simply from the relations

$$Q = C\Delta T = C(T_2 - T_1)$$

or

$$Q = \int_{T_1}^{T_2} (C \text{ as a function of } T) \, dT.$$

Experience indicates that, if we begin with the initial state shown in Figure 7.1, and if the system is left to itself, the final state will be a uniform temperature of 75°C. This is the process that will occur *spontaneously*.

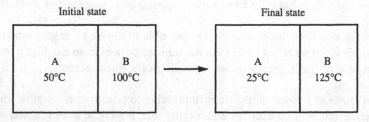

Figure 7.1 An impossible heat transfer process. The rectangles represent the amounts of materials A and B, both having the same total heat capacity

One useful way of examining a system before and after a spontaneous process is to consider its disorder or randomness. If we isolate the system boundaries, a spontaneous process will be characterized by an increase in the randomness of the isolated system. The process involves a change (sometimes obvious and sometimes more subtle) from a more ordered to a less ordered state. To illustrate the idea of an increase in randomness, let us take two containers A and B; A is full of a gas while B is a vacuum (empty). As a spontaneous process we would expect that when A and B are connected together, the gas from the container A will expand to fill the available space in B. As it expands, the system goes from a more ordered (all molecules in A) to a less ordered condition (gas molecules in both the containers). Thus, the disorder or randomness has increased. In a spontaneous process the system is more ordered in its initial state than in its final state.

More detailed analysis indicates that, if any process occurs spontaneously and the system is isolated, there is an increase in randomness. The thermodynamic function that is a direct measure of randomness is known as *entropy* and is denoted by S. The concept of entropy is not very concrete; rather it is somewhat philosophical in nature. Unfortunately, there is no better concept of entropy yet available. As a working hypothesis the existence of this function is assumed. The second law can then be formulated in terms of entropy. We can now state the second law of thermodynamics as follows: '*The entropy of an isolated assembly must increase or in the limit remain constant.*' This does not necessarily mean that the entropy of a non-isolated system can never decrease. Transfer of energy between a non-isolated system and its environment can definitely result in lowering of the entropy of the system. The above statement of the second law can be mathematically expressed as $\Delta S \geqslant 0$ for an isolated system.

7.3   SIGNIFICANCE OF ENTROPY

We have seen that the efficiency of a Carnot cycle is given by the relation

$$\frac{Q_1}{T_1} = \frac{Q_2}{T_2}$$

where Q_1 is the heat given out at temperature T_1 and Q_2 is the amount of heat absorbed at temperature T_2. Rearranging this equation, we have

$$\frac{Q_1}{T_1} - \frac{Q_2}{T_2} = 0.$$

This signifies that, for the closed cycle that brings the system back to its initial state,

$$\sum \frac{Q_{rev}}{T} = 0 \qquad\qquad (7.1)$$

This is exactly what we found in the case of internal energy U and precisely what we would demand of any other function of state. This new function of state, the left-hand side of Equation 7.1, is defined by the term entropy S by means of the equation

$$dS = \frac{Q_{rev}}{T}. \qquad\qquad (7.2)$$

Entropy will increase when heat passes into a system (for Q_{rev} is then positive), decrease when heat passes out of a system (for Q_{rev} is then negative), and thus around the closed circuit of the Carnot cycle we have

$$\Delta S = \sum \frac{Q_{rev}}{T} = 0 \qquad (7.3)$$

Now we may rightly ask: what is entropy? To answer this, let us ask: what is energy? The point is, do we exactly know what energy is? Perhaps not; we do not need to know what energy is, but we find it satisfactory to interpret internal energy, on the basis of a kinetic-molecular hypothesis, as the kinetic and potential energies of atoms and molecules. Similarly, we do not need to know what entropy is, but we find it satisfactory to interpret entropy, on the basis of a kinetic-molecular hypothesis, in terms of the randomness of the distribution of atoms and molecules in space and in energy states.

7.4   ENTROPY IS A VARIABLE OF STATE

Entropy, defined by the equations

$$\Delta S = \frac{Q_{rev}}{T} \qquad\qquad dS = \frac{dQ_{rev}}{T}$$

$\quad$ (for a finite change at $\qquad$ (for an infinitesimal change)
$\quad$ constant temperature)

is a thermodynamic function. Its changes depend only on the initial and final states of a system, and not on the paths taken between states. To prove this, let us consider a particular kind of cycle made up of isothermals and adiabatics arranged alternately as shown in Figure 7.2. The isothermals are AB at T_1, CD at T_2 and EF at T_3; the respective heat quantities are Q_1, Q_2 and Q_3. The adiabatics are BC, DE and FA. Let us extend the isothermal CD backward to meet the adiabatic FA at point G. Let Q'_2 be the heat along CG.

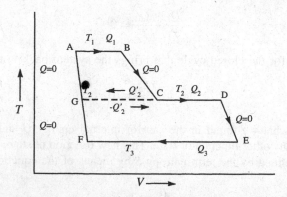

Figure 7.2   A cycle made up of isothermals and adiabatics arranged alternately

This overall cycle can be regarded as a composite of two Carnot cycles, ABCG and GDEF. From Equation 7.1 we have

$$\frac{Q_1}{T_1} + \frac{Q_2'}{T_2} = 0 \tag{7.4}$$

and

$$\frac{-Q_2' + Q_2}{T_2} + \frac{Q_3}{T_3} = 0. \tag{7.5}$$

Adding Equations 7.4 and 7.5 together we have

$$\frac{Q_1}{T_1} + \frac{Q_2}{T_2} + \frac{Q_3}{T_3} = 0. \tag{7.6}$$

This can be generalized for any reversible cycle made up of n isothermals and n adiabatics arranged alternately to give

$$\frac{Q_1}{T_1} + \frac{Q_2}{T_2} + \frac{Q_3}{T_3} + \cdots + \frac{Q_n}{T_n} = 0 \tag{7.7}$$

or

$$\sum_{i=1}^{n} \frac{Q_i}{T_i} = 0. \tag{7.8}$$

This shows that the change in entropy depends only on the initial and final states of a system, and not on the path taken between the states. The same argument can also be applied to a cycle not necessarily made up of isothermals and adiabatics.

7.5 CALCULATION OF ENTROPY CHANGES

We have already stated that it is not necessary to know what entropy is to calculate the entropy changes associated with certain processes. If the given process is irreversible (e.g. expansion of a gas into a vacuum) we obtain no value of Q_{rev} from which to determine ΔS with the help of $\Delta S = Q_{\text{rev}}/T$; we then try to imagine some path whereby the same final state is achieved reversibly and $\sum (Q_{\text{rev}}/T)$ for this gives the value of ΔS for the irreversible process. We shall discuss here the methods of calculating entropy changes under different conditions.

7.5.1 Isothermal expansion of an ideal gas into a vacuum

In passing from an initial volume V_1 to a final volume V_2, n mol of the gas change entirely irreversibly without doing any work or absorbing any heat. Nevertheless, as S is a function of state, the value of ΔS for this expansion is exactly the same as the value of ΔS for the reversible isothermal expansion of n mol of the ideal gas from V_1 to V_2. We have already deduced that

$$Q_{rev} = nRT \ln \left(\frac{V_2}{V_1} \right) = 2.3 nRT \log \left(\frac{V_2}{V_1} \right) = nRT \ln \left(\frac{P_1}{P_2} \right).$$

Then,

$$\Delta S = \frac{Q_{rev}}{T} = nR \ln \left(\frac{V_2}{V_1} \right) = nR \ln \left(\frac{P_1}{P_2} \right). \tag{7.9}$$

Where the expansion is in any degree irreversible, both the work done and the heat Q_{irrev} absorbed would to that degree be less than those for the reversible expansion and here the relation is

$$\frac{Q_{rev}}{T} = \Delta S > \frac{Q_{irrev}}{T}.$$

Example 7.1

2.0 mol of an ideal gas is expanded isothermally to five times its initial volume. Calculate the change in entropy.

Solution

Since the gas is expanded to five times its initial volume, $V_2/V_1 = 5$ and therefore $\log(V_2/V_1) = \log 5 = 0.699$. Then, given that $n = 2$ and assuming that $R = 8.314 \, \text{J} \, ^\circ\text{C}^{-1} \, \text{mol}^{-1}$ and substituting in $\Delta S = 2.303 nR \log(V_2/V_1)$, we have $\Delta S = 2.303 \times 2 \times 8.314 \times 0.699 = 26.768 \, \text{J} \, ^\circ\text{C}^{-1}$.

Note: It is not necessary to specify the temperature or the absolute values of the initial and final volumes, if the expansion is reversible.

7.5.2   Entropy changes at constant volume and at constant pressure

The equation $dU = dQ - P \, dV$ applies to any system. Hence, for a reversible change

$$dS = \frac{dQ_{rev}}{T} = \frac{dU + P \, dV}{T}.$$

If the volume is constant, $dV = 0$ and the above equation becomes

$$dS = \frac{dU}{T}$$

or

$$\left(\frac{\partial S}{\partial U}\right)_V = \frac{1}{T}.$$

Since for any system (not necessarily an ideal gas) $C_V = (\partial U/\partial T)_V$, we have $C_V = T(\partial S/\partial T)_V$. Integrating this equation between the temperature limits T_1 to T_2, we get the *change in entropy at constant volume* due to a change in temperature from T_1 to T_2. That is,

$$\Delta S = \int_{T_1}^{T_2} C_V \frac{dT}{T} = \int_{T_1}^{T_2} C_V \, d(\ln T).$$

By definition $H = U + PV$. At constant pressure, $dH = dU + P\,dV$, or $dH/T = (dU + P\,dV)/T = dS$. Hence, $(\partial S/\partial H)_P = 1/T$. Since for any system $C_P = (\partial H/\partial T)_P$, we get $C_P = T(\partial S/\partial T)_P$. For a change in temperature from T_1 to T_2 at constant pressure, the entropy change is given by

$$\Delta S = \int_{T_1}^{T_2} C_P \frac{dT}{T} = \int_{T_1}^{T_2} C_P \, d(\ln T).$$

Let us consider an interesting result. Suppose that 1 mol of an ideal gas A at standard temperature and pressure is allowed to mix isothermally with 1 mol of another ideal gas B at standard temperature and pressure, both having the same initial volume. As the gases are assumed to be ideal, during mixing they move independently of each other. Therefore, the total change in entropy is the sum of the change in entropy that each gas undergoes individually in the expansion (assuming that the gases were originally confined in two separate containers and allowed to mix together by joining the containers, thus there is an expansion). Equation 7.9 gives the change in entropy for an isothermal expansion. According to this equation, gas A has a change in entropy $R \ln 2$ and gas B has a change in entropy $R \ln 2$. Hence the total change in entropy is $2R \ln 2$. J. Willard Gibbs noticed that $\Delta S = 2R \ln 2$ irrespective of how closely identical the gasses are, provided that they are not the same. Of course, if they are the same, then ΔS equals zero. This result is often called the *Gibbs paradox*.

The above result holds for any two different ideal gases mixing together. It should be noted that the change in entropy is associated with change in partial pressure at constant total pressure. If two gases consisting of the same substance mix together, the partial pressure equals the total pressure and remains constant. Hence there is no change in entropy. Gibbs' paradox is connected with the principle of 'complete indistinguishability of identical fundamental particles' which plays an important role in quantum and statistical mechanics. From his paradox, Gibbs inferred the statistical nature of the entropy. We discuss this in Section 7.6.

7.5.3 Phase change

Phase changes occur at constant temperature and pressure. If equilibrium conditions are maintained, the process is reversible and $Q_{\text{rev}} = Q_P = (\Delta H)_{\text{phase change}}$. Then, from the relation $\Delta S = Q_{rev}/T$ we can write

$$\Delta S = \frac{(\Delta H)_{\text{phase change}}}{T}. \tag{7.10}$$

It should be noted that the applicability of this relation is restricted to phase changes under reversible conditions; it cannot be employed for an ordinary chemical reaction.

Example 7.2

The heat of vaporization of a compound at 27°C was found to be 29.288 kJ mol^{-1}. Calculate the molar entropy of vaporization at 27°C.

Solution

Given that $T = 273 + 27 = 300$ K, $(\Delta H)_{\text{vap}} = 29.288$ kJ mol^{-1}. Substituting these values into Equation 7.10, we have

$$\Delta S = \frac{29.288}{300} = 97.627 \text{ J}$$

7.5.4 Temperature changes

When the temperature of a system is changed, it leads to a change in entropy of that system. To obtain the total entropy change ΔS, we have to add an infinite number of processes, in which an amount of heat Q_{rev} has been absorbed. We know that, for a constant temperature process, $Q_P = nC_P \Delta T$ and this for each infinitesimal step of the process can be written as

$$dQ_P = nC_P \, dT = dQ_{rev}.$$

Then, from the relation $\Delta S = Q_{rev}/T$ we have

$$\Delta S = \int_{T_1}^{T_2} \frac{nC_P}{T} dT. \tag{7.11}$$

This equation can be used to calculate the absolute entropy of a substance. If C_P remains constant over the temperature range where there is no phase change, then Equation 7.11 can be written as

$$\Delta S = nC_P \int_{T_1}^{T_2} \frac{dT}{T} = nC_P \ln\left(\frac{T_2}{T_1}\right) = 2.303 nC_P \log\left(\frac{T_2}{T_1}\right). \tag{7.12}$$

Although this equation is derived for a reversible path, the same entropy change will result from the transition along any path.

Example 7.3

4 mol of water is heated at constant pressure from 30 to 40°C. Calculate the change in entropy when $C_P = 75.312 \, \text{J} \, °\text{C}^{-1} \, \text{mol.}^{-1}$

Solution

Given that $n = 4$, $T_1 = 273 + 30 = 303$ K, $T_2 = 273 + 40 = 313$ K, and $C_p = 75.312 \, \text{J} \, °\text{C}^{-1} \, \text{mol}^{-1}$. Then substituting these into Equation 7.12 we have

$$\Delta S = 2.303 \times 4 \times 75.312 \times 0.0141 \, \text{J} \, °\text{C}^{-1} = 9.782 \, \text{J} \, °\text{C}^{-1}.$$

7.5.5 Variation in C_P with temperature

If the heat capacity of a system is given as a function of temperature, namely $C = a + bT + cT^2 + \cdots$, the entropy change for heating such a system from a temperature T_1 to a temperature T_2 is given by

$$\Delta S = a \ln \left(\frac{T_2}{T_1} \right) + b(T_2 - T_1) + \frac{c}{2}(T_2^2 - T_1^2) + \cdots . \tag{7.13}$$

7.5.6 Phase changes under non-equilibrium conditions

For a transition or phase change under non-equilibrium conditions, the entropy change is not given by Equation 7.10. In this case, the entropy change must be calculated by integrating dQ/T for some reversible means of bringing about the same change in state.

To illustrate this, let us consider the isothermal freezing of supercooled water at −5°C. The overall change in state for this can be represented by

$$H_2O(\ell) \rightarrow H_2O(s)$$
$$-5°C \qquad -5°C.$$

Supercooled liquid water cannot be reversibly converted, by direct freezing, into a solid at the same temperature. We, therefore, must find some indirect reversible means of bringing about the stated change. One way of doing it is to heat the liquid to 0°C, reversibly freezing it to solid at that temperature, and finally cooling the solid back to −5°C. This can be represented as

$$
\begin{array}{lll}
\text{stage 1}: & H_2O(\ell) \rightarrow H_2O(\ell) \\
& -5°C \qquad\ \ 0°C \\
\text{stage 2}: & H_2O(\ell) \rightarrow H_2O(s) \\
& 0°C \qquad\ \ 0°C \\
\text{stage 3}: & H_2O(s) \rightarrow H_2O(s) \\
& 0°C \qquad -5°C
\end{array}
$$

Then the overall entropy change will be given by

$$\Delta S = \int_{268\,K}^{273\,K} \frac{C_\ell}{T}\,dT - \frac{(\Delta H)_f}{273} + \int_{273\,K}^{268\,K} \frac{C_s}{T}\,dT = \int_{268\,K}^{273\,K} \frac{C_\ell - C_s}{T}\,dT - \frac{\Delta H_f}{273} \qquad (7.14)$$

where C_ℓ is the heat capacity of the liquid water, C_s is the heat capacity of the solid (water) and $(\Delta H)_f$ is the heat of fusion at 0°C.

This result can also be obtained isothermally by reversibly evaporating the liquid, expanding the vapour until its pressure is equal to the vapour pressure of the solid and then condensing the vapour to solid. This can be represented as

stage 1 : $H_2O(\ell) \rightarrow H_2O(g)$
stage 2 : $H_2O(g) \rightarrow H_2O(g)$
stage 3 : $P_\ell \qquad\quad P_s$
$\qquad\qquad H_2O(g) \rightarrow H_2O(s)$
$\qquad\quad P_{(s)}$

where P_ℓ is the vapour pressure of the liquid at −5°C and P_s is the vapour pressure of the solid at −5°C.

The overall entropy change can be computed by adding all the entropy changes for all the steps. If we assume that the vapour is an ideal gas, the entropy change for stage 2 may be given by the equation

$$\Delta S = nR \ln\left(\frac{V_2}{V_1}\right) = nR \ln\left(\frac{P_1}{P_2}\right). \qquad (7.15)$$

In our cases, $P_1 = P_\ell$ and $P_2 = P_s$. The entropy change can also be written as

$$\Delta S = \frac{(\Delta H)_{vap}}{268} + R \ln\left(\frac{P_\ell}{P_s}\right) - \frac{(\Delta H)_{sub}}{268} \qquad (7.16)$$

where $(\Delta H)_{sub}$ is the heat of sublimation and $(\Delta H)_{vap}$ is the heat of vaporization. However, the heat of sublimation minus the heat of vaporization is precisely equal to the heat of fusion, i.e. $(\Delta H)_{sub} - (\Delta H)_{vap} = (\Delta H)_f$. Therefore, we can write

$$\Delta S = \frac{-(\Delta H)_f}{268} + R \ln\left(\frac{P_\ell}{P_s}\right). \qquad (7.17)$$

Note: The value of $(\Delta H)_f$ in Equation 7.17 must correspond to −5°C and will not be exactly the same as $(\Delta H)_f$ at 0°C which is used in Equation 7.14.

7.6   STATISTICAL INTERPRETATION OF ENTROPY

The ideas of disorder and entropy are clearly brought home with the use of a statistical analysis of the molecular state of a system. This approach, first used by Ludwig Boltzmann, makes a distinction between the 'microstate' and the 'macrostate' of a system. The microscopic interpretation, although not part of proper thermodynamics, is particularly helpful in understanding entropy and its behaviour. The macroscopic state or macrostate of an isolated system may be described by stating chemical composition, volume, pressure, total energy and similar macroscopic quantities. However, it is not necessary to describe the velocity and the position of all the atoms and molecules involved. A state in which all the details about atoms and molecules are specified is called a microscopic state or microstate. Usually, a large number of possible microstates correspond to a given macrostate. These microstates are indistinguishable at the macroscopic level. We shall call these microstates 'realizations' of the macrostate, and denote the number of microscopic realizations of a macroscopic state of fixed energy by the symbol Ω.

Let us suppose that an isolated system is in certain macrostate 1 with Ω_1 microscopic realizations, and that there is another macrostate 2 with Ω_2 microscopic realizations accessible to state 1. If $\Omega_2 > \Omega_1$, the system tends to assume state 2 under the randomizing influence of thermal motion. It is observed that the state of a system tends to move in the direction of increasing numbers of realizations until Ω has reached the highest possible value.

To illustrate this behaviour, let us imagine a box containing many green and yellow balls of the same size and mass. To start with, all the green balls are placed in layers at the bottom of the box, and the yellow balls at the top. If the box is then shaken vigorously, the distribution of the balls will no longer remain ordered but will be random. The microscopic realizations of the initial state may be obtained by interchanging all the green balls and separating all the yellow balls amongst themselves. Similarly the microstates that will correspond to the final macrostate may be obtained by interchanging all the balls with each other irrespective of their colour. This evidently indicates that the final state has more microscopic realizations than the initial state.

From this it is apparent that the microscopic realizations and entropy S behave in the same way. Both increase during spontaneous processes in isolated systems and reach their maximum values when equilibrium is attained. Experience shows that the same behaviour is shared by any function that increases steadily with increasing Ω. Now the question is whether it is possible to equate any such function with S. Boltzmann showed that for an isolated system the entropy must be proportional to $\ln \Omega$, i.e.

$$S = k \ln \Omega \qquad (7.18)$$

where k is the Boltzmann constant given by $k = R/N$, R and N being the gas constant and the Avogadro number, respectively. The values of k are 3.299×10^{-24} cal K^{-1} or 1.380×10^{-23} J K^{-1}. Equation 7.18 refers to the absolute entropy.

7.6.1 Justification for setting $k\ln\Omega$ equal to the entropy

Let us consider two isolated systems 1 and 2 with entropies S_1 and S_2, respectively, and realizations associated with them Ω_1 and Ω_2, respectively. If these two systems are combined into one composite system, each of the Ω_1 realizations of system 1 may be combined with any one of the Ω_2 realizations of system 2 to give a possible realization of the state of the composite system. The number Ω of such realizations is thus given by

$$\Omega = \Omega_1 \Omega_2. \tag{7.19}$$

The entropy of the combined system will be given by the sum of the entropies of the individual systems, Therefore,

$$S = S_1 + S_2. \tag{7.20}$$

However, according to Equation 7.18,

$$S_1 = k \ln \Omega_1 \qquad S_2 = k \ln \Omega_2.$$

Therefore,

$$S_1 + S_2 = k \ln \Omega_1 + k \ln \Omega_2 = k \ln(\Omega_1 \Omega_2).$$

Using Equation 7.19,

$$S_1 + S_2 = k \ln \Omega = S.$$

This justifies the appropriateness of setting $k \ln \Omega$ equal to the entropy.

We have just seen that entropy is a measure of molecular disorder. An increase in entropy indicates an increase in disorder. We can, therefore, say that the second law of thermodynamics is a matter of probability. In terms of probability, the second law, which tells us that in any process entropy increases, states that those processes occur which are most probable. However, this law in terms of probability does not exclude a decrease in entropy, but the probability is extremely low. It is to be noted that, if an increase of entropy is a probability, there is always a change that the second law might be broken. The chance that the second law is broken can be calculated. These chances are so small for any macroscopic object that the possibility can be ruled out.

7.7   CALCULATION OF ENTROPY CHANGE FOR AN ADIABATIC EXPANSION OF A PERFECT GAS ON THE BASIS OF STATISTICAL CONSIDERATIONS

Let us consider a volume V of an ideal gas and imagine that this volume is divided into a large number of very small volumes, and that $N' = nN$ molecules are distributed in these small volumes. Assume that these small volumes are all the same and constant so that the total number of such small volumes is proportional to the volume V. The total number of micro-states is then proportional to the number of different distributions in the small volumes. Since there is no interaction between the molecules of a perfect gas, each molecule may occupy any of these small volumes independently of all other molecules. Therefore, for each molecule the number of such possibilities is proportional to the volume V and consequently for N' molecules such possibilities are proportional to $V^{N'}$ so that

$$\Omega = \text{constant} \times V^{N'}. \tag{7.21}$$

If the gas expands from the volume V_1 to V_2, the ratio of the corresponding Ω-values is given by

$$\frac{\Omega_1}{\Omega_2} = \frac{\text{constant} \times V_1^{N'}}{\text{constant} \times V_2^{N'}} = \left(\frac{V_1}{V_2}\right)^{N'}. \tag{7.22}$$

Using Equation 7.18, we have

$$\Delta S = k \ln \, \Omega_2 - k \ln \, \Omega_1 = k \ln \left(\frac{\Omega_2}{\Omega_1}\right).$$

From Equation 7.22,

$$\Delta S = k \ln \left(\frac{V_2}{V_1}\right)^{N'} = k \ln \left(\frac{V_2}{V_1}\right)^{nN}$$

since $N' = nN$. Then,

$$\Delta S = nNk \ln \left(\frac{V_2}{V_1}\right) = nR \ln \left(\frac{V_2}{V_1}\right) \tag{7.23}$$

since $Nk = R$. This is the same as Equation 7.9.

Example 7.4

What would be the order of magnitude of Ω for a system having entropy $41.840 \, \text{J K}^{-1}$, given that the Boltzmann constant is $1.380 \times 10^{-23} \, \text{J K}^{-1}$

Solution

$k = 1.380 \times 10^{-23} \, \text{J K}^{-1}$ and $S = 41.840 \, \text{J K}^{-1}$. Using Equation 7.18, $41.840 = 1.380 \times 10^{-23} \times 2.303 \log \Omega$ or $\log \Omega = 13.164 \times 10^{23}$. Therefore, $\Omega = 10^{1.32 \times 10^{24}}$. This shows that in this particular case the order of Ω is almost the same as that of the Avogadro number.

7.8   THERMODYNAMIC DEFINITION OF PRESSURE

A thermodynamic definition of pressure arises from the second law. Let us consider two bodies A and B which join together to form a composite isolated assembly C as shown in Figure 7.3. Let T_A and T_B be the temperatures of A and B, respectively. A and B are free to

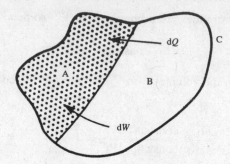

Figure 7.3   A composite assembly of two bodies A and B

exchange energy (either as heat or work). Since the composite assembly C is isolated, its energy and volume must remain constant, i.e.

$$U_C = U_A + U_B = \text{constant} \tag{7.24}$$

$$V_C = V_A + V_B = \text{constant}. \tag{7.25}$$

According to the second law, since the volume and energy of C are given in its equilibrium state, its entropy is the maximum. We also assume that A and B are both in thermodynamic states, even though they may not be in equilibrium with each other. Then according to Equation 7.20 we have

$$S_C = S_A + S_B \tag{7.26}$$

where S_A, S_B and S_C are the entropies of A, B and C, respectively. We are interested in the distributions of the internal energy and volume between A and B to achieve the largest value for S_C. The state of maximum S_C will be determined by the two conditions, namely

$$\left(\frac{\partial S_C}{\partial U_A}\right)_{V_A} = 0$$

and

$$\left(\frac{\partial S_C}{\partial V_A}\right)_{U_A} = 0.$$

The first is a condition of thermal equilibrium which requires that $T_A = T_B$, and the second is the condition of mechanical equilibrium. Keeping U_A and U_B fixed, if we differentiate S_C with respect to V_A and equate to zero, we have from Equation 7.26,

$$\left(\frac{\partial S_A}{\partial V_A}\right)_{U_A} + \left(\frac{\partial S_B}{\partial V_B}\right)_{U_B} \frac{\partial V_B}{\partial V_A} = 0. \tag{7.27}$$

However, from Equation 7.25, $dV_B = -dV_A$. Then, Equation 7.27 becomes

$$\left(\frac{\partial S_A}{\partial V_A}\right)_{U_A} = \left(\frac{\partial S_B}{\partial V_B}\right)_{U_B} \tag{7.28}$$

which is a necessary condition for mechanical equilibrium. This leads to a thermodynamic definition of pressure. *The thermodynamic pressure P of a simple compressible substance is defined by*

$$\frac{P}{T} = \left(\frac{\partial S}{\partial V}\right)_U. \tag{7.29}$$

Therefore, for the bodies A and B we have

$$\frac{P_A}{T_A} = \left(\frac{\partial S_A}{\partial V_A}\right)_{U_A}$$

$$\frac{P_B}{T_B} = \left(\frac{\partial S_B}{\partial V_B}\right)_{U_B}.$$

Hence, from Equation 7.28 for equilibrium between A and B we have

$$\frac{P_A}{T_A} = \frac{P_B}{T_B}. \tag{7.30}$$

Therefore, the necessary conditions for thermal and mechanical equilibrium between two simple compressible bodies are $T_A = T_B$ and $P_A = P_B$.

7.9 THERMODYNAMIC DEFINITION OF TEMPERATURE

A thermodynamic definition of temperature follows from a second-law analysis of thermal equilibrium. We have already given a thermodynamic definition of temperature in Section 3.5. We shall define it again from the concept of entropy of a system.

Let us consider two bodies A and B, each initially in a state of equilibrium (Figure 7.4). Their respective entropies are a function of their initial states. If we assume that both A and B are simple compressible substances, then we can write

$$S_A = f(U_A, V_A) \tag{7.31}$$

$$S_B = f(U_B, V_B) \tag{7.32}$$

where U_A, U_B and V_A, V_B are the internal energies and volumes, and S_A and S_B are the entropies of A and B, respectively.

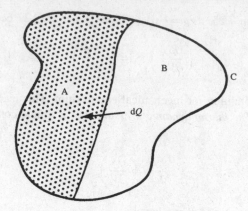

Figure 7.4 A composite assembly of two bodies A and B

Let us combine A and B to form a composite body C. A and B maintain their volumes but share their energy as heat. The composite body C forms an isolated body and attains a state in which A and B are finally in thermal equilibrium. If we assume that the possible macro-states of A and B are independent, the entropy of the composite body C is

$$S_C = S_A + S_B. \tag{7.33}$$

Application of the second law would indicate that the final entropy of the isolated body C must be greater than, or at least equal to, its initial entropy. That is, we have

$$S_{C_2} \geqslant S_{C_1} \tag{7.34}$$

where subscripts 1 and 2 denote the initial and final states, respectively. Therefore, S_{C_2} indicates the largest possible value of the initial entropy S_{C_1}. Our problem is to find the possible distribution of the total energy U between A and B so that the composite assembly C has the greatest entropy. Now combining Equations 7.31–7.33, we have

$$S_C = S_A(U_A, V_A) + S_B(U_B, V_B). \tag{7.35}$$

Since we are holding the volumes fixed, the entropy of C depends on U_A and U_B. As the total internal energy must remain constant, i.e.

$$U_C = U_A + U_B = \text{constant} \tag{7.36}$$

we cannot vary both U_A and U_B independently. If this is so, then the entropy S_C will have some maximum value for some distribution of the total energy between A and B, and this state must correspond to the state in which A and B are in thermal equilibrium. This state can be determined by the condition giving the maximum value of S_C. Now, S_C will have a maximum when the first derivative of S_C with respect to U_A (keeping V_A and V_B fixed) vanishes. Hence, differentiating Equation 7.35 with respect to U_A and equating to zero, we have

$$\frac{dS_C}{dU_A} = \left(\frac{\partial S_A}{\partial U_A}\right)_{V_A} + \left(\frac{\partial S_B}{\partial U_B}\right)_{V_B} \frac{dU_B}{dU_A} = 0. \tag{7.37}$$

From Equation 7.36, $dU_B = -dU_A$. Putting this in Equation 7.37, we obtain a necessary condition for thermal equilibrium between A and B as

$$\left(\frac{\partial S_A}{\partial U_A}\right)_{V_A} = \left(\frac{\partial S_B}{\partial U_B}\right)_{V_B}. \tag{7.38}$$

$(\partial S/\partial U)_V$ is a property which two bodies will have in common when they are in thermal equilibrium. This leads to a thermodynamic definition of temperature. We define the thermodynamic temperature T of a simple compressible substance by the relation

$$\frac{1}{T} = \left(\frac{\partial S}{\partial U}\right)_V. \tag{7.39}$$

Although we have restricted our definition of thermodynamic temperature to simple compressible substances, this definition may be extended to other classes of substance.

It is a general case that, when two bodies at different temperatures are brought into contact and allowed to share their energy, energy transfer in the form of heat occurs. The second law requires that the entropy of a combined isolated body must increase as a result of the energy transfer as heat. In the case of the composite body C in Figure 7.4, if we imagine C as isolated, and the volumes of A and B are fixed, then an amount of energy transfer as heat, dQ, from B to A will give

$$dU_A = dQ \qquad dU_B = -dQ. \tag{7.40}$$

The change in entropy S_C associated with dQ may be determined by differentiating Equation 7.35 keeping V_A and V_B constant. Then,

$$dS_C = \left(\frac{\partial S_A}{\partial U_A}\right)_{V_A} dU_A + \left(\frac{\partial S_B}{\partial U_B}\right)_{V_B} dU_B = \left(\frac{\partial S_A}{\partial U_A}\right)_{V_A} dQ - \left(\frac{\partial S_B}{\partial U_B}\right)_{V_B} dQ$$

from Equation 7.40. However, according to our thermodynamic definition of temperature

$$\left(\frac{\partial S_A}{\partial U_B}\right)_{V_A} = \frac{1}{T_A} \qquad \left(\frac{\partial S_B}{\partial U_B}\right)_{V_B} = \frac{1}{T_B}.$$

Therefore, the above equation becomes

$$dS_C = \left(\frac{1}{T_A} - \frac{1}{T_B}\right) dQ \geqslant 0. \tag{7.41}$$

The second law will be satisfied for positive dQ only if $T_B > T_A$ and for negative dQ only if $T_A > T_B$. The energy transfer as heat therefore occurs from a body at a higher temperature to body at a lower temperature.

7.10 EQUIVALENCE OF THE THERMODYNAMIC AND MECHANICAL PRESSURES

Let us consider a body of a simple compressible substance as shown in Figure 7.5. Here heat dQ is supplied to the body, work is done and the internal energy is changed. Then, from the first law, $dU = dQ + dW$.

Let us imagine that his body undergoes a reversible process, maintaining its equilibrium state. Then the change in entropy for an infinitesimal part of this reversible process can be determined by differentiating $S = f(U, V)$. From Equations 7.29 and 7.39 we can write

$$dS = \frac{1}{T}dU + \frac{P}{T}dV. \tag{7.42}$$

This equation is known as the '*Gibbs equation*'. If we denote our thermodynamic pressure by P and the mechanical pressure by P_m, then the amount of energy transfer as work to the simple compressible substance when the volume is increased by dV can be expressed as

$$dW = -P_m\, dV. \tag{7.43}$$

Putting this in $dU = dQ + dW$, we have

$$dU = dQ - P_m\, dV. \tag{7.44}$$

Combining Equations 7.42 and 7.44, we get

$$dS = \frac{dQ}{T} + \frac{P - P_m}{T}dV. \tag{7.45}$$

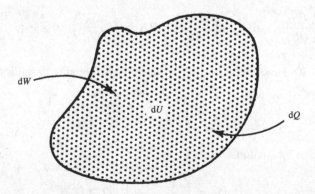

Figure 7.5 A body made up of a simple compressible substance

If we now imagine that the process is not only reversible but also adiabatic, then there will be no entropy change and no heat transfer, so that $dS = 0$ and $dQ = 0$. Hence, Equation 7.45 becomes

$$P - P_m = 0 \quad \text{or} \quad P = P_m \tag{7.46}$$

It should be noted that P and P_m are functions only of the thermodynamic state and are independent of the process which is occurring, as long as the body is in thermodynamic state. We, therefore, say that in thermodynamic states $P = P_m$. It is important to note that the mechanical and thermodynamic pressures will not be equal unless the assembly is in the equilibrium state.

7.11 EQUIVALENCE OF THE THERMODYNAMIC AND EMPIRICAL TEMPERATURE SCALES

To show that the empirical and thermodynamic temperature scales coincide, we need the experimental evidence that the gases used in empirical thermometers have the property that at very low pressures their internal energies U and the pressure–volume product PV depend entirely on temperature; conversely the temperature is a function only of the internal energy.

It is a matter of common experience that two bodies in thermal equilibrium have the same values for their empirical temperatures; also their thermodynamic temperatures will be identical. So, we can say that the thermodynamic temperature of a body is a function only of the empirical temperature. Therefore, if U is a function only of the empirical temperature, then it must also be a function only of the thermodynamic temperature. Hence, for gases at sufficiently low pressures we can write

$$T = f_1(U) \tag{7.47}$$

$$PV = f_2(T) \tag{7.48}$$

where T is the thermodynamic temperature and f_1 and f_2 are functional notations. There is considerable experimental evidence to show that these equations hold. In the equilibrium state, the entropy is a function only of the internal energy and the volume, i.e. $S = S(U, V)$. Then, differentiating

$$\left(\frac{\partial [(\partial S / \partial U)_V]}{\partial V} \right)_U = \left(\frac{\partial [(\partial S / \partial V)_U]}{\partial U} \right)_V.$$

Expressing this in terms of thermodynamic temperature and pressure (see Equations 7.29 and 7.39) we have

$$\left(\frac{\partial (1/T)}{\partial V} \right)_U = \left(\frac{\partial (P/T)}{\partial U} \right)_V. \tag{7.49}$$

Since the temperature of a gas at a low pressure is independent of volume when the internal energy is constant, the left-hand side of Equation 7.49 is zero. Therefore,

$$\left(\frac{\partial(P/T)}{\partial U}\right)_V = 0. \tag{7.50}$$

This shows that P/T is independent of U when V is constant. Now, since U depends only on T, the ratio P/T must be independent of T when V is constant. Hence, we conclude that low-density gases have the property given by Equation 7.50, or by

$$\left(\frac{\partial(P/T)}{\partial T}\right)_V = 0.$$

From Equation 7.48 we have

$$\left[\frac{\partial}{\partial T}\left(\frac{f_2(T)}{T}\right)\right]_V = \frac{d}{dT}\left[\frac{f_2(T)}{T}\right] = 0. \tag{7.51}$$

The ratio $[f_2(T)]/T$ must be a constant, which means that $f_2(T)$ must be of the form

$$PV = f_2(T) = \text{constant} \times T. \tag{7.52}$$

This is exactly the form that was arbitrarily selected for the empirical temperature. Equation 7.52 shows that, by properly choosing the constant in the defining equation for the entropy, the empirical and the thermodynamic temperature scales can be made identical.

7.12 ENTROPY CHANGE IN AN IRREVERSIBLE PROCESS

Let us consider a body A which receives energy $(dQ)_1$ and $(dQ)_2$ from the bodies B and C, and delivers energy as work to another adiabatic body D as shown in Figure 7.6. The

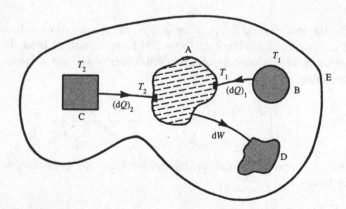

Figure 7.6 A body A receiving energies $(dQ)_1$ and $(dQ)_2$ from the bodies B and C and delivering energy as work to another adiabatic body D

combined bodies form an isolated body E. We want to investigate the effects of irreversibilities within the body A and, therefore, assume that the processes within the bodies B, D and C are reversible. By this assumption, the body D undergoes a reversible adiabatic process and therefore its entropy does not change. From Equation 7.2, the entropy changes of B and C are given by

$$(dS)_B = -\frac{(dQ)_1}{T_1} \tag{7.53}$$

$$(dS)_C = -\frac{(dQ)_2}{T_2}. \tag{7.54}$$

The minus sign is due to the convention that dQ flows out of B and C. Now application of the second law requires that the total entropy of the whole isolated system must increase or, in the limit, remain constant. Then we have,

$$(dS)_E = dS + (dS)_B + (dS)_C \geqslant 0 \tag{7.55}$$

where $(dS)_E$ = entropy of the isolated system E, and dS is the entropy change of A associated with the process. Combining Equations 7.53–7.55, we have

$$dS \geqslant \frac{(dQ)_1}{T_1} + \frac{(dQ)_2}{T_2}. \tag{7.56}$$

The equality in this equation is associated with the reversible processes within E and the inequality is associated with the irreversible process. If the temperatures T_1 and T_2 of the bodies B and C, respectively, are only infinitesimally different from the temperatures at the corresponding points on the boundary of A, the heat transfer will be reversible. In that case the inequality would be due entirely to irreversibilities within A. We may, therefore, take T_1 and T_2 as the temperatures on the boundary of A where $(dQ)_1$ and $(dQ)_2$ enter into A. Now, if the boundary of A is at a uniform temperature, say T, we may write

$$dQ = (dQ)_1 + (dQ)_2$$

where dQ is the total energy transferred to A during the infinitesimal process. Hence, Equation 7.56 becomes

$$dS \geqslant \frac{dQ}{T}. \tag{7.57}$$

We can extend this result to a body which receives energy as heat from several other bodies or sources. Then the generalized form of Equation 7.57 becomes

$$dS \geqslant \sum_{n=1}^{\infty} \frac{(dQ)_n}{T_n} \tag{7.58}$$

where $(dQ)_n$ is an infinitesimal amount of energy transferred as heat from a source to the body at a point where the boundary temperature is T_n.

We find from the above discussions that the entropy of a body undergoing an adiabatic process must increase or, in the limit of a reversible adiabatic process, remain constant. For an irreversible adiabatic process the entropy of a body always increases regardless of whether energy is transferred to or from the body as work during the process. The entropy of a body can be decreased only through energy transfer from the body as heat. Hence, for a body undergoing an adiabatic process we may write

$$\Delta S \geqslant 0. \tag{7.59}$$

This is the entropy change in an irreversible process for any substance.

Let us now consider an adiabatic process for a simple compressible substance. If we apply Equation 7.57 to Equation 7.45, we find that the thermodynamic and mechanical pressures for a simple compressible substance undergoing an irreversible process must differ, and also that

$$(P - P_m)\, dV > 0. \tag{7.60}$$

This shows that, if the volume is increasing, P must be greater than P_m, i.e. the mechanical pressure is less than the thermodynamic pressure. The term $P\, dV$ represents the work that the body would have done if the process were reversible, and $P_m\, dV$ represents the actual work done by the body in an irreversible process. Consideration of a negative volume change would similarly imply that more work would be required to compress a substance irreversibly over a given volume than if the process could be done reversibly. From Equations 7.53–7.55 we may write

$$(dS)_E = dS - \frac{(dQ)_1}{T_1} - \frac{(dQ)_2}{T_2} = dS - \left(\frac{(dQ)_1}{T_1} + \frac{(dQ)_2}{T_2} \right). \tag{7.61}$$

As before, we can generalize Equation 7.61 (see Equation 7.58)

$$(dS)_E = dS - \sum_{n=1,2,3\ldots} \frac{(dQ)_n}{T_n}. \tag{7.62}$$

The right-hand side of this equation may be regarded as 'entropy production'. The second law then sates that

$$dS - \sum_n \frac{(dQ)_n}{T_n} \begin{cases} > 0 & \text{for all real processes} \\ = 0 & \text{for reversible processes} \\ < 0 & \text{never} \end{cases}$$

7.13 DIAGRAMMATIC REPRESENTATION OF THE ENTROPY–TEMPERATURE RELATIONSHIP

We have already seen that the entropy of a substance can be regarded as a function of either T and V, or V and P, or T and P. Hence, it is possible to represent entropy by a surface plotting entropy and any pair of the state variables T, P and V along three mutually perpendicular axes. Of the three sets of such diagrams, the most useful is that in which entropy, pressure and temperature are the variables. Such entropy surfaces have the same general shape for all gases.

Instead of considering an entropy surface on a three-dimensional diagram it is often sufficient to consider only the projection of such a surface on the T–S plane. To signify the properties of such projections let us consider any arbitrary reversible process represented by the curve AB in Figure 7.7. The area represented by the vertically shaded strip is $T\,dS$, which equals the heat dQ absorbed. Therefore, the total heat Q absorbed between the states A and B is given by

$$\int dQ = \int_{S_A}^{S_B} T\,dS$$

or

$$Q = \int_{S_A}^{S_B} T\,dS.$$

This is represented by the area bounded by the curve AB, the entropy axis (i.e. the segment $S_A S_B$) and the vertical lines AS_A and BS_B (parallel to the T axis). We have seen that in a P–V diagram the area under the curve represents the work done. Analogously, the area under the curve in a temperature–entropy diagram represents heat. As a matter of convention, the heat Q is considered positive if it is absorbed by a system (with reference to Figure 7.7 this is the case when the process takes place in the direction shown by the arrows). If heat is given out (i.e. if the process takes place in the opposite direction), Q is considered as negative.

It is obvious from Figure 7.7 that isothermal processes in a T–S diagram are represented by horizontal straight lines perpendicular to the T axis (since there will be no change in temperature), and adiabatic processes are represented by vertical lines parallel to the T axis (since there will be a change in temperature).

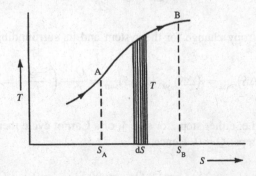

Figure 7.7 Diagrammatic representation of entropy versus temperature relationship

7.14   ENTROPY CHANGE IN THE FOUR STEPS OF A CARNOT CYCLE

We have already stated that for reversible change in state the entropy is given by the relation (see Equation 7.2)

$$dS = \frac{(dQ)_{rev}}{T}.$$

From this it is clear that the entropy change for a system for isothermal reversible changes is given by

$$(\Delta S)_{sys} = \int dS = \int \frac{dQ}{T} = \frac{Q}{T}. \tag{7.63}$$

In a Carnot cycle, the first and the third steps are isothermal steps; also these steps are reversible. If we take V_i as the initial volume and V_f as the final volume in these steps, and Q as the amount of heat involved (either absorbed or given out), then according to Equations 6.38 and 6.40 we have

$$Q = W = nRT \ln\left(\frac{V_f}{V_i}\right). \tag{7.64}$$

Note: Temperature T is a general term. We did not consider the sign of work W in writing Equation 7.64.

Hence, combining Equations 7.63 and 7.64 gives

$$(\Delta S)_{sys} = \frac{Q}{T} = nR \ln\left(\frac{V_f}{V_i}\right). \tag{7.65}$$

If Q is the heat absorbed or given out by the system in any isothermal stage of a Carnot cycle, then $-Q$ must be the heat absorbed or given out by the surroundings. Therefore,

$$(\Delta S)_{surr} = \frac{-Q}{T}. \tag{7.66}$$

Hence, the total entropy change for the system and its surroundings is

$$(\Delta S)_{total} = (\Delta S)_{sys} + (\Delta S)_{surr} = \frac{Q}{T} + \left(\frac{-Q}{T}\right) = 0. \tag{7.67}$$

In any adiabatic step, i.e. either step 2 or step 4, of a Carnot cycle there is no heat exchange. Therefore, for these steps, $Q = 0$. Hence,

$$(\Delta S)_{sys} = (\Delta S)_{surr} = (\Delta S)_{total} = 0 \tag{7.68}$$

7.14.1 Calculation of changes in pressure, temperature, volume, internal energy, enthalpy and entropy for each step of a reversible Carnot cycle for an ideal gas

Let us assume that T_1, T_3, V_1 and V_3 (see Figure 6.1) are known and that C_P and C_V are constant.

Step 1: isothermal expansion. In the case of an ideal gas as the working substance, the internal energy and the enthalpy depend only on the temperature. For an isothermal expansion, $\Delta T = 0$, $\Delta U = 0$, $\Delta H = 0$ and $PV = nRT$. Then,

$$\Delta P = nRT_1 \Delta \left(\frac{1}{V} \right) = nRT_1 \left(\frac{1}{V_2} - \frac{1}{V_1} \right)$$

$$\Delta S = \frac{Q}{T_1} = \frac{-W}{T_1} = nR \ln \left(\frac{V_2}{V_1} \right)$$

since the isothermal reversible $W = -nRT_1 \ln(V_2/V_1)$.

Step 2: adiabatic expansion. In this case, $\Delta T = T_3 - T_2 = T_3 - T_1$ and $\Delta U = nC_V(T_3 - T_2) = nC_V(T_3 - T_1)$. Since the change is reversible and adiabatic, there is no change in entropy, i.e. $\Delta S = 0$.

$$\Delta H = \Delta U + \Delta(PV) = \Delta U + nR(\Delta T) = nC_V(T_3 - T_1) + nR(T_3 - T_1)$$
$$= n(C_V + R)(T_3 - T_1) = nC_P(T_3 - T_1)$$
$$\Delta V = V_3 - V_2.$$

However, for an adiabatic reversible change, $TV^{\gamma-1} = \text{constant}$. Therefore,

$$\Delta V = V_2 \left(\frac{T_1}{T_3} \right)^{1/(\gamma-1)} - V_2 = \left[\left(\frac{T_1}{T_3} \right)^{1/(\gamma-1)} - 1 \right] V_2$$
$$\Delta P = P_3 - P_2.$$

For an adiabatic reversible change, $PV^{\gamma} = \text{constant}$. Then,

$$\Delta P = P_2 \left(\frac{V_2}{V_3} \right)^{\gamma} - P_2 = P_2 \left[\left(\frac{V_2}{V_3} \right)^{\gamma} - 1 \right] = \frac{nRT_2}{V_2} \left[\left(\frac{V_2}{V_3} \right)^{\gamma} - 1 \right]$$
$$= \frac{nRT_1}{V_2} \left[\left(\frac{T_3}{T_1} \right)^{\gamma/(1+\gamma)} - 1 \right].$$

Step 3: isothermal compression. In this case, $\Delta T = 0$, $\Delta U = 0$, $\Delta H = 0$ and $\Delta V = V_4 - V_3$. As steps 2 and 4 are adiabatic, $S_2 = S_3$ and $S_1 = S_4$, or $S_1 - S_2 = S_4 - S_3$, but $\Delta S = S_4 - S_3 = nR \ln(V_4/V_3)$. Hence,

$$\Delta S = S_4 - S_3 = S_1 - S_2 = nR \ln \left(\frac{V_4}{V_3} \right) = nR \ln \left(\frac{V_1}{V_2} \right).$$

This also gives $V_1/V_2 = V_4/V_3$, or $V_4 = V_1 V_3/V_2$. Then,

$$\Delta V = V_4 - V_3 = \frac{V_1 V_3}{V_2} - V_3 = V_3\left(\frac{V_1}{V_2} - 1\right) = \frac{V_3}{V_2}(V_1 - V_2)$$

$$= \left(\frac{T_1}{T_3}\right)^{1/(1+\gamma)}(V_1 - V_2)$$

$$\Delta P = P_4 - P_3 = \frac{nRT_4}{V_4} - \frac{nRT_3}{V_3} = nRT_3\left(\frac{1}{V_4} - \frac{1}{V_3}\right) = \frac{nRT_3}{V_1 V_3}(V_2 - V_1)$$

because $V_1/V_2 = V_4/V_3$. Hence,

$$\Delta P = \frac{nRT_3}{V_1 V_2}\left(\frac{T_3}{T_1}\right)^{1/(1+\gamma)}(V_2 - V_1).$$

Step 4: adiabatic compression. In this case,

$$\Delta T = T_1 - T_4 = T_1 - T_3$$

$$\Delta V = V_1 - V_4 = V_1\left(1 - \frac{V_3}{V_2}\right) = V_1\left[1 - \left(\frac{T_1}{T_3}\right)^{1/(1+\gamma)}\right]$$

$$\Delta P = P_1 - P_4.$$

However, P_4 can be written as $P_4 = P_1 + (P_2 - P_1) + (P_3 - P_2) + (P_4 - P_3)$ or

$$P_4 = P_1 + \left(\frac{nRT_2}{V_2} - \frac{nRT_1}{V_1}\right) + \left[\frac{nRT_1}{V_2}\left(\frac{T_3}{T_1}\right)^{\gamma/(1+\gamma)} - 1\right]$$

$$+ \frac{nRT_3}{V_1 V_2}\left(\frac{T_3}{T_1}\right)^{1/(1+\gamma)}(V_2 - V_1)$$

$$= P_1 + nRT_1\left(\frac{1}{V_2} - \frac{1}{V_1}\right) + \left[\frac{nRT_1}{V_2}\left(\frac{T_3}{T_1}\right)^{\gamma/(1+\gamma)} - 1\right]$$

$$+ \frac{nRT_3}{V_1 V_2}\left(\frac{T_3}{T_1}\right)^{1/(1+\gamma)}(V_2 - V_1).$$

Therefore,

$$\Delta P = P_1 - P_4 = -nRT_1\left(\frac{1}{V_2} - \frac{1}{V_1}\right) - \left[\frac{nRT_1}{V_2}\left(\frac{T_3}{T_1}\right)^{\gamma/(1+\gamma)} - 1\right]$$

$$- \frac{nRT_3}{V_1 V_2}\left(\frac{T_3}{T_1}\right)^{1/(1+\gamma)}(V_2 - V_1)$$

$$\Delta U = nC_V(T_1 - T_3)$$

$$\Delta H = nC_P(T_1 - T_3)$$

$$\Delta S = 0.$$

7.15   TEMPERATURE–ENTROPY DIAGRAM FOR A CARNOT CYCLE

A temperature–entropy diagram is particularly useful, as it illustrates graphically the work and heat involved in a reversible cycle. A typical temperature–entropy diagram for reversible Carnot cycle is illustrated in Figure 7.8.

Step 1. In this step there is no change in temperature (isothermal), heat Q_2 is absorbed by the system, and the entropy increases from S_1 to S_2. If T_2 is the temperature, then

$$\frac{Q_2}{T_2} = \Delta S = S_2 - S_1. \tag{7.69}$$

From this equation it is obvious that the heat Q_2 absorbed is given by the product of temperature and the entropy change. Graphically, this can be represented by the area under the line AB in Figure 7.8.

Step 2. In this step the process is adiabatic reversible expansion. The temperature decreases to T_1 but there is no heat exchange and, therefore, no change in entropy. This step is illustrated by the line BC in Figure 7.8.

Step 3. This is an isothermal reversible compression. Heat Q_1 is given out, temperature remains constant but the entropy decreases to S_1. Hence,

$$\frac{Q_1}{T_1} = \Delta S = S_1 - S_2 = -(S_2 - S_1). \tag{7.70}$$

The heat Q_1 is given by the product of temperature and the change in entropy. Graphically, this can be represented by the negative of area under the line DC in Figure 7.8.

Step 4. This is an adiabatic reversible compression. The temperature rises to T_2 but there is no entropy change, since there is no heat exchange. This change is illustrated by the line DA in Figure 7.8.

For a complete Carnot cycle, $\Delta U = 0$. Therefore, $Q_1 + Q_2 = -W$. Hence,

$$W = -[T_2(S_2 - S_1) - T_1(S_2 - S_1)] = -(T_2 - T_1)(S_2 - S_1)$$

$$= \text{negative of the area enclosed by the cycle.} \tag{7.71}$$

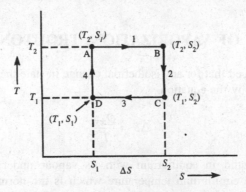

Figure 7.8   Entropy versus temperature diagram for the Carnot cycle

This shows that the work and heats involved in a Carnot cycle are clearly illustrated by a temperature–entropy diagram.

The area within the rectangle ABCD represents a cyclic process and the net heat absorbed by the system in the cycle. According to the first law of thermodynamics this area also represents the net work. *This concept applies only to closed curves. The area between a curve and the S axis will not represent work.*

The heat Q_2 absorbed at the higher temperature T_2 is equal to the area $\Delta S\, T_2$. The work W is given by the area $(T_2 - T_1)\Delta S$. Therefore, the efficiency is given by

$$\eta = \frac{W}{Q_2} = \frac{(T_2 - T_1)\Delta S}{T_2 \Delta S} = \frac{T_2 - T_1}{T_2}.$$

This is the same expression as we deduced before for the efficiency of a Carnot engine.

Example 7.5

1 mol of an ideal gas undergoes a reversible Carnot cycle with $V_1 = 20\,\ell$ and $V_2 = 40\,\ell$. Calculate the change in entropy at each step, given that $T_1 = 27°C$ and $C_V = \frac{3}{2}R$.

Solution

From the given data we can calculate V_3 and V_4, which are $74\,\ell$ and $37\,\ell$, respectively. Hence, using Equation 7.65 for the isothermal expansion

$$\Delta S = 2.30R\ \log\left(\frac{V_2}{V_1}\right) = (2.30)(8.314)\log\,2 = 5.858\,\text{J}\,°\text{C}^{-1}.$$

For isothermal compression,

$$\Delta S = 2.30R\ \log\left(\frac{V_4}{V_1}\right) = -(2.30)(8.314)\log\,2 = -5.858\,\text{J}\,°\text{C}^{-1}.$$

For adiabatic expansion and compression there is no entropy change.

7.16   ENTROPY OF VAPORIZATION: TROUTON'S RULE

We have already deduced that for any isothermal change in state the change in entropy for a finite change is given by the equation

$$\Delta S = \frac{Q_{rev}}{T}. \tag{7.72}$$

Let us consider a liquid in equilibrium with its vapour under 0.1 MPa pressure, the temperature being the equilibrium temperature which is the normal boiling point of the liquid considered. Let us imagine that this liquid is confined in a cylinder fitted with a

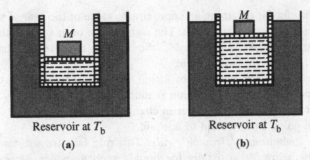

Reservoir at T_b Reservoir at T_b

(a)   (b)

Figure 7.9   A liquid confined in a cylinder fitted with a movable piston having a weight M: (a) before vaporization; (b) after vaporization

movable piston carrying a weight M which is equivalent to 0.1 MPa pressure (Figure 7.9). Let this cylinder be immersed in a reservoir at an equilibrium temperature T_b. If the temperature T_b is raised by an infinitesimal quantity, a small quantity of heat flows from the reservoir to the liquid in the cylinder. Because of this heat flow some liquid vaporizes and the mass M rises (as shown in Figure 7.9(b)). Then, if the temperature T_b is lowered by an infinitesimal quantity, the same quantity of heat will flow back to the reservoir. In that case, the vapour originally formed will condense and the weight M will drop back to its original position (as shown in Figure 7.9(a)). At the end of this step both the system and the reservoir will be restored to their initial condition. Since the temperature change was infinitesimal, the transformation is reversible. The quantity of heat required in this cycle is Q_{rev}. Since the pressure remains unchanged, $Q_p = \Delta H$. Hence for vaporization of a liquid, Equation 7.72 can be written as

$$(\Delta S)_{vap} = \frac{(\Delta H)_{vap}}{T_b}.\tag{7.73}$$

Since the transformation of liquid to its vapour requires heat, $(\Delta H)_{vap}$ and $(\Delta S)_{vap}$ are always positive. Arguing in the same way, the entropy of fusion at the melting point T_m can be written as

$$(\Delta S)_f = \frac{(\Delta H)_f}{T_m}\tag{7.74}$$

where $(\Delta H)_{vap}$ is the heat of vaporization at the boiling temperature T_b and $(\Delta H)_f$ is the heat of fusion at the melting temperature T_m. For the transformation of a solid to a liquid, ΔH is positive; therefore, the entropy of a solid increases when it melts.

For many liquids, the entropy of vaporization at the normal boiling point has approximately the same value:

$$(\Delta S)_{vap} \simeq 87.864\,\text{J}\,°\text{C}^{-1}\,\text{mol}^{-1}.\tag{7.75}$$

This is known as the rule of Pictet and Trouton. From this it follows that, for liquids which obey this rule

$$(\Delta H)_{vap}(\text{J}\,\text{mol}^{-1}) = 87.864 T_b.\tag{7.76}$$

This relation is useful for obtaining an approximate value of the heat of vaporization of a liquid from knowledge of its boiling point. The significance of this rule is that since $(\Delta S)_{vap}$ is always approximately $87.864 \, J \, °C^{-1} \, mol^{-1}$, the increase in molecular disorder on vaporization is almost the same for many liquids.

There is no general rule for entropies of fusion of solids at the melting temperature. For most solid substances the entropy of fusion is much less than the entropy of vaporization. This indicates that the increase in disorder in changing from a liquid to its vapour is larger than the change in going from a solid to a liquid. .

There are some limitations of Trouton's rule. This rule fails for substances with boiling points of 150 K or below. It also fails for liquids such as water, alcohols, amines and ammonia. Such liquids are known as 'abnormal' liquids. The 'constant' in this empirical rule varies from about 75.312 for O_2 boiling at 90 K to 97.236 for zinc boiling at 1180 K. This variation can be explained as follows.

The quantities related by Trouton's rule are those at the normal boiling point. The change in state is that of 1 mol of a liquid to its vapour at 0.1 MPa which is in equilibrium with the liquid at the normal boiling temperature. When 1 mol of a substance is vaporized at its boiling point, although we consider pressure as a uniform pressure of 0.1 MPa, the 1 mol of the substance spreads over very different volumes depending on the magnitude of the boiling point. Therefore, in addition to the entropy of vaporization which is possibly constant, Trouton's rule involves entropies of expansion which certainly differ.

7.16.1   Modification of Trouton's rule: Hildebrand's rule

Hildebrand argued that, for comparison, it would be more appropriate to evaluate $(\Delta S)_{vap}$ for a change in state chosen so that the molar volumes of different vapours are always the same, rather than comparing at a pressure of 0.1 MPa. He found that, if for any liquid T' is the boiling temperature at which the vapour occupies $22.4 \, \ell$, then the change in entropy for vaporization of that liquid may be expressed empirically by the relation

$$[(\Delta S)_{vap}]_{T'} = \frac{(\Delta H)_{vap}}{T'} = 84.935 \, J \, °C^{-1}. \tag{7.77}$$

The boiling temperature T' can be found from the following conditions.

1. The vapour pressure P' must satisfy the relation $P' = (P_{vap})_{T'}$.

2. On the assumption that the vapour behaves like an ideal gas, P' must satisfy the relation $22.4P' = RT'$.

Clearly, the combination of these two conditions will specify the temperature T'. Since the difference between the heat $(\Delta H)_{vap}$ of vaporization at the boiling temperature T_b and at the temperature T' is small, we can neglect it.

Sometimes for convenience the entropy unit is abbreviated to eU instead of $J \, °C^{-1} \, mol^{-1}$. eU stands for 'entropy unit':

$$1 \, eU = 1 \, J \, °C^{-1} mol^{-1}.$$

Example 7.6

Calculate the entropy change when 1 mol of ice at 0°C is heated to steam at 100°C at 0.1 MPa pressure. The molar heat of fusion of ice at 0°C is 6.008 kJ and the heat of vaporization of water at 100°C is 40.668 kJ. The specific heat of water is approximately $4.184 \, \mathrm{J \, °C^{-1} \, g^{-1}}$.

Solution

The total entropy change is the sum of the molar entropies of the fusion of ice and the vaporization of water together with the increase in the entropy between 0 and 100°C. Using Equation 7.74 at 0°C,

$$(\Delta S)_f = \frac{6.008}{273} = 22.007 \, \mathrm{J \, °C^{-1} mol^{-1}}.$$

Again, using Equation 7.73 at 100°C we have

$$(\Delta S)_{vap} = \frac{40.668}{373} = 109.029 \, \mathrm{J \, °C^{-1} mol^{-1}}.$$

The increase in entropy of the water between 0 and 100°C is given by the relation

$$(\Delta S)_P = \int_{T_1}^{T_2} C_P \frac{dT}{T} = C_P \ln\left(\frac{T_2}{T_1}\right).$$

The molar heat capacity of water at constant pressure is equal to its molecular weight multiplied by its specific heat. Hence, for the water this is equal to $18 \times 1 = 18$. Then,

$$(\Delta S)_P = 18 \times \ln\left(\frac{373}{273}\right) \times 4.184$$

$$= 18 \times 4.184 \times 2.303 \log\left(\frac{373}{273}\right) = 98.383 \, \mathrm{J \, °C^{-1} mol^{-1}}.$$

Therefore, the total entropy change is

$$(\Delta S)_{vap} + (\Delta S)_f + (\Delta S)_p = 109.029 + 22.007 + 98.383$$

$$= 229.419 \, \mathrm{J \, °C^{-1} mol^{-1}}.$$

7.17 UNAVAILABLE ENERGY

When heat flows from a hot body to a cold body, the entropy of the cold body increases and it goes from order to disorder. Separate hot and cold bodies can serve as the high- and low-temperature sources for a heat engine. Therefore, they can be used to obtain useful work.

However, when the two bodies are connected together and they reach the same temperature, no work can be obtained from these bodies; in this process, order has transformed to disorder.

This example illustrates another aspect of the second law of thermodynamics: that, in any natural (irreversible) process, some energy is not available to do useful work. It can be shown that the energy which is not available to do useful work is given by $T_\ell \Delta S$, where ΔS is the increase in entropy during the process and T_ℓ is the lowest available temperature. From the law of conservation of energy we know that no energy is ever lost in any process; it becomes less useful and can do less useful work. So, for the system concerned, with time energy degrades, i.e. goes from more orderly forms (such as mechanical energy) to the least orderly form (such as thermal or internal energy).

From this it can be predicted that, as time passes, the Universe will approach maximum disorder; matter will become a uniform mixture and heat will have flowed from high-temperature regions to low-temperature regions until the entire Universe reaches one temperature. Under this condition, no work can be done; all the energy of the Universe will have degraded to thermal energy and all changes will have stopped. This is known as the *heat death of the Universe*. This state appears to be an inevitable consequence of the second law of thermodynamics. However, luckily it lies infinitely far in the future but, if the Universe is finite, as is assumed by some cosmologists, this unfortunate consequence may not be that far in the future as one can think. Another point to remember is the question of whether the second law in its present state actually applies in the infinite reaches of our Universe. The answer to this question is still unknown.

EXERCISES 7

7.1 Identify whether the following statements are correct.

(a) The second law of thermodynamics is another way of stating the law of conservation of entropy.

(b) The second law of thermodynamics is only of statistical importance.

(c) At equilibrium the entropy of a system is a maximum.

(d) When a substance vaporizes, its entropy increases.

(e) It is impossible by a cyclic process to take heat from a reservoir and to convert it into work without simultaneously transferring heat from a hot to a cold reservoir.

(f) The second law is not a deduction from the first law but is a separate law of nature, referring to an aspect of nature different from that contemplated by the first law.

(g) Entropy is not a variable of state.

(h) In thermodynamic states, mechanical and thermodynamic pressures are not equivalent.

(i) The rule of Pictet and Trouton is valid for certain substances.

(j) All reversible engines working between the same temperature limits are equally efficient.

7.2 (a) What do you understand by the symbol Ω in statistical thermodynamics?

(b) Why is entropy defined in terms of logarithms?

(c) Why is the entropy constant selected as the Boltzmann constant?

(d) What does the second law of thermodynamics deal with?

(e) Can the entropy of an assembly ever decrease?

(f) Why is entropy a measure of disorder?

(g) What is the conceptual basis and definition of the thermodynamic temperature?

(h) Suppose a substance could exist in equilibrium at negative temperatures. What would this mean with regard to the energy and entropy (on a microscopic scale)?

(i) What is the thermodynamic definition of pressure?

(j) What do you understand by entropy production?

(k) Under what conditions does the entropy change of matter undergoing an adiabatic process remain constant?

(l) Is $-10°C$ a negative thermodynamic temperature?

7.3 Show that the change in entropy depends only on the initial and final states of a system, and not on the path taken between the states.

7.4 In a statistical interpretation of entropy we set $S = k \ln \Omega$. Can you justify this?

7.5 Show that for an isothermal expansion of an ideal gas into a vacuum, the change in entropy is given by $\Delta S = nR \ln(P_1/P_2)$, where P_1 and P_2 are the initial and final pressures.

7.6 An ideal heat engine operates through a reversible Carnot cycle between the temperatures $T_1 = 273$ K and T_2 and produces 418.400 J of work per cycle. The changes in entropy associated with this process are shown in the diagram.

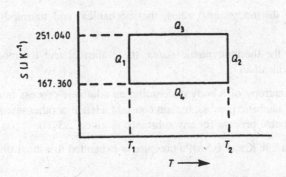

(a) Indicate the steps of the Carnot cycle in this diagram and the direction along each step.

(b) Calculate Q_1, Q_2, Q_3 and Q_4 for a complete cycle, and T_2.

7.7 The process $H_2O(\ell)$ $(-10°C) \rightarrow H_2O(s)$ $(-10°C)$, involving 1 mol of H_2O, is carried out in an isolated system consisting of a thermostat at $-10°C$. The heat of fusion of water is $334.720 \, J \, g^{-1}$ at $0°C$, the specific heats of water and ice are $4.184 \, J \, °C^{-1} \, g^{-1}$ and $2.092 \, J \, °C^{-1} \, g^{-1}$, respectively and are assumed to be constant with respect to temperature.

(a) Calculate the change in entropy for the fusion.

(b) Calculate the change in entropy of the thermostat.

(c) Is the process spontaneous? If so, discuss why.

7.8 2 mol of nitrogen at 300 K and atmospheric pressure are heated reversibly to 600 K

(a) at constant pressure and

(b) at constant volume.

Calculate the changes in entropy associated with each of these processes. C_P for nitrogen for the temperature range is given by the empirical relation $C_P = (27.0 + 0.0060 \, T) \, J \, K^{-1} \, mol^{-1}$. Assume that nitrogen behaves as an ideal gas.

7.9 Show that, on the basis of statistical consideration, the change in entropy for an adiabatic expansion of n mol of a perfect gas is given by

$$\Delta S = nR \ln\left(\frac{V_2}{V_1}\right)$$

where V_1 and V_2 are the initial and final volumes.

7.10 Derive an expression for thermodynamic pressure of a simple compressible substance in terms of its temperature and the change in entropy with respect to its volume when its internal energy is constant.

7.11 Show that the thermodynamic temperature of a simple compressible substance is given by the relation $1/T = (dS/dU)_V$.

7.12 Prove that, in thermodynamic states, the mechanical and thermodynamic pressures are equivalent.

7.13 Prove that, in the thermodynamic states, the empirical and thermodynamic temperature scales can be made identical.

7.14 Show that the entropy of a body undergoing an adiabatic process must increase or, in the limit of a reversible adiabatic process, remain constant. Hence or otherwise show that the entropy change in an irreversible process for any substance is given $\Delta S \geqslant 0$.

7.15 An ideal gas at 298 K and 0.5 MPa pressure is expanded to a final pressure of 0.1 MPa

(a) isothermally and reversibly and

(b) isothermally against a constant pressure of 0.1 MPa.

Calculate the change in entropy for each of these expansions.

7.16 1 mol of a liquid substance of molecular weight 120 is vaporized in a reversible manner at its boiling point of 60°C, vapour expanding against a pressure of 0.2 MPa. Calculate Q, W, ΔU, and ΔH and ΔS. Assume that the vapour of this liquid behaves as a perfect gas, given that the heat of vaporization of this liquid is $209.200 \, J \, g^{-1}$.

7.17 Show that for a Carnot cycle the total entropy change for the system and its surroundings is zero.

7.18 The work and heat involved in a Carnot cycle can be illustrated by a temperature–entropy diagram. Discuss and illustrate.

7.19 State the rule of Pictet and Trouton. Discuss its limitations and significance.

7.20 Discuss the basis of Hildebrand's rule. How does it modify Trouton's rule?

7.21 Discuss why the molar entropy of vaporization of a substance is greater than its molar entropy of fusion.

7.22 What do you understand by 'entropy unit'? How is it related to the general unit of entropy? Estimate the heat of vaporization of chloroform, taking its boiling point as 61°C.

7.23 Assuming that C_P and C_V are constant, sketch diagrams for the reversible Carnot cycle of an ideal gas for

(a) pressure versus entropy

(b) entropy versus volume

(c) internal energy versus entropy

(d) enthalpy versus entropy.

7.24 A block of ice at 0°C of mass 2 kg is placed inside a container which acts as a heat reservoir. The ice melts very slowly to water at 0°C. Assuming that the temperature of the container is only infinitesimally higher than 0°C, determine the change in entropy of the ice and the container, given that the heat of fusion of ice is $333.0 \, kJ \, kg^{-1}$.

7.25 An iron ball of mass 4 kg and at a temperature 607°C is thrown into a lake at a temperature 7°C. Assuming that the rise in temperature of the lake is not appreciable, determine the entropy changes of the ball and the lake, given that the specific heat of the iron is $460.240 \, J \, °C^{-1} kg^{-1}$ and does not change with temperature.

7.26 1 mol of an ideal gas at 25°C is expanded from 20 to $40 \, \ell \, mol^{-1}$

(a) isothermally and reversibly

(b) isothermally and irreversibly against a zero opposing pressure.

Calculate the changes in entropy in both cases.

7.27 A spring is placed in a large thermostat at a constant temperature of 27°C and stretched isothermally and reversibly from its equilibrium length to ten times that. During this process the spring absorbs 4.184 J of heat. It is then released without any restraining back tension and allowed to retract to its equilibrium initial length. During retraction it gives up 14.644 J of heat. Calculate

(a) the change in entropy for the stretching of the spring

(b) the change in entropy for the retraction of the spring

(c) the total change in entropy for stretching and retraction of the spring and the thermostat.

7.28 Calculate the change in entropy when 1 mol of water at $-173°C$ is heated at a constant pressure of 0.1 MPa to $227°C$. For the water the latent heat of vaporization is 40.292 kJ mol^{-1}, the latent heat of fusion is 6.004 kJ mol^{-1}, and the molar heat capacities in the solid, liquid and vapour forms are $0.50+0.03T$, 75.312 and $7.256+2.30 \times 10^{-3}T+2.83 \times 10^{-7}T^2$ J, respectively.

7.29 1 mol of an ideal gas at $0°C$ is allowed to expand isothermally from 10.0 MPa to 1.0 MPa. Calculate the change in entropy of

(a) the gas when the expansion is reversible and the gas expands freely so that no work is done

(b) the isolated system (i.e. the gas and its surroundings) when the expansion is reversible and the gas expands freely.

7.30 Calculate the entropy change per mole which occurs when cadmium vapour at $767°C$ and 0.1 MPa pressure is heated to $1027°C$ and then compressed so that its final pressure is 0.6 MPa. Assume that cadmium vapour behaves as an ideal monatomic gas.

7.31 The following data for sulphur have been obtained from K.K. Kelley, 1949, *Bulletin of US Bureau of Mines No 746*:

sulphur (rhombic) $C_P=3.58+6.24 \times 10^{-3}T$ cal K^{-1} mol^{-1} for 298–368.6 K
sulphur (monoclinic) $C_P=3.56+6.96 \times 10^{-3}T$ cal K^{-1}mol^{-1} for 368.6 to melting point
sulphur (liquid) $C_P=5.40+5.00 \times 10^{-3}T$ cal K^{-1}mol^{-1} for melting point to boiling point
Transition temperature from rhombic $\rightarrow$ monoclinic, 368.6 K
Melting point of monoclinic sulphur, 392 K
Latent heat of transition, 0.086 kcal mol^{-1}
Latent heat of fusion, 0.30 kcal mol^{-1}.

Calculate the change in entropy when 1 mol of sulphur is heated from 300 to 410 K.

7.32 State with reasons which substance in each of the following groups has the higher value of entropy:

(a) molar entropies of vaporization of benzene and of water

(b) molar entropies of water and nitrogen at 298 K

(c) molar entropies of copper and mercury at 298 K

(d) molar entropies of diamond and copper at 298 K.

7.33 Consider the following chemical reactions at 298 K and predict the sign of the entropy change in each case.

(a) $Ba(s) + \frac{1}{2}O_2(g) \rightarrow BaO(s)$

(b) $BaCO_3(s) \rightarrow BaO(s) + CO_2(g)$

(c) $H_2(g) + Br_2(\ell) \rightarrow 2HBr$

7.34 Evaluate the magnitude of Ω for a system which has an entropy $41.840\,J\,K^{-1}$, given that Boltzmann's constant is $1.38 \times 10^{-23}\,J\,K^{-1}$.

8

Some Further Applications of the Combined First and Second Laws

8.1 COMBINED FIRST AND SECOND LAWS OF THERMODYNAMICS

We know that the work dW done due to a volume change dV is given by $dW = -P\, dV$, where P is the pressure. We have also seen that, for a reversible process, the change dS in entropy is given by $dS = dQ/T$. Putting these in the first law, i.e. $dU = dQ + dW$, we have

$$dU = T\, dS - P\, dV \tag{8.1}$$

This relationship is sometimes called the combined first and second law. It applies to any system of constant composition, in which only PV work is considered. If U is a function of S and V, i.e. $U = f(S, V)$, we can write

$$dU = \left(\frac{\partial U}{\partial S}\right)_V dS + \left(\frac{\partial U}{\partial V}\right)_S dV. \tag{8.2}$$

Comparing Equations 8.1 and 8.2, we have

$$\left(\frac{\partial U}{\partial S}\right)_V = T \tag{8.3}$$

and

$$\left(\frac{\partial U}{\partial V}\right)_S = -P \tag{8.4}$$

These equations connect the intensive variables P and T with the extensive variables U, V and S of the system.

It is a matter of interest to mention that *Equation 8.1 is the fundamental equation of thermodynamics and the whole of this science grows out of its consequences.*

A large number of thermodynamic relations can now be derived by considering T, P and V in pairs as independent variables (i.e. T and V, T and P, or P and V as independent), and expressing dS in terms of the differentials of these variables and the partial derivatives of S. Since the internal energy U, the enthalpy H, etc., are functions of the state of a system, the state of a homogeneous system is completely defined by any pair of these variables. Therefore, it is evident that the partial derivative of any one variable with respect to any other (with any one of those remaining held constant) has a physical meaning. We shall now consider the individual cases, taking P, T and V in pairs as independent.

8.2   ENTROPY AS A FUNCTION OF TEMPERATURE AND VOLUME

If U is a function of T and V, we have

$$dU = \left(\frac{\partial U}{\partial T}\right)_V dT + \left(\frac{\partial U}{\partial V}\right)_T dV. \tag{8.5}$$

Combining Equations 8.1 and 8.5, we have

$$dS = \frac{1}{T}\left(\frac{\partial U}{\partial T}\right)_V dT + \frac{1}{T}\left[P + \left(\frac{\partial U}{\partial V}\right)_T\right]dV. \tag{8.6}$$

Considering S as a function of V and T, we can also write

$$dS = \left(\frac{\partial S}{\partial T}\right)_V dT + \left(\frac{\partial S}{\partial V}\right)_T dV. \tag{8.7}$$

Now, since T and V are independent, dT and dV are also independent. Therefore, the coefficients of dT and dV in Equations 8.6 and 8.7 must be equal. Then,

$$\left(\frac{\partial S}{\partial T}\right)_V = \frac{1}{T}\left(\frac{\partial U}{\partial T}\right)_V = \frac{C_V}{T} \tag{8.8}$$

and

$$\left(\frac{\partial S}{\partial V}\right)_T = \frac{1}{T}\left[P + \left(\frac{\partial U}{\partial V}\right)_T\right]. \tag{8.9}$$

Let us use the mathematical fact that the second derivative of S with respect to T and V is independent of the order of differentiation, i.e.

$$\left[\frac{\partial}{\partial V}\left(\frac{\partial S}{\partial T}\right)_V\right]_T = \left[\frac{\partial}{\partial T}\left(\frac{\partial S}{\partial V}\right)_T\right]_V = \frac{\partial^2 S}{\partial V\,\partial T} = \frac{\partial^2 S}{\partial T\,\partial V}.$$

Now, differentiating Equation 8.8 partially with respect to V, and differentiating Equation 8.9 partially with respect to T, we have

$$\frac{1}{T}\frac{\partial^2 U}{\partial V\,\partial T} = \frac{1}{T}\left[\frac{\partial^2 U}{\partial T\,\partial V} + \left(\frac{\partial P}{\partial T}\right)_V\right] - \frac{1}{T^2}\left[\left(\frac{\partial U}{\partial V}\right)_T + P\right].$$

Hence,

$$\left(\frac{\partial U}{\partial V}\right)_T + P = T\left(\frac{\partial P}{\partial T}\right)_V \tag{8.10}$$

or

$$\left(\frac{\partial U}{\partial V}\right)_T = T\left(\frac{\partial P}{\partial T}\right)_V - P. \tag{8.11}$$

We have seen that

$$C_P - C_V = \left[\left(\frac{\partial U}{\partial V}\right)_T + P\right]\left(\frac{\partial V}{\partial T}\right)_P.$$

Using Equation 8.10 we can write,

$$C_P - C_V = T\left(\frac{\partial P}{\partial T}\right)_V\left(\frac{\partial V}{\partial T}\right)_P. \tag{8.12}$$

This shows that the difference $C_P - C_V$ can be computed for any substance for which $(\partial P/\partial T)_V$ and $(\partial V/\partial T)_P$ are known.

Again, using Equations 8.10 and 8.11 and the relationship $C_V = (\partial U/\partial T)_V$ in the partial derivatives of S with respect to T (from Equation 8.8) and with respect to V (from Equation 9.9) we have

$$\left(\frac{\partial S}{\partial T}\right)_V = \frac{C_V}{T} = \frac{C_P}{T} - \left(\frac{\partial P}{\partial T}\right)_V\left(\frac{\partial V}{\partial T}\right)_P \tag{8.13}$$

and

$$\left(\frac{\partial S}{\partial V}\right)_T = \left(\frac{\partial P}{\partial T}\right)_V. \tag{8.14}$$

Putting Equations 8.13 and 8.14 into Equation 8.7 we have

$$dS = \frac{C_V}{T}dT + \left(\frac{\partial P}{\partial T}\right)_V dV \tag{8.15}$$

or

$$T \, dS = C_V \, dT + T \left(\frac{\partial P}{\partial T} \right)_V dV \tag{8.16}$$

If C_V is known as a function of T, and $(\partial P/\partial T)_V$ is known as a function of V, then Equation 8.16 can be integrated to find the change in entropy of any substance in terms of T and V. From Equation 8.16 it is also possible to compute the heat dQ (which equals $T \, dS$) absorbed by any homogeneous substance in a reversible process.

Differentiating both sides of Equation 8.13 with respect to V, we have

$$\frac{\partial^2 S}{\partial V \, \partial T} = \frac{1}{T} \left(\frac{\partial C_V}{\partial V} \right)_T . \tag{8.17}$$

Similarly, differentiating both sides of Equation 8.14 with respect to T, we have

$$\frac{\partial^2 S}{\partial T \, \partial V} = \left(\frac{\partial^2 P}{\partial T^2} \right)_V . \tag{8.18}$$

Since $\partial^2 S/(\partial T \, \partial V) = \partial^2 S/(\partial V \, \partial T)$, it follows from Equations 8.17 and 8.18 that

$$\left(\frac{\partial C_V}{\partial V} \right)_T = T \left(\frac{\partial^2 P}{\partial T^2} \right)_V \tag{8.19}$$

For an ideal gas, $P = nRT/V$; therefore, $(\partial P/\partial T)_V = nR/V$. Hence,

$$\left(\frac{\partial^2 P}{\partial T^2} \right)_V = 0. \tag{8.20}$$

For a van der Waals gas,

$$P = \frac{nRT}{V - b} - \frac{a}{V^2} .$$

Therefore, $(\partial P/\partial T)_V = nR/(V-b)$. Hence,

$$\left(\frac{\partial^2 P}{\partial T^2} \right)_V = 0. \tag{8.21}$$

From Equations 8.20 and 8.21 we conclude that, for any substance for which the pressure is a linear function of the temperature at constant volume

$$\left(\frac{\partial^2 S}{\partial T^2} \right)_V = 0.$$

Substituting this in Equation 8.19, we have

$$\left(\frac{\partial C_V}{\partial V}\right)_T = \left(\frac{\partial^2 S}{\partial T^2}\right)_V = 0. \tag{8.22}$$

This shows that C_V is independent of V. Also, C_V is independent of P, because

$$\left(\frac{\partial C_V}{\partial P}\right)_T = \left(\frac{\partial C_V}{\partial V}\right)_T \left(\frac{\partial V}{\partial P}\right)_T = 0. \tag{8.23}$$

Now, since C_V/T is always positive, Equation 8.8 indicates that at constant volume the entropy increases with increase in temperature. It is clear that the dependence of entropy on temperature is very straightforward and simple, the differential coefficient being the appropriate heat capacity divided by the temperature. For a finite change in temperature at constant volume,

$$\Delta S = \int_{T_1}^{T_2} \frac{C_V}{T} \, dT \tag{8.24}$$

and, for n mol of a gas

$$\Delta S = \int_{T_1}^{T_2} \frac{nC_V}{T} \, dT. \tag{8.25}$$

For a special case, in which C_V is independent of temperature, integration of Equation 8.25 gives

$$\Delta S = nC_V \ln\left(\frac{T_2}{T_1}\right). \tag{8.26}$$

A similar procedure for *constant-pressure* processes gives

$$\Delta S = nC_P \ln\left(\frac{T_2}{T_1}\right). \tag{8.27}$$

These integrations are conveniently carried out graphically as in Figure 8.1. Data on the heat capacity C_P of a substance as a function of T allows one to calculate ΔS by this integration process. The area under the curve gives ΔS between the initial and final states. If phase transition occurs, the corresponding $(\Delta S)_{tran} = (\Delta H)_{tran}/T_{tran}$ must be included, where $(\Delta S)_{tran}$ and $(\Delta H)_{tran}$ are the entropy and enthalpy of transition.

Equations 8.26 and 8.27 are applicable to processes in which the system is heated or cooled reversibly from T_1 to T_2. However, since the change in entropy of a system is independent of the path, these equations also give the entropy change of a system whose temperature has changed irreversibly.

In contrast with the simplicity of the temperature dependence, the volume dependence at constant temperature given by Equation 8.9 is quite complicated. Except for gases, the

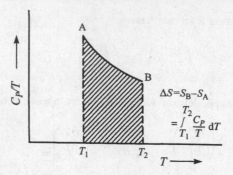

Figure 8.1   Graphical representation of $\Delta S = nC_P \ln(T_2/T_1)$

dependence of entropy on volume at constant temperature is negligibly small in most practical situations.

Example 8.1

Calculate the change in entropy when 1 mol of water is heated from 250 to 290 K, given that the molar heat capacities at constant pressure are $(C_P)_{ice} = 2.09 + 0.126T \, \text{J K}^{-1}$ and $(C_P)_{water} = 75.3 \, \text{J K}^{-1}$ and the heat of melting $(\Delta H)_m = 6000 \, \text{J mol}^{-1}$.

Solution

We have already stated that, if a phase transition occurs, then the corresponding change in entropy, i.e. $(\Delta S)_{tran} = (\Delta H)_{tran}/T_{tran}$, must be included in calculating the total entropy change. Hence, in this case the total entropy change is given by the equation

$$\Delta S = \int_{T_1}^{273\,\text{K}} (C_P)_{ice} \frac{dT}{T} + \frac{(\Delta H)_{tran}}{273} + \int_{273\,\text{K}}^{T_2} (C_P)_{water} \frac{dT}{T}.$$

In this case, $T_1 = 250\,\text{K}$, $T_2 = 290\,\text{K}$, $(\Delta H)_{tran} = 6000 \, \text{J mol}^{-1}$, $(C_P)_{ice} = 2.09 + 0.126T \, \text{J K}^{-1}$, $(C_P)_{water} = 75.3 \, \text{J K}^{-1}$. Substituting these values into the above equation, we have

$$\Delta S = \int_{250\,\text{K}}^{273\,\text{K}} (2.09 + 0.126T) \frac{dT}{T} + \frac{6000}{273} + \int_{273\,\text{K}}^{290\,\text{K}} (75.3) \frac{dT}{T}$$

$$= 2.09 \ln\left(\frac{273}{250}\right) + 5.04 + 22.0 + 75.3 \ln\left(\frac{290}{273}\right) = 31.60 \, \text{J K}^{-1}.$$

8.3   ENTROPY AS A FUNCTION OF TEMPERATURE AND PRESSURE

Let us now consider T and P as independent variables. Then the total differential is given by

$$dS = \left(\frac{\partial S}{\partial T}\right)_P dT + \left(\frac{\partial S}{\partial P}\right)_T dP. \tag{8.28}$$

Since $H = U + PV$, we have, on differentiation

$$dH = dU + P\,dV + V\,dP$$

or

$$dU + P\,dV = dH - V\,dP. \tag{8.29}$$

Substituting Equation 8.29 into 8.1 gives

$$T\,dS = dH - V\,dP. \tag{8.30}$$

Now, if H is also a function of T and P, we may write

$$dH = \left(\frac{\partial H}{\partial T}\right)_P dT + \left(\frac{\partial H}{\partial P}\right)_T dP. \tag{8.31}$$

Eliminating dH from Equations 8.30 and 8.31, we obtain

$$dS = \frac{1}{T}\left(\frac{\partial H}{\partial T}\right)_P dT + \frac{1}{T}\left[\left(\frac{\partial H}{\partial P}\right)_T - V\right]dP. \tag{8.32}$$

Now, comparing Equations 8.28 and 8.32, we have

$$\left(\frac{\partial S}{\partial T}\right)_P = \frac{1}{T}\left(\frac{\partial H}{\partial T}\right)_P = \frac{C_P}{T} \tag{8.33}$$

and

$$\left(\frac{\partial S}{\partial P}\right)_T = \frac{1}{T}\left[\left(\frac{\partial H}{\partial P}\right)_T - V\right] \tag{8.34}$$

or

$$\left(\frac{\partial H}{\partial P}\right)_T = V + T\left(\frac{\partial S}{\partial P}\right)_T. \tag{8.35}$$

Differentiating both sides of Equation 8.35 with respect to T and keeping P constant, we get

$$\frac{\partial^2 H}{\partial T\,\partial P} = T\frac{\partial^2 S}{\partial T\,\partial P} + \left(\frac{\partial S}{\partial P}\right)_T + \left(\frac{\partial V}{\partial T}\right)_P. \tag{8.36}$$

Again, differentiating both sides of Equation 8.33 with respect to P and keeping T constant, we obtain

$$\frac{\partial^2 H}{\partial P\,\partial T} = T\left(\frac{\partial^2 S}{\partial P\,\partial T}\right). \tag{8.37}$$

Now combining Equations 8.36 and 8.37 gives

$$\left(\frac{\partial S}{\partial P}\right)_T = -\left(\frac{\partial V}{\partial T}\right)_P. \tag{8.38}$$

Hence, combining Equations 8.28, 8.34 and 8.38, we obtain

$$dS = \frac{C_P}{T}dT - \left(\frac{\partial V}{\partial T}\right)_P dP. \tag{8.39}$$

It should be noted that Equations 8.15 and 8.35 are perfectly general and can be applied equally well to pure solids, liquids and gases.

Now, since C_P/T for any substance is always positive, Equation 8.33 indicates that at constant pressure the entropy increases with an increase in temperature. We find again that the dependence of entropy on temperature is simple, the differential coefficient being the appropriate heat capacity divided by the temperature. For a finite change in temperature at constant pressure,

$$\Delta S = \int_{T_1}^{T_2} \frac{C_P}{T} dT. \tag{8.40}$$

For a special case, in which C_P is independent of temperature, integration of Equation 8.40 gives

$$\Delta S = nC_P \ln\left(\frac{T_2}{T_1}\right)$$

which is the same as Equation 8.27. As mentioned before, this integration can be carried out conveniently by graphical methods as shown in Figure 8.1.

Like the volume dependence of entropy at constant temperature, the pressure dependence of the entropy at constant temperature is quite complicated but, as a matter of interest, except for gases the dependence of the entropy on pressure at constant temperature is negligibly small.

To evaluate the difference between entropy S_0 of a substance at 0 K and its entropy S at some temperature T, let us rewrite Equation 8.40 as

$$S = \int_0^T \frac{C_P}{T} dT + S_0. \tag{8.41}$$

If any changes in state occur between the temperature limits, the corresponding entropy changes must be included in S. Then, for a gas at temperature T, the general expression for the entropy becomes

$$S = \int_0^{T_m} \frac{(C_P)_{\text{crystal}}}{T} dT + \frac{(\Delta H)_m}{T_m} + \int_{T_m}^{T_b} \frac{(C_P)_{\text{liquid}}}{T} dT + \frac{(\Delta H)_V}{T_b}$$
$$+ \int_{T_b}^{T} \frac{(C_P)_{\text{gas}}}{T} dT + S_0 \tag{8.42}$$

where T_m is the melting temperature of the solid, $(C_P)_{crystal}$ is the specific heat of the crystallized substance at constant pressure, $(\Delta H)_m$ is the enthalpy at the melting temperature, T_b is the boiling temperature of the molten substance (liquid), $(C_P)_{liquid}$ is the specific heat of the liquid at constant pressure, $(\Delta H)_V$ is the enthalpy change due to vaporization and $(C_P)_{gas}$ is the specific heat of the gas at constant pressure.

8.4   ENTROPY AS A FUNCTION OF PRESSURE AND VOLUME

Equations corresponding to Equation 8.38 can be derived by considering P and V as independent variables. The procedure is the same as before. The results are

$$\left(\frac{\partial S}{\partial P}\right)_V = -\frac{C_V}{T}\left(\frac{\partial V}{\partial P}\right)_T \frac{1}{(\partial V/\partial T)_P} \tag{8.43}$$

$$\left(\frac{\partial S}{\partial V}\right)_P = \frac{C_P}{T}\frac{1}{(\partial V/\partial T)_P}. \tag{8.44}$$

We find that in the foregoing equations for the partial derivatives of S the partial derivatives $(\partial V/\partial T)_P$ and $(\partial V/\partial P)_T$ frequently occur. Naturally, to evaluate the partial derivatives of S, we must find these derivatives. It is observed that, even if the equation of state of a substance is not known or cannot be expressed in any simple analytical form, the partial derivatives $(\partial V/\partial T)_P$ and $(\partial V/\partial P)_T$ can be found from the coefficient β of cubic expansion and the compressibility Z.

The elementary definition of cubical expansion is given by

$$\bar{\beta} = \frac{V_2 - V_1}{V_1(T_2 - T_1)}$$

where V_2 and V_1 are the volumes of a substance at temperatures T_2 and T_1 respectively. In fact, $\bar{\beta}$ is the mean fractional increase in volume (per degree Celsius rise in temperature). The true coefficient β of volume expansion at any temperature T is the limiting value of the above expression when the increase in temperature and the corresponding volume change become infinitesimal. Then, for a volume V at temperature T, we can write

$$\beta = \frac{dV}{V\,dT} = \frac{1}{V}\frac{dV}{dT}.$$

Since the volume of an object depends on both pressure and temperature, it is implied that, in the above definition, pressure is considered as constant. To indicate this we write

$$\beta = \frac{1}{V}\frac{(dV)_P}{(dT)_P}.$$

The subscript P indicates that the pressure is kept constant during the changes in volume and temperature. Since V is a function of T and P, the ratio of a small change in volume to a small change in temperature (both at constant pressure) is the same thing as the partial derivative of V with respect to T at constant pressure. Therefore,

$$\frac{(dV)_P}{(dT)_P} = \left(\frac{\partial V}{\partial T}\right)_P.$$

Hence,

$$\beta = \frac{1}{V}\left(\frac{\partial V}{\partial T}\right)_P. \tag{8.45}$$

For an ideal gas, $PV = nRT$ or $V = nRT/P$. Therefore,

$$\left(\frac{\partial V}{\partial T}\right)_P = \frac{nR}{P}$$

or

$$\beta = \frac{1}{V}\left(\frac{\partial V}{\partial T}\right)_P = \frac{nR}{PV} = \frac{1}{T}. \tag{8.46}$$

To compute β in the case of a van der Waals gas, let us write $(\partial V/\partial T)_P$ in the form

$$\left(\frac{\partial V}{\partial T}\right)_P = \frac{(\partial P/\partial T)_V}{(\partial P/\partial V)_T}.$$

Now, the van der Waals equation

$$\left(P + \frac{a}{V^2}\right)(V - b) = nRT$$

can be written in the form

$$P = \frac{nRT}{V - b} - \frac{a}{V^2}.$$

Therefore,

$$\left(\frac{\partial P}{\partial T}\right)_V = \frac{nR}{V - b}$$

$$\left(\frac{\partial P}{\partial V}\right)_T = -\frac{nRT}{(V - b)^2} + \frac{2a}{V^3}.$$

Then, for a van der Waals gas

$$\beta = \frac{1}{V}\left(\frac{\partial V}{\partial T}\right)_P = \frac{nRV^2(V - b)}{nRTV^3 - 2a(V - b)^2}. \tag{8.47}$$

The mean compressibility $\bar{Z}$ of a material is defined by the equation

$$\bar{Z} = -\frac{V_2 - V_1}{V_1(P_2 - P_1)}$$

where V_1 and V_2 are the volumes at the pressures P_1 and P_2, respectively. The true compressibility Z is the limiting value of this expression when the changes in pressure and volume become infinitesimal. Then, for a volume V at a pressure P, we can write

$$Z = -\frac{dV}{V\,dP} = -\frac{1}{V}\frac{dV}{dP}.$$

The negative sign arises because an increase in pressure always results in a decrease in volume. Therefore, if dP is positive, dV is negative and Z is a positive quantity. The unit of Z is reciprocal newtons square metre ($\mathrm{N^{-1}\,m^2}$).

The change in volume due to a change in pressure depends not only on the pressure change but also on the temperature change that may occur simultaneously. The value of Z at constant temperature is called the *isothermal compressibility*. Mathematically this is written as

$$Z = -\frac{1}{V}\left(\frac{\partial V}{\partial P}\right)_T.$$

The compressibility Z is given by

$$Z = -\frac{1}{V}\frac{-nRT}{P^2} = \frac{1}{P} \tag{8.48}$$

for an ideal gas and by

$$Z = \frac{V^2(V-b)^2}{nRTV^3 - 2a(V-b)^2} \tag{8.49}$$

for a van der Waals gas. Like the coefficient of expansion, the compressibility is a function of both temperature and pressure.

8.5 ENTROPY CHANGES IN THE IDEAL GAS

The equations derived in the preceding sections are applicable to any system. For ideal gases they have a simple form because in the ideal gas the energy and the temperature are equivalent variable, i.e. $dU = nC_V\,dT$. Putting this value of dU into Equation 7.42, we have

$$dS = \frac{nC_V}{T}\,dT + \frac{P}{T}\,dV. \tag{8.50}$$

Let us now consider the total differential of S for n mol of an ideal gas, using first V and T as independent variables and then P and T.

Case (i): T and V as independent variables. From $PV = nRT$, we have $P = nRT/V$. Then, Equation 8.50 becomes

$$dS = \frac{nC_V}{T} dT + \frac{nR}{V} dV. \tag{8.51}$$

Comparison of Equations 8.7 and 8.50 gives

$$\left(\frac{\partial S}{\partial V}\right)_T = \frac{nR}{V}. \tag{8.52}$$

This derivative is always positive, since in an isothermal transformation the entropy of an ideal gas increases with increase in volume; the rate of increase is less at large volumes, as V is in the denominator. For a finite change in state, we integrate Equation 8.51:

$$\Delta S = n \int_{T_1}^{T_2} \frac{C_V}{T} dT + nR \int_{V_1}^{V_2} \frac{dV}{V}.$$

If C_V is a constant, this integrates to

$$\Delta S = nC_V \ln\left(\frac{T_2}{T_1}\right) + nR \ln\left(\frac{V_2}{V_1}\right). \tag{8.53}$$

The first term on the right-hand side of this equation gives the change in entropy of n mol of an ideal gas due to a change in temperature at constant volume. The second term gives the change in entropy of n mol of an ideal gas due to a change in volume at constant temperature. At constant temperature the first term vanishes, and at constant volume the second term vanishes. If C_V is not a constant, then we shall have to express C_V as a function of T to compute the integral.

Case (ii): P and T as independent variables. We know that the enthalpy of an ideal gas is a function of temperature only; therefore, its differential must be given by $dH = nC_P \, dT$. Putting this value of dH in Equation 8.30, we have

$$dS = \frac{nC_P}{T} dT - \frac{V}{T} dP. \tag{8.54}$$

Using $PV = nRT$ we can write $V = nRT/P$. Putting this value of V in Equation 8.54, we have

$$dS = \frac{nC_P}{T} dT - \frac{nR}{P} dP. \tag{8.55}$$

Comparison of Equation 8.28 and 8.55 gives

$$\left(\frac{\partial S}{\partial P}\right)_T = -\frac{nR}{P}. \tag{8.56}$$

This shows that, at constant temperature, the entropy decreases with increase in pressure. As before, for a finite change in state, Equation 8.55 integrates to

$$\Delta S = n \int_{T_1}^{T_2} \frac{C_P \, dT}{T} - nR \int_{P_1}^{P_2} \frac{dP}{P} .$$

If C_P is a constant, this integrates to

$$\Delta S = nC_P \ln\left(\frac{T_2}{T_1}\right) - nR \ln\left(\frac{P_2}{P_1}\right). \tag{8.57}$$

If C_P is not a constant, we shall have to express C_P as a function of T to compute the integral.

The first term on the right-hand side of Equation 8.57 gives the entropy change of n mol of an ideal gas due to the change in temperature at constant pressure, and the second term gives the entropy change due to the change in pressure at constant temperature.

Equations 8.16 and 8.39 may be used to calculate entropy changes for non-ideal gases provided that sufficient information is available to obtain values of $(\partial P/\partial T)_V$ as a function of volume, or $(\partial V/\partial T)_P$ as a function of pressure. Then these equations may be used in the same manner as for ideal gases.

Case (iii): P and V as independent variables. The equation for change in entropy for a finite change in state is given by

$$\Delta S = nC_P \ln\left(\frac{V_2}{V_1}\right) + nC_V \ln\left(\frac{P_2}{P_1}\right). \tag{8.58}$$

Equations 8.53, 8.57 and 8.58 are equivalent. Any two of these three equations can be derived from the third with the help of the equation of state.

Example 8.2

Using Equation 8.11 show that the equation

$$C_P = C_V + \left[P + \left(\frac{\partial U}{\partial V}\right)_T\right]\left(\frac{\partial V}{\partial T}\right)_P$$

reduces to

$$C_P - C_V = T\left(\frac{\partial P}{\partial T}\right)_V \left(\frac{\partial V}{\partial T}\right)_P .$$

Hence, show that

$$C_P - C_V = \frac{TV\beta^2}{Z}.$$

Solution

From Equation 8.11 we have

$$\left(\frac{\partial U}{\partial V}\right)_T + P = T\left(\frac{\partial P}{\partial T}\right)_V.$$

Substituting this into the given equation we get

$$C_P = C_V + T\left(\frac{\partial P}{\partial T}\right)_V\left(\frac{\partial V}{\partial T}\right)_P. \tag{i}$$

By definition, $\beta = (1/V)(\partial V/\partial T)_P$ and $Z = -(1/V)(\partial V/\partial P)_T$. Substituting for $(\partial V/\partial T)_P$ in (i), we get

$$C_P - C_V = TV\beta\left(\frac{\partial P}{\partial T}\right)_V. \tag{ii}$$

The pressure P is a function of T and V, i.e. $P = f(T, V)$. Hence,

$$dP = \left(\frac{\partial P}{\partial T}\right)_V dT + \left(\frac{\partial P}{\partial V}\right)_T dV.$$

If P is constant, $dP = 0$. Then the above equation becomes

$$0 = \left(\frac{\partial P}{\partial T}\right)_V dT + \left(\frac{\partial P}{\partial V}\right)_T dV.$$

Dividing both sides by dT, we get

$$\left(\frac{\partial P}{\partial T}\right)_V + \left(\frac{\partial P}{\partial V}\right)_T\left(\frac{\partial V}{\partial T}\right)_P = 0$$

or

$$\left(\frac{\partial P}{\partial T}\right)_V = -\left(\frac{\partial P}{\partial V}\right)_T\left(\frac{\partial V}{\partial T}\right)_P.$$

We know that $(\partial P/\partial V)_T = 1/(\partial V/\partial P)_T$. Using this relation and the expressions for β and Z, we obtain

$$\left(\frac{\partial P}{\partial T}\right)_V = \frac{\beta}{Z}.$$

Substituting this in (ii) and simplifying, we have

$$C_P - C_V = \frac{TV\beta^2}{Z}.$$

8.6   ENTROPY CHANGE OF A VAN DER WAALS GAS

For a van der Waals gas the equation of state is

$$\left(P + \frac{a}{V^2}\right)(V - b) = nRT$$

Solving for P and putting it into Equation 8.11, i.e.

$$\left(\frac{\partial U}{\partial V}\right)_T = T\left(\frac{\partial P}{\partial T}\right)_V - P$$

we have

$$\left(\frac{\partial U}{\partial V}\right)_T = T\left(\frac{nR}{V - b}\right) - \left(\frac{nRT}{V - b} - \frac{a}{V^2}\right)$$

or

$$\left(\frac{\partial U}{\partial V}\right)_T = \frac{a}{V^2}. \tag{8.59}$$

On integration we have,

$$U = -\frac{a}{V} + f(T) \tag{8.60}$$

where $f(T)$ is a function of temperature only. This is a function whose derivative is precisely the heat capacity at constant volume:

$$\left(\frac{\partial U}{\partial T}\right)_V = C_V = f'(T). \tag{8.61}$$

If we assume C_V to be constant, we can conclude that

$$U = -\frac{a}{V} + C_V T + \text{constant}. \tag{8.62}$$

From Equations 8.60 and 8.62 it is evident that, as the volume becomes infinite, the energy approaches that of an ideal gas. This is in line with the fact that at sufficiently low pressures all gases tend to behave ideally.

To obtain the entropy change of a van der Waals gas we write Equation 8.50 as

$$dS = \frac{nC_V}{T}dT + \frac{nR}{V - b}dV. \tag{8.63}$$

As in the case of ideal gases, we shall here consider two cases, namely in one case V and T as independent and in another case T and P as independent.

Case (i): T and V as independent. For a finite change in state, we integrate Equation 8.63.

$$\Delta S = n \int_{T_1}^{T_2} \frac{C_V}{T} \, dT + nR \int_{V_1}^{V_2} \frac{dV}{V - b}. \tag{8.64}$$

If C_V is a constant, this integrates to

$$\Delta S = nC_V \ln \left(\frac{T_2}{T_1} \right) + nR \ln \left(\frac{V_2 - b}{V_1 - b} \right). \tag{8.65}$$

If C_V is not a constant, we shall have to express C_V as a function of T to compute the integral.

Evidently, Equation 8.65 is similar to Equation 8.53 which is applicable to an ideal gas. As in the case of energy, we find that, as V becomes infinite, the entropy approaches that of the ideal gas, because b becomes negligible compared with V under such circumstances. Although the van der Waals equation has two constants that give rise to deviations from ideality, only one of them, a, affects the energy, and the other, b, modifies the entropy provided that V and T are chosen as independent variables.

The heat added during an *isothermal reversible* expansion of a van der Waals gas is given by

$$Q_{rev} = T\Delta S = nRT \ln \left(\frac{V_2 - b}{V_1 - b} \right). \tag{8.66}$$

Also, from Equation 8.60 we can conclude that the isothermal energy change for a van der Waals gas is given by

$$\Delta U = -a \left(\frac{1}{V_2} - \frac{1}{V_1} \right). \tag{8.67}$$

Therefore, putting these values of Q and U in the first law, we have

$$W_{rev} = -nRT \ln \left(\frac{V_2 - b}{V_1 - b} \right) - a \left(\frac{1}{V_2} - \frac{1}{V_1} \right). \tag{8.68}$$

This result is identical with that obtained by direct integration of $P \, dV$.

Case (ii): P and T are independent variables. Although P can be readily eliminated as a function of V and T, it is not possible to write V as a function of P and T without solving a cubic equation. To avoid the complication of solving a cubic equation we shall rearrange van der Waals equation for 1 mol of a gas in an approximate form (valid for low pressures only) as

$$PV = RT + \left(b - \frac{a}{RT} \right) P \tag{8.69}$$

or

$$V = \frac{RT}{P} + \left(b - \frac{a}{RT} \right). \tag{8.70}$$

Therefore,

$$\left(\frac{\partial V}{\partial T}\right)_P = \frac{R}{P} + \frac{a}{RT^2}. \tag{8.71}$$

Substituting the value of V from Equation 8.70 and the value of $(\partial V/\partial T)_P$ from Equation 8.71 in $(\partial H/\partial P)_T = V - T(\partial V/\partial T)_P$, we get

$$\left(\frac{\partial H}{\partial P}\right)_T = b - \frac{2a}{RT}. \tag{8.72}$$

Therefore, on integration we have

$$H = \left(b - \frac{2a}{RT}\right)P + F(T) \tag{8.73}$$

where $F(T)$ is a function of temperature T. The heat capacity at constant pressure is given by

$$C_P = \left(\frac{\partial H}{\partial T}\right)_P = \frac{2aP}{RT^2} + F'(T). \tag{8.74}$$

From the relation $(\partial S/\partial P)_T = -(\partial V/\partial T)_P$ and using Equation 8.71 we find that

$$\left(\frac{\partial S}{\partial P}\right)_T = -\frac{R}{P} - \frac{a}{RT^2}. \tag{8.75}$$

On integrating Equation 8.75 we obtain, for an isothermal expansion of a van der Waals gas between moderate pressures,

$$\Delta S = -R \ln\left(\frac{P_2}{P_1}\right) - \frac{a}{RT^2}(P_2 - P_1). \tag{8.76}$$

It can also be shown that, for a van der Waals gas,

$$C_P - C_V = R + \frac{2aP}{RT^2}. \tag{8.77}$$

Example 8.3

1 mol of helium gas, initially at 300 K and 0.2 MPa pressure, expands isothermally until the final pressure is 0.4 MPa. Calculate the change in entropy for this process, given that $R = 0.00828\,\ell\,\text{MPa}\,\text{K}^{-1}\,\text{mol}^{-1}$ and the van der Waals constant $a = 0.002979\,\text{MPa}\,\ell^2\,\text{mol}^{-2}$.

Solution

From the problem we have

$P_1 = 0.2\,\text{MPa}$
$P_2 = 0.4\,\text{MPa}$
$T = 300\,\text{K}$
$R = 0.00828\,\ell\,\text{MPa}\,\text{K}^{-1}\,\text{mol}^{-1}$
$a = 0.002979\,\text{MPa}\,\ell^2\,\text{mol}^{-2}.$

Substituting these values into Equation 8.76, we get

$$\Delta S = -0.00828\,\ell\,\text{MPa}\,\text{K}^{-1}\,\text{mol}^{-1}\ln\left(\frac{0.4\,\text{MPa}}{0.2\,\text{MPa}}\right)$$

$$-\frac{(0.002979\,\text{MPa}\,\ell^2\,\text{mol}^{-2})(0.4 - 0.2)\,\text{MPa}}{(0.00828\,\ell\,\text{MPa}\,\text{K}^{-1}\,\text{mol}^{-1}(300\,\text{K})^2}$$

$$= -0.00828 \times 0.701\,\ell\,\text{MPa}\,\text{K}^{-1}\,\text{mol}^{-1}$$

$$- 0.00799 \times 10^{-4}\,\frac{\text{MPa}^2\,\ell^2\,\text{mol}^{-2}}{\ell\,\text{MPa}\,\text{K}\,\text{mol}^{-1}}$$

$$= -5829.204 \times 10^{-4}\,\ell\,\text{MPa}\,\text{K}^{-1}\,\text{mol}^{-1}$$

$$= -5.829\,\text{J}\,\text{K}^{-1}\,\text{mol}^{-1}$$

because $1\,\ell\,\text{MPa}\,\text{K}^{-1}\,\text{mol}^{-1} = 1003.86\,\text{J}\,\text{K}^{-1}\,\text{mol}^{-1}.$

8.7   STANDARD STATE FOR THE ENTROPY OF AN IDEAL GAS

If we consider a change in state at constant temperature, we can write Equation 8.56 in the form (*for 1 mol of the gas*)

$$dS = -\frac{R}{P}dP.$$

Let us integrate this within the limits $P = 1$ atm to $P = P$ (any pressure) and let us take $S = S°$ at $P = 1$ atm where $S°$ is the value of the molar entropy at that pressure. We shall define $S°$ as the standard entropy at the particular temperature concerned. Then we have

$$S - S° = -R\,\ln\left(\frac{P}{1\,\text{atm}}\right). \tag{8.78}$$

When P is in atmospheres the logarithm will be a pure number. Since the denominator of the logarithm in Equation 8.78 is unity, we can write this equation as

$$S - S° = -R\,\ln P. \tag{8.79}$$

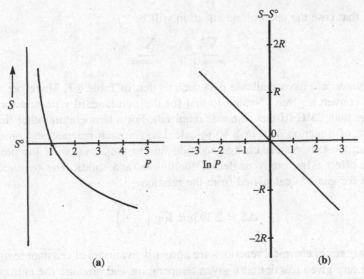

Figure 8.2 (a) $S-S°$ as a function of pressure; and (b) $S-S°$ as a function of $\ln P$

It must be remembered that P in Equation 8.79 is a pure number, obtained only by dividing the pressure in atmospheres by 1 atm.

The quantity $S-S°$ is the molar entropy at the pressure P relative to that at 1 atm. We can plot $S-S°$ as a function of pressure as shown in Figure 8.2.

It is evident from Figure 8.2(a) that the rate of entropy increase with pressure is sharp at low pressures and becomes less sharp at higher pressures. Therefore, it is more advantageous to plot $S-S°$ against $\ln P$ because such a plot is linear.

8.8   ENTROPY CHANGES IN CHEMICAL REACTIONS

If the absolute values of entropies of elements and compounds are known, it is easy to calculate the change in entropy of a system undergoing a chemical reaction. ΔS for the process will be the difference between the total entropy of the products and that of the reactants.

If we have a reaction at 298 K in which all species are present in their standard states, the change in entropy is given by

$$(\Delta S°)_{298\,K} = \sum_{products} n(S°)_{298\,K} - \sum_{reactants} n(S°)_{298\,K}. \tag{8.80}$$

In this equation the terms in each summation represent the standard absolute entropy per mole multiplied by the number of moles of that substance in the reaction. The units of $\Delta S°$ are in entropy units, i.e. joules per degree Celsius.

Equation 8.80 can also be used to calculate the change in entropy in some processes, where the species are not in their standard state, simply by using the absolute entropies in that state. In that case the appropriate equation will be

$$\Delta S = \sum_{\text{products}} nS - \sum_{\text{reactants}} nS. \tag{8.81}$$

Normally we would have available data such as that in Table 8.1. Therefore, it would be necessary to convert a given $S°$-value to that for the nonstandard state and for some temperature other than 25°C (if this also was required). For a temperature other than 25°C we can make use of Equation 8.25 or 8.40 to calculate for each reactant and product the ΔS resulting from the temperature change and add this to the $S°$-value. If the pressure is not 0.1 MPa, the effect is generally neglected for liquids and solids. The required additional entropy term for gases is calculated from the relation

$$\Delta S = 2.303nR \log\left(\frac{P_1}{P_2}\right). \tag{8.82}$$

Entropy changes of chemical reactions are normally evaluated at constant temperature and pressure. For any given reaction at a given temperature and pressure the entropy change is definite; it is just as characteristic of that reaction as the change in the internal energy or the enthalpy. To study the manner in which the entropy change of a reaction depends on the temperature at constant pressure, let us consider a general reaction

$$a\text{A} + b\text{B} + \cdots = m\text{M} + n\text{N} + \cdots.$$

The entropy change for this reaction is given by

$$\Delta S = (mS_\text{M} + nS_\text{N} + \cdots) - (aS_\text{A} + bS_\text{B} + \cdots)$$

where $S_\text{A}, S_\text{B}, S_\text{M}, S_\text{N}, \ldots$, are the entropies per mole of the various species in the reaction. Now, differentiating both sides of this equation with respect to temperature at constant pressure, we get

$$\left(\frac{\partial}{\partial T}(\Delta S)\right)_P = \left(m\frac{\partial}{\partial T}(S_\text{M})_P + n\frac{\partial}{\partial T}(S_\text{N})_P + \cdots\right)$$
$$- \left(a\frac{\partial}{\partial T}(S_\text{A})_P + b\frac{\partial}{\partial T}(S_\text{B})_P + \cdots\right).$$

However, by Equation 8.33 we have $(\partial S/\partial T)_P = C_P/T$. Hence, the above equation becomes

$$\left(\frac{\partial}{\partial T}(\Delta S)\right)_P = \frac{m(C_P)_\text{M} + n(C_P)_\text{N} + \cdots}{T} - \frac{a(C_P)_\text{A} + b(C_P)_\text{B} + \cdots}{T}$$
$$= \frac{\Delta C_P}{T} \tag{8.83}$$

where ΔC_P is the difference between the C_P-values for the products and the reactants.

Integrating Equation 8.83 between the limits of any two temperatures T_1 and T_2, we get

$$\int_{\Delta S_1}^{\Delta S_2} d(\Delta S) = \int_{T_1}^{T_2} \frac{\Delta C_P}{T} dT \quad \text{or} \quad \Delta S_2 - \Delta S_1 = \int_{T_1}^{T_2} \frac{\Delta C_P}{T} dT \tag{8.84}$$

Table 8.1 Standard absolute entropies at 298 K (the values in calories per degree Celsius per mole are taken from *National Bureau of Standards Circular* No. 500 1952 (Washington, DC: US Department of Commerce) (a complete list is given in Appendix 3)

Substance	Entropy $S°$ $(JK^{-1} mol^{-1})$	Substance	Entropy $S°$ $(JK^{-1} mol^{-1})$	Substance	Entropy $S°$ $(JK^{-1} mol^{-1})$
Ag(s)	42.702	MgO(s)	26.778	Ar(g)	154.737
AgCl(s)	96.106	α-Mn(s)	31.757	Br_2(g)	245.346
Ag_2O(s)	121.713	MnO(s)	60.249	CH_4(g)	186.188
Ag_2S(rhombic)(s)	145.603	MnO_2(s)	53.137	C_2H_2(g)	200.819
Al(s)	28.321	Na(s)	51.045	C_2H_4(g)	219.451
α-Al_2O_3(s)	51.003	NaCl(s)	72.383	C_2H_6(g)	229.492
Ba(s)	66.944	NaF(s)	58.576	Cl_2(g)	222.949
$BaCO_3$(s)	112.131	$NaNO_3$(s)	116.315	CO(g)	197.907
BaO(s)	70.291	Na_2O(s)	72.802	CO_2(g)	213.639
C(diamond)(s)	2.439	Na_2SO_4(s)	149.494	F_2(g)	203.342
C(graphite)(s)	5.690	NH_4Cl(s)	94.558	H_2(g)	130.587
Ca(s)	41.631	Ni(s)	30.125	HBr(g)	198.476
CaC_2(s)	70.291	NiO(s)	38.576	HCHO(g)	218.656
$CaCO_3$(calcite)(s)	92.885	P_4(s)	177.402	HCl(g)	186.678
$CaSO_4$(s)	106.90	Pb(s)	64.894	He(g)	120.047
CaO(s)	39.748	PbO(red)(s)	67.781	HF(g)	173.610
Co(s)	26.359	PbS(s)	91.211	HI(g)	206.329
CoO(s)	53.032	S_8(rhombic)(s)	255.057	H_2O(g)	188.734
Cu(s)	33.305	Si(s)	18.702	H_2S(g)	205.644
CuO(s)	43.014	SiO_2(quartz II)(s)	41.840	I_2(g)	260.579
Cu_2O(s)	93.834	Sn(s)	51.463	Kr(g)	163.971
CuS(s)	66.526	SnO(s)	56.484	N_2(g)	191.489
Cu_2S(s)	120.918	Sr(s)	52.392	Ne(g)	146.222
Fe(s)	27.154	SrO(s)	54.400	NH_3(g)	192.506
FeO(wurstite)(s)	53.974	$SrCO_3$(s)	97.069	NO(g)	210.618
Fe_2O_3(s)	89.956	Ti(s)	30.292	NO_2(g)	240.454
Fe_3O_4(s)	146.440	TiO_2(rutile)(s)	50.249	N_2O(g)	219.995
α-FeS(s)	67.362	Zn(s)	41.631	N_2O_4(g)	304.303
FeS_2(s)	53.137	ZnO(s)	43.932	O_2(g)	205.029
HgO(red)(s)	71.965			O_3(g)	238.78
I_2(s)	116.734	Br_2(ℓ)	152.298	P_4(g)	279.909
K(s)	63.597	CCl_4(ℓ)	216.430	PCl_3(g)	312.921
KBr(s)	96.441	CH_3OH(ℓ)	126.775	PCl_5(g)	352.711
KCl(s)	82.676	Cyclo-C_6H_{12}(ℓ)	298.236	PF_3(g)	268.278
$KClO_3$(s)	142.967	CH_3COOH(ℓ)	159.829	SiF_4(g)	284.512
$KClO_4$(s)	151.042	C_2H_5OH(ℓ)	160.666	SiH_4(g)	203.761
KF(s)	66.567	HCOOH(ℓ)	128.951	SO_2(g)	248.529
KI(s)	104.349	H_2O(ℓ)	69.873	SO_3(g)	256.228
Mg(s)	32.509	Hg(ℓ)	77.404	Xe(g)	169.578
$MgCO_3$(s)	65.689	HNO_3(ℓ)	155.603		
		$TiCl_4$(ℓ)	252.714		

where ΔS_2 and ΔS_1 are the entropy changes corresponding to T_1 and T_2 respectively. If ΔC_P is constant over the temperature change, this equation becomes

$$\Delta S_2 - \Delta S_1 = \Delta C_P \ln\left(\frac{T_2}{T_1}\right). \tag{8.85}$$

If ΔC_P is not constant over the temperature change, then the expression for ΔC_P as a function of T must be obtained and then the integration must be carried out term by term between the given temperature limits.

Example 8.4

Calculate the standard entropy change at 298 K in the reaction

$$C_2H_5OH(\ell) + O_2(g) \rightarrow CH_3COOH(\ell) + H_2O(\ell)$$

using the S°-values given in Table 8.1.

Solution

According to Equation 8.81, we have

$$(\Delta S^{\circ})_{298\,K} = [(S^{\circ}_{298\,K})_{CH_3COOH} + (S^{\circ}_{298\,K})_{H_2O}] - [(S^{\circ}_{298\,K})_{C_2H_5OH}$$
$$+ (S^{\circ}_{298\,K})_{O_2}] = 159.829 + 69.873 - 160.666 + 205.016$$
$$= -135.980\,JK^{-1}.$$

8.9   ENTROPY OF MIXING FOR IDEAL GASES

If two or more ideal gases are mixed together, there is always a change in entropy. To calculate the entropy of mixing $(\Delta S)_m$ for ideal gases, let us imagine that we mix n_1 mol of one gas at an initial pressure P_1^0 and n_2 mol of another gas at an initial pressure P_2^0. Let P_1 and P_2 be the partial pressures of the two gases, respectively, in the mixture.

At any constant temperature the change in entropy due to a change in pressure is given by the second term on the right-hand side of Equation 8.57. Therefore, because of mixing, the entropy changes for the first and second gases respectively, would be given by

$$(\Delta S)_1 = -n_1 R \ln\left(\frac{P_1}{P_1^0}\right) = n_1 R \ln\left(\frac{P_1^0}{P_1}\right) \tag{8.86}$$

$$(\Delta S)_2 = -n_2 R \ln\left(\frac{P_2}{P_2^0}\right) = n_2 R \ln\left(\frac{P_2^0}{P_2}\right). \tag{8.87}$$

The total entropy change of mixing would be obtained by adding Equations 8.86 and 8.87. Therefore,

$$(\Delta S)_m = (\Delta S)_1 + (\Delta S)_2 = R\left[n_1 \ln\left(\frac{P_1^0}{P_1}\right) + n_2 \ln\left(\frac{P_2^0}{P_2}\right)\right].$$

However, from Dalton's law of partial pressures we have $P_1 = x_1 P_t$ and $P_2 = x_2 P_t$, where x_1 and x_2 are the mole fractions of the two gases, and P_t is the total pressure of the mixture. Substituting these into the above equation, we obtain

$$(\Delta S)_m = R\left[n_1 \ln\left(\frac{P_1^0}{x_1 P_t}\right) + n_2 \ln\left(\frac{P_2^0}{x_2 P_t}\right)\right]. \qquad (8.88)$$

In the special case when $P_t = P_1^0 = P_2^0$, Equation 8.88 reduces to

$$(\Delta S)_m = -R(n_1 \ln x_1 + n_2 \ln x_2). \qquad (8.89)$$

Since x_1 and x_2 are less than unity, $(\Delta S)_m$ is positive. Therefore, the mixing results in an increase in entropy. Equation 8.89 also shows that $(\Delta S)_m$ is independent of temperature.

Example 8.5

3 mol of nitrogen gas, originally at 0.1 MPa pressure, are mixed isothermally with 5 mol of hydrogen gas, also at 0.1 MPa pressure, to give a mixture having total pressure 0.1 MPa. Assuming that the gases behave ideally, calculate the total entropy change of mixing.

Solution

For nitrogen, $n_1 = 3$, $P_1^0 = 0.1$ MPa, $x_1 = \frac{3}{8}$; for hydrogen, $n_2 = 5$, $P_2^0 = 0.1$ MPa, $x_2 = \frac{5}{8}$. The total pressure P_t of the mixture is 0.1 MPa. Substituting these values in Equation 8.88, we have

$$(\Delta S)_m = 8.314(3 \ln \tfrac{8}{3} + 5 \ln \tfrac{8}{5})$$
$$= 8.314 \times 2.303(8 \log 8 - 5 \log 5 - 3 \log 3)$$
$$= 43.852 \text{ JK}^{-1}.$$

EXERCISES 8

8.1 Derive an equation which connects the intensive variables P and T with the extensive variables H, V and S of a system.

8.2 Show that the difference $C_P - C_V$ can be computed for any substance for which $(\partial P/\partial T)_V$ and $(\partial V/\partial T)_P$ are known.

8.3 Starting with Equation 8.1 deduce the expression for the change in entropy due to heating of n mol of an ideal gas from volume V_1 at temperature T_1 to volume V_2 at temperature T_2.

8.4 The equation of state of an ideal gas is given by $PV = RT$. Show that $\beta = 1/T$ and $Z = 1/P$.

8.5 An approximate equation of state of a real gas at moderate pressure is $P(V-b) = RT$, where R and b are constants. Show that

$$\text{(a)} \quad \beta = \frac{1/T}{1 + bP/RT}$$

$$\text{(b)} \quad Z = \frac{1/P}{1 + bP/RT}.$$

8.6 An approximate equation of state of a real gas at moderate pressure is given by $PV = RT(1 + B/V)$, where R is a constant and B is a function of T only. Show that

$$\beta = \frac{1}{T} \frac{V + B + T(\mathrm{d}B/\mathrm{d}T)}{V + 2B}.$$

8.7 If S is a function of T and P, show that, in any system in which C_P is a constant, any isobaric curve in a plot of S versus $\ln T$ is a straight line whose slope is C_P.

8.8 If S is a function of T and V, show that, in any system in which C_V is constant, any isochoric curve in a plot of S versus $\ln T$ is a straight line whose slope is C_V.

8.9 For an ideal gas $C_P = \frac{5}{2} R$ JK^{-1} mol^{-1}. Calculate the change in entropy of 3 mol of this gas when it is heated from 30 to 300 K at (a) constant pressure, and (b) constant volume.

8.10 Show that, when S is expressed as a function of T and V, the change in entropy of a van der Waals gas for a finite change in state is given by

$$\Delta S = n \int_{T_1}^{T_2} \frac{C_V}{T} \mathrm{d}T + nR \int_{V_1}^{V_2} \frac{\mathrm{d}V}{V - b}.$$

8.11 Show that, when S is expressed as a function of P and T, the change in entropy for an isothermal expansion of a van der Waals gas between moderate pressures is given by

$$\Delta S = -R \ln\left(\frac{P_2}{P_1}\right) - \frac{a}{RT^2}(P_2 - P_1).$$

8.12 What do you understand by 'standard state for the entropy of an ideal gas'? Calculate the entropy change for the reaction $2C(\text{graphite}) + 2H_2(g) = C_2H_4(g)$. Use the relevant standard entropies of the substances involved from Table 8.1. Is it possible to conclude from the answer whether the reaction is spontaneous?

8.13 Using the relevant standard entropies at 298 K from Table 8.1 calculate

(a) the standard entropy of formation of 1 mol of gaseous HCl and

(b) the standard entropy change accompanying the decomposition of $CaCO_3$.

8.14 1 mol of an ideal gas at 0°C and 0.1 MPa pressure is mixed adiabatically with another mole of the same gas at 100°C and 0.1 MPa pressure to give a mixture whose pressure is also 0.1 MPa. Calculate the change in entropy for this process, given that, for this gas, $C_P = \frac{5}{2}R$ J °C^{-1} mol^{-1}.

8.15 Calculate the change in entropy in joules per kelvin which results when 4 mol of nitrogen and 1 mol of oxygen are mixed together at 298 K. Assume that the pressure is constant and the gases behave ideally.

8.16 Deduce an expression for the total change of entropy due to mixing of two or more ideal gases together at a constant temperature. Hence show that the mixing results in an increase in entropy. Is the total change in entropy in such cases independent of temperature?

8.17 If the temperature and pressure are independent variables, show that

$$\left(\frac{\partial S}{\partial T}\right)_P = \frac{1}{T}\left(\frac{\partial H}{\partial T}\right)_P = \frac{C_P}{T}$$

and

$$\left(\frac{\partial S}{\partial P}\right)_T = \frac{1}{T}\left[\left(\frac{\partial H}{\partial P}\right)_T - V\right].$$

Hence show that

$$\left(\frac{\partial H}{\partial P}\right)_T = V - T\left(\frac{\partial V}{\partial T}\right)_P.$$

8.18 Show that the Joule–Thomson coefficient $(\partial T/\partial P)_H$ can be expressed in the form

$$\left(\frac{\partial T}{\partial P}\right)_H = \frac{T^2[\partial(V/T)/\partial T]_P}{C_P} = \frac{V(\beta T - 1)}{C_P}$$

where β is the cubic or volume expansion given by $(1/V)(\partial V/\partial T)_P$. Prove that, for an ideal gas, $(\partial T/\partial P)_H = 0$.

8.19 Show that, for an ideal gas

$$\left(\frac{\partial U}{\partial P}\right)_T = -\beta VT + ZPV$$

where β is the coefficient of volume expansion and Z is the compressibility.

8.20 Assuming that, for an ideal gas,

$$C_P - C_V = T\left(\frac{\partial P}{\partial T}\right)_V\left(\frac{\partial V}{\partial T}\right)_P$$

holds, prove that

$$C_P - C_V = \frac{\beta^2 TV}{Z}.$$

8.21 3 kg of pure liquid water at 400 K are mixed adiabatically at constant pressure with 4 kg of pure liquid water at 300 K. Calculate the total entropy change for this mixing process. Assume that the heat capacity of water is constant over the given temperature range and has a value $C_P = 4180 \, \mathrm{JK^{-1} \, kg^{-1}}$.

8.22 (a) Show that for an ideal gas with constant heat capacities the entropy can be given as a function of pressure and volume by the equation $S = C_P \ln V + C_V \ln P + S'$ where S' is a constant.

(b) An ideal gas, for which $C_P = 21.0 \, \mathrm{J \, °C^{-1} \, mol^{-1}}$ and $C_V = 13.0 \, \mathrm{J \, °C^{-1} \, mol^{-1}}$, has zero molar entropy at 0°C and 10 MPa. Calculate the value and find the units of S' for this gas.

8.23 1 mol of an ideal gas expands adiabatically from 350 K and 5 MPa to double its volume. Calculate the change in entropy of the gas when the expansion is

(a) a reversible expansion and

(b) a free expansion against no opposing pressure.

9

The Third Law of Thermodynamics

9.1 INTRODUCTION

We have seen that the second law of thermodynamics does not enable the absolute value of entropy of any substance to be calculated; this law only permits evaluation of entropy changes for specified changes in physical or chemical state. More precisely, the second law provides a complete description of the equilibrium state, its temperature dependence and the quantitative relation between equilibrium states in related systems.

In 1905 Nernst stated the general principle that in any chemical reaction between solid or liquid substances $(d/dT)[(\Delta G)_T]$ approaches zero at the absolute zero of temperature, where $(\Delta G)_T$ is the change in free energy at constant temperature (we shall discuss this later). It has been shown that this rule applies to crystalline solids only, with some exceptions. According to the Nernst principle, for changes in chemical state involving only perfect crystalline substances, $\Delta S_0 = 0$. This requirement is satisfied only if for perfect crystalline substances $S_0 = 0$. More precisely, we can say that the Nernst principle is satisfied if S_0 per atom is the same for all elements when in a perfect crystalline form. If this is true, then for any change in state involving only perfect crystalline substances $\Delta S_0 = 0$. This is, in fact, the basis of the third law of thermodynamics.

The third law is a very useful principle, but an entirely satisfactory statement of it cannot be made. As we have just seen, any statement of this law will be either subjected to exception or limited to a group of substances such as 'perfect crystalline substances'. Unfortunately, the definite identification of the members of such a group is rather difficult. However, since entropy is a measure of disorder, $S_0 = 0$ will imply a highly ordered state of matter. The most highly ordered state of matter that one can conceive of is the crystalline state at the absolute zero of temperature, because at this temperature even the vibrational and rotational motions of the molecules are minimized. A glassy or amorphous solid is not completely ordered even at the absolute zero of temperature and any disorder remaining at this temperature produces a finite value of S_0.

Note: Certain crystals are also known to have a finite value of S_0, e.g. carbon monoxide $(S_0 = 4.602 \, \text{J K}^{-1} \, \text{mol}^{-1})$. This is attributed to the end-for-end randomness of carbon monoxide molecules in the lattices.

We now find that the essential criterion for application of the third law is the maximum order in the solid. Therefore, the third law can be applied to any substance which can be obtained in a perfect crystalline state at temperature near the absolute zero. Then the evaluation of the entropy content at some other temperature is possible. In some cases, one or more phase changes may occur between $0\,K$ and the temperature of interest. In such cases the appropriate entropy changes must be included. For a gas, we can write

$$S_T = \int_0^{T_1} C_s \, d(\ln T) + \frac{(\Delta H)_f}{T_1} + \int_{T_1}^{T_2} C_\ell \, d(\ln T) + \frac{(\Delta H)_{vap}}{T_2}$$
$$+ \int_{T_2}^{T} C_g \, d(\ln T) \tag{9.1}$$

where T_1 is the melting point of the substance, T_2 is the boiling point of the substance, T is the temperature of interest, and C_s, C_ℓ and C_g are the heat capacities of the solid, liquid and gas, respectively.

9.2 STATEMENTS OF THE THIRD LAW

Experiment shows that the fundamental feature of all cooling processes is that, the lower the temperature attained, the more difficult it is to cool further. For example, the colder a liquid is, the lower the vapour pressure, and the harder it is to produce further cooling by pumping away the vapour. The same is true for the magnetocaloric effect. If one demagnetization produces a temperature T_1, say one-fifth of the original temperature T, then a second demagnetization from the same original field will produce a temperature T_2 which is also approximately one-fifth of T_1. Under these circumstances, an infinite number of adiabatic demagnetizations would be required to attain absolute zero. We may generalize this experience by saying: '*By no finite series of processes is the absolute zero attainable*'.

This is known as the principle of the unattainability of absolute zero, or the unattainability statement of the third law of thermodynamics. As in the case of the second law of thermodynamics, the third law has a number of alternative or equivalent statements. Another statement of this law is the outcome of experiments leading to the calculations of the way that the entropy change of a condensed system during a reversible isothermal process behaves as the temperature approaches zero. This is known as the *Nernst–Simon statement of the third law*, which states: 'The entropy change associated with any isothermal reversible process of a condensed system approaches zero as the temperature approaches zero'.

We have given Nernst's original statement in Section 9.1. When he stated the original law, he did not think in terms of entropy. He was of the opinion that his statement and the unattainability statement could be derived from the second law with the additional assumption that the heat capacities of all materials approached zero as the temperature approached zero. He also thought that both the statements were true for all kinds of process, both reversible and irreversible. It was mainly the experiments and arguments of Simon (during 1927–37) that identified the domain of validity of the third law. To prove the equivalence of the unattainability and the Nernst–Simon statements of the third law, we shall proceed in the same manner as in the case of the Kelvin–Planck and Clausius statements of the second law.

(i) *Planck statement of the third law*: In the Nernst–Simon statement of the third law no comment was made on the value of the absolute entropy at 0 K. (It must be finite or zero, if ΔS_0 is to be finite for a reaction involving condensed phases.) Planck gave another statement by incorporating an additional postulate. According to him, 'The absolute value of the entropy of a pure solid or a pure liquid approaches zero at 0 K'. Mathematically,

$$\lim_{T \to 0} S = 0. \tag{9.2}$$

If this statement is accepted, Nernst's statement follows immediately for pure solids and liquids. Planck's statement emphasizes that S_0 is zero only for pure solids and pure liquids; on the other hand, Nernst assumed that his statement was applicable to all condensed phases (including solutions).

(ii) *Lewis and Randall statement*: According to Planck, solutions at 0 K have a positive entropy equal to the entropy of mixing. This idea was also supported by Lewis and Gibson. They also pointed out that supercooled liquids (e.g. glasses), even when composed of a single element, probably retain positive entropies even when the temperature tends to absolute zero. In view of this, Lewis and Randall proposed another statement of the third law: 'If the entropy of each element in some crystalline state be taken as zero at the absolute zero of temperature, every substance has a finite positive entropy, but at the absolute zero of temperature the entropy may become zero and does so become in the case of perfect crystalline substances'.

If the entropy of the system at the absolute zero is called the zero-point entropy, another equivalent statement of the third law can be stated: 'By no finite series of processes can the entropy of a system be reduced to its zero-point value'. This is illustrated in Figure 9.1. We find that, in the isothermal process $1 \to 2$, there is a decrease in entropy; in the isothermal process $3 \to 4$, there is another decrease, and so on. Hence, it is obvious that, to reach to zero-point entropy, we shall require an infinite number of processes. This diagram also illustrates the equivalence of all the three statements of the third law.

So far we have discussed different forms of the statement of the third law. Now we shall give a practical statement to imply that it is a working principle rather than a complete and rigorous summary of the facts with which it deals. We shall state the third law as: *'At any pressure the entropy of perfect crystalline elements and compounds may be taken to approach zero as the temperature approaches 0 K.'*

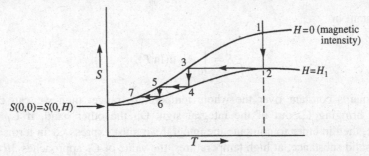

Figure 9.1 Graphical representation of the third-law statement that by no finite series of processes can the entropy of a system be reduced to its zero-point value

9.2.1   Corollaries to the practical third law

Several corollaries of the practical third law applicable to substances which are subject to it can be readily derived.

(i) The values of ΔH and ΔG (change in the Gibbs free-energy) for chemical reactions among crystalline substances approach equality (as $\Delta S \to 0$) as the temperature approaches zero.

(ii) Another corollary follows from the independence of the entropy at 0 K of the pressure. We can write

$$\left(\frac{\partial S}{\partial P} \right)_T \to 0 \quad \text{as} \quad T \to 0.$$

However, the volume may be considered as a function of P and T, and hence it follows that

$$\left(\frac{\partial S}{\partial V} \right)_T \to 0 \quad \text{as} \quad T \to 0.$$

9.2.2   Consequence of the third law

As a consequence of the third law, we can evaluate the absolute entropy S of a substance by means of the equation

$$S - S_0 = \int_0^T \frac{dQ_{rev}}{T} \qquad S_0 = 0.$$

To determine the entropy of a substance at room temperature (25°C), (assuming that there is no phase transition between $T = 0$ and $T = 298$ K) we need only substitute $dQ_{rev} = C_P\, dT$ in the above equation. Then,

$$S = \int_0^T \frac{C_P\, dT}{T}$$

if P is constant or

$$S = \int_0^T C_P\, d(\ln T). \tag{9.3}$$

If C_P remains constant over the whole temperature range, then we can compute the integral by bringing C_P out of the integral sign. On the other hand, if C_P varies with temperature, then in order to compute the integral we must express C_P in terms of temperature. For a solid substance, at high temperatures, the value of C_P approaches $3R$ per mole of atoms whereas, at low temperatures, C_P approaches zero. Unfortunately, there is no convenient and satisfactory representation of C_P over the complete temperature range.

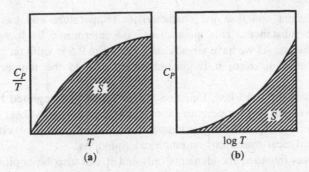

Figure 9.2 C_P versus T: (a) C_P/T as a function of T; (b) C_P as a function of log T

If C_P is expressed as a function of temperature, then Equation 9.3 can be integrated graphically either by plotting C_P/T against T or by plotting C_P against log T. In either case, the area under the curve gives the value of S. Figure 9.2 shows the curves. If a phase transition (such as melting) occurs at some temperature T_f before 298 K, we write Equation 9.3 in the form

$$S = \int_0^{T_f} \frac{C_P}{T} dT + \frac{(\Delta H)_f}{T_f} + \int_{T_f}^{T} \frac{C_P'}{T} dT \qquad (9.4)$$

where C_P is the heat capacity of the solid, T_f is the melting temperature at 0.1 MPa pressure and C_P' is the heat capacity of the liquid. Equation 9.1 is a general equation for all the phase transitions within any temperature range.

Evaluation of absolute entropy is not the only consequence of the third law. Ever since its formulation in the 1920s, it has had great influence on thermodynamics. It has enabled determination of free energies in reactions by calorimetry, computation of partition functions by the statistical spectroscopic method and from that entropies and other thermodynamic quantities. Other consequences include determination of heats of reactions, equilibrium constants, electromotive forces and heat capacities of substances down to very low temperatures.

9.3 USE OF THE THIRD LAW OF THERMODYNAMICS

The most useful application of the third law is the computation of absolute entropies of pure substances at temperatures other than 0 K from their heat capacities and heats of transition. A solid material at any temperature, say T, will have an entropy given by Equation 9.3, whereas a liquid material will have an entropy given by Equation 9.4. Under some circumstances, other transitions may also be involved. In such cases, entropy can be determined by Equation 9.1.

The heat capacities in Equations 9.1, 9.3 and 9.4 can be determined by calorimetric means; at low temperatures, the heat capacities of crystals may be determined with the help of theoretical formulae. For temperatures approaching the absolute zero, the atomic heat capacity of a crystalline substance is accurately given by the Debye equation

$$C_V = \frac{12\pi^4 R}{5}\left(\frac{T}{\theta}\right)^3 = 1943.05\left(\frac{T}{\theta}\right)^3 \text{ J K}^{-1}(\text{g-atom})^{-1} = aT^3 \qquad (9.5)$$

where a is a constant and θ is the characteristic temperature and has different values depending on the substances. This quantity can be determined by low-temperature heat capacity measurements. As we have already said, Equation 9.5 is valid only for temperatures approaching the absolute zero, it is generally restricted to the temperature range $0 < T < \theta/10$.

For application of the third law, Equation 9.5 is generally integrated from the absolute zero to some convenient low temperature, above which experimental heat capacity data can be used. Integrations at any other higher temperatures can be carried out either by numerical methods or by analytical operations on empirical equations.

Equation 9.5 was intended for elements only but it can also be applied to compounds, provided that the heat capacity given by Equation 9.5 is multiplied by the number of atoms in the molecular formula.

As we know that the difference between C_P and C_V is negligible for solid substances (especially at low temperatures), Equation 9.5 can also be used for C_P. However, for precise work of very high accuracy, corrections of an empirical nature can be used.

9.4 EXPERIMENTAL PROCEDURE FOR THE VERIFICATION OF THE THIRD LAW

Figure 9.3 is an outline showing the experimental procedure by which the third law can be verified. In this diagram $\Delta S_{\text{products}}$ is the change in entropy of the products and $\Delta S_{\text{reactants}}$ is the change in entropy of the reactants. These two thermodynamic quantities are determined by calorimetric measurements (almost always at constant pressure). The results are then treated by

$$\Delta S = \int \frac{C_P \, dT}{T}$$

for a temperature range in which there is no phase change, and by

$$\Delta S = \frac{\Delta H}{T}$$

for a phase transition and extrapolated to $0\,\text{K}$.

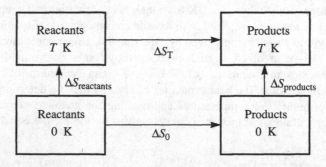

Figure 9.3 An outline of the experimental procedure by which the third law can be verified

ΔS_T is determined separately by a method derived from the second law. For a constant-temperature case,

$$\Delta S = \frac{1}{T}(\Delta H - \Delta G)$$

where ΔG is the change in the Gibbs free energy. We shall deduce this relation later.

ΔH is determined calorimetrically and ΔG is calculated through equilibrium experiments or measurements on electrochemical cells. Then, from Figure 9.3

$$\Delta S_0 = \Delta S_{\text{reactants}} + \Delta S_T - \Delta S_{\text{products}}.$$

The experiments of the type mentioned above can only show that the difference between the entropies of reactants and products at 0 K is almost always zero within the limits of accuracy of experiment. This is, in fact, the substance of Nernst's statement.

9.5 UNATTAINABILITY OF ABSOLUTE ZERO

We have already said that by no finite series of processes is the absolute zero attainable. It is a common experience in low-temperature cooling processes that, the lower the temperature attained, the harder it becomes to produce further cooling. This experience may be considered as a direct consequence of the first and second laws of thermodynamics. For example, let us consider a change in entropy at constant pressure. For this we may write

$$dS = \frac{(dQ)_P}{T} = \frac{C_P}{T} dT$$

or

$$\left(\frac{dS}{dT}\right)_P = \frac{C_P}{T}.$$

Hence, in the limiting case as $T \to 0$ we have

$$\lim_{T \to 0} \left(\frac{\partial S}{\partial T}\right)_P = \lim_{T \to 0} \left(\frac{C_P}{T}\right) = \infty. \qquad (9.6)$$

Since according to classical mechanics the specific heats of substances have a finite value at all temperatures, Equation 9.6 implies that the entropy of any substance should tend towards ∞ as $T \to 0$. Therefore, obviously the absolute zero cannot be attained by a finite series of processes.

We may now generalize the above experience by saying that no system can be reduced to absolute zero. Conversely, we may say that the assumption that at 0 K the entropy is zero leads to unattainability of the absolute zero. To prove this, let us consider a case of cooling a system below the lowest temperature attainable. Such processes may include allotropic transformation, or a change in external magnetic field, or a volume change, etc. Let the initial state of the system be denoted by i corresponding to temperature T_i, and the final state be denoted by f corresponding to temperature T_f. Obviously, the process will be an adiabatic, since no other reservoir available would be at a lower temperature than the system itself. In addition, this process may be reversible or irreversible. The entropies of the system in states i and f would be given by

$$\underset{(at\ T_i)}{S_i} = \underset{(at\ 0\ K)}{S_i} + \int_0^{T_i} \frac{C_i}{T} dT \tag{9.7}$$

$$\underset{(at\ T_f)}{S_f} = \underset{(at\ 0\ K)}{S_f} + \int_0^{T_f} \frac{C_f}{T} dT. \tag{9.8}$$

If the process is reversible, as the change in entropy is zero, we have

$$\underset{(at\ T_f)}{S_f} - \underset{(at\ T_i)}{S_i} = 0.$$

Hence, from Equations 9.7 and 9.8,

$$\underset{(at\ 0\ K)}{S_i} + \int_0^{T_i} \frac{C_i}{T} dT = \underset{(at\ 0\ K)}{S_f} + \int_0^{T_f} \frac{C_f}{T} dT. \tag{9.9}$$

However, from the third law, S_i (at 0 K) $= S_f$ (at 0 K). Hence, Equation 9.9 reduces to

$$\int_0^{T_i} \frac{C_i}{T} dT = \int_0^{T_f} \frac{C_f}{T} dT. \tag{9.10}$$

Since $T_i > 0$ and $C_i > 0$ for any non-zero temperature T_i, the left-hand side of Equation 9.10 must be positive. Therefore, the right-hand side of this equation is also positive. This is possible only when $T_f \neq 0$. This implies that the final temperature T_f cannot be equal to 0 K. Hence, we find that for a reversible cooling process the absolute zero of temperature cannot be attained.

Let us now consider the irreversible case. The process is carried out exactly in the same manner as for the reversible case. For this case $\Delta S > 0$. Therefore, from Equations 9.7 and 9.8,

$$\left[\underset{(at\ 0\ K)}{S_f} + \int_0^{T_f} \frac{C_f}{T} dT \right] > \left[\underset{(at\ 0\ K)}{S_i} + \int_0^{T_i} \frac{C_i}{T} dT \right]. \tag{9.11}$$

Again, application of the third law would give

$$S_f \text{ (at 0 K)} = S_i \text{ (at 0 K)}.$$

Therefore, Equation 9.11 reduces to

$$\int_0^{T_f} \frac{C_f}{T} dT > \int_0^{T_i} \frac{C_i}{T} dT. \tag{9.12}$$

Arguing in the same way as for the reversible case, the right-hand side of this equation is positive; therefore, the left-hand side is also positive. This is possible only when $T_f \neq 0$. It follows again that T_f cannot be equal to $0\,\text{K}$. Hence, we conclude that the absolute zero of temperature is not attainable.

9.6 STANDARD AND ABSOLUTE ENTROPIES

The concept of entropy associated with the third law is often called *absolute entropy*, but the term 'absolute' is misleading. To assign zero entropy to well-ordered crystals in their lowest states of internal motion is a convention but is by no means comparable with the absolute nature of the zero point of the thermodynamic temperature scale. Absolute entropy is denoted by the symbol S (unless it refers to the standard state and one mole of the substance concerned); standard entropy is simply denoted by the symbol $S°$ and this refers to the entropy at $298.15\,\text{K}$.

The absolute entropy must not be confused with the *standard entropy of formation* of a compound, which is denoted by $(S°)_f$. This is the entropy of forming the compound from the elements at the standard states. Therefore,

$$(S°)_f = (S°)_{\text{compound}} - \sum (S°)_{\text{elements in most stable form at standard conditions}}.$$

To find the standard entropy of a reaction, one may use absolute entropy or entropy of formation of products and reactants, or any other kind, as long as the reference state for reactants and products is the same. Therefore,

$$\Delta S° = \sum (S°)_{\text{products}} - \sum (S°)_{\text{reactants}} = \sum [(S°)_f]_{\text{products}} - \sum [(S°)_f]_{\text{reactants}}$$

because the entropies of any reference states cancel.

9.7 ENTROPIES OF SYSTEMS OTHER THAN PERFECT CRYSTALS AT ABSOLUTE ZERO

We have noticed that the statement of the third law concerns the entropy of perfect crystals at absolute zero. Now the question may arise in our mind of what would be the entropy for other systems under similar circumstances. To find the answer to this question, let us consider a solid crystalline solution at the absolute zero, and imagine that this crystal is perfectly ordered with respect to the distribution of lattice sites but contains two different kinds of atom which are distinguishable from each other. If we imagine that the two types of atom are distributed throughout the lattice sites perfectly randomly, then it is obvious that the thermodynamic probability will not be unity. Therefore, we cannot expect zero or negative entropy. This implies that the formation of a mixed crystal from perfect crystals of different types of atom is associated with an increase in entropy.

A solid solution will not have a positive entropy unless it possesses some degree of randomness. If each atom of one type bears a definite spatial relation to atoms of the other kind, the system will have a low thermodynamic probability, even if it were a mixture in composition. To illustrate, let us consider a crystal of potassium chloride. Such a crystal is made up of potassium and chlorine ions arranged in a definite spatial structure. Such a crystal should not be considered as a solid solution because both potassium and chlorine ions occur in a well-defined pattern. Therefore, a perfect crystal of KCl at the absolute zero will have a thermodynamic probability of unity, even though two species are present.

A solid solution which is nearly perfect but still has a positive entropy (even at the absolute zero) is a crystalline mixture of isotopes. In supercooled liquids or glasses at the absolute zero, one can also expect entropies greater than zero. From what we have said it appears that, since a liquid does not possess the characteristic order of a crystal, it will have a positive entropy. This entropy can be regarded as 'frozen in' when the liquid is subjected to supercooling.

9.8   MOLECULAR INTERPRETATION OF THE THIRD LAW

We have already discussed that the molecular basis for the third-law result, namely that the entropy is zero at absolute zero, is the perfectly ordered state of the crystals, with all the molecules in the same lowest energy level. The positive values for entropies of all compounds at temperatures above absolute zero result from the fact that, as the temperature is raised, more and more energy levels become available to the molecules. Since each individual molecule has its own particular pattern of energy levels, the entropies at temperatures higher than absolute zero are very much characteristic of the individual molecule.

With a few exceptions, it is observed that the entropy values calculated from the details of the molecular energies are in agreement with those obtained from calorimetric third-law measurements (within experimental error). The third-law entropies for CO and N_2O are smaller by about $4.6 \, J \, K^{-1} \, mol^{-1}$; that for H_2O is smaller by about $3.3 \, J \, K^{-1} \, mol^{-1}$. This discrepancy can be explained by stating that these materials fail to form the perfect crystalline state that is necessary at the absolute zero for the application of the third law.

To illustrate what we have just said, let us take the case of CO. The molecular arrangement in the crystal of this material should be CO CO CO CO, which is a perfect order. However, instead of a perfect crystalline order, if a crystal is initially formed in a disordered pattern as CO CO CO OC CO, then this crystal may have this disorder 'frozen in' as the temperature is lowered because there will be very little thermal energy available for the molecules to rearrange to the ordered state. In that case this randomness will provide two states to each molecule instead of a single state. Naturally, the entropy of such a crystal will be greater by a factor of $k \ln(2^{N_0}) = N_0 k \ln 2 = R \ln 2 = 5.7 \, J \, K^{-1} \, mol^{-1}$ (where N_0 is the Avogadro number) than that for a perfect crystal of this material.

Other discrepancies may be explained in the same way. A glassy material at absolute zero will not have the necessary molecular order to provide an entropy of zero at absolute zero. It is worthwhile to mention here that, of all the exceptions to the third law, hydrogen is the first to be considered. It has an entropy of $117.529 \, J \, K^{-1}$ owing to translation and $R \ln 4$, i.e. $11.527 \, J \, K^{-1}$, as the residual entropy, including the contribution from nuclear spin. This makes a total of $129.056 \, J \, K^{-1}$ at $298.15 \, K$. The entropy of hydrogen in the standard state is $130.587 \, J \, K^{-1}$, based on the third law. The discrepancy between these two values can be explained by the fact that the rotational levels that characterize the para- and ortho-molecules went over into the solid state, leaving randomness of distribution to some extent even at the lowest temperatures, at which the entropy measurement was made. Therefore, the entropy of solid hydrogen obtained from experimental measurements cannot be zero. The fact that rotation in solid hydrogen exists provides an explanation for a definite exception to the third law. Deuterium and one or two compounds of hydrogen, and one or two compounds of deuterium behave in a similar manner. However, such exceptions are very rare. The entropy of water calculated by statistical method is about $4\,J\,K^{-1}$ higher than that obtained by the third-law assumption. This may be explained by the existence of rotation of the molecule in the solid. Another suggested explanation is that the entropy of water at $0\,K$ is not zero because of the uncertainty of the location of some hydrogen bonds in the crystal lattice.

EXERCISES 9

9.1 What is the essential criterion for application of the third law? To what type of substances can the third law of thermodynamics be applied? If one or more phase changes occur between $0\,K$ and the temperature of interest, is it necessary to include the appropriate entropy of such changes in the third law?

9.2 Write the different statements of the third law of thermodynamics. Show how Nernst's statement of the third law follows from Planck's statement of the same law. How do these two statements differ fundamentally?

9.3 Discuss the consequences and use of the third law of thermodynamics. Can you state any corollary of this law?

9.4 Outline an experimental procedure for verification of the third law. What conclusion would you draw from such experiments?

9.5 What do you understand by the principle of unattainability of the absolute zero, or the unattainability statement of the third law of thermodynamics? Show analytically that the absolute zero of temperature cannot be attained.

9.6 We have proved the unattainability of the absolute zero of temperature by carrying out adiabatic transformation from initial state i to the final state f with substances for which we have taken S_i (at $0\,K$) $= 0$. Let us now imagine that the state i consists of a glassy substance for which S_i (at $0\,K$) > 0, and the state f consists of the same substance in crystalline form. Can you show that absolute zero is still unattainable by means of a transformation from the glassy to the crystalline state?

9.7 How would you differentiate between standard and absolute entropies?

9.8 Calculate the change in entropy for the change in state H_2O (solid at $0°C$) $\rightarrow H_2O$ (liquid at $25°C$), given that the molar enthalpy of fusion of ice at $0°C$ is $6.025\,kJ\,mol^{-1}$ and C_P for liquid water is $75.312\,J\,°C^{-1}\,mol^{-1}$ (assume them to be constant over the temperature range).

9.9 Calculate the change in entropy when $100\,ml$ of water at $25°C$ are mixed with $10\,g$ of ice at $-10°C$, given that the enthalpy of fusion of ice is $333.465\,J\,g^{-1}$ and the specific heat capacities for ice and liquid water are $2.092\,J\,°C^{-1}\,g^{-1}$ and $4.184\,J\,°C^{-1}\,g^{-1}$, respectively. Assume that these specific heat capacities are constant.

9.10 Using the Debye formula for heat capacities of solids, show that the entropy of a solid at low temperatures is equal to one third of its heat capacity.

9.11 The heat of transition for the reversible transformation α-AgI $\rightleftharpoons$ β-AgI at 0.1 MPa pressure and 146.5°C is 6401.520 J mol^{-1}. Calculate the entropy change involved in the transformation of 2 mol of the β form to the α form.

9.12 It is found that, for propane at 0.1 MPa pressure, $(\Delta H)_f = 3523.765$ J mol^{-1} at the melting point 85.45 K, and $(\Delta H)_{vap} = 18.773$ kJ mol^{-1} at the boiling point 231.04 K. To heat the solid from 0 to 85.45 K, $\Delta S = 41.505$ J K^{-1}; to heat the liquid from 85.45 K to the boiling point, $\Delta S = 88.115$ J K^{-1}. For propane the critical temperature and pressure are 368.8 K and 4.3 MPa, respectively. Calculate the entropy of gaseous propane at 231.04 K.

9.13 Calculate the absolute standard entropy of gaseous chlorine at 298.15 K. Assume that gaseous chlorine behaves ideally and has $C_P = 34.225$ J K^{-1} mol^{-1} which obeys the Debye cube law below 15 K. The latent heat of fusion is 6.406 kJ mol^{-1}, the latent heat of vaporization is 20.409 kJ mol^{-1}, and the temperatures are 172.12 K and 239.05 K, respectively. The following heat capacity data are taken from W.F. Giauque and T.F. Powell, 1939, *J. Am. Chem. Soc.* **61**, 1970.

T (K)	C_P (J K^{-1} mol^{-1})	T (K)	C_P (J K^{-1} mol^{-1})	T (K)	C_P (J K^{-1} mol^{-1})
15	3.724	80	38.618	172.12	55.522
20	7.740	90	40.627		
25	12.092	100	42.258	180	67.027
30	16.694	110	43.806	190	66.902
35	20.794	120	45.480	200	66.735
40	23.974	130	47.237	210	66.484
45	26.736	140	49.078	220	66.275
50	29.246	150	51.045	230	65.982
60	33.472	160	53.053	240	63.178
70	36.317	170	55.103		

9.14 Calculate the absolute standard entropy of solid silver at 298.15 K from the heat capacity data of metallic silver given below. Assume that the heat capacity obeys the Debye Cube law, i.e. it is proportional to T^3 below 15 K.

T (K)	C_P (J K^{-1} mol^{-1})	T (K)	C_P (J K^{-1} mol^{-1})	T (K)	C_P (J K^{-1} mol^{-1})
15	0.669	90	19.133	210	24.422
20	1.715	110	20.961	230	24.732
30	4.774	130	22.129	250	24.732
40	8.388	150	22.970	270	25.313
50	11.648	170	23.615	290	25.439
70	16.334	190	24.087	300	25.501

9.15 The following data are given for two crystalline forms of benzothiophene, a stable form 1 at low temperatures and a high-temperature form 2.

Crystal form	C_P at 12 K $(\mathrm{J\,K^{-1}\,mol^{-1}})$	Value of $\int_{12\,\mathrm{K}}^{261.6\,\mathrm{K}} C_P\,\mathrm{d}(\ln T)$ $(\mathrm{J\,K^{-1}\,mol^{-1}})$
1	4.469	148.105
2	6.573	152.733

Calorimetric measurements on each of these crystalline forms down to 12 K provide that, at the normal transition temperature 261.6 K, the molar enthalpy of transition $1 \rightarrow 2$ is $3012.480\,\mathrm{J\,mol^{-1}}$. Assume that below 12 K the heat capacity of benzothiophene obeys the Debye Cube law. Establish whether the high-temperature crystal form of benzothiophene is a perfect crystal at the absolute zero of temperature.

9.16 Calculate the standard entropy change at 298.15 K and 0.1 MPa for the reaction $H_2(g) + \frac{1}{2}O_2(g) \rightarrow H_2O(g)$.

9.17 Calculate the entropy change for the reaction $H_2O(\ell) \rightarrow H_2O(g)$ at 298 K and 0.1 MPa pressure.

10

Free Energy and Applications of Thermodynamic Principles

10.1 PERMITTED AND FORBIDDEN PROCESSES

We have seen that the first law of thermodynamics is concerned with the conservation of energy in physical and chemical processes; it does not provide any information regarding whether or not a specific change can occur spontaneously. On the other hand, the second law of thermodynamics does provide information as to whether or not a specified change can occur spontaneously. A spontaneous process may be fast or extremely slow. In order to characterize spontaneous changes, whether fast or vanishingly slow, we shall use the term *permitted change*. Our experience also tells us that there are classes of change which are not permitted according to our convention. In such cases we shall use the term *forbidden changes*. For example, a gas enclosed in a rigid container will not spontaneously contract unless some force is applied on it.

Forbidden changes are obviously the reverse of permitted changes but it may be more precise to say that the types of change that we have termed 'permitted' are not forbidden. There is another class of physical–chemical process which is neither forbidden nor permitted. These are reversible processes. Such processes are not observed in nature and are, therefore, hypothetical. To distinguish between permitted, forbidden and reversible processes let us consider Figure 10.1.

In this figure, path 1 represents a reversible path on a P–V diagram where P_{surr} is the pressure of the surroundings and V is the volume of the gas. Path 2 is an irreversible process representing an expansion from V_1 to V_2 against a constant external pressure $P_2(<P_1)$. This is, no doubt, a permitted process. Any path of compression which is permitted is one in which the external pressure is greater than the gas pressure. These paths are also irreversible paths (for compression). Path 3 is such a path. Compression along any path for which the external pressure is less than the gas pressure, or expansion along any path for which the external pressure is higher than the gas pressure, is a forbidden path (or process).

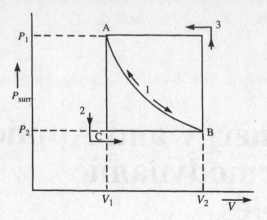

Figure 10.1 Graphical representation distinguishing between a permitted, forbidden and reversible process

10.1.1 Criterion of forbiddenness

We now know what is meant by a permitted, forbidden or reversible process. The classification of physical–chemical processes as permitted, forbidden or reversible is simply a matter of experience. To develop a more formal criterion of forbiddenness let us consider $P_{surr} \, dV$ work done by an ideal gas in the three types of expansion indicated in Figure 10.1. Since the work done is given by the equation

$$W = -\int_{V_1}^{V_2} P_{surr} \, dV$$

the maximum work is done when the gas pressure $P_g = P_{surr}$. This condition can be obtained only for a reversible case. Hence, the maximum work would be done in the reversible expansion along path 1. Hence, we have

$$W_{max} = -\int_{V_1}^{V_2} P_g \, dV.$$

The value of this integral corresponds to the area under the curve AB in Figure 10.1.

In a permitted or irreversible expansion, such as that indicated by path 2, the work done by the gas would be given by $W = -P_2(V_2 - V_1)$ ($P_{surr} = P_2 =$ constant) and represented by the area BCV_1V_2 in Figure 10.1.

In a forbidden expansion, i.e. one against an external pressure greater than the gas pressure, the work done by the gas would have to be greater than the maximum work. Of course, such an expansion is never observed.

We are now able to summarize the difference between permitted, reversible and forbidden processes (at least in the case of isothermal expansions) as follows:

$$\text{permitted process: } W < W_{max}$$
$$\text{reversible process: } W = W_{max} \qquad (10.1)$$
$$\text{forbidden process: } W > W_{max}.$$

If proper regard to sign is observed, these distinctions also apply to isothermal gas compression. In compression, work is done on the gas. We consider that the work done by the gas is negative. In a permitted (irreversible) compression the pressure on the gas is, at all times, greater than the gas pressure. Therefore, the work done on the gas is greater than the minimum required, that is to say that work done by the gas is less than W_{max}.

For any change in state at constant volume, W is necessarily zero (although the ability to produce W_{max} is not necessarily zero). Hence, for a constant volume isothermal process we have the following classification:

$$\text{permitted process: } 0 < W_{max}$$
$$\text{reversible process: } 0 = W_{max} \qquad (10.2)$$
$$\text{forbidden process: } 0 > W_{max}.$$

All the above criteria apply to expansion or contraction in a rigid container. Let us now see what happens if a gas expands or contracts in vacuum. If a gas expands isothermally into an evacuated space, $W_{max} > 0$ but no actual work is done; the expansion is permitted according to Equation 10.2. On the other hand, if it is an isothermal contraction in an evacuated space, a decrease in the volume of the gas would occur, leaving a portion of the container empty. For this contraction, $W_{max} < 0$, and this process is forbidden according to Equation 10.2.

Note: In these processes it must be remembered that no exchange of energy, other than the heat required to maintain constant temperature, is allowed.

From the above discussion it appears that we have to examine more closely the nature of the maximum work.

10.2   WORK CONTENT

We have seen that the value of W_{max} for a change in state is of general applicability for the characterization of permitted, reversible and forbidden processes. Therefore, it is appropriate to emphasize its importance by introducing the corresponding thermodynamic property of the system.

We have seen before that, for isothermal processes, $T \Delta S \geqslant Q$ (the equality holds for reversible processes, and the inequality for all other processes). Combining this with the first law of thermodynamics, we have $-W \leqslant -\Delta U + T\Delta S$. We have already sent that W can assume any real value for a given change in state depending on the path followed. Here we find that, if the change in state is brought about isothermally, there is an upper limit to W.

This upper limit is equal to the isothermal reversible work. Since ΔU and ΔS are both functions of initial and final states, this isothermal reversible work must also be a function of the initial and final states of the system. This conclusion suggests that we define a new function of the state of a system whose change will measure the reversible or maximum work associated with an isothermal process. We shall call this function 'work content' and denote it by the symbol A with the definition

$$A = U - TS. \tag{10.3}$$

Therefore, for an isothermal process

$$-\Delta A \geqslant -W_{max}. \tag{10.4}$$

This equation implies that the decrease in work content is an upper limit to the maximum work obtainable in an isothermal process. With the help of Equations 10.2 and 10.4 we can write

$$\text{permitted process: } \Delta A < 0$$
$$\text{reversible process: } \Delta A = 0 \tag{10.5}$$
$$\text{forbidden process: } \Delta A > 0.$$

If an isothermal process is carried out in such a way that $W = 0$ (which can be done by keeping the volume constant and excluding all other work), then from Equation 10.4 we have

$$\Delta A \leqslant 0. \tag{10.6}$$

Hence, we conclude that the work content of a constant-volume isothermal process tends to decrease spontaneously as the process continues, until it attains the minimum value and, once the minimum value is reached, the system will be in equilibrium state.

We have just discussed an isothermal constant-volume process. Let us now consider an isothermal constant-pressure process. In such a case the reversible pressure–volume work is given by $P\Delta V$; all other non-PV work is then given by $W_{max} - P\Delta V$. If we denote this non-PV work by $W'(=W_{max} - P\Delta V)$, we have from Equation 10.4,

$$-W' \leqslant -\Delta A - P\Delta V. \tag{10.7}$$

If we are considering a system which consists of several parts at different constant pressures, Equation 10.7 may be generalized as

$$-W' \leqslant -\Delta A - \sum_j (P_j \Delta V_j) \tag{10.8}$$

where P_j represents the constant pressure of the jth part of the system, and ΔV_j is the change in volume of that part.

Equations 10.7 and 10.8 indicate that still another function may be defined. We call this function the 'free energy'. It bears the same relationship to the work content as the enthalpy bears to the internal energy. We denote this function by G, with the definition

$$G = A + PV = (U - TS) + PV$$

from Equation 10.3 and finally

$$G = H - TS. \tag{10.9}$$

This function is also known as the *Gibbs free energy* and is often denoted by F, but there is no general agreement on nomenclature. Obviously, at constant pressure we have

$$G = A + P\Delta V. \tag{10.10}$$

Combining Equations 10.7 and 10.10, we find that

$$-W' \leqslant -\Delta G. \tag{10.11}$$

From this equation we may conclude that the upper limit of all work, other than pressure–volume work, is the decrease in free energy attending an isothermal constant-pressure process. If an isothermal process is carried out in such a way that only pressure–volume work is done, we have $W' = 0$, and therefore

$$0 \leqslant -\Delta G$$

or

$$\Delta G \leqslant 0. \tag{10.12}$$

This shows that at constant temperature and pressure the Gibbs free energy tends to decrease until it reaches the minimum value. When the minimum value is attained, the system will be in equilibrium.

It should be noted that the free energy is not a new function in our development but rather a new functional relationship between the state functions already defined. This is an invented function obtained to acquire a workable criterion of spontaneity.

Although it is not possible to know the absolute value of work content A in any case, it is possible, in principle, to calculate ΔA for any specified change.

10.3 THE HELMHOLTZ FREE ENERGY

The Helmholtz free energy of any system is denoted by A and defined by Equation 10.3. As all the variables on the right-hand side of this equation are functions of the state of the system, A must also be a function of state. Hence, if a system undergoes a change from state 1 to state 2, the change in the Helmholtz free energy of that system is given by

$$\begin{aligned}
\Delta A = A_2 - A_1 &= (U_2 - T_2 S_2) - (U_1 - T_1 S_1) \\
&= (U_2 - U_1) - (T_2 S_2 - T_1 S_1) = \Delta U - \Delta(TS)
\end{aligned} \tag{10.13}$$

This equation is the most general definition of change ΔA in the Helmholtz free energy. If the process is an isothermal process, we have $T_1 = T_2$. Then, Equation 10.13 reduces to

$$\Delta A = \Delta U - T \Delta S. \tag{10.14}$$

However, under an isothermal condition, $T \Delta S = Q_{rev}$. Therefore,

$$\Delta A = \Delta U - Q_{rev} = W_{max}. \tag{10.15}$$

This shows that at constant temperature the maximum work done by a system is obtained by the change in the Helmholtz free energy of that system. If, for any pure substance, A is a function of T and V only, we have

$$dA = \left(\frac{\partial A}{\partial T}\right)_V dT + \left(\frac{\partial A}{\partial V}\right)_T dV. \tag{10.16}$$

Now, differentiating both sides of Equation 10.3, we get

$$dA = dU - T\,dS - S\,dT = dU - S\,dT - dQ_{rev}$$

(because for an isothermal change $T\,dS = dQ_{rev}$). However, from the first law of thermodynamics

$$dU = dQ_{rev} - P\,dV.$$

Hence,

$$dA = -S\,dT - P\,dV. \tag{10.17}$$

Comparing Equations 10.16 and 10.17, we get

$$\left(\frac{\partial A}{\partial T}\right)_V = -S \tag{10.18}$$

$$\left(\frac{\partial A}{\partial V}\right)_T = -P. \tag{10.19}$$

These equations give the variation in A with respect to V and T.

As the entropy of any substance is positive, Equation 10.18 shows that the work content (or the Helmholtz free energy) of any substance decreases with increase in temperature. The rate of this decrease is greater, the greater the entropy of the substance. For gases, which have high entropies, the rate of decrease in A with increasing temperature is larger than that for liquids and solids which have comparatively small entropies.

Equation 10.19 shows that A decreases with increase in volume; the rate of decrease is greater, the higher the pressure. For any reaction of the type

$$bB + cC + \cdots = mM + nN + \cdots$$

ΔA is given by the sum of the A-values for the products minus the sum of the A-values for the reactants. Then,

$$\Delta A = (mA_M + nA_N + \cdots) - (bA_B + cA_C + \cdots) \tag{10.20}$$

where $A_M, A_N, A_B, A_C, \ldots$, are the Helmholtz free energies of the various species per mole of the substance involved.

To obtain *the variation in ΔA with temperature at constant volume*, we proceed as follows. Differentiating both sides of Equation 10.20 with respect to T, keeping V constant, gives

$$\left(\frac{\partial(\Delta A)}{\partial T}\right)_V = \left[m\left(\frac{\partial A_M}{\partial T}\right)_V + n\left(\frac{\partial A_N}{\partial T}\right)_V + \cdots\right]$$
$$- \left[b\left(\frac{\partial A_B}{\partial T}\right)_V + c\left(\frac{\partial A_C}{\partial T}\right)_V + \cdots\right].$$

From Equation 10.18, $(\partial A_M/\partial T)_V$, etc., are equal to S_M, etc. Hence, the above equation becomes

$$\left(\frac{\partial(\Delta A)}{\partial T}\right)_V = (-mS_M - nS_N - \cdots) - (-bS_B - cS_C - \cdots) = -S. \tag{10.21}$$

Combining Equation 10.14 and 10.21, we have

$$\Delta A = \Delta U + T\left(\frac{\partial(\Delta A)}{\partial T}\right)_V. \tag{10.22}$$

10.4 THE GIBBS FREE ENERGY

We have already defined the Gibbs free energy by Equation 10.9. From this equation it is obvious that, since H, T and S are all state functions, the Gibbs free energy is also a state function. For any infinitesimal change we can write

$$dG = dH - T\,dS - S\,dT. \tag{10.23}$$

As many chemical processes are isothermal, we shall restrict ourselves to constant-temperature processes. Then $dT = 0$, and Equation 10.23 becomes

$$dG = dH - T\,dS. \tag{10.24}$$

For a finite change we integrate this equation to obtain

$$\Delta G = \Delta H - T\Delta S. \tag{10.25}$$

For a reversible process, $dQ_{rev} = T dS$ and, from the first law of thermodynamics, $dU = dQ_{rev} - P\,dV$. Hence, $T\,dS = dU + P\,dV$. From $H = U + PV$ we have $dH = dU + P\,dV + V\,dP$. Substituting these into Equation 10.23, we have

$$dG = (dU + P\,dV + V\,dP) - dU - P\,dV - S\,dT$$
$$= -S\,dT + V\,dP. \tag{10.26}$$

Like H, G can also be most conveniently expressed in terms of T and P as independent variables. Then, for any substance

$$dG = \left(\frac{\partial G}{\partial T}\right)_P dT + \left(\frac{\partial G}{\partial P}\right)_T dP \tag{10.27}$$

Comparing Equations 10.26 and 10.27, we have

$$\left(\frac{\partial G}{\partial T}\right)_P = -S \tag{10.28}$$

$$\left(\frac{\partial G}{\partial P}\right)_T = V. \tag{10.29}$$

These equations give the variation in G with respect to T and P. As the entropy of any substance is positive, Equation 10.28 shows that the Gibbs free energy of any substance decreases with increase in temperature. The rate of this decrease is greater, the greater the entropy of the substance. For gases, which have high entropies, the rate of decrease in G with temperature is larger than that for liquids and solids which have comparatively small entropies. For any general reaction of the type

$$bB + cC + \cdots = mM + nN + \cdots$$

ΔG is given by the sum of the G-values for the products minus the sum of the G-values for the reactants. Then,

$$\Delta G = (mG_M + nG_N + \cdots) - (bG_B + cG_C + \cdots)$$

where G_M, G_N, ..., are the Gibbs free energies of the various species per mole of the substance involved.

10.4.1   The variation in ΔG with temperature at constant pressure

The free-energy change is a function of T and P. Then,

$$d(\Delta G) = \left(\frac{\partial(\Delta G)}{\partial T}\right)_P dT + \left(\frac{\partial(\Delta G)}{\partial P}\right)_T dP. \tag{10.31}$$

Now, differentiating Equation 10.25 with respect to T at constant P we have

$$\left(\frac{\partial(\Delta G)}{\partial T}\right)_P = \left(\frac{\partial(\Delta H)}{\partial T}\right)_P - T\left(\frac{\partial(\Delta S)}{\partial T}\right)_P - \Delta S.$$

However,

$$\left(\frac{\partial(\Delta H)}{\partial T}\right)_P = \Delta C_P \quad \text{and} \quad \left(\frac{\partial(\Delta S)}{\partial T}\right)_P = \frac{\Delta C_P}{T}.$$

Hence, the above equation becomes

$$\left(\frac{\partial(\Delta G)}{\partial T}\right)_P = \Delta C_P - T\frac{\Delta C_P}{T} - \Delta S = -\Delta S. \tag{10.32}$$

Again,

$$\left(\frac{\partial(\Delta G)}{\partial P}\right)_T = \frac{\partial}{\partial P}(G_{\text{products}} - G_{\text{reactants}})_T = \left(\frac{\partial G_{\text{products}}}{\partial P}\right)_T - \left(\frac{\partial G_{\text{reactants}}}{\partial P}\right)_T \tag{10.33}$$

where G_{products} and $G_{\text{reactants}}$ are the free energies of the products and the reactants. Using Equation 10.29 we have $(\partial G_{\text{products}}/\partial P)_T = V_{\text{products}}$ and $(\partial G_{\text{reactants}}/\partial P)_T = V_{\text{reactants}}$ where V_{products} and $V_{\text{reactants}}$ are the volumes of the products and the reactants, respectively. Hence Equation 10.33 becomes

$$\left(\frac{\partial(\Delta G)}{\partial P}\right)_T = V_{\text{products}} - V_{\text{reactants}} = \Delta V. \tag{10.34}$$

Now combining Equations 10.31, 10.32 and 10.34, we have

$$d(\Delta G) = -\Delta S\,dT + \Delta V\,dP. \tag{10.35}$$

In this equation $\Delta S\,dT$ gives the effect on ΔG due to a change in temperature at constant pressure, and the term $\Delta V\,dP$ gives the effect of a change in pressure at constant temperature. Combining Equations 10.25 and 10.32 we get the Gibbs–Helmholtz equation

$$\Delta G = \Delta H + T\left(\frac{\partial(\Delta G)}{\partial T}\right)_P. \tag{10.36}$$

Note: We shall discuss this equation later.

The Gibbs free energy is generally referred to simply as the free energy. Hereafter, we shall also follow this custom and include the terms 'Gibbs' or 'Helmholtz' only when we want to distinguish one from the other.

10.4.2 Properties of free energy

We have established that the free-energy change of any process is a function of states of the system. Hence, it is a definite quantity at any given temperature and pressure; it changes only when these two variables change. The absolute values of the free energy of any substance cannot be known; only the differences may be obtained. The free-energy changes of any process can be added and subtracted. From the definition of the free energy it is possible to obtain the following properties of this quantity.

1. Since H, U and S are all extensive quantities, the Gibbs and the Helmholtz free energies are also extensive quantities.

2. Since H, U and S are state functions, the Gibbs and the Helmholtz free energies are also state functions.

3. Since H, U and TS have the units of energy, the Gibbs and the Helmholtz free energies also have the units of energy.

4. Both the Gibbs and the Helmholtz free energies are a measure of the tendency for a reaction to take place, i.e. a criterion of spontaneous changes.

5. The Gibbs and the Helmholtz free energies are a measure of useful work of a process. They can also be used to measure the maximum work that may be obtained from any change and the maximum yields obtainable from equilibrium reactions.

6. They are a criterion for equilibrium.

10.5 STANDARD FREE ENERGY

We see that ΔG can be calculated directly from the defining Equation 10.25, if the values of enthalpy and entropy changes are available for the given conditions, but this situation is not encountered very often. Therefore, other methods must be developed for the calculation of ΔG. These methods are usually based on computation of the standard free-energy change ΔG°, the difference between the free energy of the products and that of the reactants in their standard state. Although such a method is hypothetical, this does not prevent it from being used to obtain the free-energy alteration in a real process.

If the standard enthalpy and entropy values are available, ΔG° can be evaluated from the equation

$$\Delta G^\circ = \Delta H^\circ - T\Delta S^\circ. \tag{10.37}$$

The standard free energy $\Delta G_{\mathrm{f}}^\circ$ of formation is defined as the free-energy change in the reaction, in which 1 mol of a compound is formed at constant temperature, all species being in their standard state. The standard state is a reference state that is assigned the zero free energy of formation by convention. These are exactly the same reference states that were assigned the zero standard enthalpies of formation-the elements at standard conditions in their most stable forms. The standard molar free energies G_{f}° of formation of compounds, and of elements in less stable conditions, are then their free energies of formation at standard conditions from the elements in their most stable forms. For substances in aqueous solution, standard conditions imply unit concentration (or, more precisely, activity); for elements, standard conditions imply a pressure of 0.1 MPa and a temperature of 298 K. So, by convention we shall take G_f° for elements at 298 K as zero.

For compounds, ΔG_f° has the units of joules per mole. Unless otherwise indicated, standard free energies refer to 25°C or 298 K. Occasionally, another temperature T is more useful and molar standard free energies at T are indicated by the symbol $G_f^\circ(T)$. Thus, $G_f^\circ(400\,K)$ is the free energy of formation of the substance in question from the elements at standard conditions and at 400 K. Again, the convention is to take as reference states the elements in their most stable forms at 0.1 MPa and the temperature T, to which are thus assigned the zero standard free energies of formation.

The standard free-energy change for any reaction can be found from the standard free energies of all substances involved by the relationship

$$\Delta G^\circ = \sum_{products} nG_f^\circ - \sum_{reactants} nG_f^\circ. \tag{10.38}$$

This quantity is meaningful only if the overall balanced equation of the reaction is given. It refers to the complete conversion into products of as many moles of reactants as are indicated by the coefficients in the chemical equation, all at standard conditions. A temperature of 298 K is also implied here unless another temperature is specifically mentioned.

Having evaluated the standard free-energy change, it is then necessary to make the additional calculations required to obtain the value of ΔG under the conditions of the experiment. Essentially, we must calculate the additional free-energy terms arising from the difference between G for the standard state and G for the non-standard state for each species. If a particular reaction involves only species in their standard state at 298 K, then ΔG and $(\Delta G^\circ)_{298\,K}$ calculated directly from the available data are identical. In general, however, this is not the case.

From Equation 10.37 we see that ΔG° can be calculated from the standard enthalpy and entropy values. Since

$$\Delta H^\circ = \sum_{products} H_f^\circ - \sum_{reactants} H_f^\circ$$

and

$$\Delta S^\circ = \sum_{products} S_f^\circ - \sum_{reactants} S_f^\circ$$

then, knowing the standard enthalpies of formation and standard entropies of reactants and products, we can calculate ΔG°.

Values for the standard free energy of formation of some common compounds are given in Table 10.1. These values are taken from 'Selected values of chemical thermo-

Table 10.1 Standard free energies of formation at 298 K (more data are given in Appendix 3)

Substance	ΔG_f° (kJ mol^{-1})	Substance	ΔG_f° (kJ mol^{-1})
AgCl(s)	−109.704	HCHO(g)	−109.621
Al$_2$O$_3$(s)	−1581.98	HCOOH(ℓ)	−346.017
Br(g)	82.383	HI(g)	1.697
Br$_2$(g)	3.138	H$_2$O(g)	228.614
C(g)	672.955	H$_2$O(ℓ)	−237.191
C(diamond)(s)	2.845	H$_2$O$_2$(ℓ)	−119.972
CCl$_4$(g)	−61.015	H$_2$S(ℓ)	−33.612
CCl$_4$(ℓ)	−66.618	Hg(g)	31.757
C$_2$H$_2$(g)	209.200	Hg$_2$Cl$_2$(s)	−210.664
C$_2$H$_4$(g)	68.116	I(g)	70.166
C$_2$H$_6$(g)	−32.886	I$_2$(g)	19.372
C$_2$H$_5$OH(ℓ)	−174.766	KCl(s)	−408.317
C$_3$H$_8$(g)	−23.472	KNO$_3$(s)	−393.129
C$_6$H$_6$(ℓ)	124.516	MgCl$_2$(s)	−592.329
C$_6$H$_6$(g)	129.662	MnO$_2$(s)	−464.842
Cyclo C$_6$H$_{12}$(ℓ)	26.652	N(g)	455.554
CHCl$_3$(ℓ)	−73.546	NH$_3$(g)	−16.652
CH$_3$OH(ℓ)	−166.230	NH$_4$Cl(s)	−203.886
CH$_3$COOH(ℓ)	−392.459	NO(g)	86.692
CH$_4$(g)	−50.794	NO$_2$(g)	51.839
CO(g)	−137.277	N$_2$O(g)	103.596
CO$_2$(g)	−394.384	N$_2$O$_4$(g)	98.282
CaCO$_3$(calcite)(s)	−1128.759	NaCl(s)	−384.008
CaCO$_3$(aragonite)(s)	−1127.714	NaHCO$_3$(s)	−851.862
CaO(s)	−604.169	Na$_2$CO$_3$(s)	−1047.674
Ca(OH)$_2$(s)	−896.757	O(g)	230.120
Cl(g)	105.685	O$_3$(g)	163.427
CuCl(s)	−118.826	PbCl$_2$(s)	−313.967
CuS(s)	−53.6	S(monoclinic)(s)	0.096
Fe$_2$O$_3$(s)	−740.986	SO$_2$(g)	−300.369
Fe$_3$O$_4$(s)	−1014.202	SO$_3$(g)	−370.368
H(g)	203.259	SiF$_4$(g)	−1506.240
HBr(g)	−53.220	ZnCl$_2$(s)	−369.279
HCl(g)	−95.269	ZnO(s)	−318.193

dynamic properties' 1952 *National Bureau of Standards Circular No* 500 and 'Selected values of properties of hydrocarbons' 1947 *National Bureau of Standards Circular No* C461.

Example 10.1

At 298 K the standard free energy for the formation of acetylene is 209.326 kJ mol^{-1} and that for benzene is 129.662 kJ mol^{-1}. Calculate ΔG° at 298 K for the reaction $3C_2H_2 = C_6H_6$.

Solution

The chemical equations for the formation of C_2H_2 and C_6H_6 are

$$2C + H_2 = C_2H_2 \qquad G_f^{\circ} = 209.326 \, kJ \, mol^{-1}$$
$$6C + 3H_2 = C_6H_6 \qquad G_f^{\circ} = 129.662 \, kJ \, mol^{-1}.$$

Therefore, for 3 mol of C_2H_2 we have $3G_f^{\circ} = 3 \times 209.326 = 627.977 \, kJ$. Substituting these values into Equation 10.38 we get

$$(\Delta G^{\circ})_{298 \, K} = 129.662 - 627.977 = -498.315 \, kJ.$$

10.6 SPONTANEOUS AND NON-SPONTANEOUS PROCESSES

If we consider a process for which both pressure and temperature remain constant, then ΔH for such a process is equal to Q, the actual amount of heat absorbed. $T\Delta S$ is equal to Q_{rev}. Then from Equation 10.25 we have

$$(\Delta G)_{T,P} = Q - Q_{rev}. \tag{10.39}$$

From this equation we see that, in the case of a reversible process at constant temperature and pressure, $\Delta G = 0$, or $dG = 0$, since in this case $Q = Q_{rev}$. For an irreversible process at constant temperature and pressure, $Q_{irrev} < Q_{rev}$, and therefore $\Delta G < 0$ or $dG < 0$.

These results indicate that, in a system at constant temperature and pressure, G may decrease (irreversible process) or remain constant (reversible process) but never increases. This must mean that, when such a system reaches equilibrium, its free energy is a minimum, i.e. a system not in equilibrium will tend to change irreversibly, thereby lowering its free energy until no further change is possible. At this point its free energy is a minimum, and the system is at equilibrium. The conditions of this argument are exactly those which are commonly used in chemical processes. For this reason the Gibbs free energy provides the most convenient criterion for equilibrium in chemical systems.

From Equation 10.25 we find that, the more negative ΔH, the more negative ΔG will be. So, processes in which the system goes from a high to a low energy state tend to proceed spontaneously. We also see that, the more positive ΔS, the more negative ΔG will be. Therefore, a process in which the system changes from a state of low randomness to one of higher randomness also tends to proceed spontaneously (as we would expect). Hence, we can say that there are two driving forces which govern the behaviour of systems at constant temperature and pressure, namely firstly the tendency towards minimum enthalpy and secondly the tendency towards maximum entropy. ΔG is a measure of the tendency for a reaction to take place and is known as the driving force of reaction. We have summarized these facts in Table 10.2.

From Equation 10.39 and from the established characteristics of Q_{rev} we can arrive at a quantitative distinction between the three types of thermodynamic process.

(a) *Spontaneous process.* An actual spontaneous process is an irreversible process. For the same change in state, $Q_{irr} < Q_{rev}$. Hence, from Equation 10.39 we find that, for a

Table 10.2   Conditions for a reaction to take place

ΔH	ΔS	ΔG	
$-$	$+$	$-$	Reaction favoured at all temperatures (spontaneous reaction)
$+$	$-$	$+$	Reaction favoured at no temperature (never spontaneous)
$-$	$-$	?	Reaction favoured at low temperatures (conditional on temperature)
$+$	$+$	?	Reaction favoured at high temperatures (conditional on temperature)

spontaneous process,

$$(\Delta G)_{T,P} = Q_{irr} - Q_{rev}$$

or

$$(\Delta G)_{T,P} < 0. \tag{10.40}$$

Hence, we can say that a process for which $(\Delta G)_{T,P}$ is negative is a spontaneous process. This indicates that a spontaneous process at constant temperature and pressure is associated with the decrease in the free energy of the system.

(b) *Non-spontaneous process*. If a particular process shows a positive free-energy change, then from Equation 10.25 it can be shown that such a process is thermodynamically impossible. It follows from Equation 10.39 that, if $(\Delta G)_{T,P}$ is positive, the heat actually absorbed would have to be greater than Q_{rev}. This is impossible simply because $Q_{rev} = Q_{max}$. Therefore, we can say that a process for which $(\Delta G)_{T,P}$ is positive is thermodynamically impossible.

(c) *Reversible process*. For a reversible process the heat actually absorbed is equal to Q_{rev}. Then, from Equation 10.39, $(\Delta G)_{T,P} = 0$. Hence, we can say that a process for which $(\Delta G)_{T,P} = 0$ is thermodynamically reversible. We have seen that a system undergoing a reversible process is in a state of continuous equilibrium. Therefore, we can restate that a system is in equilibrium if there is zero free-energy difference between the initial and final states.

Another important conclusion is that the free energy per mole of any substance in one phase is the same as that of the same substance in any other phase in equilibrium with the former phase.

10.7   CONDITIONS FOR EQUILIBRIUM AND SPONTANEITY

Although equilibrium and spontaneity have been used in our discussions, no thermodynamic criteria for these have been given so far. Quite a number of such criteria can be deduced in terms of various thermodynamic functions and the conditions under which the equilibrium is attained. Here we shall discuss the different cases.

Case (i): equilibrium in a reversible process. In such a process, at every step the system departs from the equilibrium state only infinitesimally. Therefore, it is logical to assume that the system is transformed and yet remains effectively at equilibrium throughout a reversible change. Hence, the condition for reversibility is a condition for equilibrium.

Since the condition of reversibility is $T\,dS = dQ_{rev}$ (from the defining equation), the condition of equilibrium is $T\,dS = dQ_{rev}$. For an irreversible change in state, $T\,dS > dQ$. This is a condition of spontaneity. Combining these two together, we may write

$$T\,dS \geqslant dQ \qquad (10.41)$$

where the equality sign implies a reversible value of dQ.

Combining the first law, i.e. $dU = dQ + dW$, with Equation 10.41 we have

$$T\,dS \geqslant dU - dW$$

or

$$T\,dS - dU + dW \geqslant 0. \qquad (10.42)$$

Since the work dW includes all kinds of work ($P\,dV$ work, electrical work, mechanical work, etc.), we may rewrite Equation 10.42 as

$$T\,dS - dU - P\,dV + dW' \geqslant 0 \qquad (10.43)$$

where dW' is any other type of work except $P\,dV$ work.

Equations 10.42 and 10.43 provide the condition of equilibrium by the equality sign, and of spontaneity by the inequality sign, for a transformation in terms of changes in properties dU, dV and dS of the system and the amount of work $P\,dV$ and dW' associated with the transformation.

If for such a transformation we keep the volume and the internal energy constant, then $dU = 0$ and $dV = 0$. Since no other type of work can be obtained, $dW' = 0$. Then, substituting these values in Equations 10.42 and 10.43, we have $dS \geqslant 0$. Therefore, we may conclude that for equilibrium in a system at constant volume and internal energy, $(dS)_{U,V} = 0$ for an infinitesimal change, and $(\Delta S)_{U,V} = 0$ for an finite change.

Case (ii): equilibrium in an isolated system. For such a system, by definition, we have $dU = 0$, $dW = 0$ and $dQ = 0$. Then, Equation 10.42 becomes (taking only the equality)

$$T\,dS = 0 \text{ or } dS = 0. \qquad (10.44)$$

This implies that, for a transformation in an isolated system, dS must be positive; the entropy must increase. The entropy of any isolated system continues to increase as long as changes occur within the system. When there is no more change, the system has attained the equilibrium state and the entropy has reached the maximum value. Therefore, we conclude that the condition of equilibrium in an isolated system is that the system has the maximum entropy.

Case (iii): equilibrium at constant temperature and volume. Differentiating Equation 10.3, we have

$$T\,dS - dU = -dA - S\,dT.$$

Substituting this into Equation 10.42, we get

$$dA + S\,dT - dW = 0$$

Table 10.3 Conditions of spontaneity and equilibrium

Constraint	Condition for spontaneity		Condition for equilibrium	
	Infinitesimal change	Finite change	Infinitesimal change	Finite change
Reversible process	$(T\,dS - dU - P\,dV + dW') > 0$		$(T\,dS - dU - P\,dV + dW') = 0$	
Isolated system	$dS > 0$	$\Delta S > 0$	$dS = 0$	$\Delta S = 0$
T and V constant; $W' = 0$	$dA < 0$	$\Delta A < 0$	$dA = 0$	$\Delta A = 0$
T and P constant; $W' = 0$	$dG < 0$	$\Delta G < 0$	$dG = 0$	$\Delta G = 0$

or

$$dA + S \, dT + P \, dV = 0. \tag{10.45}$$

If the temperature and volume are constant, $dT = 0$ and $dV = 0$. Then Equation 10.45 becomes $(dA)_{T,V} = 0$ for an infinitesimal change, and $(\Delta A)_{T,V} = 0$ for a finite change. This corresponds to a minimum value of A for the system, as ΔA is negative for a spontaneous change and positive for a non-spontaneous change.

Case (iv): equilibrium at constant temperature and pressure. Differentiating Equation 10.9, we have

$$dG = dA + P \, dV + V \, dP$$

or

$$dA + P \, dV = dG - V \, dP.$$

Substituting this in Equation 10.45, we get

$$dG + S \, dT - V \, dP = 0. \tag{10.46}$$

If the temperature and pressure are constant, $dP = 0$ and $dT = 0$. Then, Equation 10.46 becomes $(dG)_{T,P} = 0$ for an infinitesimal change and $(\Delta G)_{T,P} = 0$ for a finite change. Arguing in the same way as for the Helmholtz free energy, we can say that the Gibbs free energy of a system is minimum at the equilibrium state.

Any spontaneous transformation may occur in such a way as to produce some other type of work in addition to pressure–volume work, but it need not necessarily be so. We are interested in those transformations which do not produce any work other than PV work. Therefore, $dW' = 0$. Spontaneous changes can continue to occur in such a system as long as the free energy of the system decreases, i.e. until the free energy of the system reaches the minimum value.

We can summarize the conditions of spontaneity and equilibrium as in Table 10.3.

10.8 EQUILIBRIUM INVOLVING PURE SUBSTANCES: THE CLAPEYRON EQUATION

Any variation in the free energy for any pure substance in a single phase is given by Equation 10.26. At constant temperature and pressure, for equilibrium in the phase, dG must be equal to zero.

Let us consider the transformation of any pure substance from one phase to another. For all such transformations the change in the Gibbs free energy is given by

$$\Delta G = G_2 - G_1 \tag{10.47}$$

where G_1 and G_2 are the molar free energies of the substance in the initial and final states. As we have already said, such a transformation will reach equilibrium when $\Delta G = 0$ at constant T and P. Then from Equation 10.47 we have $G_1 = G_2$. Hence, we can say that, at constant T

and P, such a transformation will be in equilibrium when the molar free energies of the substance are the same in both phases.

When two phases are in equilibrium, then, if we want to change the pressure of the system by an infinitesimal quantity dP, we must also change the temperature of the system by an infinitesimal quantity dT to maintain the equilibrium. In such a case the relation between dP and dT can be expressed as follows.

From Equation 10.26 we have $(dG)_1 = -S_1 \, dT + V_1 \, dP$ and $(dG)_2 = -S_2 \, dT + V_2 \, dP$. Since, at the equilibrium state, $(dG)_1 = (dG)_2$, we have

$$-S_1 \, dT + V_1 \, dP = -S_2 \, dT + V_2 \, dP$$

or

$$\frac{dP}{dT} = \frac{S_2 - S_1}{V_2 - V_1} = \frac{\Delta S}{\Delta V}. \tag{10.48}$$

From Equation 10.25 we have at the equilibrium state $(\Delta G = 0)$, $\Delta H/T = \Delta S$. Substituting this in Equation 10.48, we get

$$\frac{dP}{dT} = \frac{\Delta H}{T \Delta V}. \tag{10.49}$$

This equation is known as the *Clausius–Clapeyron equation* or simply the *Clapeyron equation*. It relates the change in temperature necessarily accompanying a change in pressure occurring in a system containing two phases of a pure substance in equilibrium.

To evaluate Equation 10.49 it is necessary to know ΔH and ΔV as functions of temperature or pressure. As such information is not usually available, Equation 10.49 is used in the form

$$\frac{P_2 - P_1}{T_2 - T_1} = \frac{\Delta H}{T \Delta V} \tag{10.50}$$

where $T = (T_2 + T_1)/2$.

Alternatively, Equation 10.49 may be integrated, assuming that ΔH and ΔV are constant. Then,

$$\int_{P_1}^{P_2} dP = \frac{\Delta H}{\Delta V} \int_{T_1}^{T_2} \frac{dT}{T}$$

or

$$P_2 - P_1 = \frac{\Delta H}{\Delta V} \ln \left(\frac{T_2}{T_1} \right). \tag{10.51}$$

The Clapeyron equation is used for calculations involving equilibria in solids and liquids (condensed-phase equilibria). If one solid form of a pure substance transforms to another solid form, the variation in the transition temperature T with pressure P is given by the

Clapeyron equation. ΔH and ΔV in this case are the enthalpy change and volume change accompanying the transformation. For such a transformation, ΔH is always positive if the transformation is from a form stable at a lower temperature to a form stable at a higher temperature; ΔH is negative for the reverse process.

If P is in megapascals, ΔV in litres and ΔH is in joules, then Equation 10.50 will take the form

$$\frac{P_2 - P_1}{T_2 - T_1} = \frac{99.610 \times 10^{-5}\Delta H}{T\Delta V} \tag{10.52}$$

where 99.610×10^{-5} is a conversion factor for joules to litre megapascals.

Example 10.2

Calculate the melting point of a pure substance at $2\,\text{MPa}$ pressure, if the melting point at $0.1\,\text{MPa}$ pressure is $33.22°\text{C}$, given that the changes in enthalpy and volume accompanying the transformation at $0.1\,\text{MPa}$ and $33.22°\text{C}$ are $8368.00\,\text{J}\,\text{mol}^{-1}$ and $0.01\,\ell\,\text{mol}^{-1}$, respectively.

Solution

Let T_1 be the melting temperature at $0.1\,\text{MPa}$ and T_2 the melting temperature at $2\,\text{MPa}$. Then, $T_1 = 306.37\,\text{K}$, $P_1 = 0.1\,\text{MPa}$, $P_2 = 2\,\text{Mpa}$, $\Delta H = 8368.00\,\text{J}\,\text{mol}^{-1}$ and $\Delta V = 0.01\,\ell\,\text{mol}^{-1}$. Substituting these values into Equation 10.52, we get

$$T_2 = \frac{(306.37)(2 - 0.1)(0.01)}{(8368.0)(99.610 \times 10^{-5})} + 306.37\,\text{K} = 307.06\,\text{K}.$$

10.9   THE EHRENFEST EQUATIONS

For first-order phase changes we have

$$G_1 = G_2 \qquad V_1 = \frac{\partial G_1}{\partial P} \qquad V_2 = \frac{\partial G_2}{\partial P} \qquad \frac{\partial G_1}{\partial P} \neq \frac{\partial G_2}{\partial P}$$

$$S_1 = -\frac{\partial G_1}{\partial T} \qquad S_2 = -\frac{\partial G_2}{\partial T} \qquad \frac{\partial G_1}{\partial T} \neq \frac{\partial G_2}{\partial T}.$$

In a higher-order phase change these are replaced by equalities. The simplest view is that of the Ehrenfest classification, namely that in *second-order transition the second derivatives of G with respect to P and T are discontinuous, that in the third-order transition the third derivatives of G are discontinuous, and so on.* The derivatives are as follows; the first

derivatives are

$$\frac{\partial G}{\partial P} = V$$

$$\frac{\partial G}{\partial T} = -S$$

and the second derivatives are

$$\frac{\partial^2 G}{\partial P^2} = \left(\frac{\partial V}{\partial P}\right)_T = -V\left[-\frac{1}{V}\left(\frac{\partial V}{\partial P}\right)_T\right] = -VZ$$

where Z is the isothermal compressibility

$$\frac{\partial^2 G}{\partial P \partial T} = \left(\frac{\partial V}{\partial T}\right)_P = V\left[\frac{1}{V}\left(\frac{\partial V}{\partial T}\right)_P\right] = \beta V$$

where β is the isobaric coefficient of volume or cubic expansion, and

$$\frac{\partial^2 G}{\partial T^2} = -\left(\frac{\partial S}{\partial T}\right)_P = \frac{1}{T}\left[-T\left(\frac{\partial S}{\partial T}\right)_P\right] = \frac{-C_P}{T}.$$

Higher derivatives of G are expressed in terms of derivatives of β, Z and C_P. However, not all higher-order phase transitions fit the Ehrenfest classification. The λ point in liquid helium is an example of misfit as shown in Figure 10.2.

If we apply the Clapeyron equation to transitions of order higher than first, we get an intermediate result because both the numerator and the denominator of the Clapeyron equation are zero. We may obtain analogous equations for second-order transitions by using the equality of the entropies or the volumes at the transition, i.e. $S_1 = S_2$ or $V_1 = V_2$.

Let V be a function of P and T, i.e. $V = f(P, T)$. Then,

$$dV = \left(\frac{\partial V}{\partial T}\right)_P dT + \left(\frac{\partial V}{\partial P}\right)_T dP.$$

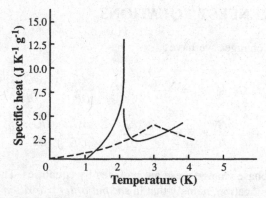

Figure 10.2   Specific heat versus temperature of liquid ^{4}He(——). Specific heat of ideal Bose–Einstein gas having the same density as liquid ^{4}He(– – – –).

Since $V_1 = V_2$, we have

$$\left(\frac{\partial V_1}{\partial T}\right)_P dT + \left(\frac{\partial V_1}{\partial P}\right)_T dP = \left(\frac{\partial V_2}{\partial T}\right)_P dT + \left(\frac{\partial V_2}{\partial P}\right)_T dP$$

or

$$\frac{dP}{dT} = -\frac{(\partial V_2/\partial T)_P - (\partial V_1/\partial T)_P}{(\partial V_2/\partial P)_T - (\partial V_1/\partial P)_T} = \frac{(1/V)(\partial V_2/\partial T)_P - (1/V)(\partial V_1/\partial T)_P}{(1/V)(\partial V_2/\partial P)_T - (1/V)(\partial V_1/\partial P)_T}.$$

According to the definitions of cubic (or volume) expansion β and compressibility Z, we can write

$$\beta = \frac{1}{V}\left(\frac{\partial V}{\partial T}\right)_P \qquad Z = -\frac{1}{V}\left(\frac{\partial V}{\partial P}\right)_T.$$

Hence, the above equation becomes

$$\frac{dP}{dT} = \frac{\beta_2 - \beta_1}{Z_2 - Z_1}. \tag{10.53}$$

If we now consider the entropy S as a function of P and T, we have $S = f(P, T)$. Proceeding the same way as above and using the equality $S_1 = S_2$ and $C_P = T(\partial S/\partial T)_P$, we obtain

$$\frac{dP}{dT} = \frac{1}{VT}\frac{(C_P)_2 + (C_P)_1}{\beta_2 - \beta_1} = \frac{1}{VT}\frac{\Delta C_P}{\Delta \beta}. \tag{10.54}$$

Equations 10.53 and 10.54 are the Ehrenfest equations. It should be remembered that, with a higher-order transition, there is just one Gibbs surface without any possibility of superheating or supercooling.

10.10   TEMPERATURE DEPENDENCE OF THE FREE ENERGY: THE GIBBS–HELMHOLTZ EQUATION

The mechanical properties P and V, the three fundamental properties U, S and T, and the three composite properties H, G and A of a system are all connected by the equations

$$G = H - TS$$
$$A = U - TS$$
$$H = U + PV.$$

Hence,

$$G = U + PV - TS.$$

The differential forms of these potential functions are

$$dH = dU + P\,dV + V\,dP$$
$$dA = dU + T\,dS - S\,dT$$
$$dG = dU + P\,dV + V\,dP - T\,dS - S\,dT.$$

However, from the combined first and second laws we have

$$dU = T\,dS - P\,dV.$$

So replacing dU in each of the above equations by its value, we get

$$dU = T\,dS - P\,dV$$
$$dH = T\,dS + V\,dP \tag{10.55}$$

$$dA = -S\,dT - P\,dV \tag{10.56}$$

$$dG = -S\,dT + V\,dP. \tag{10.57}$$

It appears that the differentials dS and dV in the equation $dU = T\,dS - P\,dV$ have changed to dS and dP in Equation 10.55. This transformation is known as the *Legendre transformation*. These four equations are sometimes called the four fundamental equations of thermodynamics. In fact, they are simply four different ways of expressing the same fundamental equation, i.e. $dU = T\,dS - P\,dV$.

If we are considering a process where P and T are independent variables, then G can be expressed as

$$dG = \left(\frac{\partial G}{\partial T}\right)_P dT + \left(\frac{\partial G}{\partial P}\right)_T dP. \tag{10.58}$$

Comparing Equations 10.57 and 10.58, we have

$$\left(\frac{\partial G}{\partial T}\right)_P = -S$$
$$\left(\frac{\partial G}{\partial P}\right)_T = V$$

(see Equations 10.28 and 10.29). From the definition, $G = H - TS$; so we have

$$-S = \frac{G - H}{T}.$$

Then from above we have

$$\left(\frac{\partial G}{\partial T}\right)_P = \frac{G - H}{T} = \frac{G}{T} - \frac{H}{T}. \tag{10.59}$$

Differentiating G/T with respect to T, we have

$$\left[\frac{\partial}{\partial T}\left(\frac{G}{T}\right)\right]_P = \frac{1}{T}\left(\frac{\partial G}{\partial T}\right)_P - \frac{1}{T^2}G.$$

Using Equation 10.28 in this equation, we get

$$\left[\frac{\partial}{\partial T}\left(\frac{G}{T}\right)\right]_P = -\frac{TS+G}{T^2}.$$

Since $TS + G = H$, we have

$$\left[\frac{\partial}{\partial T}\left(\frac{G}{T}\right)\right]_P = -\frac{H}{T^2}. \tag{10.60}$$

This is the Gibbs–Helmholtz equation (see Equation 10.36). Again, since $d(1/T) = -(1/T^2)\,dT$, we can replace ∂T in Equation 10.36 by $-T^2\,\partial(1/T)$. Then,

$$\left[\frac{\partial(G/T)}{\partial(1/T)}\right]_P = H. \tag{10.61}$$

Any of the Equations 10.59, 10.60 and 10.61 is simply a different form of the fundamental Equation 10.28. We shall call them the first, second, third and fourth forms of the Gibbs–Helmholtz equation. These equations show the temperature dependence of the free energy. For a process or for a chemical reaction, the free-energy change as a function of T is given by

$$\Delta G = \Delta H + T\left(\frac{\partial}{\partial T}(\Delta G)\right)_P$$

from Equation 10.36 and

$$\left(\frac{\partial}{\partial T}(\Delta G)\right)_P = -\Delta S = \frac{\Delta G - \Delta H}{T}$$

from Equation 10.32. Therefore,

$$\left(\frac{\partial}{\partial T}(\Delta G)\right)_P - \frac{\Delta G}{T} = \frac{-\Delta H}{T} \tag{10.62}$$

$$\left(\frac{\partial}{\partial T}\left(\frac{\Delta G}{T}\right)\right)_P = \frac{-\Delta H}{T^2} \tag{10.63}$$

$$\left(\frac{\partial(\Delta G/T)}{\partial(1/T)}\right)_P = \Delta H. \tag{10.64}$$

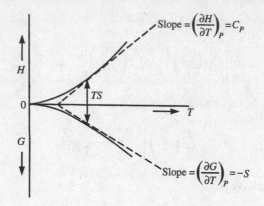

Figure 10.3 Variation in the free energy and enthalpy of a pure substance with temperature at constant pressure

Graphical representation of Equation 10.59 in Figure 10.3 shows the variation in G and H of a pure substance with T at constant P. Since $G = H - TS$, at 0 K, $G_0 = H_0 = 0$.

The Gibbs–Helmholtz equation in the form of Equation 10.64 also enables us to obtain ΔH of a reaction or a process, since ΔH is the slope of the plot of $\Delta G/T$ versus $1/T$.

Example 10.3

Given that for the formation of silver chloride from the elements the free-energy change is -110.039 kJ and the enthalpy change is -126.775 kJ mol^{-1} at 18°C. Calculate the entropy change at 18°C and the free energy change at 25°C, assuming that the rate of change in ΔG with temperature remains constant.

Solution

To calculate entropy change ΔS, we use Equation 10.25, i.e. $\Delta G = \Delta H - T\Delta S$. Then,

$$-110.039 = -126.775 - (273 + 18)\Delta S$$

or

$$\Delta S = -57.51 \ \text{J}°\text{C}^{-1}\text{mol}^{-1}$$

To calculate ΔG at 25°C we again use that same equation but keep in mind that

$$\left(\frac{\mathrm{d}}{\mathrm{d}T}(\Delta G)\right)_P = -\Delta S = 57.51.$$

Therefore, $\Delta G = \Delta H - T\Delta S$ will become

$$\Delta G = -126775 - (273 + 25)(-57.51) = -109.637 \ \text{kJ mol}^{-1}$$

10.11   THE MAXWELL RELATIONSHIPS

For a system with two degrees of freedom, there are four thermodynamic potentials, namely U, $H(=U+PV)$, the Helmholtz function $A(=U-TS)$ and the Gibbs function $G(=U-TS+PV)$. We already know the differential form of these potential functions, i.e.

$$dU = T\,dS - P\,dV$$
$$dH = T\,dS + V\,dP$$
$$dA = -S\,dT - P\,dV$$
$$dG = -S\,dT + V\,dP.$$

Let us consider $dU = T\,dS - P\,dV$. From this we can write

$$\left(\frac{\partial U}{\partial S}\right)_V = T \qquad \left(\frac{\partial U}{\partial V}\right)_S = -P.$$

If we differentiate the first with respect to V and the second with respect to S, we get

$$\frac{\partial^2 U}{\partial V\,\partial S} = \left(\frac{\partial T}{\partial V}\right)_S \qquad \frac{\partial^2 U}{\partial S\,\partial V} = -\left(\frac{\partial P}{\partial S}\right)_V.$$

However,

$$\frac{\partial^2 U}{\partial V\,\partial S} = \frac{\partial^2 U}{\partial S\,\partial V}.$$

Therefore, from the above two equations, we get

$$\left(\frac{\partial T}{\partial V}\right)_S = -\left(\frac{\partial P}{\partial S}\right)_V. \tag{10.65}$$

Let us consider $dH = T\,dS + V\,dP$. From this we can write

$$\left(\frac{\partial H}{\partial S}\right)_P = T \quad \text{and} \quad \left(\frac{\partial H}{\partial P}\right)_S = V.$$

Differentiating the first with respect to P and the second with respect to S, we get

$$\frac{\partial^2 H}{\partial P\,\partial S} = \left(\frac{\partial T}{\partial P}\right)_S \qquad \frac{\partial^2 H}{\partial S\,\partial P} = \left(\frac{\partial V}{\partial S}\right)_P.$$

However,

$$\frac{\partial^2 H}{\partial P\,\partial S} = \frac{\partial^2 H}{\partial S\,\partial P}.$$

Therefore, from the above two equations we get

$$\left(\frac{\partial T}{\partial P}\right)_S = \left(\frac{\partial V}{\partial S}\right)_P. \tag{10.66}$$

Similarly, taking $dA = -S\,dT - P\,dV$ and $dG = -S\,dT + V\,dP$ in turn and following the above procedures, it can be shown that

$$\left(\frac{\partial T}{\partial V}\right)_P = -\left(\frac{\partial P}{\partial S}\right)_T \tag{10.67}$$

and

$$\left(\frac{\partial T}{\partial P}\right)_V = \left(\frac{\partial V}{\partial S}\right)_T. \tag{10.68}$$

Equations 10.65, 10.66, 10.67 and 10.68 are known as the Maxwell relationships.

10.11.1 Applications of the Maxwell relationships

There are many applications of the Maxwell relationships in thermodynamics. Let us consider one or two examples. We know that the entropy of an ideal gas depends on any two of the independent variables P, V and T. If we take the entropy as a function of P and T, we can write $S = f(P, T)$. Then, differentiating gives

$$dS = \left(\frac{\partial S}{\partial P}\right)_T dP + \left(\frac{\partial S}{\partial T}\right)_P dT = -\left(\frac{\partial V}{\partial T}\right)_P dP + \left(\frac{\partial S}{\partial T}\right)_P dT$$

(using Equation 10.67. For 1 mol of an ideal gas, $PV = RT$. Then $P(\partial V/\partial T)_P = R$. By definition $T(\partial S/\partial T)_P = C_P$. Then the above equation becomes

$$dS = -\frac{R}{P}\,dP + \frac{C_P}{T}\,dT. \tag{10.69}$$

This equation relates the entropy of an ideal gas to its pressure, temperature and C_P. Integrating Equation 10.69 we get

$$\int_{S_1}^{S_2} dS = -R \int_{P_1}^{P_2} \frac{dP}{P} + C_P \int_{T_1}^{T_2} \frac{dT}{T} \tag{10.70}$$

or

$$\Delta S = -R \ln P + C_P \ln T + S_0' \tag{10.71}$$

where S_0' is a constant of integration. Equation 10.70 is the same as Equation 8.57.

As another application of the Maxwell relationships, let us consider $dU = T\,dS - P\,dV$. This gives

$$\left(\frac{\partial U}{\partial V}\right)_T = T\left(\frac{\partial S}{\partial V}\right)_T - P$$

(dividing by dV). Equation 10.68 gives $(\partial S/\partial V)_T = (\partial P/\partial T)_V$. Then the above equation becomes

$$\left(\frac{\partial U}{\partial V}\right)_T = T\left(\frac{\partial P}{\partial T}\right)_V - P.$$

For 1 mol of an ideal gas, $PV = RT$, or $P = RT/V$. This gives $(\partial P/\partial T)_V = R/V$. Therefore, the above equation becomes

$$\left(\frac{\partial U}{\partial V}\right)_T = \frac{RT}{V} - P = 0.$$

This shows that *the internal energy of an ideal gas is a function of the temperature only.*

10.12   INTEGRATION OF THE GIBBS–HELMHOLTZ EQUATION

We have already deduced the following relation (see Equation 10.63)

$$\left[\frac{d}{dT}\left(\frac{\Delta G}{T}\right)\right]_P = -\frac{\Delta H}{T^2}.$$

Since ΔH (or alternatively ΔC_P) can be expressed as a function of temperature, this equation can be integrated directly. Then,

$$\frac{\Delta G}{T} = \int\left(-\frac{\Delta H}{T^2}\right)dT + D \qquad (10.72)$$

where D is an integration constant. From the Kirchoff equation we have

$$\frac{d}{dT}(\Delta H) = \Delta C_P.$$

If ΔC_P is not constant over the temperature range concerned, then this can be expressed as a function of T in the familiar form $\Delta C_P = a + bT + cT^2 + \cdots$. Substituting this value of ΔC_P in the above equation and integrating give

$$\Delta H = aT + \frac{bT^2}{2} + \frac{cT^3}{3} + \cdots + I$$

where I is an integration constant. At $T=0$, $\Delta H = \Delta H_0$ and $I = \Delta H_0$. Then,

$$\Delta H = aT + \frac{bT^2}{2} + \frac{cT^3}{3} + \cdots + \Delta H_0. \tag{10.73}$$

Putting Equation 10.73 in Equation 10.64 and integrating, we obtain

$$\frac{\Delta G}{T} = \frac{\Delta H_0}{T} - a \ln T - \frac{bT}{2} - \frac{cT^2}{6} - \cdots + D$$

or

$$\Delta G = \Delta H_0 - aT \ln T - \frac{bT^2}{2} - \frac{cT^3}{6} - \cdots + DT. \tag{10.74}$$

The value of the integration constant D can be calculated from the free-energy change at a known temperature. The value of ΔH_0 can be calculated, if one value of the heat of reaction is known. Equation 10.74 can also be used to show the variation in the standard free energy change with temperature, i.e.

$$\Delta G^\circ = \Delta H_0 - aT \ln T - \frac{bT^2}{2} - \frac{cT^3}{6} - \cdots + DT.$$

Example 10.4

Calculate the free-energy change at 773 K for the formation of ammonia from gaseous nitrogen and hydrogen, given that, for the reaction

$$(\Delta H)_{298\,K} = -91\ 629.60\ \text{J} \qquad (\Delta G)_{298\,K} = -32\ 718.88\ \text{J}$$
$$(C_P)_{N_2} = 6.50 + 0.0010T\ \text{J K}^{-1}\text{mol}^{-1}$$
$$(C_P)_{H_2} = 6.50 + 0.0009T\ \text{J K}^{-1}\text{mol}^{-1}$$
$$(C_P)_{NH_3} = 8.40 + 0.0007T + 0.0000051T^2\ \text{J K}^{-1}\text{mol}^{-1}.$$

Solution

The reaction is $N_2(g) + 3H_2(g) = 2NH_3(g)$; also

$$\Delta C_P = \sum_{\text{products}} C_P - \sum_{\text{reactants}} C_P.$$

Therefore,

$$\Delta C_P = 2(8.04 + 0.0007T + 0.0000051T^2) - 3(6.50 + 0.0009T) - (6.50 + 0.0010T)$$
$$= -9.92 - 0.0023T + 0.0000102T^2.$$

Substituting this value of ΔC_P into Equation 10.73, we have

$$(\Delta H)_{298\,K} = (\Delta H_0)_{298\,K} - 9.92T - \frac{0.0023T^2}{2} + \frac{0.0000102T^3}{3}$$

or

$$-91\ 629.60 = (\Delta H_0)_{298\,K} - (9.92)(298) - \frac{0.0023}{2}(298)^2$$
$$+ (0.000003\ 4)(298)^3$$

or

$$(\Delta H_0)_{298\,K} = -88661.30 \text{ J}.$$

By substituting this value of ΔH_0 and the value of $(\Delta G)_{298\,K}$ into Equation 10.74, we find that

$$-32\ 718.88 = -88\ 661.300 - (-9.92)(298)(2.303)\log 298$$
$$+ \left(\tfrac{1}{2}\right)(0.0023)(298)^2 - \left(\tfrac{1}{6}\right)(0.0000102)(298)^3 + 298D$$

or

$$D = 131.01.$$

Knowing the value of the integration constant D, we can calculate the value of ΔG at 773 K by using Equation 10.74 again. Then,

$$(\Delta G)_{733\,K} = -88\ 661.300 + (9.92)(773)(2.303)\log 773$$
$$+ (0.00115)(773)^2 - (0.0000017)(773)^3 - (131.01)(773)$$
$$= -139.017 \text{ kJ}.$$

10.13 PRESSURE DEPENDENCE OF THE FREE ENERGY

We have already deduced the net relation $dG = V\ dP - S\ dT$ for a pure substance (see Equation 10.57). This equation can be extended to a change in state such as vaporization, or to a chemical reaction such as $aA = bB$. Let

$$\Delta V = bV_B - aV_A$$
$$\Delta S = bS_B - aS_A$$

where V_A, V_B, S_A and S_B are the molar volume and entropies of A and B. Then, for the reactant and the product,

$$(dG)_{\text{reactant}} = aV_A\ dP - aS_A\ dT$$
$$(dG)_{\text{product}} = bV_B\ dP - bS_B\ dT.$$

Subtracting gives

$$d(\Delta G) = \Delta V \, dP - \Delta S \, dT.$$

For pressure changes at constant temperature, $dT = 0$ and then Equation 10.57 becomes $dG = V \, dP$. For a change in volume of an ideal gas this relation can be integrated:

$$\Delta G = \int_{P_1}^{P_2} V \, dP. \tag{10.75}$$

However, from the ideal-gas law, $V = nRT/P$. Therefore,

$$\Delta G = \int_{P_1}^{P_2} \frac{nRT}{P} \, dP = nRT \ln \left(\frac{P_2}{P_1} \right). \tag{10.76}$$

If we take state 1 as the standard state, then $P_1 = 1$ atm and $G_1 = G^\circ$; for state 2, if we take $P_2 = P$ atm and $G_2 = G$, then Equation 10.76 becomes

$$G - G^\circ = nRT \ln \left(\frac{P}{1} \right) = nRT \ln P. \tag{10.77}$$

For 1 mol of the gas,

$$G = G^\circ + RT \ln P. \tag{10.78}$$

The pressure P is expressed in atmospheres, and T is assumed constant in deriving Equation 10.78. This equation is strictly applicable to ideal gases, since $PV = nRT$ was used for the P–V relation in the integration of Equation 10.76. However, if the details of non-ideal behaviour are ignored, it can be used, and assumed to apply approximately, for all gases. Equation 10.78 shows that the free energy of a gas at pressure P is equal to the free energy that it has at 1 atm plus an additional term that is positive for P larger than 1 atm and negative for P less than 1 atm.

10.14   FUGACITY AND ACTIVITY

It is observed that reproducible results for the change in free energy for real gases, especially at higher pressures, cannot be obtained from Equation 10.76. The reason for this is that in the case of non-ideal gases the relation $PV = nRT$ does not hold. In such cases the exact dependence of the volume on the pressure must be known to integrate Equation 10.75. Since this dependence of volume on pressure cannot be generalized for every non-ideal gas, the change ΔG in free energy is given by different equations in different cases, making it impossible to obtain a general equation. This difficulty was first overcome by G. N. Lewis who introduced two new quantities called *fugacity* and *activity*.

It is a general experience that any substance in a particular state has a tendency to escape from that state. Fugacity, denoted by f, is a measure of this escaping tendency; it is related to the free-energy content of the substance (per mole) by

$$G = RT \ln f + D \tag{10.79}$$

where D is a constant dependent on the substance and the temperature.

As we have already said that the absolute values of the free energy cannot be determined, the constant D cannot, therefore, be evaluated. To avoid this difficulty we refer to a standard reference point for all substances. If G° is the free energy at this standard state and f° is the corresponding fugacity at this state, then by Equation 10.79 we have

$$G^\circ = RT \ln f^\circ + D. \tag{10.80}$$

Then the change in free energy from the standard state to any other state, in which the free energy is G, is given by

$$G - G^\circ = RT \ln \left(\frac{f}{f^\circ} \right) \tag{10.81}$$

or

$$G = G^\circ + RT \ln \left(\frac{f}{f^\circ} \right). \tag{10.82}$$

Equation 10.82 gives the free-energy content of any substance in any state in terms of the free energy in the standard state and the fugacities. If we write

$$\frac{f}{f^\circ} = a \tag{10.83}$$

then Equation 10.82 becomes

$$G = G^\circ + RT \ln a. \tag{10.84}$$

The quantity $f/f^\circ = a$ is called the *activity*. From Equation 10.84 it is obvious that, in the standard state, $G = G^\circ$ and $RT \ln a = 0$, i.e. $a = 1$. So, we conclude that in the standard state the activity of any substance is unity; in any other state the value of the activity depends on the difference between the free energies at the standard state and the particular state concerned. From Equation 10.84 we can write, for any two states,

$$G_1 = G^\circ + RT \ln a_1$$
$$G_2 = G^\circ + RT \ln a_2.$$

Hence, $\Delta G = G_2 - G_1 = RT \ln(a_2/a_1)$. If we are considering n mol of a gas

$$\Delta G = nRT \ln \left(\frac{a_2}{a_1} \right). \tag{10.85}$$

From Equations 10.76 and 10.85 we can say that the activity is the thermodynamic counter-part of gas pressure (also of the concentration in the case of solutions).

In order to evaluate the activity it is necessary to define the standard state. This definition is based on certain conventions. Here we shall confine ourselves only to the conventions applicable to pure gases, solids and liquids.

10.14.1   Standard state for gases

The standard state for a gas at any given temperature is a state in which the fugacity is unity, i.e. $f° = 1$. Then from Equation 10.83 we have

$$a = \frac{f}{f°} = f. \tag{10.86}$$

Substituting this into Equation 10.84, we get

$$G = G° + RT \ln f. \tag{10.87}$$

For an ideal gas $f = P$, i.e. the fugacity is equal to the gas pressure. As we know that at very low pressures all gases behave ideally, any gas can be brought into an ideal state by reducing its pressure to zero. Then, we can generalize the definition of fugacity by saying that

$$f = P \text{ as } P \rightarrow 0. \tag{10.88}$$

Alternatively,

$$\lim_{P \to 0} \left(\frac{f}{P} \right) = 1. \tag{10.89}$$

The ratio f/P is a measure of ideality of any gas. As long as a gas behaves ideally, the value of this ratio remains unity. As soon as a gas deviates from ideality, the value of f/P no longer remains unity. The ratio f/P is called the *activity coefficient* of a gas and is denoted by γ. The value of γ gives a direct measure of the extent to which any real gas deviates from ideality at any given temperature and pressure.

10.14.2   Standard states for solids and liquids

We have already defined the standard state of a pure solid or liquid as the solid or liquid at 1 atm pressure at each temperature. In this state of the solid or liquid, activity $a = 1$ and $G = G°$. As the free energies of solids and liquids are not very dependent on pressure, for these substances $a = 1$ (approximately) at all temperatures and for wide ranges of pressure.

10.15 PRESSURE DEPENDENCE OF THE FREE ENERGY OF NON-IDEAL GASES

Equations 10.76 and 10.78 show the pressure dependence of the free energy of ideal gases. To show the pressure dependence of the free energy of non-ideal gases, equations analogous to Equations 10.76 and 10.78 would be necessary. Although such equations can be derived by using suitable equations of state, such as the van der Waals equations, such a procedure would produce a complicated expression for ΔG. To avoid this, a satisfactory procedure is to use fugacity. If G_1 and G_2 are the molar free energies of a gas at pressures P_1 and P_2, respectively, and f_1 and f_2 are the fugacities of the gas at these pressures, then from what we have said in the previous section we may write

$$G_2 - G_1 = RT \ln \left(\frac{f_2}{f_1} \right). \tag{10.90}$$

Equation 10.75 may be rewritten in the form

$$G_2 - G_1 = \int_{P_1}^{P_2} \left[\frac{RT}{P} + \left(V - \frac{RT}{P} \right) \right] dP.$$

The volume V in this equation is molar volume, since Equation 10.75 was deduced for 1 mol of a gas. Then we have

$$G_2 - G_1 = \int_{P_1}^{P_2} \frac{RT}{P} dP + \int_{P_1}^{P_2} \left(V - \frac{RT}{P} \right) dP$$

$$= RT \ln \left(\frac{P_2}{P_1} \right) + \int_{P_1}^{P_2} \left(V - \frac{RT}{P} \right) dP. \tag{10.91}$$

Combining Equations 10.90 and 10.91, we get

$$RT \ln \left(\frac{f_2}{f_1} \right) = RT \ln \left(\frac{P_2}{P_1} \right) + \int_{P_1}^{P_2} \left(V - \frac{RT}{P} \right) dP$$

or

$$RT \ln \left(\frac{f_2/P_2}{f_1/P_1} \right) = \int_{P_1}^{P_2} \left(V - \frac{RT}{P} \right) dP. \tag{10.92}$$

As $P_1 \to 0$, the ratio f_1/P_1 will also tend to unity. If we replace P_2 and f_2 by simply P and f, respectively for the arbitrary pressure and fugacity in Equation 10.92, we get

$$RT \ln \left(\frac{f}{P} \right) = \int_{P=0}^{P=P} \left(V - \frac{RT}{P} \right) dP. \tag{10.93}$$

If V can be expressed as a function of P at some temperature, value of f/P, and therefore of f, at some pressure P can be obtained from Equation 10.93. Substituting γ for f/P in

Equation 10.93, we have

$$RT \ln \gamma = \int_{P=0}^{P=P} \left(V - \frac{RT}{P} \right) dP. \tag{10.94}$$

This equation may be used to obtain γ. In order to do this, when an equation of state for a gas is available, $V - RT/P$ is substituted as a function of P and then integrated between $P = 0$ and any desired pressure. Alternatively, if $P-V$ data are available at a given temperature, then $V - RT/P$ is plotted against P, and the integration is carried out graphically.

10.16   ADDITION OF FREE-ENERGY EQUATIONS

We know how thermochemical equations could be added and subtracted (Hess's law) to obtain the value of ΔH for one of the reacting substances. Since Hess's law is also applicable to free energy, this quantity can be treated in exactly the same way. To illustrate, let us consider the combustion of graphite in oxygen. In the presence of excess oxygen, complete combustion of graphite into carbon dioxide is ensured, i.e.

$$C + O_2 = CO_2 \qquad (\Delta G)_{298\,K} = -394.384 \, kJ.$$

In a limited supply of oxygen, carbon monoxide will be formed such that

$$C + \tfrac{1}{2} O_2 = CO \qquad (\Delta G)_{298\,K} = -136.022 \, kJ.$$

By subtracting one from the other we can find the free-energy change for further oxidation of the monoxide to the dioxide, i.e.

$$CO + \tfrac{1}{2} O_2 = CO_2 \qquad (\Delta G)_{298\,K} = -258.362 \, kJ.$$

When each reacting substance is in its standard state, this procedure gives the standard free energy of the reaction. Standard free-energy values of individual substances may thus be determined in the cases where the experimental methods are not applicable

10.17   THE REACTION ISOTHERM

We know how to calculate the change in free energy in a reaction under standard conditions, and also ΔG for the transition from the standard to a non-standard state by individual species. The combination of these techniques is the final step to develop relationships for the free-energy change involved in any type of transformation, whether physical or chemical. The overall calculation involves the separation of products and the reaction on the absolute free-energy scale by adding to $\Delta G°$ for the reaction the difference between the standard-to-non-standard-state ΔG terms for the components. Let us consider a general equation of the type

$$bB + cC + \cdots = mM + nN + \cdots.$$

For this reaction, the free-energy change is given by

$$\Delta G = (mG_M + nG_N + \cdots) - (bG_B + cG_C + \cdots).$$

With the help of Equation 10.84 the above equation may be written as

$$\Delta G = [m(G_M^\circ + RT \ln a_M) + n(G_N^\circ + RT \ln a_N) + \cdots]$$
$$- [b(G_B^\circ + RT \ln a_B) + c(G_C^\circ + RT \ln a_C) + \cdots]$$

where a_M, a_N, a_B, a_C, ..., are the activities of the components M, N, B, C, ... Rearranging, we get

$$\Delta G = [(mG_M^\circ + nG_N^\circ + \cdots) - (bG_B^\circ + cG_C^\circ + \cdots)]$$
$$+ RT[(m \ln a_M + n \ln a_N + \cdots) - (b \ln a_B + c \ln a_C + \cdots)].$$

The terms within the first set of parentheses on the right-hand side of this equation represent the free-energy change in the process when all the species are present in their standard states, and the terms in the second set of parentheses represent ΔG for the standard-to-non-standard-state transition for products and reactants. Therefore, we can rewrite the above equation as

$$\Delta G = \Delta G^\circ + RT \ln \left(\frac{a_M^m a_N^n \cdots}{a_B^b a_C^c \cdots} \right). \tag{10.95}$$

The quantity $a_M^m a_N^n \cdots / a_B^b a_C^c \cdots$ is often known as the *activity quotient* and denoted by Q. Substituting Q for the ratio of activities in Equation 10.95, we get

$$\Delta G = \Delta G^\circ + RT \ln Q. \tag{10.96}$$

This equation has the same form as the relationships deduced for the individual components, where the free-energy change is obtained by the addition of a correction term to the value applicable to the system in the standard state.

It must be carefully noted that the temperature in the second term on the right-hand side of Equations 10.95 and 10.96 must be the same as that at which ΔG° is calculated.

Equation 10.95 gives the free-energy change of a reaction in terms of the free-energy change in the standard state and the activities of the reactants and the products at any constant temperature. Hence, this equation is called the *reaction isotherm*. If $Q = 1$ or, in other words, the activities of the reactants and the products are all unity, the second term on the right-hand side of Equation 10.95 disappears, and $\Delta G = \Delta G^\circ$. For any reaction, ΔG° is constant at any given temperature and entirely independent of pressure.

Note: Although Q in Equation 10.96 has the same form as the general mass action law equilibrium constant, the two are not equivalent. Equilibrium constant and Q are closely related.

Example 10.5

Calculate the free-energy change involved in the following reaction carried out at 298 K and 0.5 MPa pressure:

$$2CO(g) + O_2(g) = 2CO_2(g).$$

Solution

To apply Equation 10.96 we must calculate ΔG° first. From Equation 10.38,

$$\Delta G^\circ = 2(\Delta G_f^\circ)_{CO_2(g)} - 2(\Delta G_f^\circ)_{CO(g)}.$$

Substituting the values of ΔG_f° for $CO_2(g)$ and $CO(g)$ from Table 10.1, we have

$$\Delta G^\circ = (2)(-394.384) - (2)(-137.277) = -514.214 \, kJ$$

$$Q = \frac{a_{CO_2}^2}{a_{CO}^2 a_{O_2}} = \frac{P_{CO_2}^2}{P_{CO}^2 P_{O_2}}.$$

Since the reactants and the products are in gaseous form, the activity of these substances are the corresponding pressure. Therefore,

$$Q = \frac{(0.5)^2}{(0.5)^2(0.5)} = \frac{1}{0.5} = 2.$$

Substituting the values of ΔG° and Q in Equation 10.96 gives

$$\Delta G = \Delta G^\circ + RT \ln Q = -514 \, 214 + (2.303)(8.314)(298)\log 2$$
$$= -514 \, 214 + (2.303)(8.314)(298)(0.301)$$
$$= -512496 \, J.$$

EXERCISES 10

10.1 What do you understand by the terms *permitted processes* and *forbidden processes*? Discuss the criteria of forbiddenness.

10.2 Discuss the importance of work content in connection with the classification of permitted, reversible and forbidden processes.

10.3 Show that the work content of a constant-volume isothermal process tends to decrease spontaneously until it attains the minimum, value.

10.4 Define the Helmholtz and the Gibbs free energies. Is it possible to calculate the absolute values of work content and the Gibbs free energy? Show that at constant temperature and pressure the Gibbs free energy tends to decrease until it reaches the minimum value.

10.5 Show that at constant temperature the maximum work done by a system may be obtained by a change in the Helmholtz free energy of that system.

10.6 Deduce suitable relationships to show the variations in the Helmholtz free energy with temperature and volume. Hence show that the Helmholtz free energy of any substance decreases with increase in temperature.

10.7 Deduce suitable relationships to show the variations in the Gibbs free energy with temperature and pressure. Hence show that the Gibbs free energy of any substance decreases with increase in temperature.

10.8 Discuss the fundamental properties of the thermodynamic quantities A and G. What do you understand by standard free-energy change? Knowing the values of the standard enthalpies and entropies of the substances involved in a reaction, is it possible to calculate the standard free-energy change for that reaction?

10.9 Calculate the values of ΔG and ΔA for the phase change of 1 mol of liquid water at 373 K. and $101.3 \, kN \, m^{-2}$ ($= 0.1 \, MPa$) to gaseous water: $H_2O(\ell) \rightarrow H_2O(g)$. The molar volumes of liquid water and steam at 373 K are $18.8 \times 10^{-6} \, m^3$ and $31.0 \times 10^{-3} \, m^3$, respectively.

10.10 Calculate the Gibbs free energy for the reduction of 1 mol of acetone at 298 K and 0.1 MPa pressure to isopropyl alcohol under the same conditions. The standard free energies $(\Delta G°)_{298 \, K}$ for formation of $CH_3CHOHCH_3(\ell)$ and $CH_3COCH_3(\ell)$ are $-184.096 \, kJ \, mol^{-1}$ and $-155.728 \, kJ \, mol^{-1}$, respectively.

10.11 Calculate ΔA and ΔG for the vaporization of 2 mol of benzene at 80.2°C (boiling point), given that the vapour behaves ideally, and the latent heat of vaporization of benzene is $422.584 \, J \, g^{-1}$. Also calculate the new values of ΔA and ΔG, if the final pressure is 0.05 MPa at 80.2°C.

10.12 For the vaporization of water at temperatures near its boiling point, $H_2O(\ell) \rightarrow H_2O(g)$, and at 0.1 MPa pressure, $\Delta H = 40.125 \, kJ \, mol^{-1}$ and $\Delta S = 107.529 \, J \, K^{-1} \, mol^{-1}$. Calculate the temperatures for the formation of water vapour at 0.1 MPa when the system is

(a) at equilibrium

(b) spontaneous

(c) non-spontaneous

10.13 Define and distinguish spontaneous, non-spontaneous and reversible processes. 'The free energy per mole of any substance in one phase is the same as that of the same substance in any other phase in equilibrium with it.' Discuss this statement.

10.14 Show that for a reversible process the condition of equilibrium is given by the equality sign and that of spontaneity by the inequality sign of the expression $T \, dS - dU + dW \geqslant 0$, where the work term dW includes all kinds of work.

10.15 Is it necessarily true that the condition of equilibrium in an isolated system is that the system has the maximum entropy? Discuss.

10.16 For a certain process the change in free energy is expressed by the relation

$$\Delta G = 13 \, 580 + 16.1T \, \log T - 72.59T \, J.$$

Calculate ΔS and ΔH for this process at 298.15 K.

10.17 Deduce the criteria of equilibrium in a reversible process for the following conditions:

(a) constant S and P

(b) constant H and P

(c) constant S and H

(d) constant S and V

(e) constant S and U.

10.18 Which of the following reactions is spontaneous in the standard state?

$$C_2H_5OH(\ell) + O_2(g) = CH_3COOH(\ell) + H_2O(\ell)$$
$$2CO_2(g) = 2CO(g) + O_2(g)$$

Use necessary data from Table 10.1.

10.19 Deduce an equation showing the change in temperature necessarily accompanying a change in pressure occurring in a system containing two phases of a pure substance in equilibrium. Discuss the significance of this equation.

10.20 At 97.8°C and normal pressure, metallic sodium melts with a latent heat of fusion 2635.920 J mol^{-1} and an increase in specific volume $27.90 \times 10^{-6} \ell g^{-1}$. Calculate the melting temperature of metallic sodium.

10.21 For the reaction $FeCO_3(s) = FeO(s) + CO_2(g)$, it is given that $\Delta G° = 18\,660 - 14.42T \log T - 6.07T + 8.24 \times 10^{-3}T^2$ J. Find $\Delta H°$ and $\Delta S°$ for the reaction at 25°C.

10.22 Assuming that the vapour of carbon tetrachloride (CCl$_4$) behaves ideally, calculate the latent heat of vaporization of CCl$_4$ at 77°C (boiling point of CCl$_4$). The vapour pressure P of liquid CCl$_4$ is given by the equation $\log P = -2400/T - 2.30 \log T + 23.6$, where T is temperature.

10.23 An ideal gas at 298 K and 0.5 MPa pressure is expanded to a final pressure of 0.1 MPa (a) isothermally and reversibly and (b) isothermally against a constant pressure of 0.1 MPa. Calculate ΔA for each of these expansions.

10.24 (a) What kinds of reaction can occur in a system at constant temperature and pressure without any external energy?

(b) What kinds of reaction require external energy to make them occur at constant temperature and pressure?

(c) The reaction $H_2 + Cl_2 + 2H_2O = 2H_3O^+ + 2Cl^-$ was carried out reversibly under standard conditions.

(i) If under the specified conditions $\Delta G = -262.0$ kJ, calculate the maximum useful work that can be obtained from the reaction.

(ii) Calculate the range of values for $W_{irreversible}$, If the reaction is carried out irreversibly.

(iii) Given that $W_{irrev} = -182$ kJ, what happens to the difference $\Delta G - W_{irrev} = 80$ kJ?

10.25 1 mol of an ideal gas at 0°C is allowed to expand isothermally from 10.0 to 1.0 MPa pressure. Calculate ΔG and ΔA

(a) of the gas, if the expansion is reversible

(b) of the entire isolated system (gas and its surroundings), if the expansion is reversible

(c) of the gas, if it is allowed to expand freely so that no work is done by it

(d) of the entire isolated system, if the gas expands freely

10.26 Calculate $(\Delta G°)_{298 K}$ for the reaction $H_2O(g) + CO(g) = H_2(g) + CO_2(g)$.

10.27 For the reaction $Ag_2O(s) \rightleftarrows 2Ag(s) + \frac{1}{2}O_2(g)$, $\Delta G° = 7700 + 4.2T \log T - 27.80T$. Express $\Delta S°$ as a function of temperature.

10.28 (a) What quantity measures the total energy to do mechanical work from the reaction $C_3H_8(g) + 5O_2(g) = 3CO_2(g) + 4H_2O(g)$?

(b) Name a process for which $\Delta U = 0$, $\Delta S = 0$ and $\Delta G = 0$. State all the necessary conditions or restrictions for each of the chosen processes.

(c) Discuss whether gaseous cadmium sulphide can be reduced to metallic cadmium with hydrogen at 1100°C. At 1100°C,

$$H_2(g) + \frac{1}{2}S_2(g) = H_2S(g) \quad (\Delta G°)_{1100°C} = -48.953 \text{ kJ mol}^{-1}$$
$$Cd(g) + \frac{1}{2}S_2(g) = CdS(g) \quad (\Delta G°)_{1100°C} = -127.194 \text{ kJ mol}^{-1}.$$

10.29 For the conversion of 1-butene into 1,3-butadiene by elimination of hydrogen, i.e.

$$CH_2 = CH - CH_2.CH_3 \rightarrow CH_2 = CH - CH = CH_2 + H_2$$

the following data are available for 298 K.

	$S°$ (J mol^{-1})	$\Delta H_f°$ (kJ)
1-Butene	307.106	1.117 152 0
1,3-butadiene	278.738	111.922
Hydrogen	130.583	

Calculate $\Delta H_f°$, $\Delta S°$ and $\Delta G°$ for the reaction.

10.30 Which does entropy change favour, product or reactant? Is the product or the reactant energetically more stable? Does the free-energy change favour a reaction?

10.31 Calculate the free energy change for the irreversible process $H_2O(1100°C, 700 \text{ mm}) = H_2O(100°C, 700 \text{ mm})$, given that the density of water is 1 g cm^{-3}.

10.32 Derive an equation in terms of temperature for the free-energy change for the reaction $CO(g) + 2H_2(g) = CH_3OH(g)$, given that the latent heat of vaporization of CH_3OH is $37.405\,kJ\,mol^{-1}$ at 298 K and the variation of heat capacity with temperature at constant pressure is $C_P = a + bT + cT^2$ where the values of a, b and c are as follows.

Substance	a $(J\,K^{-1}\,mol^{-1})$	$b \times 10^3$ $(J\,K^{-1}\,mol^{-1})$	$c \times 10^7$ $(J\,K^{-1}\,mol^{-1})$
$CO(g)$	26.861	6.966	−8.201
$H_2(g)$	29.041	−0.837	20.117
$CH_3OH(g)$	18.401	101.546	−286.813

10.33 What is the free energy change when 4.50 g of water is converted from liquid at 25°C to vapour at 25°C and 2×10^{-5} atm pressure? The equilibrium vapour pressure of water at 25°C is 0.0313 atm. Assume that this vapour behaves ideally.

10.34 Deduce the Gibbs–Helmholtz equation. Starting from this equation deduce the relation

$$\Delta G° = \Delta H_0 - aT \ln T - \frac{bT^2}{2} - \frac{cT^3}{6} - \cdots + AT$$

where A is a constant of integration.

10.35 Show that for ideal gases the pressure dependence of the free energy is given by $G = G° + RT \ln P$.

10.36 Can you use the relation $G = G° + RT \ln P$ to show the pressure dependence of the free energy in the case of non-ideal gases? If not, discuss how this problem can be overcome.

10.37 Discuss the standard states for solids, liquids and gases.

10.38 A certain gas at temperature T obeys the relation $PV = RT + BP + CP^2$. Deduce an expression for $\ln \gamma$ of the gas as a function of P at this temperature. Calculate the activity coefficient of the gas at 10.0 MPa pressure when $T = 223.2\,K$, $B = -3.69 \times 10^{-2}$ and $C = 1.79 \times 10^{-4}$ for volume in litres and pressure in atmospheres.

10.39 At moderately high pressures the PV behaviour of gases conforms to the relation $PV = RT + BP$. For 1 mol of oxygen at 298.15 K and for pressures up to about 1 atm, this expression becomes $PV = 298.15R - 0.0211P\,\ell$ atm. Calculate the fugacity of oxygen at 1 atm.

10.40 Show that the pressure dependence of the free energy of a non-ideal gas may be expressed by the relation

$$\Delta G = RT \ln\left(\frac{P_2}{P_1}\right) + \int_{P_1}^{P_2} \left(V - \frac{RT}{P}\right) dP.$$

Hence, show that the activity coefficient γ is given by

$$RT \ln \gamma = \int_{P=0}^{P=P} \left(V - \frac{RT}{P}\right) dP.$$

10.41 Show that the change in any type of transformation, whether physical or chemical, is given by $\Delta G = \Delta G° + RT \ln Q$, where Q is the activity quotient of that transformation.

10.42 The standard molar free energies of formation at 25°C and 0.1 MPa of the compounds $Ag_2S(s)$, $H_2O(\ell)$, $H_2S(g)$, $Ag(s)$ and $O_2(g)$ are $-40.7\,kJ\,mol^{-1}$, $-237.19\,kJ\,mol^{-1}$, $-33.59\,kJ\,mol^{-1}$, $0\,kJ\,mol^{-1}$ and $0\,kJ\,mol^{-1}$, respectively. Calculate the free-energy change $\Delta G°$ for the reaction of Ag_2S with $H_2O(\ell)$.

10.43 Calculate the standard free-energy change for the reaction $CaO(s) + SO_3(g) \rightarrow CaSO_4(s)$ at 298 K.

10.44 Calculate the free-energy change for the reaction $CaO(s) + SO_3(g) \rightarrow CaSO_4(s)$ at 298 K when the pressure of SO_3 is 0.02 MPa.

10.45 Calculate the temperature at which liquid and gaseous water are in equilibrium with each other at 0.1 MPa pressure. Assume that $\Delta H°$ and $\Delta S°$ are independent of temperature and comment on this assumption.

10.46 For the reaction $CuS(s) + H_2(g) \rightarrow Cu(s) + H_2S(g)$, calculate

(a) $\Delta G°$ at 298.15 K and 0.1 MPa pressure

(b) $\Delta G°$ at 800 K and 0.1 MPa pressure

(c) the temperature at which $\Delta G°$ is zero at 0.1 MPa pressure (here assume that there is no significant change of $\Delta H°$ and $\Delta S°$ as the temperature increases).

11

Free Energy and Chemical Equilibrium

11.1 INTRODUCTION

Many of the chemical reactions that we come across are known to occur, to some extent, in both directions. In such systems it is possible to observe experimentally the conversion of reactants to products and the reverse process. Such systems attain a state of dynamic equilibrium, in which there is no cessation of reaction, but the system undergoes no further alteration unless there is a change in conditions. It is observed that when a system, say $A + B \rightleftharpoons C + D$, comes to equilibrium at some temperature, the ratio of the concentrations of the products and the reactants has a unique value. This ratio is the value of the term known as the *equilibrium constant*, K. Irrespective of the proportions of species present initially, the same numerical value of K characterizes the system at that temperature.

This is valuable information for determining equilibrium. For example, if a chemical reaction reaches an equilibrium state, we can use this property to determine whether true equilibrium is attained. The reaction can be studied starting with several sets of initial concentrations at the same temperature. When there is no change in composition, the ratio of the concentration of products to the concentration of reactants should remain the same in every case. Although the concentration of individual reactants and products may vary, the value of their ratio is fixed at equilibrium. From the viewpoint of the law of mass action, the equilibrium state can be regarded as the situation where the rates of the overall forward and backward reactions are equal. Both processes continue to occur, but no further change in the gross composition of the system occurs. Here we try to understand the relation between the thermodynamic properties and the equilibrium constant.

11.2 EXPRESSIONS FOR THE EQUILIBRIUM CONSTANTS

For molecular systems, equilibrium constants can be expressed using nominal concentrations, namely pressure, mole fraction or molarity. For ionic systems, equating activity with nominal concentration is frequently invalid, and the correction terms for non-ideal behaviour become significant. According to the law of mass action, the rate of a chemical reaction is proportional to the active masses of the reacting substances. The equilibrium constant K for the reaction

$$bB + dD + \cdots \rightleftharpoons mM + nN + \cdots \qquad (11.1)$$

is given by

$$K = \frac{a_M^m a_N^n \cdots}{a_B^b a_D^d \cdots} \qquad (11.2)$$

where a is the activity of the components B, D, M, N, ... in the equilibrium mixture, and b, d, m, n, ... are the numbers of molecules of B, D, M, N, ..., respectively.

The equilibrium constant K has a large value when the forward reaction from left to right is favoured and products are readily formed. On the other hand, a small value of K indicates that the reverse reaction is favoured. The equilibrium constant has different values depending on the temperature at which the reaction is taking place. As long as the activity concept is utilized, Equation 11.2 can be applied to reversible reactions involving gases, liquids or solids. We know that the activity is related to molar concentration by the relation $a = \gamma c$, where γ is the activity coefficient and c is the concentration expressed in moles per litre. For ideal conditions, e.g. for perfect gases or very dilute solutions, the activity coefficient γ is unity and, therefore, $a = c$. The standard state of unit activity is a hypothetical solution of 1 mol ℓ^{-1} which possesses the properties of a very dilute solution.

For many practical purposes, the equilibrium constant can be expressed in molar concentration with sufficient accuracy. In such cases, Equation 11.2 takes the form

$$K_C = \frac{[M]^m [N]^n \cdots}{[B]^b [D]^d \cdots} \qquad (11.3)$$

or

$$K_C = \frac{C_M^m C_N^n \cdots}{C_B^b C_D^d \cdots} \qquad (11.4)$$

where C_M, C_N, C_B, C_D, ... are the molar concentrations of M, N, B, D, ... respectively. The square brackets in Equation 11.3 are used conventionally to represent concentrations of the components in moles per litre, and the subscript C to K indicates that the numerical value of K_C quoted is for the case of molar concentrations.

For homogeneous gaseous systems the equilibrium state can be conveniently stated in terms of the partial pressures of the reactants and the products, each being raised to a power corresponding to the coefficient of the particular gaseous component in the balanced equation. Therefore, we have

$$K_P = \frac{P_M^m P_N^n \cdots}{P_B^b P_D^d \cdots} \qquad (11.5)$$

where P is the partial pressure of b molecules of B, d molecules of D and so on. The subscript P to K indicates that the value of K_P quoted is for the case of partial pressures of the reactants and the products.

For a heterogeneous equilibrium (one which occurs in two phases, e.g. at a gas–solid interface) the concentration of any pure species in the condensed phase is constant and by convention is included in the K-value. Thus, for a gas–solid reaction

$$b B(s) + d D(g) + \cdots \rightleftharpoons m M(s) + n N(g) + \cdots$$

the equilibrium constant would be simply

$$K = \frac{P_N^n \cdots}{P_D^d \cdots}. \qquad (11.6)$$

At a given temperature, the system has a unique value of K_C or K_P, whose units depend on the stoichiometry. If all the coefficients in the general reaction are unity, then K_C and K_P both would be dimensionless. Also, for this special case they would have the same numerical value, since the factor for conversion from atmospheres to moles per litre to be applied to each term would cancel. In any other case where the coefficients are not unity, or where there is a change in the number of moles in going from reactants to products, K_C and K_P will have different numerical values, even though they describe the same equilibrium state.

For a process such as $B + D \rightleftharpoons M$, substitution of concentration in moles per litre in the equilibrium expression shows that K_C has units of litres per mole, while K_P, with gas pressures in atmospheres (or megapascals), has units of reciprocal atmospheres (or reciprocal megapascals). Equilibrium constant values are frequently quoted without units. This usually does not create any problem (although not strictly correct), since in most problems the units of K do not enter into the calculation directly. However, it must be specified whether the equilibrium constant is K_C or K_P.

11.2.1   Properties of equilibrium constants

Needless to say, K_P or K_C for a reaction has to be taken as a true constant.

Equilibrium constants have the following properties.

(i) The equilibrium constant principle is valid only at the equilibrium condition. Unless true equilibrium concentrations or pressures are substituted in Equations 11.4 and 11.5, these equations will not produce a constant value.

(ii) The equilibrium constant of a reaction varies with temperature but, at any particular temperature, it must be a constant independent of concentration or pressure at all concentrations and pressures.

(iii) The equilibrium constant quantitatively indicates the effect of concentrations of reactants and products on the extent of reaction.

(iv) The magnitude of the equilibrium constant determines the extent of any particular reaction under any given condition. If the concentrations of the products are greater than those of the reactants, the values of the equilibrium constant will be large and the reaction will favour the formation of the products, and vice versa.

Example 11.1

For the reaction $N_2(g) + 3H_2(g) \rightleftharpoons 2NH_3(g)$, the partial pressures of gases at equilibrium at a certain temperature are $P_{N_2} = 6.84$ atm, $P_{H_2} = 4.12$ atm and $P_{NH_3} = 7.25$ atm. Calculate the equilibrium constant K_P.

Solution

$$K_P = \frac{P_{NH_3}^2}{P_{N_2} P_{H_2}^3} = \frac{(7.25)^2}{(6.84)(4.12)^3} = 0.110 \, \text{atm}^{-2}.$$

11.3.   RELATIONSHIP BETWEEN K_C AND K_P

The concentration of a substance, say B in Equation 11.1, is the number of moles of B in 1 ℓ. Therefore, $C_B = 1/V_B$ where V_B is the volume in litres of the gaseous mixture which contains 1 mol of B. Now, if P_B is the partial pressure of B, which is a perfect gas and obeys the ideal gas equation $PV = RT$, we have

$$C_B = \frac{1}{V_B} = \frac{P_B}{RT}.$$

Similarly,

$$C_D = \frac{1}{V_D} = \frac{P_D}{RT}$$

$$C_M = \frac{1}{V_M} = \frac{P_M}{RT}$$

$$C_N = \frac{1}{V_N} = \frac{P_N}{RT}$$

and so on.

Therefore, from Equation 11.4,

$$K_C = \frac{(P_M/RT)^m (P_N/RT)^n \cdots}{(P_B/RT)^b (P_D/RT)^d \cdots} = \frac{P_M^m P_N^n \cdots}{P_B^b P_D^d \cdots} (RT)^{(b+d+\cdots)-(m+n+\cdots)}$$

or

$$K_C = K_P (RT)^{\Sigma n} \tag{11.7}$$

where Σn is the algebraic sum of the number of molecules taking part in the reaction. It should be noted that the numbers of molecules of the reactants are taken as positive while those of the products are taken as negative. Alternatively, Equation 11.7 can be written as

$$K_C = K_P(RT)^{\Delta n} \qquad (11.8)$$

where Δn is the difference between the sum of the number of molecules of the products and that of the number of molecules of the reactants. To use Equation 11.8 it is not necessary to consider the sign of the number of molecules of the products as negative and that of the reactants as positive.

When the number of molecules of reactants is equal to the number of molecules of the products, $\sum n$ or Δn is necessarily equal to zero; so $K_P = K_C$. This implies that there is no volume change on reaction. If there is a volume change, $K_P \neq K_C$. For an increase in volume on reaction, $\sum n$ or Δn is positive. Then K_P is numerically greater than K_C. On the other hand, if there is a decrease in volume, $\sum n$ or Δn is negative; therefore K_P is less than K_C. It should be remembered that, in using Equations 11.7 and 11.8, R must be expressed in the same units as for pressures and volumes in K_P and K_C.

Example 11.2

Calculate the value of K_C for the reaction $\frac{1}{2}O_2 + SO_2 \rightleftharpoons SO_3$, for which K_P is 1.85 at 727°C.

Solution

$$K_C = \frac{C_{SO_3}}{C_{SO_2} C_{O_2}^{1/2}} = \frac{P_{SO_3}/RT}{[(P_{SO_2}/RT)(P_{O_2}/RT)]^{1/2}} = K_P(RT)^{1/2}$$

$$= (1.85)[(0.0821)(1000)]^{1/2} = 16.75 \text{ at } 727°C.$$

This problem can be worked out by direct application of Equation 11.7 where $\sum n = 1.5 - 1.0 = 0.5$.

11.4   EQUILIBRIUM AND FREE ENERGY

In all the processes considered in the discussion of thermodynamics we have invariably considered state-to-state transitions in one direction only. The change in the various state functions is calculated as the system is altered from the reactants to products. To apply thermodynamics to chemical reactions that can occur to appreciable extents in both forward and reverse directions, we now examine how the thermodynamic functions change as the system goes in the reverse direction from the *final* state (products) to the *initial* state (reactants). We can apply an approach of this type to the relationship between equilibrium and free energy. A reaction reactants→products with a positive ΔG under some specified conditions is thermodynamically non-spontaneous. In such cases this must mean that the reverse process products→reactants has a negative free energy change and is, therefore, spontaneous. So, if the reaction is chemically reversible to an observable extent, a positive ΔG indicates that the process will occur in the reverse direction but does not necessarily mean that 'nothing will happen'.

If the value of ΔG is not zero, this indicates the possibility that the system can attain a state of lower free energy through the formation of more reactants or more products (as the case may be). If we consider the free energy decrease as the driving force for chemical reactions, the system moves in a direction consistent with this tendency. Then, there will be the normal reactant-to-product or product-to-reactant conversion. Nevertheless, if the free-energy difference is zero, there is no thermodynamic drive for further change, i.e. there is no driving force for either the forward or the reverse reactions. Consequently, we shall say that the system has reached equilibrium. To establish a relationship between the equilibrium constant and the free-energy change, let us consider the general Equation 11.1.

Here we shall have to assume that the reactants B, D, ... and the products M, N, ... are present in any state that we may choose instead of in the equilibrium state. Expressing Equation 11.1 in terms of molar free energies G of the substances taking part in the reaction, we may write

$$b\mathrm{G_B} + d\mathrm{G_D} + \cdots \rightleftharpoons m\mathrm{G_M} + n\mathrm{G_N} + \cdots.$$

Then the change in free energy resulting from the reaction is given by

$$\Delta G = (m\mathrm{G_M} + n\mathrm{G_N} + \cdots) - (b\mathrm{G_B} + d\mathrm{G_D} + \cdots). \tag{11.8a}$$

We have seen that the molar free energy G of any substance in any state may be expressed in terms of its activity a by the equation

$$G - G° = RT \ln a$$

or

$$G = G° + RT \ln a$$

where $G°$ is the molar free energy in the standard state of unit activity. We can then express Equation 11.8a in terms of activity:

$$\Delta G = [m(G_M° + RT \ln a_M) + n(G_N° + RT \ln a_N) + \cdots]$$
$$- [b(G_B° + RT \ln a_B) + d(G_D° + RT \ln a_D) + \cdots]$$

where $a_M, a_N, a_B, a_D, \ldots$, are the activities of the reactants and the products in the state we have chosen for them. Rearranging this equation, we have

$$\Delta G = \Delta G° + RT \ln \left(\frac{a_M^m a_N^n \cdots}{a_B^b a_D^d \cdots} \right) \tag{11.9}$$

where $\Delta G°$ is the standard free-energy change for the reaction, i.e. when the reactants and the products are all in their standard states of unit activity. Equation 11.9 does not necessarily imply that conditions of equilibrium are present. If the arbitrarily chosen state should also be the equilibrium state, then Equation 11.9 becomes

$$\Delta G^{\circ} = -RT \ln \left(\frac{a_M^m a_N^n \cdots}{a_B^b a_D^d \cdots} \right)_{equ} \tag{11.10}$$

because, when a system is in equilibrium and the temperature and pressure are constant, $\Delta G = 0$. If we now substitute the appropriate concentration terms for activities, then it is apparent that $(a_M^m a_N^n \cdots / a_B^b a_D^d \cdots)_{equ}$ is the same expression as that for the equilibrium constant K. Then Equation 11.10 can be written

$$\Delta G^{\circ} = -RT \ln K. \tag{11.11}$$

Since the equilibrium constant of a reaction can be determined from laboratory experiments, the standard free-energy change for the same reaction can also be calculated from Equation 11.11. If the calculated value of ΔG° is inserted into Equation 11.9, we can calculate the free-energy change for the reaction with the reactants and products in any chosen state. With the help of Equation 11.11, Equation 11.9 can be written

$$\Delta G = -RT \ln K + RT \ln \left(\frac{a_M^m a_N^n \cdots}{a_B^b a_D^d \cdots} \right). \tag{11.12}$$

As before, by replacing $a_M^m a_N^n \cdots / a_B^b a_D^d \cdots$ by the symbol Q, Equation 11.12 becomes

$$\Delta G = -RT \ln K + RT \ln Q. \tag{11.13}$$

This is the *reaction isotherm of van't Hoff.* When a system is at equilibrium, the arbitrary reaction quotient Q becomes equal to K so that $\Delta G = 0$. Since both Q and K are represented by identical formal expressions, it is important to distinguish between them. The distinction between Q and K is that Q is the ratio of the activities of the products to the activities of the reactants under conditions of interest, whereas K is the specific value of Q when the system is in equilibrium at a specified temperature.

In a molecular system we use nominal concentrations rather than activities. The expressions for Q in such systems will be designated by the subscripts P, C and x for concentrations expressed as pressures, molarities and mole fractions. Then, for general purposes we shall have three alternative expressions,

$$Q_P = \frac{P_M^m P_N^n \cdots}{P_B^b P_D^d \cdots} \qquad Q_C = \frac{[M]^m [N]^n \cdots}{[B]^b [D]^d \cdots} \qquad Q_x = \frac{x_M^m x_N^n \cdots}{x_B^b x_D^d \cdots}.$$

At any given temperature, Q will have a specific numerical value for the system at equilibrium, namely $Q_P = K_P$, $Q_C = K_C$ and $Q_x = K_x$. Then, Equation 11.13 will become

$$\Delta G = -RT \ln K_P + RT \ln Q_P \tag{11.14}$$

$$\Delta G = -RT \ln K_C + RT \ln Q_C \tag{11.15}$$

$$\Delta G = -RT \ln K_x + RT \ln Q_x. \tag{11.16}$$

Equations 11.14, 11.15 and 11.16 can also be written as

$$\Delta G = RT \ln \left(\frac{Q_P}{K_P} \right) \tag{11.17}$$

$$\Delta G = RT \ln \left(\frac{Q_C}{K_C} \right) \tag{11.18}$$

$$\Delta G = RT \ln \left(\frac{Q_x}{K_x} \right). \tag{11.19}$$

Each of these equations uses its appropriate units and gives the same value of ΔG. Thus, ΔG is determined by the ratio of Q to K, i.e. by the ratio of the chosen or arbitrary reaction quotient to the equilibrium reaction quotient. However, $\Delta G°$ is related only to the equilibrium reaction quotient.

If we combine Equations 10.37 and 11.11, we get $\Delta G° = \Delta H° - T\Delta S° = -RT \ln K$. This relationship is developed for conditions of constant pressure and temperature. We have already stated that $\Delta H°$ and $\Delta S°$ are approximately independent of temperature. However, if we assume that they are really independent of temperature, it is then possible to obtain a simple relationship between the equilibrium constant K and temperature T.

Let us consider an equilibrium reaction at two different temperatures T_1 and T_2 with equilibrium constants K_1 and K_2 respectively. Then we have

$$\Delta H° - T_1 \Delta S° = -RT_1 \ln K_1$$

and

$$\Delta H° - T_2 \Delta S° = -RT_2 \ln K_2.$$

Rearranging these equations we get

$$\frac{\Delta H°}{T_1} - \Delta S° = -R \ln K_1 \tag{11.19a}$$

$$\frac{\Delta H°}{T_2} - \Delta S° = -R \ln K_2. \tag{11.19b}$$

Subtracting Equation 11.19b from Equation 11.19a and rearranging gives

$$\Delta H° \left(\frac{1}{T_1} - \frac{1}{T_2} \right) = R \ln \left(\frac{K_2}{K_1} \right)$$

or

$$\frac{\Delta H° (T_2 - T_1)}{RT_1 T_2} = \ln \left(\frac{K_2}{K_1} \right). \tag{11.19c}$$

Equations 11.11, 11.19a, 11.19b and 11.19c are useful for calculating the equilibrium constant. For example, if we know $\Delta G°$ for 298 K, we can with the help of Equation 11.11 calculate the equilibrium constant at 298 K. Then, if we know $\Delta H°$ for 298 K (which, we have said, is virtually independent of temperature), we can calculate the value of K for any other temperature using Equation 11.19c (assuming that both $\Delta S°$ and $\Delta H°$ are independent of temperature). The value of K for temperatures other than 298 K can also be obtained from Equation 11.11, provided that we know the value of $\Delta G°$ for the corresponding temperature. However, it should be noted that values for $\Delta G°$ at temperatures other than 298 K are not readily available. This makes it difficult to obtain K-values for temperatures other than 298 K by using $\Delta G°$ alone. Another point to remember is that, unlike $\Delta H°$ and $\Delta S°$, $\Delta G°$-values significantly change with temperature.

It is important to note that care must be taken to ensure that the standard states chosen to calculate $\Delta G°$ are consistent. For example, if we are using Q_P (i.e. the gas pressure in atmospheres or megapascals), the standard free energies of formation used to obtain $\Delta G°$ must be those in the gas phase at 1 atm. Particular attention to this requirement is necessary for reactions in solution where various standard state definitions may have been used for tabulated $\Delta G_f°$ values.

The sign of ΔG at a particular temperature depends on the relative magnitudes of Q and K. Two cases may arise. From Equation 11.13 it appears that, if $K > Q$, ΔG is negative; the process will occur spontaneously from left to right. The physical significance of this situation is that the relative concentrations of product species present are smaller than those at equilibrium. As the reaction proceeds, large amounts of products are formed and the value of Q increases until $Q = K$ and equilibrium is established.

If $K < Q$, ΔG is positive. Therefore, the reverse reaction will be thermodynamically spontaneous. Consequently, there will be conversion of some products with an associated reduction in the value of Q.

Although a negative value for ΔG makes a reaction theoretically possible, it bears no relationship to the speed of the reaction. A reaction might only become spontaneous and proceed at an observable speed when the pressure is favourable or perhaps in the presence of a specific catalyst. The fact that there is a quantitative relationship between ΔG and K allows us not only to calculate the value of one from the other but also to establish a qualitative relationship between the magnitude of $\Delta G°$ and the general characteristics of the system.

Example 11.3

The free-energy change at 298 K in the process

$$2NO_2(g)(5\ atm) \rightleftharpoons N_2O_4(g)(5\ atm)$$

is −9380.528 J. Calculate the equilibrium constant at this temperature.

Solution

We shall use the equation $\Delta G = \Delta G° + RT \ln Q$:

$$\log Q = \log \left(\frac{P_{N_2O_4}}{P_{NO_2}^2} \right) = \log \left(\frac{5}{25} \right) = -0.699.$$

Substituting this value into the above equation, we have

$$-9380.528 = \Delta G^\circ + (2.303)(8.314)(298)(-0.699)$$

or

$$\Delta G^\circ = -5393.176\,\text{J}.$$

Substituting this value of ΔG° in $\Delta G^\circ = -RT\ln K$, we have

$$-5393.176 = -8.314\,(298)(2.303)\log K$$

or

$$\log K = 0.945$$

or

$$K = 8.81.$$

11.5 THE VAN'T HOFF EQUATION

We can write the van't Hoff isotherm for standard free-energy change as

$$\Delta G^\circ = -RT \ln K_P.$$

Differentiating this equation with respect to temperature, the pressure remaining constant, we have

$$\frac{\text{d}}{\text{d}T}(\Delta G^\circ)_P = -RT \ln K_P - RT \frac{\text{d}}{\text{d}T}(\ln K_P).$$

Multiplying both sides by T,

$$T\frac{\text{d}}{\text{d}T}(\Delta G^\circ)_P = -RT \ln K_P - RT^2 \frac{\text{d}}{\text{d}T}(\ln K_P) = \Delta G^\circ - RT^2 \frac{\text{d}}{\text{d}T}(\ln K_P)$$

(see Equation 11.11). However, from the Gibbs–Helmholtz equation,

$$\Delta G^\circ = \Delta H^\circ + T\frac{\text{d}}{\text{d}T}(\Delta G^\circ)_P.$$

This equation is modified to refer all the substances involved to their standard states. Combining these two equations together gives

$$\Delta H^\circ = RT^2 \frac{\text{d}}{\text{d}T}(\ln K_P)$$

or

$$\frac{\text{d}}{\text{d}T}(\ln K_P) = \frac{\Delta H^\circ}{RT^2}. \tag{11.20}$$

This equation is known as the *van't Hoff equation*. $\Delta H°$ is the change in the heat of reaction at constant pressure when all reactants and products are in their standard states. Equation 11.20 is sometimes more conveniently written as follows:

$$\frac{d(\ln K_P)}{d(1/T)} = \frac{-\Delta H°}{R} \tag{11.21}$$

or

$$\frac{d(\log K_P)}{d(1/T)} = \frac{-\Delta H°}{2.303R} \tag{11.22}$$

Note: In practice, $\Delta H°$ can be replaced by ΔH with only slight loss of accuracy so that the equation can be of much greater application.

On the assumption that $\Delta H°$ is essentially independent of temperature, these equations may be integrated to give

$$\log K_P = \left(-\frac{\Delta H°}{2.303R} \right) \frac{1}{T} + \text{constant}. \tag{11.23}$$

Equations 11.20–11.23 show that a plot of $\log K_P$ versus $1/T$ should give a straight line with a slope equal to $-\Delta H°/2.303R$.

For exact integration of Equation 11.20, $\Delta H°$ must be known as a function of temperature. If the temperature interval is not large, $\Delta H°$ may be considered as constant over the temperature range concerned, and in that case integration may be performed within the temperature range as follows:

$$\int_{(K_P)_1}^{(K_P)_2} d(\ln K) = \int_{T_1}^{T_2} \frac{\Delta H°}{RT^2} dT$$

or

$$\ln \left(\frac{(K_P)_2}{(K_P)_1} \right) = \frac{\Delta H°}{R} \frac{T_2 - T_1}{T_1 T_2}. \tag{11.24}$$

This equation allows the calculation of $(K_P)_2$ at a temperature T_2, if $(K_P)_1$ at another temperature T_1 and $\Delta H°$ are known. If the equilibrium constants at two different temperatures are known, the average heat of the reaction over that temperature range can be calculated from Equation 11.24.

11.6 TEMPERATURE DEPENDENCE OF THE EQUILIBRIUM CONSTANT

We have already deduced the equations $\Delta G° = -RT \ln K$ and $\Delta G° = \Delta H° - T\Delta S°$. Combining these two equations we have

$$\ln K = \frac{-\Delta H°}{RT} + \frac{\Delta S°}{R}. \tag{11.25}$$

This equation shows that, if $\Delta H°$ and $\Delta S°$ are constants independent of temperature, then $\ln K$ is a simple linear function of the reciprocal of temperature. Now the question is: when are $\Delta H°$ and $\Delta S°$ independent of temperature? We have seen earlier that, for two temperatures T_1 and T_2, the change in enthalpy is given by

$$(\Delta H)_2 = (\Delta H)_1 + \Delta C_P(T_2 - T_1).$$

This equation shows that $\Delta H°$ will be independent of temperature if $\Delta C_P = 0$, i.e. if the heat capacities of reactants and products are the same. Again, we also know that the entropy of any substance at a temperature T is given by

$$S_T° = S_{298\,K}° + \int_{298\,K}^{T} \Delta C_P \frac{dT}{T}.$$

Using such an equation for both reactants and products and then subtracting one from the other, we have

$$\Delta S_T° = \Delta S_{298\,K}° + \int_{298\,K}^{T} \Delta C_P \frac{dT}{T}$$

where ΔC_P is the difference between the heat capacities of reactants and products. This equation shows that, if $\Delta C_P = 0$, $\Delta S°$ is independent of temperature. Now we can say that, if the difference in heat capacities of reactants and products is zero, $\ln K$ is a linear function of $1/T$.

Differentiating Equation 11.25 with respect to $1/T$, we have

$$\frac{d(\ln K)}{d(1/T)} = -\frac{\Delta H°}{R}. \tag{11.26}$$

Since $d(1/T) = -dT/T^2$, we have

$$\frac{d}{dT}(\ln K) = \frac{\Delta H°}{RT^2}.$$

This is the same as Equation 11.20. These equations show that $\Delta H°$ of a reaction determines the effect of temperature on its equilibrium constant. If $\Delta H°$ is positive, then Equation 11.20 shows that $d(\ln K)$ is positive provided that dT is positive. Therefore, for an endothermic reaction, K is proportional to T i.e. K increases as T increases. If $\Delta H°$ is negative, K is inversely proportional to T, i.e. K decreases as T increases. Rewriting Equation 11.26 as

$$d(\ln K) = \frac{-\Delta H°}{R}d\left(\frac{1}{T}\right)$$

and integrating between the temperatures T_1 and T_2, at which the equilibrium constant has the values K_1 and K_2, respectively, we have

$$\ln\left(\frac{K_2}{K_1}\right) = -\frac{\Delta H°}{R}\left(\frac{1}{T_2} - \frac{1}{T_1}\right). \tag{11.27}$$

This equation basically gives the temperature dependence of the equilibrium constant.

Equation 11.25 shows that a plot of $\log K$ versus $1/T$ will be a straight line with the slope $\Delta H^\circ/2.303\,R$ and the intercept $\Delta S^\circ/2.303\,R$. It is not convenient to calculate ΔS° from the intercept of such plots; it can be calculated directly, once ΔH° is known. An advantage of the graphical approach is that any significant temperature dependence of the enthalpy change will be immediately apparent in a curvature of the plot. A true straight line will be obtained only if ΔH° is rigorously independent of temperature. If curvature is present, we can obtain an average value and, at the same time, the amount of variation in ΔH can be estimated.

An equation similar to Equation 11.27 may be derived for K_C, but as a general rule such an equation will yield different results from those of Equation 11.27. They will be identical only when $K_C = K_P$, and $\sum n$ or Δn is equal to zero, i.e. there is no volume change.

Example 11.4

The values of equilibrium constant at the temperatures 834 K and 921 K for the reaction $SO_2 + \frac{1}{2}O_2 \rightleftharpoons SO_3$ are 18.0 and 4.96, respectively. Calculate the average value of ΔH° over the temperature range.

Solution

$K_1 = 18.0$, $T_1 = 834\,K$; $K_2 = 4.96$, $T_2 = 921$ K. Substituting these values into Equation 11.27, we have

$$\log \left(\frac{4.96}{18.0} \right) = \frac{-\Delta H^\circ}{(2.303)(8.314)} \left(\frac{1}{921} - \frac{1}{834} \right)$$

or

$$\Delta H^\circ = -94.140\,kJ.$$

11.7 PRESSURE DEPENDENCE OF THE EQUILIBRIUM CONSTANT

We have said that for molecular systems the equilibrium constant can be expressed using nominal concentrations, i.e. pressure, mole fraction or molarity. Equations 11.7 and 11.8 give a relationship between K_C and K_P. To establish a relation between K_x and K_P, let us consider x mol of the substances B, D, M, N, ... in our general Equation 11.1 i.e.

$$b_B + d_D + \cdots \rightleftharpoons m_M + n_N + \cdots.$$

From Dalton's law of partial pressure (which states that the partial pressure of each gas in an ideal mixture of ideal gases is equal to its mole fraction multiplied by the total pressure) we have

$$x_B = \frac{P_B}{P} \qquad x_M = \frac{P_M}{P}$$

$$x_D = \frac{P_D}{P} \qquad x_N = \frac{P_N}{P}$$

where P_B, P_D, P_M, P_N, ... are the partial pressures of B, D, M, N, ..., and P is the total pressure. The equilibrium constant in mole fractions is given by

$$K_x = \frac{x_M^m x_N^n \cdots}{x_B^b x_D^d \cdots} = \frac{P_M^m P_N^n \cdots}{P_B^b P_D^d \cdots} P^{(b+d+\cdots)-(m+n+\cdots)} = K_P P^{\Delta n} \tag{11.28}$$

where Δn is the difference between the number of moles of the reactants and the products. Since K_P for ideal gases is independent of pressure, it is evident that K_x is a function of pressure except when $\Delta n = 0$. It is thus constant only with respect to variations in the x at constant temperature and pressure. Taking logarithms of both sides of Equation 11.28, we have

$$\ln K_x = \ln K_P + \Delta n \ln P.$$

Differentiating both sides with respect to P, we get

$$\frac{d}{dP}(\ln K_x) = \frac{\Delta n}{P} = \frac{\Delta V}{RT} \tag{11.29}$$

(since, at constant pressure, $P\Delta V = \Delta nRT$).

When a reaction occurs without any change in the total number of moles of the gas in the system, $\Delta n = 0$. In such instances, the constant K_P is the same as K_x or K_C and, for ideal gases, the position of equilibrium does not depend on the total pressure. When $\Delta n \neq 0$, the pressure dependence of K_x is given by Equation 11.29. When there is a decrease in the number of moles ($\Delta n < 0$) and, therefore, a decrease in the volume, K_x increases with increasing pressure. If there is an increase in the number of moles ($\Delta n > 0$) and, therefore, an increase in the volume, K_x decreases with increasing pressure. An important class of reactions for which $\Delta n \neq 0$ is that of molecular dissociation.

11.8   PARTIAL MOLAR QUANTITIES

From our previous discussion it should be obvious now that any state of a system may be defined by the extensive or intensive properties of the system. Actually, few of the many extensive or intensive properties which we can determine are required completely to specify the state of any system. The properties which are generally required for such specification are P, V, T and the numbers, say N_i, N_j, ..., of the various types of molecule present. Let us consider an extensive property, say X, of a system, and assume that this property is a function of T, P and composition of the system. Then,

$$X = f(T, P, N_i, N_j, \ldots).$$

From this we may write, for any change in X,

$$dX = \left(\frac{\partial X}{\partial T}\right)_{P,N_1,N_2,N_3,\ldots} dT + \left(\frac{\partial X}{\partial P}\right)_{T,N_1,N_2,N_3,\ldots} dP$$
$$+ \left(\frac{\partial X}{\partial N_1}\right)_{T,P,N_2,N_3,\ldots} dN_1 + \left(\frac{\partial x}{\partial N_2}\right)_{T,P,N_1,N_3,\ldots} dT + \cdots$$

In particular, if we are considering a system at constant T and P, then the first two terms in the above equation will disappear, and we shall have

$$(dX)_{T,P} = \left(\frac{\partial X}{\partial N_1}\right)_{T,P,N_2,N_3,\ldots} dN_1 + \left(\frac{\partial X}{\partial N_2}\right)_{T,P,N_1,N_3,\ldots} dN_2 + \cdots$$
$$= X_1 dN_1 + X_2 dN_2 + \cdots \tag{11.30}$$

where

$$X_1 = \left(\frac{\partial X}{\partial N_1}\right)_{T,P,N_2,N_3,\ldots}$$

$$X_2 = \left(\frac{\partial X}{\partial N_2}\right)_{T,P,N_1,N_3,\ldots}$$

and so on. Hence, Equation 11.30 may be written as

$$dX = X_1 dN_1 + X_2 dN_2 + \cdots = \sum_i (X_i dN_i). \tag{11.31}$$

Here, although we have omitted the subscripts on the left-hand side, it must be remembered that Equation 11.31 is valid for a system at constant temperature and pressure. The implication of this equation is that a system has been built up at a constant temperature and pressure by simultaneous addition of small quantities dN_1, $dN_2, \ldots,$ in the constant proportions finally required, so that the final composition of the system contains N_1 molecules of type 1, N_2 molecules of type 2 and so on. Since the final composition of the system is constant, the value of $X_1, X_2, \ldots,$ must remain constant during the building up of the system. Accepting this implication, Equation 11.31 may then be directly integrated to obtain

$$X = X_1 N_1 + X_2 N_2 + \cdots = \sum_i X_i N_i. \tag{11.32}$$

This equation shows that X_i is the magnitude of the property X for the molecule of type i so that the total magnitude of X for the system may be expressed on a molecular additivity basis.

In the special case, if we take N_1 as the number of moles instead of the number of molecules, then X_i will be called the partial molar value of X for species i. Partial molar quantities are homogeneous functions of degree zero in the numbers of moles of substances that make up the system.

There are several partial molar quantities, all of which are not equally important in thermodynamics. Here we shall discuss some of these partial molar quantities.

11.8.1   Partial molar volumes

These molar quantities provide examples for illustrating partial molar concepts. There are several methods for the determination of partial molar volumes, but we shall not discuss the volume of a mixture on composition. Let us consider a system consisting of two components with mole numbers n_1 and n_2, and mole fractions x_1 and x_2. Then the average molar volume V_m can be expressed as

$$V_m = \frac{V}{n_1 + n_2} = x_1 \bar{V}_1 + x_2 \bar{V}_2 \tag{11.33}$$

where $\bar{V}_1$ and $\bar{V}_2$ are the partial molar volumes of components 1 and 2, respectively. Since $x_2 = 1 - x_1$, this equation becomes

$$V_m = x_1 \bar{V}_1 + (1 - x_1)\bar{V}_2.$$

Now, partially differentiating both sides with respect to x_1, we have

$$\frac{\partial V_m}{\partial x_1} = \bar{V}_1 - \bar{V}_2 + x_1 \frac{\partial \bar{V}_1}{\partial x_1} + (1 - x_1)\frac{\partial \bar{V}_2}{\partial x_1} = \bar{V}_1 - \bar{V}_2 + x_1 \frac{\partial \bar{V}_1}{\partial x_1} + x_2 \frac{\partial \bar{V}_2}{\partial x_1}. \tag{11.34}$$

We shall show later that $\sum_i N_i \, dX_i = 0$, where X_i is any partial molar quantity. In view of this, the last two terms of Equation 11.34 disappear and we get

$$\frac{\partial V_m}{\partial x_1} = \bar{V}_1 - \bar{V}_2$$

or

$$\bar{V}_1 = \bar{V}_2 + \frac{\partial V_m}{\partial x_1}$$

or

$$x_2 \bar{V}_1 = x_2 \bar{V}_2 + x_2 \frac{\partial V_m}{\partial x_1}$$

or

$$(1 - x_1)\bar{V}_1 = x_2 \bar{V}_2 + x_2 \frac{\partial V_m}{\partial x_1}$$

or

$$\bar{V}_1 = (x_1 \bar{V}_1 + x_2 \bar{V}_2) + x_2 \frac{\partial V_m}{\partial x_1} = V_m + x_2 \frac{\partial V_m}{\partial x_1} \tag{11.35}$$

Similarly,

$$\bar{V}_2 = V_m - x_1 \frac{\partial V_m}{\partial x_1}. \tag{11.36}$$

From Equations 11.35 and 11.36 it is clear that partial molar volumes can be obtained from the average molar volume. These two equations also show the dependence of partial molar volumes on composition.

11.8.2   Partial molar enthalpy

We have seen that the absolute value of enthalpy in any system is an indeterminate quantity, but the change in enthalpy of a system associated with a change in state of that system may be determined. Likewise, the partial molar enthalpies are indeterminate but the determination of change in such quantities is possible.

Let us consider a homogeneous solution formed by isothermal mixing of several pure substances of mole numbers $n_1, n_2, \ldots$. This mixing process will be accompanied by an enthalpy change $(\Delta H)_{\text{mix}}$, which will generally be a function of the composition. If we denote the molar enthalpies of pure substances by $\bar{H}'_1, \bar{H}'_2, \ldots$, then the total enthalpy of the mixture will be

$$H = (\Delta H)_{\text{mix}} + n_1\bar{H}'_1 + n_2\bar{H}'_2 + \cdots. \tag{11.37}$$

To obtain the partial molar enthalpies of the various substances present in the solution we partially differentiate Equation 11.37 with respect to the number of moles of the substance concerned. Then,

$$H_i = \left(\frac{\partial H}{\partial n_i}\right)_{T,P,n_1,\ldots} = \frac{\partial}{\partial n_i}[(\Delta H)_{\text{mix}}]_{T,P,n_1,\ldots} + \bar{H}'_1. \tag{11.38}$$

The change in partial molar enthalpy associated with the transformation of pure substances to a solution will then be given by

$$\Delta\bar{H}_i = \bar{H}_i - \bar{H}'_i = \frac{\partial}{\partial n_i}[(\Delta H)_{\text{mix}}]_{T,P,n_1,\ldots}. \tag{11.39}$$

11.8.3   Partial molar energies

Although these quantities are seldom used, they can be obtained by combination of partial molar enthalpies and partial molar volumes.

The most important partial molar quantity is the *partial molar free energy*. This is also known as the *chemical potential*. We shall denote this quantity by the symbol μ_i, with the definition

$$\mu_i = \left(\frac{\partial G}{\partial n_i}\right)_{T,P,n_1,n_2,\ldots}. \tag{11.40}$$

The chemical potential is an intensive quantity that characterizes the thermodynamic behaviour of any chemical species when reacting with other chemical species, both as to the direction in which reactions are permitted to occur, and as to the position of chemical equilibria. This applies to both homogeneous reactions (reactions that occur in a single phase) and heterogeneous reactions (reactions at the interfaces between different phases). Chemical potentials also apply to phase transformations and equilibria, irrespective of whether they occur in pure substances or in a mixture. Application of the chemical potential leads to such well known laws as Henry's law, Raoult's law and the osmotic pressure law.

The free energy of a mixture of n_i moles of i species is equal to the sum of the chemical potentials μ_i of each species in the mixture multiplied by the respective number n_i of moles. Mathematically,

$$G = \sum_i n_i \mu_i. \tag{11.41}$$

The chemical potential depends on the composition of the mixture, the temperature T and the pressure P. For a pure substance, its chemical potential is equal to its free energy G_f of formation, i.e. $\mu = G_f$. In a mixture of perfect gases the chemical potential of the species i ($i = 1, 2, 3, \ldots$) is given by

$$\mu_i = (\mu_i^\circ)_T + RT \ln P_i \tag{11.42}$$

where P_i is the partial pressure (in atmospheres or megapascals) of the species i and $(\mu_i^\circ)_T$ is the chemical potential of the same species at a standard pressure of 1 atm (0.1 MPa) and a temperature T. This temperature T is chosen as 298 K (or 25°C) unless another temperature is specifically mentioned. The quantity $(\mu_i^\circ)_T$ may be equated to $(G_f^\circ)_T$, the standard molar free energy of formation for the pure gas i.

It is to be noted that only the logarithm of the numerical part of P_i is to be taken; the term $RT \ln(\text{atm})$ is automatically compensated by a term $-RT \ln(\text{atm})$ in μ_i° that is usually not mentioned explicitly, Thus, from Equation 11.42 it is obvious that, when $P_i = 1$ atm, $\mu_i = \mu_i^\circ$, as it should do. The same result follows if P_i is in megapascals.

For dilute solutions, the chemical potential of the solute species i is given by

$$(\mu_i)_{\text{soln}} = (\mu_i^\circ)_{\text{soln}} + RT \ln C_i \tag{11.43}$$

where $(\mu_i^\circ)_{\text{soln}}$ is the chemical potential of the species i in a solution at a standard concentration 1 M, at temperature T, usually 298 K, and at 1 atm; C_i is the concentration in moles per litre of the species i. We may set $(\mu_i^\circ)_{\text{soln}}$ equal to $(G_f^\circ)_T$, the standard molar free energy of formation for the species i. $(\mu_i)_{\text{soln}}$ depends on the temperature T and the total pressure P.

It should be noted that, although the units of C_i are moles per litre, the logarithm $\ln C_i$ relates only to the numerical part of C_i. The term $RT \ln (\text{mol } \ell^{-1})$ is automatically compensated by a term $-RT \ln(\text{mol } \ell^{-1})$ in $(\mu_i^\circ)_{\text{soln}}$ that is usually not specifically mentioned. Thus, from Equation 11.43 it is obvious that, when $C_i = 1$ mol ℓ^{-1}, $(\mu_i^\circ)_{\text{soln}} = (\mu_i)_{\text{soln}}$ as it should do.

For other than dilute solutions (about 0.01 M), Equation 11.43 is valid if the concentrations are replaced by activity a_i. Since activity $a_i = \gamma_i C_i$ where γ_i is the activity coefficient, Equation 11.42 becomes

$$(\mu_i)_{\text{soln}} = (\mu_i^\circ)_{\text{soln}} + RT \ln a_i \tag{11.44}$$

or

$$(\mu_i)_{\text{soln}} = (\mu_i^\circ)_{\text{soln}} + RT \ln (\gamma_i C_i). \qquad (11.45)$$

Similarly, for a mixture of real gases, Equation 11.42 is valid, if the partial pressures P_i are replaced by corrected partial pressure (fugacities).

The chemical potential of a pure solid, liquid or gas may be set equal to the molar free energy of formation of that substance.

11.9   PARTIAL DERIVATIVES OF THE CHEMICAL POTENTIAL

Choosing $T, P, n_1, n_2, \ldots$ as the independent variables for the free energy of a system, we may write

$$dG = \left(\frac{\partial G}{\partial T}\right)_{P,n_1,n_2,\ldots} dT + \left(\frac{\partial G}{\partial P}\right)_{T,n_1,n_2,\ldots} dP + \left(\frac{\partial G}{\partial n_1}\right)_{T,P,n_2,n_3\ldots} dn_1$$

$$+ \left(\frac{\partial G}{\partial n_2}\right)_{T,P,n_1,n_3\ldots} dn_2 + \cdots. \qquad (11.46)$$

Using Equations 10.28, 10.29 and 11.40, the above equation can be written as

$$dG = -S\,dT + V\,dP + \mu_1\,dn_1 + \mu_2\,dn_2 + \cdots$$

$$= -S\,dT + V\,dP + \sum_i (\mu_i\,dn_i). \qquad (11.47)$$

From Equation 10.10 we get $dA = dG - P\,dV - V\,dP$. Combining this with Equation 11.47, we obtain

$$dA = -S\,dT - P\,dV + \mu_1\,dn_1 + \mu_2\,dn_2 + \cdots$$

$$= -S\,dT - P\,dV + \sum_i (\mu_i\,dn_i). \qquad (11.48)$$

From Equation 10.10, $dH = dG + T\,dS + S\,dT$. Combining this with Equation 11.47 we can write

$$dH = T\,dS + V\,dP + \mu_1\,dn_1 + \mu_2\,dn_2 + \cdots = T\,dS + V\,dP + \sum_i (\mu_i\,dn_i). \qquad (11.49)$$

Finally, from $H = U + PV$, we have $dH = dU + P\,dV + V\,dP$. With the help of this, Equation 11.49 becomes

$$dU = T\,dS - P\,dV + \mu_1\,dn_1 + \mu_2\,dn_2 + \cdots = T\,dS - P\,dV + \sum_i (\mu_i\,dn_i). \qquad (11.50)$$

From Equations 11.49 and 11.50 we can write

$$dS = \frac{1}{T}dH - \frac{V}{T}dP - \frac{\mu_1}{T}dn_1 - \frac{\mu_2}{T}dn_2 - \cdots$$
$$= \frac{1}{T}dH - \frac{V}{T}dP - \frac{1}{T}\sum_i(\mu_i \, dn_i) \tag{11.51}$$

$$dS = \frac{1}{T}dU + \frac{P}{T}dV - \frac{\mu_1}{T}dn_1 - \frac{\mu_2}{T}dn_2 - \cdots$$
$$= \frac{1}{T}dU + \frac{P}{T}dV - \frac{1}{T}\sum_i(\mu_i \, dn_i). \tag{11.52}$$

From the foregoing equations it is apparent that the chemical potential can be defined in any of the following six different ways:

$$\mu_i = \left(\frac{\partial G}{\partial n_i}\right)_{T,P,n_1,n_2,\ldots} \tag{11.53a}$$

$$\mu_i = \left(\frac{\partial A}{\partial n_i}\right)_{T,V,n_1,n_2,\ldots} \tag{11.53b}$$

$$\mu_i = \left(\frac{\partial U}{\partial n_i}\right)_{S,V,n_1,n_2,\ldots} \tag{11.53c}$$

$$\mu_i = \left(\frac{\partial H}{\partial n_i}\right)_{S,P,n_1,n_2,\ldots} \tag{11.53d}$$

$$\mu_i = -T\left(\frac{\partial S}{\partial n_i}\right)_{U,V,n_1,n_2,\ldots} \tag{11.53e}$$

$$\mu_i = -T\left(\frac{\partial S}{\partial n_i}\right)_{H,P,n_1,n_2,\ldots} \tag{11.53f}$$

From Equation 11.47 we get (since dG is an exact differential)

$$\frac{\partial^2 G}{\partial T \, \partial n_i} = -\left(\frac{\partial S}{\partial n_i}\right)_{T,P,n_1,n_2,\ldots} = \left(\frac{\partial \mu_i}{\partial T}\right)_{P,n_1,n_2,\ldots} \tag{11.54}$$

and also

$$\frac{\partial^2 G}{\partial P \, \partial n_i} = \left(\frac{\partial V}{\partial n_i}\right)_{T,P,n_1,n_2,\ldots} = \left(\frac{\partial \mu_i}{\partial P}\right)_{T,n_1,n_2,\ldots} \tag{11.55}$$

Similarly, from Equation 11.48 we get (since dA is an exact differential)

$$\left(\frac{\partial \mu_i}{\partial T}\right)_{V,n_1,n_2,\ldots} = -\left(\frac{\partial S}{\partial n_i}\right)_{T,V,n_1,n_2,\ldots} \tag{11.56}$$

$$\left(\frac{\partial \mu_i}{\partial V}\right)_{T,n_1,n_2,\ldots} = -\left(\frac{\partial P}{\partial n_i}\right)_{T,V,n_1,n_2,\ldots}. \tag{11.57}$$

It should be carefully noted that the quantities on the right-hand side of Equations 11.56 and 11.57 are not partial molar quantities. These two equations are not used very frequently.

11.10   CHEMICAL POTENTIAL OF AN IDEAL GAS IN A MIXTURE OF IDEAL GASES

Equation 11.42 gives the chemical potential of the species i ($i = 1, 2, 3, \ldots$) in a mixture of i perfect gases. If P is the total pressure of such a mixture of gases, then $P_i = x_i P$, where x_i is the mole fraction of the ith species in the mixture. Substituting this into Equation 11.42, we have

$$\mu_i = (\mu_i^\circ)_T + RT \ln(x_i P) = (\mu_i^\circ)_T + RT \ln P + RT \ln x_i. \tag{11.58}$$

Since the first two terms on the right-hand side of this equation are the chemical potential of the pure ith species under the pressure P, we can write

$$\mu_i = (\mu_{i(\text{pure})})_{T,P} + RT \ln x_i. \tag{11.59}$$

Since x_i is a fraction, obviously $\ln x_i$ is negative. Therefore, Equation 11.59 shows that *under the same total pressure the chemical potential of any gas in a mixture is always less than the chemical potential of that gas in a pure state.*

11.11   FREE ENERGY, ENTROPY, ENTHALPY AND VOLUME OF MIXING OF IDEAL GASES

Let us consider a mixture of three ideal gases with n_1 mol, n_2 mol and n_3 mol and chemical potentials μ_1°, μ_2° and μ_3°, respectively. Let us assume that the gases are mixed together at a constant temperature and pressure, i.e. the initial and the final temperatures and pressures are the same. Then, for the pure individual gases the respective free energies are

$$G_1 = n_1\mu_1^\circ$$
$$G_2 = n_2\mu_2^\circ$$
$$G_3 = n_3\mu_3^\circ.$$

So the free energy of the initial state will be

$$G_{\text{initial}} = G_1 + G_2 + G_3 = n_1\mu_1^\circ + n_2\mu_2^\circ + n_3\mu_3^\circ = \sum_i n_i\mu_i^\circ \tag{11.60}$$

where $i = 1, 2, 3, \dots$.

The free energy of the final state will be give by Equation 11.41, i.e.

$$G_{\text{final}} = \sum_i n_i\mu_i.$$

The free energy of mixing will be given by

$$(\Delta G)_{\text{mix}} = G_{\text{final}} - G_{\text{initial}} = n_1(\mu_1 - \mu_1^\circ) + n_2(\mu_2 - \mu_2^\circ) + n_3(\mu_3 - \mu_3^\circ)$$
$$= \sum_i n_i(\mu_i - \mu_i^\circ). \tag{11.61}$$

Now, substituting for $\mu_i - \mu_i^\circ$ from Equation 11.59, we have

$$(\Delta G)_{\text{mix}} = RT \sum_i (n_i \ln x_i). \tag{11.62}$$

If N is the total number of moles of all the gases in the mixture, and x_i is the mole fraction of the ith component, then $n_i = x_iN$. Substituting this into Equation 11.62, we get

$$(\Delta G)_{\text{mix}} = RT(Nx_1 \ln x_1 + Nx_2 \ln x_2 + Nx_3 \ln x_3) = NRT \sum_i (x_i \ln x_i). \tag{11.63}$$

Since every term on the right-hand side of Equation 11.63 is negative (as x_i is a fraction), the entire sum is negative, and hence the free energy of mixing of ideal gases is negative.

To obtain the entropy $(\Delta S)_{\text{mix}}$, of mixing, let us differentiate $(\Delta G)_{\text{mix}}$ with respect to T. Then,

$$\frac{\partial}{\partial T}[(\Delta G)_{\text{mix}}]_{P,n_i} = \frac{\partial}{\partial T}(G_{\text{final}})_{P,n_i} - \frac{\partial}{\partial T}(G_{\text{initial}})_{P,n_i}$$

However, from Equation 10.28

$$\frac{\partial}{\partial T}(G_{\text{final}})_{P,n_i} = -S_{\text{final}}$$
$$\frac{\partial}{\partial T}(G_{\text{initial}})_{P,n_i} = -S_{\text{initial}}.$$

Therefore, we have

$$\frac{\partial}{\partial T}[(\Delta G)_{\text{mix}}]_{P,n_i} = -(S_{\text{final}} - S_{\text{initial}}) = -(\Delta S)_{\text{mix}}. \tag{11.64}$$

Again, differentiating both sides of Equation 11.63 with respect to T, we get

$$\frac{\partial}{\partial T}[(\Delta G)_{\text{mix}}]_{P,n_i} = NR\sum_i (x_i \ln x_i).$$

Substituting this in Equation 11.64, we have

$$(\Delta S)_{\text{mix}} = -NR\sum_i (x_i \ln x_i). \tag{11.65}$$

The negative sign on the right-hand side of this equation implies that the entropy of mixing is always positive (because $\ln x_i$ is always negative). The positive entropy corresponds to the increase in randomness that occurs on mixing.

Substituting the values of $(\Delta G)_{\text{mix}}$ and $(\Delta S)_{\text{mix}}$ from Equations 11.63 and 11.65 in $(\Delta G)_{\text{mix}} = (\Delta H)_{\text{mix}} - T(\Delta S)_{\text{mix}}$, we get

$$(\Delta H)_{\text{mix}} = NRT\sum_i (x_i \ln x_i) - NRT\sum_i (x_i \ln x_i) = 0. \tag{11.66}$$

This shows that *there is no change in the enthalpy associated with the formation of an ideal mixture.* Hence, we conclude that the driving force $-(\Delta G)_{\text{mix}}(=T(\Delta S_{\text{mix}}))$ is an entropy effect.

The volume of mixing can be obtained by differentiating the free energy $(\Delta G)_{\text{mix}}$ of mixing with respect to pressure, keeping the temperature and the composition constant. Therefore,

$$(\Delta V)_{\text{mix}} = \frac{\partial}{\partial P}[(\Delta G)_{\text{mix}}]_{T,n_i}.$$

However, since $(\Delta G)_{\text{mix}}$ is independent of P,

$$\frac{\partial}{\partial P}[(\Delta G)_{\text{mix}}]_{T,n_i} = 0$$

and therefore $(\Delta V)_{\text{mix}} = 0$. So, we conclude that *there is no change in volume for an ideal mixing.*

11.12   THE GIBBS–DUHEM EQUATION

We have seen that the free energy of a mixture of n_i mol of i species is given by Equation 11.41, i.e.

$$G = \sum_i n_i \mu_i.$$

Although this equation was used for the particular case when the temperature and pressure are constant, it must be generally true because G is a thermodynamic function. The effect of variation in T and P on G will be reflected by the changes in the μ- and n-values. Therefore, we can differentiate Equation 11.41 without any restriction regarding the type of process. Hence, on differentiating, we get

$$dG = \sum_i (\mu_i \, dn_i) + \sum_i (n_i \, d\mu_i). \tag{11.67}$$

From what we have said above, it is clear that this equation is generally true; the effects of all changes of conditions will be reflected in the μ- and n-values. Now, comparing Equation 11.47 with Equation 11.67, we get

$$S \, dT - V \, dP + \sum_i (n_i \, d\mu_i) = 0. \tag{11.68}$$

In this equation the values of S and V refer to all the moles of the mixture, i.e., to $\sum_i n_i$. So, for 1 mol we shall have to divide throughout by $\sum_i n_i$. Hence, for 1 mol we have

$$S \, dT - V \, dP + \sum_i (x_i \, d\mu_i) = 0 \tag{11.69}$$

where $x_i = n_i / \sum_i n_i$ which equals the mole fraction of species i.

We have said that Equation 11.68 and therefore Equation 11.69 are applicable to any general case without any restriction regarding the type of process. In particular, if we are considering a system at constant temperature and pressure, then $dP = 0$ and $dT = 0$. In that case, Equations 11.68 and 11.69 become

$$\sum_i (n_i \, d\mu_i) = n_1 \, d\mu_1 + n_2 \, d\mu_2 + \cdots = 0 \tag{11.70}$$

and

$$\sum_i (x_i \, d\mu_i) = x_1 \, d\mu_1 + x_2 \, d\mu_2 + \cdots = 0. \tag{11.71}$$

These two equations are the general relationships for the variation in the chemical potentials of the various species present in a phase in internal equilibrium but the composition of the phase is varied at constant temperature and pressure. These equations are known as the *Gibbs–Duhem equations*.

Equation 11.70 shows that, if the composition of a mixture varies, the chemical potentials do not change independently. This can be better illustrated by considering a system of only two constituents. Then Equation 11.70 becomes

$$n_1 \, d\mu_1 + n_2 \, d\mu_2 = 0 \, (T, P \, \text{constant})$$

or

$$d\mu_2 = -\left(\frac{n_1}{n_2} \right) d\mu_1. \tag{11.72}$$

This equation shows that, if $d\mu_2$ is positive, i.e. μ_2 increases, then $d\mu_1$ must be negative and μ_1 must decrease at the same time, and vice versa. The significance of this behaviour is discussed later in connection with solutions.

11.13 THE DUHEM–MARGULES EQUATION

Equation 11.72 is the Gibbs–Duhem equation for a system of two components. Now, dividing both sides of this equation by $n_1 + n_2$ and rearranging, we get

$$x_1 \, d\mu_1 = -x_2 \, d\mu_2 \tag{11.73}$$

where x_1 and x_2 are the mole fractions of the components.

Differentiating Equation 11.42 at constant temperature and constant total pressure (not the partial pressure of any component), we get

$$d\mu_i = RT \, d(\ln P_i). \tag{11.74}$$

With the help of Equation 11.74, Equation 11.73 can be written as

$$x_1 RT \, d(\ln P_1) = -x_2 RT \, d(\ln P_2)$$

or

$$x_1 \, d(\ln P_1) = -x_2 \, d(\ln P_2) \tag{11.75}$$

where P_1 and P_2 are the partial pressures of the components with mole fractions x_1 and x_2, respectively. Dividing both sides of Equation 11.75 by dx_1 we have

$$\frac{x_1 \, d(\ln P_1)}{dx_1} = \frac{-x_2 \, d(\ln P_2)}{dx_1}$$

but $x_1 + x_2 = 1$. Therefore, $dx_1 = -dx_2$. Hence, the above equation becomes

$$x_1 \frac{d(\ln P_1)}{dx_1} = x_2 \frac{d(\ln P_2)}{2 \, dx_2} \tag{11.76}$$

or

$$\frac{d(\ln P_1)}{d(\ln x_1)} = \frac{d(\ln P_2)}{d(\ln x_2)}. \tag{11.77}$$

These two equations are called the *Duhem–Margules equations*. It should be noted that, although in the case of a gaseous mixture the assumption involved in the deduction of the Duhem–Margules equation is that the gases behave ideally, no assumption is involved as to the ideality of a liquid mixture.

Equation 11.76 was first deduced by J. Willard Gibbs (1876), and thereafter independently by P. Duhem (1886), M. Margules (1895) and R. A. Lehfeldt (1895).

The actual relationship between P_1 and P_2 cannot be derived from Equation 11.77 without an arbitrary assumption that permits integration of the equation. Margules integrated the equation as

$$P_1 = P_1^0 x_1 \exp(\alpha x_2^2)$$
$$P_2 = P_2^0 x_2 \exp(\alpha x_1^2)$$

(11.78)

where α is a constant which is the same for both components, and P_1^0 and P_2^0 are the vapour pressures of the pure components.

If P_1^0 and P_2^0 are known and P_1 (or P_2) is measured, then α can be determined from Equation 11.78, and hence P_2 (or P_1) can be determined. This shows that the two partial pressures are directly related.

11.14 LAW OF MASS ACTION

We have now come to a stage where the condition for equilibrium in a reaction taking place in a homogeneous mixture at constant temperature and pressure can be expressed in terms of chemical potentials. Let us consider a general stoichiometric reaction $B + D = M + N$. For an infinitesimal amount of reaction there will be infinitesimal amounts of decrease in B and D, say dn_B mol and dn_D mol, respectively. At the same time dn_M mol of M and dn_N mol of N will appear. Therefore, for this infinitesimal reaction we may write

$$dn_M = dn_N = -dn_B = -dn_D = dn \text{ (say).}$$

(11.79)

The change in free energy for this reaction is given by (see Equation 11.30)

$$dG = \left(\frac{\partial G}{\partial n_B}\right)_{T,P,n_D,n_M,n_N} dn_B + \left(\frac{\partial G}{\partial n_D}\right)_{T,P,n_B,n_M,n_N} dn_D$$
$$+ \left(\frac{\partial G}{\partial n_M}\right)_{T,P,n_B,n_D,n_N} dn_M + \left(\frac{\partial G}{\partial n_N}\right)_{T,P,n_B,n_D,n_M} dn_N$$

or

$$dG = \mu_B \, dn_B + \mu_D \, dn_D + \mu_M \, dn_M + \mu_N \, dn_N.$$

With the help of Equation 11.79 this equation can be written as

$$dG = (\mu_M + \mu_N - \mu_B - \mu_D)dn.$$

(11.80)

We have already seen that the condition for equilibrium for changes (in a closed system) at constant temperature and pressure is $dG = 0$. Therefore, if our reaction is at equilibrium, Equation 11.80 becomes

$$(\mu_M + \mu_N - \mu_B - \mu_D) \, dn = 0.$$

Hence,

$$\mu_M + \mu_N = \mu_B + \mu_D. \tag{11.81}$$

This is known as the *law of mass action* for the type of reaction considered. If we are considering a general equation of the type

$$bB + dD + \cdots = mM + nN + \cdots$$

then the general form of the law of mass action for any reaction becomes

$$m\mu_M + n\mu_N + \cdots = b\mu_B + d\mu_D + \cdots \tag{11.82}$$

We can deduce the same equation for a reaction at constant temperature and volume by using the condition for equilibrium, $dA = 0$, for such cases.

It should be remembered that *the law of mass action does not tell us anything about the rate of any reaction*.

11.15   EXCESS FUNCTIONS

In Section 11.11 we have derived different expressions for different thermodynamic quantities for mixing of several pure substances together. In all those cases we assumed that the mixture behaves ideally. In actual systems, the observed values of the thermodynamic quantities for mixing may differ from the ideal values. We shall define these differences as the excess functions of mixing. These functions are defined as follows for a two-component system but they can be extended to any number of components.

(i) The excess free energy is

$$\begin{aligned}(\Delta G)_{mix}^E &= (\Delta G)_{mix} - [(\Delta G)_{mix}]^{ideal} \\ &= (\Delta G)_{mix} - x_1 RT \ln x_1 - x_2 RT \ln x_2. \end{aligned} \tag{11.83}$$

(ii) The excess entropy is

$$\begin{aligned}(\Delta S)_{mix}^E &= (\Delta S)_{mix} - [(\Delta S)_{mix}]^{ideal} \\ &= (\Delta S)_{mix} + x_1 R \ln x_1 + x_2 R \ln x_2. \end{aligned} \tag{11.84}$$

(iii) The excess enthalpy is

$$(\Delta H)_{mix}^E = (\Delta H)_{mix} - [(\Delta H)_{mix}]^{ideal} = (\Delta H)_{mix} - 0 = (\Delta H)_{mix}. \tag{11.85}$$

(iv) The excess volume is

$$(\Delta V)_{mix}^E = (\Delta V)_{mix} - [(\Delta V)_{mix}]^{ideal} = (\Delta V)_{mix} - 0 = (\Delta V)_{mix}. \tag{11.86}$$

(v) The excess chemical potential is

$$(\Delta \mu_i)^E = \mu_i - \mu_i^{ideal} = \mu_i - \mu_i^\circ - RT \ln x_i. \tag{11.87}$$

It can easily be shown that the usual relationships between temperature coefficients of the free energy, enthalpy or entropy also apply, if stated for the excess functions. Hence we have

$$(\Delta S)_{mix}^E = -\frac{\partial}{\partial T}[(\Delta G)_{mix}^E] \tag{11.88}$$

$$(\Delta H)_{mix}^E = (\Delta G)_{mix}^E - T\frac{\partial}{\partial T}[(\Delta G)_{mix}^E]. \tag{11.89}$$

The excess functions written above may be considered as excess changes in property on mixing. These are sometimes simply denoted by G^E, S^E, etc., of the mixture.

Combining Equations 11.45 and 11.87 we can obtain the relationship between excess chemical potential and activity coefficient as

$$\mu_i^E = RT \ln \gamma_i. \tag{11.90}$$

From Equation 11.71 we can write for a binary mixture

$$x_1 \frac{\partial \mu_1}{\partial x_1} + x_2 \frac{\partial \mu_2}{\partial x_1} = 0$$

or

$$x_1 \left(\frac{\partial}{\partial x_1} (\mu_1^E + RT \ln x_1) \right)_{T,P} + x_2 \left(\frac{\partial}{\partial x_1} (\mu_2^E + RT \ln x_2) \right)_{T,P} = 0. \tag{11.91}$$

Since $x_1 + x_2 = 1$, the above equation may be simplified to

$$x_1 \left(\frac{\partial}{\partial x_1} (\mu_1^E + RT \ln x_1) \right) - x_2 \left(\frac{\partial}{\partial x_2} (\mu_2^E + RT \ln x_2) \right) = 0$$

or

$$x_1 \frac{\partial \mu_1^E}{\partial x_1} - x_2 \frac{\partial \mu_2^E}{\partial x_2} = 0. \tag{11.92}$$

Using Equation 11.90 the above equation can be written as

$$x_1 \frac{\partial}{\partial x_1} (\ln \gamma_1) - x_2 \frac{\partial}{\partial x_2} (\ln \gamma_2) = 0. \tag{11.93}$$

Equation 11.92 may be written in the integral form as

$$\int_{x_1=0}^{x_1} d\mu_1^E = -\int_{x_1=0}^{x_1} \frac{x_2}{x_1} d\mu_2^E. \tag{11.94}$$

This equation enables the excess chemical potential of one component to be determined from knowledge of the excess chemical potential of the other component as a function of composition. Now, integrating Equation 11.92, we get

$$\int_{x_1=0}^{x_1=1} x_1 \frac{\partial \mu_1^E}{\partial x_1} dx_1 = \int_{x_2=0}^{x_2=1} x_2 \frac{\partial \mu_2^E}{\partial x_2} dx_2 \tag{11.95}$$

or

$$[x_1 \mu_1^E]_0^1 - \int_0^1 \mu_1^E dx_1 = [x_2 \mu_2^E]_0^1 - \int_0^1 \mu_2^E dx_2. \tag{11.96}$$

Since μ_1^E is finite at $x_1 = 0$ and zero at $x_1 = 1$, and also μ_2^E is finite at $x_2 = 0$ and zero at $x_2 = 1$, Equation 11.96 becomes

$$\int_{x_1=0}^{x_1=1} \mu_1^E dx_1 = \int_{x_2=0}^{x_2=1} \mu_2^E dx_2 \tag{11.97}$$

for constant T and P. This implies that, if thermodynamic consistency prevails, the areas under the graphs of μ_1 and μ_2 between $x_1 = 0$ and $x_1 = 1$ are equal. Another way of expressing this condition is

$$\int_{x_1=0}^{x_1=1} (\mu_1^E - \mu_2^E) dx_1 = 0. \tag{11.98}$$

With the help of Equation 11.90, Equation 11.98 may be written as

$$\int_{x_1=0}^{x_1=1} (RT \ln \gamma_1 - RT \ln \gamma_2) dx_1 = 0$$

or

$$\int_{x_1=0}^{x_1=1} \ln\left(\frac{\gamma_1}{\gamma_2}\right) dx_1 = 0. \tag{11.99}$$

Within the ambit of restrictions provided by the Gibbs–Duhem equation the chemical potential (i.e. vapour pressures) of the individual components in a mixture may vary in many ways, although the commonest observed behaviour of non-electrolyte mixtures is that the sign of deviations from the ideal behaviour is always the same for both components, but this is not the only kind. From Equation 11.98 it can be seen that the deviations from the ideal behaviour of both components cannot possibly be of opposite signs over the entire composition range concerned. In such cases, $\mu_1^E - \mu_2^E$ would have the same sign over the entire composition range; therefore, Equation 11.98 cannot hold. However, an interesting point is that Equation 11.98 does not say that the two components must not deviate from ideality in opposite directions over any part of the composition range.

EXERCISES 11

11.1 State the important properties of equilibrium constants. Assuming ideal gas behaviour, deduce a mathematical relationship between K_C and K_P. Using this relationship discuss the effect of volume change of reaction on K_C and K_P.

11.2 Calculate the equilibrium constant at 298.15 K for the reaction $CO(g) + \frac{1}{2}O_2(g) = CO_2(g)$.

11.3 Calculate the equilibrium constant at 500°C for the formation of benzene from acetylene by the reaction $3C_2H_2(g) \rightarrow C_6H_6(\ell)$. Make any assumption, if necessary.

11.4 Deduce a relationship between the standard free-energy change and the activities of the substances in a reaction. Hence show that the relationship between the standard free energy change and the equilibrium constant K can be expressed as $\Delta G° = -RT \ln K$.

11.5 Using problem 11.4 above, deduce the reaction isotherm of van't Hoff for a reaction. Hence deduce the different forms of this equation connecting K_P, K_C, K_x, Q_P, Q_C and Q_x, where these symbols have their usual meanings. Do all these relationships give the same value of ΔG?

11.6 Discuss the change of sign of ΔG for $K \gtrless Q$. What does a negative value of ΔG imply?

11.7 The equilibrium constant for the reaction $\frac{1}{2}I_2(g) + \frac{1}{2}Br_2(g) = IBr(g)$ is given by the equation $(\log K)_T = 277.4/T + 0.3811$ J. Calculate $\Delta G°$ for this reaction.

11.8 At 3227°C the equilibrium constant for the reaction $CO_2(g) + H_2(g) = CO(g) + H_2O(g)$ is 8.28. Calculate $\Delta G°$ at 3227°C for this reaction. Calculate the value of ΔG for transforming at 3227°C 1 mol each of CO_2 and H_2 both held at 0.10 atm to 1 mol each of CO and H_2O both held at 2 atm.

11.9 At high temperatures the decomposition of carbon dioxide is given by the reaction $2CO_2 = 2CO + O_2$. At 1 atm pressure the percentage decomposition is 2×10^{-5} at 727°C and 1.27×10^{-2} at 1127°C. Calculate

(a) the equilibrium constants at 727 and 1127°C

(b) the standard free-energy change at 727°C

(c) the standard entropy change at 727°C, assuming that ΔH is independent of temperature.

11.10 The equilibrium constant at 444°C for the reaction $2HI(g) = H_2(g) + I_2(g)$ is given as 0.0198. Calculate the free energy of formation of HI(g) at this temperature.

11.11 Starting from the van't Hoff isotherm, deduce the van't Hoff equation. Assuming that $\Delta H°$ of this equation is independent of temperature, show that $\log K_P$ is directly proportional to $1/T$.

11.12 Show that, if ΔH° and ΔS° are constants independent of temperature, then $\ln K$ is a simple linear function of the reciprocal of temperature. Hence show that, if $\Delta C_P = 0$, ΔS° is independent of temperature.

11.13 Deduce a relationship between K_x and K_P for an ideal gas. Hence show that, when there is a change in the total number of moles of a gas in a system, the pressure dependence of K_x is given by

$$\frac{d}{dP}(\ln K_x) = \frac{\Delta n}{P} = \frac{\Delta V}{RT}$$

where Δn and ΔV are the changes in the number of moles and in the volume, respectively. Using this equation discuss the effect of change in the number of moles of a gas in a system on K_x when P increases.

11.14 Calculate the ratio of K_P to K_C at 300 K for the reaction $C_2H_6(g) = C_2H_4(g) + H_2(g)$.

11.15 Calculate the equilibrium constant for the reaction $CH_3COOH(aq) \rightarrow H^+(aq) + CH_3COO^-(aq)$ at 25°C, given that the standard state free energy change for this reaction is 27.0 kJ.

11.16 Calculate the equilibrium constant at 298.15 K for the reaction $H_2(g) + \frac{1}{2}O_2(g) = H_2O(g)$.

11.17 For the reaction $CuS(s) + H_2(g) \rightarrow Cu(s) + H_2S(g)$, calculate (a) the equilibrium constant at 298.15 K and 0.1 MPa pressure, and (b) the equilibrium constant at 798 K and 0.1 MPa pressure.

11.18 Discuss the various partial molar quantities. Show that partial molar volumes of the constituents of a mixture can be obtained from the average molar volume.

11.19 Show that the free energy of mixing of ideal gases is negative. Also show that there is no change in enthalpy associated with the formation of an ideal mixture.

11.20 Show that there is no change in volume for an ideal mixing. Deduce the general relationships for the variation of the chemical potentials of the various species present in a phase in internal equilibrium when the composition of the phase is varied at constant temperature and pressure.

11.21 Using the Gibbs–Duhem equation for a mixture of two components, show that at constant T and P, if the chemical potential of one component increases, then the chemical potential of the other component decreases and vice versa.

11.22 Starting from the Gibbs–Duhem equation, deduce the Duhem–Margules equation. Is it necessary to assume that the gases behave ideally in the mixture for deducing this equation? Can you apply the Duhem–Margules equation to a liquid mixture without making any assumption as to the ideality?

11.23 Deduce the condition for equilibrium in a reaction taking place in a homogeneous mixture at constant temperature and pressure in terms of chemical potential. Is it possible to deduce the condition for equilibrium in a reaction taking place in a homogeneous mixture at constant temperature and volume in terms of chemical potential?

11.24 Define the excess functions of mixing. Deduce an equation for a binary mixture which enables the excess chemical potential of one component to be determined from the knowledge of the excess chemical potential of the other component as a function of composition.

11.25 Derive an expression for the equilibrium constants K_P as a function of temperature for the reaction $CO(g) + H_2O(g) \rightarrow H_2(g) + CO_2(g)$, given that the heat capacity of the substances involved is $C_P = a + bT + cT^2$ J K^{-1} mol^{-1} for the temperature range 273.15–1500 K with the following values of a, b, and c.

	a	$b \times 10^{-3}$	$c \times 10^{-7}$
CO	6.420	1.665	−1.96
CO_2	6.214	10.396	−35.45
H_2	6.947	−0.200	4.808
H_2O	7.256	2.298	−2.83

11.26 For the reaction CH_3NH_2(ideal gas, 1 atm) + HCl(ideal gas, 1 atm) = CH_3NH_3Cl(s) at 298.15 K, $\Delta H^\circ = -182.506$ kJ. Calculate the equilibrium constant for this reaction at 298.15 K, given that the absolute entropies of CH_3NH_2(ideal gas, 1 atm), HCl(ideal gas, 1 atm) and CH_3NH_3Cl(s) are 243.592 J K^{-1} mol^{-1}, 186.690 J K^{-1} mol^{-1} and 138.616 J K^{-1} mol^{-1}, respectively, at 298.15 K.

11.27 For the reaction $Ag_2O(s) \rightleftharpoons 2Ag(s) + \frac{1}{2}O_2(g)$, $\Delta G^\circ = 7700 + 4.2T$ log $T - 27.80T$ J. Express the equilibrium constant K_P as a function of temperature.

11.28 Consider the reaction $CO(g) + H_2(g) \rightleftharpoons CH_2O(g)$. Given that, at 298.15 K, $\Delta G^\circ = 27.196$ kJ, $\Delta H^\circ = -5.439$ kJ, and the heat capacities are

$C_P(CH_2O(g)) = 4.490 + 13.950 \times 10^{-3}T - 3.730 \times 10^{-6}T^2$ J K^{-1} mol^{-1}
$C_P(H_2(g)) = 6.940 - 0.20 \times 10^{-3}T + 4.80 \times 10^{-7}T^2$ J K^{-1} mol^{-1}
$C_P(CO(g)) = 6.340 + 1.84 \times 10^{-3}T - 2.80 \times 10^{-7}T^2$ J K^{-1} mol^{-1}
Calculate K_P at 1000 K (a) when ΔH° is independent of temperature, and (b) when ΔH° is dependent on temperature.

11.29 It is required to pass carbon monoxide at 1 MPa and water vapour at 0.5 MPa pressure into a reaction chamber at 700°C, and to withdraw carbon dioxide and hydrogen at partial pressures of 0.15 MPa. Is it theoretically possible? The reaction occurring in the chamber is $CO(g) + H_2O(g) = CO_2(g) + H_2(g)$ with an equilibrium constant of 0.71.

11.30 For the reaction $\frac{1}{2}N_2 + \frac{1}{2}O_2 \rightarrow NO$ the following data are available at 298 K

$$S^\circ(NO) = 210.623 \text{ J K}^{-1} \text{ mol}^{-1}$$
$$S^\circ(N_2) = 191.502 \text{ J K}^{-1} \text{ mol}^{-1}$$
$$S^\circ(O_2) = 205.058 \text{ J K}^{-1} \text{ mol}^{-1}$$

The enthalpy of formation of gaseous NO is 90.374 kJ mol^{-1}. Calculate the equilibrium constant for the formation of NO at 298 K. If the sign of the free-energy change is reversed, what would be the implication of this on the equilibrium constant?

11.31 Consider problem 10.29 and the data given there. Calculate

(a) the equilibrium constant for 298 K

(b) ΔG° at 500 K, stating any assumption, and

(c) the equilibrium constant at 1000 K, stating any necessary assumptions.

11.32 If ammonia is 98% dissociated at 400°C and a total pressure of 1 MPa, calculate the free-energy change for the reaction $2NH_3(g) \rightleftharpoons N_2(g) + 3H_2(g)$.

11.33 The standard free-energy change for the vaporization of molten aluminium at 1850 K is 50.836 kJ mol^{-1}. Calculate the vapour pressure of molten aluminium at 1850 K.

11.34 Calculate the standard enthalpy change for the reaction $N_2(g) + O_2(g) = 2NO(g)$.

The following data are given.

Temperature (K)	1900	2000	2100	2200	2300	2400	2500	2600
$K_P \times 10^4$	2.31	4.08	6.86	11.0	16.9	25.1	36.0	50.3

Is the sign of $\Delta H°$ for this reaction in agreement with that predicted by Le Chatelier's principle?

11.35 Consider problem 10.32. Using the expression for ΔG_T° obtained in that problem, find an expression for $\ln K_P$ at T.

12
Fundamental Concepts of Statistical Mechanics

12.1 INTRODUCTION

The entire world that we know consists of objects which are large compared to atoms or molecules. While atoms and molecules are *microscopic* in size, the objects of our world are *macroscopic*. In fact, the world which we know of is quite complex and varied, consisting of solids, liquids and gases of the most diverse compositions. Naturally, we would like to know how the concepts of atomic theory can be used to understand the observed behaviour of macroscopic systems, how quantities describing the directly measurable properties of macroscopic systems are interrelated and how all these measurable properties can be derived from atomic behaviours. Classical equilibrium thermodynamics developed as a series of laws relating the macroscopic state variables of a system, such as pressure, volume and temperature. This was done mainly in the Nineteenth Century. Long after this, in the Twentieth Century, detailed atomic models were developed.

In classical equilibrium thermodynamics, there is no need for any assumptions about atomic structure. The laws of thermodynamics, which we studied in the earlier chapters, lead to general relations between all thermodynamic variables which, in turn, enable us to predict the value of one variable from the values of other variables. However, the laws of thermodynamics are unable to predict the actual magnitudes of individual quantities directly from an atomic model. Therefore, we intend to explore whether possibilities exist to interpret the laws of classical thermodynamics using a microscopic theory. Here lies the application of statistical mechanics. Statistical mechanics can also be used to predict the values of individual macroscopic quantities.

The object of statistical mechanics is the study of the properties of physical systems containing a very large number of particles (of the order of 10^{23}). This number is an important one, and is known as *Avogadro's constant*. In fact, this constant relates microscopic quantities to macroscopic quantities. It is, indeed, an extremely large number. It would be neither possible nor desirable to study the individual properties of each of the huge number of particles that comprise a physical system. Therefore, we shall be interested in the statistical properties of systems rather than the particles that form the system.

As we did in the earlier chapters, in the next few chapters we shall limit ourselves to the study of only those systems which are in thermodynamic equilibrium. In fact, most physical systems are either in thermodynamic equilibrium, or they tend to achieve an equilibrium state very quickly. It is interesting to note that statistical mechanics extends and supplements the science of thermodynamics in two very distinct ways. As said earlier, thermodynamics is not able to determine the values of the thermodynamic properties of a particular physical system. Such properties actually depend on the properties of the individual molecules that constitute the system. The number of such molecules being extremely large, we may apply the laws of probability to their motions and configurations. This is the domain of statistical mechanics. The other application of the science of statistical mechanics extends much deeper. Probabilistic methods applied to the study of systems consisting of large numbers of molecules can throw much light on the nature of the fundamental laws of thermodynamics.

It should be noted that the above discussion does not imply that the movement from statistical mechanics to thermodynamics is a one-way route. It is also possible, via a suitable and natural hypothesis, to move from thermodynamics to statistical mechanics. The laws and procedures of statistical mechanics are highly important in the study of certain processes such as chemical reactions (theory of rate processes), diffusion, etc. However, the principal aim of statistical mechanics is to relate the properties of bulk matter with those of individual atoms and molecules.

12.2 SUMMARY OF DEFINITIONS

The following terms are commonly used in statistical mechanics. Therefore, it is convenient to know their definition.

Configuration. Each way in which the molecules of a substance (particularly those of a gas inside a container) can be distributed is called a configuration. There are two types of distribution in statistical mechanics.

(i) *Ordered distribution.* In such a case the number of possible configurations of the molecules is the minimum.

(ii) *Disordered or random distribution.* In this case the number of possible configurations is the maximum.

Microstate of a system. The microstate of a system at any time is given by specifying the maximum possible information about the system (say, a gas) molecules at this time, e.g. the position and velocity of each molecule. It is a particular quantum state of a system.

Macrostate of a system. A complete description of a system in terms of parameters which are macroscopically measurable.

Equilibrium state of a system. A system of many particles, such as a gas, whose macroscopic (i.e. large scale) state does not tend to change in time is said to be in equilibrium.

Accessible microstate. Any microstate of a system which is its quantum state in which the system can be found without breaking any conditions imposed by the macroscopic information available about the system.

Degeneracy. This is the number of accessible microstates.

Nondegenerate. A system is said to be nondegenerate if no two particles in the system have the same energy.

Relaxation time. If a system is far removed from equilibrium, it tends to attain equilibrium over a period of time. The approximate time required for the system to revert to the equilibrium situation is called relaxation time.

Phase space. To describe both the position and the state of motion of a point particle in space, it is customary to set up a six-dimensional space, called the phase space, in which the six co-ordinates x, y, z, p_x, p_y, p_z are marked out along six mutually perpendicular axes (p_x, p_y and p_z are momenta).

Reduced mass of a diatomic molecule. If a diatomic molecule is made up of two atoms a and b of mass m_a and m_b, respectively, then the reduced mass of the molecule is given by $m_a m_b / (m_a + m_b)$.

Quantum numbers. For a particle in a space defined by the co-ordinate system (x, y, z), the quantum numbers are given by (n_x, n_y, n_z) which are positive, and for each different set of values of (n_x, n_y, n_z) we say that the particle is in a different quantum state.

There are three types of quantum number, *viz.* *translational*, *rotational* and *vibrational* quantum numbers associated, respectively, with translation, rotation and vibration of the particle.

Canonical ensemble. An ensemble of systems, all of which are in contact with a heat reservoir or other system of known temperature can exchange energy with the reservoir or the other system, so that the energy of the ensemble is no longer constant (all in thermal equilibrium with reservoirs at the same temperature). An ensemble of such systems is called a canonical ensemble. Although the system exchanges energy with the reservoir, there is no exchange of particles (the number of particles remains fixed). This is a *closed system*.

Microcanonical ensemble. An ensemble of identical isolated systems is called a microcanonical ensemble. In such an ensemble the energy is constant and all the microstates with this energy are equally probable. The prefix 'micro' indicates the fact that, because the energies of all the systems are equal and constant, this ensemble is actually a smaller part of a canonical ensemble.

Fermions. Particles which obey Fermi–Dirac statistics are called fermions. These particles have half-integer spins. Examples of fermions are electron, proton, neutron, etc.

Bosons. These particles obey Bose–Einstein statistics and have spins of integral numbers. Examples are photons, π and K mesons, etc.

Grand canonical ensemble. This is a collection of systems which are all in thermal equilibrium with a heat reservoir or other systems but exchange particles with one another. Such systems are called *open systems*.

Energy level. The collection of all the states with the same energy is often called an energy level. It is composed of all states having the same energy.

Temperature parameter. The parameter $\beta = 1/kT$ is called the temperature parameter of a system. It is an extremely useful abbreviation to use in place of $1/kT$, where k is the Boltzmann constant and T is the temperature of the system.

Density of states. It is the number of microstates (i.e. the number of independent quantum states) of an N particle system per unit energy range (see Equation 13.91). Macroscopically, the density of states can be treated as a continuous function of the internal energy of the system.

12.3 MICROSTATES, MACROSTATES AND STATISTICAL ENSEMBLES

The *macrostate* of a system is a complete description of the system itself in terms of macroscopically measurable parameters. On the other hand, the *microstate* of a system is a particular quantum state of that system. At any time, it is given by specifying the maximum possible information about the system (say, a gas) molecules at this time, e.g. the position and velocity of each molecule.

We shall call the *accessible microstates* of a system those of its quantum states in which the system can be found without contradicting any conditions imposed by the information available about the system. It follows then that an isolated system (which is isolated in the sense that it does not interact with any other systems so as to result in interchange of energy with them) is not in equilibrium if it is not found with equal probability in each one of its accessible microstates. Such an isolated system tends to change in time until it eventually reaches the equilibrium state. Upon reaching the equilibrium situation, the system is found with equal probability in each one of its accessible microstates.

The number of accessible microstates is often called the *degeneracy* or *statistical weight*. Any system which is not isolated can be considered as a part of a bigger system which is isolated.

In principle, a precise knowledge of the particular macroscopic state, in which a system of particles is found at any one time, will permit us to use the laws of mechanics to compute in the greatest possible detail all of the properties of the system at any given time. Generally, such precise microscopic knowledge about a macroscopic system is not available. Besides, we are not interested in such detailed description. Therefore, we shall discuss the system in terms of probability, and instead of considering the single macroscopic system of interest we consider a system or ensemble consisting of a very large number of such macroscopic systems, all of which satisfy the same conditions as those which are satisfied by the system under consideration. However, a description of the microstate of a system of many particles is based entirely on the specification of macroscopically measurable quantities alone; it does not provide much information about the particles in the system. If the system is in a given macrostate, the number of quantum states accessible to the system is very large, because the number of particles in the system is very large.

Consider an ensemble of N particles in a cubical box. The energy ε_i of the ith state of the N-particle system is a function of both the volume V of the box and the number N of the particles inside the box. ε_i is the sum of the total kinetic and potential energies of all the N particles in the system. In some special cases, if the interactions between the particles are small, as a first approximation the particles can be treated as independent particles. This is true of an ideal gas. If there is no interaction between the particles, the total energy of the N-particle system is given by

$$\varepsilon_i(N, V) = E_1 + E_2 + E_3 + \cdots + E_N = \sum_{i=1}^{N} E_i \qquad (12.1)$$

where E_i is the energy of the ith independent particle.

One could not distinguish between non-localized identical particles, such as the atoms of an ideal monatomic gas inside a box. Later, we shall discuss the implications of the distinguishable and indistinguishable particles. Particles which obey the Fermi–Dirac statistics, i.e. fermions, have half-integral spin $\left(\frac{1}{2}, \frac{3}{2}, \ldots\right)$. On the other hand, particles which obey the Bose–Einstein statistics, i.e. bosons, have integral spin $(0, 1, \ldots)$. The wave function of an N-particle system of identical fermions is antisymmetric; the wave function changes sign if the co-ordinates of any two fermions are interchanged. The consequence of this is that only one of the fermions in the N-particle system can be in any particular single particle quantum state with a given set of single particle quantum numbers. However, it is not possible to say which of the fermions is in which particular single particle quantum state. This leads to the Pauli exclusion principle. Contrary to the fermion, the wave function of an N-particle system of identical bosons is symmetric, i.e. if the co-ordinates of any two bosons are interchanged, the wave function of the N-particle system remains unchanged. In this case, there is no limit to the number of bosons that can be in a particular quantum state with a given set of single particle quantum numbers. Here also, one cannot say which of the bosons is in which particular single particle state.

If U units of energy are distributed among N distinguishable particles of zero spin, the total number of accessible microstates Ω is given by

$$\Omega = (N + U - 1)!/(N - 1)!U! \qquad (12.2)$$

We shall verify this equation with an example.

Example 12.1

Three units of energy have been distributed among 3 distinguishable particles of zero spin. Calculate the total number of accessible microstates in the system.

Solution

Here, the number of particles $N = 3$ and energy $U = 3$. The total number of accessible microstates Ω can be calculated using Equation 12.2. Hence,

$$\Omega = (3 + 3 - 1)!/(3 - 1)!(3!) = 10$$

Let us solve this problem analytically and see whether we get a value of Ω which is consistent with the value obtained above.

Let us denote the three distinguishable particles as A, B and C. We shall consider the different ways in which 3 units of energy can be distributed among A, B and C. One possible configuration is one in which one of the particles has 3 energy units and the other two have none (zero energy). In this configuration, either A, B or C could have the 3 units of energy. Thus, there are three microstates in the $3;0;0$ configuration; these can be written as $3,0,0;0,$ $3,0$ and $0,0,3$, where the first, second and third numbers refer to the energies of A, B and C, respectively. Consistent with the condition that we have only 3 units of energy, there are two other possible configurations, *viz.* $2;1;0$ and $1;1;1$. The different microstates associated with these configurations are:

$$2;1;0 \quad 2,1,0 \quad 2,0,1 \quad 1,2,0$$
$$1,0,2 \quad 0,2,1 \quad 0,1,2$$

$$1;1;1 \quad 1,1,1$$

Thus, we find that there are altogether 10 accessible microstates, which is in agreement with the result obtained above using Equation 12.2.

It can easily be shown that the number of accessible microstates Ω increases rapidly as the energy (U) of the system is increased. If the volume of a system is changed, the energies of the particles comprising the system also change. The energy eigenvalues for a particle of mass m contained in a cubical box of volume V and side L are given by

$$\varepsilon = (h^2/2mV^{2/3})(n_x^2 + n_y^2 + n_z^2) \tag{12.3}$$

where n_x, n_y and n_z are the quantum numbers which can have the positive integer values $1, 2, 3, \ldots$. According to quantum mechanics, a measurement of the energy of the particle will give one of the energy eigenvalues given by Equation 12.3. In the ground state, which is the lowest energy value, the quantum numbers are $n_x = n_y = n_z = 1$, so the energy of this state is

$$\varepsilon_{1,1,1} = (h^2/2mV^{2/3})(1 + 1 + 1) = 3h^2/2mV^{2/3}.$$

According to Equation 12.3, for a particle in a cubical box the energies are inversely proportional to $V^{2/3}$. As the volume is increased, the energies are decreased. In Example 12.1 we have calculated the number of accessible microstates $\Omega = 10$ for $U = 3$ distributed among three distinguishable particles of zero spin. To illustrate that Ω depends on N, we use $U = 3$ and $N = 4$ in the same example. A simple calculation shows that increasing N from 3 to 4 increases Ω from 10 to 20. This shows that, in general, the number of accessible microstates (i.e. the degeneracy or statistical weight of the macrostate) is a function of U, N and V.

12.4 EQUILIBRIUM AND FLUCTUATIONS

Many of the chemical reactions that we come across are known to occur, to some extent, in both directions, i.e. reactants $\rightleftharpoons$ products. In such systems it is possible to observe experimentally the conversion of reactants to products and the reverse process. Such systems attain a state of dynamic equilibrium in which there is no cessation of reaction, but the system undergoes no further alteration unless there is a change in conditions.

However, the situation is quite different in a system of many particles such as a gas of identical molecules. In such a gaseous system two situations may occur. (i) if the total number of molecules per unit volume is small, the inter-molecular distances are large, i.e. the molecules are widely separated, and consequently their mutual interaction is small, and (ii) if the total number of molecules per unit volume is high, the inter-molecular distances are small and their mutual interaction is high. The former is said to be a dilute gas, and it is ideal if it is so dilute that the molecular interaction is almost negligible. A situation of nearly negligible molecular interaction occurs when the total potential energy of interaction is negligible compared with their total kinetic energy. However, the total potential energy should be sufficiently large for the molecules to be able to interact and exchange energy with one another.

In an ideal gas, each molecule can move like a free particle, uninfluenced by the surrounding molecules or the walls of the container. An interaction with another molecule or a collision with the container walls occurs only rarely when it comes sufficiently near to the other molecules or the walls. The average inter-molecular distance is comparatively larger than the average de Broglie wavelength of a molecule, so that the quantum-mechanical effects are of very little importance. Therefore, it is possible to treat the molecules as distinguishable particles moving along classical paths. The validity of this classical approximation will become apparent when we consider the Maxwell Velocity Distribution.

With this background information, let us now consider an ideal gas with N molecules in a container, as shown in Figure 12.1. Imagine that the container is divided into two equal parts by an imaginary partition. We also assume that the entire system is isolated and left undisturbed for a long time. Each molecule inside the container moves strictly according to the laws of motion. So, any given molecule moves along a straight line until it collides with another molecule or the walls of the container. Once a collision occurs, it will then move along some other straight line until another collision occurs. The process continues for each of the molecules in the system. It is quite easily comprehensible that, unless N is very small,

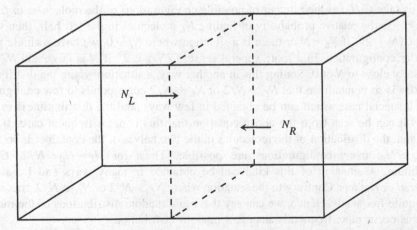

Figure 12.1   A container having an ideal gas of N molecules which are in random motion colliding with one another and the walls

N molecules moving randomly in the container and colliding with the walls and one another present a complex and chaotic situation.

Next we consider the positions of the molecules and their distribution within the container. Let there be N_L molecules in the left half of the container and N_R number of molecules in the right half. Thus, $N_L + N_R = N$. Approximately, $N/2$ molecules are in each half of the container, i.e. $N_L \simeq N_R \simeq N/2$. During their motion, colliding with one another or the walls, some of the molecules leave the left half of the container, while others enter it from the right half. In the process, N_L fluctuates constantly. However, these fluctuations are generally not large; therefore, N_L is not much different from $N/2$. Another situation may also occur. All of the N molecules may stay in the left half, in which case $N_L = N$ and $N_R = 0$. Let us consider this situation in detail.

There are many ways in which the molecules can be distributed in the two halves of the container. Let us call each way in which the molecules can be distributed in each half a *configuration*. If the total volume of all the N molecules is very much smaller than the total volume of the container, it is reasonable to assure that the possibility of finding a particular molecule in any one half of the container is not affected by the presence of the other molecules. Thus, a single molecule could be either in the left half or in the right half. Since the volumes of both halves are equal in size and are otherwise equivalent, there is an equal possibility that the molecule can be found in either half. For two molecules, each one can be found in either half, i.e. the total number of configurations is $2 \times 2 = 4 = 2^2$. This is because for each possible configuration of one molecule there are two possible configurations of the other. Similarly, for three molecules the total number of their possible configurations will be $2 \times 2 \times 2 = 8 = 2^3$, because for each of the 2^2 possible configurations of the first two molecules there are two possible configurations of the third one. Thus, by generalizing this result we can say that for N molecules the total number of possible configurations will be 2^N.

However, of the 2^N possible configurations of the N molecules there are two special cases: (i) all N of them could be in the left half, or (ii) all N of them could be in the right half. Now, if we take $\mathscr{P}_N$ as the relative *probability* of finding all the N molecules in the left half, then $\mathscr{P}_N = 1/2^N$. Similarly, if $\mathscr{P}_0$ denotes the probability of finding all the N molecules in the right half (i.e. no molecule in the left half), then $\mathscr{P}_0 = 1/2^N$.

We can now consider a situation in which N_L molecules are in the left half of the container. Let us take $C(N_L)$ as the number of possible configurations of the molecules in this case, and if $\mathscr{P}_{N_L}$ is the relative probability of finding N_L molecules in the left half, then we have $\mathscr{P}_{N_L} = C(N_L)/2^N$. If $N_L = N$ or there is a situation where $N_L = 0$, we have a single possible molecular configuration. In a more general sense, $C(N_L) \ll 2^N$ if N is large and N_L is even moderately close to N or 0. Stating this in another way, a situation where the distribution of molecules is so nonuniform that $N_L \gg N/2$ or $N_L \ll N/2$ corresponds to few configurations. This is a special case which can be obtained in few ways, and the distribution is said to be *orderly*. It can be seen from the above equation that this is not a frequent case. If, on the other hand, the distribution of the molecules in the two halves of the container is so uniform that $N_L \simeq N_R$, many configurations are possible. Thus, for $N_L = N_R = N/2$, $C(N_L)$ is a maximum. A situation of this kind can be obtained in many ways and is said to be *disordered* or *random*. Contrary to the situation where $N_L \gg N/2$ or $N_L \ll N/2$, this situation occurs quite frequently. Hence, we can say that more random distributions of the molecules in the gas occur more frequently than less random distributions.

From the above discussion, our general conclusion is that if the total number of particles is large, fluctuations corresponding to a nonuniform distribution of the molecules very rarely occur. We summarize by saying that the number N_L of molecules in one half of the container fluctuates in time about the value $N/2$ (a constant) which occurs most frequently, and the frequency of occurrence of a particular value of N_L fast decreases as the difference $N_L - N/2$ becomes greater. If N is large, only values of N_L with $N_L - N/2 \ll N$ occur significantly frequently. However, positive and negative values of $N_L - N/2$ occur equally frequently.

In the equilibrium condition, it is generally expected that N_L is always close to $N/2$. However, it is possible to have values of N_L far from $N/2$, but such values are rare. If we watch the gas for a long time, we might be able to see at some instant of time t a value of N_L which is different from $N/2$. If we do get a large spontaneous fluctuation of $N_L - N/2$, for example N_L at some particular time t takes a value N_1 where $N_1 \gg N/2$, then we can say that to the extent that $N_1 - N_2$ is big, the value N_1 corresponds to a highly nonuniform distribution of the molecules. This occurs very rarely in equilibrium. As time passes, N_L must decrease with accompanying small fluctuations until it goes back to the equilibrium state. Upon attaining the equilibrium situation it will no longer tend to change other than fluctuate about the constant value $N/2$. Thus, we see that there is a time lapse for the decay of this fluctuation. The time required for the large fluctuation to revert to the usual equilibrium situation is called the *relaxation* time. As we said earlier, if N_L at some particular time t assumes a value N_1, it corresponds to a very nonuniform distribution of the molecules. In this state, the molecules would have to travel in a very special manner to maintain the non-uniformity. Thus, the continuous motion of the molecules almost always mixes them up so thoroughly that all the molecules are distributed over the whole container in the most uniform or random way. This argument is equally applicable whether the fluctuation $N_L - N/2$ is positive or negative. If it is positive, the value of N_1 corresponds to the maximum of a fluctuation in N_L. On the other hand, if it is negative, it will correspond to the minimum of a fluctuation in N_L. It is interesting to note that this statement is equally valid whether the change in time is in the forward or backward direction.

The gas which we have considered (or any other gas) can be described in greatest detail by specifying its *microstate* (e.g. the position and velocity of every molecule) at any time. As the molecules randomly move, the microstate of the gas changes in a very complicated way. However, from macroscopic or large-scale point of view, we are more interested in a much less detailed description of the gas than the behaviour of individual molecules. The macrostate of the gas at any time may be sufficiently described by specifying the number of molecules in any part of the container at that time. The macrostate of the isolated gas which has been left undisturbed for a long time does not tend to change in time. A system of many particles, such as a gas, whose macroscopic state does not tend to change in time is said to be in *equilibrium*. If a system is not in equilibrium, it fluctuates about the average value $N/2$.

Before we finish this section, let us consider the probability equation $\mathscr{P}_{N_L} = C(N_L)/2^N$ again. We already know that 2^N represents the total number of microstates. The probability $\mathscr{P}_{N_L}$ that, out of the total N particles, N_L number of particles are in the left half of the container is given by the binomial distribution (see Appendix 5) as follows.

$$\mathscr{P}_{N_L} = \frac{N!}{N_L!(N-N_L)!}(p)^{N_L}(1-p)^{(N-N_L)} \tag{12.4}$$

where p is the probability that any particular one of the molecules is in the left half. It is assumed that $p = \frac{1}{2}$. Then, $1 - p = \frac{1}{2}$ as well. As there are a total of 2^N molecules, it follows that the number of microstates $C(N_L)$, in which N_L molecules are in the left half, is given by

$$
\begin{aligned}
C(N_L) &= (2^N)(\mathscr{P}_{N_L}) \\
&= (2^N)\left[\frac{N!}{N_L!(N - N_L)!}\right]\left(\frac{1}{2}\right)^{N_L}\left(\frac{1}{2}\right)^{(N-N_L)} \\
&= \frac{N!}{N_L!(N - N_L)!}.
\end{aligned}
\tag{12.5}
$$

For the purposes of discussion it may be assumed that each particle in a system changes its state every 10^{-12} seconds, i.e. 10^{12} times per second. Each change of state of *one* of the N particles in the system changes the microstate of the entire N-particle system. For a system of 6.02×10^{23} particles there would be $6.02 \times 10^{23} \times 10^{12} = 6.02 \times 10^{35}$ changes of microstates per second.

Generally, systems which are not in equilibrium are much more difficult to treat than those in equilibrium. In the case of non-equilibrium systems, it is necessary to deal with processes that change in time. Naturally, we want to know how quickly or slowly they change. The answers to these questions demand a detailed analysis of how molecules interact with each other and this type of analysis could be quite complicated. Consider the chemical reaction $2AB_2 \rightleftharpoons 2AB + B_2$ where AB_2, AB and B_2 are all gases. From the given equation we see that both forward and backward reactions occur. If we want to know how long would it take to achieve the equilibrium concentration of AB, we would like to calculate the rate at which the reaction proceeds from left to right. Consider another case where two bodies 1 and 2, which are at temperatures T_1 and T_2, respectively, are connected by a rod. Clearly, the situation here is not one of equilibrium, because T_1 and T_2 are different, so heat will flow from one body to another through the connecting rod. The question is: how long does it take for a given quantity of heat to flow from one to the other? This depends on the thermal conductivity of the rod. If the rod has a high thermal conductivity, the heat flow will occur more readily than when its thermal conductivity is low. To obtain the answer to the above question we need to calculate the thermal conductivity of the rod.

The above examples give us some indication of the range of macroscopic natural phenomena which we may wish to treat quantitatively on the basis of microscopic considerations. From now on, we shall turn the above discussion into a more systematic quantitative discussion of the properties of macroscopic systems to answer the most basic questions, such as those mentioned above.

12.4.1 Some properties of equilibrium

From our discussion in the previous sections we can say that the equilibrium situation of a macroscopic system is quite simple. The reasons for this are as follows.

(a) Fluctuations are ever-present in a system. Except for the presence of fluctuations, the macroscopic state of a system in equilibrium is independent of time. In general, the macrostate of a system can be described by those parameters which characterize the properties of the system on a large scale. Such parameters are called *macroscopic parameters*. In the equilibrium condition, the average values of all macroscopic parameters of a system remain constant in time; however, the parameters themselves may fluctuate about their average value, and if they do, such fluctuations are quite small. In contrast to an equilibrium system, the macroscopic parameters of nonequilibrium systems tend to change in time. Therefore, the equilibrium situation of a system is easier to treat than a nonequilibrium situation.

(b) We have seen earlier that, except for fluctuations, under the specified conditions the microstate of a system in equilibrium is the most random. This has many implications: (i) the equilibrium macrostate of a system can be completely defined by very few macroscopic parameters, and (ii) the equilibrium macrostate of a system is independent of its past. Let us illustrate these with examples.

(i) Consider an isolated gas of N identical molecules confined in a container of volume V. We assume that the total energy of all the molecules is U, which is constant. If the gas is in equilibrium, then: it is in its most random state; all the molecules must be uniformly distributed throughout the volume V; and the molecules, on the average, equally share the total energy U. This information enables us to conclude that the average number (N_s) of molecules in any part (V_s) of the volume V of the container is given by $N_s = NV_s/V$, and the average energy $(\bar{U})$ per molecule is $\bar{U} = U/N$.

If the gas were not in equilibrium, the situation would be much more complicated, because in that case the distribution of molecules would generally be very nonuniform, and the knowledge of the total number of molecules (N) in the container would be inadequate to determine N_s of molecules in any given subvolume (V_s) of the container.

(ii) Consider the isolated gas of N molecules in a container. We may imagine that these molecules might originally have been confined by an imaginary partition to one half of the container or to one-quarter or one-eighth of the container. We also assume that the total energy of all the molecules remains the same in each case. When the partition is removed, the equilibrium is attained and the macrostate of the gas is the same in each case. The final macrostate corresponds merely to the uniform distribution of all the molecules throughout the container.

12.4.2　Irreversibility

Consider again an isolated gas of a large number N of molecules in a box. If the fluctuations occurring in this gas in equilibrium are such that the difference $N_L - N/2$ is very appreciable, we call it a *large fluctuation*. Such a situation can occur in two different ways.

Generally, in the gas in equilibrium, N_L is always close to $N/2$. However, although it is rare, N_L values much bigger than N can occur. Let us suppose that a large spontaneous fluctuation has occurred in the gas. We explain the situation by saying that N_L at some particular time t_1 takes a value N_1 which is much bigger than $N/2$ (see Figure 12.2). We wish to examine the probable behaviour of N_L as time passes. As we discussed in Section 12.4, for $|N_1 - N_2|$ very large, the particular value of N_1 corresponds to a highly nonuniform distribution of the molecules in the two halves of the box. Such a situation occurs very seldom in equilibrium. As shown in Figure 12.2, as time proceeds N_L must decrease (with accompanying small fluctuations) until it goes back to the usual equilibrium situation where it fluctuates close to $N/2$. The approximate time required for the large fluctuation to revert back to the equilibrium situation, i.e. $N_L \simeq N/2$, is called the *relaxation time*.

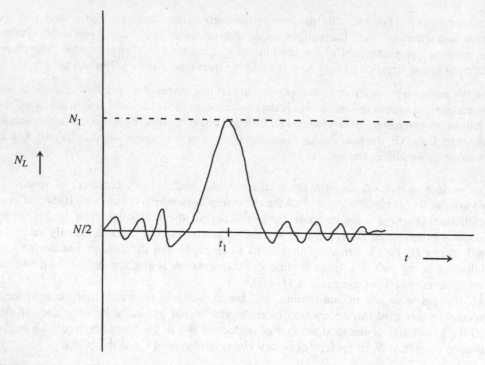

Figure 12.2   An illustration where the number N_L of molecules in one half of a box shows large fluctuations about its equilibrium value $N/2$

We can conclude that if N_L takes a value N_1 which is appreciably different from its equilibrium value $N/2$, then N_L will always tend to change in such a direction that it approaches the equilibrium value $N/2$. Since a value of N_L far from $N/2$ corresponds to a very nonuniform distribution of the molecules, they have to move in a very special way to retain this nonuniformity. However, the ceaseless motion of the molecules will most certainly result in mixing them up thoroughly, thus distributing them throughout the box in the most uniform (or random) way. This situation is equally applicable for positive or negative large fluctuation. If the fluctuation is positive, the value of N_1 will almost always correspond to the maximum of a fluctuation in N_L. On the other hand, if it is negative, it will almost always correspond to the minimum of a fluctuation in N_L.

The above statements are equally valid whether the change in time is in the future (forward) or the past (backward). If N_1 corresponds to a maximum fluctuation, then N_L must decrease for both $t > t_1$ and $t < t_1$, i.e. in either direction of t_1. A large fluctuation occurs very rarely, so it would almost never be observed in reality. Most macroscopic systems with which we deal are not in equilibrium because these have not remained isolated and undisturbed for very long periods of time. Nonrandom situations are quite common because of interactions that affected the system in the recent past, and it is quite easy to effect a nonrandom situation of a system through external intervention.

From what we have said above we may conclude that an isolated system in an appreciably nonrandom situation will change over time to approach in the end its most random situation, which is its equilibrium situation. This implies that an isolated macroscopic system tends to

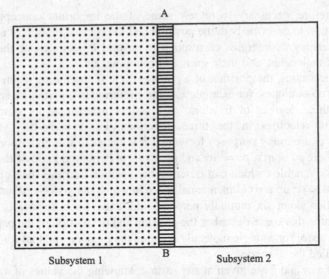

Subsystem 1   B   Subsystem 2

Figure 12.3 Example of an irreversible process in a composite system

change in time from a less random to a more random situation. An irreversible process is one which would almost never occur in the reverse direction. Irreversibility does not imply that it is impossible to force a process in the reverse direction, but only indicates that such a reversal cannot be achieved simply by changing parameters by infinitesimal amounts.

As an example of an irreversible process, let us consider two subsystems of equal volumes V, as shown in Figure 12.3, where the subsystem 1 consists of a mole of ideal gas molecules and the subsystem 2 is a vacuum, separated by the partition AB. If the partition is removed, the gas will expand to fill the entire volume 2V and the process is irreversible, because once the partition is removed it will never happen that all the gas molecules will go back to subsystem 1 in the composite system. It follows from classical thermodynamics or statistical mechanics that, in the above example, the increase in the entropy per mole is

$$\Delta S = R \ln 2 \tag{12.6}$$

The direction of the irreversible process is determined by the fact that the total entropy of any isolated system increases in an irreversible process. By using the general form of Clausius' theorem, it can be shown that for the general case of a system interacting with its surroundings, we have

$$(\Delta S_{system} + \Delta S_{surroundings}) \geq 0. \tag{12.7}$$

Note that the equality holds for a reversible change.

12.4.3 The phase spaces for a single molecule and an assembly of molecules

A knowledge of the properties of individual atoms and molecules is a prerequisite to understand statistical mechanics, simply because one of the *prima facie* aims of the subject is to correlate the properties of atoms and molecules with those of bulk matter. Bearing this in mind, it is, therefore, necessary to review some of the important concepts of statistical mechanics, e.g. how to describe both the position and the state of motion of a particle inside an atom, what energy or energies correspond to several trajectories of the particles, the energy levels of molecules, and their spins and rotations, etc.

In classical mechanics, the position of a point particle in three dimensions is determined by a set of three co-ordinates, for example x, y and z referred to a particular set of axes. Such a particle has three degrees of freedom. In classical mechanics, its state of motion is described by its velocities in the three directions, i.e. $\dot{x}(= \mathrm{d}x/\mathrm{d}t)$, $\dot{y}(= \mathrm{d}y/\mathrm{d}t)$ and $\dot{z}(= \mathrm{d}z/\mathrm{d}t)$. There are many purposes for which it is better to use corresponding momenta p_x, p_y and p_z, where $p_x = m\dot{x}$, $p_y = m\dot{y}$ and $p_z = m\dot{z}$, m being the mass of the particle. Now that there are six variables which can describe the position and the state of motion of the particle, we need to set up a six-dimensional space in which our six co-ordinates x, y, z, p_x, p_y and p_z are labelled along six mutually perpendicular axes. Such a space is called a *phase space* and a point in this space describes the position and the motion of the particle at a given time. The phase space for a single molecule is called the *μ-space* and that for an assembly of molecules is called the *γ-space*.

The values of $\dot{x}$, $\dot{y}$ and $\dot{z}$ are given at any instant, knowing the values of x, y and z at that instant. If we know the force acting upon the particle as a function of x, y and z, then the values of p_x, p_y and p_z are also known. Therefore, it is possible to know both the position of the point and the rate of change of this position in the six-dimensional phase space. Hence, every new position after an instant is known and so is its entire trajectory. This idea of phase space can be extended to an assembly of molecules. Obviously, if an assembly contains, say, N_1 molecules with n_1 atoms, N_2 molecules with n_2 atoms and so on, then the total number of rectangular co-ordinates in the phase space will be $6N_1n_1 + 6N_2n_2 + 6N_3n_3 + \cdots$.

The above discussion is not restricted to Cartesian co-ordinates. In fact, we can use polar or cylindrical co-ordinates, or any suitably defined set of co-ordinates. Let us use a general set of co-ordinates r_1, r_2 and r_3 so that the corresponding velocities are $\dot{r}_1$, $\dot{r}_2$ and $\dot{r}_3$. Then, the kinetic energy of the system, ε_k, can be expressed as a function of these co-ordinates and the corresponding velocities. It will always remain quadratic in $\dot{r}_i$. If we assume that $r_i = f(x, y, z)$, then

$$\dot{r}_i = \frac{\partial r_i}{\partial x}\frac{\mathrm{d}x}{\mathrm{d}t} + \frac{\partial r_i}{\partial y}\frac{\mathrm{d}y}{\mathrm{d}t} + \frac{\partial r_i}{\partial z}\frac{\mathrm{d}z}{\mathrm{d}t} \tag{12.8}$$

Here, all the partial derivatives are functions of x, y and z, or r_1, r_2 and r_3 (by inversion of the equation $r_i = f(x, y, z)$). The values of $\dot{x}$, $\dot{y}$ and $\dot{z}$ in Equation 12.8 may be obtained in terms of $\dot{r}_1$, $\dot{r}_2$ and $\dot{r}_3$; these will be linear functions of the $\dot{r}_i$ with coefficients as functions of x, y and z, or of r_1, r_2 and r_3.

The original kinetic energy equation of a particle of mass m, as given in Equation 12.9, will then be of the form of Equation 12.10.

$$\varepsilon_k = \frac{1}{2}m(\dot{x}^2 + \dot{y}^2 + \dot{z}^2) \tag{12.9}$$

$$\varepsilon_k = \frac{1}{2}\sum \dot{r}_i\left(\frac{\partial \varepsilon_k}{\partial \dot{r}_i}\right) \tag{12.10}$$

where $i = 1$, 2 and 3 and all the r_i are constant.

The transformation of Equation 12.9 to Equation 12.10 occurred through the introduction of the newly found expressions for $\dot{x}$, $\dot{y}$ and $\dot{z}$ which gives a form which is quadratic in $\dot{r}_1$, $\dot{r}_2$ and $\dot{r}_3$ with terms such as $\dot{r}_1\dot{r}_2$, $\dot{r}_2\dot{r}_3$, etc., and coefficients which are functions of r_1, r_2 and r_3. The quantity $\partial\varepsilon_k/\partial\dot{r}_i$ is the generalized momentum, and if we write this as p_i Equation 12.10 becomes

$$\varepsilon_k = \frac{1}{2}\sum_i p_i\dot{r}_i. \tag{12.11}$$

This a general form of the kinetic energy which is also valid for Cartesian co-ordinates. Clearly, we can conclude from this equation that a six-dimensional volume in phase space has the dimensions of (time × energy)^3 which hold irrespective of the type of co-ordinates used.

It is interesting to note that upon change of variable any given six-dimensional volume becomes a six-dimensional volume of equal size in the new phase space, and a trajectory of motion of a given particle in one phase space becomes a corresponding trajectory in the new phase space.

We conclude this discussion by generalizing what we have said above. Everything we have said about transformations of co-ordinates of one particle is also applicable to many atoms. In that case, $p_i = \partial\varepsilon_k/\partial\dot{r}_i$ can be used to determine the generalized momentum, and in general the new co-ordinates may depend on the co-ordinates of all the atoms in the assembly, i.e. $r_i = r_i(x_1, y_1, z_1, \ldots, x_n, y_n, z_n)$. Therefore, any molecule which has n atoms requires $3n$ co-ordinates to determine its position; it has $3n$ degrees of freedom and its motion is determined by $3n$ velocities or momenta. If the positions of all the atoms are known, then the forces acting between the atoms can be determined. Then, we can set up a $6n$-dimensional phase space with the help of $3n$ co-ordinates and $3n$ momenta as rectangular co-ordinates of the phase space.

From the above discussion we can see that the motion of a particle or a system of particles can be described in classical mechanics by a trajectory in the phase space. Such trajectories are arbitrarily densely distributed. In the next section we discuss how only certain trajectories are possible, so that they are not densely spaced, and how the possible trajectories correspond to certain allowed energy levels. This is an effect of quantization.

12.4.4 Spacing of energy levels of molecules: quantization

In the previous section we noted that a six-dimensional volume in the phase space has the dimensions of (energy × time)^3 and these dimensions hold irrespective of the co-ordinate system used. Therefore, if the phase space has n co-ordinates and n momentum co-ordinates, the dimensionality is (energy × time)^n. Then the density of the trajectories of a particle or a system of particles is such that each trajectory has a six-dimensional volume which is equal to h^n where h is Planck's constant.

There are situations when this information is sufficient, but sometimes we need to know the energies corresponding to several trajectories. The uncertainty principle states that it is not possible simultaneously to ascertain precisely both the position and the momentum of a particle or a system of particles. Therefore, information about a trajectory in phase space does not allow us to follow the motion of a system, because this would then mean that both position and momentum are known. If ∂r is the uncertainty about our knowledge of a co-ordinate being used and the corresponding uncertainty in the momentum is ∂p, then according to the uncertainty principle

$$\partial r \partial p \gtrsim h/2\pi. \tag{12.12}$$

It is possible to assign to each quantum state a region in phase space equal in size to h^n, and it is large enough to give the necessary uncertainty. The uncertainty principle states that it is not possible to determine precisely both the energy and the time. So, we can write an equation similar to Equation 12.12 as

$$\partial \varepsilon \partial t \gtrsim \frac{h}{2\pi} \tag{12.13}$$

where ε is the energy and t is the time. This equation signifies that if a system is disturbed after a short time ∂t, the energy level broadens and it cannot be definitely known.

If $n = 1$ (i.e. one degree of freedom), the quantum region becomes equal to h. If the *range of the co-ordinate* is Δr and the average difference between the momentum of two adjacent quantum states is Δp, then

$$\Delta r \Delta p \approx h. \tag{12.14}$$

A point to note is that in the case of usual periodic motion Δr becomes the total distance travelled along the co-ordinate axis in both directions in a complete period.

If Δt is the *range of time*, i.e. the time required for the particle to complete a cycle in its motion (which is its period, assuming it to be bound in a force field), and $\Delta \varepsilon$ is the energy between two adjacent energy levels, then for one degree of freedom an equation similar to Equation 12.14 can be written

$$\Delta \varepsilon \Delta t \approx h. \tag{12.15}$$

Since $1/\Delta t = \nu$ (frequency), it follows from Equation 12.15 that

$$\Delta \varepsilon \approx h\nu \tag{12.16}$$

which is Planck's equation.

The magnitude of ∂t in Equation 12.13 is often much larger than that of Δt in Equation 12.15. This is why energies can be well defined. Consider a gas which is not too dense, and whose molecules are occasionally colliding with one another. For such a gas, the time interval between collisions will be long enough not to broaden the energy levels appreciably. In that case, we can consider the molecules to be in definite energy levels. This will change only upon collisions, so we can ignore the effect of the collisions on energy level broadening. If the molecules in a system interact appreciably, we must consider energy levels of the entire system.

Another point to note is that there may be more than one characteristic time in the motion of a system. It is possible that a particle can oscillate with a certain frequency in a certain *state* and may make a transition to another *state* which might be at a different position in space but has the same energy. This can happen in quantum mechanics and it is a quantum-mechanical tunnelling phenomenon through energy barriers. In classical mechanics, a system with constant energy cannot make a transition from one state to another with the same energy if it has to pass through a situation. Where the total energy is less than the potential energy. If it did that, the kinetic energy would be negative. These situations are much more complicated than those considered in formulating Equation 12.16. In such cases, the spacing of the energy levels may seem superficially to be not so regular; energy differences between successive levels of varying magnitudes will exist, corresponding to the various characteristic times.

12.5 MODELS OF REAL SYSTEMS

The real world is complicated. Because of this, we rather naturally attempt to make a simplified model of a real system which does not lose the salient physical features of the system being considered. Historically, there are separate models for gases and solids. Here, we shall consider some of the models of macroscopic systems which have been developed from a long history of trial and error. They are all classical models; they have limitations but are very useful for developing a simplified picture of various systems. These models help in our understanding of nature.

(A) *Models for gases*:   A gas is considered a system in which the molecules are quite widely separated, except for occasional collisions between them. So, it can be assumed that the interaction energy between the molecules is insignificant; thus the total energy of the system is the sum of the energies of all the molecules. Such a model represents a perfect gas and we have

$$E = \sum_i \varepsilon_i \tag{12.17}$$

where E is the total energy, ε_i is the energy of the ith molecule and $i = 1, 2, \ldots, N$. A perfect gas may be *monatomic* (molecules composed of a single atom), *diatomic* (composed of molecules which are made up of two atoms) and *polyatomic* (made up of molecules consisting of many atoms). We shall consider these separately.

(i) *Monatomic gas*:   The energy of an atom consists of several parts. The most significant of these are the translational or kinetic energy of the atom $\varepsilon_k = \frac{1}{2}mv^2$ where m is the mass and v is the velocity of the atom. Let $\phi(\mathbf{r})$ be the potential energy due to external forces and let ε_e be the energy of the electrons bound to the nucleus. So, if we add up all these, we get the total energy of the ith molecule (monatomic) as

$$\varepsilon_i = \frac{1}{2}m_i v_i^2 + \phi(\mathbf{r}_i) + (\varepsilon_e)_i. \tag{12.18}$$

The total energy of the system is then given by Equation 12.17.

The motion of the electrons is so fast compared with that of the nuclei that in considering electronic energy states we can ignore nuclear motion. The energy of the electrons depends on the positions of the nuclei, and the mutual potential energy of the nuclei must be added to this electronic energy to get the potential energy surfaces. Then the total energy is separated into the *kinetic energy* of the nuclei and the *potential energy* with respect to the motion of the nuclei. The motions of the nuclei and those of the electrons are quantized. However, the translational energy of the molecule in its entirety can be considered and quantized separately from the internal motions of the molecule. In many cases the energy of the electrons (ε_e) remains constant and it can be ignored. However, if it changes, then a quantum-mechanical expression for ε_e must be used.

(ii) *Diatomic molecule*:   A diatomic molecule is made up of two atoms bound together by the interatomic force. A simple way to visualize a diatomic molecule is to imagine two balls attached to the two ends of a spring; the spring represents the interatomic force. In this situation, the two balls can *vibrate* along the spring (i.e. along the line connecting them), *rotate* about the spring (the axis of rotation) and *translate* through the space. Identical situations are all possible in the case of a diatomic molecule. The total energy of such a molecule is, therefore, given by the sum of the energies derived from these motions:

$$\varepsilon = \varepsilon_t + \varepsilon_r + \varepsilon_v + \varepsilon_p \tag{12.19}$$

where ε_t, ε_r, ε_v and ε_p are the translational, rotational, vibrational and potential energies, respectively. Remember, ε_p represents the total potential energy due to external forces and we should add to it the electron energy. However, for simplicity we shall omit it in our discussion.

If the atoms do not move far from their equilibrium positions because of the restrictions of their motion, the internal energy may be divided into parts which are quantized separately to a first approximation. These parts are the rotation and the vibration of different normal modes of vibration.

Classically, the position of a diatomic molecule can be fully described by assigning the three co-ordinates of its centre of gravity, the two polar angles θ and ϕ (which determine the direction of the line joining the two atoms) as well as the distance r between the two atoms. So, a diatomic molecule has six degrees of freedom of nuclear motion. The motion of the molecule is given by the momenta corresponding to the assigned co-ordinates. Hence, there are three degrees of freedom for translation motion (involving the motion of the centre of gravity), two degrees of freedom for rotational motion involving changes in θ and ϕ, and one degree of freedom for vibration involving r. To a good approximation, all of these can be quantized separately.

To derive the classical energy of a diatomic molecule we have to know the relevant expressions for ε_t, ε_r, ε_v and ε_p. Consider the spring model of a diatomic molecule as mentioned earlier (Figure 12.4). Here, a and b are two atoms. If we consider these atoms as points and ignore their electronic structure and spatial extension, then the energy of the molecule can be written as

$$\varepsilon = \frac{1}{2}m_a\mathbf{v}_a^2 + \frac{1}{2}m_b\mathbf{v}_b^2 + \Phi(r_{ab}) + \phi(\mathbf{r}_a) + \phi(\mathbf{r}_b) \tag{12.20}$$

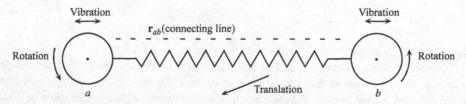

Figure 12.4 Schematic diagram of translational, rotational and vibrational motion of atoms in a diatomic molecule

where m_a and m_b are the masses and $\mathbf{v}_a$ and $\mathbf{v}_b$ are the velocities of a and b, respectively, and $\mathbf{r}$ represents the position of the particle. The quantity r_{ab} is equal to $|\mathbf{r}_a - \mathbf{r}_b|$ and $\phi_a(\mathbf{r}_a) + \phi_b(\mathbf{r}_b) = \varepsilon_p$, the total potential energy due to external forces. We need to simplify $\Phi(r_{ab})$. $\Phi(r)$ is the intermolecular potential energy called the Lennard–Jones potential which has the form shown in Figure 12.5 and a mathematical expression

$$\Phi(r) = \varepsilon_0\left\{ \left(\frac{r_0}{r}\right)^{12} - 2\left(\frac{r_0}{r}\right)^{6} \right\} \tag{12.21}$$

where r_0 is the equilibrium separation and ε_0 is the binding energy, i.e. the energy required to pull the atoms apart; both r_0 and ε_0 are constants and depend on the type of molecule. The force between the molecules is along their connecting line and its magnitude is given by

$$-\frac{d\Phi}{dr} = \frac{12\varepsilon_0}{r_0}\left\{ \left(\frac{r_0}{r}\right)^{13} - \left(\frac{r_0}{r}\right)^{7} \right\} \tag{12.22}$$

which vanishes when the molecules are separated by a distance $r = r_0$. At this distance the potential energy becomes $-\varepsilon_0$ from Equation 12.21.

In a diatomic molecule the atoms are bound together. They are separated by a distance approximately equal to r_0, and we can make the following approximation in the Taylor expansion (neglecting the higher order terms in the expansion)

$$\Phi(r) \equiv \Phi[r_0 + (r - r_0)]$$
$$\simeq \Phi(r_0) + (r - r_0)\left(\frac{d\Phi}{dr}\right)_{r_0} + \frac{1}{2}(r - r_0)^2\left(\frac{d^2\Phi}{dr^2}\right)_{r_0} \tag{12.23}$$

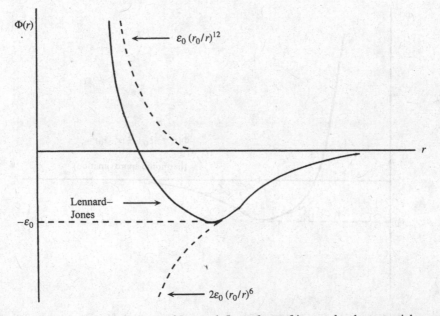

Figure 12.5 Schematic diagram of Lennard–Jones form of intermolecular potential energy

This approximation is reasonable, provided that $(r-r_0)$ is not too large. At the equilibrium point we have $(d\Phi/dr)_{r_0} = 0$ and from Equation 12.21 we have $\Phi(r_0) = -\varepsilon_0$. Therefore, Equation 12.23 reduces to

$$\Phi(r) \simeq -\varepsilon_0 + \frac{\kappa}{2}(r - r_0)^2 \tag{12.24}$$

where $\kappa = (d^2\Phi/dr^2)_{r_0}$ is the *force constant* and $-\varepsilon_0$ represents the constant binding energy of the molecule. This approximation of $\Phi(r)$ is called the *harmonic approximation* (represented graphically as shown in Figure 12.6), since the resulting force $-d\Phi/dr = -(r-r_0)(d^2\Phi/dr^2)$ is proportional to the displacement from the equilibrium separation r_0.

Using the harmonic approximation, the classical expression for the energy given by Equation 12.20 can be written as

$$\varepsilon = \frac{1}{2}m_a\mathbf{v}_a^2 + \frac{1}{2}m_b\mathbf{v}_b^2 + \frac{\kappa}{2}(r_{ab} - r_0)^2 - \varepsilon_0 + \varepsilon_p. \tag{12.25}$$

If we take the centre of mass as

$$\mathbf{R} = \frac{m_a\mathbf{r}_a + m_b\mathbf{r}_b}{m_a + m_b},$$

its velocity is given by

$$\mathbf{V} = \frac{d\mathbf{R}}{dt} = \left[m_a\left(\frac{d\mathbf{r}_a}{dt}\right) + m_b\left(\frac{d\mathbf{r}_b}{dt}\right) \right] \Big/ (m_a + m_b)$$

$$= \frac{m_a\mathbf{v}_a + m_b\mathbf{v}_b}{m_a + m_b} \tag{12.26}$$

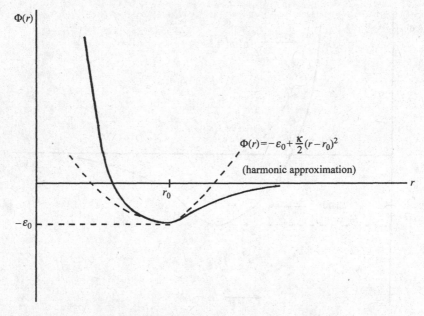

Figure 12.6   Harmonic approximation of $\Phi(r)$

Then,

$$
\begin{aligned}
\frac{1}{2}m_a\mathbf{v}_a^2 + \frac{1}{2}m_b\mathbf{v}_b^2 &= \frac{(m_a + m_b)(m_a\mathbf{v}_a^2)}{2(m_a + m_b)} + \frac{(m_a + m_b)(m_b\mathbf{v}_b^2)}{2(m_a + m_b)} \\
&= \frac{m_a^2\mathbf{v}_a^2 + m_am_b\mathbf{v}_a^2 + m_am_b\mathbf{v}_b^2 + m_b^2\mathbf{v}_b^2}{2(m_a + m_b)} \\
&= \frac{m_a^2\mathbf{v}_a^2 + m_b^2\mathbf{v}_b^2}{2(m_a + m_b)} + \frac{m_am_b(\mathbf{v}_a^2 + \mathbf{v}_b^2)}{2(m_a + m_b)} \\
&= \left(\frac{m_a + m_b}{2}\right)\left[\frac{m_a^2\mathbf{v}_a^2 + 2m_am_b\mathbf{v}_a\mathbf{v}_b + m_b^2\mathbf{v}_b^2}{(m_a + m_b)^2}\right] \\
&\quad + \frac{m_am_b(\mathbf{v}_a^2 + \mathbf{v}_b^2)}{2(m_a + m_b)} - \frac{2m_am_b\mathbf{v}_a\mathbf{v}_b}{2(m_a + m_b)} \\
&= \left(\frac{m_a + m_b}{2}\right)\left(\frac{m_a\mathbf{v}_a + m_b\mathbf{v}_b}{m_a + m_b}\right)^2 + \frac{m_am_b(\mathbf{v}_a^2 - 2\mathbf{v}_a\mathbf{v}_b + \mathbf{v}_b^2)}{2(m_a + m_b)} \\
&= \frac{1}{2}(m_a + m_b)\mathbf{V}^2 + \frac{1}{2}\left(\frac{m_am_b}{m_a + m_b}\right)(\mathbf{v}_a - \mathbf{v}_b)^2. \qquad (12.27)
\end{aligned}
$$

In this equation $m_a + m_b = M$ is the total mass and $m_am_b/(m_a + m_b) = \mu$ is the reduced mass of the molecule.

Therefore, the total energy of the molecule can now be written as

$$
\begin{aligned}
\varepsilon &= \frac{1}{2}M\mathbf{V}^2 + \frac{1}{2}\mu(\mathbf{v}_a - \mathbf{v}_b)^2 + \frac{\kappa}{2}(r_{ab} - r_0)^2 - \varepsilon_0 + \varepsilon_p \\
&= \frac{1}{2}M\mathbf{V}^2 + \frac{1}{2}\mu\left(\frac{dr_{ab}}{dt}\right)^2 + \frac{\kappa}{2}(r_{ab} - r_0)^2 - \varepsilon_0 + \varepsilon_p
\end{aligned} \qquad (12.28)
$$

where $\mathbf{v}_a - \mathbf{v}_b = d\mathbf{r}_{ab}/dt$. In this equation, $\frac{1}{2}M\mathbf{V}^2$ represents the translational (kinetic) energy of the molecule, so we write $\varepsilon_t = \frac{1}{2}M\mathbf{V}^2$. According to Figure 12.4, the vibrational motion is represented by the change in $\mathbf{r}_{ab}$ along the connecting line and the rotational motion is due to the change in $\mathbf{r}_{ab}$ in a direction perpendicular to itself. Therefore, we see that rotational motion changes the direction and vibrational motion changes the magnitude of $\mathbf{r}_{ab}$. To a good approximation, we can assume that during the rotational motion $\mathbf{r}_{ab}$ has a fixed magnitude r_0.

If ω_1 and ω_2 are the angular velocities in the two directions perpendicular to the connecting line ($\mathbf{r}_{ab}$), $r_0(\omega_1 + \omega_2) = (d\mathbf{r}_{ab}/dt)_{\text{rotation}}$. As vibrational motion changes $\mathbf{r}_{ab}$ along the connecting line, we can write

$$
(d\mathbf{r}_{ab}/dt)_{\text{vib}} = (dr_{ab}/dt)(\mathbf{r}_{ab}/r_{ab})
$$

Here, dr_{ab}/dt represents the change in magnitude and $\mathbf{r}_{ab}/r_{ab}$ is a unit vector. The total rate of change of $\mathbf{r}_{ab}$ is then

$$
d\mathbf{r}_{ab}/dt = (d\mathbf{r}_{ab}/dt)_{\text{rotation}} + (d\mathbf{r}_{ab}/dt)_{\text{vibration}}
$$

Clearly, ω_1, ω_2 and $\mathbf{r}_{ab}$ are perpendicular, so we have

$$(\mathrm{d}\mathbf{r}_{ab}/\mathrm{d}t)^2 = \left(\frac{\mathrm{d}r_{ab}}{\mathrm{d}t}\right)^2 + r_0^2(\omega_1^2 + \omega_2^2). \tag{12.29}$$

Substituting Equation 12.29 into Equation 12.28, we obtain the classical energy of a diatomic molecule

$$\varepsilon = \frac{1}{2}M\mathbf{V}^2 + \frac{1}{2}\mu r_0^2(\omega_1^2 + \omega_2^2) + \frac{1}{2}\mu\left(\frac{\mathrm{d}r_{ab}}{\mathrm{d}t}\right)^2$$
$$+ \frac{\kappa}{2}(r_{ab} - r_0)^2 + \varepsilon_p - \varepsilon_0. \tag{12.30}$$

In Equation 12.30, $\frac{1}{2}\mu r_0^2(\omega_1^2 + \omega_2^2)$ is the rotational energy and

$$\frac{\mu}{2}(\mathrm{d}r_{ab}/\mathrm{d}t)^2 + \frac{\kappa}{2}(r_{ab} - r_0)^2 = \varepsilon_v,$$

the vibrational energy. The binding energy ε_0, which is a constant, can be ignored in most cases. To find the total energy of a perfect diatomic gas, simply substitute Equation 12.30 into Equation 12.19.

Equation 12.30 is called the *harmonic model*. In deriving this model we assumed that in a diatomic molecule the two atoms are attached at the two ends of a spring. There is another model of a diatomic molecule where it is taken that the two atoms are connected by a rigid rod. This is similar to a dumbbell and called the *dumbbell model*. Since the connecting rod is rigid, there is no vibrational motion and we have zero vibrational energy. Thus, in this case Equation 12.30 becomes

$$\varepsilon = \frac{1}{2}M\mathbf{V}^2 + \frac{I}{2}(\omega_1^2 + \omega_2^2) + \varepsilon_p \tag{12.31}$$

where we have ignored ε_0 and used $I = \mu r_0^2$, which is the *moment of inertia* of the molecule.

Example 12.2

Calculate the moment of inertia of HF, given that the distance of separation between H and F atoms is 0.921×10^{-8} cm and the atomic mass unit is 1.66×10^{-24} g.

Solution

The atomic weights of H and F are 1 and 18.998, respectively. The reduced mass $\mu = (18.998/19.998) \times 1.66 \times 10^{-24} = 1.58 \times 10^{-24}$ g. The distance between H and F atoms $r_0 = 0.921 \times 10^{-8}$ cm. Moment of inertia $I = \mu r_0^2 = (1.58 \times 10^{-24}\text{ g})(0.921 \times 10^{-8}\text{ cm})^2 = 1.455 \times 10^{-40}$ g cm^2.

Tables 12.1 and 12.2 contain the values of the equilibrium distance of separation (r_0), the binding energy (ε_0) between two molecules (e.g. between one Ne and another Ne atom, or

Table 12.1   Values of ε_0 and r_0

Molecule	$\varepsilon_0 \times 10^8$ J	$r_0 \times 10^{-8}$ cm
HCl	50.0	3.73
CO	14.0	4.24
O_2	16.5	3.90
He	1.42	2.88
Ne	4.84	3.12
Ar	16.7	3.83
Kr	24.9	4.05
Xe	30.7	4.61

Table 12.2   Values of moment of inertia

Molecule	Moment of inertia ($\times 10^{33}$ kg m^2)
H_2	0.460
Cl_2	115.00
I_2	745.00
CO	14.49
NO	16.50
HF	1.35
HCl	2.66
CsCl	393.01

between one HCl molecule and another HCl molecule), and the moment of inertia (I) of some molecules. Note that in Table 12.1, the values of r_0 and ε_0 quoted for HCl, O_2 and CO are given for the temperature range 100–300 K, because these values are temperature-dependent.

Note: For the benefit of those readers who are not familiar with the harmonic oscillator, let us define this term. Imagine that a particle is connected to a fixed point, $x = 0$, by a spring as shown in Figure 12.7(a) and that the length of the spring in the undisturbed condition is x_0. If x is the position of the particle, then $(x - x_0)$ represents the extent to which the spring is stretched (or compressed). If the spring obeys Hooke's law, then the force it exerts on the

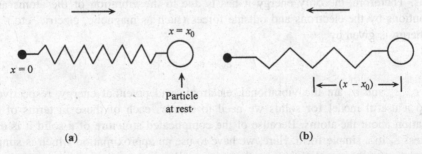

(a) (b)

Figure 12.7   A particle is connected to a fixed point by a spring: (a) the equilibrium condition and (b) when acted upon by a force

particle should be proportional to the extension of the spring from its equilibrium length (undisturbed). Then,

$$F_x = -\kappa(x - x_0)$$

where κ is the *force constant*. The force is always such as to restore the particle to the point $x = x_0$, because it is clear from the above equation that when $x > x_0$ we have $F_x < 0$, and when $x < x_0$ we have $F_x > 0$. However, when the particle arrives at the point x_0 it has acquired a large velocity, and as a result overshoots the point; it oscillates back and forth past $x = x_0$. This is a case of *harmonic oscillation*. The motion is periodic with a period $2\pi/\omega$ where ω is the angular velocity ($\omega^2 = \kappa/m$). If we take $\phi(x_0) = 0$, then the potential energy is

$$\phi = \frac{\kappa}{2}(x - x_0)^2$$

and the kinetic energy is $\frac{1}{2}mv^2$, where m and v are the mass and velocity. The frequency $\nu = 1/\text{period} = (1/2\pi)(\kappa/m)^{1/2}$. The main feature of a harmonic potential is that it is a quadratic function of $(x - x_0)$. The total energy is

$$\varepsilon = \frac{mv^2}{2} + \frac{\kappa}{2}(x - x_0)$$

and it is constant in time.

If we plot $\phi(x)$ against x, we shall get a curve which is effectively a parabola facing upward and touching the x-axis at some point x_0. A potential of this shape is called a *potential well* or *potential valley*. At the bottom of the well (point x_0), $\mathrm{d}\phi/\mathrm{d}x = 0$ and there is force on the particle. A particle placed at x_0 remains there and it is called the point of "mechanical equilibrium". On the right of this equilibrium point, the force is towards the left, $F_x = -\mathrm{d}\phi/\mathrm{d}x$, and on the left of the point x_0 the force is towards the right, $\mathrm{d}\phi/\mathrm{d}x < 0$.

(B) *Models for solids*:   Unlike gases and liquids, in any solid the atoms are arranged in some regular spatial order because they are so closely packed that the interactions between them make each atom stay in its position. Some solids have a simple cubic lattice (e.g. NaCl), some have a body-centred cubic lattice (e.g. CsCl) or face-centred cubic lattice (e.g. Cu and Ag), while there are solids which have fairly complicated structures (e.g. diamond, SiC). The object of a model is to avoid this complication. If we consider a solid as a large molecule, then we can assume that because of its size it has no translational or rotational motions. Therefore, the only energy it has is due to the vibration of the atoms and any contributions by the electrons and outside forces (such as magnetic, electric, etc.). So the total energy is given by

$$E = \varepsilon_v + \varepsilon_e + \varepsilon_p \tag{12.32}$$

where ε_v, ε_e and ε_p are the vibrational, electronic and potential energy, respectively. To develop a useful model for solids we need to express each of these in terms of known information about the atoms. Because of the complicated structure of a solid it is difficult to express ε_v in a simple form. Here, we have to use an approximation which is simple but contains most of the essential features that we need. We assume that each atom in the lattice can move in any of the three directions about the point of its minimum potential energy (which is its equilibrium position). If the atom does not move too far from this position, then there is a linear relationship between the force acting on the atom and its displacement from the equilibrium position; these are proportional to each other. Thus, the motion of the atom will be that of a three-dimensional harmonic oscillator. The energy of such an oscillator is

$$\varepsilon = \frac{1}{2}m\mathbf{v}^2 + \frac{1}{2}\kappa\mathbf{r}^2 \tag{12.33}$$

where $\mathbf{v}$ and $\mathbf{r}$ are the velocity and displacement from the equilibrium, respectively, and m is the mass of the atom.

In this equation, κ is the force constant which may be different for different atoms, and ε is the vibrational energy of one atom. So the total vibrational energy of the solid will be the sum of the vibrational energies of each atom. Thus,

$$\varepsilon_v = \sum_i \frac{1}{2}m_i\mathbf{v}_i^2 + \frac{1}{2}\kappa_i\mathbf{r}_i^2 \tag{12.34}$$

where $i = 1, 2, \ldots, N$. This is called *harmonic lattice model*. According to this model, the system behaves like an aggregate of $3N$ independent harmonic oscillators. An important point to note is that in the classical case there is no need to evaluate the frequencies given by

$$\nu_i = (1/2\pi)(\kappa_i/m_i)^{1/2}. \tag{12.35}$$

These can be arbitrary. However, in the next section we shall see that in the case of quantum mechanics we need to make some assumptions about their values.

Let us now consider the contribution of the electrons to the total energy of a solid. Unfortunately, this cannot be sufficiently represented by any classical model. One good way to circumvent the problem is to assume that ε_e is constant and can be ignored, but this cannot be justified by any classical argument. It is well known that in metals there are electrons which are fairly free to move through the entire lattice; these electrons do not belong to any one atom. Each one of these would have a kinetic energy $\frac{1}{2}m_e\mathbf{v}^2$, where m_e is the mass and $\mathbf{v}$ is the velocity of the electron. Then the total contribution of all the free electrons is the sum of the kinetic energy of each free electron. Thus,

$$\varepsilon_e = \sum_i \frac{1}{2}m_e\mathbf{v}_i^2 \tag{12.36}$$

where $i = 1, 2, \ldots\ldots$. This is a *classical model*.

The contribution of potential energy ε_p depends on the type of material. Molecules may have a dipole moment $\mathbf{p}$ or a magnetic moment μ, and when an electric or magnetic field is applied their potential energy is given by

potential energy $\phi = -\mathbf{p} \cdot \mathbf{\varepsilon}$ (due to electric field $\mathbf{\varepsilon}$)

potential energy $\phi = -\mu \cdot \mathbf{B}$ (due to magnetic field $\mathbf{B}$).

Therefore, if a solid (liquid or gas) is made up of such molecules, it will have a potential energy depending on the nature of the molecules and the applied field. If an external electric or magnetic field is applied to a simple dielectric or paramagnetic solid (in which the interaction energy between the dipole moments or the permanent magnetic moments, as the case may be, is very much smaller than their potential energy due to the applied field, so it can be ignored), its total potential energy is equal to the sum of the potential energies of each molecule. Thus,

$$\left.\begin{array}{l} \text{for dielectric solids,} \quad \varepsilon_p = -\sum_i \mathbf{p}_i \cdot \boldsymbol{\varepsilon} \\[2mm] \text{for paramagnetic solids,} \quad \varepsilon_p = -\sum_i \boldsymbol{\mu}_i \cdot \mathbf{B} \end{array}\right\} \tag{12.37}$$

where $i = 1, 2, 3, \ldots$. However, in ferromagnetic solids the magnetic moments interact sufficiently strongly that the interaction energy becomes significant. This energy causes all the magnetic moments to line up in parallel to one another; therefore, the substance has a permanent magnetic field. Modelling of such systems is complicated.

12.6   QUANTUM RESTRICTIONS ON TRANSLATIONAL, ROTATIONAL AND VIBRATIONAL FORMS OF THE ENERGY

Before the Twentieth Century it was assumed that Newton's laws of classical mechanics are applicable to all mechanical phenomena, including planetary motions. However, at the beginning of the Twentieth Century it became known that some experimental results could not be explained by classical mechanics. At first, these problems were limited to the interaction between matter and radiation. This led Max Planck to the idea of the quantization of energy. Later Ludwig Boltzmann developed the statistical theory of gases on the basis of some assumptions. Still the problems remained and some scientists began to doubt whether Boltzmann's assumptions in developing the statistical theory of gases were right. Through further painstaking work of many people it became clear that there was nothing wrong with the statistical concepts; rather that the concepts of classical mechanics were at fault. Consequently, Planck modified the theory which he developed to explain the observed radiation from systems in thermal equilibrium and concluded that energy could only be exchanged in certain finite or quantized amounts. The idea of quantization of energy led to the discovery of the theory of quantum mechanics.

The distinction between the quantum theory and the classical theory becomes important when an exchange of a small amount of energy takes place with the system. According to the classical theory it is always possible to exchange a small amount of energy with a system. On the other hand, quantum mechanics requires that we must have a certain minimum amount of energy. This minimum amount is the energy difference between two of the quantum states of the system. If energy is added to a system, say by heating to raise the temperature, it becomes rich in energy and can easily make the quantum jumps in energy. However, at lower temperatures these transitions may become difficult. Therefore, at lower temperatures the energy is relatively scanty and the quantum restrictions become rather important. Here, we shall consider the quantum restrictions of the translational, vibrational and rotational energies.

12.6.1 Translational energy

We have seen in the previous sections that a molecule has rotational, vibrational and electronic energy, as well as the translational energy of the centre of gravity which is also quantized. The energy levels in translational energy are very closely spaced. Often it is convenient to assume that the molecule is inside a cubical box of sides L_x, L_y and L_z. In this model, within the box the potential energy due to external forces acting on the molecule is constant; it may be taken as zero. However, at the walls of the box it suddenly becomes infinite. In such a case the translational energy (ε_t) can be expressed as

$$\varepsilon_t = \varepsilon_x + \varepsilon_y + \varepsilon_z \tag{12.38}$$

where ε_x, ε_y and ε_z are the kinetic energies of the motions in the x, y and z directions when the box is aligned parallel to the x, y and z co-ordinate axes, and are given by

$$\varepsilon_x = \frac{n_x^2 h^2}{8mL_x^2}$$

$$\varepsilon_y = \frac{n_y^2 h^2}{8mL_y^2} \tag{12.39}$$

$$\varepsilon_z = \frac{n_z^2 h^2}{8mL_z^2}$$

where h is Planck's constant and m is the mass of the molecule. n_x, n_y and n_z are called *translational quantum numbers*. These are all positive integers, which may be the same or different, and can have values $1, 2, 3, \ldots$, but not zero. For each set of values (n_x, n_y, n_z) the molecule is said to be in a different quantum state. Combining Equations 12.38 and 12.39, we get

$$\varepsilon_t = \frac{h^2}{8m}\left(\frac{n_x^2}{L_x^2} + \frac{n_y^2}{L_y^2} + \frac{n_z^2}{L_z^2}\right) \tag{12.40}$$

where n_x, n_y and n_z are all positive integers.

When n_x, n_y and n_z are large, ε_t may be considered as a quasi-continuous function of these numbers. Constant energy occurs on a surface in a space formed by n_x, n_y and n_z marked off along the three axes, and Equation 12.40 represents this surface. Actually, Equation 12.40 is the equation of an ellipsoid which has the volume in the n_x, n_y and n_z space equal to

$$\frac{4}{3}\pi(8m\varepsilon_t)^{3/2}\,V/h^3$$

where $V = L_x L_y L_z$ is the volume of the box.

We wish to know what will be the number of energy levels below ε_t, and between ε_t and $\varepsilon_t + d\varepsilon_t$, where $d\varepsilon_t$ is an infinitesimal change in ε_t. The volume in the n_x, n_y and n_z space is equal to the number of combinations of integral values included in the volume, because this will be the number of unit cubes inside that volume. Therefore, the number of energy levels below ε_t is the volume of the part of the ellipsoid, where n_x, n_y and n_z are all positive, or

$$\left(\frac{1}{8}\right)\left(\frac{4\pi}{3}\right)(8m\varepsilon_t)^{3/2} V/h^3 = \frac{4\pi}{3}(2m\varepsilon_t)^{3/2} V/h^3. \tag{12.41}$$

The number of levels between ε_t and $\varepsilon_t + d\varepsilon_t$ is

$$n_{\varepsilon_t} = [2\pi(2m)^{3/2}\varepsilon_t^{1/2} V/h^3]d\varepsilon_t. \tag{12.42}$$

Quantization of the translational energy for a polyatomic molecule is carried out exactly in the same way. Each vibrational mode is quantized, in first approximation, by using its particular frequency, but there is no simple reduced mass.

The quantum expression for the translational (kinetic) energy (Equation 12.40) is very similar to the classical expression

$$\varepsilon_t = \frac{1}{2}m(v_x^2 + v_y^2 + v_z^2)$$
$$\equiv \frac{1}{2m}[(mv_x)^2 + (mv_y)^2 + (mv_z)^2].$$

However, the allowed values of the components of the momentum (mv) are discrete. If L_x, L_y and L_z are large, the values of ε_t will be very close together. So, the energy can be treated as continuous.

Example 12.3

An atom of mass 1.66×10^{-24} g in a cubical box of $1\,cm^3$ is in a state of translational motion described by the quantum numbers $n_x = 3 \times 10^7$, $n_y = 0$ and $n_z = 0$. Calculate its momentum, velocity and translational energy.

Solution

The momenta in the x, y and z directions are given by

$$\frac{n_x h}{2L_x}, \frac{n_y h}{2L_y} \quad \text{and} \quad \frac{n_z h}{2L_z}$$

respectively, where h is Planck's constant $= 6.626 \times 10^{-27}$ erg-s, and L_x, L_y and L_z are the lengths of the box. Here, $L_x = L_y = L_z = 1$ cm and $n_y = n_z = 0$. Therefore, the momentum of the atom has value only in the x direction, which is

$$\frac{3 \times 10^7 \times 6.626 \times 10^{-27} \text{ erg-s}}{2 \times 1 \text{ cm}} = 0.994 \times 10^{-19} \text{ erg-s/cm}$$

Since 1 erg $= \frac{1 \text{ g} \times 1 \text{ cm}^2}{1 \text{ s}^2}$, this momentum $= 0.994 \times 10^{-19}$ g cm/s.

$$\text{Velocity} = \frac{\text{momentum}}{\text{mass}} = \frac{0.994 \times 10^{-19} \text{ g cm/s}}{1.66 \times 10^{-24} \text{ g}} = 0.60 \times 10^5 \text{ cm/s}$$

Translational energy

$$\varepsilon_t = (momentum)^2/2(mass)$$
$$= (0.994 \times 10^{-19} \text{ g cm/s})^2/2(1.66 \times 10^{-24} \text{ g})$$
$$= 0.30 \times 10^{-14} \text{ g cm}^2/\text{s}^2 = 0.30 \times 10^{-14} \text{ ergs}$$
$$= 0.30 \times 10^{-21} \text{ J}$$

From the above example it can be shown by changing the values of n_x by 1 that the translational energy difference between different quantum states is less than 10^{-27} J. Hence, it can be considered as continuous.

12.6.2 Rotational energy

Compared with the translational energy of a diatomic molecule, the rotational energy of such a molecule is more important, due to the quantum effects. The rotational quantum states of molecules are quite involved. According to quantum mechanics the allowed values of the rotational energy of a diatomic molecule are given by

$$\varepsilon_r = \left(\frac{h^2}{8\pi^2 I} \right) J(J+1) \tag{12.43}$$

where I is the moment of inertia of the molecule, h is Planck's constant and J is known as the *rotational quantum number*, which has values $0, 1, 2, \ldots$. The classical equation for the rotational energy of a diatomic molecule is

$$\varepsilon_r = \frac{I}{2}\omega^2 = \frac{1}{2I}(I\omega)^2 \tag{12.44}$$

where $\omega^2 = \omega_1^2 + \omega_2^2$ (the sum of the square of the angular velocities which are in two directions perpendicular to the line joining the atoms (see Section 15.5).

Comparing Equations 12.43 and 12.44, we see that the quantum theory requires that the total angular momentum $I\omega$ can have only values $(h/2\pi)[J(J+1)]^{1/2}$. The rotational quantum state of a molecule is not specified by the value of the rotational quantum number, because there are a number of quantum states which all have the same value of J. According to quantum mechanics, for any fixed value of J the angular momentum along any fixed direction can have the values

$$Jh/2\pi, \ (J-1)h/2\pi, \ (J-2)h/2\pi, \ldots, -Jh/2\pi.$$

This implies that when the total angular momentum is

$$\left(\frac{h}{2\pi} \right)[J(J+1)]^{1/2}$$

there are $(2J+1)$ possible orientations of the angular momentum.

Example 12.4

Show that the quantum effects are much more important in restricting the rotational motion than in restricting the translational motion.

Solution

To prove this let us consider the case of HCl which has a moment of inertia 2.648×10^{-40} g cm^2. Consider the rotational energies in the quantum states $J = 0$ and $J = 1$. For $J = 0$ the rotational energy is zero. Therefore, the difference in energy between these two states is

$$\Delta \varepsilon_r = 2h^2/8\pi I$$
$$= 2 \times (6.626 \times 10^{-27} \text{ erg s})^2/(8 \times 9.872 \times 2.648 \times 10^{-40} \text{ g cm}^2)$$
$$= 4.198 \times 10^{-15} \text{ erg}^2 \text{ s}^2/\text{g cm}^2 = 4.2 \times 10^{-15} \text{ ergs}$$

(since 1 erg = g cm^2/s^2). The translational energy difference between different quantum states is characteristically less than 10^{-27} J. In the above example we see that the jump in energy is 4.2×10^{-22} J which is 10^5 times larger than that in the translational case. Hence, it is obvious that the quantum effects are much more important in restricting the rotational motion than in restricting the translational motion.

We shall see later that, in general, the energy difference between two vibrational states is one-hundred times larger than that between two rotational states. So, the vibrational motion of a molecule is significantly more restricted than the rotational motion. Let us now consider the restriction which quantum mechanics places on the vibrational motion of atoms.

12.6.3 Vibrational energy

The possible energies of a harmonic oscillator can be calculated by Schrödinger's equation, and they are given by

$$\varepsilon_\nu = \left(n + \frac{1}{2}\right)h\nu \tag{12.45}$$

where h is Planck's constant, and ν is the frequency of the oscillator, which is a fixed number depending on the physical situation. Classically,

$$\nu = \left(\frac{1}{2\pi}\right)(\kappa/\mu)^{1/2}$$

where μ is the reduced mass of the molecule, κ is the harmonic force constant, and n is the vibrational quantum number which can have values $0, 1, 2, \ldots$. The values of ν (in s^{-1}) can be determined experimentally if the vibrational motion of diatomic molecules is treated in the harmonic approximation.

The minimum value of the energy of a harmonic oscillator is that one which corresponds to the value for $n = 0$, i.e. $\frac{1}{2}h\nu$ (by Equation 12.45). This value is called the *zero-point energy*. Since both h and ν are constants, the zero-point energy is also a constant. So, this constant term can be neglected in the energy (because we can measure only the difference in energies) if we define the state of zero energy to be that state in which the oscillator is at rest. This has been incorporated into Equation 12.45. The significance of the zero-point energy is that it indicates that even when the oscillator has its lowest values of the energy it cannot be at rest. Thus, the zero-point energy has no effect on the difference in vibrational energy between two quantum states ($\Delta\varepsilon_v$).

Let us calculate the difference in the vibrational energy of the HCl molecule for $n = 0$ and $n = 1$. An HCl molecule has a frequency of $8.65 \times 10^{13}\,\text{s}^{-1}$. Then,

$$\Delta\varepsilon_v = \left[\left(n + \frac{1}{2}\right)h\nu\right]_{n=1} - \left[\left(n + \frac{1}{2}\right)h\nu\right]_{n=0}$$

$$= h\nu = 8.65 \times 10^{13} \times 6.626 \times 10^{-27}\,\text{erg}$$

$$= 5.73 \times 10^{-20}\,\text{J}.$$

We see that this value is 10^2 times bigger than the $\Delta\varepsilon_r$ value calculated in Example 12.4. Therefore, we can conclude that the vibrational motion of a molecule is more restricted than the rotational motion, which is more restricted than the translational motion.

According to the model for solids which we developed in Section 12.5, a solid may be considered to be made up of $3N$ independent harmonic oscillators (where N is the number of atoms). If we consider the ith oscillator, then the vibrational energy of this will be

$$(\varepsilon_v)_i = \left(n_i + \frac{1}{2}\right)h\nu_i$$

where ν_i is the frequency of the ith oscillation. The total vibrational energy of the solid is then obtained by summing $(\varepsilon_v)_i$ over all the $3N$ oscillators. Thus,

$$(\varepsilon_v)_{\text{total}} = \sum_{i=1}^{3N} \left(n_i + \frac{1}{2}\right)h\nu_i. \tag{12.46}$$

To specify the quantum state of a solid it is necessary to assign the values of the $3N$ vibrational quantum numbers, i.e. $n_i = 0, 1, 2, \ldots$ for all the $3N$ independent oscillators. As for the frequencies, there are two methods for assigning their values. These lead to slightly different models of solids in terms of quantum mechanics.

(a) The first method is to assume that all the ν_i are the same and have a general value equal to ν which depends on the type of solid in some way. This simplifies Equation 12.46 to the following form, which is the *Einstein model*.

$$(\varepsilon_v)_{\text{total}} = \sum_{i=1}^{3N} \left(n_i + \frac{1}{2}\right)h\nu. \tag{12.47}$$

Since this model describes the solid in terms of one frequency only, it is more appropriate for monatomic solids than for diatomic solids.

(b) A solid is composed of many atoms. These atoms do not really vibrate independently of one another for the simple reason that the motion of one atom affects that of the others. Actually, the motion of all the atoms inside a solid can be imagined as a collection of $3N$ standing waves. However, as in the case of a harmonic oscillator, the amplitude of each wave varies periodically with time. So, the solid can be imagined as a collection of $3N$ harmonic oscillators, except that the frequencies of these oscillators are determined by the frequencies of the $3N$ standing waves. This model is called the *Debye model*.

If v is the velocity of a wave of wavelength λ and frequency ν, then $v = \lambda\nu$. If v is common to all the waves, then according to this equation the shorter wavelengths correspond to higher frequencies. We need to determine two things: (i) the number of waves (i.e. oscillators) which have a frequency in the range $d\nu$ of frequency ν and (ii) the number of waves with a wave number in the range $d(1/\lambda)$ of $1/\lambda$.

(i) Let us imagine that a standing wave is confined in a space bounded by three sides of length L_x, L_y and L_z in the x, y and z directions. Such a wave is mathematically described by

$$A \sin\left(\frac{n_x \pi x}{L_x}\right) \sin\left(\frac{n_y \pi y}{L_y}\right) \sin\left(\frac{n_z \pi z}{L_z}\right) \sin\ (2\pi\nu t)$$

where t is the time and n_x, n_y and n_z are all positive integers. It is obvious that the wave disappears at $x = 0$, L_x, because then

$$\sin\left(\frac{n_x \pi x}{L_x}\right) = 0.$$

Similarly, it also vanishes at $y = 0$, L_y and $z = 0$, L_z. The terms $n_x/2L_x$, $n_y/2L_y$ and $n_z/2L_z$ are called the components of the wave number which is given by

$$[(n_x/2L_x)^2 + (n_y/2L_y)^2 + (n_z/2L_z)^2]^{1/2} = 1/\lambda.$$

Also,

$$\nu = v\left[\left(\frac{n_x}{2L_x}\right)^2 + \left(\frac{n_y}{2L_y}\right)^2 + \left(\frac{n_z}{2L_z}\right)^2\right]^{1/2} = v(1/\lambda). \qquad (12.48)$$

Since the velocity is the same for all waves, we can say that

$$d\nu = v d\left[\left(\frac{n_x}{2L_x}\right)^2 + \left(\frac{n_y}{2L_y}\right)^2 + \left(\frac{n_z}{2L_z}\right)^2\right]^{1/2} = v d(1/\lambda)$$

and the number of these waves with a frequency in the range $d\nu$ of ν is related to the number of waves with a wave number in the range $d(1/\lambda)$ of $1/\lambda$.

(ii) To find the number of waves with a wave number in the range $d(1/\lambda)$ of $1/\lambda$, we consider a space defined by the co-ordinates

$$\left(\frac{n_x}{2L_x}\right), \left(\frac{n_y}{2L_y}\right) \quad \text{and} \quad \left(\frac{n_z}{2L_z}\right)$$

The permissible values of these co-ordinates can be represented by points in this space, which form a simple cubic-type lattice. The volume of each of these is equal to

$$(1/2L_x)(1/2L_y)(1/2L_z) = 1/8L_xL_yL_z = 1/8V$$

where V is the volume enclosed by L_x, L_y and L_z. We can see that there is only one cubic-type lattice for each point in this space. So, $1/8\,V$ is the volume per point (a point is a wave).

The space defined by the components of the wave number i.e. $n_x/2L_x$, $n_y/2L_y$ and $n_z/2L_z$ is a spherical shell of radius

$$\left[\left(\frac{n_x}{2L_x}\right)^2 + \left(\frac{n_y}{2L_y}\right)^2 + \left(\frac{n_z}{2L_z}\right)^2\right]^{1/2} = 1/\lambda$$

which has a thickness $d(1/\lambda)$. The volume of this spherical shell is $\frac{1}{8}$ of a complete spherical shell, because each component of the wave number is positive. Thus, the volume of the spherical shell of radius $1/\lambda$ and thickness $d(1/\lambda)$ is simply $\frac{1}{8}(4\pi)(1/\lambda)^2\,d(1/\lambda)$. The number of waves in this shell is then given by dividing this volume by the volume of one cubic-type lattice, i.e. $1/8\,V$, which gives $4\pi V(1/\lambda)^2\,d(1/\lambda)$. Since the velocity v of the wave is related to $1/\lambda$ and the frequency ν by $\nu = v/\lambda$, we find that the number of waves with a frequency in the range $d\nu$ of ν is

$$4\pi V\left(\frac{\nu}{v}\right)^2 d\left(\frac{\nu}{v}\right) = (4\pi V/v^3)\nu^2\,d\nu.$$

Since v is constant,

$$d(\nu/v) = \frac{1}{v}d\nu.$$

A point to remember is that the above result is true only if the frequencies are continuous. However, it is a reasonable approximation if the frequencies are discrete but very close together.

We must also remember that in a solid both longitudinal and transverse waves exist. There are two perpendicular transverse waves and one longitudinal wave. If v_t is the velocity of the transverse waves and v_l is that of the longitudinal wave, then the number of modes with a frequency in the range $d\nu$ of ν is given by

$$dN(\nu) = 4\pi V\left(\frac{1}{v_l^3} + \frac{2}{v_t^3}\right)\nu^2\,d\nu = 4\pi V(v_l^{-3} + 2v_t^{-3})\nu^2\,d\nu.$$

If we make the approximation $v_l^{-3} + 2v_t^{-3} \simeq 3/v^3$, then we get

$$4\pi V(v_l^{-3} + 2v_t^{-3})\nu^2\,d\nu \simeq \left(\frac{12\pi V}{v^3}\right)\nu^2\,d\nu.$$

If a solid contains N number of atoms, it has $3N$ degrees of freedom so it can be described in terms of $3N$ waves. Evidently, there is some maximum frequency ν_m for all of these $3N$

waves. This maximum frequency is given by

$$3N = \int_0^{\nu_m} \left(\frac{12\pi V}{v^3} \right) \nu^2 \, d\nu = \frac{12\pi V}{v^3} \int_0^{\nu_m} \nu^2 \, d\nu$$

$$= \left(\frac{4\pi V}{v^3} \right) \nu_m^3$$

or

$$\nu_m = (3N v^3 / 4\pi V)^{1/3}. \tag{12.49}$$

This shows that the Debye model treats the possible frequencies as continuously distributed in the range $\nu_m \geqslant \nu \geqslant 0$, and that all summations of the form $\sum f(\nu_i)$ (where $f(\nu_i)$ is any function of the frequencies, and $i = 1, 2, \ldots, 3N$) tend to

$$\int_0^{\nu_m} f(\nu) \, dN(\nu) = \frac{12\pi V}{v^3} \int_0^{\nu_m} f(\nu) \nu^2 \, d\nu.$$

However, the energy of each wave or oscillator is still given by $(n + \frac{1}{2})h\nu$.

12.6.4   Electronic energy

The relative separations in the energy levels for ε_t, ε_r and ε_v in a diatomic molecule are of the order of $\Delta \varepsilon_t < 10^{-27}$, $\Delta \varepsilon_r \simeq 10^{-22}$ and $\Delta \varepsilon_v \simeq 10^{-20}$ J. The difference in energies between the electronic levels in hydrogen atom is $\Delta \varepsilon_e \simeq 10^{-19}$ J. The quantum-mechanical equation for these energy levels is

$$\varepsilon_e = \frac{-21.76 \times 10^{-19}}{n^2} \text{ J} = \frac{-13.6}{n^2} \text{ eV} \tag{12.50}$$

where n is a number which can have values $1, 2, 3, \ldots \ldots$.

When an electron travels across a potential drop of 1 volt it gains a kinetic energy equivalent to $1 \, \text{eV} = 1.6 \times 10^{-19}$ J. Thus, the energy difference $\Delta \varepsilon_e$ between different electronic states is approximately 10^{-19} to 10^{-18} J which is quite high. So, $\Delta \varepsilon_e$ tends to be constant and can be ignored, provided there is no large quantity of energy in the system.

12.7   THE BASIC ASSUMPTION OF STATISTICAL MECHANICS

We know that all macroscopic properties of a system are, in fact, averages of some microscopic behaviour within the system. Which macroscopic property should be identified with which average property or properties is a question answered by statistical mechanics. However, before we calculate any averages, we must first find the probability of certain *events*. When these probabilities are obtained, calculations of the desired averages are simple. The question is: What is an event?

Consider a system of N particles. In relation to this system, we may define a physical event as the detailed microscopic description of the system at some instant of time. By classical mechanics, the most detailed information that we can obtain about a system is the velocity and the position of each of the N particles, i.e.

$$\left.\begin{array}{c} r_1, r_2, r_3, \ldots, r_N \\ v_1, v_2, v_3, \ldots, v_N \end{array}\right\} \qquad (12.51)$$

Let us first consider the case of $N = 1$, which can then be generalized for all the N particles. The position of a particle and its velocity in the xyz-space are given by

$$x, y, z \quad \text{and} \quad v_x, v_y, v_z. \qquad (12.52)$$

Since the particle is moving all the time, these values can vary continuously. Therefore, it is necessary to consider the following infinitesimal ranges to discuss the probability of these events.

$$dx, dy, dz \quad \text{and} \quad dv_x, dv_y, dv_z. \qquad (12.53)$$

According to classical mechanics these values could be arbitrarily small, but quantum mechanics demands that the accuracy cannot be higher than the value given by the Heisenberg uncertainty principle.

As in the case of continuous variables, the usual procedure to describe the probability of these events is to introduce the distribution function $f(x, y, z, v_x, v_y, v_z)$. Then, the probability that the particle is situated between x and $x + dx$, y and $y + dy$, z and $z + dz$, with a velocity between v_x and $v_x + dv_x$, v_y and $v_y + dv_y$, and v_z and $v_z + dv_z$ is given by

$$f(x, y, z, v_x, v_y, v_z) \, dx \, dy \, dz \, dv_x \, dv_y \, dv_z. \qquad (12.54)$$

This can be abbreviated to the form

$$f(\mathbf{r}, \mathbf{v}) \, d^3 r \, d^3 v \qquad (12.55)$$

where $d^3 r = dx \, dy \, dz$ and $d^3 v = dv_x \, dv_y \, dv_z$. Then, Equation 12.55 represents the probability that the particle is in the space $d^3 r$ of $\mathbf{r}$ having a velocity in the range $d^3 v$ of $\mathbf{v}$.

For the original N particles there will be $6N$ variables and we can generalize Expression 12.55 to write

$$f(\mathbf{r}_1, \mathbf{r}_2, \ldots, \mathbf{r}_N, \, \mathbf{v}_1, \mathbf{v}_2, \ldots, \mathbf{v}_N) \, d^3 r_1 \cdots d^3 r_N \, d^3 v_1 \cdots d^3 v_N \qquad (12.56)$$

which represents the probability that N particles are in the regions $d^3 r_1$ of $\mathbf{r}_1$, $d^3 r_2$ of $\mathbf{r}_2, \ldots, d^3 v_N$ of $\mathbf{v}_N$ all at the same time. This can be abbreviated to

$$f(\mathbf{r}_1, \cdots, \mathbf{v}_N) \, d^3 r_1 \cdots d^3 v_N. \qquad (12.57)$$

In statistical mechanics, these events are represented by microstates. In classical mechanics, a microstate is specified by assigning values to the variables $\mathbf{r}_1, \mathbf{r}_2, \ldots, \mathbf{v}_N$ within an accuracy determined by $d^3r_1, \ldots, d^3v_N$ and the Expression 12.57 gives the probability of this microstate. On the other hand, in quantum mechanics it is not possible to determine the exact position and velocity $\mathbf{r}_1, \ldots, \mathbf{v}_N$ of the particles, because the Heisenberg uncertainty principle stipulates that $d^3r \, d^3v > h^3/m^3$ for each of the N particles, where m is the mass of the particle. We can, however, describe the system in terms of certain quantum numbers, and these can only have discrete values and they depend on the system. Each set of values represents a different microstate. If E_k is the energy of a microstate, the subscript k indicates which microstate we are referring to. As the classical energy $E(\mathbf{r}_1, \ldots, \mathbf{v}_N)$ may have the same value for different values of the variables, so also the value of E_k may be the same for different microstates (for different k).

In reality, the number of microstates is so high that it is not possible to have sufficient data to accurately determine the probabilities of the microstates. So, we make some assumptions about the probabilities. Let us consider an isolated system of known total energy and volume V. The law of conservation of energy tells us that the energy of an isolated system does not change in time (because it does not exchange energy with other systems). Contrary to this, the microstate of the system is constantly changing with time, so it is not possible to know which microstate will occur at some instant of time. We generally know that the atoms of the system are located within V in a definite region of space. Given only the above information, the most reasonable assumption to make is that all microstates which satisfy these conditions are equally probable. The most reasonable assumption is the basic assumption of statistical mechanics, which is: *all microstates of a system which have the same energy are assumed to be equally probable.*

This assumption is the basis of entire statistical mechanics and so far nobody has proved it to be incorrect.

We wish to define two simple but often used terms: *microcanonical ensemble* and *canonical ensemble*. The literary meaning of *canonical* is *accepted* or *authoritative*. In the scientific sense it refers to the notion that the ensemble is the most significant type. We already know that in any isolated system the energy is conserved, i.e. it does not change, so according to the above assumption all the microstates with this energy are equally probable. An ensemble or collection of such identical isolated systems is called a microcanonical ensemble. On the other hand, if a system is in thermal contact with other systems of known temperature, the energy of the system is no longer constant because it then exchanges energy with the system or systems in contact with it. An ensemble or collection of such systems which are in thermal equilibrium with other system or systems at the same temperature is called a canonical ensemble.

From the above discussion we are confronted with a question: if a system exchanges energy with another system, what is the probability of finding it in some particular microstate? According to the basic assumption of statistical mechanics the probability of a microstate for a classical system given by the Expression 12.57 must be the same for all microstates with the same energy $E(\mathbf{r}_1, \ldots, \mathbf{v}_N)$, provided the distribution function $f(\mathbf{r}_1, \ldots, \mathbf{v}_N)$ depends on the energy $E(\mathbf{r}_1, \ldots, \mathbf{v}_N)$ only. So, the basic assumption of statistical mechanics means that the probability of a classical microstate can be written as

$$f[E(\mathbf{r}_1, \ldots, \mathbf{v}_N)] \; d^3r_1 \cdots d^3v_N. \tag{12.58}$$

Likewise, if the energy is discrete, the probability $\mathscr{P}$ of a microstate must depend on E_k of that microstate, i.e. $\mathscr{P}$ is a function of E_k and we can write

$$
\begin{aligned}
\mathscr{P}(E_k) = \text{ the probability that the system} \\
\text{can be found in a particular} \\
\text{microstate with an energy } E_k.
\end{aligned}
\tag{12.59}
$$

Remember, the probability given by Expressions 12.58 and 12.59 is for a particular microstate, which may be just one of many microstates with the same energy. At first it may appear to be a reasonable assumption that all values of the energy are equally probable. If this is true, both $\mathscr{P}(E_k)$ and $f(E)$ would have to be independent of the energy. This is not possible, as it would not be possible to normalize $\mathscr{P}(E_k)$ and $f(E)$. We know that the sum of the probabilities must equal one, i.e.

$$
\int_V d^3 r_1 \cdots \int_{-\infty}^{\infty} d^3 v_N \, f[E(\mathbf{r}_1, \ldots, \mathbf{v}_N)] = 1
\tag{12.60}
$$

where $\int_V$ indicates that the spatial integrals extend over the volume of the system. Now, if $f(E)$ is independent of E, we can take it out of the integral sign and find that the second integral equals ∞. Thus, Equation 12.60 cannot be satisfied. So, $f(E)$ and $\mathscr{P}(E_k)$ cannot be independent of E. Later we shall consider how to determine $\mathscr{P}(E_k)$ and $f(E)$.

12.8 SOME USEFUL POSTULATES

For theoretical predictions of various probabilities we need some statistical postulates. In this section we shall consider these postulates and how we can formulate them. We shall consider the simple case of an isolated system which has energy in a small range between E and $E + \partial E$. Such a system can be found in any one of a large number of accessible microstates (degeneracy). We wish to find out what we know about the probability of finding the system in any one of these microstates.

Postulate 1: Consider an ensemble of isolated systems which is known at some time to be uniformly distributed over all the accessible microstates. This system in a given state will not remain there forever, but will continually make transitions between the different accessible microstates without giving preference to any one of the accessible microstates over another. So, as time passes it is not expected that the number of systems in some particular subset of the accessible microstates will be greater, while that in some other subset will become less. Hence, we can say that if the systems are at the beginning uniformly distributed over all their accessible microstates, then they will remain unchanged in time. This means that if an isolated system has equal probability of being found in each one of its accessible microstates, then we can say that the probability of finding the system in each one of its microstates is independent of time. This statement implies that such an ensemble is in equilibrium. Hence, we can combine these two statements together to a single statement: *If an isolated system is in equilibrium, it has equal probability of being found in each one of its accessible microstates.*

Postulate 2: Let us consider an ensemble of isolated systems which are such that at some initial time each system is in some subset of the states which are actually accessible to it. Clearly, this ensemble at this time will contain many systems in some subset of their accessible states, and there will be no systems in the remaining accessible microstates. The laws of mechanics can neither favour some of the accessible microstates over the others, nor prevent the system from being found in any one of these microstates. Thus, as time goes on, it is not at all likely that a system in the ensemble will stay indefinitely in the subset of the microstates in which it was at the initial time and will not make a transition to the other microstates that are equally accessible to it.

The situation now is similar to the case of an ideal gas in a box. If the system is left alone over a period, it will continually make transitions, through interactions between its particles, between all of its accessible microstates. Consequently, there will be a mixing such that each system in the ensemble will eventually go through virtually all the microstates where it can possibly be found, and a uniform (random) distribution will occur. According to *Postulate 1* this distribution will remain uniform, which is actually an equilibrium state independent of time. Therefore, we can conclude: *An isolated system is not in equilibrium if it is not found with equal probability in each one of its accessible states.*

There are a number of implication of this postulate. *Firstly*, if the isolated system is not in equilibrium, it will tend to change in time until it reaches the equilibrium position. Then, it will be found with equal probability in each one of its accessible microstates. *Secondly*, this postulate provides an indication about the direction in which the system will change; this is the direction in which the system changes towards the equilibrium position, where it will exist in a uniform statistical distribution over all its accessible microstates. *Thirdly*, the postulate does not give any indication as to the actual time that will be necessary for it to attain the final equilibrium condition. The time required to achieve the final equilibrium condition is called *relaxation time*. This time may be quite long (many years) or very short (a microsecond) depending on a number of factors such as (i) the frequency with which transitions occur between the accessible microstates of the system, and (ii) the nature of the interactions between the particles of the system.

Postulate 3: The converse of *Postulate 2* leads to another postulate: *An isolated system in equilibrium can be found with equal probability in each one of its accessible microstates.* It is valid by virtue of what we said in *Postulate 2*. This postulate is the fundamental postulate of equilibrium statistical mechanics, and is referred to as the *postulate of equal a priori probabilities*.

According to the postulate of equal *a priori* probabilities a closed system which is in internal thermodynamic equilibrium is likely to be found in any one of the microstates accessible to it. A closed system is not under any external influence; also, its total volume, total energy and total number of particles are all constant. Therefore, there is no reason for us to believe that if such a system is at thermodynamic equilibrium, any particular one of the microstates accessible to the system is more probable than any of the other microstates accessible to it.

Postulate 4: It follows that due to the sharpness of the distributions of probability for macroscopic systems it can be assumed to a good approximation that the equilibrium values of the macroscopic variables of classical thermodynamics are given by the ensemble averages of the corresponding microscopic quantities.

Actually, for macroscopic systems, the ensemble averages are given to a very good approximation by the most probable values. However, it is often easier to estimate the most probable value than to find the ensemble average. This postulate relates the macroscopic variables of classical equilibrium thermodynamics to microscopic theory.

A quantitative description of systems which are not in equilibrium is generally much more difficult than that of equilibrium systems. Here, it is necessary to deal with processes which change in time and require a detailed analysis of (i) how effectively molecules interact with each other, (ii) how the probability of finding a system in each of its states changes with time, and (iii) how fast or slowly the processes change. Except for simple cases such as dilute gases, a quantitative description of a nonequilibrium system can become fairly complicated. On the contrary, such a description of a system in equilibrium only requires use of the postulate of *equal a priori probabilities*.

12.9 DISTINGUISHABLE AND INDISTINGUISHABLE PARTICLES

In classical mechanics there is no need to distinguish between the particles which form the system being considered, but in quantum mechanics we must distinguish between two possible situations. If two particles have some different measurable property (say, mass, frequency, etc.), then we say the two particles are *distinguishable*. On the other hand, if no experiment can distinguish one particle from another, then we say they are *indistinguishable*. For example, two H atoms are indistinguishable, but a H atom and a He atom are distinguishable.

In classical mechanics it is always assumed that particles are distinguishable even when their physical properties are the same. This helps to identify *which is which* by measuring their velocity and position at small intervals of time. However, in quantum mechanics there are restrictions on the accuracy of measurements of position and velocity at small intervals of time. So, it becomes impossible to identify one particle from another unless they have different physical properties.

(a) *Distinguishable particles*: A gas with only one type of gas molecule is not a case of distinguishable particles. As a common example, we can refer to a situation involving N distinguishable particles is sound waves in solid or electromagnetic waves, where each wave (may be considered as *particle*) is distinguishable from another by its direction of propagation, wavelength or frequency. Consider a system of N distinguishable particles. Since the particles are distinguishable, it is possible to label them as $i = 1, 2, 3, \ldots, N$ and we can easily identify any particle by the value of i. In such a system, if ε_i is the energy of the ith particle, then the total energy (E) of the N noninteracting particles is given by

$$E = \sum_i \varepsilon_i \tag{12.61}$$

where $i = 1, 2, 3, \ldots, N$ and the partition function is given by

$$Z = \sum_{ms} \exp\left(-\beta \sum \varepsilon_i \right) \tag{12.62}$$

where ms indicates microstates and β is a constant. We shall later discuss the partition function and its significance, and also identify β in section 13.3. The summation of the microstates is now the sum over all possible values of the energies of each particle, i.e. $\varepsilon_1, \varepsilon_2, \ldots, \varepsilon_N$. Hence, Equation 12.62 takes the form

$$Z = \left(\sum_{\varepsilon_1} \exp(-\beta\varepsilon_1) \right) \left(\sum_{\varepsilon_2} \exp(-\beta\varepsilon_2) \right) \cdots \left(\sum_{\varepsilon_N} \exp(-\beta\varepsilon_N) \right). \tag{12.63}$$

This equation can be written in a condensed form as

$$Z = \prod_{i=1}^{N} \left(\sum_{\varepsilon_i} \exp(-\beta\varepsilon_i) \right). \tag{12.64}$$

The symbol $\prod$ represents the product.

(b) *Indistinguishable particles*: In this case, our problem is: how do we characterize a microstate of a system. In the case of distinguishable particles, ε_i represented the energy of the ith particle. In the case of indistinguishable particles, let $\varepsilon(s)$ represent the possible values of the energy of any one of the indistinguishable particle in its sth quantum state. Here, s refers to the particle's quantum state, not to the particle. Thus, each particle can have energy $\varepsilon(1), \varepsilon(2), \ldots$. Let

$$n_s = \text{the number of particles in the state } s. \tag{12.65}$$

Two microstates in the system can be distinguished if n_s (generally referred to as the *occupation number*) values are different in the two cases. Therefore, we can say that *a microstate of a system of noninteracting indistinguishable particles can be specified by* $n_1, n_2, n_3, \ldots$. In this respect, the occupation numbers are important.

In terms of n_s, the total energy of the system of noninteracting particles can be written as

$$E = \sum_s n_s \varepsilon(s) \tag{12.66}$$

where the summation is for $s = 1, 2, 3, \ldots, \infty$. In words, the total energy is equal to the number of particles in the particle state s, multiplied by the energy of that state and then summed over all the states. For N particles, we have

$$N = \sum_s n_s \tag{12.67}$$

because each particle is in some particle state s. Logically, we should now substitute the value of E given by Equation 12.66 into Equation 12.62 and then sum over all allowed values of n_s which satisfy Equation 12.67. According to what we said above, this summation represents the summation over all microstates. However, this operation is not easy.

Here, we make use of an approximation to avoid the difficulty in computing the summation. The simplification occurs in the case of a *nondegenerate system*. We have defined the number of accessible microstates as the degeneracy or *statistical weight*. A microstate is said to be nondegenerate if no two particles have the same energy. Another way of defining a nondegenerate system is to say that

$$(\text{The number of particle states with } \varepsilon(s) < kT) \gg N \tag{12.68}$$

where k is the Boltzmann constant, T is the temperature and N is the number of particles. We shall prove this inequality at a later stage in this section. However, clearly this equation can always be satisfied if T is large and N is not too big.

The probability of finding a particle in a particle state with energy $\varepsilon(s)$ is proportional to $\exp(-\beta\varepsilon(s))$. When Equation 12.68 is satisfied, each particle will have a choice of a large number of states, each with about the same probability. If many states with energy $\varepsilon(s) < kT$ exist, then a particle is equally likely to be found in any of these states. However, remember that the number of particles is much less than that of these states. So, it is not likely that two particles will be in the same state. This implies that for most microstates either $n_s = 0$ or $n_s = 1$ for all values of s.

For $N = 1$, the microstates are $(1, 0, 0, \ldots), (0, 1, 0, \ldots), (0, 0, 1, \ldots), \ldots$, corresponding to $n_1 = 1, n_2 = 1, n_3 = 1, \ldots$. Thus, the partition function for a system with only one particle, i.e. $N = 1$ (which is known as the *particle partition function*) is

$$z = \exp(-\beta\varepsilon(1)) + \exp(-\beta\varepsilon(2)) + \exp(-\beta\varepsilon(3)) + \cdots$$
$$= \sum_s \exp(-\beta\varepsilon(s)) \tag{12.69}$$

where $s = 1, 2, \ldots, \infty$. We can extend the same procedure for any number of particles.

Consider the case for $N = 2$. Here, some microstates will correspond to the existence of both particles in the same particle state ($n_s = 2$). However, if the system satisfies Equation 12.68, i.e. if it is nondegenerate, then it is quite logical to imagine that most of the microstates will correspond to the existence of the two particles in different particle states. This means that $n_s = 0$ or $n_s = 1$, and if we disregard the microstates corresponding to $n_s = 2$, then the partition function Z is approximately given by

$$Z \simeq \exp[-\beta[\varepsilon(1) + \varepsilon(2)]] + \exp[-\beta[\varepsilon(1) + \varepsilon(3)]]$$
$$+ \exp[-\beta[\varepsilon(1) + \varepsilon(4)]] + \cdots + \exp[-\beta[\varepsilon(2) + \varepsilon(3)]]$$
$$+ \exp[-\beta[\varepsilon(2) + \varepsilon(4)]] + \cdots. \tag{12.70}$$

This equation is very similar to the square of Equation 12.69, i.e.

$$z^2 = [\exp[-\beta\varepsilon(1)] + \exp[-\beta\varepsilon(2)] + \exp[-\beta\varepsilon(3)] + \cdots]^2$$
$$= \exp[-\beta2\varepsilon(1)] + 2\exp[-\beta[\varepsilon(1) + \varepsilon(2)]] + 2\exp[-\beta[\varepsilon(1) + \varepsilon(3)]]$$
$$+ \cdots + \exp[-\beta2\varepsilon(2)] + 2\exp[-\beta[\varepsilon(2) + \varepsilon(3)]]$$
$$+ 2\exp[-\beta[\varepsilon(2) + \varepsilon(4)]] + \cdots + \exp[-\beta2\varepsilon(3)] + \cdots.$$

In this equation the terms $\exp[-\beta2\varepsilon(1)]$, $\exp[-\beta2\varepsilon(2)]$, $\exp[-\beta2\varepsilon(3)]$, etc., correspond to $n_s = 2$. If we disregard these terms, we find that the above equation and Equation 12.70 differ only by a factor of 2, i.e.

$$\text{for } N = 2, \quad Z \simeq z^2/2 \tag{12.71}$$

For $N = 3$, we shall have terms such as $\exp[-\beta[\varepsilon(1) + \varepsilon(2) + \varepsilon(3)]]$, $\exp[-\beta[\varepsilon(2) + \varepsilon(3) + \varepsilon(4)]]$, etc., whereas the square of Equation 12.69 will give terms of the form

$3! \exp[-\beta[\varepsilon(1)+\varepsilon(2)+\varepsilon(3)]]$. Thus, to make Z equal to z^2 we have to divide z^2 by $3!$, which will give

$$\text{for } N = 3, \quad Z \simeq z^3/3!. \tag{12.72}$$

We can now extend the above argument to a *nondegenerate system* of higher values of N and find that for indistinguishable particles

$$Z \simeq z^N/N!. \tag{12.73}$$

Consider again Equation 12.64. If the possible values of ε_i for all the particles are the same [say, $\varepsilon(1)$, $\varepsilon(2), \ldots$], then each of the sums in this equation would be equal to

$$\sum_{s=1}^{\infty} \exp[-\beta\varepsilon(s)].$$

Because Equation 12.64 is a product of N number of these sums, we get

$$Z = \left(\sum_{s=1}^{N} \exp[-\beta\varepsilon(s)] \right)^N = z^N. \tag{12.74}$$

Comparing Equations 12.73 and 12.74, we see that the result for distinguishable particles differs from the approximate result for indistinguishable particles by a factor of $1/N!$. This difference can be attributed to the difference in the number of microstates in the two cases.

Finally, we now attempt to prove the inequality, Equation 12.68, and determine under what conditions the inequality holds for perfect gases. In Section 12.6.1 we have seen that the quantum equation for the translational energy (ε_t) for a molecule in a cubic box of sides L_x, L_y and L_z is given by Equation 12.40. We also know that the difference between the energy levels is usually very small. Equation 12.68 suggests that the system will not be a degenerate system if the number of states with translational energy $\varepsilon_t < kT$ is much bigger than N. To find the maximum number of translational states which satisfy this condition we rewrite Equation 12.40 as

$$(\varepsilon_t)_{max} = \left(\frac{h^2}{2mL^2} \right) n_{max}^2 = kT \tag{12.75}$$

and

$$n_{max} = (L/h)(2mkT)^{1/2} \tag{12.76}$$

where $L_x = L_y = L_z = L$, and $(\varepsilon_t)_{max}$ and n_{max}^2 represent the maximum values of ε_t and n. The possible values of the translational quantum numbers (n_x, n_y, n_z) are all positive integers. In the co-ordinate system (n_x, n_y, n_z) the allowed values of these translational quantum numbers are given by points in this space (see Figure 12.8) and these points are arranged in a cubic lattice. Since each cube is of side unity, it has a volume of one. For each allowed point there is only one such cube.

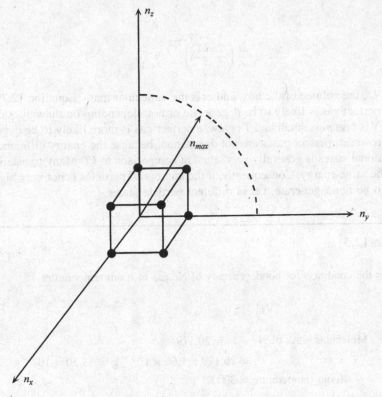

Figure 12.8 The points given by the allowed values of n_x, n_y and n_z, and the spherical region generated by n_{max}

Now, imagine that n_{max} generates a spherical region of radius n_{max}. The number of states which satisfy $\varepsilon_t \leqslant kT$ are equal to the number of points which fall within this spherical region. In Section 12.6.1 we said that the volume of this spherical region is $\frac{1}{8}$ of that of a complete sphere ($\frac{4}{3}\pi n_{max}^3$). We also said earlier that the volume per point is 1. Therefore, the number of states inside this spherical region is approximately

$$\left[\left(\frac{1}{8}\right)\left(\frac{4}{3}\pi n_{max}^3\right)\right]\Big/ 1.$$

Hence, Equation 12.68 becomes

$$\left(\frac{1}{8}\right)\left(\frac{4}{3}\right)\pi n_{max}^3 \gg N. \tag{12.77}$$

Combining Equations 12.76 and 12.77, we get, for a nondegenerate perfect gas,

$$\frac{\pi L^3}{6}\left(\frac{2mkT}{h^2}\right)^{3/2} \gg N \tag{12.78}$$

or

$$\frac{\pi}{6}\left(\frac{2mkT}{h^2}\right)^{3/2} \gg \frac{N}{V} \tag{12.79}$$

where $V = L^3$, the volume of the box, and m is the molecular mass. Equation 12.79 indicates whether a perfect gas is likely to be degenerate or not, depending on the values of m, T and N/V. If N/V is big, m is small and T is low, a perfect gas is more likely to be degenerate. At ordinary pressures, perfect gases are not degenerate, because the energy difference between the translational states is generally very small in comparison to kT. Many translational states can have the same energy. Consequently, if the number of particles is not very high, they are all likely to be nondegenerate, i.e. in different particle states.

Example 12.5

Determine the condition for nondegeneracy of Ne gas at room temperature.

Solution

$$\text{Molecular mass of Ne} = m = 20.179$$
$$= 20.179 \times 1.66 \times 10^{-24}\,\text{g} = 33.50 \times 10^{-24}\,\text{g}$$
$$\text{Room temperature} = 300\,\text{K}$$
$$\text{Boltzmann constant} = 1.38 \times 10^{-23}\,\text{JK}^{-1}$$
$$\text{Planck constant} = 6.63 \times 10^{-34}\,\text{J s}$$

Then,

$$\frac{2mkT}{h^2} = \frac{2(33.50 \times 10^{-24}\,\text{g})(1.38 \times 10^{-23}\,\text{JK}^{-1})(300\,\text{K})}{(6.63 \times 10^{-34}\,\text{J s})^2}$$
$$= 630.98 \times 10^{21}\,\text{g/J s}^2$$
$$= (630.98 \times 10^{21})\frac{\text{g}}{\text{s}^2}\left(\frac{\text{s}^2}{1000\,\text{g} \times 10^4\,\text{cm}^2}\right) \quad \left(\text{as } 1\,\text{J} = \frac{\text{kg m}^2}{\text{s}^2}\right)$$
$$= 630.98 \times 10^{14}\,\text{cm}^{-2}$$

Therefore,

$$\frac{\pi}{6}\left(\frac{2mkT}{h^2}\right)^{3/2} = \frac{\pi}{6}(630.98 \times 10^{14}\,\text{cm}^{-2})^{3/2}$$
$$= 8.29 \times 10^{24}\,\text{cm}^{-3}.$$

Hence, the gas is nondegenerate if $8.29 \times 10^{24}\,\text{cm}^{-3} \gg N/V$. At room temperature, a density of this size would correspond to a pressure $\simeq 3000\,\text{MPa}$. So at ordinary pressures, perfect gases are not degenerate.

12.10 PARTICLES DEFINED BY SYMMETRIC AND ANTISYMMETRIC WAVE FUNCTIONS

We have considered gases consisting of identical particles. Among identical particles, we shall have to make distinction between them; this is what we have discussed in Section 12.9. The distinction is defined as: *distinguishable* particles which are described by classical mechanics, and *indistinguishable* particles which are described by quantum mechanics. The indistinguishable particles will further have to be classified to study many of the thermodynamic properties of gases. These are classified as bosons or fermions.

Consider a gas with N indistinguishable particles. Let $\psi(1, 2, \ldots, N)$ be the wave function of the system, obeying the Schrödinger equation

$$H\psi(1, 2, \ldots, N) = E\psi(1, 2, \ldots, N) \qquad (12.80)$$

where E is the energy.

By the numbers $1, 2, \ldots, N$ we represent the totality of the properties that characterize the particles numbered from 1 to N. However, the numbering of the particles is wholely artificial, as they are indistinguishable. To avoid this artificiality we introduce an operator P which permutes some arbitrary sub-group of the N indistinguishable particles; there are $N!$ possible permutations.

As all the particles of the system are identical, the Hamiltonian H is invariant, i.e. it is the same for all the particles, under any such permutation P, and H and P are commutative. Thus, $PH = HP$. Applying P to Equation 12.80, we get

$$\begin{aligned} P[H\psi(1, 2, \ldots, N)] &= H[P\psi(1, 2, \ldots, N)] \\ &= E[P\psi(1, 2, \ldots, N)]. \end{aligned} \qquad (12.81)$$

This equation implies that $P\psi(1, 2, \ldots, N)$ is an eigenfunction of H corresponding to E. Therefore, we have $N!$ linear combinations of these wave functions which are able to describe a system of N indistinguishable particles with an energy E. However, there are only two different types of particle which are described by two very particular linear combinations of these wave functions. These two categories of particles are: particles described by a wave function that is totally *symmetric* with respect to the interchange of any pair of particles and is given by

$$\psi_S(1, 2, \ldots, N) = (1/N!) \sum_P P\psi(1, 2, \ldots, N) \qquad (12.82)$$

where the summation is over the $N!$ possible permutations of all the particles; and particles which are described by a wave function that is totally *antisymmetric* with respect to the interchange of any pair of particles and is given by

$$\psi_{A \cdot S}(1, 2, \ldots, N) = (1/N!) \sum_P \alpha_P \psi(1, 2, \ldots, N). \qquad (12.83)$$

α_P can be $+1$ or -1, depending on whether P is an even permutation of $1, 2, \ldots, N$, or an odd permutation of $1, 2, \ldots, N$.

Any system of particles described by Equation 12.82 (i.e. a totally symmetric wave function) is referred to as a system of *bosons*, and the system obeys *Bose–Einstein statistics*. On the other hand, a system of particles described by Equation 12.83 (i.e. a totally anti-symmetric wave function) is called a system of *fermions*, and the system obeys *Fermi–Dirac statistics*. All particles with spins of integral numbers are bosons (for example, photons, π and κ mesons, etc.) and those with half-integer spins are fermions (for example, electron, proton, neutron and μ mesons). Later, we shall present a comprehensive discussion of bosons and fermions.

In the case of free particles, the wave function $\psi(1, 2, \ldots, N)$ can be represented by a product of 1-particle wave functions as

$$\psi(1, 2, \ldots, N) = \psi_i(1)\psi_j(2)\cdots\psi_r(N) \tag{12.84}$$

where $i, j, \ldots, r$ represent the energy states of the particles. With the help of this equation we can now write Equation 12.83 in the following form:

$$\psi_{A \cdot S}(1, 2, \ldots, N) = \begin{vmatrix} \psi_i(1) & \psi_i(2) & \cdots & \psi_i(N) \\ \psi_j(1) & \psi_j(2) & \cdots & \psi_j(N) \\ \cdots & \cdots & \cdots & \cdots \\ \cdots & \cdots & \cdots & \cdots \\ \psi_r(1) & \psi_r(2) & \cdots & \psi_r(N) \end{vmatrix}. \tag{12.85}$$

This determinant is totally antisymmetrical; the interchange of two particles can be done by interchanging two columns, and this will involve a change of sign. The determinant becomes zero if $\psi_i = \psi_j$, i.e. two particles are in the same state. This leads to a useful postulate: *if a system of particles is described by a totally antisymmetric wave function, no two particles can occupy the same state* (Pauli exclusion principle).

We now see that the condition that a system of bosons or fermions has a wave function with specified symmetry properties implies that there are some correlations among the particles even if the particles are not interacting.

In Chapter 14, Fermi–Dirac and Bose–Einstein statistics will be considered in some detail to bring out the inherent distinctions between boson and fermion.

12.11 ENERGY BETWEEN MACROSCOPIC SYSTEMS

Let us consider two macroscopic systems A and B with energies U_A and U_B. Suppose that the energy scale in each system is subdivided into very small but equal intervals of size ∂U, where the size of ∂U is large enough to contain many states. Let Ω_A be the number of states accessible to A when its energy is between U_A and $U_A + \partial U$, and let Ω_B be the number of states accessible to B when its energy is between U_B and $U_B + \partial U$. As, to a good approximation, we can treat all energies in a way such that they could take only discrete values separated by ∂U, the counting of states is simplified.

Both A and B are assumed to be free to exchange energy in the form of heat, and the energy of each system separately is not constant. However, the combined system $A + B$ is isolated so that its total energy U_{A+B} must remain constant and is given by

$$U_A + U_B = U_{A+B} + U_{int} \tag{12.86}$$

where U_{int} is the interaction energy. This interaction energy depends on both A and B and on the amount of work required to bring the systems together. We know from our previous discussion that thermal interaction is weak enough to be ignored. Since $U_{int} \ll U_A$ or U_B, Equation 12.86 becomes

$$U_A + U_B = U_{A+B} = \text{constant.} \qquad (12.87)$$

Let us now consider the situation where the combined system $A + B$ is in equilibrium. In this situation the energy of A can take many possible values; however, we want to know what is the probability $\mathscr{P}(U_A)$ that the energy of A lies in between U_A and $U_A + \partial U$, where U_A has any specified value. As for B, once the energy U_A of A has been specified, the energy U_B of B can be found from Equation 12.87.

According to *Postulate 3* (see Section 12.8), the combined system can be found with equal probability in each one of its accessible microstates.

Let $\Omega_{(A+B)}$ be the total number of states accessible to the total system $(A + B)$. We want to find out the number $\Omega(U_A)$ of states of $A + B$ which are such that the subsystem A has an energy equal to U_A. Also, we want to see whether the number $\Omega(U_A)$ can be expressed in terms of the numbers of states separately accessible to A and B. According to our fundamental postulate the system has equal probability of being found in each one of its accessible states. Through this simple argument, the desired probability $\mathscr{P}(U_A)$ is given by

$$\mathscr{P}(U_A) = \frac{\Omega(U_A)}{\Omega_{(A+B)}}.$$

Since $\Omega_{(A+B)}$ is merely a constant independent of U_A, we can write the above equation as

$$\mathscr{P}(U_A) = \Omega_{(A+B)}^{-1} \Omega(U_A).$$

Now, when A has an energy U_A, it can be in any one of its Ω_A possible states. The system B will then have an energy U_B given by $U_B = U_{A+B} - U_A$ (from Equation 12.87). This follows from the conservation of energy. Hence, the system B could be in any of the $\Omega_B(U_B) = \Omega_B(U_{A+B} - U_A)$ states which are accessible to it under these conditions.

We can combine every possible state of A with every possible state of B to get a different possible state of the combined system $A + B$. Therefore, the number of distinct states which are accessible to the total system $A + B$ when A has an energy U_A is given by

$$\Omega(U_A) = \Omega_A \Omega_B(U_{A+B} - U_A).$$

Substituting this into the above equation for $\mathscr{P}(U_A)$ we find that the probability that the system A has an energy U_A is given by

$$\mathscr{P}(U_A) = \Omega_{(A+B)}^{-1} \Omega_A \Omega_B(U_{A+B} - U_A).$$

It can be easily seen from this equation that if U_A increases, Ω_A increases very rapidly, but $\Omega_B(U_{A+B} - U_A)$ decreases very rapidly. Thus, $\mathscr{P}(U_A)$ exhibits a sharp maximum for some particular energy U_A' of U_A.

12.12   CLOSED AND OPEN SYSTEMS

In an isolated system the energy is constant, and according to the basic assumption of statistical mechanics (i.e. all microstates of a system that have the same energy are assumed to be equally probable) all microstates with this energy are equally probable. A collection or ensemble of such isolated systems (which are all identical) is often called a *microcanonical ensemble*. However, we are interested not only in isolated systems, but also in those systems which are in thermal contact with heat reservoirs or other systems. In this case, the system can exchange energy with the heat reservoir or other systems; therefore, the energy of the system is no longer constant. A collection or ensemble of such systems which are all in thermal equilibrium with reservoirs at the same temperature is called a *canonical ensemble*. Here, although the system exchanges energy with the reservoir, there is no exchange of particles; the number of particles remains fixed. Hence, they are *closed systems*.

On the other hand, there are systems in which an exchange of particles occurs; thus the number of particles will vary. Such systems are called *open systems*. A collection of such systems which are all in thermal equilibrium with heat reservoirs or other systems at the same temperature is referred to as a *grand canonical ensemble*. An example of such a situation is a liquid in equilibrium with its vapour. In this case, both the energy and the number of atoms in the liquid are continually changing. So, we can define an open system as one which can exchange particles with a particle reservoir and energy with a thermal reservoir (that may be distinct from the particle reservoir). However, the exchange of energy and particle is quite independent. Therefore, the number of particles N and the energy E are both independent variables of the system.

Both 'canonical' and 'grand canonical' ensembles are rather pompous names for simple concepts. It is much easier to comprehend what are *closed* and *open* systems than *canonical* and *grand canonical* ensembles. Actually, the epithet *canonical* refers to the fact that the ensemble is the simplest type. The prefix *micro* simply means a small part of a canonical ensemble. We shall use the concept of a grand canonical ensemble in dealing with distribution functions in Chapter 14.

The fact that an open system can exchange both energy and particles leads to the question: how do we treat these systems? The microstates of an open system are defined in the same way as for closed systems, except that now a system is also considered to be in a different microstate if the number of particles N has changed. Therefore, we can apply the same reasoning as used for closed systems to extend the basic assumption to open systems and write: *All (micro) states with the same number of particles N and the same amount of energy E are assumed to be equally probable.* We represent the probability of a particular microstate by $\mathscr{P}(E, N)$ and determine the dependence of $\mathscr{P}(E, N)$ on E and N by a method similar to that used in the case for closed systems.

Let us consider a composite system $X + Y$ which is made up of the macroscopic systems X and Y having the number of particles N_X and N_Y, and the energy E_X and E_Y, respectively. Then, $N = N_X + N_Y$, and since X and Y are macroscopic systems, $E_{X+Y} \simeq E_X + E_Y$. We assume that these systems are free to exchange both energy and particles with each other, as well as with a common reservoir (thermal and particle). We represent the probability that the composite system $X + Y$ is in a state with energy E_{X+Y} and number of particles N by $\mathscr{P}_{X+Y}(E_{X+Y}, N)$.

As in the case of closed systems, here we consider that the energies of X and Y are independent of each other, and also that the number of particles in X is independent of that in Y (because both X and Y are free to exchange particles with the particle reservoir). Thus, we say that the microstate (E_X, N_X) is not dependent on the microstate (E_Y, N_Y). So it follows that

$$\mathscr{P}_{X+Y}(E_X + E_Y, N_X + N_Y) = \mathscr{P}_X(E_X, N_X)\mathscr{P}_Y(E_Y, N_Y). \tag{12.88}$$

It can be seen from Equation 12.88 that

$$\mathscr{P}_X(E_X, N_X)\frac{\partial}{\partial E_Y}[\mathscr{P}_Y(E_Y, N_Y)] = \mathscr{P}_Y(E_Y, N_Y)\frac{\partial}{\partial E_X}[\mathscr{P}(E_X, N_X)] \tag{12.89}$$

and

$$\mathscr{P}_X(E_X, N_X)\frac{\partial}{\partial N_Y}[\mathscr{P}(E_Y, N_Y)] = \mathscr{P}_Y(E_Y, N_Y)\frac{\partial}{\partial N_X}[\mathscr{P}_X(E_X, N_X)]. \tag{12.90}$$

Note: Since E_X and E_Y are independent variables, we can differentiate Equation 12.88 with respect to either E_X or E_Y, holding the other variable constant. For any function $f(X+Y)$ which depends on the sum $X+Y$ we have $\partial f(X+Y)/\partial X = \partial f(X+Y)/\partial Y$.

$$\frac{\partial f(X+Y)}{\partial X} = \left[\frac{\mathrm{d}f(X+Y)}{\mathrm{d}(X+Y)}\right]\frac{\partial(X+Y)}{\partial X} = \frac{\mathrm{d}f(X+Y)}{\mathrm{d}(X+Y)}$$

$$\frac{\partial f(X+Y)}{\partial Y} = \left[\frac{\mathrm{d}f(X+Y)}{\mathrm{d}(X+Y)}\right]\frac{\partial(X+Y)}{\partial Y} = \frac{\mathrm{d}f(X+Y)}{\mathrm{d}(X+Y)}.$$

Hence the result.

From Equation 12.89 we get

$$\frac{\partial}{\partial E_X}[\ln \mathscr{P}_X(E_X, N_X)] = \frac{\partial}{\partial E_Y}[\ln \mathscr{P}(E_Y, N_Y)]. \tag{12.91}$$

Therefore, each side must be equal to some constant. We take this constant as $-\beta$ (β depends on the composition of the system). Integrating the left-hand side of Equation 12.91, we get

$$\ln \mathscr{P}_X(E_X, N_X) = -\beta E_X + C_1 \tag{12.92}$$

where C_1 is a constant of integration dependent on N_X. Hence,

$$\mathscr{P}_X(E_X, N_X) = \exp(-\beta E_X)\exp C_1. \tag{12.93}$$

Similarly, we get an equation for Y which is

$$\mathscr{P}_Y(E_Y, N_Y) = \exp(-\beta E_Y)\exp C_2 \tag{12.94}$$

where C_2 is a constant of integration dependent on N_Y.

Next, we determine the integration constant. From Equation 12.90 we get

$$\frac{\partial}{\partial N_X}[\ln \mathscr{P}_X(E_X, N_X)] = \frac{\partial}{\partial N_Y}[\ln \mathscr{P}(E_Y, N_Y)].$$

Since the left and the right-hand sides depend on different independent variables, they must be equal to the same constant. We take this constant as $\mu\beta$, where the constant μ does not depend on the composition of the system. The distinction between μ and β is that, while the former characterizes the particle reservoir, the latter actually represents the thermal reservoir in the same way. We now have

$$\frac{\partial}{\partial N_X}[\ln \mathscr{P}_X(E_X, N_X)] = \mu\beta.$$

With the help of Equation 12.93 the above equation becomes

$$\frac{dC_1}{dN_X} = \mu\beta$$

and integrating

$$C_1 = \mu\beta N_X + d \tag{12.95}$$

where d is a constant of integration independent of E and N. Then, by combining Equations 12.93 and 12.94 we get

$$\mathscr{P}_X(E_X, N_X) = \exp(d) \exp(-\beta E_X + \mu\beta N_X). \tag{12.96}$$

We can now generalize Equation 12.96 to write the probability that the system is in a particular state with energy E and number of particles N as

$$\mathscr{P}(E, N) = Z^{-1} \exp(-\beta E + \mu\beta N) \tag{12.97}$$

where $Z^{-1} = \exp(d)$. In this equation the energy E is the energy of the N particles and Z is the *grand canonical partition function*. The value of Z is determined by the condition

$$\sum_{N=0}^{\infty} \sum_k \mathscr{P}(E_k, N) = 1 \quad \text{(normalization)} \tag{12.98}$$

where k is the number of states. Combining Equations 12.97 and 12.98, we get

$$Z = \sum_{N=0}^{\infty} \sum_k \exp(\mu\beta N) \exp(-\beta E_k) = \sum_{N=0}^{\infty} \exp(\mu\beta N) Z_N. \tag{12.99}$$

In this equation, the new constant μ is just the Gibbs free energy per particle. The energy sum is over all the k states of the system with N particles and includes any repetition due to different states with the same energy (degenerate states). The energy sum is followed by another sum over all allowed values of N. The first sum, i.e. the energy sum

$$\sum_k \exp(-\beta E_k)$$

is simply the partition function for a system of fixed value of N particles, and we denote it by Z_N.

Equations 12.97 and 12.99 are both equally valid for the classical case where the energy varies continuously, except that one must use the distribution function

$$f[E(\mathbf{r}_1, \mathbf{r}_2, \ldots, \mathbf{v}_N), N] = Z^{-1} \exp(\mu \beta N) \exp[-\beta E(\mathbf{r}_1, \ldots, \mathbf{v}_N)] \tag{12.100}$$

and

$$Z = \sum_{N=0}^{\infty} \exp(\mu \beta N) \int_V \cdots \int_{-\infty}^{\infty} \exp[-\beta E(\mathbf{r}_1, \ldots, \mathbf{v}_N)] d^3 r_1 \cdots d^3 v_N \tag{12.101}$$

The distributions Equations 12.97 and 12.100 are called *grand canonical distributions*.

It can be shown that the grand canonical partition function Z can be written as (see Equation 12.113)

$$Z = \exp(PV/KT) = \exp(\beta PV) \tag{12.102}$$

Hence, the grand canonical distributions 12.97 and 12.100 can be written

$$\mathscr{P}(E, N) = \exp[-\beta(E + PV - \mu N)] \tag{12.103}$$

and

$$f[E(\mathbf{r}_1, \ldots, \mathbf{v}_N), N] = \exp[-\beta\{E(\mathbf{r}_1, \ldots, \mathbf{v}_N) - \mu N + PV\}]. \tag{12.104}$$

12.12.1   The grand partition function and thermodynamics

Here an outline is given of the approach to statistical thermodynamics based on the grand partition function Z which is given by Equation 12.99 and is a function of the absolute temperature T of the reservoir, the chemical potential μ of the particle reservoir and the volume V of the small system as shown in Figure 12.9.

The small system and the heat and particle reservoir are separated by a rigid diathermic partition with holes in it which allow the exchange of energy and particles. Both the small system and the heat and particle reservoir are surrounded by rigid, adiabatic, outer walls which form a closed system of constant total energy, constant total number of particles and constant total volume.

In classical thermodynamics, the grand potential Ω is defined by

$$\Omega = U - TS - \mu N. \tag{12.105}$$

Considering the variables on the right-hand side of this equation, we can immediately say that Ω is an extensive variable and a function of the thermodynamic state of the system. For infinitesimal changes

$$d\Omega = dU - T\,dS - S\,dT - \mu\,dN - N\,d\mu. \tag{12.106}$$

However, $dU = T\,dS - P\,dV + \mu\,dN$. So, Equation 12.106 becomes

$$d\Omega = -S\,dT - N\,d\mu - P\,dV. \tag{12.107}$$

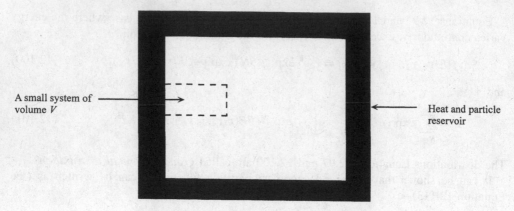

Figure 12.9 The grand canonical distribution

Since $d\Omega$ is a function of T, V, μ, i.e. $d\Omega = f(T, V, \mu)$, we can write

$$d\Omega = \left(\frac{\partial \Omega}{\partial T}\right)_{V, \mu} dT + \left(\frac{\partial \Omega}{\partial V}\right)_{T, \mu} dV + \left(\frac{\partial \Omega}{\partial \mu}\right)_{T, V} d\mu. \tag{12.108}$$

Comparing the coefficients of dT, dV and $d\mu$ in Equations 12.107 and 12.108, we get

$$-\left(\frac{\partial \Omega}{\partial T}\right)_{V, \mu} = S; \quad -\left(\frac{\partial \Omega}{\partial V}\right)_{T, \mu} = P; \quad -\left(\frac{\partial \Omega}{\partial \mu}\right)_{T, V} = N. \tag{12.109}$$

If we are considering a one-component system, we get

$$G = \mu N = U - TS + PV$$

or

$$-PV = U - TS - \mu N. \tag{12.110}$$

By combining Equations 12.105 and 12.110, we get

$$\Omega = -PV. \tag{12.111}$$

However, in statistical mechanics

$$\Omega = -kT \ln Z. \tag{12.112}$$

Then, by using Equation 12.111, Equation 12.112 becomes

$$-\Omega = PV = kT \ln Z. \tag{12.113}$$

It can be easily seen from Equations 12.109 and 12.112 that

$$-\left(\frac{\partial \Omega}{\partial T}\right)_{V,\mu} = S = \left[\frac{\partial(kT \ln Z)}{\partial T}\right]_{V,\mu} \tag{12.114}$$

$$-\left(\frac{\partial \mu}{\partial V}\right)_{T,\mu} = P = kT\left[\frac{\partial}{\partial V}(\ln Z)\right]_{T,\mu} \tag{12.115}$$

$$-\left(\frac{\partial \Omega}{\partial \mu}\right)_{V,T} = N = kT\left[\frac{\partial}{\partial \mu}(\ln Z)\right]_{V,T}. \tag{12.116}$$

From Equation 12.110 we get $U = TS - PV + \mu N$. If we now substitute the values of PV, S and N from Equations 12.111, 12.114 and 12.116, respectively, into this equation, we get

$$U = \mu kT\left[\frac{\partial}{\partial \mu}(\ln Z)\right]_{T,V} + kT^2\left[\frac{\partial}{\partial T}(\ln Z)\right]_{V,\mu}. \tag{12.117}$$

We conclude that if the grand partition function Z of the small system (in Figure 12.9) is known as a function of T, V and μ, then we can calculate the thermodynamic quantities P, U, S and N using Equations 12.115, 12.117, 12.114 and 12.116, respectively.

EXERCISES 12

12.1 Consider a system consisting of 4 distinguishable particles which are designated as 1, 2, 3 and 4 inside a box which is divided into two halves. Determine the number of possible ways in which all the 4 molecules can be distributed between the two halves of the box.

12.2 Calculate the total number of microstates of a system consisting of 6.02×10^{23} particles.

12.3 Consider the composite system consisting of the subsystems 1 and 2 being separated by a diathermic partition, as shown in the following figure. Let the initial temperatures of the two

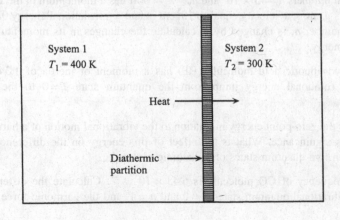

systems be $T_1 = 400\,K$ and $T_2 = 300\,K$. Assume that 1 J of heat has been transferred from system 1 to system 2 through the diathermic partition and then this partition has been immediately replaced with an adiabatic one. Calculate the changes in the entropy of the systems. Discuss the conditions for equilibrium, and reversible and irreversible changes in this case.

12.4 Define phase space, μ-space and γ-space. Show that the general form of the kinetic energy of a particle of mass m moving in a three-dimensional space can be expressed as $\varepsilon_k = \frac{1}{2}\sum p_i \dot{r}_i$ where ε_k is the kinetic energy, p_i is the generalized momentum $\dot{r}_i$ are the velocities corresponding to a set of co-ordinates r_i, $i = 1, 2, 3$.

12.5 Imagine that in classical mechanics a system with constant energy makes a transition from one state to another with the same energy to pass through a situation where the potential energy is higher than the total energy. Discuss whether this transition is possible. If so, what will be the resultant consequences?

12.6 Calculate the total number of accessible microstates in a system where 10^2, 10^3, 10^4, 10^5 and 10^6 units of energy have been distributed between 3 distinguishable particles of zero spin.

12.7 If the moment of inertia of HCl is $2.65 \times 10^{-40}\,\mathrm{g\,cm^2}$, calculate the distance of separation between H and Cl.

12.8 Determine the order of magnitude of the dipole moment for a molecule in which the distance of separation between an electron and a proton is 1 angstrom.

12.9 Prove that the quantum expression for the translational kinetic energy of a molecule inside a rectangular box is very similar to the classical expression $\frac{1}{2}m(v_x^2 + v_y^2 + v_z^2)$, except that the allowed values of the components of the momentum mv are discrete. Here, m is the mass, and v_x, v_y and v_z are the velocities in the x, y and z directions.

12.10 The translational kinetic energy of a molecule inside a rectangular box is given by Equation 12.40. Deduce by explaining logically that the number of energy levels below ε_t is

$$\frac{4\pi}{3}(2M\varepsilon_t)^{3/2} V/h^3$$

where all the symbols have their usual meanings.

12.11 A hydrogen atom inside a cubical box of volume $1\,\mathrm{cm^3}$ is in a translational state described by the quantum numbers $n_x = 5 \times 10^7$ and $n_y = n_z = 0$. It has a momentum in the x direction equal to $1.66 \times 10^{-19}\,\mathrm{g\,cm/s}$, a velocity equal to $10^5\,\mathrm{cm \cdot s}$ and a translational energy $0.83 \times 10^{-21}\,\mathrm{J}$. If the quantum number n_x is changed by 1, calculate the changes in its momentum, velocity and translational energy.

12.12 The hydrofluoric acid molecule (HF) has a moment of inertia of $1.34 \times 10^{-40}\,\mathrm{g\,cm^2}$. Calculate the rotational energy jump from the quantum state $J = 0$ to the quantum state $J = 1$.

12.13 Define the zero-point energy in relation to the vibrational motion of a harmonic oscillator and discuss its significance. What is the effect of this energy on the difference in vibrational energy between two quantum states of a diatomic molecule.

12.14 The frequency of CO molecule is $6.45 \times 10^{13}\,\mathrm{s^{-1}}$. Calculate the difference in energy between the vibrational quantum states $n = 0$ and $n = 1$, and the harmonic force constant.

12.15 Compare the energy difference $\Delta\varepsilon_v$ between the vibrational quantum states $n=0$ and $n=1$ of a chlorine molecule with that of CO molecule obtained in Exercise 12.14. The frequency of the Cl_2 molecule is $1.68 \times 10^{13}\,s^{-1}$. Calculate the angular velocity and the period of motion.

12.16 Discuss the interaction energy between two physically identical macroscopic systems.

12.17 In statistical mechanics the term *microstate* represents the statistical events that describe the microscopic state of the system. Schematically show the microstate of a hypothetical one-dimensional system in its phase space and discuss how one microstate changes into another.

12.18 Calculate the number of ways in which X indistinguishable balls can be put into N distinguishable holes. Assume there is no restriction on how many of the X balls are in each box.

12.19 A perfect gas of molecular mass $6.64 \times 10^{-24}\,g$ is confined in a cubical box of sides 2 cm at 300 K. What should be the value of the number density for this gas to be nondegenerate? Hence, estimate the number of particles in the total volume of the gas. Comment on the magnitude of this number.

12.20 Discuss the concept of *canonical* and *grand canonical* ensembles. What are the basic differences between the two?

12.21 In the case of a grand canonical ensemble, can we apply the basic assumption that all (micro)states with the same number of particles N and the same amount of energy E are equally probable?

12.22 Show that in the case of a grand canonical ensemble the probability that the system is in a particular state with energy E and number of particles N depends on E and N.

12.23 If the probability derived in Exercise 12.22 is to be valid for the classical case where the energy varies continuously, what is the other condition required?

12.24 Given that the grand partition function of a small system Z is a function of the absolute temperature T of the heat reservoir, the chemical potential μ of the particle reservoir and the volume V of the small system. Derive appropriate equations for the thermodynamic variables S, P, N and U in terms of Z, where all the symbols have their usual meanings.

12.25 One mole of a monatomic helium gas is confined in a cubical box of side $28.2 \times 10^{-2}\,m$ and volume $22.4 \times 10^{-3}\,m^3$ at 0.1 MPa pressure and 273 K. The mass of a helium atom is $6.65 \times 10^{-27}\,kg$.

(a) Calculate the energy of this gas in the ground state.

(b) Show that the next two quantum states above the ground state are threefold degenerate and have energies equal to 2 and 3 times the ground state energy.

(c) Prove that the mean energy of this gas is about 1.81×10^{19} times larger than the energy in the ground state.

(d) Show that a helium gas atom in the cubical box described above at 273 K is likely to be in a single particle quantum state with quantum numbers in the range 10^9 to 10^{10}.

(e) Discuss what happens to the energies of helium if the volume of the box is increased or decreased at STP (0.1 MPa pressure and 273 K temperature).

13

Statistical Mechanics Applied to Classical Thermodynamics

13.1 INTRODUCTION

In the previous chapter we have discussed various fundamental and basic principles of statistical mechanics. Here, we intend to use this knowledge to interprete the thermodynamic quantities and functions such as heat, temperature, pressure, enthalpy, entropy, etc., as well as to develop the laws of classical equilibrium thermodynamics. This will provide an opportunity to see the inter-relationships between statistical mechanics and classical thermo-dynamics. In Section 12.7 of Chapter 12 we discussed the basic assumption of statistical mechanics and introduced the idea of probability $\mathscr{P}(E_k)$ and energy $f(E)$. Before we go any further let us first see how we can determine these two quantities.

13.2 DEPENDENCE OF $\mathscr{P}(E_k)$ AND $f(E)$ ON THE ENERGY

In Section 12.7 of Chapter 12 we introduced the functions $\mathscr{P}(E_k)$ and $f(E)$. Here, we consider how these functions depend on the energy. Let us imagine that we have two systems A and B which are in thermal equilibrium with another systems, say a heat reservoir. Both A and B are free to exchange energy with this reservoir, so they interact with the reservoir indepen-dently of one another. This implies that at any instant of time the state of system A is independent of the state of system B. However, the probability of each system may depend on the nature of the system itself, i.e. the types of atom in it, etc. By definition the probability of finding the system A in a particular microstate with energy E_A is $\mathscr{P}_A(E_A)$ and that of finding the system B in a particular state with energy E_B is $\mathscr{P}_B(E_B)$. Similarly, the probability of finding the composite system $A + B$ made up of the systems A and B in a particular state with energy E_{A+B} is $\mathscr{P}_{(A+B)}(E_{A+B})$.

We already know that the interaction energy between two macroscopic systems is very small, so we may assume that both E_A and E_B are comparatively larger than the interaction energy of the systems A and B. This means that the total energy of the composite system $A + B$ is equal to the energies of A and B added together, i.e. $E_A + E_B = E_{A+B}$. Thus, the probability of finding the composite system $A + B$ in a particular state with energy E_{A+B} is given by

$$\mathscr{P}_{(A+B)}(E_{A+B}) = \mathscr{P}_{(A+B)}(E_A + E_B). \tag{13.1}$$

This equation may be viewed from another angle as the intersection of two states: (i) the existence of the system A in some particular state with energy E_A, and (ii) the existence of the system B in some particular state with energy E_B. Mathematically, this may be written as $\mathscr{P}[A(E_A)B(E_B)]$. In view of what we said at the beginning, the two events (i) and (ii) are independent. Also, we know that for two independent events the probability of their intersection is equal to the product of the probabilities of the individual events. Thus, in our case we may write $\mathscr{P}(AB) = \mathscr{P}(A)\mathscr{P}(B)$. Following this argument, Equation 13.1 becomes

$$\mathscr{P}_{(A+B)}(E_A + E_B) = \mathscr{P}[A(E_A)B(E_B)]$$
$$= \mathscr{P}_A(E_A)\mathscr{P}_B(E_B). \tag{13.2}$$

In this equation, E_A and E_B are independent variables. Also, if we have a function $f(x+y)$ which depends on x and y only through $x + y$, then

$$\frac{\partial f}{\partial x} = \left[\frac{df}{d(x+y)}\right]\left[\frac{\partial(x+y)}{\partial x}\right] = \frac{df}{d(x+y)}$$

$$\frac{\partial f}{\partial y} = \left[\frac{df}{d(x+y)}\right]\left[\frac{\partial(x+y)}{\partial y}\right] = \frac{df}{d(x+y)}.$$

By virtue of these results, we can show by differentiating Equation 13.2 with respect to E_A keeping E_B constant, and also differentiating with respect to E_B keeping E_A constant, that

$$\left[\frac{d\mathscr{P}_A(E_A)}{dE_A}\right]\left[\frac{1}{\mathscr{P}_A(E_A)}\right] = \left[\frac{d\mathscr{P}_B(E_B)}{dE_B}\right]\left[\frac{1}{\mathscr{P}_B(E_B)}\right]. \tag{13.3}$$

Since E_A and E_B can take any value, and as the left-hand side of Equation 13.3 is a function of E_A only and the right-hand side is a function of E_B only, then the only way this equation can be satisfied is for both the sides to be equal to a quantity which is not dependent on E_A or E_B. If we denote this quantity by $-\beta$, then from Equation 13.3 we get

$$\frac{d\mathscr{P}_A(E_A)}{dE_A} = -\beta\mathscr{P}_A(E_A)$$

or

$$\mathscr{P}_A(E_A) = C_A \exp\,(-\beta E_A) \tag{13.4}$$

and

$$\frac{d\mathscr{P}_B(E_B)}{dE_B} = -\beta\mathscr{P}_B(E_B)$$

or

$$\mathcal{P}_B(E_B) = C_B \exp\ (-\beta E_B). \tag{13.5}$$

In Equations 13.4 and 13.5, the factors C_A and C_B depend on the types of atoms of A and B, respectively. However, β does not depend on the composition of A or B, but it is common to both A and B, because both the left and right sides of Equation 13.3 equal $-\beta$. The question is: how is β common to both A and B?

Remember that the systems A and B are in thermal equilibrium with the heat reservoir, i.e. the only quantity which is common to both A and B is the temperature of the reservoir. Since β is common to both A and B, this quantity must be related to the temperature of the reservoir. Hence, we can generalize Equations 13.4 and 13.5 to write

$$\mathcal{P}(E) = C_1 \exp\ (-\beta E). \tag{13.6}$$

This equation gives the probability that a system which is in thermal equilibrium with a reservoir will be found in a particular state with energy E.

The quantity β does not depend on the composition of the system; it is related to the temperature of the reservoir. On the other hand, the factor C_1 is dependent on the composition of the system. Where the energy can vary continuously, we can write for the energy distribution function $f(E)$ (see Equation 12.58)

$$f[E(\mathbf{r}_1, \ldots, \mathbf{v}_N)] = C_2 \exp\ (-\beta E(\mathbf{r}_1, \ldots, \mathbf{v}_N)).$$

Ignoring the terms within the parentheses, we shall write this as

$$f(E) = C_2 \exp\ (-\beta E). \tag{13.7}$$

Equations 13.6 and 13.7 are often written in a different form, with C_1 and C_2 replaced by some expressions. These are derived as follows.

If the probabilities of all the events are added together, the sum must equal 1. So, for Equation 13.6, summing over all the microstates, we get

$$\sum_{ms} \mathcal{P}(E_k) = \sum_k C_1 \exp\ (-\beta E_k) = C_1 \sum_k \exp\ (-\beta E_k) = 1 \tag{13.8}$$

where ms represents *microstate*. From this equation we get

$$C_1 = \frac{1}{\sum_k \exp\ (-\beta E_k)} = Z^{-1} \tag{13.9}$$

where

$$Z = \sum_k \exp\ (-\beta E_k)$$

which is a dimensionless quantity. For discrete energies we use this value of C_1 and rewrite Equation 13.6 as

$$\mathscr{P}(E_k) = C_1 \exp \, (-\beta E_k) = Z^{-1} \exp \, (-\beta E_k). \tag{13.10}$$

Now we need a similar dimensionless quantity for the classical case. The basic assumption of statistical mechanics implies that the probability of a classical microstate can be written as in Equation 12.58. Since the probabilities of all the events added together must equal 1, we have

$$\int_V d^3 r_1 \cdots \int_{-\infty}^{\infty} f[E(\mathbf{r}_1, \ldots, \mathbf{v}_N)] d^3 v_N$$

$$= \int_V d^3 r_1 \cdots \int_{-\infty}^{\infty} C_2 \exp \, (-\beta E) d^3 v_N \quad [\text{using equation (13.7)}]$$

$$= C_2 \int_V d^3 r_1 \cdots \int_{-\infty}^{\infty} \exp \, (-\beta E) d^3 v_N = 1 \tag{13.11}$$

where the first integral extends over the volume of the system (see Section 13.3). From this, $C_2 = CZ^{-1}$, where

$$C = \frac{1}{\int_V d^3 r_1 \cdots \int_{-\infty}^{\infty} d^3 V_N}$$

and

$$Z^{-1} = \frac{1}{\int_{-\infty}^{\infty} \exp \, (-\beta E)}$$

Thus, Equation 13.7 becomes

$$f(E) = C Z^{-1} \exp \, (-\beta E). \tag{13.12}$$

It can be seen that C cancels out of the expression for $f(E)$. Its numerical value cannot be measured because the value of $f(E)$ does not depend on it. Furthermore, Z in Equation 13.12 is dimensionless if C has the dimensions (length-velocity)$^{-3N}$. From the point of view of classical mechanics, it is quite appropriate to use the simplest value of C as

$$C = 1 \, (\text{length-velocity})^{-3N}. \tag{13.13}$$

Equation 13.10 is the basic equation used for all calculations in statistical mechanics. The average microscopic properties of a system can be calculated only after the probabilities of the various microstates have been determined. Since the macroscopic properties of a system are related to its average microscopic properties, once the latter are known then the former could be determined. However, we have to be certain about which averages are related to which macroscopic properties before we perform a computation.

We have introduced the factor Z in Equations 13.10 and 13.12. The reader may be wondering what is the significance of it. For now, Z is only a normalizing factor which makes sure that the sum of the probabilities (Equations 13.10 and 13.12) is equal to 1. In fact, Z is a function of β and the volume of the system, and it is called the *partition function*.

It is easy to see that when $E = 0$, $\exp(-\beta E) = 1$ which is the largest value, but this does not mean that the system is most likely to be found with zero energy because, in general, there are many states of the system with the same energy. We can write

$$\begin{pmatrix} \text{Probability that the} \\ \text{system has an} \\ \text{energy } E \end{pmatrix} = \begin{pmatrix} \text{number of states} \\ \text{with energy } E \end{pmatrix} \mathscr{P}(E). \qquad (13.14)$$

The left-hand side does not mean that the system is in a particular state with energy E. The quantity in the parentheses on the right-hand side is called an *energy level*. It would, therefore, appear that an energy level is a compound event. Then, following the definition of the probability of such an event, we can write

$$\mathscr{P}_l(E) = \sum_{E_k} \mathscr{P}(E_k) = \sum \mathscr{P}(E) = \Omega(E)\mathscr{P}(E) \qquad (13.15)$$

where $\mathscr{P}_l(E)$ is the probability of an energy level and $\Omega(E)$ is the number of states with energy E.

$\Omega(E)$ is called the *degeneracy* or the *weighting factor* of the energy level E. It generally increases with increasing E; this means that there are more states with higher energy.

13.3   IDENTIFICATION OF β

In the previous section we said that the partition function Z is a function of β. In the derivation of Equation 13.10 we showed that β is independent of the composition of the system, i.e. the types of atom, forces, etc. However, so far we have not identified β in the real sense. Here we intend to investigate the true nature of β. We shall do this for a perfect gas and then generalize the result for any system.

For a rarefied real gas, the atoms in the gas rarely collide with one another. Therefore, the potential energy due to the interatomic forces is virtually nil most of the time, and the energy of the gas is essentially due to the sum of the kinetic energies of the atoms. Real gases, when rarefied by reduction of pressure (and hence the density), behave like perfect gases. If we have a perfect monatomic gas of N atoms, all of mass m and having a velocity $\mathbf{v}$, then the energy of this gas can be represented by

$$E = \sum_i \varepsilon_i = \sum_i \frac{1}{2} m \mathbf{v}_i^2 \qquad (13.16)$$

where ε_i is the energy of the ith atom and $i = 1, 2, \ldots, N$. However, for a polyatomic gas, we must consider additional terms representing the potential energy between the atoms in each molecule. If we combine Equations 13.12, 13.13 and 13.16, we have for a system of monatomic gas

$$f(E) = Z^{-1} \exp \ (-\beta E) = Z^{-1} \exp \ (-\beta E) = Z^{-1} \exp \ \left(-\beta \sum_i \varepsilon_i \right)$$

$$= Z^{-1} \exp \ \left(-\beta \sum_i \frac{1}{2} m \mathbf{v}_i^2 \right). \tag{13.17}$$

We have seen above that the energy is a sum of all the individual energies. So Equation 13.17 can be written as

$$f(E) = Z^{-1} [\exp \ (-\beta \varepsilon_1)][\exp \ (-\beta \varepsilon_2)] \cdots [\exp \ (-\beta \varepsilon_N)]. \tag{13.18}$$

With reference to Equation 13.12 we find that

$$Z = \int_V d^3 r_1 \cdots \int_{-\infty}^{\infty} [\exp \ (-\beta \varepsilon_1)][\exp \ (-\beta \varepsilon_2)] \cdots [\exp \ (-\beta \varepsilon_N)] d^3 v_N \tag{13.19}$$

where the spatial integrals extend over the volume of the system. As the energy does not depend on the position of the atoms, the volume integral can be computed, and for a system of N particles each integral of a variable $\mathbf{r}_i$ will give the volume V so that

$$Z = V^N \int_{-\infty}^{\infty} dv_1 \ \exp \ (-\beta \varepsilon_1) \int_{-\infty}^{\infty} dv_2 \ \exp \ (-\beta \varepsilon_2) \cdots \int_{-\infty}^{\infty} dv_N \ \exp \ (-\beta \varepsilon_N).$$

These integrals are all identical and are equal to

$$\int_{-\infty}^{\infty} d^3 v \ \exp \ (-\beta \varepsilon).$$

So, we can write $Z = z^N$, where z is the particle (or molecular) partition function. It can be easily seen that this form for the particle function is valid whenever Equation 13.19 can be written (irrespective of the form of the energy ε of the particle). For a monatomic gas, where $\varepsilon = \frac{1}{2} m \mathbf{v}^2$, the particle partition function is

$$z = V \int_{-\infty}^{\infty} d^3 v \ \exp \ \left[(-\beta) \left(\frac{1}{2} m v^2 \right) \right] \tag{13.20}$$

where V is the volume of the system. Combining Equations 13.18 and $Z = z^N$, we have the distribution function for the whole system as

$$f(E) = [z^{-1} \exp \ (-\beta \varepsilon_1)][z^{-1} \exp \ (-\beta \varepsilon_2)] \cdots [z^{-1} \exp \ (-\beta \varepsilon_N)]. \tag{13.21}$$

The right-hand side contains the distribution functions for individual atoms, and the terms are simply $f(\varepsilon_1), f(\varepsilon_2) \ldots, f(\varepsilon_N)$. Hence, Equation 13.21 becomes

$$f(E) = f(\varepsilon_1) f(\varepsilon_2) \ldots f(\varepsilon_N). \tag{13.22}$$

Statistically, each atom in a perfect gas behaves as an independent system, because they do not have any interaction energy. Then, the probability of finding any atom in the region d^3r of r with a velocity in the range d^3v of $\mathbf{v}$ can be written as

$$f(\varepsilon)\,d^3r\,d^3v = [z^{-1}\,\exp\,(-\beta\varepsilon)]\,d^3r\,d^3v \quad [\text{as} f(\varepsilon) = z^{-1}\,\exp\,(-\beta\varepsilon)]$$

$$= z^{-1}\,\exp\,\left[(-\beta)\left(\frac{1}{2}mv^2\right)\right]d^3r\,d^3v \qquad (13.23)$$

where ε and z are given by Equations 13.16 and 13.20, respectively.

The velocity $\mathbf{v}$ of a particle in a gas can be resolved along the x, y and z directions as v_x, v_y and v_z. So, Equation 13.20 can be written in terms of these components as

$$z = V \int_{-\infty}^{\infty} dv_x \int_{-\infty}^{\infty} dv_y \int_{-\infty}^{\infty} dv_z \,\exp\,\left[(-\beta)\frac{1}{2}m(v_x^2 + v_y^2 + v_z^2)\right].$$

Here, each integral has the same value, because the value of the integral is not affected by the variable of integration. So the above equation can be written as

$$z = V\left[\int_{-\infty}^{\infty} dv \,\exp\,(-\beta mv^2/2)\right]^3. \qquad (13.24)$$

The integral is Gaussian and has the value $(2\pi/\beta m)^{1/2}$. Thus,

$$z = V(2\pi/\beta m)^{3/2}. \qquad (13.25)$$

Combining Equation 13.25 and $Z = z^N$, we get

$$Z = V^N(2\pi/\beta m)^{3N/2}. \qquad (13.26)$$

We shall show in the next section that the internal energy of a system is given by (see Equation 13.36)

$$U = -\frac{\partial(\ln z)}{\partial \beta}.$$

Taking the logarithm of both sides of Equation 13.26, we get

$$\ln Z = N \ln V + \frac{3N}{2} \ln\,(2\pi/\beta m).$$

Then,

$$U = -\frac{\partial}{\partial \beta}(\ln Z) = -\frac{3N}{2}\frac{\partial}{\partial \beta}[\ln\,(2\pi/\beta m)]$$

$$= \frac{3N}{2}\frac{\partial}{\partial \beta}\left[\ln\,\left(\frac{\beta m}{2\pi}\right)\right] = \frac{3N}{2\beta}. \qquad (13.27)$$

As both m and π are constant,

$$\frac{\partial}{\partial \beta}\left[\ln\left(\frac{\beta m}{2\pi}\right)\right] = \frac{1}{\beta}.$$

This is the internal energy of a perfect monatomic gas. We know that the heat capacity at constant volume (C_V) of a monatomic gas is given by $3nR/2 = 3Nk/2$, where n is the number of moles, R is the universal gas constant, N is the number of atoms and k is the Boltzmann constant. We also know that $C_V = (\partial U/\partial T)_V$. Then,

$$\left(\frac{\partial U}{\partial T}\right)_V = C_V = \frac{3Nk}{2}.$$

Separating the variables and integrating, we obtain

$$U = (3Nk/2)T = \frac{3N}{2\beta}$$

using Equation 13.27. Hence,

$$\beta = \frac{1}{kT}. \tag{13.28}$$

Although Equation 13.28 has been derived for a perfect gas, it is entirely general; it holds for all systems. The simple reason for this is that in the derivation of Equation 13.10 we showed that β does not depend on the type of atom, force, etc., in the system. Thus, β can be determined using any system. Suffice to say, we have chosen a perfect monatomic gas because its partition function can be easily evaluated.

Example 13.1

A system consists of 10^{24} atoms and is at a temperature of 300 K. Assuming that there is no interatomic energy in the system, calculate its total internal energy.

Solution

The internal energy of a system is given by $U = 3N/2\beta$ where N is the number of atoms and $\beta = 1/kT$ where k is the Boltzmann constant $1.380 \times 10^{-23}\,\mathrm{JK^{-1}}$ and T is the temperature 300 K. Then,

$$\beta = \frac{1}{1.38 \times 10^{-23}\,\mathrm{JK^{-1}} \times 300\,\mathrm{K}} = 2.42 \times 10^{20}\,\mathrm{J^{-1}}$$

Substituting this value of β into $U = 3N/2\beta$, we get

$$U = \frac{3 \times 10^{24}}{2.42 \times 10^{20}\,\mathrm{J^{-1}}} = 12.4\,\mathrm{kJ}$$

13.4 STATISTICAL APPROACH TO THERMODYNAMIC QUANTITIES

When chemical and physical changes take place, a change in energy occurs. The study of these energy transformations is known as thermodynamics. In the early chapters we have considered thermodynamic quantities, such as internal energy, enthalpy, work, heat, entropy, etc., on purely macroscopic basis. Here, we wish to consider these quantities from the statistical mechanics point of view, with the aim of establishing interrelationships between thermodynamics and statistical mechanics. We shall also use statistical mechanics to explain adiabatic and isothermal changes, as well as the laws of thermodynamics.

13.4.1 Statistical approach to thermodynamic pressure

Let us consider an assembly of definite numbers of various species of molecules, contained in a cylinder fitted with a weighted piston (see Figure 13.1). Let P be the pressure on the piston due to the force of gravity (not due to the pressure of the substance inside the cylinder, except at equilibrium). So P is constant. The total energy of the system is the sum of the internal energy U of the collection of molecules and PV, V being the volume of the assembly. Let us imagine that the system is isolated, so that $U + PV$ is constant. However, if the piston moves, the volume changes and also U, because energy is transferred to the piston. Some of this transferred energy may be, temporarily, kinetic energy of the molecules and of the piston. Because of any frictional forces within the assembly, this kinetic energy will ultimately become internal energy, i.e. heat energy. The system will have a tendency to attain that volume for which the number of energy levels available to the system or the entropy is a maximum under the constraint $dU = -P\,dV$.

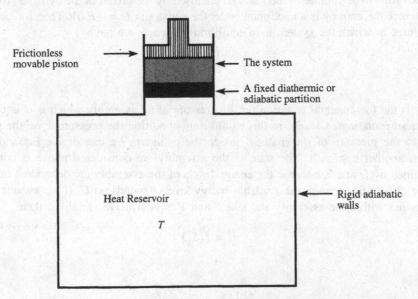

Frictionless movable piston

The system

A fixed diathermic or adiabatic partition

Heat Reservoir

T

Rigid adiabatic walls

Figure 13.1 A system in contact with a heat reservoir through a fixed rigid diathermic or adiabatic partition

It is almost impossible to prevent some exchange of energy with the piston and cylinder. However, the actual composition of the cylinder and piston has no decisive influence in this system. Their effect as heat reservoirs can be neglected, because it is always possible to make their mass negligible with respect to that of the substance in the assembly. In a gravitational field, however, a large mass could be required to produce a large force, but such a mass can be insulated thermally. A point to note is that even if the mass of the cylinder and piston were considered negligible relative to that of the material in the assembly, the piston would always exchange kinetic energy with the assembly, and would have an energy of the order of kT. However, energy exchange with a relatively small number of molecules of the cylinder and piston would not affect the energy to a significant extent.

If we consider the entire system, cylinder, piston and assembly of molecules, and if this system is truly isolated, then, irrespective of the position of the piston, it must be in a state of definite total energy. Generally, the system would tend to reach an equilibrium position, since, even with energy fixed, the equilibrium configurations actually occupy a very large fraction of the accessible phase space. From the quantum mechanical consideration, states in which the energy is divided in different ways between the piston and the assembly of molecules (in some quantum state of its own) may be considered unperturbed quantum states. Under the perturbation of the interaction between the piston and the assembly these quantum states combine linearly to produce the stationary states of the whole system. Generally, in these linear combinations the unperturbed states with the properties of the equilibrium state of the system would be very high because of their high numbers. A system in almost any one of these stationary states of the entire system (piston, cylinder and assembly) would have virtually the equilibrium properties.

It must be assumed that in any statistical theory there is some random process, whatever the nature of it may be. So, in our case, it will be enough to assume only a random exchange of energy between the piston and the assembly of molecules. We stated earlier that there is a tendency for V (the volume of the material enclosed by the piston in the cylinder) to attain a value where the entropy is a maximum under the constraint $U = -P\,dV$. Therefore, $dS = 0$ at the volume at which the system is in equilibrium. Hence, we have

$$P = -\left(\frac{\partial U}{\partial V}\right)_S.$$

(13.29)

This is the fundamental equation for the pressure of an assembly which is in equilibrium with its surroundings. Clearly, in this equilibrium condition the pressure P on the piston is equal to the pressure of the material inside the cylinder. We can derive Equation 13.29 through another approach. The state of the assembly we considered above is completely determined by U and V, because the energy levels of the assembly are dependent only on V, and the distribution among the available energy levels depends on U. If we assume internal equilibrium within the assembly and take S and V as alternative variables, then

$$U = f(S, V)$$

or

$$dU = \left(\frac{\partial U}{\partial S}\right)_V dS + \left(\frac{\partial U}{\partial V}\right)_S dV.$$

If we hold both P and $U + PV$ constant, we can divide both sides of the above equation by ∂V and get

$$\left(\frac{\partial U}{\partial V}\right)_{(U+PV)} = -P = \left(\frac{\partial U}{\partial V}\right)_{S} + \left(\frac{\partial U}{\partial S}\right)_{V}\left(\frac{\partial S}{\partial V}\right)_{(U+PV)}.$$

When the system is at equilibrium

$$\left(\frac{\partial S}{\partial V}\right)_{(U+PV)} = 0$$

and the above equation reduces to

$$P = -\left(\frac{\partial U}{\partial V}\right)_{S}$$

which is Equation 13.29.

13.4.2   Statistical approach to thermodynamic temperature

Since the state of the assembly is determined by U and V, and as S is assumed to be independent of the surroundings of the system (i.e. whether or not the system is in thermal contact with its surroundings), we have defined the temperature by

$$\frac{1}{T} = \left(\frac{\partial S}{\partial U}\right)_{V}. \qquad \text{See Equation 7.39.} \qquad (13.30)$$

Let us now see whether $(\partial S/\partial U)_V$ has the properties expected of temperature. Consider two systems A and B of any kind at constant volume V_A and V_B, respectively. We place them in thermal contact to allow energy to flow from one to the other. If the composite system is isolated, the energy dU_A gained by A will be equal to the energy $-dU_B$ lost by system B. So, the total change of entropy of the composite system is

$$dS = \left(\frac{\partial S_A}{\partial U_A}\right)_{V_A} dU_A + \left(\frac{\partial S_B}{\partial U_B}\right)_{V_B} dU_B$$

$$= \left[\left(\frac{\partial S_A}{\partial U_A}\right)_{V_A} - \left(\frac{\partial S_B}{\partial U_B}\right)_{V_B}\right] dU_A \qquad (13.31)$$

since $dU_A = -dU_B$. At the equilibrium state of the composite system we have $dS = 0$. Therefore,

$$\left[\left(\frac{\partial S_A}{\partial U_A}\right)_{V_A} - \left(\frac{\partial S_B}{\partial U_B}\right)_{V_B}\right] dU_A = 0.$$

Since $dU_A \neq 0$, we have

$$\left(\frac{\partial S_A}{\partial U_A}\right)_{V_A} = \left(\frac{\partial S_B}{\partial U_B}\right)_{V_B} \tag{13.32}$$

or

$$\frac{1}{T_A} = \frac{1}{T_B}, \text{ i.e} \quad T_A = T_B.$$

This was expected, because the systems A and B are in thermal equilibrium with each other.

At the equilibrium condition, the entropy must be at a maximum, so $(\partial^2 S/\partial U_A^2)_{V_A, V_B}$ must be negative. This holds for any system. The two portions in thermal contact could be identical. In this case, when Equation 13.32 is valid, we have

$$\left(\frac{\partial^2 S_A}{\partial U_A^2}\right)_{V_A} = \left(\frac{\partial^2 S_B}{\partial U_B^2}\right)_{V_B} \tag{13.33}$$

Also,

$$\left(\frac{\partial^2 S}{\partial U_A^2}\right)_{V_A, V_B} = \left(\frac{\partial^2 S_A}{\partial U_A^2}\right)_{V_A} + \left(\frac{\partial^2 S_B}{\partial U_B^2}\right)_{V_B}$$

$$= 2\left(\frac{\partial^2 S_A}{\partial U_A^2}\right)_{V_A} \tag{13.34}$$

by virtue of Equation 13.33. Hence, $(\partial^2 S_A/\partial U_A^2)_{V_A}$ must be negative. This means that $1/T_A$ decreases with U_A. This is expected.

13.4.3   Statistical approach to internal energy

It is quite apparent that the internal energy of a system is related, in some way, to the *energy* inside a system when it is in contact with a heat reservoir. We already know that when a system is in contact with a reservoir it continually exchanges energy with the reservoir; therefore, the energy of the system fluctuates. If that is the case, what is then the internal energy? The instruments we use in the laboratory to measure the change in the internal energy are not sensitive enough to measure all the minute changes which occur at the microscopic level of the system. Therefore, such instruments do not really indicate the actual energy exchange between systems at a particular instant; rather they measure the average energy over some interval of time. In other words, we can say that our instrument, in effect, has averaged the results of a large number of separate measurements of the instantaneous energy of the system. This problem has been extensively discussed in classical statistical mechanics in terms of the so-called *ergodic theorem*. What we said above leads us to conclude that the internal energy is really the statistical average energy E_{av}. Then,

$$U = E_{av} = \sum_k E_k \mathscr{P}(E_k).$$

Using Equation 13.10, we get

$$U = Z^{-1} \sum_k E_k \exp(-\beta E_k) \tag{13.35}$$

where

$$Z = \sum_k \exp(-\beta E_k)$$

is the partition function which is a function of β. Differentiating Z with respect to β, we get

$$\frac{\partial Z}{\partial \beta} = \sum_k \frac{\partial}{\partial \beta}[\exp(-\beta E_k)] = -\sum_k E_k \exp(-\beta E_k).$$

Multiplying both sides of this equation by $1/Z$ and then comparing with Equation 13.35, we get

$$\frac{1}{Z}\left(\frac{\partial Z}{\partial \beta}\right) = -Z^{-1} \sum_k E_k \exp(-\beta E_k) = -U.$$

Thus,

$$U = -\frac{\partial}{\partial \beta}(\ln Z) \tag{13.36}$$

since

$$\frac{1}{Z}\left(\frac{\partial Z}{\partial \beta}\right) = \frac{\partial}{\partial \beta}(\ln Z).$$

Equation 13.36 enables us to find the internal energy of a system if the partition function and β are known. In Section 13.3 we proved that β depends only on the temperature and not on the composition of the system. Therefore, to find how β is related to the temperature of a simple system, we require to evaluate only Z and use Equation 13.36.

Example 13.2

A simple system is found to be in a state with the energy 1×10^{-20} J at a temperature $300\,K$. Assuming that there is no atomic interaction energy, calculate the value of $\exp(-\beta E)$ and the internal energy of the system.

Solution

Given that $E = 1 \times 10^{-20}$ J and $T = 300$ K. We know that $\beta = 1/kT$, where k is the Boltzmann constant 1.38×10^{-23} JK^{-1}. Substituting these values, we obtain

$$\beta = \frac{1}{(1.38 \times 10^{-23}\, \text{JK}^{-1}(300\, \text{K})} = 2.42 \times 10^{20}\, \text{J}^{-1}$$

Then,

$$\exp(-\beta E) = \exp[(-2.42 \times 10^{20}\, \text{J}^{-1})(1 \times 10^{-20}\, \text{J})]$$
$$= \exp(-2.42) = 0.0889.$$

Internal energy

$$U = -\frac{\partial}{\partial \beta}(\ln Z) = -\frac{1}{Z}\frac{\partial Z}{\partial \beta}.$$

Now, $Z = \exp(-\beta E)$, so $\partial Z/\partial \beta = -E \exp(-\beta E)$. Substituting this into the above equation, we get

$$U = -\frac{1}{Z}[-E \exp(-\beta E)] = [\exp(\beta(1 \times 10^{-20}))][(10^{-20}\, \text{J}) \exp(-\beta(1 \times 10^{-20}))]$$
$$= 10^{-20}\, \text{J}.$$

13.4.4   Statistical approach to heat and work

In Chapter 2 we have explained heat and work on the basis of classical thermodynamics, i.e. there we considered the macrostates of a system. Here, we proceed to explain the nature of heat and thermodynamic work on the basis of a statistical approach.

Let us consider the system shown in Figure 13.1. At one end, the system is in thermal contact with a heat reservoir at a temperature T through a fixed, rigid and diathermic partition. The reservoir has rigid outer adiabatic walls. At the other end of the system, the reservoir is fitted with an adiabatic frictionless and movable piston. Let the system have a volume V when it is in thermal equilibrium with the reservoir at an absolute temperature T. The ensemble average energy of the system is given by

$$U_{av} = \sum_{states} \mathscr{P}_i U_i \tag{13.37}$$

where U_{av} is the average energy, $\mathscr{P}_i$ is the probability that the system is in the ith microstate with energy U_i when it is in thermal equilibrium with the reservoir.

Remember, the volume V is constant and the thermal capacity of the system is very much less than that of the heat reservoir. The system and the reservoir can exchange energy through the rigid diathermic partition, but they do not interchange particles, so that the number of particles in the system remains constant. The reservoir and the system together can be considered as a *closed isolated* composite system of fixed total energy, fixed total number of particles and fixed total volume. Thus, the total energy U_T of the reservoir and the system is constant. Then, when the system is in its ith microstate with energy U_i, we have

$$U_R = U_T - U_i \qquad (13.38)$$

where U_R is the energy of the reservoir.

Note that U_i is the total energy of the system, which includes both the kinetic energy and the potential energy of all the particles that constitute the system. Actually, the value of U_i depends on the volume of the system and the number of particles in it. According to the postulate of *equal a priori probabilities*, when the composite system in Figure 13.1 has reached thermal equilibrium, it is equally possible that the closed composite system may be in any one of the microstates that are accessible to it, provided that the energy is conserved. It is interesting to note that the number of microstates accessible to the reservoir depends only on its energy, because its volume and the number of particles in it are maintained constant by the rigid diathermic partition and the rigid adiabatic outer walls.

Let us consider two subsystems 1 and 2 (see Figure 12.3) which are in thermal equilibrium with each other. Let $\Omega_1(U_1')$ be the number of microstates accessible to system 1 when it has an instantaneous energy U_1', and let $\Omega_2(U_2')$ be the number of microstates accessible to system 2 when it has an instantaneous energy U_2'. Then, the total number of microstates accessible to the composite system is

$$\Omega_T(U_1', U_2') = \Omega_1(U_1')\Omega_2(U_2'). \qquad (13.39)$$

This equation requires a little explanation. We have said that system 1 and system 2 are in thermal equilibrium, but we did not specify the type of partition between them (partition AB in Figure 12.3). If the partition between them is fixed, rigid and adiabatic, it is an internal constraint which prevents the flow of heat from one system to another, and keeps the total energies of subsystems 1 and 2 constant and equal to U_1' and U_2', respectively. Under this condition, at any instant, system 1 can be in any one of its $\Omega_1(U_1')$ accessible microstate and system 2 could be in any one of its $\Omega_2(U_2')$ accessible microstates. The total number of microstates accessible to the composite system is given by Equation 13.39. On the other hand, if the partition is a fixed, rigid and diathermic one, then the volumes V_1 and V_2 of systems 1 and 2, respectively, remain constant and the energy of the single particle quantum states of systems 1 and 2 are unchanged. The particles cannot go from one system to the other through the partition, but energy, in the form of heat, can flow through the diathermic partition, subject to the condition that the total energy U_T of the subsystems 1 and 2 is constant, i.e.

$$U_T = U_1' + U_2'. \qquad (13.40)$$

In this case, too, Equation 13.39 holds. However, the value of the total number of microstates accessible to the composite system (Ω_T) will not be the same as that obtained in the case with an adiabatic partition, even if $U_1' + U_2'$ remains the same. In this case, the total number of microstates (Ω_T) accessible to the composite system is obtained by adding all $\Omega_1(U_1')\Omega_2(U_2')$ values over all possible distributions of energy between subsystem 1 and subsystem 2, provided that $U_1' + U_2'$ does not change. Then,

$$\Omega_T = \sum_{U_1} \Omega_1(U_1')\Omega_2(U_2')$$
$$= \sum_{U_1} \Omega_1(U_1')\Omega_2(U_T - U_1') \tag{13.41}$$

since $U_T = U_1' + U_2'$. The alternative form of Equation 13.41 is

$$\Omega_T = \sum_{U_2} \Omega_2(U_2')\Omega_1(U_1') = \sum_{U_2} \Omega_2(U_2')\Omega(U_T - U_2').$$

With the help of Equation 13.41 it can be easily shown that if the partition is a diathermic partition so that heat can flow from one subsystem to the other, the number of microstates accessible to the composite system is higher, in some cases many times higher, than that for an adiabatic partition. The composite system in Figure 13.1 (or Figure 12.3) is an isolated system of constant total energy, constant total volume and constant total number of particles; it is a closed system. According to the principle of *equal a priori probabilities*, when a closed system is in internal thermodynamic equilibrium it is equally likely to find the closed system in any one of the microstates accessible to it. Therefore, when the closed system in Figure 13.1 has had time to reach internal thermal equilibrium, the composite closed system is equally likely to be found in any one of the accessible microstates of the composite closed system given by Equation 13.41.

A closed system is an isolated system of constant total energy, constant total volume and constant total number of particles. Such a system is not under any external influence, so there is no reason to think that at thermodynamic equilibrium any particular one of the microstates accessible to it is any more probable than any of the remaining accessible microstates.

Example 13.3

(a) Consider two systems A and B, where A consists of two distinguishable particles of zero spin and B is made up of two distinguishable particles (different from those two in A) of zero spin. Let A and B initially have energies 5 and 1, respectively, and be in thermal equilibrium with each other through a rigid, fixed and adiabatic partition between them. Calculate the total energy of the composite system and the total number of microstates accessible to it. Write these accessible microstates.

(b) The rigid, fixed and adiabatic partition above is replaced by a rigid, fixed and diathermic partition. Show that for a distribution of energy where $U_A = 4$ and $U_B = 2$ units of energy the total number of microstates accessible to the composite system is higher than that obtained above.

Solution

(a) The initial energies of A and B are $U_A^0 = 5$ energy units and $U_B^0 = 1$ energy units. The total energy U_T of the composite system is $U_T = U_A^0 + U_B^0 = 5 + 1 = 6$. The initial 5 units of energy of system A can be distributed between its two distinguishable particles in six different ways: 5,0; 0,5; 4,1; 1,4; 3,2; and 2,3. Initially, system A has $\Omega_A^0 = 6$ accessible microstates consistent with $U_A^0 = 5$ units of energy. For system B, initially it has $\Omega_B^0 = 2$ accessible microstates consistent with $U_B^0 = 1$ units of energy; these microstates are 1,0 and 0,1.

Systems A and B together form an isolated closed composite system. The adiabatic partition acts as an internal constraint which prevents the flow of heat from one system to the other, and keeps the total energies of systems A and B constant with A and B having 5 and 1 units of energy, respectively. At any instant, system A could be in any one of its 6 accessible microstates mentioned above and system B could be in any one of its 2 accessible microstates given above. However, the total number of microstates Ω_T accessible to the composite system is (using Equation 13.39)

$$\Omega_T = \Omega_A^0 \times \Omega_B^0 = 6 \times 2 = 12.$$

These microstates are

$$5,0,1,0; \quad 5,0,0,1; \quad 0,5,1,0; \quad 0,5,0,1; \quad 4,1,1,0; \quad 4,1,0,1;$$
$$1,4,1,0; \quad 1,4,0,1; \quad 3,2,1,0; \quad 3,2,0,1; \quad 2,3,1,0; \quad 2,3,0,1.$$

(b) Turning to the second part of the example, the rigid, fixed and adiabatic partition has been replaced by a rigid, fixed and diathermic partition. The diathermic partition keeps the volumes V_A and V_B of systems A and B constant; there is no exchange of particles between A and B, but energy, in the form of heat, can flow through the diathermic partition, provided that $U_T = U_A^0 + U_B^0 = 4 + 2 = 6$. The 4 units of energy of system A can be distributed between its two distinguishable particles in 5 different ways: 4,0; 0,4; 3,1; 1,3; 2,2. Similarly, the 2 units of energy of system B can be distributed between its two distinguishable particles in 3 different ways: 2,0; 0,2; 1,1. Therefore, initially system A has 5 accessible microstates consistent with $U_A^0 = 4$ units of energy, and system B has 3 accessible microstates consistent with $U_B^0 = 2$ units of energy. Then, using Equation 13.39, the total number of different microstates accessible to the composite system is

$$\Omega_T = \Omega_A \times \Omega_B = 5 \times 3 = 15$$

which is higher than that obtained above.

At thermal equilibrium (with a diathermic partition), if U_2' is the energy of subsystem 2 then $U_T - U_2' = U_1'$ is the energy of subsystem 1 (Figure 12.3). Let $\mathscr{P}_2(U_2')$ be the probability that at thermal equilibrium subsystem 2 has energy U_2'. Then, subsystem 2 has energy U_2' in $[\Omega_2(U_2')\Omega_1(U_T - U_2')]$ of the total Ω_T microstates accessible to the composite system.

At thermal equilibrium, if we are equally likely to find the composite system in any one of its Ω_T accessible microstates, then the probability of finding the composite system in any one of its $\Omega_2(U_2')\Omega_1(U_T - U_2')$ accessible microstates where subsystem 2 has energy U_2' is given by

$$\mathscr{P}_2(U_2') = [\Omega_2(U_2')\Omega_1(U_T - U_2')]/\Omega_T$$
$$= [\Omega_1(U_1')\Omega_2(U_2')]/\Omega_T. \tag{13.42}$$

Again, if $\mathscr{P}_1(U_1')$ is the probability that at thermal equilibrium subsystem 1 has energy U_1', then we have

$$\mathscr{P}_1(U_1') = [\Omega_1(U_1')\Omega_2(U_T - U_1')]/\Omega_T$$
$$= [\Omega_1(U_1')\Omega_2(U_2')]/\Omega_T. \tag{13.43}$$

In both Equations 13.42 and 13.43, Ω_T is given by Equation 13.41.

We can generalize what we said above. It $\mathscr{P}_{2i}(U_{2i})$ is the probability that at thermal equilibrium subsystem 2 is in its ith microstate which has an energy eigenvalue U_{2i}', then

$$\mathscr{P}_{2i}(U_{2i}') = \Omega_1(U_T - U_{2i}')/\Omega_T. \tag{13.44}$$

In this equation $\Omega_1(U_T - U_{2i}')$ is the number of microstates accessible to subsystem 1 (Figure 12.3) when it has energy $U_1' = (U_T - U_{2i}')$. According to Equation 13.44, we can write

$$\mathscr{P}_i = \Omega_R(U_T - U_i)/\Omega_T \tag{13.45}$$

where $\Omega_R(U_T - U_i)$ is the number of microstates accessible to the heat reservoir when the reservoir has energy $(U_T - U_i)$, U_T is the total energy of the system and the reservoir (see Figure 13.1) and Ω_T is the total number of microstates accessible to the closed composite system (see Equation 13.41) made up of the reservoir and the system in thermal equilibrium with it.

Going back to Figure 13.1, let us replace the fixed rigid diathermic partition between the system and the heat reservoir with a fixed rigid adiabatic partition, and move the adiabatic piston inwards very slowly until the volume of the system has changed to $V + dV$, where dV is negative. Under these conditions, let dU_1' be the change in the ensemble average energy of the system. Now, we rigidly fix the piston to its new position. By the adiabatic approximation of quantum mechanics, if the system is compressed *quasistatically* under adiabatic conditions, we can assume that the probability that the system makes a transition to a different quantum state is negligible. If $\mathscr{P}_i$ is constant, then all the $d\mathscr{P}_i$ are zero. According to Equation 13.37, the change in the ensemble average energy (dU_1') of the system in the reversible adiabatic change is given by

$$dU_1' = \sum_{\text{states}} \mathscr{P}_i dU_i. \tag{13.46}$$

Let us assume that in its ith microstate the system exerts a pressure P_i on the piston. If the piston were given a virtual displacement dx inwards, the change in the volume of the system would be equal to $A \, dx$ where A is the area of the piston, and the work done on the system would be pressure × volume change $= P_i A \, dx = -P_i \, dV$ where dV is the increase in the volume of the system. If all the work is used to increase the energy of the microstate from U_i to $U_i + dU_i$, then

$$dU_i = -P_i \, dV. \tag{13.47}$$

Combining Equations 13.46 and 13.47, we get

$$dU'_1 = -\sum_{\text{states}} \mathscr{P}_i P_i \, dV = -dV \sum_{\text{states}} \mathscr{P}_i P_i.$$

However

$$\sum_{\text{states}} \mathscr{P}_i P_i = P = \text{the ensemble average pressure.}$$

Hence, for an infinitesimal reversible adiabatic change we have

$$dU'_1 = \sum_{\text{states}} \mathscr{P}_i dU_i = -P \, dV = dW \tag{13.48}$$

that is, $\sum \mathscr{P}_i \, dU_i$ is equal to the mechanical work dW done on the system in a reversible adiabatic change. We conclude that *the values of the energy U_i change in a reversible adiabatic change of volume, but the probability $\mathscr{P}_i$ that the system is in a particular microstate does not change.*

Consider again the assembly in Figure 13.1. We now replace the fixed, rigid, adiabatic partition with a fixed, rigid, diathermic partition, and raise the temperature of the heat reservoir to $T + dT$, keeping the volume of the system constant at $V + dV$. Under this condition, heat will flow from the reservoir to the system. This flow will continue until thermal equilibrium is attained and the temperature of the system is increased to $T + dT$. Because of the heat that flowed from the reservoir to the system, the ensemble average energy will change. Let dU'_2 represent this change.

Assuming that the number of particles in the system remains constant, and if dU is the total change in the ensemble average energy of the system because of the addition of heat to the system at constant volume as well as the reversible adiabatic change of volume, then by virtue of Equation 13.37 we can write

$$dU = dU'_1 + dU'_2 = \sum \mathscr{P}_i dU_i + \sum U_i \, d\mathscr{P}_i. \tag{13.49}$$

According to Equation 13.48, $dU'_1 = \sum \mathscr{P}_i dU_i$. Therefore, from Equation 13.49 we conclude that

$$dU'_2 = \sum U_i d\mathscr{P}_i \tag{13.50}$$

that is, $\sum U_i d\mathscr{P}_i$ is equal to the heat added to the system to raise the temperature from T to $T + dT$ (keeping the volume of the system constant). Hence, we can arrive at the following conclusions.

(i) In a reversible adiabatic change of volume, the values of the energy eigenvalues U_i change, but the probability $\mathscr{P}_i$ that the system is in its ith microstate remains unchanged.

(ii) When heat is added to a system at constant volume, the values of the energy eigenvalues of the system do not change, but the probability $\mathscr{P}_i$ that the system is in its ith microstate changes.

It is this difference between the effect of addition of heat to and work done on a system that leads to the second law of thermodynamics: *No process is possible whose sole result is the absorption of heat from a reservoir and the conversion of this heat into work (the Kelvin–Planck principle).*

We know that $dU = TdS - PdV + \mu N$ (see Equation 13.93) where N is the number of particles in the system and μ is its chemical potential. If the volume V of the system and N are constant, the increase in the (internal) energy of a system is then equal to $T\,dS$. Hence, Equation 13.50 can be rewritten as

$$dU_2' = T\,dS = \sum U_i\,d\mathscr{P}_i \tag{13.51}$$

This equation suggests that there is a connection between the entropy S of the system and the probabilities $\mathscr{P}_i$. Let us see whether this is true.

Consider the system given in Figure 13.1. Let us assume that the heat capacity of the reservoir is large enough for us to apply the Boltzmann distribution to the system. According to Equation 13.10 we have

$$\mathscr{P}_i = \frac{\exp\left(-U_i/kT\right)}{Z}$$

where Z is the partition function. Since Z is a function of T, V and N, it is a constant for any given values of these parameters. Taking the log of both sides and rearranging, we get

$$U_i = -kT(\ln\mathscr{P}_i + \ln Z).$$

Hence, Equation 13.51 becomes

$$dS = -k\left[\sum(\ln\mathscr{P}_i)d\mathscr{P}_i + \ln Z\sum d\mathscr{P}_i\right]$$

Remember that $\sum\mathscr{P}_i = 1$, so $\sum d\mathscr{P}_i = 0$. Hence,

$$dS = -k\left[\sum(\ln\mathscr{P}_i)d\mathscr{P}_i\right].$$

However, $\sum\mathscr{P}_i d(\ln\mathscr{P}_i) = \sum\mathscr{P}_i(d\mathscr{P}_i/\mathscr{P}_i) = \sum d\mathscr{P}_i = 0$. So,

$$dS = d\left[-k\sum\mathscr{P}_i\ln\mathscr{P}_i\right].$$

By integrating both sides, we get

$$S = -k\sum\mathscr{P}_i\ln\mathscr{P}_i + A$$

where A is an integration constant.

At $T = 0$, the system is in its ground state. Under this condition all the probabilities $\mathscr{P}_i$ are zero except $\mathscr{P}_0$. If the ground state is not degenerate, $\mathscr{P}_0 = 1$ which gives $\ln \mathscr{P}_0 = 0$ and $\sum \mathscr{P}_i \ln \mathscr{P}_i = 0$. The third law of thermodynamics requires that at $T = 0$, the entropy S must be zero. Hence, $A = 0$. Thus, at a finite absolute temperature, the entropy of the small system in Figure 13.1 is related to the probability by

$$S = -k \sum_{\text{states}} \mathscr{P}_i \ln \mathscr{P}_i. \tag{13.52}$$

This is the Boltzmann definition of the entropy. It can also be applied to other cases. For example, take the case of a closed system. According to the postulate of *equal a priori probabilities*, if the closed system is in internal thermodynamic equilibrium, all the Ω microstates accessible to the system are equally probable. So, all the $\mathscr{P}_i$ will be equal to $1/\Omega$ and there will be Ω terms of the type $\mathscr{P}_i \ln \mathscr{P}_i$ in Equation 13.52. Each of these terms is equal to $(-1/\Omega) \ln \Omega$, so that the entropy given by Equation 13.52 is $S = -k \ln \Omega$ which is the same as Equation 7.18.

13.4.5 Statistical approach to entropy

In Sections 7.6, 7.6.1 and 7.6.2 we have developed the concept of entropy on the basis of statistical mechanics, and defined entropy by the equation

$$S = k \ln \Omega \tag{13.53}$$

(see Equation 7.18). There we have also justified for setting $k \ln \Omega$ equal to the entropy S of a system. k is the Boltzmann constant. Thus, in the microscopic approach, the entropy is a logarithmic measure of the total number of accessible microstates Ω. In fact, Ω is a function of the volume V, the internal energy U and the number of particles in a system. So, the entropy S of a system is also a function of U, V and N, i.e. $S = f(U, V, N)$.

We have already shown that $\beta = 1/kT$ (see Equation 13.28). Therefore, we can write

$$T = 1/k\beta. \tag{13.54}$$

β is defined in the general case for any system, which has $\Omega(U, V, N)$ accessible microstates, energy U, volume V and N particles, by the equation

$$\beta = \left(\frac{\partial \ln \Omega}{\partial U} \right)_{V,N} = \frac{1}{\Omega} \left(\frac{\partial \Omega}{\partial U} \right)_{V,N} \tag{13.55}$$

Using this equation, Equation 13.54 becomes

$$T = \frac{\Omega}{k(\partial \Omega / \partial U)_{V,N}} = \frac{1}{k} \left(\frac{\partial U}{\partial \ln \Omega} \right)_{V,N}. \tag{13.56}$$

By combining Equations 13.53 and 13.56, we get

$$T = \left(\frac{\partial U}{\partial S}\right)_{V,N}. \tag{13.57}$$

This equation is the same as Equation 7.39 developed from classical equilibrium thermodynamics.

The precise variation of $\ln \Omega (= S/k)$ with energy U depends on the quantum mechanical properties of the system. A schematic of the variation of $\ln \Omega$ with internal energy U for a typical thermodynamic system is shown in Figure 13.2a (on the basis of statistical mechanics) and that of entropy S is shown in Figure 13.2b (on the basis of classical equilibrium thermodynamics).

The slope of the curve in Figure 13.2a is equal to β. U_0 is the ground state energy (for statistical consideration) or initial energy (for classical consideration). When the energy of the system is close to U_0, the *temperature parameter* $\beta = (\partial \ln \Omega / \partial U)_{V,N}$ tends to infinity. Since $\beta = 1/kT$, T (absolute temperature) tends to zero as U tends to U_0. On the other hand, β is inversely proportional to the energy U, i.e. the slope of the curve decreases as U increases. Hence, the absolute temperature T increases as U increases, so $\partial U / \partial T$ is positive. This implies that the molar heat capacity at constant volume given by the equation

$$C_V = (\partial U / \partial T)_V \tag{13.58}$$

must be positive for a normal thermodynamic system. For such systems, the variation of $\ln \Omega$ with U is shown in Figure 13.2a.

Note that there is a class of system which has a spin system 1 in a magnetic field. For such a system there is an upper limit to the energy the system can have. In these cases, the variation of the entropy with the internal energy has a maximum at a particular value of U;

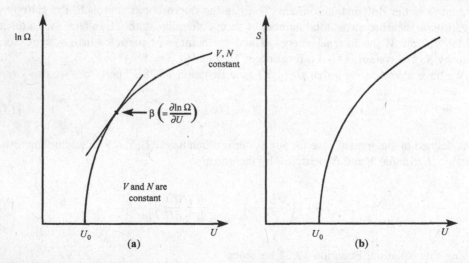

Figure 13.2   (a) Schematic diagram of the variation of $\ln \Omega$ with the internal energy U for a normal thermodynamic system (statistical mechanical consideration). (b) Schematic diagram of the variation of the entropy S with the internal energy U for a normal thermodynamic system (by classical equilibrium thermodynamics)

for higher values of U, $(\partial U/\partial S)_{V,N}$ becomes negative and corresponds to negative absolute temperature.

In classical equilibrium thermodynamics, if a quantity of heat dQ is added to a system, which is at the absolute temperature T, the change in the entropy of the system is given by

$$dS = dQ/T \qquad (13.59)$$

(see Equation 7.2). Let us develop this equation from statistical mechanics.

Consider a macroscopic system which has a volume V, energy U and N particles. Let $\Omega(U, V, N)$ be the number of accessible microstates when the system is in a macrostate specified by U, V and N. We add a small amount of heat dQ to the system by putting it in thermal contact with a heat reservoir for a short time, keeping both V and N constant. We have $S = k \ln \Omega$ and S is a function of U, V and N, i.e.

$$S = f(U, V, N).$$

Then,

$$dS = \left(\frac{\partial S}{\partial U}\right)_{V,N} dU + \left(\frac{\partial S}{\partial V}\right)_{U,N} dV + \left(\frac{\partial S}{\partial N}\right)_{U,V} dN. \qquad (13.60)$$

If V and N are constant, this equation becomes

$$dS = \left(\frac{\partial S}{\partial U}\right)_{V,N} dU.$$

By virtue of Equation 13.57 and putting $dU = dQ$, we get

$$dS = dQ/T. \qquad (13.61)$$

This is the same as Equation 7.2 or Equation 13.59 of classical equilibrium thermodynamics. This can be extended to the general case where V and N can also vary. Equation 13.61 gives the increase in the entropy of a system when a small quantity of heat dQ is added to the system, keeping both V and N constant.

Example 13.4

In statistical méchanics, the entropy S of a system is set equal to $k \ln \Omega$ where k is the Boltzmann constant and Ω is the total number of microstate accessible to the system. Can you justify this equality?

Solution

Consider two isolated systems 1 and 2 with entropies S_1 and S_2, respectively, and microstates accessible to them being Ω_1 and Ω_2, respectively. If these two systems are combined into one composite system, each of Ω_1 microstates of system 1 may be combined with any one of the Ω_2 microstates of system 2 to give a possible microstate of the composite system. Then the number Ω of such microstates is given by $\Omega = \Omega_1 \Omega_2$. However, the entropy of the composite system will be given by $S = S_1 + S_2$, but according to Equation 13.53 we have $S_1 = k \ln \Omega_1$ and $S_2 = k \ln \Omega_2$. Therefore,

$$S = S_1 + S_2 = k \ln \Omega_1 + k \ln \Omega_2 = k \ln (\Omega_1 \Omega_2).$$

Since $\Omega = \Omega_1 \Omega_2$, we get $S = k \ln \Omega$. This justifies the appropriateness of setting $k \ln \Omega$ equal to S.

13.5   REVERSIBLE ADIABATIC CHANGE

Consider the system in Figure 13.1. The heat reservoir has an absolute temperature T, and a diathermic partition separates the system from the reservoir. When the system is in thermal equilibrium with the reservoir, the diathermic partition is changed to an adiabatic one. Since the other walls of the system and the reservoir are adiabatic, no heat can now flow into or out of the system in any expansion or compression. We shall consider changes due to both expansion and compression.

(A) *Changes due to a reversible expansion*: The adiabatic piston is allowed to move freely and the external pressure is varied very slowly so that the difference between the external pressure and the pressure of the system is infinitesimally small. Under these conditions, work is done by the system on its surroundings.

When the volume of the system is increased, the energy eigenvalues of the N particle system are decreased, but in the transition the energies of all the representative energy eigenvalues remain closer than they were before the transition, and the mean energy of the system is less than the mean energy value of the system before the transition.

Note: According to quantum mechanics, a measurement of the total energy of the N particle system will show that the system is in one of a set of N particle quantum eigenstates. This can be determined by solving Schrödinger's time-independent equation, Equation 13.62, using the appropriate boundary conditions.

$$H\psi_i = U_i \psi_i. \tag{13.62}$$

In this equation, ψ_i is the wave function of the system when it is in the ith microstate, H is the Hamiltonian operator and U_i is the energy corresponding to ψ_i. For given boundary conditions, Equation 13.62 can only be solved for certain values of the energy U_i, which are the *energy eigenvalues*.

Sometimes it is possible to have more than one independent solution of Equation 13.62 for the same energy eigenvalue (each solution gives a wave function representing a quantum state). The energy level is then said to be *degenerate*, and the degeneracy is equal to the number of independent solutions of Equation 13.62 for the given value of energy.

We have earlier defined each independent N particle quantum state as a *microstate* of the system made of N particles. Although a microstate is specified by an appropriate wave function, in practice it can often be specified by a set of quantum numbers. Another point to note is that in classical mechanics the Hamiltonian is the expression for the sum of the total potential and kinetic energies of the N particles given in terms of their positions and momenta. Thus, the Hamiltonian operator is obtained from the Hamiltonian by replacing the position and momentum variables with the appropriate operators.

It was shown in Equation 13.47 that if the system is in its ith microstate with energy U_i, we have

$$P_i = -(\partial U_i / \partial V)$$

where P_i is the pressure of the system. This shows that P_i would not be positive unless U_i decreased as V increased. Thus, for a hydrostatic system, if the system expands, the energies of the N particle system must decrease.

According to Ehrenfest's principle, i.e. the adiabatic approximation of quantum mechanics, if a gas is allowed to expand very slowly under adiabatic conditions, the probability $\mathscr{P}_i$ that the system is in its ith microstate does not change. However, the mean energy of the system decreases in a reversible adiabatic expansion. As according to Ehrenfest's principle all the $\mathscr{P}_i$ are constant in a reversible adiabatic expansion, the entropy given by Equation 13.52 is also constant, but the absolute temperature decreases.

(B) *Changes due to a reversible compression*: In a reversible adiabatic compression, work is done on the system. In such cases, the values of the microstate energy are increased; however, the probability that the system is in a particular microstate remains unchanged. The most probable energy is increased, the entropy remains unchanged, but the absolute temperature increases.

13.6   ISOTHERMAL EXPANSION

Consider Figure 13.1. We now use a gas as our system, and a diathermic partition is used between the gaseous system and the heat reservoir. To constitute a reversible process, let us imagine that the gas is allowed to expand isothermally and reversibly very slowly by releasing the piston in such a way that the internal pressure of the gas is balanced by the external pressure on the piston at every step; in other words, the difference between the internal and the external pressures is infinitesimally small. During the expansion from the initial state 1 to the final state 2, the heat reservoir maintains the gas at a constant absolute temperature T. This situation is schematically represented in Figure 13.3.

In the special case of an ideal monatomic gas, the ensemble average energy is $\frac{3}{2}(NkT)$ (see Equation 13.28). This shows that the ensemble average energy is proportional to T. If T is kept constant, i.e. for an isothermal case, the ensemble average energy is independent of the volume of the gas. However, the attractions between the molecules of a real gas depend on the inter-molecular distances, so that the internal energy of a real gas depends on its volume, even when the absolute temperature of the gas is maintained constant.

Let us now consider the change of entropy of the system in the finite isothermal change from the initial state 1 to the final state 2, as shown in Figure 13.3. We do this in several steps. Consider first an infinitesimal isothermal expansion. This can be achieved by an infinitesimal adiabatic expansion, then adding an infinitesimal quantity of heat from the heat reservoir of

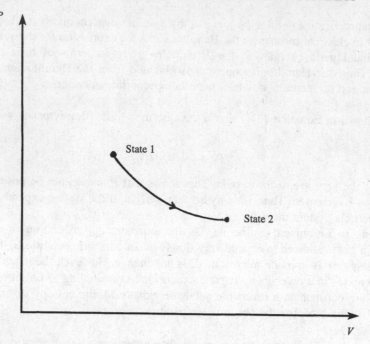

Figure 13.3   Reversible isothermal expansion from the initial state 1 to the final state 2 at a constant absolute temperature T

absolute temperature T to convert the infinitesimal adiabatic change into the equivalent of an infinitesimal isothermal change (as is done for a finite isothermal change in Figure 13.3).

Equation 13.61 was developed from statistical mechanics, which gives the increase in the entropy of a system at constant volume (V) and number of particles (N) when an infinitesimal amount of heat dQ is added to the system. Therefore, in our case the total change of entropy during the infinitesimal isothermal expansion is given by

$$dS = dQ/T$$

because in the infinitesimal reversible adiabatic expansion there was no change in the entropy. Following this argument, the finite isothermal expansion in Figure 13.3 can be divided into a number of such infinitesimal isothermal expansions carried out at the fixed absolute temperature T. The total change of entropy is then given by

$$\int_1^2 dS = \frac{1}{T} \int dQ$$

or

$$S_2 - S_1 = \Delta S = Q/T \tag{13.63}$$

where Q is the total heat added to the system from the reservoir, and S_1 and S_2 are the entropies of the system in state 1 and state 2, respectively.

13.7   STATISTICAL APPROACH TO THE LAWS OF CLASSICAL THERMODYNAMICS

We have now essentially completed our basic discussions of the thermal and mechanical interaction between macroscopic systems; our discussion has produced all the fundamental statements of the theory of *statistical thermodynamics*. The first four of these statements are the laws of thermodynamics. In Sections 3.2, 4.3, 7.1 and 9.2 we have listed the statements of the zeroth, first, second and third laws of thermodynamics. These are very general statements which are completely macroscopic and do not make any explicit reference to the atoms composing the system. They are, therefore, independent of any detailed microscopic model. A complete discussion of these laws constitutes the subject of thermodynamics. The subject can be broadened by adding another statement and then it becomes statistical thermodynamics.

We have already established that the entropy S of a system is given by $S = k \ln \Omega$, where Ω is the number of microstates accessible to a system and k is the Boltzmann constant. The fundamental postulate of *equal a priori probabilities* allows the statistical calculation of all time-independent properties of any system in equilibrium. According to this postulate, the probability of finding an isolated system in equilibrium in each one of its accessible microstates Ω is the same and is simply equal to $1/\Omega$. However, the probability of finding such a system in a microstate which is not accessible to it is zero. Suppose that we are interested in some parameter x (say, the pressure exerted by it). When the system is in a particular microstate, x will have some definite value corresponding to this state. The parameter can have many possible values such as $x_1, x_2, \ldots x_i, \ldots x_n$. So, if Ω is the number of microstates accessible to the system, then there will be some microstate Ω_i in which x will have the particular value x_i. Then, the probability $\mathscr{P}_i$ that the parameter x has the value x_i is the probability that the system is found among the Ω_i microstates given by the value x_i. Hence, $\mathscr{P}_i$ can be found by adding the probability of finding the system in each one of its accessible microstates, (i.e. $1/\Omega$), over the Ω_i microstates, where x is x_i. Thus,

$$\mathscr{P}_i = \Omega_i/\Omega. \qquad (13.64)$$

By using this equation we can calculate probabilities of an isolated system in equilibrium. It is apparent that the probability of finding the system in a situation defined by particular values of its parameters is proportional to Ω (the number of microstates accessible to the isolated system in equilibrium), i.e.

$$\mathscr{P} \propto \Omega.$$

Since $S = k \ln \Omega$, $\Omega = \exp(S/k)$, so we can write

$$\mathscr{P} \propto \exp(S/k). \qquad (13.65)$$

This is *the fifth statement of thermodynamics*. This statement implies that if an isolated system is in equilibrium, the probability of finding it in a macrostate defined by an entropy S is proportional to $\exp(S/k)$.

The statements of the four laws of classical thermodynamics are very general statements. Since they do not make any explicit reference to the atoms of the system concerned, they are totally independent of any detailed microscopic models that might exist about the atoms or molecules of the system. Therefore, these statements can be used without any knowledge about the atoms of the system. This subject can be further broadened without changing its completely macroscopic content by adding the statement given by Equation 13.65. It then becomes what is known as *statistical thermodynamics*. If we combine statistical ideas with the *microscopic* concept of the atoms or molecules in a system, we have the subject of statistical mechanics, which includes $S = k \ln \Omega$ as well. Then it is possible to calculate S of a system from first principles and use it in Equation 13.65 to make probability statements. Thus, we are in a position to evaluate the properties of macroscopic systems on the basis of microscopic information.

13.7.1   The zeroth law of thermodynamics

Let us consider two subsystems 1 and 2 which make up a composite closed system of fixed total energy, fixed total volume and fixed total number of particles. This is shown in Figure 13.4. It is assumed that subsystems 1 and 2 have discrete energy levels and that each energy level has an extremely large degeneracy (of the order of exp N, where N is the number of particles in the subsystem).

The composite system is surrounded by fixed, rigid, impermeable, adiabatic walls, and the two subsystems are separated by a fixed, rigid, impermeable, diathermic partition. Let U_1 and U_2 be the internal energies of system 1 and system 2, respectively, before they were placed in thermal contact. The systems would start with the most probable division of energy (the mean value); on average, no heat would flow from one system to the other through the diathermic partition when they were placed in thermal contact. In terms of classical equilibrium thermodynamics, we can say that the two subsystems were in thermal equilibrium as soon as they were placed in thermal contact. For macroscopic systems, there will be small fluctuations in the energies of both subsystems even at thermal equilibrium. However, the energies of both subsystems will remain extremely close to the most probable energies.

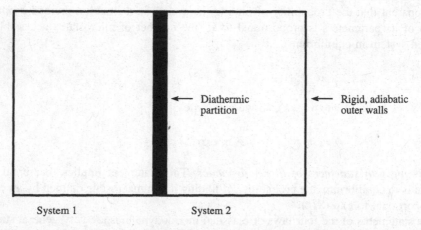

System 1   System 2

Figure 13.4   A composite system surrounded by rigid, impermeable, adiabatic walls

When subsystems 1 and 2 are in thermal equilibrium, if U_1' and U_2' are their instantaneous energies, respectively, and if energy is conserved, then the total energy U_T of the composite system is

$$U_T = U_1' + U_2'. \tag{13.66}$$

Equation 13.55 implies that for U_1' and U_2' to be equal to U_1 and U_2, respectively, before the two subsystems were placed in thermal contact, they needed to be in such a state before thermal contact was made that

$$\beta_1 = \beta_2. \tag{13.67}$$

Under this condition, the subsystems 1 and 2 would be in thermal equilibrium instantaneously after they were placed in thermal contact.

Let us place another macroscopic system, system 3, in thermal contact with system 2. If, on average, no heat is transferred between them, then the initial division of energy between them would be the most probable one; systems 2 and 3 would be in thermal equilibrium immediately after they were placed in thermal contact. The required condition that characterizes the most probable division of energy between system 2 and system 3, when they are in thermal contact and in thermal equilibrium, is

$$\beta_2 = \beta_3. \tag{13.68}$$

If systems 2 and 3 were in such states before they were put in thermal contact that the β-values for both were equal, they would be in thermal equilibrium immediately after they were put in thermal contact.

Combining Equations 13.67 and 13.68, we can immediately write

$$\beta_1 = \beta_2 = \beta_3. \tag{13.69}$$

This implies that if systems 1 and 3 were placed in thermal contact, they would start with the most probable division of energy, because they would start with the same value of β. When they are placed in thermal contact they would be in thermal equilibrium immediately, and, on average, no energy would be transferred from one system to the other. Equation 13.69 also implies that if two systems are in thermal equilibrium with another system, then all of them are in thermal equilibrium with one another. This is the *zeroth law* of thermodynamics, as stated in Section 3.2.

The parameter β is quite useful as an indicator of whether or not thermodynamic systems would be in thermal equilibrium when they were placed in thermal contact. On average, heat flows from a system with a low value of β to one with a higher value of β.

13.7.2 The first law of thermodynamics

The first law of thermodynamics incorporates both the law of conservation of energy and the concept of internal energy. In the microscopic approach, the law of conservation of energy is taken as axiomatic and can be considered as one of the postulates of statistical thermodynamics. We can use the principle of conservation of energy to define a function called the internal energy (U) which is assumed to have a precise value, can be treated as an extensive variable, and is a function of the thermodynamic macrostate of the system. Let us imagine that a closed system undergoes a process by which it passes from a state A to another state B. If the only interaction of the system with its surroundings is in the form of transfer of heat Q to the system, or performance of work W on the system, the change in U will be

$$\Delta U = U_B - U_A = Q + W. \tag{13.70}$$

The first law of thermodynamics states that this energy difference ΔU depends only on the initial and final states, and not on the path followed between them. Both Q and W have many possible values, depending on exactly how the system passes from A to B, but $Q + W = \Delta U$ is invariable and independent of the path. If this were not true, it would be possible, by passing from A to B along one path and then returning from B to A along another, to get a net change in the energy of the closed system in contradiction to the principle of conservation of energy. Therefore, the change (ΔU) in internal energy must be equal to the energy absorbed in the process from the surroundings in the form of heat (Q) plus work (W) done on the system.

Let us consider again systems 1 and 2 in Figure 13.4. When these two systems are in thermal equilibrium, system 2 can be in any one of a very large number of its microstates, which can have different energy eigenvalues. When system 2 is in a particular microstate, the energy eigenvalue of that microstate of system 2 is equal to the sum of the kinetic and potential energies of all the particles in system 2.

When system 2 is in thermal equilibrium with system 1, the ensemble average energy of system 2 is

$$U_2 = \sum_{\text{states}} \mathscr{P}_{2i} U_{2i} \tag{13.71}$$

(see Equation 13.37) where $\mathscr{P}_{2i}$ is the probability that at thermal equilibrium system 2 is in its ith microstate with energy eigenvalue U_{2i}; it is given by Equation 13.44.

If we are to ignore surface energy effects in order to treat the internal energy U of classical equilibrium thermodynamics as an extensive variable, then we must take the thermodynamic limit. In this limit, the volume V and the number of particles N of the system tend to infinity, but N/V is kept constant. For system 2, in the thermodynamic limit, the number of particles N_2 would tend to infinity and the energy per unit volume at thermal equilibrium would tend to a definite value. For finite macroscopic systems 1 and 2 (and for that matter, any finite macroscopic systems), our microscopic approach can give the most probable value of the energy of system 2 (which is the same as the mean value) as well as the magnitude of possible fluctuations in the energy of system 2 when it is in thermal equilibrium with system 1.

13.7.3 The second law of thermodynamics

Here we shall derive the second law of classical thermodynamics, first from the concept of heat flow based on statistical mechanics, and then from the concept of entropy.

Consider the composite system given in Figure 13.4 made up of system 1 and system 2 which are separated by an adiabatic partition. Let U_1^0 and U_2^0 be the initial energies of system 1 and system 2, respectively. We assume that at the start the division of energy between system 1 and system 2 is such that U_2^0 is considerably smaller than U_2 (the most probable value of U_2' when systems 1 and 2 are in thermal equilibrium, where U_2' is the instantaneous value of the energy of system 2). In that case, on average, in the transition from the initial state to the final equilibrium state, energy in the form of heat will flow from system 1 to system 2. For normal thermodynamic systems, we have $C_V = (\partial U / \partial T)_V$ (see Equation 5.6) which is positive. Therefore, the heat capacity of system 2 is positive. So, as the initial energy U_2^0 of system 2 increases to a value close to the most probable value U_2 of the instantaneous energy U_2', the temperature of system 2 must increase until it becomes equal to the final equilibrium temperature of system 1. On the other hand, during this process, system 1 loses energy to system 2, and its temperature decreases until it becomes equal to the final increased temperature of system 2; now the two systems are in thermal equilibrium.

From the above discussion we conclude that during the transition from the initial state of $U_2^0 < U_2$ to the final state of thermal equilibrium, system 1 was at a higher temperature than system 2, and in the approach to thermal equilibrium heat flows, on average, from the system at higher temperature to a system at a lower temperature. This is the *Clausius statement of the second law of thermodynamics*. In terms of the temperature parameter $\beta = 1/kT$, we can say that, on average, heat flows from a system with the smaller value of β to a system with the higher value of β.

The second law of thermodynamics can be formulated in terms of entropy as: *the entropy of an isolated assembly must increase or in the limit remain constant*. This can be mathematically expressed as

$$\Delta S \geqslant 0 \qquad \text{for an isolated system.} \tag{13.72}$$

Note that this does not necessarily mean that the entropy of a non-isolated system can never decrease. Transfer of energy between a non-isolated system and its surroundings can definitely result in lowering of the entropy of the system. We intend to derive Equation 13.72 on the basis of statistical consideration.

Consider again the composite system given in Figure 13.4. Let the initial energies of systems 1 and 2 be U_1^0 and U_2^0, respectively, and the partition between them is an adiabatic one. Let the number of microstates accessible to system 1 when it has energy U_1^0 be designated by $\Omega_1(U_1^0)$ and that accessible to system 2 when it has energy U_2^0 be denoted by $\Omega_2(U_2^0)$. Since system 2 can be in any one of its $\Omega_2(U_2^0)$ accessible microstates when system 1 is in any one of its $\Omega_1(U_1^0)$ accessible microstates, for our initial condition the total number of microstates accessible to the composite system in Figure 13.4 is given by

$$\Omega_T^0 = \Omega_1(U_1^0)\Omega_2(U_2^0) \tag{13.73}$$

where Ω_T^0 is the total number of microstates accessible to the composite system under the initial condition, and its total energy is given by

$$U_T = U_1^0 + U_2^0. \tag{13.74}$$

If S_i is the total entropy of the composite system in the initial state, then according to Equation 13.53

$$S_i = k \ln \Omega_T^0 = k \ln [\Omega_1(U_1^0)\Omega_2(U_2^0)]$$
$$= k \ln [\Omega_1(U_1^0)] + k [\ln \Omega_2 (U_2^0)]$$
$$= S_1(U_1^0) + S_2(U_2^0) \tag{13.75}$$

where $S_1(U_1^0)$ and $S_2(U_2^0)$ are the initial entropies of system 1 and system 2, respectively, when they are considered as two isolated systems with energies U_1^0 and U_2^0, respectively.

Now the adiabatic partition is replaced with a diathermic one, so that heat can flow from one system to the other. According to the principle of *equal a priori probabilities* (Postulate 3), at thermal equilibrium, the composite system is equally likely to be found in any one of the total microstates Ω_T accessible to it. Then, according to Equation 13.41

$$\Omega_T = \sum \Omega_1(U_1')\Omega_2(U_2') \tag{13.76}$$

where $U_1' + U_2' = U_T$ (the law of conservation of energy) and the summation is to be carried out over all possible values of instantaneous energies U_1' and U_2'. Equation 13.76 includes the right-hand side of Equation 13.73 for one of the possible values of U_2', that is $U_2' = U_2^0$. Hence, clearly Ω_T must be greater than Ω_T^0. Noting that the number of microstates accessible to the composite system does not decrease even when an internal constraint is removed (in our case, replacement of the adiabatic partition with a diathermic one), because all of the original possibilities remain, we conclude that Ω_T^0 must be always smaller than or equal to Ω_T, i.e. $\Omega_T \geqslant \Omega_T^0$. This means

$$k \ln \Omega_T \geqslant k \ln \Omega_T^0. \tag{13.77}$$

Since $k \ln \Omega_T = S_f$, the entropy of the composite system in the final state when systems 1 and 2 are in thermal equilibrium, combining Equations 13.75 and 13.77, we get

$$S_f \geqslant S_i. \tag{13.78}$$

It can be proved, to an excellent approximation, that Equation 13.76 can be written as

$$\ln \Omega_T = \ln \Omega_1(U_1) + \ln \Omega_2(U_2) \tag{13.79}$$

and equation $S_f = k \ln \Omega_T$ can be written as

$$S_f = k \ln [\Omega_1(U_1)] + k \ln [\Omega_2(U_2)]. \tag{13.80}$$

However, $k \ln [\Omega_1(U_1)] = S_1(U_1)$ and $k \ln [\Omega_2(U_2)] = S_2(U_2)$, which are the entropies system 1 and system 2 would have if they were two separate isolated systems with the most probable energies U_1 and U_2, respectively. Therefore,

$$S_f = S_1(U_1) + S_2(U_2). \tag{13.81}$$

Actually, S_f of the composite system should be greater than the sum of $S_1(U_1)$ and $S_2(U_2)$ because $\ln \Omega_T > \ln (\Omega_1\Omega_2)_{maximum}$. However, because of the sharpness of the probability distribution we can ignore the difference for a macroscopic system.

As $\Omega_1(U_1)\Omega_2(U_2)$ is the maximum value of the product $\Omega_1(U_1')\Omega_2(U_2')$ corresponding to the maximum value of the probability $\mathscr{P}_2(U_2') = \Omega_1\Omega_2/\Omega_T$, we shall always have

$$\Omega_1(U_1)\Omega_2(U_2) \geqslant \Omega_1(U_1^0)\Omega_2(U_2^0) \tag{13.82}$$

or

$$\ln [\Omega_1(U_1)\Omega_2(U_2)] \geqslant \ln [\Omega_1(U_1^0)\Omega_2(U_2^0)]$$

or

$$k \ln [\Omega_1(U_1)\Omega_2(U_2)] \geqslant k \ln [\Omega_1(U_1^0)\Omega_2(U_2^0)]. \tag{13.83}$$

By virtue of Equations 13.75, 13.80 and 13.81 the above equation becomes

$$S_f = [S_1(U_1) + S_2(U_2)] \geqslant S_i = [S_1(U_1^0) + S_2(U_2^0)]. \tag{13.84}$$

This equation shows that the total entropy of the composite system in the final state, i.e. when system 1 and system 2 are separated by a diathermic partition and in thermal equilibrium, is greater than or equal to the total entropy of the composite system at the initial state when systems 1 and 2 were separated by an adiabatic partition.

In Equation 13.84 we have two signs, $>$ and $=$. We should be clear about the circumstances when the left-hand side of this equation is equal to or greater than the right-hand side. If the left-hand side is to be equal to the right-hand side, then we must have $U_1 = U_1^0$ and $U_2 = U_2^0$. In that case, systems 1 and 2, when separated by a diathermic partition, would start with the most probable division of energy and at the same absolute temperature $(T_1 = T_2)$. According to the criterion of reversibility in classical thermodynamics, when systems 1 and 2 are at the same absolute temperature, any infinitesimal transfer of energy would be a reversible change. Therefore, we conclude that in Equation 13.84 the left-hand side will be equal to the right-hand side only for a reversible change.

Equation 13.84 is often written in the following form in the change from the initial state of the composite system (when subsystems 1 and 2 are separated by an adiabatic partition) to the final equilibrium state of the composite system (when subsystems 1 and 2 are separated by a diathermic partition)

$$\Delta S_{total} \geqslant 0 \tag{13.85}$$

where ΔS_{total} is the change in the total entropy of the composite system. This is the same equation as Equation 13.72 which is the mathematical representation of the second law of thermodynamics. The equality sign in Equation 13.85 holds only for a reversible change.

13.7.4 The third law of thermodynamics

The second law of thermodynamics does not enable the absolute value of entropy of any substance to be calculated. This law only permits evaluation of entropy changes for specified changes in physical or chemical states. In 1905 Nernst stated the general principle that in any chemical reaction between solid or liquid substances

$$\frac{d}{dT}[(\Delta G)_T] \rightarrow 0 \tag{13.86}$$

at the absolute zero of temperature. It has been proved that this rule applies to crystalline solids only, with some exceptions.

According to this principle, for changes in chemical state involving only perfect crystalline substances, $\Delta S_0 = 0$. This requirement is satisfied only if for perfect crystalline substances the entropy at the absolute zero of temperature $S_0 = 0$. More precisely, the Nernst principle is satisfied if S_0 per atom is the same for all elements when in a perfect crystalline form. If this is true, then for any change in state involving only perfect crystalline substances $\Delta S_0 = 0$. In fact, this is the basis of the third law of thermodynamics.

Since entropy is a measure of disorder or randomness in a substance, $S_0 = 0$ will imply a highly ordered state of matter. The most highly ordered state of matter that we can imagine is the crystalline state at the absolute zero of temperature, because at this temperature even the rotational and vibrational motions of the molecules are minimized. A glassy or amorphous solid is not completely ordered even at the absolute zero of temperature and any disorder remaining at this temperature produces a finite value of S_0.

Of course, no substance is a gas at sufficiently low temperatures, and we must take into account the interaction between the molecules. However, usually we can assume that there is one lowest energy level for an assembly. If this is true, then at the equilibrium state the assembly will certainly be in this lowest energy level at absolute zero of temperature. If E_1 is the energy of this energy level, then at a low enough temperature the canonical partition function for the assembly reduces to $\exp[-E_1/kT]$, because all the higher terms will be negligible compared with the first term. It can be shown statistically that for any assembly the entropy S is given by

$$S = \frac{E}{T} + k \ln (\text{p.f.}) \tag{13.87}$$

where E is the energy of the lowest energy level at absolute temperature T, and p.f. is the canonical partition function. By substituting $E = E_1$ and p.f. $= \exp[-E_1/kT]$ into Equation 13.87, we have

$$S = \frac{E_1}{T} + k \ln \exp[-E_1/kT]$$
$$= \frac{E_1}{T} - \frac{kE_1}{kT} = 0.$$

This shows that at the zero of absolute temperature the entropy S_0 is equal to zero.

We have noticed that the statement of the third law concerns the entropy of perfect crystals at absolute zero. The question may arise of what would be the entropy for other systems under similar circumstances. To find the answer, let us consider a solid crystalline solution at the absolute zero of temperature, and imagine that this crystal is perfectly ordered with respect to the distribution of lattice sites, but contains two different kinds of atom which are distinguishable from each other. If we imagine that the two types of atom are distributed throughout the lattice sites perfectly randomly, then it is obvious that the thermodynamic probability will not be unity. Therefore, we cannot expect zero or negative entropy. This implies that the formation of a mixed crystal from perfect crystals of different types of atom is associated with an increase in entropy.

We have already discussed that the molecular basis for the third law result, i.e. that the entropy is zero at absolute zero, is the perfectly ordered state of the crystals, with all the molecules in the same lowest energy level. The positive values for entropies of all compounds at temperatures above absolute zero result from the fact that, as the temperature is raised, more and more energy levels become available to the molecules. Since each individual molecule has its own particular pattern of energy levels, the entropies at temperatures higher than absolute zero are very much characteristic of the individual molecule. With a few exceptions, it is observed that the entropy values calculated from the details of the molecular energies are in agreement with those obtained from calorimetric third law measurements (within experimental error).

Experiments show that the fundamental feature of all cooling processes is that the lower the temperature attained, the more difficult it is to cool further. For example, the colder a liquid is, the lower the vapour pressure, and the harder it is to produce further cooling by pumping away the vapour. We can generalize this experience by saying: *By no finite series of processes is the absolute zero attainable.* This is known as the *principle of the unattainability of absolute zero, or the unattainability statement of the third law of thermodynamics.*

In the microscopic approach, we have defined the entropy of a system by

$$S = k \ln \Omega \tag{13.88}$$

where Ω is the total number of accessible microstates (degeneracy) of the system consistent with its macrostate. The microstates of a system can be specified by the values of the volume V and the energy U of the system together with the number of particles N in it. It can be shown that to allow for the splitting of the degeneracies of energy levels it is better in the general case to use the density of state rather than the degeneracies of individual energy levels. So, Equation 13.88 should be replaced by

$$S = k \ln g \tag{13.89}$$

where g is the density of state. If an N-particle system has a total of n independent microstates with energy eigenvalues in the range U to $U + \Delta U$ ($\Delta U \ll U$ but ΔU is sufficiently bigger than the energy separation of neighbouring microstates), then the density of state is given by

$$g = n/\Delta U \tag{13.90}$$

Therefore, the density of state is the number of microstates of the N-particle system per unit energy range. In a macroscopic concept, it can be treated as a continuous function of the internal energy of the system.

Schematic of the variation of $\ln \Omega$ with U for a thermodynamic system is shown in Figure 13.2. There we noted that as $U \rightarrow U_0$ (the lowest energy eigenvalue of the N-particle system, the slope β of the curve tends to infinity, so that the absolute temperature $T(= 1/\beta k)$ tends to zero. This means that the system would be in the lowest energy level (the ground state) if it were at the absolute zero of temperature. If the ground state of the system were non-degenerate, we have

$$S = k \ln 1 = 0.$$

This shows that if the system were in a non-degenerate ground state at the absolute zero, its entropy would be zero. Although it has not yet been definitely proved, it is accepted that the ground state of a system is non-degenerate.

Because at room temperatures the entropy S as given by Equation 13.88 is $\approx Nk$, for it to be negligible at the absolute zero we must have $\ln \Omega_T^0 \ll N$, where Ω_T^0 is the degeneracy of the lowest energy level (ground state). Therefore, for the third law to be valid for a macroscopic system at $T = 0$, we must have $\Omega_T^0 \ll \exp(N)$.

Example 13.5

The molar heat capacity of a metal at very low temperatures is given by $7 \times 10^{-4} T$ JK^{-1} mol^{-1}, where T is the absolute temperature. Determine the rise in its temperature and the increase of its entropy when 1×10^{-7} J of heat is add to it at constant volume. Assume that the entropy of this metal is zero at the absolute zero. Calculate the total number of microstates accessible to it and comment on your answer.

Solution

The specific heat at constant volume is given by $dU = nC_V dT$, where n is the number of moles. In this case, $C_V = 7 \times 10^{-4} T$ J-K^{-1} mol^{-1}, so $n = 1$, and $\Delta U = 1 \times 10^{-7}$ J. Therefore,

$$\int dU = \int_0^{T_1} C_V dT = 7 \times 10^{-4} \int_0^{T_1} T \, dT$$

or

$$\Delta U = 7 \times 10^{-4} [T/2]_0^{T_1} = 3.5 T_1^2$$

or 1×10^{-7} J $= 3.5 T_1^2$, i.e. $T_1 = 0.0169$ K which is the rise in temperature from the absolute zero.

The change in entropy from the absolute zero to T_1 is

$$\Delta S = \int_0^{T_1} C_V (dT/T) = 7 \times 10^{-4} \int_0^{T_1} T(dT/T)$$
$$= 7 \times 10^{-4} [T]_0^{T_1} = 1.18 \times 10^{-5} \text{ JK}^{-1}.$$

The number of microstates accessible to the metal under the given condition can be calculated using Equation 13.88. Therefore,

$$\Omega = \exp(S/k)$$

where k is the Boltzmann constant which is $1.38 \times 10^{-23} \, JK^{-1}$. Then,

$$\Omega = \exp\,(8.57 \times 10^{17}) \simeq 10^{37 \times 10^{16}}.$$

This is a very big number. With this number of microstates accessible to the metal, we cannot say that at $0.0169 \, K$ its properties are dominated by the properties of its ground state and the first excited state of the N-particle system. From this example it may be concluded that for macroscopic systems the properties of the ground state do not normally determine those properties of materials which are in agreement with the predictions of the Nernst form of the third law at temperatures as high as $1 \, K$.

Actually, contributions to the entropy come from several sources which are more or less independent, e.g. the nuclear spins, lattice vibrations, etc. To allow for such nearly independent contributions to the total entropy of a system, the third law of thermodynamics is expressed in a form given by Nernst–Simon (see Section 9.2).

There are substances which are not in internal thermodynamic equilibrium, e.g. glasses. The third law should not be applied to such substances. Besides, many substances can have significant residual entropies at low temperatures because of the entropy of mixing, for example the mixing of the various isotopes making up the substance. The statement of Nernst–Simon makes it possible to apply the third law to other aspects of such systems, for example the paramagnetic behaviour of the ions at low temperatures.

We noted earlier that one form of the third law which is often used in classical equilibrium thermodynamics is the law of the unattainability of the absolute zero of temperature, which states: 'By no finite series of processes is the absolute zero attainable'. Consider the system given in Figure 13.4 and assume that subsystem 1 and subsystem 2 are separated by a fixed rigid diathermic partition. It has been shown in section 13.4.4 that when subsystem 2 is in thermal equilibrium with subsystem 1, the probability $\mathscr{P}_{2i}(U'_{2i})$ that subsystem 2 is in its ith microstate with energy eigenvalue U'_{2i} is given by

$$\mathscr{P}_{2i}(U'_{2i}) = \Omega_1(U_T - U'_{2i})/\Omega_T = \Omega_1(U'_1)/\Omega_T$$

(see Equation 13.44). Since for normal thermodynamic systems, Ω_1 increases as U'_1 increases, $\Omega_1(U_T - U'_{2i})$ decreases as U'_{2i} increases. Thus, the probability $\mathscr{P}_{2i}(U'_{2i})$ has the maximum value when U'_{2i} is at a minimum. U'_{2i} is at a minimum when subsystem 2 is in its ground state. When subsystems 1 and 2 are in thermal equilibrium (when separated by a diathermic partition as in Figure 13.4), subsystem 2 is more likely to be in its ground state than in any other microstate. This means that it would be at the absolute zero. This is true irrespective of the relative sizes of subsystems 1 and 2.

In order to find the probability $\mathscr{P}_2(U'_2)$ that at thermal equilibrium subsystem 2 has energy $U'_2 = U_{2i}$, we have to multiply $\mathscr{P}_{2i}(U'_{2i})$ by the number of microstates, $\Omega_2(U'_2)$, accessible to it when it has energy U'_2 to obtain Equation 13.42. It can be shown that this product has a very sharp maximum. For systems with particles $N > 10^{19}$, i.e. macroscopic systems, it is highly probable that at thermal equilibrium subsystem 2 would be found in a microstate with an energy eigenvalue within 1 part in 10^7 of the value of U_1, where $\mathscr{P}_2(U'_2)$ is at a maximum.

Therefore, for macroscopic systems, the possibility of finding subsystem 2 in its ground state is extremely remote. Thus, if subsystem 2 is a macroscopic system the probability of finding it in its ground state can be neglected. On the other hand, if subsystem 2 is a microscopic system this probability could be very significant.

Let us assume that our subsystem 1 in Figure 13.4 is very large so that its absolute temperature T_1 can be considered to be constant for values of U_{2i} in the range of our interest. According to Equation 13.56 we have

$$\frac{\partial}{\partial U_1'} (\ln \Omega_1) = \beta_1 = 1/kT_1$$

for constant V_1 and N_1. If T_1 is constant, then integrating between the limits $U_1' = U_T$ and $U_1' = U_T - U_{2i}$, we have

$$\ln \left[\Omega_1(U_T - U_{2i})\right] - \ln \left[\Omega_1(U_T)\right] = -U_{2i}/kT_1$$

or

$$\Omega_1(U_T - U_{2i}) = \Omega_1(U_T) \exp \left[-U_{2i}'/kT_1\right]. \tag{13.91}$$

Combining equations 13.44 and 13.91, we get

$$\mathscr{P}_{2i} \propto \exp \left[-U_{2i}'/kT_1\right] \tag{13.92}$$

since both Ω_T and $\Omega_1(U_T)$ are constant. This is the Boltzmann distribution, which will be discussed later. Note that $\mathscr{P}_{2i}$ decreases exponentially with increasing U_{2i}' only when subsystem 1 is big enough to be considered as a heat reservoir. If subsystem 1 is not large enough to be treated as a heat reservoir, $\mathscr{P}_{2i}$ will still decrease with increasing U_{2i}' but it will not be exponential because of the finite change in T_1.

13.8   THE HELMHOLTZ FREE ENERGY AND CHEMICAL POTENTIAL

Let us imagine that a small system is in thermal equilibrium with a heat reservoir. Before we proceed further we need to develop a number of equations. These will be necessary to deduce the Helmholtz free energy as well as the chemical potential. Imagine that the small system is in internal thermodynamic equilibrium before it is placed in contact with the heat reservoir. If U is the internal energy, V is the volume and N is the number of particles of this system, then the number of microstates Ω accessible to the system is a function of U, V and N; that is, $\Omega = f(U, V, N)$. Then, we can write

$$S(U, V, N) = k \ln \Omega(U, V, N)$$

where $S(U, V, N)$ is the entropy of the system. If the system is placed in contact with a heat reservoir, the energy of the system is increased by dU, the volume by dV and the number of particles by dN, and the system is allowed to reach a new state of internal thermodynamic equilibrium. Since S is a function of U, V and N, that is $S = f(U, V, N)$, we can write

$$dS = \left(\frac{\partial S}{\partial U}\right)_{V,N} dU + \left(\frac{\partial S}{\partial V}\right)_{U,N} dV + \left(\frac{\partial S}{\partial N}\right)_{U,V} dN$$

(see Equation 13.60). We have already proved that $(\partial S/\partial U)_{V,N} = 1/T$ (see Equation 13.57), $(\partial S/\partial V)_{U,N} = P/T$ and $(\partial S/\partial N)_{U,V} = -\mu/T$, where μ is the chemical potential. Then, the above equation becomes

$$dS = \frac{dU}{T} + (P/T)dV - (\mu/T)dN \tag{13.93}$$

or

$$dU = T\, dS - P\, dV + \mu\, dN \tag{13.94}$$

and for constant N

$$dU = T\, dS - P\, dV \tag{13.95}$$

(see Equation 8.1). If the system has more than one chemical component, the entropy can be expressed as a function of $U, V, N_1, N_2 \ldots$. Then Equation 13.94 can be generalised to

$$dU = T\, dS - P\, dV + \sum \mu_i\, dN_i \tag{13.96}$$

where N_i is the number of particles of the ith chemical component and

$$\mu_i = -T\left(\frac{\partial S}{\partial N_i}\right)_{U,V,N_1,\ldots,N_n(\text{except } N_i)} \tag{13.97}$$

is the chemical potential of the ith chemical component.

The number of particles N in our small system is assumed to be constant. Since $\ln Z$ is a function of V, N and β, we have

$$d(\ln Z) = \left(\frac{\partial \ln Z}{\partial V}\right)_{N,\beta} dV + \left(\frac{\partial \ln Z}{\partial \beta}\right)_{V,N} d\beta.$$

We have shown that the pressure and the internal energy of a small system in thermal equilibrium with a heat reservoir are given by

$$P = \frac{1}{\beta}\left(\frac{\partial \ln Z}{\partial V}\right)_{\beta,N} = kT\left(\frac{\partial \ln Z}{\partial V}\right)_{T,N} \tag{13.98}$$

and

$$U = -\left(\frac{\partial \ln Z}{\partial \beta}\right)_{V,N} = kT^2\left(\frac{\partial \ln Z}{\partial T}\right)_{V,N} \tag{13.99}$$

Then,

$$d(\ln Z) = -U \, d\beta + P\beta \, dV. \tag{13.100}$$

However, $d(\beta U) = U \, d\beta + \beta \, dU$. Adding this to Equation 13.100 and rearranging, we get

$$d(\ln Z + \beta U) = \beta(dU + P \, dV). \tag{13.101}$$

Since $\beta = 1/kT$, Equation 13.101 becomes

$$dU + P \, dV = T \, d[k \ln Z + U/T]. \tag{13.102}$$

Comparing Equations 13.95 and 13.102, we get

$$dS = d[k \ln Z + U/T]$$

and integrating

$$S = k \ln Z + U/T + \text{constant}. \tag{13.103}$$

We have to evaluate the constant of integration. Let this constant be S_0 and let the energy eigenvalues of our small system be $\varepsilon_0, \varepsilon_1, \varepsilon_2, \varepsilon_3, \ldots$ in order of increasing magnitude. Then, the partition function Z is

$$
\begin{aligned}
Z &= \exp\,(-\varepsilon_0/kT) + \exp\,(-\varepsilon_1/kT) + \exp\,(-\varepsilon_2/kT) + \cdots \\
&= \exp\,(-\varepsilon_0/kT)[1 + \exp\,\{-(\varepsilon_1 - \varepsilon_0)/kT\} + \cdots].
\end{aligned}
$$

If the absolute temperature T of the heat reservoir tends to zero, $\exp\,\{-(\varepsilon_1-\varepsilon_0)/kT\} \ll 1$, $Z \to \exp\,(-\varepsilon_0/kT)$ and $\ln Z \to (-\varepsilon_0/kT)$. When $T \to 0$, $U \to \varepsilon_0$. Hence, in the limit when $T \to 0$, Equation 13.103 becomes

$$S = -k\varepsilon_0/kT + \varepsilon_0/T + S_0 = S_0. \tag{13.104}$$

Thus, the integration constant S_0 is equal to the entropy which the small system would have at the absolute zero. Consistent with the third law of thermodynamics, if we assume that $S = 0$ at $T = 0$, then Equation 13.103 becomes

$$S = k \ln Z + U/T = k[\ln Z + \beta U]. \tag{13.105}$$

Rewriting this equation, we get

$$U - TS = -kT \ln Z \tag{13.106}$$

and comparing it with Equation 10.3, we find that the *Helmholtz free energy* A becomes

$$A = -kT \ln Z \tag{13.107}$$

or

$$Z = \exp(-A/kT). \tag{13.108}$$

Differentiating both sides of Equation 13.106, we get

$$-d(kT \ln Z) = dU - T \, dS - S \, dT.$$

However, from Equation 13.94 $dU - T \, dS = -P \, dV + \mu N$. Hence, the above equation becomes

$$-d(kT \ln Z) = -P \, dV - S \, dT + \mu N. \tag{13.109}$$

If V and T are constant, then $dV = 0$ and $dT = 0$ and we get

$$\mu = -kT \left(\frac{\partial \ln Z}{\partial N} \right)_{V,T} = -\frac{1}{\beta} \left(\frac{\partial \ln Z}{\partial N} \right)_{V,\beta}. \tag{13.110}$$

If there is more than one chemical component in the system consisting of particles $N_1, N_2, \ldots$, etc., of type $1, 2, \ldots$, etc., then by using Equation 13.96 it follows that if T, V and all the N_i (except for N_j, the number of particles of the jth component present) are constant, we have the chemical potential of the jth component as

$$\mu_j = -kT \left(\frac{\partial \ln Z}{\partial N_j} \right)_{T,V,N_1,N_2,\ldots(\text{except } N_j)}. \tag{13.111}$$

13.9   THE GIBBS FREE ENERGY

In Sections 10.4 to 10.10 an extensive discussion of various aspects of the Gibbs free energy, including the conditions for equilibrium, have been presented. Here we intend to derive an equation for the Gibbs free energy (G), using an approach based on statistical mechanics. We have already defined G by the equations

$$G = U + PV - TS \tag{13.112}$$

$$G = H - TS \tag{13.113}$$

(since $H = U + PV$). U, T, S, V and P are all functions of the thermodynamic state of a system. So, G is also a function of state, as well as an extensive variable. By differentiating Equation 13.112, we get

$$\begin{aligned} dG &= dU + P \, dV + V \, dP - T \, dS - S \, dT \\ &= (dU - T \, dS + P \, dV) + V \, dP - S \, dT \\ &= \mu \, dN + V \, dP - S \, dT \end{aligned} \tag{13.114}$$

(see Equation 13.94). Equation 13.114 shows that G is a function of N, V and P, that is $G = f(N, T, P)$. So, by differentiating G with respect to each of these variables, keeping others constant, we have

$$dG = \left(\frac{\partial G}{\partial N}\right)_{T,P} dN + \left(\frac{\partial G}{\partial T}\right)_{N,P} dT + \left(\frac{\partial G}{\partial P}\right)_{N,T} dP. \qquad (13.115)$$

By comparing Equations 13.114 and 13.115, we have

$$\mu = \left(\frac{\partial G}{\partial N}\right)_{T,P} \qquad (13.116)$$

$$S = -\left(\frac{\partial G}{\partial T}\right)_{N,P} \qquad (13.117)$$

$$V = \left(\frac{\partial G}{\partial P}\right)_{N,T} \qquad (13.118)$$

$$H = G - T\left(\frac{\partial G}{\partial T}\right)_{N,P} \qquad (13.119)$$

(using Equation 13.113). Equation 13.119 is known as the *Gibbs–Helmholtz equation*.

Since $G(N, T, P)$ and N are extensive variables, then for a single component system G is proportional to N if T and P are constant. Therefore,

$$G = Nf(T, P) \qquad (13.120)$$

where f depends on T and P. Differentiating both sides with respect to N, keeping both T and P constant, we have

$$\left(\frac{\partial G}{\partial N}\right)_{T,P} = f$$

and by comparing it with Equation 13.116, we have $f = \mu$ the chemical potential of a one-component system. Thus, Equation 13.120 becomes

$$G = N\mu(T, P). \qquad (13.121)$$

Equations 13.116 and 13.121 show that the chemical potential of a one-component system is equal to the increase in G with respect to N if T and P are constant. According to Equation 13.96, for an n component system, we have

$$dG = V\,dP - S\,dT + \sum_i \mu_i\,dN_i \qquad (13.122)$$

where $i = 1, 2, \ldots, n$, so we can write

$$\mu_i = \left(\frac{\partial G}{\partial N_i} \right)_{T, P, N_1, \ldots, N_n (\text{except } N_i)}. \tag{13.123}$$

In the general case of a multicomponent system, μ_i of the ith component depends on the temperature, pressure and the concentrations of all the other components of the system. However, in the case of a mixture of ideal gases, each component can be considered as completely independent, occupying the whole volume V. So, μ_i is independent of the concentrations of the other components, and for each component Equation 13.110 becomes

$$\mu = -kT \ln (Z_1/N) \tag{13.124}$$

where Z_1 is the single particle partition function. If T and P are constant, both dT and dP are zero and Equation 13.122 becomes

$$dG = \sum_i \mu_i \, dN_i \tag{13.125}$$

where $i = 1, 2, \ldots, n$. Remember that the higher the concentration, the higher the chemical potential. Particles tend to go from areas of high chemical potential to regions of lower chemical potential. For an ideal diatomic gas, this means that particles tend to move from regions of high concentration to regions of lower concentration.

The Gibbs free energy is an extensive variable, i.e. it is proportional to the amount of matter present in the system. Therefore, if $N_1, N_2, N_3, \ldots, N_i$ are all increased by a factor, keeping T and P constant, then G is also increased by that factor.

13.10   THE GIBBS PARADOX

Let us consider a mixture of ideal gases containing n_1 moles of A_1, n_2 moles of A_2, n_3 moles of $A_3, \ldots, n_m$ moles of A_m, all at pressure P and temperature T. Then, the equation of state is

$$PV = n_1 RT + n_2 RT + \cdots + n_m RT = \sum_{i=1}^{m} n_i RT. \tag{13.126}$$

The partial pressure P_i of the ith component is

$$P_i = n_i RT/V \tag{13.127}$$

and the total pressure of the mixture is the sum of the partial pressures, that is

$$P = \sum_{i=1}^{m} P_i. \tag{13.128}$$

According to Equation 13.122, the change in the Gibbs free energy for an arbitrary change in the pressure (P), volume (V) and number of moles ($n_1, n_2, \ldots, n_m$) is

$$dG = V \, dP - S \, dT + \sum_{i=1}^{m} \mu_i \, dn_i \tag{13.129}$$

where S, V and the chemical potential μ_i are given by Equations 13.117, 13.118 and 13.123, respectively, the number of particles N_i being replaced by the number of moles n_i.

Now, let us first find the Gibbs free energy for a single component before mixing. If $PV = nRT$ is the equation of state and $C_V = \alpha nR$ (where $\alpha = \frac{3}{2}$ for monatomic gases, $\frac{5}{2}$ for diatomic gases and 3 for polyatomic gases) is the heat capacity at constant volume when $\alpha = \frac{3}{2}$, the Gibbs free energy can be written in the form

$$G(P, T, n) = nRT \ln P + nRT f(T) \tag{13.130}$$

where $f(T) = \ln (T/T_0)^{\alpha+1}$ is a function of temperature only, with T_0 being constant. Since, upon mixing, the Gibbs free energy of the mixture will be the sum of the free energies of each component, we can write

$$G_i(P, T, n_1, n_2, \ldots, n_m) = \sum_{i=1}^{m} n_i RT [\ln P + f(T) + \ln x_i] \tag{13.131}$$

where x_i is the mole fraction given by

$$x_i = \frac{n_i}{\sum_{i=1}^{m} n_i} = \frac{P_i}{P}. \tag{13.132}$$

Thus, the change in the Gibbs free energy during mixing is given by

$$G_{final} - G_{initial} = \sum_{i=1}^{m} n_i RT \ln x_i. \tag{13.133}$$

From Equations 13.117 and 13.133, it can be shown that the increase in entropy due to mixing is

$$\Delta S_{mix} = - \sum_{i=1}^{m} n_i R \ln x_i. \tag{13.134}$$

Since

$$G = \sum_i n_i \mu_i,$$

the chemical potential of the ith component is given by

$$\mu_i = RT [\ln P + f(T) + \ln x_i]. \tag{13.135}$$

For a single component, $x_i = 1$ and $\Delta S_{mix} = 0$. However, if there are two components, each containing one mole, then $x_1 = 1/2 = x_2$. So, $\Delta S_{mix} = 2R \ln 2$ and the entropy will increase upon mixing.

We have derived Equation 13.134 without any reference to the type of particle present in the different components. Quite obviously, the mixing process will increase the entropy so long as the particles are different. However, according to Equation 13.134, if all the particles are identical, there will still be an increase in entropy upon mixing, although the concept of mixing loses its meaning. Under this condition, Equation 13.134 does not work, i.e. it fails for identical particles.

This anomaly was first observed by Gibbs; hence it is called the *Gibbs paradox*. The resolution of this paradox lies in quantum mechanics. Identical or indistinguishable particles have different '*statistics*', so they must be counted in a different way from distinguishable particles. The difference between distinguishable and indistinguishable particles exists even in the classical limit; this leads to a resolution of the Gibbs paradox.

13.11   CONDITIONS FOR EQUILIBRIUM

In Section 10.7 we have discussed various thermodynamic criteria for conditions for equilibrium, but so far have not done this in terms of statistical mechanics. Actually, quite a number of statistical mechanical criteria can be deduced in terms of various thermodynamic functions and conditions under which the equilibrium is attained. Here we shall discuss the different cases.

Consider the composite system shown in Figure 13.4. Let subsystem 1 and subsystem 2 be separated by a diathermic partition which allows heat energy to flow from one subsystem to another. Let U'_1 and U'_2 be the instantaneous energies, V_1 and V_2 be the volumes, and N_1 and N_2 be the number of particles of subsystem 1 and subsystem 2, respectively. Under the condition stated above, V_1, V_2, N_1 and N_2 are all constant. Equation 13.76 gives the total number of microstates Ω_T accessible to the composite system.

Assume that subsystem 1 and subsystem 2 are left in contact with each other for sufficient time and they have reached thermal equilibrium. Under this condition, according to the principle of *equal a priori probabilities*, the closed isolated composite system is equally likely to be found experimentally in any one of the Ω_T microstates accessible to it. The probability $\mathscr{P}_2(U'_2)$ that subsystem 2 has energy U'_2 (at thermal equilibrium) is given by the ratio of the number of microstates $[\Omega_1(U_T - U'_2)\Omega_2(U'_2)]$ accessible to the closed composite system (in which subsystem 2 has energy U'_2) to the total number Ω_T of microstates accessible to the closed composite system, that is

$$\mathscr{P}_2(U'_2) = [\Omega_1(U_T - U'_2)][\Omega_2(U'_2)]/\Omega_T \qquad (13.136)$$

where Ω_T is a constant independent of U'_2. This is the same as Equation 13.42. Since according to the law of conservation of energy $U_T = U'_1 + U'_2$, $U'_1 = U_T - U'_2$ and $\mathscr{P}_2(U'_2)$ can be expressed as a function only of U'_2. The probability $\mathscr{P}_1(U'_1)$ that at thermal equilibrium subsystem 1 has energy $U'_1 = U_T - U'_2$ is equal to $\mathscr{P}_2(U'_2)$.

It can be seen from a plot of $\mathscr{P}_2(U'_2)$ against U'_2 (Figure 13.5) that for a macroscopic system, the number $\Omega_2(U'_2)$ of microstates accessible to subsystem 2 when it has energy U'_2 increases very rapidly with the increase of U'_2. Since $U'_1 + U'_2 = U_T$ (constant), as U'_2 increases,

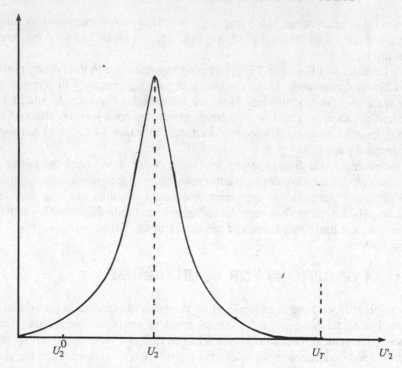

Figure 13.5   Schematic of variation of the probability $\mathscr{P}_2(U_2) = \Omega_1(U_T - U_2)\Omega_2(U_2)$ with U_2

U_1' decreases so that the number of microstates $\Omega_1(U_T - U_2')$ accessible to subsystem 1 when it has energy $U_1' = U_T - U_2'$ decreases rapidly. The product $\Omega_1(U_T - U_2')\Omega_2(U_2')$ has a very sharp maximum at $U_2' = U_2$ (a particular value; see Figure 13.5).

The most probable division of energy between subsystems 1 and 2 under the condition stated at the beginning of this section is found above by finding the value of U_2' at which $\mathscr{P}_2(U_2')$ given by Equation 13.136 is at a maximum. This leads us to the condition that the probability $\mathscr{P}_2(U_2')$ is at a maximum when $\beta_1 (= 1/kT_1) = \beta_2 (= 1/kT_2)$, i.e. $T_1 = T_2$, where T_1 and T_2 are the temperatures of subsystem 1 and subsystem 2, respectively. From Equation 13.136 we have

$$\ln \left[\mathscr{P}_2(U_2')\right] = \ln \left[\Omega_1(U_T - U_2')\right] + \ln \left[\Omega_2(U_2')\right] - \ln \Omega_T. \qquad (13.137)$$

It is obvious that when $\mathscr{P}_2(U_2')$ is a maximum, $\ln[\mathscr{P}_2(U_2')]$ is also a maximum. Therefore, we can say that the most probable value of U_2' is that vlaue of it for which $\ln[\mathscr{P}_2(U_2')]$ is at a maximum. As can be seen in Figure 13.5, $\mathscr{P}_2(U_2')$ is at a maximum at a particular value of $U_2' = U_2$. So, from Equation 13.137 the condition for U_2' to be equal to U_2 is

$$\ln \left[\Omega_1(U_T - U_2')\right] + \ln \left[\Omega_2(U_2')\right] - \ln \Omega_T \text{ is at a maximum.} \qquad (13.138)$$

The total number of microstates Ω_T accessible to the composite system is independent of the instantaneous value of U_2'. Therefore, $\ln \Omega_T$ is a constant, and from Equation 13.138

we have

$$k \ln [\Omega_1(U_T - U_2')] + k \ln [\Omega_2(U_2')] \text{ is at a maximum} \qquad (13.139)$$

where k is the Boltzmann constant. According to the equation $S = k \ln \Omega$ (Equation 13.88) where S is the entropy, we can write

$$S_1(U_1', V_1, N_1) = k \ln [\Omega_1(U_T - U_2')] \qquad (13.140)$$

$$S_2(U_2', V_2, N_2) = k \ln [\Omega_2(U_2')] \qquad (13.141)$$

which are the entropies that subsystems 1 and 2 would have if they were isolated systems consisting of N_1 and N_2 particles and with volumes V_1 and V_2, and energies U_1' and U_2', respectively. By combining Equations 13.139, 13.140 and 13.141, we get

$$S_1(U_1', V_1, N_1) + S_2(U_2', V_2, N_2) \text{ is at a maximum.} \qquad (13.142)$$

According to this equation the most probable division of energy between subsystem 1 and subsystem 2 when they are in thermal equilibrium is that division of energy which makes $S_1 + S_2$ a maximum, provided that V_1, V_2, N_1, N_2 and $U_T = U_1' + U_2'$ are all constant.

We have now developed the background information and the necessary mathematical tools to discuss conditions for equilibrium on the basis of a statistical mechanical approach.

13.11.1 Conditions for the equilibrium of a system of constant volume in contact with a heat reservoir

Here, we shall derive, on the basis of statistical mechanics, the fact that the Helmholtz free energy of a small system of constant volume and temperature is at a minimum at thermodynamic equilibrium. The heat reservoir and the small system in Figure 13.6 form a closed composite system. T_R is the absolute temperature of the heat reservoir; it keeps the temperature of the small system constant. The system has rigid diathermic walls and heat can flow across these walls. Let U_R, S_R, V_R and N_R be the energy, entropy, volume and number of particles of the heat reservoir, and let U, S, V and N be those of the small system if they were isolated systems. Then, according to Equation 13.142 we have

$$S_R(U_R, V_R, N_R) + S(U, V, N) \text{ is at a maximum} \qquad (13.143)$$

provided that V_R and V are constant, $U_R + U = U_T$ (the total energy) and there is no exchange of particles between the small system and the heat reservoir. Equation 13.143 corresponds to the most probable division of energy between the small system and the heat reservoir. If $\Omega_R(U_R, V_R, N_R)$ and $\Omega_S(U, V, N)$ are the numbers of microstates accessible to the heat reservoir and the small system, respectively, then it follows from Equation 13.136 that the probability $\mathscr{P}_R(U_R, V_R, N_R)$ that at thermodynamic equilibrium the heat reservoir has energy U_R, volume V_R and number of particles N_R is

$$\mathscr{P}_R(U_R, V_R, N_R) = [\Omega_R(U_R, V_R, N_R)][\Omega_S(U, V, N)]/\Omega_T \qquad (13.143)$$

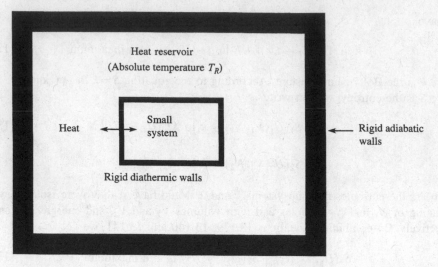

Figure 13.6   A small system is in thermal equilibrium with a heat reservoir of absolute temperature T_R

where Ω_T is a constant which is equal to the total number of accessible microstates. According to this equation the most probable division of energy between the heat reservoir and the small system is when the product

$$[\Omega_R(U_R, V_R, N_R)][\Omega_S(U, V, N)] \text{ is at a maximum.} \qquad (13.144)$$

If Ω_R^* is the number of microstates accessible to the heat reservoir when it has the energy $U_T = U_R + U$, then according to Equation 13.10 we have

$$\Omega_R = \Omega_R^*[\exp\,(-U/kT_R)] \qquad (13.145)$$

assuming that the heat reservoir is large enough to have the same absolute temperature T_R irrespective of the energy U of the small system. Substituting the value of Ω_R from Equation 13.145 into Equation 13.144, we have the condition for equilibrium

$$\Omega_R^*[\exp\,(-U/kT_R)][\Omega_S(U, V, N)] \text{ is at a maximum} \qquad (13.146)$$

Ω_R^* is a constant. So taking the logarithm, we have

$$(-U/kT_R) + \ln\,[\Omega_S(U, V, N)] \text{ is at a maximum.}$$

However, according to Equation 13.88, $\ln\,[\Omega_S(U, V, N)] = \left(\frac{1}{k}\right)S(U, V, N)$, the entropy of the small system. Hence, the condition for equilibrium of the small system is

$$(S - U/T_R) \text{ is at a maximum} \qquad (13.147)$$

or

$$U - T_R S \text{ is at a minimum.} \tag{13.148}$$

This implies that

$$dU - T_R\, dS = 0.$$

However, for an isolated system, $dU = 0$, and T_R is constant for constant V_R and N_R. Hence,

$$T_R\, dS = 0 \quad \text{or} \quad dS = 0 \tag{13.149}$$

which is the same as Equation 10.44 developed for classical thermodynamics.

Equation 13.149 implies that, for a transformation in an isolated system, dS must be positive; the entropy must increase. The entropy of any isolated system continues to increase so long as changes occur within the system. When there is no more change, the system has attained the equilibrium state and the entropy has reached the maximum value. Therefore, we conclude that *the condition of equilibrium in an isolated system is that the system has the maximum entropy.*

For a small system consisting of only one component, differentiating Equation 13.148 with respect to U, for constant V and N (T_R is a constant) we get

$$1 - T_R(\partial S/\partial U)_{V,N} = 0.$$

However, $T = (\partial U/\partial S)_{V,N}$. Therefore, the above equation gives $T = T_R$. This means that at equilibrium the absolute temperature of the small macroscopic system is equal to that of the heat reservoir. By definition, the Helmholtz free energy $A = U - TS$. Hence, by virtue of Equation 13.148 we have

$$A = U - TS \text{ is at a minimum} \tag{13.150}$$

provided that V and N are constant, and $T_R = T$. In this equation all the variables are for the small system.

From Equation 13.107, the Helmholtz free energy is given by $A = -kT \ln Z$ for constant T. This means that when A is at a minimum, $\ln Z$ is a maximum. So, the partition function Z of the small N particle system is at a maximum when $\ln Z$ is a maximum. Thus, another way of expressing the condition for the equilibrium of the small system in Figure 13.6 is

$$\left. \begin{array}{l} \ln Z \quad \text{is at a maximum} \\ Z \quad \text{is at a maximum} \end{array} \right\} \tag{13.151}$$

provided that V and N are constant, and $T_R = T$.

Let us assume that the small system in Figure 13.6 is a multicomponent system and there is an exothermic chemical reaction taking place between the components. We wish to examine the effect of the exothermic chemical reaction. In an exothermic reaction, energy is released. So, if the temperature of the small system is to remain equal to that of the heat reservoir, heat must flow from the small system to the heat reservoir. Consequently, the energy U of the small system will decrease and the energy U_R of the reservoir will increase. The latter effect will increase the value of Ω_R in Equation 13.144, so the value of the product of $\Omega_R \Omega_s$ will tend to increase, making it a more probable situation.

The exothermic chemical reaction will change the amounts of the different chemical constituents. If this effect and the transfer of energy from the small system to the heat reservoir due to the exothermic reaction reduce Ω_s, then the product $\Omega_R \Omega_s$ in Equation 13.144 will tend to decrease. An exothermic chemical reaction can go ahead only if the increase in Ω_R more than compensates for any reduction in Ω_s in the product $\Omega_R \Omega_s$ in Equation 13.144.

On the other hand, some endothermic reactions proceed, though in such cases energy has to be transferred from the heat reservoir to the small system so that the temperature of the latter remains constant. However, if this happens, Ω_R will decrease but there will be an increase in the number Ω_s of microstates accessible to the small system due to the energy transfer from the heat reservoir and the changes in the quantities of the different chemical constituents through the endothermic reaction. So that the endothermic reaction can proceed, the increase in Ω_s must be big enough to compensate for the reduction in Ω_R, so that $\Omega_R \Omega_s$ is increased following the reaction. Thermodynamically, this means that U is increased in Equation 13.150 with increasing A, but the increase in the entropy S of the small system makes the $-TS$ term in this equation more negative, which is enough to effect an overall reduction in the Helmholtz free energy A of the small system.

13.11.2   Conditions for equilibrium of a system in contact with a heat and pressure reservoir

Consider the closed system shown in Figure 13.7. The small system is fitted with a piston which can move forward and backward to keep the pressure of the small system equal to that of the reservoir P_R. The small system has diathermic walls which keep its temperature equal

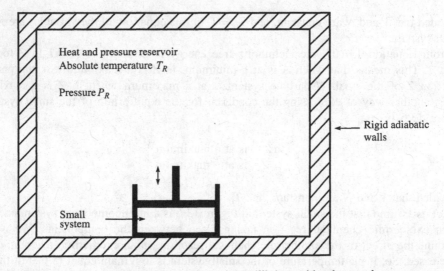

Figure 13.7   A small system is in thermodynamic equilibrium with a heat and pressure reservoir

to that of the reservoir T_R. Both the heat and pressure reservoir and the small system are in thermodynamic equilibrium.

Let U_R, S_R, V_R and N_R be the energy, entropy, volume and number of particles, respectively, of the heat and pressure reservoir, and let U, S, V and N be those of the small system. According to Equation 13.142, the condition for equilibrium between the small system and the reservoir in Figure 13.7 is

$$S_R(U_R, V_R, N_R) + S(U, V, N) \text{ is at a maximum} \qquad (13.152)$$

provided that $U + U_R = U_T$ and $V + V_R = V_T$, where U_T is the total energy and V_T is the total volume of the closed composite system, and there is no exchange of particles between the small system and the reservoir. Equation 13.152 implies that

$$[\Omega_R(U_R, V_R, N_R)][\Omega_s(U, V, N)] \text{ is at a maximum} \qquad (13.153)$$

where Ω_R is the number of microstates accessible to the reservoir and Ω_s is that accessible to the small system. In the case shown in Figure 13.7, both the energy and the volume of the small system can vary; thus those of the reservoir can also vary. However, it is assumed that the changes in these parameters in the reservoir are small, so that its temperature and pressure can remain constant.

Now, $\Omega_R(U_R, V_R)$ can be written as $\Omega_R(U_T - U, V_T - V)$. Taking the natural logarithm of it and expanding in a Taylor series, we obtain

$$\ln[\Omega_R(U_T - U, V_T - V)] \simeq \ln[\Omega_R(U_T, V_T)]$$
$$- U\left[\frac{\partial}{\partial U_R}(\ln \Omega_R)\right] - V\left[\frac{\partial}{\partial V_R}(\ln \Omega_R)\right]. \qquad (13.154)$$

By Equation 13.55 we can write

$$\frac{\partial \ln \Omega_R}{\partial U_R} = \beta_R = \frac{1}{kT_R}.$$

From Equation 13.88 we have $S_R = k \ln \Omega_R$, or $S_R/k = \ln \Omega_R$. By differentiating with respect to V_R, we have

$$\frac{\partial}{\partial V_R}(\ln \Omega_R)_{U,N} = \frac{1}{k}(\partial S_R/\partial V_R)_{U,N}.$$

However, $(\partial S/\partial V)_{U,N} = P/T$ (see Equation 7.29). So,

$$\frac{\partial}{\partial V_R}(\ln \Omega_R) = P_R/kT_R.$$

Substituting these values of

$$\frac{\partial}{\partial U_R}(\ln \Omega_R) \qquad \text{and} \qquad \frac{\partial}{\partial V_R}(\ln \Omega_R)$$

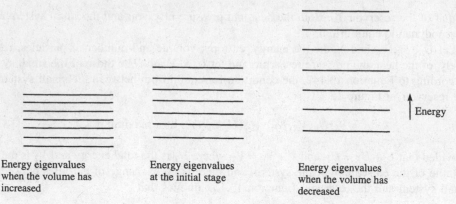

Energy eigenvalues when the volume has increased

Energy eigenvalues at the initial stage

Energy eigenvalues when the volume has decreased

$\uparrow$ Energy

Figure 13.8   The effect of volume change of a system on the energy eigenvalues

into Equation 13.154 and taking exponentials, we have

$$\Omega_R = \Omega_R^*(U_T, V_T) \ \exp \ [-(U + P_R V)/kT_R] \tag{13.155}$$

where Ω_T^* is the number of microstates accessible to the reservoir when it has the entire volume V_T and all the energy U_T. It is a constant.

If the volume of the small system in Figure 13.7 is kept constant and under this condition if its energy U increases due to heat transfer from the reservoir, the energy U_R of the reservoir will decrease and so will Ω_R. The change of Ω_R with decreasing U_R (increasing U) at constant volume is proportional to $\exp(-U/kT_R)$, where k is the Boltzmann constant. It is true that if the volume of a system is increased, the energy eigenvalues are all decreased, whereas if the volume is decreased, the energy eigenvalues are all increased and further separated (see Figure 13.8). If the volume of the small system increases when its energy U remains constant, the volume V_R of the reservoir will decrease and its energy eigenvalues will increase. The change of Ω_R with decreasing V_R (increasing V) for constant U_R is proportional to $\exp(-P_R V/kT_R)$, where k is the Boltzmann constant. Combining Equations 13.153 and 13.155, we have

$$\Omega(U, V, N) \ \exp \ [-(U + P_R V)/kT_R] \text{ is a at maximum.} \tag{13.156}$$

Using $\ln \Omega = S/k$ (see Equation 13.88) and taking the natural logarithm of both sides of Equation 13.156, we obtain

$$S - (U/T_R) - (P_R V/T_R) \text{ is at a maximum} \tag{13.157}$$

which is the condition for equilibrium between the small system and the reservoir in Figure 13.7.

By differentiating Equation 13.157 partially with respect to U, keeping V and N constant, we obtain

$$\left(\frac{\partial S}{\partial U}\right)_{V,N} - 1/T_R = 0.$$

However, $(\partial S/\partial U)_{V,N} = 1/T$. Substituting this into the above equation and simplifying, we have $T = T_R$. Again, differentiating Equation 13.157 partially with respect to V, keeping U and N constant, we get

$$\left(\frac{\partial S}{\partial V}\right)_{U,N} - P_R/T_R = 0.$$

However, $(\partial S/\partial V)_{U,N} = P/T$. Substituting this into the above equation, we obtain $P/T - P_R/T_R = 0$. Since we have already proved that $T = T_R$, this equation gives $P = P_R$. Hence, Equation 13.157 becomes

$$TS - PV - U \text{ is a at maximum} \tag{13.158}$$

provided that $T = T_R$ and $P = P_R$, where the variables T, S, P, V and U all belong to the small system. The Gibbs free energy is given by $G = U - TS + PV$. Therefore, we conclude that the condition for the small system in Figure 13.7 to be in thermodynamic equilibrium is

$$G \text{ is at a minimum} \tag{13.159}$$

where G is the Gibbs free energy of the small system. This conclusion agrees with that derived from classical thermodynamics (see Section 10.7).

13.12 EQUILIBRIUM CONSTANT

Many of the chemical reactions occur, to some extent, in both directions. Such systems attain a state of dynamic equilibrium, in which there is no cessation of reaction, but the system undergoes no further change unless conditions are altered. When a system, say $A + B \rightleftarrows C + D$, comes to equilibrium at some temperature, the ratio of the concentrations of the products and the reactants has a unique value. This ratio is the value of the term known as the *equilibrium constant K*. In Chapter 11 we have studied the relation between the equilibrium constant and the thermodynamic properties. Here, we shall try to understand equilibrium in a system from statistical point of view and derive equations for the equilibrium constant of the system.

Let us consider the equilibrium between two molecules, say A and B, to form a molecule AB, i.e. $A + B \rightleftarrows AB$. We assume that all species in this reaction are ideal gases and the reaction takes place in both directions. We already know how to express the parts of the thermodynamic functions that arise from the internal motions of molecules in terms of the corresponding partition functions. To study the equilibrium constant we need to know the partition function of the energy of the system. Equation 12.39 suggests that we treat the equilibrium between molecules in the various translational levels in the same way as we treated the equilibrium between molecules in the various internal levels, and set up a partition function for the translational motion. This will give the correct result if we follow certain precautions. However, a point to remember is that such a procedure does not have such a close relation to the methods of thermodynamics; in thermodynamics we do not deal with anything analogous to translational energy levels. This is particularly true when we consider that normally the number of molecules is far less than the translational levels.

Consequently, only occasionally is a level occupied, almost invariably by only a single molecule. It would be difficult to find a rationale for calculating the entropy, one of the most important thermodynamic properties, of such a system. Therefore, we need to proceed in a way which can be used where there is appreciable interaction between the molecules.

Now, let us assume that we have an assembly of molecules of any type: solid, liquid or gas. We also assume that we have many exactly identical assemblies with the same number of molecules and the same fixed volume, and all these together form an *ensemble of assemblies*. We take this ensemble to be a canonical ensemble, i.e. all the assemblies are in thermal contact with one another, so that they are all in thermal equilibrium. Each assembly can be considered as a big molecule with the total energy of an enormous molecule. The translational energy of such a big molecule involves only three degrees of freedom compared with the huge number of internal degrees of freedom. So, we can neglect these.

The entire energy of such an assembly is quantized, but the gap between the quantum levels are not wide. We denote such a quantum level by L and the energy by U_L. It is interesting to note that equilibrium between the quantum levels of the big molecules, i.e. the assemblies, will be established in the same way as equilibrium was established between the internal levels of ordinary molecules. Then the probability that the assembly will be in one particular quantum level is given by

$$\mathscr{P}_L = \frac{\exp\left(-U_L/kT\right)}{\sum_L \exp\left(-U_L/kT\right)}. \tag{13.160}$$

Note that we have used the Boltzmann constant k instead of the universal gas constant R. This is because U_L represents the energy of a single assembly or big molecule, not the ensemble of the assemblies (U_L includes all the energy of the assembly). The canonical partition function or the assembly partition function is given by

$$\text{P.F.} = \sum_L \exp\left(-U_L/kT\right). \tag{13.161}$$

We have started with the assumption that we have an assembly of molecules of any kind, solid, liquid or gas. However, if the assembly is a perfect gas, there will be little interaction between the individual molecules. In such a case, U_L can be taken as a sum of energies of the individual molecules which comprise the assembly, so that

$$U_L = \sum_n \varepsilon_n \tag{13.162}$$

where ε_n is the energy of an individual molecule n. The sum is to be carried out over all the molecules in the assembly. We have said that the assembly is in a particular quantum state L. This implies that the particular molecule ε_n will be in an individual quantum state different from the state L. Let us denote this quantum state by $l_{n,L}$. Then we can replace ε_n by $\varepsilon_{l_{n,L}}$ to specify the state more completely. Thus, Equation 13.162 becomes

$$U_L = \sum_n \varepsilon_{l_{n,L}} \tag{13.163}$$

Combining Equations 13.161 and 13.163, we get

$$\text{P.F.} = \sum_L \exp\left\{\sum\left(\frac{-\varepsilon_{l_{n,L}}}{kT}\right)\right\}. \tag{13.164}$$

By virtue of the property of the exponentials, i.e. a sum in the exponent is equivalent to a product of exponentials, Equation 13.164 can be written as

$$\text{P.F.} = \sum_L \prod_n \exp\left(-\varepsilon_{l_{n,L}}/kT\right). \tag{13.165}$$

If N is the total number of molecules, then each term in the sum over L is the product of N factors. If the summation is carried out over all states L, the quantum number $l_{n,L}$ for a particular molecule n will take every possible value, provided that the number of molecules occupying the energy levels in any given energy range is small compared with the number of energy levels. (If this condition is not satisfied, we use the Bose–Einstein statistics.) This is applicable to every one of the N factors, corresponding to each of the N molecules. Furthermore, every possible combination of $l_{n,L}$ values will appear in one term or another of the sum. On the right-hand side of Equation 13.165 if we interchange the $\sum$ and $\prod$ signs, this will also give every possible combination of N factors. Therefore, we can write

$$\text{P.F.} = \prod_n \sum_{l_n} \exp\left(-\varepsilon_{l_n}/kT\right). \tag{13.166}$$

Since the quantum numbers $l_{n,L}$ (for simplicity we shall write this as l) are the same for all molecules n (which means that the allowed energies are the same for all the molecules, provided that they are all of the same species), we can replace the product over all molecules by the Nth power. Then, Equation 13.166 becomes

$$\text{P.F.} = \left[\sum_l \exp\left(-\varepsilon_l/kT\right)\right]^N. \tag{13.167}$$

The right-hand side is a partition function but is slightly different from those we have written above, since it contains all energy levels (including all of those which differ only in the translational part) in the summation. Hence, we write

$$\text{P.F.} = (\text{p.f})^N. \tag{13.168}$$

Remember, we are dealing with N identical molecules which can move freely and be exchanged in positions (these being included). If two identical particles or two identical systems of particles (e.g. two like atoms or molecules) exchange their state, i.e. the first one goes into the state of the second and *vice versa*, the final state of the whole assembly becomes indistinguishable from the initial state. Then the quantum state of the entire assembly must be considered identical in the final and initial states; in fact, there is only one quantum state. The exchange of state of a pair of molecules actually corresponds to a shift of the point which represents the assembly in the phase space. This is due to the fact that each molecule has a set of co-ordinates and momenta (which generally include spin co-ordinates and momenta), and each of these provides a co-ordinate in the phase space for the assembly of molecules (commonly called γ-space). If a pair of molecules is exchanged, a phase volume of h^f will shift to another phase volume of the same size, and each quantum state will be counted more than once if we add over the entire phase space. If we have an assembly of N molecules which can exchange states with one another, then each quantum state of the entire assembly will be counted $N!$ times, provided that no two molecules are in exactly the same state, including spin state, internal quantum state and translational state. A particular state of the entire assembly implies that the states of each molecule in the assembly are given. We have N molecules, each one of which is in its particular state. However, of the N molecules it is possible to select one molecule to place in one of the states. Let us call this state the first state. We can then choose another molecule from the remaining $N-1$ to place it in the second state, then another from $N-2$ for the third state, and so on, and finally the last 1 molecule for the Nth state. Thus, clearly there will be $N(N-1)(N-2)\ldots 1 = N!$ ways of setting up the state of the N molecules in the assembly. All of these will correspond to the same quantum state. Now, if we integrate over phase space in such a way so that we duplicate possible states of the system, we have to divide by the number of duplications (in our case by $N!$) to avoid counting parts of the phase space which are not distinct. Only this procedure gives results which have been experimentally verified.

In view of the above discussion, Equation 13.168 needs to be corrected by dividing it by $N!$ because we are dealing with N identical molecules. Thus, we obtain

$$\text{P.F.} = \left(\frac{1}{N!}\right)\left[\sum_l \exp\left(\varepsilon_l/kT\right)\right]^N = \left(\frac{1}{N!}\right)(\text{p.f.})^N. \tag{13.169}$$

Instead of writing it in terms of power of N, (p.f.) can be expressed in terms of products. If we take $(\text{p.f.})_0$ as the partition function of internal energy states for the whole system, i.e. $(\text{p.f.})_0 = \sum_i \exp\left(-U_i/RT\right)$ (see Equation 13.206) where U_i represents the energy of the assembly when it is in a particular internal quantum state, then rigorously speaking

$$(\text{p.f.}) = (\text{p.f.})_0 \, (\text{p.f.})_{translation} \tag{13.170}$$

because the energies of the translational (or external) levels strictly do not depend on the internal levels.

Now let us go back to the equilibrium between two molecules A and B to form AB, i.e. $A + B \rightleftarrows AB$ as mentioned earlier. We shall use the canonical partition function given by Equation 13.161. If N_A, N_B and N_{AB} represent the numbers of molecules of A, B and AB, respectively, then by virtue of Equation 13.163 we can write

$$U_L = \sum_{n_1} \varepsilon_{k_{n_1},L} + \sum_{n_2} \varepsilon_{l_{n_2},L} + \sum_{n_3} \varepsilon_{m_{n_3},L} \tag{13.171}$$

where $\varepsilon_{k_{n_1},L}$ is the energy of the n_1th molecule in the quantum state k when the state of the entire assembly is L, and $\sum_{n_1}$ is the sum over the N_A molecules of A; similarly, $\sum_{n_2}$ and $\sum_{n_3}$ are the sums over the N_B molecules of B and N_{AB} molecules of AB, respectively, and

$\varepsilon_{l_{n_2},L}$ and $\varepsilon_{m_{n_3},L}$ are the energies of the n_2th and n_3th molecules in the quantum states l and m, respectively, when the state of the whole assembly is L. Using the same reasons which were used to derive Equation 13.169, we now have

$$\text{P.F.} = \left[\frac{(\text{p.f.})_A^{N_A}}{N_A!} \right] \left[\frac{(\text{p.f.})_B^{N_B}}{N_B!} \right] \left[\frac{(\text{p.f.})_{AB}^{N_{AB}}}{N_{AB}!} \right]. \tag{13.172}$$

An important point to remember is that Equation 13.172 is applicable for a fixed set of values of N_A, N_B and N_{AB}. If the values are changed to another set, we shall have a different set of energy levels U_L for the entire system. Since N_A, N_B and N_{AB} can have many possible values, the complete partition function for all possible values of these is given by

$$\text{P.F.} = \sum_{N_A} \sum_{N_B} \sum_{N_{AB}} \left[\frac{(\text{p.f.})_A^{N_A}}{N_A!} \right] \left[\frac{(\text{p.f.})_B^{N_B}}{N_B!} \right] \left[\frac{(\text{p.f.})_{AB}^{N_{AB}}}{N_{AB}!} \right]. \tag{13.173}$$

As said before, the sums are over all the possible values of N_A, N_B and N_{AB}, and consistent with

$$\left. \begin{array}{l} N_{AB} + N_A = N_{A,\,total} = \text{constant} \\ N_{AB} + N_B = N_{B,\,total} = \text{constant} \end{array} \right\}. \tag{13.174}$$

It does not matter which molecules are combined and which are separate; only the values of N_A, N_B and N_{AB} do matter, because exchanging two N_A (whether separate or combined) does not result in distinct energy levels. The partition functions must be referred to consistent energy levels.

The partition function given by Equation 13.173 may look formidable having so many summations, factorials, powers, etc. However, in fact, it is not necessary to evaluate all the sums which appear in it; we need take only the largest term in Equation 13.173. To find the largest term in the sums, we need the values of N_A, N_B and N_{AB} which make the term a maximum. We may just as well make the logarithm of the largest term a maximum. The condition for this is to set the variation of the logarithm of such a term equal to zero. Thus,

$$\begin{aligned} \partial \ln(\text{P.F.}) = \partial \ln &\left[\frac{(\text{p.f.})_A^{N_A}}{N_A!} \frac{(\text{p.f.})_B^{N_B}}{N_B!} \frac{(\text{p.f.})_{AB}^{N_{AB}}}{N_{AB}!} \right] = 0 \\ = &\,\partial N_A \ln(\text{p.f.})_A + \partial N_B \ln(\text{p.f.})_B + \partial N_{AB} \ln(\text{p.f.})_{AB} \\ &- \partial N_A \ln N_A - \partial N_B \ln N_B - \partial N_{AB} \ln N_{AB} \\ = &\,0. \end{aligned} \tag{13.175}$$

By differentiating Equations 13.174 we have

$$\partial N_{AB} + \partial N_A = 0 \tag{13.176}$$

$$\partial N_{AB} + \partial N_B = 0. \tag{13.177}$$

We multiply Equations 13.176 and 13.177 by the constants c_1 and c_2, respectively, and then add to Equation 13.175. Thus, we have

$$\partial N_A[\ln(\text{p.f.})_A - \ln N_A + c_1] + \partial N_B[\ln(\text{p.f.})_B - \ln N_B + c_2]$$
$$+ \partial N_{AB}[\ln(\text{p.f.})_{AB} - \ln N_{AB} + c_1 + c_2] = 0. \tag{13.178}$$

In this equation, if we put

$$c_1 = \ln N_A - \ln(\text{p.f.})_A \tag{13.179}$$

$$c_2 = \ln N_B - \ln(\text{p.f.})_B, \tag{13.180}$$

then the first two expressions in parenthesises in Equation 13.178 become zero and consequently the last expression in parentheses also becomes zero. This gives

$$\ln(\text{p.f.})_{AB} - \ln N_{AB} + c_1 + c_2 = 0$$

or

$$c_1 + c_2 = \ln N_{AB} - \ln(\text{p.f.})_{AB}. \tag{13.181}$$

Clearly, by definition of the chemical potential μ in the usual way (i.e. the expression for A for a mixture of N_A molecules of A, for B for a mixture of N_B molecules of B, and for N_{AB} for a mixture of N_{AB} molecules of AB), we have

$$c_1 = \mu_A/kT$$
$$c_2 = \mu_B/kT$$
$$c_1 + c_2 = \mu_{AB} + kT.$$

Therefore, for the maximum term, which we assume represents a state of equilibrium, we have

$$\mu_A + \mu_B = \mu_{AB}. \tag{13.182}$$

Also, from Equations 13.179, 13.180, 13.181 and 13.182 we have

$$\ln\left[\frac{N_A}{(\text{p.f.})_A}\right] = \mu_A/kT$$

$$\ln\left[\frac{N_B}{(\text{p.f.})_B}\right] = \mu_B/kT$$

$$\ln\left[\frac{N_{AB}}{(\text{p.f.})_{AB}}\right] = \mu_{AB}/kT.$$

This gives

$$\frac{N_{AB}}{N_A N_B} = \frac{(\text{p.f.})_{AB}}{(\text{p.f.})_A (\text{p.f.})_B}. \tag{13.183}$$

The equilibrium constant K of the system is given by

$$K = \frac{N_{AB}/V}{(N_A/V)(N_B/V)} = \frac{(p.f.)_{AB}/V}{[(p.f.)_A/V][(p.f.)_B/V]} \qquad (13.184)$$

where V is the volume. The thermodynamic functions of an assembly are just the same as the thermodynamic functions, per molecule, of the ensemble of assemblies mentioned in our discussion above. Therefore, the total thermodynamic functions U, S and A for arbitrary N (not necessarily one mole) are given by

$$U = RT^2 \left[\frac{\partial \ln (\text{P.F.})}{\partial T} \right] \qquad (13.185)$$

$$S = \frac{U}{T} + R \ln (\text{P.F.}) \qquad (13.186)$$

$$A = -RT \ln (\text{P.F.}) \qquad (13.187)$$

where P.F. is the partition function appropriate to the respective thermodynamic function. If we are dealing with one molecule of an ensemble, then R will be replaced by the Boltzmann constant k.

Equations 13.185, 13.186 and 13.187 are perfectly general and applicable to any assembly. However, if the assembly is an ideal gas, then by virtue of Equations 13.169 and 13.170, the translational contributions to U, S and A will be

$$U_{trans} = NkT^2 \left[\frac{\partial \ln (p.f.)_{trans}}{\partial T} \right]_V \qquad (13.188)$$

$$S_{trans} = \frac{U_{trans}}{T} + kN \ln (p.f.)_{trans} - k \ln N! \qquad (13.189)$$

$$A_{trans} = -NkT \ln (p.f.)_{trans} + kT \ln N!. \qquad (13.190)$$

Remember that $(p.f.)_{trans}$ is a function of the volume V.

For any system, the Gibbs free energy is given by $G = A + PV$. Since the pressure $P = -(\partial A/\partial V)_T$, it is possible to obtain an equation for G from Equation 13.187. However, for an ideal gas we use Equation 13.190 and $PV = NkT$. In this case, the Gibbs free energy is

$$G_{trans} = -NkT \ln (p.f.)_{trans} + kT \ln N! + NkT.$$

By the Stirling approximation, $\ln N! = N \ln N - N$. Hence,

$$G_{trans} = -NkT \ln \left[\frac{(p.f.)_{trans}}{N} \right] \qquad (13.191)$$

or

$$\frac{(p.f.)_{trans}}{V} = \frac{N}{V} \exp\left(\frac{-G_{trans}}{NkT}\right). \tag{13.192}$$

If we compare the terms $(p.f.)_A/V$ etc., in Equation 13.184 with Equation 13.192, we can see that

$$\frac{(p.f.)_A}{V} = \exp\left(-G_A^0/RT\right), \text{ etc.} \tag{13.193}$$

where G_A^0 is the molal Gibbs free energy of A when $N/V = 1$, i.e. A is at unit concentration. The same argument applies to $(p.f.)_B/V$ and $(p.f.)_{AB}/V$. Hence, Equation 13.184 can be written as

$$K = \exp\left(\frac{-\Delta G^0}{RT}\right) \tag{13.194}$$

where $\Delta G^0 = G_{AB}^0 - G_A^0 - G_B^0$. Here, we have expressed K in terms of the concentrations of A, B and AB. Hence, the standard state is the condition of unit concentration.

Taking the logarithm of both sides of Equation 13.184 and then differentiating with respect to T, we have

$$\frac{d \ln K}{dT} = \frac{d}{dT} \ln\left[\frac{(p.f.)_{AB}}{V}\right] - \frac{d}{dT} \ln\left[\frac{(p.f.)_A}{V}\right] - \frac{d}{dT} \ln\left[\frac{(p.f.)_B}{V}\right]$$

$$= U_{AB}/RT^2 - U_A/RT^2 - U_B/RT^2 \qquad \text{(see Equation 13.185)}$$

$$= \frac{1}{RT^2}(U_{AB} - U_A - U_B)$$

$$= \Delta U/RT^2 \tag{13.195}$$

where $\Delta U = U_{AB} - U_A - U_B$ is the molal energy change. This equation is valid only when K is expressed in terms of concentration. If K is expressed in terms of partial pressures of the reactants and products, we have the van't Hoff equation

$$\frac{d \ln K_P}{dT} = \frac{\Delta H}{RT^2} \tag{13.196}$$

(see Equation 11.20), where ΔH is the molal enthalpy change.

In the above discussion we have taken the statistical approach to derive the equilibrium constant of a system. It is interesting to know that a reverse approach is also possible, i.e. we can approach statistical mechanics through the use of the ideas of thermodynamics. We start with the well known law for the equilibrium constant (see Equation 11.11)

$$\Delta G^0 = -RT \ln K. \tag{13.197}$$

When this equation is used for a mixture of ideal gases, it is a general practice to express K in terms of pressure, taking as the standard state the condition at unit pressure (commonly 0.1 MPa or 1 atmosphere). However, for our discussion we shall express K in terms of concentrations. Provided that we use the condition in which the concentration is 1 as the standard state, Equation 13.197 will apply unchanged. This equation is applicable to any kind of equilibrium. If it is applied to an equilibrium between two isomers, for example, the two kinds of molecules involved in the equilibrium can be as closely related as we want, and the equation will apply all the same if the two species of molecules are of the same chemical type but in different internal quantum states (rotational, vibrational or electronic). Let us call a molecule in a particular internal quantum state a *quantum species*. Consider two quantum species, say the ith and jth quantum species, with concentrations c_i and c_j, respectively, which are in equilibrium. For this system, it follows from Equation 13.197 that

$$K_{ij} = \frac{c_j}{c_i} = \exp\left(\frac{-\Delta G_{ij}^0}{RT}\right)$$

$$= \exp\left(\frac{-\Delta H_{ij}^0 + T\Delta S_{ij}^0}{RT}\right)$$

$$= \exp\left(\frac{-\Delta H_{ij}^0}{RT}\right)\exp\left(\frac{\Delta S_{ij}^0}{R}\right) \tag{13.198}$$

where ΔG_{ij}^0 is the total molal free energy of species j minus that of species i, ΔH_{ij}^0 is the total molal enthalpy of species j minus that of species i, and $\Delta S_{ij}^0 =$ total molal entropy of species j minus that of species i. The superscript zero implies the standard state. However, we are dealing with an ideal gas mixture and there is no interaction between the molecules. Therefore, the changes in the enthalpy (ΔH) and the energy (ΔU) do not depend on the concentration. Hence, we can omit the superscript from these symbols in Equation 13.198.

By virtue of the kinetic theory of gases we may assume that the average translational energy of the molecules in any quantum state is the same as that of those in any other quantum state. This implies that the quantity PV of one quantum species is equal to that of any other quantum species. Therefore, we have, since $\Delta H = \Delta U + PV$,

$$\Delta H_{ij} = \Delta U_{ij}. \tag{13.199}$$

If U_i and U_j are the internal energies of ith and jth quantum species, respectively, then

$$\Delta U_{ij} = U_j - U_i \tag{13.200}$$

because the average translational energies cancel each other in the difference.

We use a hypothesis that assumes that *under standard conditions all quantum species have the same entropy per mole*. In other words, this assumption is the same as saying that the *a priori* probability of any quantum state is the same as that of any other. Then, under any special set of conditions, the probability of finding a molecule in any particular quantum state will depend on the energy. By virtue of our assumption we have

$$\Delta S_{ij}^0 = 0 \tag{13.201}$$

and by combining Equations 13.198, 13.199 and 13.201, we have

$$K_{ij} = \frac{c_j}{c_i} = \exp\left(\frac{-U_j + U_i}{RT}\right). \tag{13.202}$$

This equation holds for any two quantum states, and the concentration for any quantum species in proportional to the quantity $\exp(-U/RT)$, where U represents the energy of that quantum species. It follows, therefore, that the sum of the concentrations of several quantum species will be proportional to the sum of all such terms. If $\sum_i {}_1 c_i$ represents the sum of concentrations of quantum species over a particular set of quantum states, and $\sum_i {}_2 c_i$ represents the sum over another set of quantum states which may or may not overlap the former set, then we can write

$$\frac{\sum_i {}_2 c_i}{\sum_i {}_1 c_i} = \frac{\sum_i {}_2 \exp(-U_i/RT)}{\sum_i {}_1 \exp(-U_i/RT)}. \tag{13.203}$$

According to Equation 11.4, the left-hand side of Equation 13.203 is an equilibrium constant; it describes the equilibrium between two sets of quantum states. An important point to remember is that here the two sets of quantum species (or two sets of quantum states) are in equilibrium, and at the same time they are also in equilibrium within themselves. Since free energies are generally defined only for systems which are in internal equilibrium, we may write (by Equation 13.197)

$$\frac{\sum_i {}_2 c_i}{\sum_i {}_1 c_i} = \exp\left(\frac{-G_2^0 + G_1^0}{RT}\right) = \frac{\exp(-G_2^0/RT)}{\exp(-G_1^0/RT)} \tag{13.204}$$

where G_1^0 and G_2^0 are the standard molal free energies of the first and second set, respectively. This equation indicates that there is an interrelationship between G_1^0 and the sum $\sum_1 \exp(-U_i/RT)$, and G_2^0 and $\sum_2 \exp(U_i/RT)$. The difference $G_2^0 - G_1^0$ is, in fact, equal to $A_2^0 - A_1^0$, i.e. the difference between the Helmholtz free energies of two sets of quantum species. Terms such as $\sum \exp(-U/RT)$ are called the partition functions (p.f.). Equation 13.187 shows the simplest relation between A and (p.f.).

The partition function of molecules in any given set of quantum states is proportional to the corresponding total concentration of the molecules. If $c = \sum_i c_i$ represents the summation carried out over all quantum states (and so represents the total concentration), then we can write

$$\frac{c_i}{c} = \frac{\exp(-U_i/RT)}{\sum_i \exp(-U_i/RT)}. \tag{13.205}$$

In this equation $\sum_i \exp(-U_i/RT)$ is the partition function of internal energy states for the whole system, and we write

$$(\text{p.f.})_0 = \sum_i \exp(-U_i/RT) \tag{13.206}$$

If n_i is the number of molecules in the ith quantum state and N is the total number of molecules, then Equation 13.205 becomes

$$\frac{n_i}{N} = \frac{\exp\left(-U_i/RT\right)}{\sum\limits_i \exp\left(-U_i/RT\right)} = \frac{\exp\left(-U_i/RT\right)}{(\text{p.f.})_0}. \tag{13.207}$$

EXERCISES 13

13.1 A composite system is made up of two systems A and B which have 1.60 and 1.95 kJ of energy, respectively, and are in thermal equilibrium with a heat reservoir at 70°C. Assume that the interaction energy between A and B is negligible.

(a) Comment on the ability of the systems A and B as part of the composite system to exchange energy with the reservoir.

(b) Explain the relationship between the state that A is in to the state that B is in at any instant of time.

(c) What is the total energy of the composite system?

13.2 If two systems X and Y have energies E_X and E_Y at an instant of time and there is no interaction energy, show analytically that the probability of finding the composite system $X + Y$ in a particular state with energy $E_X + E_Y$ is given by $\mathscr{P}_{(X+Y)}(E_X + E_Y) = \mathscr{P}_X(E_X)\mathscr{P}_Y(E_Y)$.

13.3 Consider a system which has two microstates A and B and is in thermal equilibrium with a heat reservoir. The energy of the system in state A is -1 and that of the system in state B is $+1$ (measured in units of energy). If the probabilities of finding the system in state A and state B are $\mathscr{P}(A) = 0.70$ and $\mathscr{P}(B) = 0.30$, respectively, calculate the values of C and β in the equation $\mathscr{P}(E) = C \exp\left(-\beta E\right)$.

13.4 A hypothetical system at 290 K contains three different molecules with energies 2.4×10^{-21}, 4.8×10^{-21} and 7.2×10^{-21} J. Assuming that there is no interaction energy between these molecules, calculate the total energy of the system and the value of $\exp\left(-\beta E\right)$.

13.5 Show that the classical partition function for a perfect monatomic gas is given by $Z = V^N \left[(2\pi/\beta m)^{3N/2}\right]$ where the symbols have their usual meanings. Hence, show that $\beta = 1/kT$, where T is the temperature and k is the Boltzmann constant. Does this hold for all systems?

13.6 Consider two systems A and B surrounded by rigid, impermeable, adiabatic outer walls and separated by a rigid, impermeable diathermic partition. Let system A consist of two distinguishable particles of zero spin and have an initial energy 5 units of energy, and system B consist of two other distinguishable particles of zero spin and have an initial energy 1 unit of energy. Let the systems be in thermal contact with each other. Can the composite system be considered a closed composite system? Prepare a table of all the microstates accessible to the composite system, over all the possible distributions of energy between system A and system B consistent with the total energy of the composite system. Using this table, comment on whether, according to statistical mechanics it is possible for heat to flow from a hot system to a cold system, and whether it is possible that the mean energies $\bar{U}_A$ and $\bar{U}_B$ of systems A and B, respectively, are equal at thermal equilibrium.

13.7 Using the information given in Exercise 13.6, prepare a table of all the microstates accessible to systems A and B over all possible distributions of energy between system A and system B consistent with the total energy of the composite system. Hence, calculate the probability $\mathscr{P}_B(U_B)$ that at thermal equilibrium system B has energy U_B.

13.8 Discuss the variation of $S/k = \ln \Omega$ with respect to the internal energy of a system on the basis of (i) statistical mechanics and (ii) classical equilibrium thermodynamics. Sketch this variation and discuss the relationship of the slope of this curve with the temperature parameter β.

13.9 In classical equilibrium thermodynamics, if a small quantity of heat dQ is added to a system which is at the absolute temperature T, the entropy change of the system is given by $dS = dQ/T$. Derive this equation from statistical mechanics.

13.10

(a) Discuss the meaning of the symbol Ω in statistical thermodynamics.

(b) Explain why entropy is defined in terms of logarithms.

(c) Why is the entropy constant chosen as the Boltzmann constant?

(d) If a substance exists in equilibrium at negative temperatures, what would this mean with regard to the entropy on a microscopic scale?

13.11 Prove that two systems which are both in thermal equilibrium with a third system are in thermal equilibrium with each other.

13.12 Discuss the principle of the unattainability of absolute zero.

13.13 Prove analytically that the absolute zero of temperature cannot be attained.

13.14 Show that for the third law to be valid for a macroscopic system at $T = 0$, $\Omega_0 \ll \exp N$.

13.15 The specific heat at constant volume of an ideal gas is given by $C_V = \alpha nR$, where α is equal to 1.5, 2.5 and 3 for monatomic, diatomic and polyatomic gases, respectively. $(\partial U/\partial T)_{V,N} = C_V$ holds for ideal gases. Using these equations show that the internal energy depends only on the temperature.

13.16 With the help of the result obtained in Exercise 13.15 for an ideal gas, derive the fundamental equations for

(a) the entropy as a function of T and V

(b) the internal energy as a function of S and V

(c) the enthalpy as a function of S and P

(d) the Helmholtz free energy as a function of T and V

(e) the Gibbs free energy as a function of T and P

13.17 Sketch the variation of the Gibbs free energy with the temperature and pressure for which no phase transition occurs. What do the slopes of these curves mean? Show the point $G(P_0, T_0)$ on the figures. Sketch the variation of $(\partial G/\partial P)_T$ with P and that of $-(\partial G/\partial T)_P$ with T.

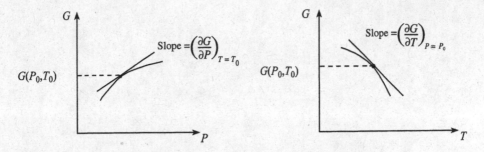

13.18 The partition function of an assembly of N molecules of a perfect gas is given by

$$(\text{p.f.})^N = \left[\sum_l \exp\left(\varepsilon_l / kT\right) \right]^N$$

which contains all energy levels, including all of those which differ only in the translational part, ε_l being the energy of an individual molecule in a special individual quantum state (see Equation 13.167). Is this partition function correct for the given system? Discuss.

14

Phase Equilibria
and Phase Transition

14.1 EQUILIBRIUM AMONG THE PHASES OF A PURE SUBSTANCE

So far our discussion has been mainly about chemical processes. We now discuss processes involving physical changes. Both physical and chemical processes can equally be treated in terms of fundamental thermodynamic principles. Phase equilibria provide some demonstration of the applicability of thermodynamics to physical processes. Any discussion on phase equilibria will provide the necessary background for discussion of solutions and their properties.

We have already deduced that, for any pure substance, $(\partial \mu / \partial T)_P = -\bar{S}$, where $\bar{S}$ is the partial molar entropy and μ is the chemical potential. Then, for any pure substance in the three phases (solid, liquid and gas), we have

$$
\left(\frac{\partial \mu_s}{\partial T}\right)_P = -\bar{S}_s
$$

$$
\left(\frac{\partial \mu_\ell}{\partial T}\right)_P = -\bar{S}_\ell \tag{14.1}
$$

$$
\left(\frac{\partial \mu_g}{\partial T}\right)_P = -\bar{S}_g.
$$

Since at any temperature, $\bar{S}_s < \bar{S}_\ell \ll \bar{S}_g$, if we plot μ-values against T at constant pressure for all the three phases of a pure substance, we shall obtain Figure 14.1.

A solid and its liquid phase will coexist in equilibrium, if $\mu_s = \mu_\ell$. This condition implies that the equilibrium temperature is given by the intersection of μ–T plots for the solid and liquid. This temperature is the melting temperature T_m of the solid substance. Similarly, a liquid and its vapour coexist in equilibrium at the temperature given by the intersection of μ–T plots for the liquid and vapour. This temperature is the boiling temperature T_b of the liquid substance. Below T_m the solid has the lowest chemical potential; between T_m and T_b the liquid has the lowest chemical potential, and above T_b the gas has the lowest chemical potential. Since a phase with the lowest value of the chemical potential is the stable phase,

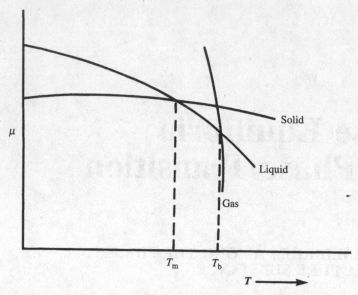

Figure 14.1 Schematic diagram of variation in chemical potential μ with temperature at constant pressure

then at any temperature $T < T_m$ the solid is stable, at $T_m < T < T_b$ the liquid is stable, and at $T > T_b$ the vapour is stable. Figure 14.1 illustrates the sequence of phase changes when a solid is heated at constant pressure. This sequence of phases is the result of the sequence of entropy values.

Melting, boiling and the change in a crystal from one lattice structure to another all involve phase transitions. Variations in these temperatures with pressure and concentration can readily be represented by curves. These and other plots of values of a state variable at which two phases can coexist are commonly known as *phase diagrams*. Such diagrams convey more information than μ–T plots. A phase diagram shows at a glance the properties of the substance, e.g. melting point, boiling point, triple point and transition point.

In a phase diagram the areas are of interest; the curves are only the boundaries. These curves divide the phase diagram into regions of solid, liquid and gas. If any point which describes a system falls within the solid region, then that implies that the system is a solid. Similarly, if a point which describes a system falls within the liquid region (or within the gas region), then the substance is a liquid (or a gas). Let us study the implication of a phase diagram with reference to the vapour pressure diagram of water as schematically shown in Figure 14.2.

If we choose any point B on the vapour pressure curve, then this point will represent the equilibrium pressure of water vapour above the liquid at that particular temperature (corresponding to B). The area above the curve is called the *liquid-phase area*. This is because, at the temperature and pressure corresponding to any point in that area, only the liquid water is present at equilibrium, provided that water is the only substance present.

Similarly, if the vapour pressure is reduced to any point below B, say C, at a given temperature, the vapour expands and the liquid evaporates to re-establish equilibrium; it

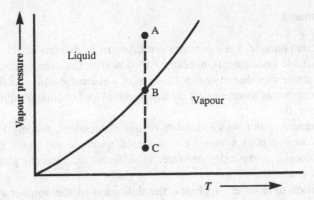

Figure 14.2 Schematic diagram of variation in vapour pressure of water with temperature

will finally disappear if the pressure is maintained at that point. The area below the vapour pressure curve is called the *vapour-phase area*. Within this area, only vapour can exist. The curve in Figure 14.2 is the common boundary of liquid- and vapour-phase regions; it is the equilibrium vapour pressure curve, along which liquid and vapour can exist in equilibrium with each other.

The number of phases that can coexist at equilibrium under a given condition is related by the phase rule to the number of components present and the number of variables. Before we proceed to derive this rule we intend to define the terms *phase*, *components* and *degree of freedom*, because a clear understanding of the significance of these terms will be necessary for derivation and application of the phase rule.

14.2 DEFINITIONS OF PHASE, COMPONENT AND DEGREE OF FREEDOM

14.2.1 Phase

This is defined as any homogeneous and physically distinct part of a system which is separated from the other parts of the system by a definite boundary. A phase may consist of any amount, large or small, of material. Ice, liquid water and water vapour are three phases. Each one is physically distinct and homogeneous, and there are clear boundaries between ice and liquid water, between liquid water and vapour, and between vapour and ice. It is to be remembered that each crystalline form of ice constitutes a separate phase, because it is clearly distinct from the other forms. In general, it can be said that every solid in a system is an individual phase.

A solid solution, being homogeneous, is a single phase irrespective of how many chemical compounds it contains. Similarly, in a liquid solution, one liquid layer is one phase whether it consists of a mixture or a pure substance as long as it is homogeneous. If there are two layers, there are two phases (e.g. water and alcohol). A gas or mixture of gases is homogeneous and constitutes only one phase.

The number of phases is denoted by the letter *P. Readers should be careful not to confuse P for number of phases with P for pressure.*

14.2.2   Component

The number of components of a system at equilibrium is defined as the least number of independently variable constituents necessary to describe the composition of each phase present in the system, either directly or in the form of a chemical equation. This definition of the number of components assumes that all the physical and chemical equilibria existing in the system are operative.

A simple example is water which consists of one component, namely H_2O. Each of the solid, liquid and vapour phases may be considered as made up of the component H_2O. Although the molecular complexity of water is different in different phases, the number of components is unaltered.

Like the definition of number of phases, the definition of the number of components is entirely definable only with reference to the phase rule for a particular equilibrium state. Unless the relationship of the number of components and the phase rule is clearly understood, any definition of the number of components will appear vague. The number of components is denoted by C.

14.2.3   Degree of freedom (or variance)

The number of degrees of freedom (or variance) of a system is defined as the number of intensive variables, such as temperature and pressure, which need to be fixed so that the condition of a system at equilibrium may be completely defined. As an example, a system consisting of one phase only of water has two degrees of freedom, because it is necessary to specify both temperature and pressure to define completely the state of that system. A system with one degree of freedom is usually called a *univariant system*, a system with two degrees of freedom is called a *bivariant system* and so on. An *invariant system* is one which has no degree of freedom.

An alternative statement of the definition of the degree of freedom is the number of intensive variables that can be independently varied without changing the number of phases present in the system. The degree of freedom is denoted by the symbol F.

14.3   PHASE RULE

In 1876 it was shown by J. W. Gibbs that there is a definite relationship in a system between the number of degrees of freedom, the number of components and the number of phases present. This relation is known as the *phase rule*. For any system at equilibrium at a definite pressure and temperature, provided that the equilibrium between any number of phases is not influenced by gravity, electrical or magnetic forces or by surface action, the number F of degrees of freedom of the system is related to the number P of phases and the number C of components by the equation

$$F = C - P + 2. \tag{14.2}$$

This rule can be deduced analytically by applying thermodynamic principles and arguments. Let us consider a system of C components and P different phases. If these C

components are distributed between the P different phases, the composition of each phase is completely defined by $C-1$ concentration terms because, if the concentrations of all the components except one are known, then the concentration of the last one must be equal to the remaining one. Therefore, in order to define the compositions of the P phases, $P(C-1)$ concentration terms are necessary. Hence, obviously the total number of concentration variables will be given by $P(C-1)$. The other intensive properties that may vary are temperature and pressure. If we now assume that the equilibrium of our system is not influenced by any other forces, then we may write

$$\text{total number of variables} = P(C-1) + 2. \tag{14.3}$$

Our problem now reduces to finding out the number of variables stipulated by the condition that the phases of the system are in equilibrium. We have already established that, for any homogeneous system in equilibrium at a definite temperature and pressure, $\sum_i(\mu_i \, dn) = 0$, where μ_i is the chemical potential and n is the number of moles. This implies that, when two phases are in equilibrium at a particular temperature and pressure, the chemical potential of a given component will be the same in each phase. Since this is true for each homogeneous phase, it must also apply to any system in equilibrium and consisting of any number of phases. For example, if a system consists of three phases, a, b and c say, then the chemical potential of a particular component, say 1, will be the same in phases a and b, as well as in b and c, so that we have

$$(\mu_1)_a = (\mu_1)_b = (\mu_1)_c$$

where the subscripts denote the particular phases in question. Two independent equations will determine the equilibrium between the three phases.

We can generalize this argument for any system in equilibrium and consisting of P phases and C components. Then, by analogy, we have

$$(\mu_1)_a = (\mu_1)_b = (\mu_1)_c = \cdots = (\mu_1)_P$$
$$(\mu_2)_a = (\mu_2)_b = (\mu_2)_c = \cdots = (\mu_2)_P$$
$$(\mu_3)_a = (\mu_3)_b = (\mu_3)_c = \cdots = (\mu_3)_P$$
$$\vdots$$
$$(\mu_C)_a = (\mu_C)_b = (\mu_C)_c = \cdots = (\mu_C)_P.$$

All these equations will constitute $C(P-1)$ independent equations and thereby $C(P-1)$ variables are automatically fixed. So the number of undetermined variables will be given by the difference between the total number of variables and $C(P-1)$ variables. Therefore, the number of undetermined variables will be (from Equation 14.3)

$$[P(C-1)+2] - [C(P-1)] = C - P + 2.$$

Now, in order to define the system in question completely, this number of undetermined variables must be fixed arbitrarily. Therefore, they must be equal to the number of degrees of freedom. Hence, we have

$$F = C - P + 2.$$

This is the phase rule given by Equation 14.2.

14.4 ONE-COMPONENT SYSTEMS

As a system with one component involves no concentration variable but usually involves two other variables, e.g. temperature and pressure, it is possible to draw a complete phase diagram in two dimensions for such a system. Let us consider the phase diagram for water at moderate pressures. Figure 14.3 schematically represents such a phase diagram.

The line BC represents the vapour pressure of liquid water up to the critical point C corresponding to a pressure of 220 atm and a temperature of 374°C, above which only one fluid phase exists. On any point along the line BC, liquid and vapour exist in equilibrium. The vapour phase area lies below BC and extends under AB. Liquid- and solid-phase areas lie above the respective vapour pressure curve.

As seen in Figure 14.3, the liquid-phase area is bounded on the left by the freezing-point curve BD. This curve defines for each pressure the lowest temperature at which liquid is the stable phase. At any temperature lower than this, solid is the stable phase. The solid-phase area is bounded on the right by the melting-point curve BD. This is identical with the freezing-point curve for all pure substances.

The line BA represents the vapour pressure of solid; this line represents the temperatures and pressures at which the solid and vapour are in equilibrium. The line BD represents the melting point of ice as a function of pressure; this line defines the temperatures and pressure at which ice and liquid water are in equilibrium.

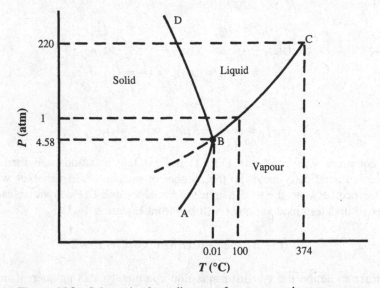

Figure 14.3 Schematic phase diagram of water at moderate pressures

If the temperature and pressure of an isolated sample of water are such that the system falls within any phase area in Figure 14.3, water will exist in the single phase corresponding to that area. Within that area, both temperature and pressure can be changed without changing the number of phases present. So the degrees of freedom within that area will be $F = 2$. This exactly corresponds to the value given by the phase rule. Because in this case $C = 1$ and $P = 1$, therefore from $F = C - P + 2$ we get $F = 2$. Hence within that area the system is bivariant.

If any point defined by a set of temperature and pressure lies on one of the three curves (each of which is the common boundary of two areas), the system may be in either area or in both areas. Experimentally one can be sure that the point lies exactly on the curve only if two phases are present. Then $P = 2$ and $C = 1$. Therefore, applying the phase rule, we have $F = 1$, i.e. the system has only one degree of freedom. Hence it is *univariant*. This implies that only one of the two variables T and P can be changed without changing the phase present. The extent of change of the one variable is limited by the amount of each phase initially present. If two phases are present, any attempt to place arbitrary values on both T and P will result in the disappearance of one of the phases.

The common intersection of the three curves, i.e. point B, lies on the boundaries of all the three areas. This point B is called the *triple point* of water. All the three phases can coexist at equilibrium only at the temperature and pressure defined by this point. In this case, $P = 3$ and $C = 1$. Then, from the phase rule, $F = 0$, i.e. the system has zero degrees of freedom. Such a system is called an *invariant system*; all the variables in the system are determined by the thermodynamic properties.

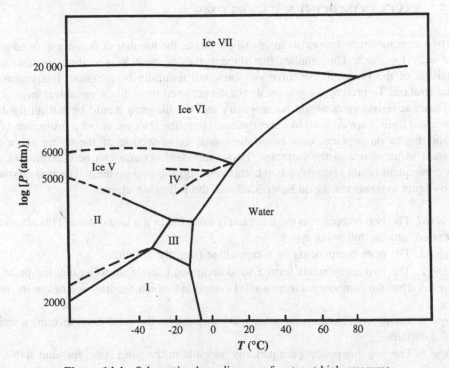

Figure 14.4 Schematic phase diagram of water at high pressures

It is observed that very often a solid phase, which is thermodynamically stable at a given temperature and pressure, forms slowly from a phase that has greater molar free energy. In this case the phase with greater molar free energy retains its existence for some time. This type of phase is called a *metastable phase*. It transforms spontaneously into the stable phase only when influenced by a mechanism for the transformation.

A schematic phase diagram for water at high pressures is shown in Figure 14.4. There are six stable forms and one metastable form of ice. These forms are denoted by I, II, III, IV, V, VI and VII.

Ice IV is metastable but may form briefly in the phase area for ice V. The transition curve for the metastable equilibrium between ice I and ice III has been found to extend into the area of ice II below $-34.7°C$ (shown by the broken curve). If the temperature is reduced rapidly below $-155°C$, ices II, III, IV, V and VI can be brought to atmospheric pressure without phase transition. When these metastable phases are heated at $0.1\,MPa$, they transform into cubic ice Ic. This form is stable below $-120°C$ and metastable between -120 and $-80°C$; above $-80°C$, ice Ic is irreversibly transformed into the hexagonal ice I. It is interesting to note that ice Ic has not yet been found to form from ice I by cooling below $-120°C$; it has been formed by condensation from the vapour phase.

The principles applied to the phase diagram for water at moderate pressures and temperatures can also be applied to the solid-phase areas and boundaries in the phase diagram for water at high pressures, and also to all other one-component phase diagrams.

Examples of other one-component systems are the sulphur system, the iron system, the tin system, etc.

14.5   TWO-COMPONENT SYSTEMS

If a two-component system exists in one single phase, the number of degrees of freedom will be $F = 2 - 1 + 2 = 3$. This implies that three variables must be specified to describe the condition of the phase. These three variables will naturally be pressure, temperature and concentration. To present the graphical relation between these three variables, three coordinate axes at right angles would be necessary and the diagram would be a solid figure. As these solid figures are difficult to draw, the usual practice is to use either a projection of such a solid figure on a plane, or a two-dimensional cross-section of the figure for a given constant value of one of the variables. Two-component systems can be classified as either two-component liquid systems or two-component solid–liquid systems. The two-component solid–liquid systems can again be divided into the following classes.

Class I. The two components are completely miscible in the liquid state. This class can be separated into the following types.

Type 1. The pure components only crystallize from the solution.

Type 2. The two components form a solid compound stable up to its melting point.

Type 3. The two components form a solid compound which decomposes below its melting point.

Type 4. The two components are completely miscible in the solid state, forming a series of solid solutions.

Type 5. The two components are partially miscible in the solid state, forming stable solid solutions.

Type 6. The two components form solid solutions that are stable only up to a transition temperature.

Class II. The two components are partially miscible in the liquid state.

Class III. The two components are immiscible in the liquid state.

14.6   THREE-COMPONENT SYSTEMS

As the name suggests, a three-component system naturally contains three independent components. In such a case, any single phase possesses four degrees of freedom, namely temperature, pressure and the two compositions of two components out of three. Owing to the greater number of variables involved in such cases it is difficult to represent graphically the phase relationships of such systems. To avoid this difficulty, data in a ternary system are usually presented at some convenient fixed pressure and at various constant temperatures. It is then possible to show the relationships between the concentrations of the components at any given temperature on a two-dimensional diagram; by joining such planar diagrams together it is possible to construct a solid model.

Since, in a three-component system, $C = 3$, then from the phase rule $F = 5 - P$. Now, if we fix the temperature and pressure, then $F = 2$ and the number P of phases becomes 3. This is the same as in the case of a two-component system at a constant pressure.

Before we discuss three-component systems let us first try to understand the method of graphical presentation of such systems. There are several methods for plotting planar equilibrium diagrams for ternary systems. Of these, the method suggested by Stokes and Roozeboom is generally used. In this method an equilateral triangle is used to plot the concentrations of the three components at any given temperature and pressure. A general diagram is schematically shown in Figure 14.5.

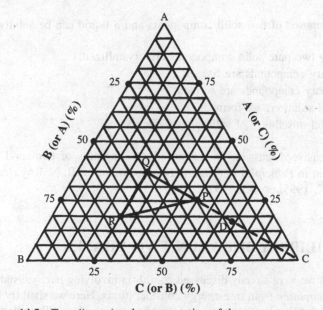

Figure 14.5   Two-dimensional representation of three-component systems

The three corners of the equilateral triangle in Figure 14.5 represent 100% of the components (apex A as 100% A, apex B as 100% B and apex C as 100% C). The lines parallel to BC give the percentages of A other than 100%. Similarly, lines parallel to AC and dividing AB and BC give the percentages of B other than 100%, and the lines parallel to AB and dividing the lines AC and BC give the percentages of C other than 100%. Any point on the side AB gives the concentration relationships of the binary system A–B; any point on the side BC gives the concentration relationships of the binary system B–C, and similarly any point on the side AC will give the concentration relationships of the binary system A–C. Any mixture composed of A, B and C must lie within the triangle. This argument is also applicable to any line or triangle within the diagram. If any mixture is prepared from P and Q, the composition of such a mixture will lie on the line PQ. Similarly, those prepared from Q and R will lie on QR, and those prepared from R and P will lie on RP. Any mixture made up of P, Q and R will lie within the triangle PQR. By similar consideration, any mixture such as D composed of P and C will lie on the line PC and the P : C composition ratio will be given by the length ratio, i.e. CD : PD.

In order to plot any point on the diagram such as P having a composition 25% A, 25% B and 50% C, we first locate the line representing 25% A and then locate the line representing 25% B. The intersection of these two lines will give the point P which will also lie on the line representing 50% C.

Any ternary system can be classified as either systems composed of three liquid components which show partial miscibility, or systems composed of two solid components and a liquid component. Systems composed of three liquid components which exhibit partial miscibility can be subdivided as follows.

Type 1. One pair of partially miscible liquids is formed.
Type 2. Two pairs of partially miscible liquids are formed.
Type 3. Three pairs of partially miscible liquids are formed.

Systems composed of two solid components and a liquid can be subdivided as follows.

Type 1. Only two pure solid components are crystallized.
Type 2. Binary compounds are formed.
Type 3. Ternary compounds are formed.
Type 4. Solid solutions are formed.
Type 5. Partial miscibility of solid phases occurs.

A comprehensive discussion of various phase diagrams of two- and three-component systems is given in *Principles of Modern Thermodynamics*; B. N. Roy; Institute of Physics Publishing, UK, 1995; pp. 355–397.

14.7 EQUILIBRIA BETWEEN TWO PHASES

In Section 10.8 we have already discussed equilibria involving pure substances and deduced the *Clapeyron equation* from free-energy considerations. Here we shall try to understand the significance of the Clapeyron equation to the phenomena of equilibria between two phases.

We discuss this under three different headings, namely liquid–vapour equilibrium, solid–vapour equilibrium and solid–liquid equilibrium.

14.7.1   Liquid–vapour equilibrium

The vapour pressure of liquids is a property which is most directly related to phase equilibria. To understand what the vapour pressure of a liquid is, let us imagine that a liquid is confined in a closed container and undergoing a process of vaporization at a constant temperature. All the molecules in the liquid state have a distribution of energies (given by the Boltzmann distribution curve). If any molecule has sufficient energy to overcome the intermolecular forces, it will escape into the vapour phase above the liquid. As the vaporization continues inside the closed container at a constant temperature, the concentration of the substance in the vapour phase will increase. At the same time the rate at which molecules will return to the liquid phase through condensation will also increase. After some time a stage will come when the rate of evaporation will be constant. The system attains equilibrium. At this stage the pressure exerted by this equilibrium vapour concentration is called the *vapour pressure*.

The vapour pressure of a liquid is a characteristic of its own at a particular temperature; it increases with increasing temperature as the average kinetic energy increases. As the temperature increases, a greater number of molecules acquire sufficient energy to escape from the liquid. Therefore, a higher pressure will be necessary to maintain equilibrium between vapour and liquid. Above the critical temperature the escaping tendency of the molecules is so great that no applied pressure is adequate to keep the molecules in the liquid state. Therefore, obviously the dependence of vapour pressure on temperature is not linear. The general nature of the variation in vapour with temperature is shown in Figure 14.6.

At lower temperatures it is observed that the vapour pressure increases slowly and then sharply. This variation in vapour pressure with temperature can be expressed mathematically. Let P be the vapour pressure for the transition of liquid to vapour at temperature T, $(\Delta H)_{vap}$ be the heat of vaporization of a given weight of liquid, V_ℓ be the volume of the liquid and V_g be the volume of the vapour. Using the Clapeyron equation for vaporization (see Equation 10.49)

$$\frac{\mathrm{d}P}{\mathrm{d}T} = \frac{(\Delta H)_{vap}}{T(V_g - V_\ell)}. \tag{14.4}$$

At temperatures not too near the critical temperature, the volume of a liquid is very small compared with that of its vapour. So V_ℓ may be neglected. If we assume that the vapour behaves ideally, then $V_g = nRT/P$; therefore, for 1 mol of vapour $V_g = RT/P$. Substituting this into Equation 14.4, we have

$$\frac{\mathrm{d}P}{\mathrm{d}T} = \frac{(\Delta H)_{vap}}{TV_g} = \frac{P(\Delta H)_{vap}}{RT^2}$$

or

$$\frac{1}{P}\frac{\mathrm{d}P}{\mathrm{d}T} = \frac{(\Delta H)_{vap}}{RT^2}$$

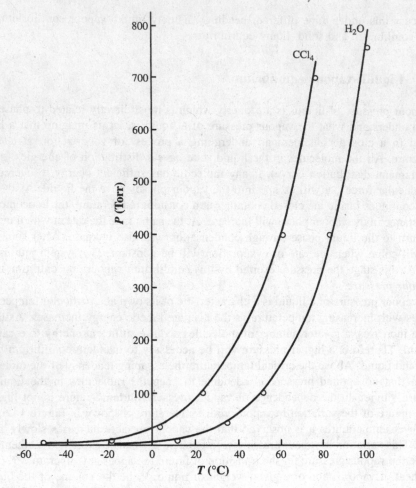

Figure 14.6 Schematic diagram of the variation in the vapour pressures of H_2O and CCl_4 with temperature

or

$$\frac{d}{dT}[\ln P] = \frac{(\Delta H)_{vap}}{RT^2} \qquad (14.5)$$

This equation is known as the *Clausius–Clapeyron equation*. In order to integrate this equation it is necessary to know the heat $(\Delta H)_{vap}$ of vaporization as a function of temperature. If it is assumed that $(\Delta H)_{vap}$ remains constant over the interval of temperature concerned, then on integration of Equation 14.5 we obtain

$$\ln P = \left(\frac{(\Delta H)_{vap}}{R}\right)\int \frac{dT}{T^2} + K = -\frac{(\Delta H)_{vap}}{RT} + K \qquad (14.6)$$

where K is an integration constant.

Equation 14.6 shows the variation in the vapour pressure with temperature. A plot of log P against $1/T$ would give a straight line with slope $-(\Delta H)_{vap}/2.303\,R$. Knowing the value of the slope, $(\Delta H)_{vap}$ can be calculated in joules per mole. This value of $(\Delta H)_{vap}$ will not be the exact value; it will be the mean value over the temperature range concerned. The integration constant can be calculated either from the intercept of a graph of log P versus $1/T$, or by substituting in the equation

$$\log P = -\frac{(\Delta H)_{vap}}{2.303\,RT} + K' \tag{14.7}$$

the calculated value of $(\Delta H)_{vap}$ and any value of log P and $1/T$ corresponding to a point on the graph. Knowing $(\Delta H)_{vap}$ and K' for any liquid, the vapour pressure of that liquid at any temperature can be calculated.

It should be noted that the magnitude of the integration constant K' will depend on the units of P. So the units of P must be stated clearly in setting up an equation for the vapour pressure.

Alternative forms of Equations 14.6 and 14.7 may be obtained by integrating Equation 14.5 over the limits P_1 and P_2 corresponding to T_1 and T_2. Then we have,

$$\int_{P_1}^{P_2} d(\ln P) = \left(\frac{(\Delta H)_{vap}}{R}\right) \int \frac{dT}{T}$$

or

$$\ln\left(\frac{P_2}{P_1}\right) = \frac{(\Delta H)_{vap}}{R}\left(\frac{1}{T_1} - \frac{1}{T_2}\right) \tag{14.8}$$

or

$$\log\left(\frac{P_2}{P_1}\right) = \frac{(\Delta H)_{vap}}{2.303\,R}\left(\frac{1}{T_1} - \frac{1}{T_2}\right). \tag{14.9}$$

Example 14.1

Given that the heat of vaporization of ethanol is 38.576 kJ mol^{-1} and that its vapour pressure at 98°C is 0.5 atm. Calculate the normal boiling point of ethanol.

Solution

The normal boiling point T_2 of ethanol would be the temperature at which the vapour pressure of ethanol equals $P_2 = 1$ atm. Then using Equation 14.9 we have

$$\log\left(\frac{1}{0.5}\right) = \frac{-38.576}{(2.303) \times (8.314)}\left(\frac{1}{T_2} - \frac{1}{371.15}\right)$$

or

$$T_2 = 78.6°C$$

14.7.2   Solid–vapour equilibrium

A solid, like a liquid, has a definite vapour pressure at each temperature, although this pressure may be extremely small. Some solids cannot exist in contact with vapour. Like the vapour pressure of a liquid, the vapour pressure of a solid increases with increasing temperature, and the variation may be represented by a curve similar to that for a liquid. This curve is generally called a *sublimation curve*. The transformation of a solid to vapour is accompanied by an absorption of heat. This is latent heat of sublimation or heat of vaporization; its units are joules per mole.

As in the case of a liquid, the variation in the vapour pressure of a solid can be expressed mathematically. If P is the vapour pressure for the transition of solid to vapour at temperature T, $(\Delta H)_{sub}$ is the heat of sublimation for the process of a given weight of solid, V_s is the volume of the solid and V_g is the volume of the vapour, then from the Clapeyron equation (Equation 10.49) we get

$$\frac{dP}{dT} = \frac{(\Delta H)_{sub}}{T(V_g - V_s)}. \tag{14.10}$$

By neglecting the volume of the solid with respect to that of the vapour and assuming that the vapour behaves ideally, an equation analogous to Equation 14.5 can be obtained:

$$\frac{d}{dT}(\ln P) = \frac{(\Delta H)_{sub}}{RT^2}. \tag{14.11}$$

This equation can be integrated to obtain equations analogous to those already obtained for liquids, namely Equations 14.6–14.9.

14.7.3   Solid–liquid equilibrium

When a pure crystalline solid is heated, a temperature is reached at which it changes sharply into liquid. This process of heating is known as *fusion*. The temperature at which transformation of solid to liquid occurs is known as the *melting point* of the solid, and it has a definite value, depending on the external pressure. If the molten solid is cooled, solidification will occur at the same temperature if the pressure is the same. This means that for a pure substance the freezing and melting points are identical, i.e. for every pressure there is a definite temperature at which a pure solid and liquid can be in equilibrium. It is thus possible to plot a melting-point curve representing the pressure and temperature conditions for equilibrium between the solid and liquid. Such a plot has virtually negligible curvature; for practical purposes it can be considered a straight line.

Like the process of evaporation and sublimation, the process of fusion is accompanied by an absorption of heat $(\Delta H)_f$. As in the case of liquid–vapour and solid–vapour equilibria, the variation in melting temperature with pressure can be mathematically expressed using the Clapeyron equation. This gives

$$\frac{dT}{dP} = \frac{T(V_\ell - V_s)}{(\Delta H)_f}. \tag{14.12}$$

By neglecting the volume V_s of the solid with respect to the volume V_ℓ of the liquid, Equation 14.12 can be reduced to

$$\frac{dT}{dP} = \frac{TV_\ell}{(\Delta H)_f}. \tag{14.13}$$

This equation is, in fact, the Clausius–Clapeyron equation written in the inverted form just to represent the influence of external pressure P on the melting of the solid. The pressure P in this equation is not vapour pressure but just the external pressure. In using Equation 14.13 for calculations, the value of $(\Delta H)_f$ must be converted from joules per mole to litre atmospheres per mole to obtain a value for $\Delta T/\Delta P$ in atmospheres per kelvin.

Example 14.2

Calculate the rate of change in melting point per atmosphere increase of pressure for the ice–water system at 0°C, given that at 0°C the volume of water is $1.0\,cm^3$, the volume of ice is $1.0907\,cm^3$ and the latent heat of fusion of ice is $334.887\,J\,g^{-1}$.

Solution

Substituting the given values into Equation 14.12 we have

$$\frac{dT}{dP} = \frac{(273)(1.00 - 1.0907)(1.013 \times 10^{-3})}{(334.887)(9.863 \times 10^{-3})} = -0.0075\,K\,atm^{-1}.$$

This shows that for every 1 atm pressure the melting point is decreased by 0.0075 K.

14.8   EQUILIBRIA BETWEEN THREE PHASES

We have seen that for a solid or liquid substance, as the temperature increases, the vapour pressure also increases. For every substance there is a characteristic temperature at which the vapour pressures of both the solid and liquid phases are the same. This is known as the *triple point*. The value of the temperature and the corresponding vapour pressure at the triple point for a particular system can be obtained by considering the temperature dependence of the vapour pressure of both phases. Therefore, a Clausius–Clapeyron plot of log P versus $1/T$ gives two straight lines that will intersect when the vapour pressure of the solid is equal to the vapour pressure of the liquid. This provides a method to determine the triple point. Alternatively, the triple point can be determined by solving simultaneously two Clausius–Clapeyron equations for the solid and liquid. If the heats of vaporization and sublimation for the substances and also the vapour pressure at one temperature for each phase are known, then the two unknown quantities that will satisfy both the equations are the temperature and the pressure at the triple point.

The triple point of any substance is a useful reference state which permits correlation of the separate graphs for the various two-phase equilibria into single pressure–temperature diagrams, such as shown in Figure 14.3 for the triple point of water. The considerations discussed in Section 14.4 are applicable to the triple point of any substance.

14.9   PHASE TRANSITIONS

The preliminary principles and concepts of phase transitions have been introduced in Sections 7.5.3, 7.5.6, 7.16 and 10.9, which included a discussion of the Clausius–Clapeyron and the Ehrenfest equations. We have seen somewhat detailed discussions of the phase rule and phase equilibria between solid–liquid, liquid–liquid, liquid–vapour and solid–vapour. It has been illustrated how thermodynamics can be used to explain the equilibria that exist between various phases of material under different conditions. A thermodynamic system can exist in various phases, each of which can exhibit different behaviours. It is well known that at high temperatures systems are highly disordered with a high degree of entropy. Generally, systems tend to become more ordered as temperature is decreased, and at the absolute zero of the temperature a system is most ordered. At high temperatures, thermal motions of molecules dominate over the forces of cohesion between them, but as temperature decreases the latter overcome the former and the atoms rearrange themselves in a more ordered state. This rearrangement suggests a phase change.

Phase changes do not occur smoothly; these are rather abrupt at some critical temperature. A phase change is not a large-scale manifestation; any evidence of such changes can only be found on a macroscopic scale as the critical temperature is approached. The study of the transition region between phases is one of the important and fascinating areas of statistical mechanics. At a transition point, two or more phases exist in equilibrium with each other, and the thermodynamic conditions for equilibrium between them can be found from the equilibrium conditions discussed earlier. From the equilibrium conditions, it is possible to find the maximum number of phases which can coexist, and we can find equations (such as the Clausius–Clapeyron equation) for the regions where phases coexist. Exchange of matter can take place between phases, so equilibrium occurs between them when their chemical potentials are equal for given values of pressure and temperature. When a phase transition occurs, the chemical potentials of the participating phases must change continuously. This requirement also necessitates the continuous change in the Gibbs free energy of the phases.

Phase changes can be accompanied by a discontinuous change of state or a continuous change of state. The former is called *first-order phase transition*, while the latter is called *continuous phase transition*. In the case of first-order phase transitions, only the first derivatives of the Gibbs free energy are discontinuous. However, for the continuous phase transitions, higher-order derivatives of the Gibbs free energy will be discontinuous. The liquid–solid, vapour–liquid and vapour–solid transitions discussed earlier all represent first-order. The Clausius–Clapeyron equation can be used to find appropriate equations for the coexistence curves in the case of vapour–solid and vapour–liquid transitions. In a binary mixture of molecules, a physical separation of the mixture can occur below a certain critical temperature; each separated part is rich in one of the two kinds of molecules. According to the *Hildebrand theory* (see Section 14.9.4) the Gibbs free energy of the mixture can be written by a simple expression which has the necessary features of the phase transition.

Although a critical temperature does exist in most phase transitions, it is interesting to note that the liquid–solid transition is one of the few exceptions. In this case, above a well defined temperature the liquid phase exists, and as the temperature is decreased the solid phase appears. The symmetry properties of liquids and solids are quite different, so when a solid phase appears as the temperature is lowered, it has symmetry properties different to those of the liquid and some new variable. This variable is called the *order parameter* and it actually characterizes the new phase. The critical point is the point at which the order parameter of a new phase starts to grow continuously from zero. In the case of first-order transitions, there may not be interrelationships between the symmetries of the high- and low-temperature phases. However, in continuous transitions well-defined relationships between symmetry properties of the two phases are generally observed. This happens due to the simple reason that the state changes continuously rather than abruptly. Ginzburg and Landau proposed a general theory which can account for continuous symmetry-breaking phase transitions. This theory involves a series expansion of the free energy in terms of the order parameter. We shall discuss this theory later and use it to explain the transitions in magnetic and superconducting substances, as well as liquid helium.

Superconductors are especially interesting from the thermodynamics point of view because these materials provide a different application of the Clausius–Clapeyron equation and a very distinct application of the Ginzburg–Landau expansion. Many features of the Ginzburg–Landau expansion are applicable to superfluid He^4 and He^3. Liquid He^3 and He^4 are quantum fluids and a good example of the third law of thermodynamics.

In many ways, the critical point is important in the study of phase transitions. On a microscopic level, a system undergoes an adjustment as it approaches its critical point from high temperatures. During this adjustment large fluctuations occur which indicate the appearance of a new order parameter. Some thermodynamic quantities can become infinite at the critical point. It is interesting to note that, irrespective of the particular substance involved, a great similarity appears to exist in the behaviour of all systems as they tend to their critical points.

Most systems can exist in a number of different phases, and each of these phases can show quite different macroscopic behaviours. It is possible that, for certain values of the independent variables, two or more phases of a system can coexist. The Gibbs phase rule (see Equation 14.2) tells us the number of phases that can coexist under a given condition. In general, coexisting phases are in mechanical as well as thermal equilibrium; they can also exchange matter. As stated earlier, when two or more phases of a system coexist, the chemical potentials and the temperatures of the participating phases must be equal. There will also be another condition between mechanical variables to show mechanical equilibrium. For example, in the case of a simple PVT system the pressure of the two coexisting phases may be equal unless they are separated by a rigid (i.e. not free to move) surface.

14.9.1   First- and higher-order phase transition

In the previous section we qualitatively defined continuous and discontinuous phase changes. Let us consider these in a quantitative way. We know that if the independent intensive variables (such as pressure, temperature and mole fraction) of a system are changed, the values of the variables can reach a point where a phase change can occur. At these points, the chemical potentials of the participating phases must be equal (because they are functions only of intensive variables) and they can coexist. The Gibbs free energy is related to the chemical potential by the equation

$$G = \sum_i \mu_i n_i \tag{14.14}$$

(see Equation 11.41) where n_i is the number of moles. For a constant pressure and temperature process, we can write

$$(dG)_{P,T} = \sum_i \mu_i dn_i$$

or

$$\mu_i = \left(\frac{\partial G}{\partial n_i}\right)_{P,T,n_1,n_2,\dots} \tag{14.15}$$

where $i \neq j$, i.e. $n_i \neq n_j$.

Equation 14.15 implies that, at a phase transition, the Gibbs free energy of each phase must have the same value and $(\partial G/\partial n_i)_{P,T,n_1,n_2,\dots}$ must be equal. Now the question is: what happens to the derivatives $V = (\partial G/\partial P)_{T,n_i}$ and $S = -(\partial G/\partial T)_{P,n_i}$? We shall use the behaviour of these derivatives to classify phase transitions.

(i) *First-order transition*: If the derivatives $(\partial G/\partial P)_{T,n_i}$ and $(\partial G/\partial T)_{P,n_i}$ are discontinuous at the transition point, i.e. if the extensive variable V and the entropy S are different in the two phases, the transition is said to be first-order. The Gibbs free energy and its *first-order* derivatives in a pure PVT system are plotted in Figure 14.7. For such a system, the Gibbs free energy must be a concave function of the temperature and pressure.

If $(\partial G/\partial P)_{T,n_i}$ is discontinuous, this implies that there is a discontinuity in the volume of the phases. If I and II represent the two phases in the system, then

$$\Delta V = V^{\mathrm{I}} - V^{\mathrm{II}} = \left(\frac{\partial G}{\partial P}\right)^{\mathrm{I}}_{T,n_i} - \left(\frac{\partial G}{\partial P}\right)^{\mathrm{II}}_{T,n_i}. \tag{14.16}$$

If $(\partial G/\partial T)_{P,n_i}$ is discontinuous, there is a discontinuity in the entropy of the two phases. We can write

$$\Delta S = S^{\mathrm{I}} - S^{\mathrm{II}} = -\left[\left(\frac{\partial G}{\partial T}\right)^{\mathrm{I}}_{P,n_i} - \left(\frac{\partial G}{\partial T}\right)^{\mathrm{II}}_{P,n_i}\right]. \tag{14.17}$$

The enthalpies of the two phases are given by

$$H^{\mathrm{I}} = G^{\mathrm{I}} + (TS)^{\mathrm{I}}$$
$$H^{\mathrm{II}} = G^{\mathrm{II}} + (TS)^{\mathrm{II}}.$$

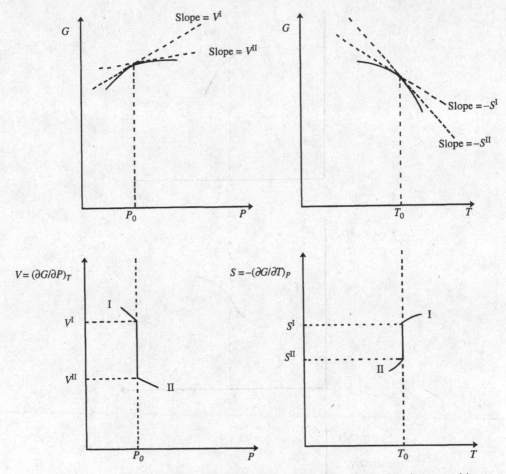

Figure 14.7   Schematic diagram of the Gibbs free energy at a first-order phase transition

Since the Gibbs free energy and the temperature are the same for both phases at the transition, for a first-order transition the difference in enthalpy of the two phases is

$$\Delta H = H^{\text{I}} - H^{\text{II}} = T(S^{\text{I}} - S^{\text{II}}) = T\Delta S. \tag{14.18}$$

In this case, ΔH is called *latent heat of transition*.

(ii) *Higher-order transitions*: If the higher order derivatives of G with respect to P and T are discontinuous, then the phase change is continuous. For such a phase transition the Gibbs free energy is continuous but its slope changes rapidly, which produces a peak in the heat capacity, $C_P/T = (\partial S/\partial T)_P = (\partial^2 G/\partial T^2)_P$, at the transition point. This is illustrated in Figure 14.8. As can be seen in this figure, for a continuous transition there is no abrupt

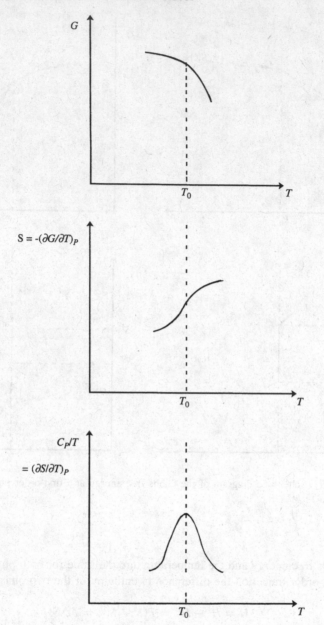

Figure 14.8 Schematic diagram of the Gibbs free energy of a continuous phase transition

change in the entropy at the transition. This is also true for the extensive variable as a function of P and T.

An *nth-order phase transition* is said to be that one for which the nth derivative of G is the first discontinuous derivative. This term was used by Ehrenfest. However, his theory fails for some systems for which higher-order derivatives are infinite.

14.9.2 Clausius–Clapeyron equations for vapour–liquid, liquid–solid and vapour–solid coexistence curves for a pure PVT system

When two phases of a pure PVT system are in equilibrium then, if we want to change the pressure of the system by an infinitesimal quantity dP, we must also change the temperature of the system by an infinitesimal quantity dT to maintain the equilibrium. The Clausius–Clapeyron equation relates the change in temperature necessarily accompanying a change in pressure occurring in a system containing two phases of a pure substance in equilibrium. We have derived this in Equations 10.48 and 10.49. The Clapeyron equation is used for calculations involving equilibria in solids and liquids (condensed-phase equilibria). If one solid form of a pure substance transforms into another solid form, the variation in the transition temperature T with pressure P is given by the Clausius–Clapeyron equation. ΔH (the latent heat) and ΔV in this case are the enthalpy and volume changes accompanying the transformation. For such a transformation, ΔH is always positive, i.e. heat must be absorbed if the transformation is from a form stable at a lower temperature to a form stable at a higher temperature. On the other hand, ΔH is always negative, i.e. heat must be given up for the reverse process.

To prove what we said above, let us consider the transformation between two phases I and II of a pure substance, and assume that phase I is the high-temperature phase and phase II is the low-temperature phase. Above the transition temperature, phase I is stable has free energy G^I, whereas below the transition temperature phase II is stable and has free energy G^{II}. Since at the equilibrium state free energy is at a minimum, $G^I < G^{II}$ above the transition temperature and $G^I > G^{II}$ below the transition temperature. So we can say $(\partial G^I/\partial T)_{P,n_i} < (\partial G^{II}/\partial T)_{P,n_i}$ for both above and below the transition temperature. Hence, $S^I = -(\partial G^I/\partial T)_{P,n_i} > S^{II} = -(\partial G^{II}/\partial T)_{P,n_i}$, and the difference in the entropy $\Delta S = \Delta H/T$ is always positive if the transformation is from the low-temperature phase to the high-temperature phase.

As a system with one component involves no concentration variable but usually involves two other variables, e.g. temperature and pressure, it is possible to draw a complete phase diagram in two dimensions for such a system. Figure 14.3 schematically represents such a phase diagram. Although it is drawn for water, it is typical of most other pure substances. However, this figure does not represent the phase diagrams of the isotopes of helium (He^3 and He^4) which have superfluid phases. Later, we shall discuss the phase diagrams of liquid helium.

The point B on the diagram is the triple point, i.e. the point at which the vapour, liquid and solid phases can coexist. The line BC represents the vapour pressure of the liquid phase up to the *critical point C* (at this point the vapour pressure curve terminates). The fact that the vapour pressure curve has this point implies that it is possible to go continuously from a vapour to a liquid without passing through a phase transition, provided that we choose the correct path. The melting curve BD does not have a critical point; this has never been found. We cannot go from the liquid to the solid state without going through a phase transition. Solids show spatial ordering, but vapours and liquids do not. This fundamental difference is a factor for the difference between the vapour–liquid and liquid–solid transitions. Let us examine the nature of the Clasius–Clapeyron equation for the three coexistence curves in Figure 14.3.

(a) *Melting curve*: This curve can have a positive slope (shown by the solid line) or a negative slope (shown by the broken line). A positive slope means that if the pressure is increased at a fixed temperature, the system is simply pushed deeper into the solid phase. On the other hand, if the slope is negative, then increasing the pressure at a fixed temperature will push the system into the liquid phase. The melting curve of water has a negative slope. The Clausius–Clapeyron equation for the liquid–solid transition is given by

$$\frac{dP}{dT} = \frac{(\Delta \bar{H})_{ls}}{T(\Delta \bar{V})_{ls}} \tag{14.19}$$

where $(\Delta \bar{H})_{ls}$ is the change in latent heat of melting per mole in going from the solid to the liquid phase, and $(\Delta \bar{V})_{ls}$ is the corresponding change in molar volume.

Note that if the solid has a bigger volume than that of the liquid, then the change in molar volume, $(\Delta \bar{V})_{ls}$, will be negative, so the slope $(dP/dT)_V$ will be negative.

(b) *Sublimation curve*: The line BA represents the vapour pressure of solid; this line represents the temperatures and pressures at which the solid and vapour are in equilibrium. The difference between the enthalpy of the solid at zero temperature and pressure and that of a point on the sublimation curve is given by

$$\bar{H}_s - (\bar{H}_s)_0 = \int_0^T (\bar{C}_P)_s \, dT \tag{14.20}$$

where $\bar{H}_s$ is the molar enthalpy of the solid at a point on the sublimation curve with temperature T, $(\bar{H}_s)_0$ is the molar enthalpy of the solid at zero temperature and pressure, and $(\bar{C}_P)_s$ is the molar specific heat of the solid. Similarly, the molar enthalpy of the vapour can be written as

$$\bar{H}_v - (\bar{H}_v)_0 = \int_0^T (\bar{C}_P)_v \, dT \tag{14.21}$$

where $\bar{H}_v$ is the molar enthalpy of the vapour at a point on the sublimation curve with temperature T, $(\bar{H}_v)_0$ is the molar enthalpy of the vapour at zero temperature and pressure, and $(\bar{C}_P)_v$ is the molar specific heat of the vapour.

Assuming that the vapour behaves ideally so that it can be described by the ideal gas law, and by using Equations 14.20 and 14.21, the Clausius–Clapeyron equation for the sublimation curve can be written as

$$\frac{dP}{dT} = \left(\frac{P}{RT^2} \right) (\Delta \bar{H})_{vs} \tag{14.22}$$

where $(\Delta \bar{H})_{vs} \simeq \bar{H}_v - \bar{H}_s$, and it has been assumed that volume changes of the solid are very much smaller than those of the vapour and can be ignored.

(c) *Vapour pressure curve*: The line BC represents the vapour pressure of the liquid up to the *critical point* C. Along this line the vapour and liquid phases coexist. For a given temperature, the pressure of the liquid and vapour is fixed: this pressure is called the *saturated vapour pressure*. If we change the temperature of the system, its vapour pressure will also change.

To obtain the Clausius–Clapeyron equation for the vapour–liquid transition, we make three assumptions:

1. The vapour behaves ideally and obeys the ideal gas law.

2. The volume changes of the liquid are very much smaller than those of the vapour so that the former can be neglected as we move along the curve BC.

3. The latent heat of vaporization is approximately constant over the temperature range considered.

Thus,

$$\Delta V \simeq V_v = \frac{nRT}{P} \tag{14.23}$$

where V_v is the volume of the vapour. Then, the Clausius–Clapeyron equation for the vapour pressure curve is given by

$$\frac{dP}{dT} = \left(\frac{P}{RT^2}\right)(\Delta \bar{H})_{vl} \tag{14.24}$$

where $(\Delta \bar{H})_{vl}$ is the change in the latent heat of vaporization per mole in going from the liquid to the vapour. Integrating this equation, we have

$$P = P_0 \exp\left[\frac{(-\Delta \bar{H})_{vl}}{RT}\right] \tag{14.25}$$

where P_0 is a constant (see Figure 14.9a).

Equation 14.25 shows that the vapour pressure increases exponentially as the temperature is increased. On the other hand, the boiling point, i.e. the temperature of coexistence, increases as the vapour pressure is increased.

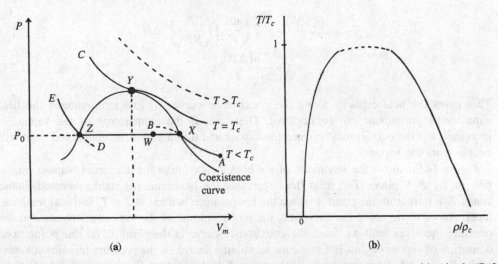

Figure 14.9 (a) Schematic diagram of the coexistence curve for liquid–vapour transition in the P–V plane (b) Schematic diagram of liquid–vapour coexistence curve in the T–V plane

Along the vapour pressure curve, temperature is the only independent variable and the pressure is related to the temperature by the Clausius–Clapeyron equation. The entropy of a vapour is a function of both T and P, i.e. $S = f(T, P)$. We define the heat capacity of the vapour as

$$C_{coexistence} = \left(\frac{dQ}{dT}\right)_{coexistence} = T\left(\frac{dS}{dT}\right)_{coexistence}. \tag{14.26}$$

If we have four state variables w, x, y and z such that $f(x, y, z) = 0$ and w is a function of any two of the variables x, y and z, then we can write

$$\left(\frac{\partial x}{\partial y}\right)_z = \left(\frac{\partial x}{\partial y}\right)_w + \left(\frac{\partial x}{\partial w}\right)_y \left(\frac{\partial w}{\partial y}\right)_z. \tag{14.27}$$

Following this rule and assuming that w, x, y and z correspond to P, S, T and coexistence, respectively, Equation 14.26 becomes

$$C_{coexistence} = T\left(\frac{\partial S}{\partial T}\right)_P + T\left(\frac{\partial S}{\partial P}\right)_T \left(\frac{dP}{dT}\right)_{coexistence}.$$

However, $T(\partial S/\partial T)_P = C_P$ and $(\partial S/\partial P)_T = -(\partial V/\partial T)_P$ (see Equation 10.67). Then we can write

$$C_{coexistence} = C_P - T\left(\frac{\partial V}{\partial T}\right)_P \left(\frac{dP}{dT}\right)_{coexistence}. \tag{14.28}$$

According to our first assumption above, we can use the ideal gas law $PV = nRT$ to the vapour. This gives $(\partial V/\partial T)_P = nR/P$. Substituting this and the value of dP/dT from Equation 14.24 into Equation 14.28, we have

$$C_{coexistence} = C_P - T\left(\frac{nR}{P}\right)\left(\frac{P}{RT^2}\right)(\Delta \bar{H})_{vl}$$

$$= C_P - \frac{n(\Delta \bar{H})_{vl}}{T}. \tag{14.29}$$

This gives the heat capacity along the coexistence curve. At low temperatures the heat capacity of the vapour can be negative. Therefore, if the temperature of the vapour is increased and the equilibrium between the vapour and liquid phases is maintained, heat will evolve from the vapour.

Figure 14.9a shows the schematic of the coexistence curve for the liquid–vapour transition in the P–V plane. The solid lines represent the isotherms for stable thermodynamic states. We start from the point A where the temperature is fixed at $T < T_c$ (critical temperature). As we move along the curve AX, the molar volume of the vapour decreases and the pressure increases until we reach the coexistence curve at the point X. At this point, condensation of vapour begins, but as the molar volume decreases the pressure remains constant until all vapour has become liquid at the point Z. If the molar volume is decreased further, the pressure begins to increase along ZE. For mechanical stability of the system we require that the response function (i.e. the isothermal compressibility) $(\partial V/\partial P)_T < 0$. If we continue the isotherm of $T < T_c$ beyond the points X and Z (indicated by the broken lines XB and ZD), we get mechanically stable curves but these no longer correspond to a minimum of free energy.

Note the two broken lines at points X and Z: XB and ZD. States along the line XB correspond to super-cooled vapour states and the states along the broken line ZD correspond to super-heated liquid states. Such states are metastable. It is possible that the super-heated liquid curve ZD can extend into the region of negative pressure. The region of metastable states decreases as the critical temperature is approached and vanishes at T_c. The other interesting thing that happens as the critical temperature is approached is that the molar volumes of the liquid and vapour approach each other and become equal at T_c.

Let us have a look at the liquid–vapour coexistence curve in the $T–V$ plane. The actual shape was given by Guggenheim (J. Chem. Phys., **13** (1945) 253) for a number of pure substances. He plotted the coexistence curves in terms of the reduced temperature (T/T_c) and the reduced density (ρ/ρ_c), ρ_c being the critical density. T_c and ρ_c are different for different substances. Guggenheim found that the plots of T/T_c versus ρ/ρ_c for most substances have approximately the same shape (see Figure 14.9b).

This is an example of the *law of corresponding states* which was developed by de Boer and Michels (*Physica*, **5** (1938) 945) and states that all pure classical fluids obey the same equation of state if described in terms of reduced quantities. According to Guggenheim, ρ/ρ_c of liquid and vapour phases along the coexistence curves follow the equations:

This is an example of the *law of corresponding states* which was developed by de Boer and Michels (*Physica*, **5** (1938) 945) and states that all pure classical fluids obey the same equation of state if described in terms of reduced quantities. According to Guggenheim, ρ/ρ_c of liquid and vapour phases along the coexistence curves follow the equations:

$$\frac{\rho_\ell + \rho_v}{2\rho_c} = 1 + \frac{3}{4}(1 - T/T_c) \tag{14.30}$$

$$\frac{\rho_\ell - \rho_v}{\rho_c} = \frac{7}{2}(1 - T/T_c)^{1/3} \tag{14.31}$$

14.9.3   Van der Waals equation and Maxwell construction for the liquid–vapour phase transition

In 1873, van der Waals derived a simple equation of state to describe many of the essential features of the liquid–vapour phase transition. It is written in the form

$$V_m^3 - \left(b + \frac{RT}{P}\right)V_m^2 + \left(\frac{a}{P}\right)V_m - \frac{ab}{P} = 0 \tag{14.32}$$

where V_m is the molar volume, P is the pressure, R is the universal gas constant, T is the temperature, and a and b are the van der Waals constants. This equation has three distinct roots, i.e. values of V_m, for every value of T and P, provided that both T and P are small. As

$T \to \infty$, this equation reduces to the equation of state for ideal gases, $PV_m = RT$. There is a critical temperature T_c at which all the roots of Equation 14.32 coalesce, and above this temperature two of the roots are imaginary. At T_c the critical isotherm $T = T_c$ has a slope $(\partial P / \partial V_m)_{T=T_c} = 0$, as well as an inflexion point $(\partial^2 P / \partial V_m^2)_{T=T_c} = 0$. An inflexion point is defined as a point on a curve where the shape of the curve changes from convex to concave and the second derivative changes sign.

Using the zero values of $(\partial P / \partial V_m)_{T=T_c}$ and $(\partial^2 P / \partial V_m^2)_{T=T_c}$ at the critical point, we have the critical temperature, pressure and molar volume as

$$\left. \begin{array}{rcl} T_c & = & \dfrac{8a}{27bR} \\[2mm] P_c & = & \dfrac{a}{27b^2} \\[2mm] (V_m)_c & = & \dfrac{a}{3b} \end{array} \right\}. \tag{14.33}$$

If we substitute the reduced state variables $\bar{T} = T/T_c$, $\bar{P} = P/P_c$ and $\bar{V} = V_m/(V_m)_c$ in Equation 14.32, we get

$$(\bar{P} + 3/\bar{V}^2)(3\bar{V} - 1) = 8\bar{T}. \tag{14.34}$$

Comparing Equations 14.32 and 14.34, we see that the van der Waals constants do not appear in the latter. The temperature, pressure and volume in Equation 14.34 are now being measured in terms of their respective distance from the critical point. Although different gases will have different values of T_c, P_c and $(V_m)_c$, all gases will follow the same equation of state if the values of $\bar{T}$, $\bar{P}$ and $\bar{V}$ are the same. This is the *law of corresponding states* (mentioned in Section 14.9.2). Figure 14.10a shows the schematic of the variation of the molar volume with pressure along a typical van der Waals isotherm and Figure 14.10b shows the schematic of the molar Gibbs free energy as a function of pressure for the same isotherm. The area under the curve in Figure 14.10a between any two points is equal to the difference in molar Gibbs free energy between those two points.

Van der Waals equation predicts a positive slope $(\partial P / \partial V_m)_T$ for certain segments of the isotherms below the critical temperature. This aspect can be seen for the segment DEF in Figure 14.10a. The line from D to F corresponds to mechanically unstable thermodynamic states. Later, we shall discuss how to remove the unphysical sections of the $P-V_m$ curve by a technique known as *Maxwell construction*.

Differential changes in the entropy are related to differential changes in the extensive state variables through the combined first and second laws of thermodynamics:

$$T \, dS \geqslant dU + P \, dV - \sum_i \mu_i \, dn_i \quad \text{[see Equation 11.52]}$$

or

$$dU \leqslant T \, dS - P \, dV + \sum_i \mu_i \, dn_i. \tag{14.35}$$

The equality holds if changes in the thermodynamic state are reversible and the inequality holds for changes which are irreversible or spontaneous. For processes carried out at

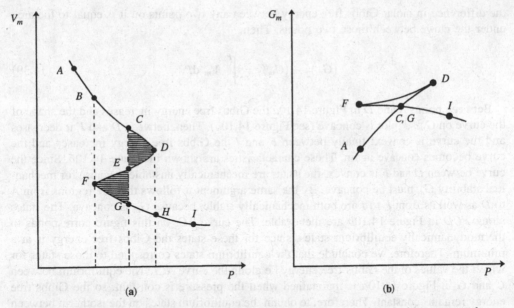

Figure 14.10 (a) Schematic diagram of the variation of the molar volume with pressure along a typical van der Waals isotherm (b) Schematic diagram of the variation of the molar Gibbs free energy as a function of pressure for the isotherm in (a)

constant P, T and n_i, the Gibbs free energy corresponds to the thermodynamic potential. Such a process is linked both mechanically and thermally to the outside world and the Gibbs free energy is obtained from the internal energy by adding terms due to the mechanical and thermal linking. Thus, we change from independent variables S, V, n_i to variables T, P, n_i. For such processes the Gibbs free energy is given by

$$G = U - TS + PV = \sum_i \mu_i n_i$$

or

$$dG = dU - T\,dS - S\,dT + P\,dV + V\,dP. \tag{14.36}$$

Substituting the value of dU from Equation 14.35, we have

$$dG \leqslant -S\,dT + V\,dP + \sum_i \mu_i\,dn_i. \tag{14.37}$$

From Equation 14.37 the equation for infinitesimal changes in the molar Gibbs free energy can be written as

$$dG_m = -S_m\,dT + V_m\,dP. \tag{14.38}$$

If we consider one of the van der Waals isotherms for which $dT = 0$, we can find how G_m varies with pressure along that isotherm. Along the isotherm $ABCDEFGHI$ in Figure 14.10a the difference in molar Gibbs free energy between any two points on it is equal to the area under the curve between those two points. Then,

$$(G_m)_2 - (G_m)_1 = \int_{P_1}^{P_2} V_m \, dP. \tag{14.39}$$

Between points A and D in Figure 14.10a the Gibbs free energy increases and the shape of the curve on $G_m{\sim}P$ plot is concave (see Figure 14.10a). Then, between D and F it decreases and the curve is convex; finally between F and I the Gibbs free energy increases and the curve becomes concave again. These characteristics are shown in Figure 14.10b. Since the curve between D and F is convex, the states are mechanically unstable, because for mechanical stability G_m must be concave. By the same argument it follows that the regions from A to D as well as from F to I are both mechanically stable, because G_m is concave. The states along FCD in Figure 14.10b are metastable. The curve ACI on this figure corresponds to thermodynamically equilibrium states, since for these states the Gibbs free energy is at a minimum. Therefore, we conclude that the equilibrium states correspond to those states for which the values of the Gibbs free energy lie along the curve ACI. The equilibrium between C and G in Figure 14.10a is maintained when the pressure is constant, so the Gibbs free energy remains constant. Therefore, to obtain the equilibrium states on the isotherm between these two points, we have to join them by a vertical line, i.e. a line of constant pressure. Then the isotherm containing equilibrium states is given by the curve $ABCEGHI$.

The points C and G have equal molar Gibbs free energy, as shown in Figure 14.10b. So the difference between their molar free energies is zero, and we can write

$$(G_m)_G - (G_m)_C = 0 = \int_{P_C}^{P_G} V_m \, dP.$$

The line CG can be divided into four segments: CD, DE, EF and FG. Hence the above integral becomes

$$\int_{P_C}^{P_D} V_m \, dP + \int_{P_D}^{P_E} V_m \, dP + \int_{P_E}^{P_F} V_m \, dP + \int_{P_F}^{P_G} V_m \, dP = 0 \tag{14.40}$$

or

$$\int_{P_C}^{P_D} V_m \, dP - \int_{P_E}^{P_D} V_m \, dP = \int_{P_F}^{P_E} V_m \, dP - \int_{P_F}^{P_G} V_m \, dP. \tag{14.41}$$

Referring to Figure 14.10a, we find that the left-hand side of the above equation is equal to the area bounded by $CDEC$ and the right-hand side is equal to the area bounded by $EFGE$. Thus, these two areas are equal, so the curve $ACEGI$ gives the equilibrium states of the system. The condition

$$\text{Area } CDEC = \text{Area } EFGE$$

is called the *Maxwell construction*. This enables us to obtain the equilibrium isotherms from the van der Waals equation and also the curves for metastable states.

14.9.4   Regular binary solutions: the Hildebrand theory

Many of the properties observed in physico-chemical systems are dependent on the interactions between the molecules and are exhibited by large collections of particles such as ions, atoms or molecules. For example, an individual molecule does not have a melting or boiling point; we cannot say that an isolated molecule is a solid, liquid or gas. Therefore, we can easily distinguish between *bulk properties* of a system which are due to large collections of particles and *molecular properties* which are exhibited by individual particles. Except for the very smallest observable systems, the bulk behaviour of systems at equilibrium does not depend on the size of the system and the time of the observation. Therefore, it is possible to express bulk behaviour in terms of some kind of time-independent average of the behaviour of the individual particles comprising the system. Bulk behaviours of mixing of solids, liquids and vapours are already known from the thermodynamics of solutions, and in Section 14.9.2 we considered phase transitions in pure PVT systems. In this section we shall consider mixtures of different types of particle. In such a system, a phase transition can occur to produce a physical separation within the system into regions of various concentrations of the different types of particle present. Examples of this type of phase transition can be found in binary mixtures.

In a mixture of two types of particle, say A and B, two situations can arise: (i) the particles do not interact with one another, and (ii) the particles do interact with one another. In the first case, the total Gibbs free energy of the system is given by

$$G = n_A(\mu_A^0) + n_B(\mu_B^0) + RTn_A \ln x_A + RTn_B \ln x_B \qquad (14.42)$$

where n_A and n_B are the number of moles of A and B, respectively, $n_A + n_B = n$ (the total number of moles in the system), μ_A^0 and μ_B^0 are the chemical potentials per mole of A and B, respectively, in the absence of mixing, and x_A and x_B are the mole fractions of A and B, respectively. Note that both μ_A^0 and μ_B^0 are functions of the pressure and temperature of the system. On the other hand, if the particles interact with one another, then the interaction will add a term to the internal energy which is proportional to the number of moles of each type of particle. If we take λ as the proportionality constant, then $U = (\lambda n_A n_B)/n$, and the Gibbs free energy per mole is given by

$$G_m = \frac{G}{n} = x_A(\mu_A^0) + x_B(\mu_B^0) + RTx_A \ln x_A + RTx_B \ln x_B + \lambda x_A x_B. \qquad (14.43)$$

A binary system described by Equation 14.43 is called a *regular binary solution*. The constant λ is a measure of the strength of interaction between the particles A and B. λ could be positive or negative. If it is negative, then A and B will attract. On the other hand, A and B will repel if λ is positive. It is the thermal energy which tends to mix the particles. At sufficiently low temperature, if λ is positive and sufficiently large, then the repulsive interaction energy will dominate over the thermal energy and the system will separate into two phases: one rich in A and the other rich in B. Equation 14.43 suggests that a phase transition can occur.

The molar Gibbs free energy depends on the independent variables P, T and the mole fractions of the particles. Since $x_A + x_B = 1$ or $x_B = 1 - x_A$, the molar Gibbs free energy then depends on P, T and x_A. Let us partially differentiate equation 14.43 with respect to x_A. Then,

$$\left(\frac{\partial G_m}{\partial x_A}\right)_{P,T} = \mu_A^0 + \mu_B^0 \frac{\partial x_B}{\partial x_A} + RT \ln x_A + RT$$

$$+ RT\left(\frac{\partial x_B}{\partial x_A}\right) \ln x_B + RT\left(\frac{\partial x_B}{\partial x_A}\right)\frac{x_B}{x_B} + \lambda x_B + \lambda x_A\left(\frac{\partial x_B}{\partial x_A}\right).$$

Since $x_B = 1 - x_A$, $\partial x_B/\partial x_A = -1$. Substituting this into the above equation, we have

$$\left(\frac{\partial G_m}{\partial x_A}\right)_{P,T} = \mu_A^0 - \mu_B^0 + RT \ln x_A + RT - RT \ln x_B - RT$$

$$+ \lambda(1 - x_A) - \lambda x_A$$

or

$$\left(\frac{\partial G_m}{\partial x_A}\right)_{P,T} = \mu_A^0 - \mu_B^0 + \lambda(1 - 2x_A) + RT \ln \left(\frac{x_A}{1 - x_A}\right). \qquad (14.44)$$

Differentiating again, we obtain

$$\left(\frac{\partial^2 G_m}{\partial x_A^2}\right)_{P,T} = -2\lambda + RT/[x_A(1 - x_A)] \geqslant 4RT - 2\lambda. \qquad (14.45)$$

Equation 14.45 suggests that for temperatures $T < (\lambda/2R)$, a range of values of x_A centred at $x_A = \frac{1}{2}$ exists where G_m can be a concave function of x_A, and outside this range G_m is a convex function of x_A. Therefore, for $T < (\lambda/2R)$, G_m can have one maximum and two minima, as shown in Figure 14.11. Each minimum corresponds to a stable thermodynamic state. The maximum and the minima occur for those values of x_A for which

$$\lambda(1 - 2x_A) + RT \ln \left(\frac{x_A}{1 - x_A}\right) = 0. \qquad (14.46)$$

The maximum is at $x_A = \frac{1}{2}$.

The stability of the binary solution is given by the condition

$$(1 - x_A)x_A \leqslant \frac{RT}{2\lambda}. \qquad (14.47)$$

A graphical representation of this equation in a T–x_A plane is shown in Figure 14.12. The curve is symmetric about the point $x_A = 1/2$ and has a maximum value at $T_c = \lambda/2R$. $x_A = 1/2$ is the critical point. At this point we have

$$\left(\frac{\partial^2 G_m}{\partial x_A^2}\right)_{P,T} = \left(\frac{\partial^3 G_m}{\partial x_A^3}\right)_{P,T} = 0 \qquad (14.48)$$

and

$$\left(\frac{\partial^4 G_m}{\partial x_A^4}\right)_{P,T} > 0 \qquad (14.49)$$

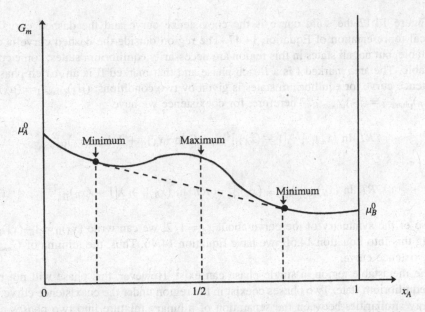

Figure 14.11 Schematic diagram of the variation of the molar Gibbs free energy of a binary mixture as a function of mole fraction x_A for constant P and $T < \lambda/2R$

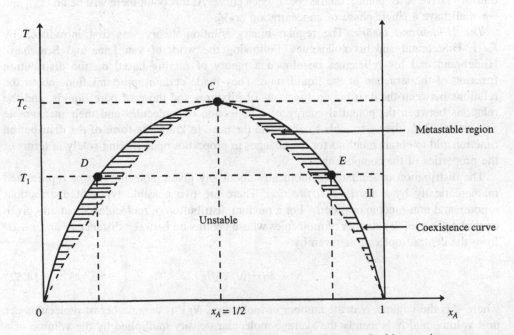

Figure 14.12 Schematic diagram of stability curve for a regular binary mixture

In Figure 14.12 the solid curve is the coexistence curve and the dashed curve is the graphical representation of Equation 14.47. The region outside the dashed curve is chemically stable, but not all states in this region are necessarily equilibrium states; some could be metastable. The area marked I is a B-rich phase and that marked II is an A-rich phase. The coexistence curve for equilibrium states is given by two conditions: $(\mu_A)_{\text{phase I}} = (\mu_A)_{\text{phase II}}$ and $(\mu_B)_{\text{phase I}} = (\mu_B)_{\text{phase II}}$. Therefore, for coexistence we have

$$RT \ln (x_A)_{\text{I}} + \lambda[1 - (x_A)_{\text{I}}]^2 = RT \ln (x_A)_{\text{II}} + \lambda[1 - (x_A)_{\text{II}}]^2 \qquad (14.50)$$

$$RT \ln (x_B)_{\text{I}} + \lambda[1 - (x_B)_{\text{I}}]^2 = RT \ln (x_B)_{\text{II}} + \lambda[1 - (x_B)_{\text{II}}]^2. \qquad (14.51)$$

Because of the symmetry of the curve about $x_A = 1/2$, we can write $(x_A)_{\text{II}} = 1 - (x_A)_{\text{I}}$. Substituting this into Equation 14.50, we have Equation 14.46. Thus, the minima of G_m fall on the coexistence curve.

In the metastable region, a single phase can exist. However, this phase will not be in a stable equilibrium state. Two phases coexist in the region under the coexistence curve. There are many similarities between the separation of a binary mixture into two phases and the liquid–vapour transition. In Figure 14.12, at temperature T_1 the concentration of x_A is zero (100% B). Here, we have a system made of only B particles. Let us now start adding A particles. Along the horizontal line T_1D, the concentration of A increases until we reach the coexistence curve at D. At this point the system separates into two phases: one in which A has concentration $(x_A)_{\text{I}}$, and another in which A has concentration $(x_A)_{\text{II}}$. As we keep on increasing A relative to B, the amount of phase $(x_A)_{\text{I}}$ decreases and that of $(x_A)_{\text{II}}$ increases until we arrive at the point E on the coexistence curve. At this point there will be no $(x_A)_{\text{I}}$ and we shall have a single phase of concentration $(x_A)_{\text{II}}$.

The Hildebrand theory: The regular binary solution theory was first introduced by J. H. Hildebrand and his colleagues. Following the work of van Laar and Scatchard, Hildebrand and his colleagues developed a theory of mixing based on the distribution function of the structure of the liquid state. They made certain approximations about the relations between the distribution functions of mixtures and those of pure liquids, and the relations between the potential energy of a collection of molecules and their measurable properties. Thus, they were able to eliminate the need to know the form of the distribution function and to obtain relations for the changes in properties upon mixing solely in terms of the properties of the components.

The distribution of similar size molecules about any given molecule can be represented mathematically by a *distribution function*. There are two possible types of distribution: random and non-random or orderly. For a random distribution of molecules about any given central molecule, the number of molecules whose centres lie between distances r and $r + \partial r$ from the central molecule is given by

$$\bar{N} = 4\pi r^2 (N/V) \partial r \qquad (14.52)$$

where V is the volume, N is the number of molecules, N/V is the number of molecules per unit volume and $\bar{N}$ is merely the average molecular density multiplied by the volume of a spherical shell between r and $r + \partial r$. If the distribution is non-random, the deviation from randomness is given by the distribution function

$$g(r) = N(r)/\bar{N} \tag{14.53}$$

where $N(r)$ is the number of molecules between r and $r + \partial r$ from the central molecule. By virtue of Equation 14.52, Equation 14.53 becomes

$$N(r) = 4\pi r^2 (N/V)g(r)\partial r. \tag{14.54}$$

Solids have long range order, so $g(r)$ will have a number of sharp intensity maxima in X-ray diffraction at values of r corresponding to the equilibrium separations of the nearest neighbours, next nearest neighbours, and so on. On the other hand, in liquids $g(r)$ has a maximum at a value of r close to that for the first maximum in the solid, and subsequently only small fluctuations from 1. In the case of a monatomic solid such as metals, inert gases, etc., if the atoms are confined exactly to their positions in the lattice, then $g(r)$ would become discontinuous with finite values at certain values of r and is zero at all other values of r.

In the Hildebrand theory, it has been assumed that the changes in energy upon mixing is entirely due to the energy of interaction of the molecules. Then,

$$(\Delta U)_{mix} = \text{the potential energy } (E) \text{ of mixture}$$
$$\text{minus the potential energy of components}$$
$$= E(\text{mixture}) - E(\text{components}). \tag{14.55}$$

Since it is assumed that the contributions to the liquid state behaviour of the motions of the molecules do not change in going from pure liquid to mixture, there is no need to consider such contributions. Hildebrand also assumed that the potential energy functions of the molecules are spherically symmetrical, so that (i) the potential energies of interaction between the molecules are dependent only on their separation, and (ii) the potential energy of a large assembly of molecules is given correctly by the sum of the potential energies of all possible pairs of molecules (this means that the potential energy of any pair of molecules is not affected by the presence of other molecules in the liquid).

Using the above assumptions and the expression for the distribution function for a pure liquid, the potential energy E of interaction of one molecule with all other in the liquid can be written as

$$E = \int_0^\infty 4\pi r^2 (N/V)g(r)u(r) \, dr \tag{14.56}$$

where $u(r)$ is the potential energy of a pair of molecules at r and $4\pi r^2(N/V)g(r)$ is the number of molecules in a spherical shell of radius r measured from the central molecule and thickness dr.

The potential energy E of a pure liquid can be found by taking the sum of potential energies of interaction of each molecule with all other molecules. So, for n_i moles of pure liquid i, we can write

$$E = \frac{N_0 n_i}{2} \int_0^\infty 4\pi r^2 \left(\frac{N_0}{V_m}\right) g_{ii}(r)u_{ii}(r) \, dr \tag{14.57}$$

where N_0 is Avogadro's number, V_m is the molar volume and the $1/2$ is due to the fact that the sum over all molecules of the interactions of each molecule with all other molecules includes each contribution to the potential energy twice.

As a simple example, let us consider two liquids of molecule type A and molecule type B mixed together. First, we take each type A molecule and count its interactions with all other type A molecules and then with all type B molecules. Then we take each type B molecule and count its interaction with all other type B molecules. Finally, we take each type A molecule and count its interactions with all type B molecules. In this counting, we have indeed counted each interaction twice. Hence, we have to divide the total count by 2 to obtain the actual sum over all the molecules of their potential energies of interaction with all other molecules. However, in this calculation we have to add a new potential energy function $u_{AB}(r)$ for the interactions between unlike molecules. We also have to allow for the changes in distribution of molecules around a given central molecule that arises due to the fact that we now have two types of molecule to be distributed.

The potential energy of a mixture of n_A moles of A and n_B moles of B would be analogous to Equation 14.57. Then,

$$
\begin{aligned}
E_{\text{mix}} = 2N_0 \Bigg[n_A \int_0^\infty &\left\{ \frac{N_0 n_A}{n_A (V_m)_A + n_B (V_m)_B} \right\} g_{AA}(r) u_{AA}(r) \pi r^2 \, dr \\
+ n_A \int_0^\infty &\left\{ \frac{N_0 n_B}{n_A (V_m)_A + n_B (V_m)_B} \right\} g_{AB}(r) u_{BB}(r) \pi r^2 \, dr \\
+ n_B \int_0^\infty &\left\{ \frac{N_0 n_A}{n_A (V_m)_A + n_B (V_m)_B} \right\} g_{BA}(r) u_{BB}(r) \pi r^2 \, dr \\
+ n_B \int_0^\infty &\left\{ \frac{N_0 n_B}{n_A (V_m)_A + n_B (V_m)_B} \right\} g_{BB}(r) u_{BB}(r) \pi r^2 \, dr \Bigg]
\end{aligned} \tag{14.58}
$$

where $(V_m)_A$ and $(V_m)_B$ are the molar volumes of A and B, respectively. Assuming that the volume of the mixture is equal to the sum of the volumes of its components, $(N_0 n_i)/\sum_i n_i (V_m)_i$ are the number of molecules of type i per unit volume, and $g_{AA}(r)$, $g_{AB}(r)$, $g_{BA}(r)$ and $g_{BB}(r)$ are distribution functions for A–A, A–B, B–A and B–B pairs in the mixture.

The four integrals in Equation 14.58 cannot be evaluated unless the functional forms of the $g_{ij}(r)$ and $u_{ij}(r)$ are known. To circumvent this situation, Hildebrand assumed that, similarly to the situation that occurs in pure monatomic fluids, one can write

$$
g_{ij}(r) = f(r/\sigma_{ij}) = f(y) \tag{14.59}
$$

$$
u_{ij}(r) = \varepsilon_{ij} F(r/\sigma_{ij}) = \varepsilon_{ij} F(y) \tag{14.60}
$$

where $y = r/\sigma_{ij}$ and ε and σ are the characteristic energy and distance apart of the (supposedly spherical) molecules at the minimum in the potential energy against separation curve shown in Figure 14.13. The Lennard–Jones potential energy function $u(r) = \varepsilon\{(\sigma/r)^{12} - 2(\sigma/r)^6\}$. The overall potential energy is taken as the sum of a repulsive term in r^{-12} and an attractive term in r^{-6}. One interesting feature of this potential function as well as of all other potential functions with only two parameters is that if for each compound we

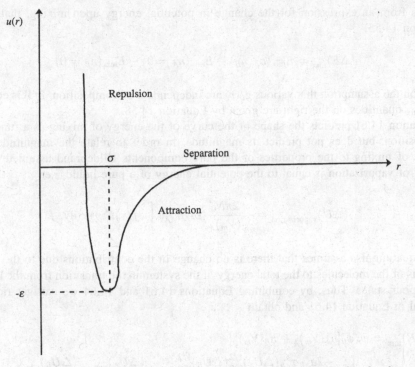

Figure 14.13 Lennard–Jones potential energy function for a pair of molecules

choose ε and σ as the units of energy and distance, respectively, then the plots of the *reduced potential* $u(r)/\varepsilon$ versus the *reduced separation* r/σ lie on the same curve. This behaviour is common to all two-parameter potential functions and leads to simple relations between the calculated properties of substances to which such a form of potential is applicable. The properties calculated for such compounds can all be represented by single curves of *reduced properties*. This is the *principle of corresponding states* mentioned before. This principle has also been applied to the correlation of the properties of mixtures with those of their pure components.

The assumptions 14.59 and 14.60 are closely equivalent to assuming a principle of corresponding states. So we may replace the terms $g_{ij}(r)u_{ij}(r)r^2\,dr$ in Equation 14.58 by $\varepsilon_{ij}(\sigma_{ij})^3 f(y)F(y)y^2\,dy$ and obtain

$$(\Delta E)_{\text{mix}} = 2\pi N_0^2 [n_A(V_m)_A + n_B(V_m)_B]\phi_A\phi_B$$
$$\times \left[\frac{2\varepsilon_{AB}\sigma_{AB}^3}{(V_m)_A(V_m)_B} - \frac{\varepsilon_{AA}\sigma_{AA}^3}{(V_m)_A^2} - \frac{\varepsilon_{BB}\sigma_{BB}^3}{(V_m)_B^2} \right] \int_0^\infty f(y)F(y)y^2\,dy \qquad (14.61)$$

$$= [n_A(V_m)_A + n_B(V_m)_B]\phi_A\phi_B K \qquad (14.62)$$

where ϕ_A and ϕ_B are volume fractions of A and B, respectively, and K is a constant for a given mixture. Equation 14.61 is the *basic equation for regular solutions*. This equation follows from an expression for the change in potential energy upon mixing, that is (see Equation 14.55)

$$(\Delta E)_{\text{mix}} = E_{\text{mix}}(n_A, n_B) - E_{\text{mix}}(n_A = 0) - E_{\text{mix}}(n_B = 0) \qquad (14.63)$$

based on the assumption that various $g_{ij}(r)$ are independent of composition. In this equation, the E_{mix} quantities on the right are given by Equation 14.58.

Equation 14.61 predicts the shape of the curve of the energy of mixing as a function of composition, but does not predict its magnitude. In order to relate the magnitude of the energy of mixing to the properties of the pure components Hildebrand assumed that the energy of vaporization is equal to the potential energy of a pure liquid, i.e.

$$(\Delta U)_{\text{vaporization}} = -\frac{2\pi N_0}{(V_m)_A} (\varepsilon_{AA}\sigma_{AA}^3) \int_0^{\infty} f(y)F(y)y^2 \, dy. \qquad (14.64)$$

This equation also assumes that there is no change in the contributions due to the internal motions of the molecules to the total energy of the system in the transition from the liquid to the vapour states. Thus, by combining Equations 14.61 and 14.64 we can get rid of the integral in Equation 14.61 and obtain

$$(\Delta E)_{\text{mix}} = \phi_A\phi_B[n_A(V_m)_A + n_B(V_m)_B]$$
$$\times \left[\frac{-2\varepsilon_{AB}\sigma_{AB}^3(\Delta U_A)_{\text{vap}}^{1/2}(\Delta U_B)_{\text{vap}}^{1/2}}{(V_m)_A(V_m)_B(\varepsilon_{AA}\varepsilon_{BB})^{1/2}(\sigma_{AA}\sigma_{BB})^{3/2}} + \frac{(\Delta U_A)_{\text{vap}}}{(V_m)_A} + \frac{(\Delta U_B)_{\text{vap}}}{(V_m)_B} \right] \qquad (14.65)$$

where $(\Delta U_A)_{\text{vap}}$ and $(\Delta U_B)_{\text{vap}}$ are the vaporization energies of A and B, respectively. This equation can be written in a more elegant form if we write

$$\frac{(\Delta U_A)_{\text{vap}}}{(V_m)_A} = \delta_A^2 \quad \text{and} \quad \frac{(\Delta U_B)_{\text{vap}}}{(V_m)_B} = \delta_B^2 \qquad (14.66)$$

Then, Equation 14.65 becomes

$$(\Delta E)_{\text{mix}} = \phi_A\phi_B[n_A(V_m)_A + n_B(V_m)_B]\left[\delta_A^2 + \delta_B^2 - \frac{2\varepsilon_{AB}\sigma_{AB}^3\delta_A\delta_B}{(\varepsilon_{AA}\varepsilon_{BB})^{1/2}(\sigma_{AA}\sigma_{BB})^{3/2}} \right] \qquad (14.67)$$

Further, we assume that the difference between the geometric mean and the arithmetic mean is insignificant for the calculation of σ_{AB} in terms of σ_{AA} and σ_{BB}, and that $\varepsilon_{AB} = (\varepsilon_{AA}\varepsilon_{BB})^{1/2}$. Then, Equation 14.67 simplifies to

$$(\Delta E)_{\text{mix}} = \phi_A\phi_B[n_A(V_m)_A + n_B(V_m)_B](\delta_A - \delta_B)^2 \qquad (14.68)$$

This is the basic equation of the *solubility parameter theory* which is used for the treatment of solutions. Both Equation 14.61 and Equation 14.68 are often combined with the assumption of ideal entropy of mixing to derive equations for the excess free energy of mixing.

14.9.5   Ginzburg–Landau theory

The van der Waals equation and the equation for regular binary mixtures describe the behaviour of systems around the critical point. The Ginzburg–Landau theory describes all transitions which involve a broken symmetry and a continuous change in the slope of the free energy curve as a function of P and T at the transition. All these theories are known as *mean field theories* as they can be derived by assuming that each particle in the system moves in the mean field of all other particles. Although mean field theories are not truly perfect because these do not correctly take into account short-ranged correlations which are important near the critical point, they do give qualitatively correct behaviour of systems near critical points. Hence, these theories are useful.

The transitions from vapour to liquid phase, from liquid to solid phase, and from vapour to solid phase in a pure PVT system are all first-order transitions. The transition from vapour to liquid phase is quite different from the liquid–solid and vapour–solid transitions, because in the former no symmetry of the system is broken at the transition point, whereas in the liquid–solid and vapour–solid transitions the translational symmetry of the high-temperature phase is broken at the transition point. This difference can be attributed to the average density of solid, liquid and vapour. This parameter of both the liquid and the vapour is independent of position and uniform throughout the system, so it is invariant under all elements of the translational group. On the other hand, the average density of the solid is a periodic one and is invariant only with respect to a subgroup of the translational group. Therefore, translational symmetry is broken in the transition from vapour or liquid phase to the solid phase.

In a first-order transition the slope of the free energy curve changes discontinuously as a function of P and T at the transition. In a first-order symmetry-breaking transition there occurs a discontinuous jump in the density, entropy or in any state of the system at the transition. Therefore, no relation between the symmetry properties of the two phases is needed. We can also have symmetry-breaking transitions where the slope of the free energy curve changes continuously as a function of P and T. In such cases, there is a continuous change in the state of the system, and the symmetry properties of the two phases are closely related. In a continuous transition, the lower temperature phase usually has a lower symmetry than that of the higher temperature phase. However, this may not always be true. Continuous transitions are called λ-*points*, because the heat capacity always shows a λ-shaped peak at the transition point.

A new macroscopic parameter is always associated with the lower symmetry phase whenever there is a symmetry-breaking transition. This parameter is called the *order parameter*. It could be a complex number, a scalar, a vector, a tensor or any other quantity. There are different types of symmetry that can be broken at a phase transition. Similarly, there are different types of order parameter. The types of symmetry that can be broken at a phase transition are: (i) translational symmetry for the liquid–solid transition, (ii) rotational symmetry for the transition from a paramagnetic to a ferromagnetic system (here a spontaneous magnetization takes place that defines a unique direction in space), and (iii) gauge symmetry for the transition from normal liquid He^4 to superfluid liquid He^4. The Ginzburg–Landau theory alone can describe all transitions which involve a continuous change in the slope of the free energy curve and a broken symmetry.

We said above that the order parameter could be a complex number, a scalar, a vector, a tensor or any other quantity. Let us assume that it is a vector denoted by α. We want to discuss the form of the free energy in the vicinity of the phase transition. For $\alpha = 0$ the free energy must be at a minimum above the transition point and for $\alpha \neq 0$ it must be at a minimum below the transition. The free energy is actually a number and hence it is a scalar function of the order parameter. Therefore, if according to our assumption the order parameter α is a vector, then the free energy depends on scalar products of the order parameter. The particular free energy of a system will depend on the system itself. Let us consider a free energy denoted by $G(T, \alpha)$. Generally, in the vicinity of the transition, we can expand the free energy to write

$$G(T, \alpha) = G_0(T) + \beta_2(T)|\alpha|^2 + \beta_4(T)|\alpha|^4 + \cdots \qquad (14.69)$$

where $\alpha \cdot \alpha = |\alpha|^2$. Note that only the even-order terms appear in the expansion. This is because a scalar cannot be constructed from the odd-order terms.

We can choose the form of $\beta_2(T)$ such that the free energy will only be minimized for $|\alpha| = 0$ at and above the critical temperature, and it will be minimized below the critical temperature for $|\alpha| > 0$. Mathematically, minimum free energy means $(\partial G/\partial \alpha)_{P,T} = 0$ and $(\partial^2 G/\partial \alpha^2)_{P,T} > 0$. In our case, the above condition is satisfied if $\beta_2(T) > 0$ for $T > T_c$ and $\beta_2(T) < 0$ for $T < T_c$. If $\alpha = 0$, then $G(T, \alpha)$ will be at a minimum when $T > T_c$. On the other hand, $G(T, \alpha)$ will be at a minimum for $\alpha \neq 0$ when $T < T_c$. Then, the question arises: what happens at $T = T_c$? The answer is that at this point $\beta_2(T) = 0$, because the free energy must vary continuously through the transition point. All of this information can be put together if we simply write

$$\beta_2(T) = \beta_0(T - T_c) \qquad (14.70)$$

where we have introduced a new quantity β_0 which is a function of the temperature; $\beta_0 = f(T)$. Now for total stability in the system we must have

$$\beta_4(T) > 0. \qquad (14.71)$$

The above condition will make sure that the free energy will increase if the scalar product $(\alpha \cdot \alpha)^{1/2} = |\alpha|$ increases to very large values. From Equation 14.69 we see that the free energy has extrema when $\partial G/\partial \alpha = 0$, that is

$$\frac{\partial G}{\partial \alpha} = 2\beta_2 \alpha + 4\beta_4 \alpha |\alpha|^2 = 0. \qquad (14.72)$$

This gives

$$\alpha = 0, \quad \text{or} \quad \alpha = \pm \left(\frac{-\beta_2}{2\beta_4} \right)^{1/2} \hat{\alpha} \qquad (14.73)$$

where $\hat{\alpha}$ is a unit vector. The minimum occurs for $\alpha = 0$ when $\beta_2 > 0$ and for

$$\alpha = \pm \left(\frac{\beta_2}{2\beta_4} \right)^{1/2} \hat{\alpha}$$

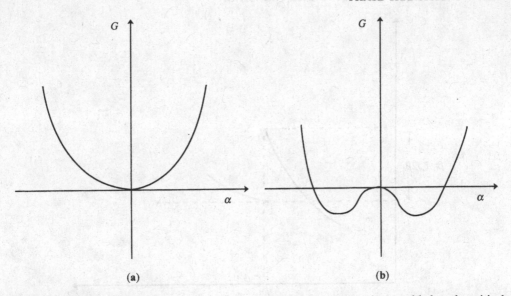

Figure 14.14 Schematic diagram of the Ginzburg–Landau free energy above and below the critical point for (a) $\beta_2 > 0$ and (b) $\beta_2 < 0$

when $\beta_2 < 0$. This is shown in Figure 14.14. Thus, for $T < T_c$, the order parameter is not zero but increases as $(|T - T_c|)^{1/2}$. The order parameter vector α can be pointed in the positive or the negative direction along any given axis. We have the following mathematical forms for Ginzburg–Landau free energy:

$$G(T, \alpha) = G_0(T) \quad \text{for} \quad T > T_c \tag{14.74}$$

and

$$G(T, \alpha) = G_0(T) - [\beta_0^2(T - T_c)^2/4\beta_4] \quad \text{for} \quad T < T_c. \tag{14.75}$$

Differentiating Equations 14.74 and 14.75 twice with respect to T and then subtracting the first from the second, we have

$$\frac{\partial^2 G}{\partial T^2}(T < T_c) - \frac{\partial^2 G}{\partial T^2}(T > T_c) = \left(\frac{\beta_0^2}{2\beta_4}\right)T_c \tag{14.76}$$

where we have ignored the derivatives of both β_0 and β_4, assuming that they change slowly with temperature. However, we know that (see Section 10.9)

$$\frac{\partial^2 G}{\partial T^2} = -\left(\frac{\partial S}{\partial T}\right)_P = \frac{1}{T}\left[-T\left(\frac{\partial S}{\partial T}\right)_P\right] = \frac{-C_P}{T}. \tag{14.77}$$

Comparing Equations 14.76 and 14.77, we can write

$$C(T < T_c) - C(T > T_c) = \left(\frac{\beta_0^2}{2\beta_4}\right)T_c. \tag{14.78}$$

Thus, the jump in the heat capacity has a shape of λ (see Figure 14.15).

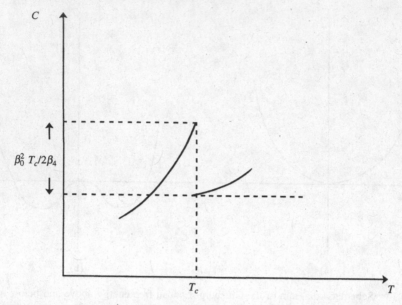

Figure 14.15 The λ-point jump in the heat capacity as predicted by the Ginzburg–Landau theory

Ginzburg–Landau theory of paramagnetic to ferromagnetic transition

The Ginzburg–Landau theory is applicable to all continuous symmetry-breaking transitions such as magnetic systems, superconductors, transition from normal liquid He^4 to superfluid liquid He^4. *The transition from a paramagnetic to a ferromagnetic system* is an example of a λ-point. A paramagnetic system contains particles which have magnetic moment. Above the Curie temperature there is random orientation of the magnetic moments but no net magnetization. As the temperature approaches the Curie temperature, magnetic interaction energy between lattice sites wins over randomizing thermal energy, and this effect persists up to the Curie temperature. Above this point, the paramagnetic system has an invariant rotational symmetry. Below this temperature the magnetic moments become ordered and at the same time a spontaneous magnetization occurs which selects a preferred direction in space. Thus, the rotation symmetry is broken. Assuming that there is negligible volume change in the system at the transition point, the Helmholtz free energy of an isotropic system is given by

$$A(\mathbf{M}, T) = A(0, T) + \beta_2(T)|M|^2 + \beta_4(T)|M|^4 + \cdots \tag{14.79}$$

where $\mathbf{M}$ is the average magnetization of the system, $\mathbf{M} \cdot \mathbf{M} = |M|^2$ ($\mathbf{M}$ being magnetization = magnetic moment per unit volume), $A(\mathbf{M}, T)$ is the Helmholtz free energy and $A(0, T)$ is a contribution, independent of magnetization, to the free energy above the Curie temperature. The Gibbs free energy can be obtained by applying an external magnetic field, $\mathbf{H}$, to the system. Then,

$$G(\mathbf{H}, T) = A(\mathbf{M}, T) - \mathbf{H} \cdot \mathbf{M}$$
$$= G_0(\mathbf{H}, T) + \beta_2(T, \mathbf{H})|M|^2 + \beta_4(T, \mathbf{H})|M|^4 + \cdots \quad (14.80)$$

where

$$G_0(\mathbf{H}, T) = A(0, T) - \mathbf{H} \cdot \mathbf{M}. \quad (14.81)$$

If $\mathbf{H}$ is fixed, then the Gibbs free energy given by Equation 14.80 will be a minimum, and also if $\mathbf{H}$ is sufficiently small, then the dependence of β_2 and β_4 on it can be ignored.

Now, from the Ginzburg–Landau theory we have Equations 14.70 and 14.71. The Gibbs free energy is minimized for lowest $\mathbf{M}$ when we have

$$\left(\frac{\partial G}{\partial \mathbf{M}}\right)_{\mathbf{H}, T} = -\mathbf{H} + 2\beta_2(T)\mathbf{M} = 0. \quad (14.82)$$

Hence,

$$\mathbf{M} = \left(\frac{1}{2\beta_2}\right)\mathbf{H}. \quad (14.83)$$

This shows that in the presence of an external field a non-zero magnetization is possible above the Curie temperature. However, in order to obtain a meaningful result we have to retain higher-order terms below the Curie temperature.

The magnetic susceptibility of a paramagnetic substance is given by

$$\chi = \mathbf{M}/\mathbf{H} \quad (14.84)$$

where $\mathbf{M}$ is magnetic moment per unit volume and $\mathbf{H}$ is the magnetic field intensity. Then, from Equation 14.83 the isothermal susceptibility of the substance is given by

$$\chi_T = \left(\frac{\partial \mathbf{M}}{\partial \mathbf{H}}\right)_T = \frac{1}{2\beta_2(T)}.$$

Substituting the value of $\beta_2(T)$ from Equation 14.70, we have

$$\chi_T = 1/[2\beta_0(T - T_c)] \quad (14.85)$$

Here we are considering an isotropic system. Therefore, the susceptibility is a scalar quantity.

Equation 14.85 shows that the susceptibility becomes infinite when $T = T_c$, i.e. at the critical temperature. Hence, at T_c a very small $\mathbf{H}$ can produce a large magnetization. In the absence of an external field, the magnetization behaviour below T_c is described by

$$\mathbf{M} = \left[\frac{\beta_0(T_c - T)}{2\beta_4}\right]^{1/2}. \quad (14.86)$$

Differentiating this with respect to T at constant $\mathbf{H}$, we obtain

$$\left(\frac{\partial \mathbf{M}}{\partial T}\right)_{\mathbf{H}} = -(1/2)\left[\frac{\beta_0}{2\beta_4(T_c - T)}\right]^{1/2}. \tag{14.87}$$

Thus, the magnetization increases abruptly for $T_c > T$.

Ginzburg–Landau theory of superconductors

There are many applications of the *Ginzburg–Landau theory of superconductors*. It is known that superconductors are perfect diamagnets. If a superconductor is cooled below its transition temperature in the presence of a magnetic field, then currents are generated on its surface in such a way that the magnetic fields created by the currents neutralize any magnetic fields which were initially inside the superconductor. Therefore, the magnetic field $\mathbf{B} = 0$ inside a superconductor irrespective of how it was prepared.

A magnetic field can cause a current to flow in a superconductor; there is no need for an electric field. In a superconductor, an applied electric field accelerates some of the electrons; there is no significant frictional force to retard the electrons. This behaviour is analogous to the frictionless superflow in liquid He^4 below $2.19\,\text{K}$. In the next section we shall discuss the helium liquids.

The apparent frictionless flow in superconductors and liquid He^4 has its origin in quantum mechanics. It is known that the electrons in a superconductor can experience an attractive interaction through interaction with lattice phonons. Due to this effect some electrons (it is not known which ones) can form *bound pairs*. In the state of minimum free energy all bound pairs have the same quantum numbers and they form a single macroscopically occupied quantum state. This state is a condensed phase and acts as one coherent state; it contains many bound pairs and has a big mass. Therefore, any friction arising from impurities in the lattice must act on the whole phase, not on some pairs only. Hence, the condensed phase moves as a whole under an applied electric field and there is no significant retardation of its motion by frictional effects. One of the most interesting applications of the Ginzburg–Landau theory is to the case of a superconductor that may contain both normal and condensed regions or is of finite size. In such cases we must allow the order parameter to vary in position. Actually, the order parameter of the condensed phase behaves like an effective wave function of the pairs.

The condensed phase moves steadily in the presence of a magnetic field but is accelerated under an applied electric field. Just below T_c only a tiny fraction of the electrons are condensed and take part in superflow. Thermal effects actually tend to destroy the condensed phase. So, as the temperature goes down below T_c the thermal effect decreases and the condensation of a large fraction of the electron occurs. However, a sufficiently large external magnetic field can destroy the superconducting state. The system will behave as a perfect diamagnet under an applied external magnetic field $\mathbf{H}$ which has a value less than some critical value $\mathbf{H}_c(T)$, its permeability $\mu = 0$ and the internal magnetic field $\mathbf{B} = 0$. If $\mathbf{H} > \mathbf{H}_c(T)$, the system is normal and $\mathbf{B} = \mu\mathbf{H}$. For normal metals $\mu \simeq \mu_0$, the permeability of the vacuum. We can summarize as

$$\begin{aligned} B &= 0 &&\text{for} \quad \mathbf{H} < \mathbf{H}_c(T) \\ &= \mu_0\mathbf{H} &&\text{for} \quad \mathbf{H} > \mathbf{H}_c(T) \end{aligned} \Biggr\}. \tag{14.88}$$

The condensed phase in a superconductor is described by a macroscopic *wave function* ψ. This is the order parameter of the system, and it is generally a complex function. The phase transition from normal to superconducting phase involves the breaking of *gauge symmetry*. A gauge transformation is one which changes the phase of all wave functions in the system. It is given by the number operator. Under this transformation the order parameter ψ changes its phase, but the free energy must not change. So, in the absence of magnetic fields the Ginzburg–Landau equation for the Helmholtz free energy per unit volume must be of the form

$$a_s(T) = a_n(T) + \beta_2(T)|\psi|^2 + \beta_4(T)|\psi|^4 \tag{14.89}$$

where $a_s(T)$ and $a_n(T)$ are the Helmholtz free energies per unit volume for the superconducting and normal phases, respectively, $\psi^*\psi = |\psi|^2$, $\beta_2 = \beta_0(T-T_c)$ and $\beta_4(T) > 0$. We interpret $|\psi|^2 = n_s$ as the number density of bound-pairs of electrons in the condensed phase. Equation 14.89 assumes an infinitely large superconductor so that the order parameter is independent of position. For temperatures lower than the critical temperature, $|\psi| = \left[|\beta_2|/2\beta_4\right]^{1/2}$ and

$$a_s(T) = a_n(T) - \frac{\beta_2^2}{4\beta_4}. \tag{14.90}$$

The variation of the critical external magnetic field $\mathbf{H}_c(T)$ with the temperature is approximately the same for most metals. This is schematically shown in Figure 14.16. The coexistence curve for the superconducting and normal phases is approximately given by the equation

$$H_c(T) = H_0(1 - T^2/T_c^2). \tag{14.91}$$

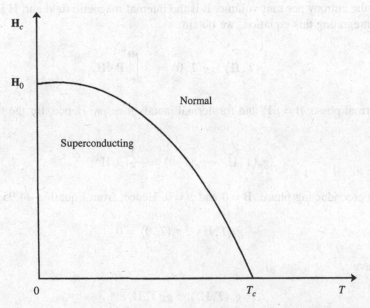

Figure 14.16 Schematic diagram of the coexistence curve for normal and superconducting phases

Here, T_c is the critical temperature in the absence of external fields. The slope of the curve, dH_c/dT, is equal to 0 at $T = 0$; it is negative at $T = T_c$. There are similarities between the phase diagram of a superconducting metal and the vapour–liquid transition in a pure PVT system if H_c replaces the specific volume.

The chemical potentials of the normal and superconducting phases must be equal anywhere on the coexistence curve. So, any changes in these must also be equal. If B_n and B_s are the internal magnetic fields of normal and superconducting phases, respectively, and s_n and s_s are the entropy per unit volume of the normal and superconducting phases, then along the coexistence curve the following condition will satisfy

$$-s_n\, dT - B_n\, dH = -s_s\, dT - B_s\, dH. \tag{14.92}$$

If we now substitute $B_n = \mu_0 H_c(T)$ and $B_s = 0$ in the above equation, we obtain

$$s_n - s_s = -\mu_0 H_c(T)\left(\frac{dH}{dT}\right)_{\text{coexistence}}. \tag{14.93}$$

In the absence of external magnetic fields, the transition is second order. However, in the presence of a magnetic field, it is first order, and has a latent heat for all temperatures but $T = T_c$ (where $H_c = 0$).

Next we consider the difference between the Gibbs free energies of the normal and superconducting phases for $H = 0$, and the change in the heat capacity per unit volume at the transition. The differential of the Gibbs free energy per unit volume (g) is given by

$$dg = -s\, dT - B\, dH \tag{14.94}$$

where s is the entropy per unit volume, B is the internal magnetic field and H is the applied field. By integrating this equation, we obtain

$$g(T, H) - g(T, 0) = -\int_0^H B\, dH. \tag{14.95}$$

For the normal phase, $B = \mu H$ and for normal metals $\mu \simeq \mu_0$. Hence, for the normal phase we have

$$g_n(T, H) - g_n(T, 0) = -\frac{1}{2}(\mu_0 H^2) \tag{14.96}$$

For the superconducting phase, $B = 0$ and $\mu = 0$. Hence, from Equation 14.95 we have

$$g_s(T, H) - g_s(T, 0) = 0 \tag{14.97}$$

We also have

$$g_n(T, H_c) = g_s(T, H_c). \tag{14.98}$$

Then, by subtracting Equation 14.97 from Equation 14.96 and using Equation 14.98, we have for the critical applied field $\mathbf{H} = 0$

$$g_s(T,0) = g_n(T,0) - \left(\frac{1}{2}\right)\mu_0 \mathbf{H}_c^2(T). \tag{14.99}$$

This equation shows that in the absence of a magnetic field the Gibbs free energy of the normal phase is bigger than that of the superconducting phase by $\left(\frac{1}{2}\right)\mu_0 \mathbf{H}_c^2(T)$ at a temperature T. The quantity $\left(\frac{1}{2}\right)\mu_0 \mathbf{H}_c^2(T)$ is called the *condensation energy*. Because the Gibbs free energy is a minimum for fixed temperature and applied field, the condensed phase is the physically realized state. Note also that at $T = T_c$ and $\mathbf{H} = 0$ we have

$$g_s(T_c,0) = g_n(T_c,0). \tag{14.100}$$

By using Equation 14.99 in Equation 14.90 we have the relation

$$\beta_2^2/4\beta_4 = \left(\frac{1}{2}\right)\mu_0 \mathbf{H}_c^2(T). \tag{14.101}$$

This relates the critical field and the Ginzburg–Landau parameters.

The change in the heat capacity per unit volume at the transition is given by

$$(c_n - c_s)_{\text{coex}} = \left[T\frac{\partial}{\partial T}(s_n - s_s)\right]_{\text{coex}} \tag{14.102}$$

where c_n and c_s are heat capacities, per unit volume, of the normal and superconducting phases. Substituting the value of $s_n - s_s$ from Equation 14.93, we obtain

$$\left[T\frac{\partial}{\partial T}(s_n - s_s)\right]_{\text{coex}} = T\frac{\partial}{\partial T}\left[-\mu_0 \mathbf{H}_c(T)\left(\frac{d\mathbf{H}}{dT}\right)_{\text{coex}}\right].$$

From Equation 14.91 we obtain the values of $\mathbf{H}_c(T)$ and $(d\mathbf{H}/dT)$, and then substitute these into the above equation to obtain

$$
\begin{aligned}
(c_n - c_s)_{\text{coex}} &= T\frac{\partial}{\partial T}\left[-\mu_0 \mathbf{H}_0\left(1 - \frac{T^2}{T_c^2}\right)\left(\frac{-2T\mathbf{H}_0}{T_c^2}\right)\right] \\
&= (2\mu_0 \mathbf{H}_0^2)\left(\frac{T}{T_c^2}\right)\frac{\partial}{\partial T}\left(T - \frac{T^3}{T_c^2}\right) \\
&= (2\mu_0 \mathbf{H}_0^2)\left(\frac{T}{T_c^2}\right)\left(1 - \frac{3T^2}{T_c^2}\right) \\
&= \left(\frac{2\mu_0 \mathbf{H}_0^2}{T_c}\right)\left(\frac{T}{T_c} - \frac{3T^3}{T_c^3}\right).
\end{aligned}
\tag{14.103}
$$

At low temperatures the heat capacity of the superconducting phase is lower than that of the normal phase, whereas at $T = T_c$ the heat capacity of the superconducting phase is higher. At T_c, c_s has a finite jump $(c_s - c_n)_{T=T_c} = (4\mu_0/T_c)\mathbf{H}_0^2$. Hence, in the transition from the normal to a superconducting state, the critical point seems to be a λ-point. As $T \to 0$, $(\mathrm{d}\mathbf{H}_c/\mathrm{d}T) \to 0$. Then it follows from Equation 14.93 that $s_s - s_n$ should also tend to zero. This is consistent with the third law of thermodynamics.

14.10   THERMODYNAMICS OF LIQUID HELIUM

The helium atom occurs in nature in two stable isotopic forms, He^3 and He^4. The former, with nuclear spin 1/2, obeys Fermi–Dirac statistics, while the latter, with nuclear spin 0, obeys Bose–Einstein statistics. At very low temperatures, where quantum effects become important, these two isotopes provide two of the very few examples in nature of quantum liquids. Because of its small atomic mass and weak attractive interaction, helium remains in the liquid state over a wide range of pressures up to 0 K. In terms of statistical physics, helium has proven to be one of the most unique and interesting elements.

In fact, chemically He^3 and He^4 are virtually identical; they differ only in mass. As they obey two different statistics, at low temperatures they exhibit very different behaviours. Liquid He^4 exhibits a rather straightforward transition to a superfluid state at 2.19 K. This happens due to a condensation of particles into a single quantum state. On the other hand, liquid He^3 undergoes a transition to a superfluid state at a much lower temperature of 2.7×10^{-3} K, and the mechanism of its superfluid transition is very different from that of liquid He^4.

In liquid He^3, it is more accurate to call the particles as *quasi-particles*. They form bound pairs with a spin $s = 1$ and relative angular momentum $l = 1$ by a mechanism that is similar to the formation of bound pairs in a superconductor, except that the pairs in a superconductor are formed with $s = 0$ and $l = 0$. Because of these differences, the bound pairs in liquid He^3 are flatter along one axis, have angular momentum and a net magnetic moment, whereas those in a superconductor are spherical in shape and have no magnetic moment. Since these systems present such a contrast to those of classical fluids, and as they tend to conform to the third law of thermodynamics, it is worthwhile to consider their phase diagrams.

(A) *Liquid He^3*: The He^3 atom is the rarer of the two isotopes; its relative abundance in natural helium gas is 1 p.p.m. In order to obtain He^3 in large quantities it must be prepared artificially from tritium solutions through β-decay of the tritium atom. It was first liquefied in 1948 by Sydoriack *et al.* (*Phys. Rev.* **75** (1949) 303). The mass of He^3 is only $\frac{3}{4}$ of that of a He^4 atom, so it has a larger zero point energy than the He^4 atom. Because of this, the boiling temperature of He^3 is about 25% lower than that of He^4, and also it requires about 25% more pressure than He^4 to solidify. The phase diagram of He^3 is shown schematically in Figure 14.17.

The diagram does not show a transition to a superfluid state; however, there is a minimum in the liquid–solid coexistence curve, which is due to the spin of the He^3 atom. At low temperatures the spin lattice of the He^3 solid has a higher entropy than the liquid system. In this temperature range if the third law of thermodynamics is to be satisfied, the slope of the liquid–solid curve has to be flat as $T \to 0$. At high temperatures, $\Delta S = (S)_{\text{liquid}} - (S)_{\text{solid}}$ is positive, it vanishes at about $T = 0.3$ K, and becomes negative for $T < 0.3$ K. However, as the volume differences remain practically unaltered, $\mathrm{d}P/\mathrm{d}T = \Delta S/\Delta V$ (the Clausius–Clapeyron law) gives a positive slope at high temperatures and a negative slope at low temperatures.

(B) *Liquid He^4*: Helium gas consists mainly of the isotope He^4 which has a resultant spin $s = 0$ and is a boson. The critical temperature of helium is about 5.2 K; above this

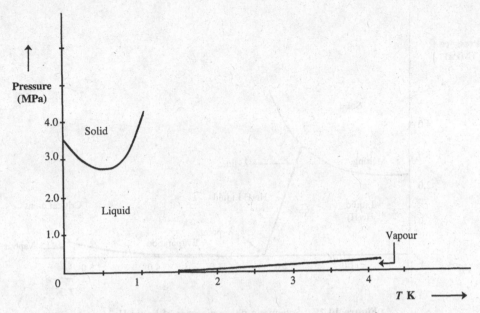

Figure 14.17   Schematic phase diagram of liquid He^3

temperature helium cannot exist as a liquid irrespective of the external pressure. He^4 was first liquefied by H. K. Onnes in 1908 (*Leiden Comm.*, 122b, 124c (1911)) at a temperature of 4.215 K and a pressure of 0.1 MPa. Actually, the interatomic (van der Waals) forces are too weak for liquid He^4 to solidify at 0.1 MPa pressure even at the lowest temperatures which have been attained. Unlike the classical liquids, liquid He^4 has two triple points. The phase diagram of He^4 is shown schematically in Figure 14.18. In naturally occurring helium, He^3 occupies only about 1.3–1.4 p.p.m. So the properties of liquid helium are dominated by those of He^4.

At low temperatures, He^4 has four phases. The solid phase cannot exist as $T \rightarrow 0$; it appears only for pressures above 2.5 MPa and the transition between the liquid and solid phases is first order. The liquid phase continues down to $T = 0$ K. However, as can be seen in the diagram, there are two liquid phases. The transition from liquid He(I) to liquid He(II) is quite interesting. As the normal liquid He(I) is cooled, a line of λ-*points* appears at about 2 K; actually, the exact temperature at which this line occurs depends on the pressure. The appearance of the line of λ-*points* is indicative of a continuous symmetry-breaking phase transition that occurs. The symmetry that is broken is gauge symmetry. The condensed phase of He^4 corresponds to a macroscopically occupied quantum state and can be described by a complex macroscopic *wave function* which is the order parameter of the system. In a gauge transformation, the phase of all wave functions in the system changes. There is a triple point at each end of the λ-line. Apart from its very low mass density, liquid He(I) is a fairly normal liquid.

The liquid phase below the λ-line is called liquid He(II); this condensed phase is a highly coherent macroscopic quantum state. It has some remarkable properties. For example, it has effectively zero viscosity. The apparently frictionless flow of liquid He(II) is analogous to the apparent frictionless flow of the condensed phase in superconductors. There are reasons to believe that the order parameter for the condensed phase in liquid He^4 is a macroscopic

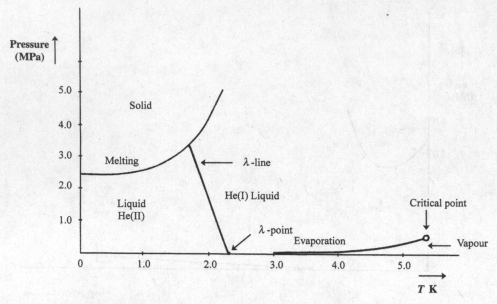

Figure 14.18 Schematic phase diagram of liquid He^4

wave function. The Ginzburg–Landau theory for the condensed phase of He^4 is very similar to that of the condensed phase in superconductors, with the exception that the particles in liquid He^4 are not charged.

According to the two-fluid theory of liquid He(II), it is assumed that, at temperatures below the λ-point, liquid He(II) consists of two fluids: a normal fluid which behaves like any other normal liquid with viscosity, and a superfluid which has no viscosity. It is this superfluid which imparts the superfluid properties of liquid He(II). It is assumed that the amount of the superfluid component is zero at the λ-point and increases to unity at the absolute zero of temperature.

We have noted that unlike liquid He(II), liquid He^3 does not have a λ-point and the same superfluid properties as liquid He(II). However, liquid He^3 does show superfluidity below 0.003 K which is attributed to the formation of Cooper-type pairs of He^3 atoms that behave as bosons. According to the ideal boson gas model, a considerable number of the He^4 atoms in liquid helium should be in the single particle ground state below the Bose temperature $T_b = 3.1$ K (see Exercise 15.25). These particles may constitute the superfluid component of liquid He(II), while the He^4 atoms in excited single particle states could make up the normal component of liquid He(II). In view of this explanation, it is reasonable to think that the superfluid properties of liquid He(II) could be due, at least partly, to a Bose–Einstein type condensation at the λ-point.

The phase diagram of He^4 is a good example of the third law of thermodynamics. The solid–liquid and vapour–liquid equilibrium curves approach the pressure axis with zero slope. This is, in fact, a consequence of the third law.

The observed heat capacity of He^4 is shown schematically in Figure 14.19. Although it does bear a vague resemlance to Figure 15.21, there are very significant differences between

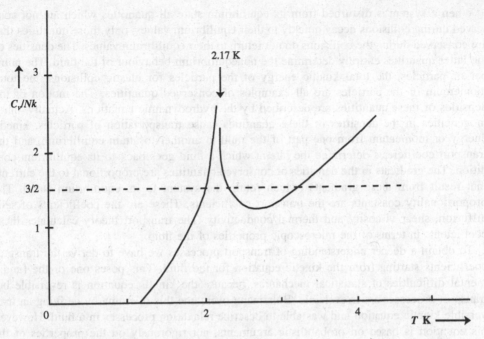

Figure 14.19   Schematic diagram of the variation of C_V of liquid He^4 with temperature

the properties of liquid He^4 and the ideal boson gas model. The heat capacity of liquid He^4 exhibits a more spectacular rise at the transition point than is produced by the perfect Bose–Einstein gas. However, the critical temperature obtained for the ideal boson gas, $T_b = 3.1$ K (see Exercise 15.25), is fairly close to the λ-point temperature, 2.17 K, of He^4. Actually, the term λ-point comes from the shape of Figure 14.19.

Another reason for the belief that the λ-point in liquid He^4 is related in some way to a Bose–Einstein condensation arises from the fact that He^3 does not show such a transition. The interatomic forces within liquid He^3 and He^4 should be quite similar. So the only significant difference between these two systems is that He^3 consists of Fermi–Dirac particles, whereas He^4 consists of Bose–Einstein particles. This difference is a strong indication that the λ-point in liquid He^4 is related to a Bose–Einstein type condensation.

14.11   KINETICS OF CHEMICAL REACTIONS

Kinetic theory is traditionally considered as one part of a trilogy, the other members being thermodynamics and statistical mechanics. In view of this, it is worthwhile to discuss, at least on a preliminary level, how kinetics is related to thermodynamics and statistical mechanics, bearing in mind that all three are primarily concerned with heat transfer as the state of a system changes in some manner determined by restrictions that are imposed on it. The discussion of chemical equilibrium presented in Section 13.12 lays the foundation for a study of chemical kinetics.

When a system is disturbed from its equilibrium state all quantities which are not conserved during collisions decay quickly to their equilibrium values; only those quantities that are conserved during the collisions do not return to their equilibrium values. The densities of the latter quantities entirely determine the nonequilibrium behaviour of the fluid. The number of particles, the total kinetic energy of the particles for elastic collisions, the total momentum of the particles are all examples of conserved quantities. The motion of the densities of these quantities are described by the hydrodynamic equations. Actually, inhomogeneities in the densities of these quantities cause transportation of particles, kinetic energy or momentum from one part of the fluid to another to attain equilibrium, and the transport coefficients determine the rate at which a fluid goes back to its equilibrium condition. The gradients in the densities of conserved quantities are proportional to the currents that result from those gradients (which return the system to its equilibrium state). The proportionality constants are the transport coefficients. These are the coefficients of self-diffusion, shear viscosity and thermal conductivity. The transport theory calculates these coefficients in terms of the microscopic properties of the fluid.

To obtain a deeper understanding of transport processes we have to derive the transport coefficients starting from the kinetic equation for the fluid. This poses one of the fundamental difficulties of statistical mechanics, because the kinetic equation is reversible but transport processes are irreversible. Boltzmann overcame this dilemma by deriving an irreversible kinetic equation and was able to describe relaxation processes in a fluid. However, his equation is based on probabilistic arguments, not rigorously on the properties of the dynamics of the system. We shall not discuss the Boltzmann equation or his H-theorem, as these are beyond the scope of this book.

His equation describes the time evolution of the distribution of particles in six-dimensional phase space for a dilute gas with inhomogeneities which are slowly varying in position space. If no external fields act on the system, it should return to equilibrium after a long time. Boltzmann's equation describes such behaviour. To show this, he introduced a function $H(t)$ defined by

$$H(t) = \int \int d\mathbf{q}_1 \, d\mathbf{p}_1 \, f(\mathbf{p}_1, \mathbf{q}_1, t) \, \ln\left[f(\mathbf{p}_1, \mathbf{q}_1, t)\right] \tag{14.104}$$

where $\mathbf{p}$ is the momentum, $\mathbf{q}$ is the position of the particle and t is the time. He then showed that, if $f(\mathbf{p}_1, \mathbf{q}_1, t)$ satisfies the Boltzmann equation, $H(t)$ always decreases because of the effect of collisions. From the H-theorem, he was able to obtain a microscopic expression for the entropy of a system near equilibrium which has the proper behaviour as the system approaches the equilibrium state. Since $H(t)$ always decreases with time, the negative of $H(t)$ will always increase with time. Boltzmann identified the following quantity as the nonequilibrium entropy:

$$S(t) = -k_B H(t) = -k_B \int \int d\mathbf{q}_1 \, d\mathbf{p}_1 \, f(\mathbf{p}_1, \mathbf{q}_1, t) \, \ln\left[f(\mathbf{p}_1, \mathbf{q}_1, t)\right]. \tag{14.105}$$

The difference between the Gibbs entropy and the Boltzmann entropy is that the former depends on the full distribution function, whereas the latter depends only on a reduced distribution function.

Having got some intuition about the Boltzmann equation and his H-theorem, we now discuss transport processes from a phenomenological point of view and obtain very simple expressions for the rate of reactions, using elementary kinetic theory. Let us start with a simple case where two atoms A and B react to form a diatomic molecule $A + B = AB$, and consider the rate at which the forward and reverse reactions take place in a state of equilibrium. If A collides with B, they will inevitably separate again, because the total energy is above the dissociation energy. Only the presence of a third body in the neighbourhood can remove some of the excess energy and prevent the separation of A and B. For our present discussion we shall not consider whether A and B actually stick together upon collision. We simply assume that upon collision there is an association, and a dissociation occurs whenever they separate. Although it is true that an association is almost always directly followed by a dissociation, this does not matter.

With this definition of dissociation, we shall find it easy to calculate the rate constant for dissociation (which is the number of dissociations per unit time divided by the number of pairs of AB, N_{AB}, when equilibrium is reached). We may imagine that A and B are attached to each other by a weightless and inextensible string so that they do not separate, but if they move away from each other they are pulled back together. This is equivalent to considering an inter-atomic potential energy curve for A and B which extends to infinity at some distance r_∞ (see Figure 14.20). We have drawn the potential energy curve with a maximum at a distance r_m. Although such a maximum is not common in the interaction of two atoms, it does occur in many more complicated reactions.

The pairs which have energy greater than ε_m and cross the distance r_m per unit time are outgoing pairs. Their number is equal to the equilibrium number in the energy level above ε_m (above the top broken line in Figure 14.20) multiplied by the specific rate at which they

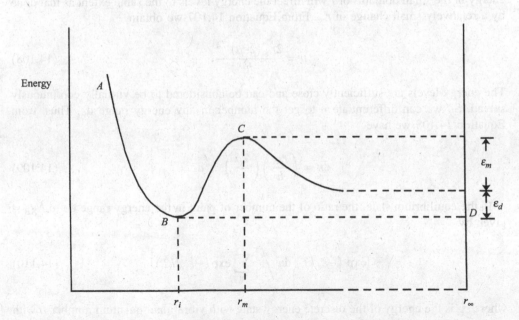

Figure 14.20 Schematic diagram of potential energy curve for collision between two atoms

move across the top of the curve (point C). This number clearly should not depend on details of the potential energy curve at very long distances. Therefore, it is quite acceptable to allow the potential energy to become infinite at an infinite distance r_∞. The rate constant for dissociation is given by

$$\begin{pmatrix} \text{rate constant} \\ \text{for dissociation} \end{pmatrix} = (\text{the rate at which pairs}$$

$$\text{with energy greater than}$$
$$\varepsilon_m \text{ cross the distance}$$
$$r_m \text{ going outward divided}$$
$$\text{by the number of}$$
$$\text{combined pairs } N_{AB}) \tag{14.106}$$

where ε_m is the energy of the maximum.

By allowing the potential energy to become infinite at the infinite distance r_∞, we allow the energy to have discrete levels even for those pairs which are dissociated. If we replace m by the reduced mass μ of A and B, and L_x by r_∞ in the equation for ε_x (see Equation 12.39), then we can express the energies of the levels for the region outside r_m approximately by the usual formula for a one-dimensional particle in a box as

$$\varepsilon = \frac{n^2 h^2}{8 \mu r_\infty^2}. \tag{14.107}$$

In this equation ε is the relative energy of the pair AB, measured from the asymptote of the potential energy curve which is taken as the zero energy. The irregularity of the potential energy in the small domains of r will affect the energy levels to the same extent as that done by a relatively small change in r_∞. From Equation 14.107 we obtain

$$n = \frac{2 r_\infty (2 \varepsilon \mu)^{1/2}}{h}. \tag{14.108}$$

The energy levels are sufficiently close and can be considered to be virtually continuously spread. So we can differentiate n to get the number in any energy range $d\varepsilon$. Thus, from Equation 14.108 we have

$$dn = \left(\frac{r_\infty}{h} \right) \left(\frac{2\mu}{\varepsilon} \right)^{1/2} d\varepsilon. \tag{14.109}$$

In the equilibrium state, the ratio of the number of pairs in the energy range $d\varepsilon$ to N_{AB} is given by

$$[\exp(-\varepsilon/kT) \, dn] \bigg/ \left[\sum_v \exp(-\varepsilon_v'/kT) \right] \tag{14.110}$$

where ε_v' is the energy of the discrete energy state with vibrational quantum number v with respect to the asymptote of the potential energy curve as the zero of energy. All the discrete levels correspond to the species AB, so the sum in expression 14.110 is taken over all the discrete levels. Now, $\varepsilon_d + \varepsilon_v' = \varepsilon_v$ where ε_v represents the energy of the state having quantum number v with respect to the minimum as the zero of energy. Then, expression 14.110 can be written as

$$[\exp\,(-\varepsilon/kT)\,dn]\bigg/\bigg[\exp\,(\varepsilon_d/kT)\sum_v \exp\,(-\varepsilon_v/kT)\bigg]$$

$$= [\exp\,(-\varepsilon/kT)\,dn]/[\exp\,(\varepsilon_d/kT)(\text{p.f.})_{\text{vib}}] \tag{14.111}$$

where

$$\sum_v \exp\,(-\varepsilon_v/kT) = (\text{p.f.})_{\text{vib}}$$

is the vibrational partition function. By substituting the value of dn from Equation 14.109 into Equation 14.111, we have

$$\bigg[\exp\,(-\varepsilon/kT)\bigg(\frac{r_\infty}{h}\bigg)\bigg(\frac{2\mu}{\varepsilon}\bigg)^{1/2}\,d\varepsilon\bigg]\bigg/[\exp\,(\varepsilon_d/kT)(\text{p.f.})_{\text{vib}}]. \tag{14.112}$$

$(\text{p.f.})_{\text{vib}}$ for a diatomic molecule such as AB can easily be calculated. If the temperature is not so high so that the vibrational energy of the molecule should become too high, then, to a good approximation, the vibrations of the molecule are harmonic and the allowed vibrational energy levels are given by

$$\varepsilon_v = \bigg(v + \frac{1}{2}\bigg)h\nu \tag{14.113}$$

where v can take the values $0, 1, 2, 3, \ldots$, and ν is the frequency of the oscillator. Then,

$$\begin{aligned}
(\text{p.f.})_{\text{vib}} &= \sum_{v=0}^{\infty} \exp\,(-\varepsilon_v/kT) \\
&= \sum_{v=0}^{\infty} \exp\,[-h\nu(v+1/2)/kT] \\
&= \exp\,\bigg(\frac{-h\nu}{2}/kT\bigg)\sum_{v=0}^{\infty}\exp\,\bigg(\frac{-vh\nu}{kT}\bigg).
\end{aligned} \tag{14.114}$$

The sum on the right of this equation is a geometric series and is equal to

$$[1 - \exp\,(-h\nu/kT)]^{-1} = \exp\,(h\nu/kT)\bigg/\bigg[\exp\,\bigg(\frac{h\nu}{kT}\bigg) - 1\bigg].$$

Hence,

$$(\text{p.f.})_{\text{vib}} = \bigg[\exp\,\bigg(\frac{-h\nu}{2kT}\bigg)\exp\,(h\nu/kT)\bigg]\bigg/[\exp\,(h\nu/kT) - 1] \tag{14.115}$$

If a pair AB has relative energy $\varepsilon > \varepsilon_m$ and moves towards or back from the infinite distance r_∞, it will have a velocity $(2\varepsilon/\mu)^{1/2}$ for most of its travel. The time needed to travel to r_∞ and back from it will be twice distance divided by the velocity, i.e. $2r_\infty/(2\varepsilon/\mu)^{1/2}$. It is easy to see that in this time the pair will travel the distance r_m once, moving away from each other. The reciprocal of this time gives the number of outward travel of the pair per unit time; this is the rate constant k_ε for the energy level with energy ε. The number of outward travel of pairs per unit time in the energy range between ε and $\varepsilon + d\varepsilon$ divided by N_{AB} is given by the product of k_ε and Equation 14.112. The number of dissociations per unit time divided by N_{AB}, i.e. the total rate constant k_d for dissociation, can then be found by integrating the product between the limits $\varepsilon = \varepsilon_m$ and $\varepsilon = \infty$. Thus,

$$k_D = \frac{\int_{\varepsilon_m}^{\infty} \exp\left(-\varepsilon/kT\right) k_\varepsilon \, dn}{\exp\left(\varepsilon_d/kT\right)(\text{p.f.})_{\text{vib}}}$$

$$= \frac{\int_{\varepsilon_m}^{\infty} \exp\left(-\varepsilon/kT\right) d\varepsilon}{h \exp\left(\varepsilon_d/kT\right)(\text{p.f.})_{\text{vib}}} \tag{14.116}$$

where we have used the definition of k_ε given above. Further simplification gives

$$k_D = \frac{kT}{h(\text{p.f.})_{\text{vib}}} \exp\left[-(\varepsilon_d + \varepsilon_m)/kT\right]. \tag{14.117}$$

Note that in this equation r_∞ does not appear. Actually it should not appear, as all dependence on r_∞ cancels. Let us see what happens at high and low temperatures. At high temperatures, the vibrational (p.f.) is given by

$$(\text{p.f.})_{\text{vib}} = kT/h\nu_0 \tag{14.118}$$

and at very low temperatures

$$(\text{p.f.})_{\text{vib}} = \exp\left(-h\nu_0/2kT\right) \tag{14.119}$$

where ν_0 is the vibrational frequency (it is a constant and has a simple form). Substituting Equations 14.118 and 14.119 into Equation 14.117, we have

$$k_D = \nu_0 \exp\left[-(\varepsilon_d + \varepsilon_m)/kT\right] \tag{14.120}$$

$$k_D = \left(\frac{kT}{h}\right) \exp\left[-(\varepsilon_d + \varepsilon_m - h\nu_0/2)/kT\right]. \tag{14.121}$$

The pre-exponential or frequency factor does not differ very much from one case to another. The factor kT/h varies from about 10^{12} to 10^{13} between room temperature and 1300 K. The frequencies of molecules are of the same order of magnitude as kT/h, i.e. around 10^{13}. This means neither of Equations 14.120 and 14.121 is exactly applicable, so the more general Equation 14.117 will have to be used. In general, 10^{13} is often considered to be the normal value for dissociations.

The above discussion has been based on the assumption that a collision is not usually followed by actual recombination of the atoms; however, a collision is usually immediately followed by a dissociation. We calculated only those number of dissociations which followed immediately after a collision. These processes are not really closely related to chemical reactions where we want to know the number of collisions that are followed by a sticking together of the reacting species, or we want to know the number of dissociations of reacting species which were joined together.

Upon collision, the atoms A and B will remain stuck together if their mutual energy can be reduced below the energy required for dissociation. This can be achieved through a *three-body collision* in the presence of another particle so that the excess energy is carried away by it. The situation is completely different in the reverse reaction. In this case, dissociation of the molecule will occur only if a third particle collides with AB to supply sufficient energy needed for dissociation. However, in the last collision, only a little amount of this energy is needed.

On the other hand, suppose that A and B are not monatomic and have a certain number of internal degrees of freedom. In such cases, the excess energy produced upon collision may be removed by the other degrees of freedom. Then the pair of molecules AB will remain stuck together until this energy is restored to the bond to break it. The bond-breaking energy may come through fluctuations in those degrees of freedom which removed the excess energy. However, before such fluctuations occur, if the pressure of the assembly is increased so that the pair AB collide with other molecules, then it is quite likely that the excess energy will be removed. Thus, the pair will remain stuck together as a single molecule. Each collision of A and B that produces a pair AB can be considered as a chemical reaction.

The dissociation of AB under high pressures is closely related to the combination of A with B (which could be radicals or polyatomic molecules) under similar conditions. The dissociation can only occur after sufficient energy has been imparted to the molecule AB after a series of favourable collisions with other molecules. This energy goes in the weak bond. However, if the pressure is high, there are possibilities of losing this energy before it goes in the weak bond. Reaction can take place only when this concentration of energy occurs. At the equilibrium state, the overall rate of association of A and B must equal that of dissociation of AB. Hence, we can write

$$A + B \rightleftarrows AB^* \tag{14.122a}$$

$$AB^* + C \rightleftarrows AB + C. \tag{14.122b}$$

In this mechanism, it is assumed that equilibrium is maintained in every step. C is another molecule which could be another AB, and AB^* is an intermediate molecule with sufficient energy to decompose.

In any real chemical reaction, there are no products at the start and no equilibrium either. So we modify the above reactions to write

$$A + B \rightarrow AB^* \tag{14.123a}$$

$$AB^* + C \rightarrow AB + C. \tag{14.123b}$$

In this scheme of reaction, very rarely the reverse of equation 14.123a will occur. If the pressure remains high, almost all of activated molecules AB^* that are formed will be deactivated through collisions according to Equation 14.123b. Therefore, the rate of appearance of AB^* will actually determine the rate of formation of AB, and we can write

$$d[AB]/dt = k_a[A][B] \tag{14.124}$$

where k_a is the association constant, $[AB]$ is the concentration of AB, and $[A] = N_A/V$ and $[B] = N_B/V$ are the concentrations of A and B, respectively, N_A and N_B are the numbers of A and B, and V is the total volume. We assume that the reaction in Equation 14.123a occurs at the same rate as the appearance of AB^* in Equation 14.122a, remembering that the reverse reaction does not occur in the second mechanism.

We have considered the association of A and B above. Let us now consider the dissociation of AB. In this case, at the beginning there are no A or B. So the reaction mechanism is

$$AB + C \rightleftarrows AB^* + C \tag{14.125a}$$

$$AB^* \rightarrow A + B. \tag{14.125b}$$

Here, we have assumed that the forward reaction in Equation 14.125a is just the reverse of reaction in Equation 14.123b and that to a good approximation equilibrium has been established in each case at high pressures of C, because usually AB^* (whatever way it is formed) is deactivated through collision. The reason for this is that at high pressures the rate of reaction in either direction in Equation 14.122b is much faster than that in Equation 14.123a or Equation 14.125b. It can also be assumed that the rate of reaction in Equation 14.125b is just the same as that of dissociation of AB, because AB^* is present in equilibrium numbers. Therefore, we can write

$$-\frac{d[AB]}{dt} = k_d^*[AB^*] = k_d[AB] \tag{14.126}$$

where k_d is the dissociation constant.

If AB is present at the same concentration in the two cases, then we may assume that the rate in Equation 14.125b is just the same as that of dissociation of AB^* in Equation 14.122a, remembering that in this occasion the reverse reaction does not take place. If it is true that reactions in Equations 14.123a and 14.125b take place at the same rate as the corresponding reactions at equilibrium, then for the equilibrium case

$$k_d[AB] = k_a[A][B] \tag{14.127}$$

or

$$\frac{k_a}{k_d} = \frac{[AB]}{[A][B]} = K \tag{14.128}$$

where k_a and k_d are the same constants as in Equations 14.124 and 14.126, and K is the equilibrium constant.

As we are calculating the equilibrium rate of association of A and B, and the association is not necessarily of permanent nature, we will actually be calculating the rate of collision with sufficient relative kinetic energy to cross over the peak at C (see Figure 14.20). The number of collisions per unit volume per unit time between A's and B's will be proportional to the number N_A/V of A's in unit volume and the number N_B/V of B's in unit volume. The rate of association (or of collision) per unit volume must be equal to the rate of dissociation, i.e.

$$k_d\left(\frac{N_{AB}}{V}\right) = k_a\left(\frac{N_A N_B}{V^2}\right). \tag{14.129}$$

Comparing this with Equation 13.184, we have

$$K = \frac{k_a}{k_d}. \tag{14.130}$$

Equation 14.117 was derived on the assumption that the molecule is in its lowest rotational state. We now consider the *effect of rotation on the rate of dissociation*. This effect can be taken care of by adding the rotational potential to the potential energy curve. The rotational potential of the rotational state that has quantum number j (where j can take the integral values $0, 1, 2, 3, \ldots$) is given by $[j(j+1)h^2]/8\pi^2\mu r^2$. The consequence of adding this potential to the potential energy curve (see Figure 14.20) will be to shift the positions of the minimum and maximum. Nevertheless, such a shift will be so small that we can neglect it. The actual main effect will be the increase of the minimum by an amount $[j(j+1)h^2]/8\pi^2\mu r_i^2$ and of the maximum by $[j(j+1)h^2]/8\pi^2\mu r_m^2$. Thus, $\varepsilon_d + \varepsilon_m$ will be changed by the difference between these quantities, and we can rewrite Equation 14.117 for this case as

$$k_{d,j} = [kT/h(\text{p.f.})_{\text{vib}}] \exp\left[-\{\varepsilon_d + \varepsilon_m + j(j+1)c^2(r_m^{-2} - r_i^{-2})\}/kT\right] \tag{14.131}$$

where $c^2 = h^2/8\pi^2\mu$.

The moment of inertia of the diatomic molecule is μr_i^2. There are actually $2j+1$ states with quantum number j. So the probability that the diatomic molecule will be in the rotational state j is taken as equal to

$$[(2j+1) \exp\{-j(j+1)c^2/r_i^2 kT\}]\bigg/\sum_j(2j+1) \exp\{-j(j+1)c^2/r_i^2 kT\}. \tag{14.132}$$

Although this procedure ignores symmetry effects, it is satisfactory for the simple reason that we are interested only in the classical limit. So, the average value of k_d for all rotational states can be found by multiplying Equation 14.131 by the probability 14.132. Hence,

$$k_d = \left[\frac{kT}{h(\text{p.f.})_{\text{vib}}}\right] \exp\{-(\varepsilon_d + \varepsilon_m)/kT\} \frac{\sum_j(2j+1) \exp[-j(j+1)c^2/r_m^2 kT]}{\sum_j(2j+1) \exp[-j(j+1)c^2/r_i^2 kT]}. \tag{14.133}$$

It is easily recognizable that the sum in the numerator is the rotational partition function, $(\text{p.f.})_{\text{rot}}$ with r_m, and the sum in the denominator is $(\text{p.f.})_{\text{rot}}$ with r_i. As said earlier, we are interested in the classical limit, where the ratio of these sums becomes r_m^2/r_i^2. Hence,

$$k_d = \left[\frac{kT}{h(\text{p.f.})_{\text{vib}}} \right] \exp \left[-(\varepsilon_d + \varepsilon_m)/kT \right] (r_m^2/r_i^2). \tag{14.134}$$

Generally, in the recombination of atoms, $\varepsilon_m = 0$, and then the potential energy has no maximum for $j = 0$. Although for all other j greater than zero there is a slight maximum due to the rotational potential, the position of this maximum actually depends on j. This leads to some ambiguity in r_m. However, if we assume that r_m is somewhere near that value of r for which the potential energy curve for $j = 0$ tends to its asymptotic value within about kT (more precisely $kT/3$), then r_m may be reasonably well approximated.

14.12   PRACTICAL APPLICATIONS OF REACTION RATE THEORY

In this section we intend to discuss a more specific area of kinetics and thermodynamics. A mathematical model is developed which enables calculations of kinetic parameters and thermodynamic properties in diffusion-controlled solid-state reactions such as crystallization of many inorganic substances from melts. Another model is introduced for the determination of the exact distance of separation between a diffusing particle and its host in the melt immediately prior to a successful diffusion encounter.

The rate and the rate constant (k) of a reaction generally increases with increasing temperature. This dependence of k on temperature can be represented fairly accurately by the empirical Arrhenius equation

$$k = k_0 \exp \left(-E/RT \right) \tag{14.135}$$

where E is the activation energy, R is the universal gas constant and k_0 is a pre-exponential factor. The method of determination of kinetic parameters in chemical reactions using this equation is well established. Such data are available in the literature for many chemical reactions where the products are formed by chemical bonding of the reacting molecules.

On the other hand, the reaction kinetics in a solid-state reaction depends on many factors, e.g. the temperature of the reaction, reaction order, reaction mechanism, activation energy, etc. The correct determination of the actual mechanism is an important step in the study of kinetics of solid-state reactions, because the activation energy and the pre-exponential factor in Equation 14.135 depend on the rate-controlling process.

If the nature of the rate-controlling process and the relevant rate constant of a solid-state reaction are correctly determined, then Equation 14.135 can be modified by replacing k with the appropriate rate constant of that reaction to give a new equation. The author has developed a unique mathematical model (B. N. Roy, *J. Am. Ceram. Soc.* **63** (1980) 10) for diffusion-controlled solid-state reactions and successfully used it to estimate the energy (E), the enthalpy (ΔH_a), the entropy (ΔS_a), and the free energy (ΔG_a) of activation, and also the pre-exponential factor (k_0) and the distance of separation (d_{12}) between a diffusing particle and its host crystal for a successful diffusion for diffusion-controlled crystal growth of alkaline-earth metal salts from solutions in metal chloride melts.

If a solid-state reaction is diffusion rate-controlled and k_D is its rate constant, then to use Equation 14.135 in this case we modify it to write

$$k_D = k_0 \exp(-E/RT) \tag{14.136}$$

where E is the activation energy and T is the temperature of the reaction. As mentioned earlier, the reaction kinetics in a solid-state reaction depends on the temperature. So it is possible to obtain several values of k_D at different reaction temperatures for the same reaction. Then Equation 14.136 can be used to estimate E and k_0 from a graph of $\ln k_D$ vs. $1/T$. Such a graph will be a straight line. E and k_0 can then be estimated from the slope and intercept of the line, respectively.

Equation 14.136 postulates that a molecule must attain an activated state by acquiring an energy E in excess of its normal energy before it can diffuse into its host. The simplest interpretation of the factor k_0 is that it is equal to the number of diffusion encounters (Z) between a diffusing particle and its host. Generally, Z is of the order 10^{11}, but it could be smaller than this by as much as 10^{10}. Therefore, it would appear that a discrepancy may occur. This discrepancy is dealt with in the collision theory by taking $k_0 = \mathscr{P}Z$, where $\mathscr{P}$ is a probability factor. This factor may be less than 1 and can be interpreted in various ways. Equation 14.136 then becomes

$$k_D = (\mathscr{P}Z) \exp(-E/RT). \tag{14.137}$$

To understand the idea of activation as a prerequisite, let us consider the transition state theory. According to this theory, in a chemical reaction there is a state, known as the transition state (say X), which represents the top of the energy barrier, i.e. the most activated state. A reacting system composed of one, two or more molecules must pass across this state in the course of a reaction. This state may be represented by a simple equation

$$\text{Reactants} \underset{k_2}{\overset{k_1}{\rightleftharpoons}} X \xrightarrow{k_3} \text{Products} \tag{14.138}$$

where k_1 and k_2 are the rate constants, respectively, for the forward and reverse reaction between the reactants and X, and k_3 is the rate constant for conversion of X into the products. The ascent of the energy barrier at X is treated as a thermodynamic equilibrium. This is assigned an equilibrium constant (K_e) defined by the ratio k_1/k_2. This equilibrium constant depends on the temperature according to

$$RT \log K_e = \Delta G_a = \Delta H_a - T\Delta S_a \tag{14.139}$$

where ΔG_a, ΔH_a and ΔS_a are as defined above. The overall rate constant (k) of a reaction will then be given by

$$k = K_e k_3. \tag{14.140}$$

It can be shown by statistical mechanics that k_3 has the universal value given by RT/Nh (see Equation 14.121), where N is the Avogadro number and h is Planck's constant ($R = Nk$, R is the universal gas constant and k is the Boltzmann constant). Substituting these into Equation 14.139, we have the overall rate constant

$$k = \exp\,(\Delta S_a/R)(RT/Nh)\,\exp\,(-\Delta H_a/RT). \tag{14.141}$$

If we now replace k by the rate constant, k_D, for diffusion rate-controlled solid-state reactions, we obtain

$$k_D = \exp\,(\Delta S_a/R)(RT/Nh)\,\exp\,(-\Delta H_a/RT). \tag{14.142}$$

The activation energy ΔH_a in this equation is related to the experimental activation energy E by the simple relationship

$$E = RT + \Delta H_a. \tag{14.143}$$

By comparing Equations 14.137 and 14.142, we have

$$Z = RT/Nh \tag{14.144}$$

$$\mathscr{P} = \exp\,(\Delta S_a/R). \tag{14.145}$$

Hence, if k_D and E are known, it is possible to calculate ΔH_a, ΔS_a, ΔG_a and $\mathscr{P}$ for a diffusion-controlled solid-state reaction using Equations 14.137 and Equations 14.142–14.145. Many experimental values for these kinetic and thermodynamic parameters for diffusion-controlled crystal growth of alkaline-earth metal tungstate from solutions in lithium chloride and sodium tungstate melts have been determined by the author (*Crystal Growth From Melts*; B. N. Roy; John Wiley & Sons, UK (1991); Ch. 14).

Goddard (*Phys. Rev. Lett.* **49** (1983) 1847) has introduced an equation identical to Equation 14.142 to express the rate of thermal desorption of atoms and molecules from a metal surface. His equation is

$$R = (\Omega_0/2\pi)f(T)\,\exp\,(-D_e/kT). \tag{14.146}$$

In this equation, R (the rate of desorption) corresponds to k_D in Equation 14.142, $\Omega_0/2\pi$ corresponds to $Z\,(= RT/Nh$, the number of diffusion encounters between a diffusing particle and its host), and $f(T)$, which has been defined by Goddard as the 'frustrated rotational motion' necessary for the successful desorption of a molecule from a metal surface, has exactly the same meaning as $\exp\,(\Delta S_a/R)$, which is a probability factor determined by the steric requirement of the diffusing particle in diffusion-controlled reactions, e.g. in diffusion-controlled crystal growth. The diffusion process of an ion into a solid phase in a melt at high temperatures could be considered as the reverse of the process of thermal desorption of an atom or molecule from a metal surface (which is exactly the case where Goddard used his equation). Therefore, the Goddard equation, i.e. Equation 14.146, cannot be said to be new and original, as reported by Baum (*Chem. Eng. News* February, 21 (1983) p.24). Goddard acknowledged that his equation and Equation 14.142 are identical in all respects but are applied to different circumstances.

In a diffusion-controlled solid-state reaction, when a diffusing particle and its host come near to each other within a critical distance, say d_{12}, it is assumed that they stick together and diffusion occurs whereby the particle penetrates the host, allowing it to grow. However, it must be remembered that strongly repulsive 'overlap' forces exist which prevent all except the most energetic particles from approaching to within d_{12}. Therefore, there exists a potential energy barrier. This must be surmounted before any diffusion can occur. The height of this potential energy barrier, which is equal to the minimum energy required for diffusion, is taken as the activation energy. From the kinetic theory it can be shown that

$$k_D = 2(d_{12})^2 (2\pi RT/M^*) \ \exp \ (-E/RT) \tag{14.147}$$

where M^* is the reduced mass given by

$$M^* = \frac{M_1 M_2}{M_1 + M_2}$$

and M_1 and M_2 are the molecular weights of the reacting partners. So, if the diffusion rate-constant k_D and the activation energy E are known, then the critical distance of separation d_{12} between a diffusing particle and its host in melts immediately before a successful diffusion encounter can be evaluated from Equation 14.147. Such values are generally of the order of molecular dimensions. However, higher values may be obtained. Experimental values of d_{12} for diffusion-controlled crystal growth of alkaline-earth metal tungstates from solutions in lithium chloride and sodium tungstate melts have been determined by the author (*Crystal Growth From Melts*; B. N. Roy; John Wiley & Sons, UK (1991); Ch. 14).

EXERCISES 14

14.1 Define the terms phase, component and degree of freedom. State the number of components present in the systems: $H_2(g) + O_2(g)$, and NH_3.

14.2 State the number of degrees of freedom that each of the following systems possesses.

(a) $KCl(s)$ in equilibrium with its saturated solution at standard temperature and pressure

(b) $HCl(g)$ and $NH_3(g)$ in equilibrium with $NH_4Cl(s)$ when the equilibrium is attained starting with the two gases.

14.3 A one-component system exhibits gas and liquid phases and two solid modifications. Determine the numbers of one-, two-, three- and four-phase equilibria possible in this system.

14.4 A system in equilibrium consists of two solid components existing in three phases. Predict the number of degrees of freedom of this system.

14.5 Show that, for a system in equilibrium consisting of P phases and C components, the number F of degrees of freedom is given by $F = C - P + 2$.

14.6   Using the following data for the vapour pressure of water, determine graphically the heat of vaporization of water.

T (°C)	20	30	40	50	60	75
P (Torr)	17.54	31.83	55.32	92.51	149.4	289.1

14.7   The boiling point of iodine is 183.0°C and its vapour pressure at 116.5°C is 100 Torr. Calculate the temperature and pressure at the triple point, if the heat of fusion is $15.648 \text{ kJ mol}^{-1}$ and the vapour pressure of the solid is 1 Torr at 38.7°C.

14.8   Liquid water under an air pressure of 1 atm at 298.15 K has a larger vapour pressure than in the absence of air pressure. Calculate the increase in vapour pressure produced by the pressure of the atmosphere on the water, given that the density of water is 1 g cm^{-1} and the vapour pressure (in the absence of air pressure) is 23.76 Torr.

14.9   A substance melts at 125°C and has the heat of fusion 6276 J mol^{-1}. Calculate the change in volume on melting when the melting point is increased by 0.47°C by changing the applied pressure from 1 to 175 atm.

14.10   Prove that, at a phase transition, the Gibbs free energy of each phase must be equal.

14.11   In Figure 14.16a, W represents a system with a temperature $T < T_c$, pressure P_0 and total molar volume v_W. This system is in a state in which the liquid and vapour phases coexist. Prove that the ratio of the mole fractions of liquid to vapour at the point W is equal to the inverse ratio of the distance between the points W and X.

14.12   Find an expression for the heat capacity at constant volume per mole inside the coexistence curve given in Figure 14.16a. How does the heat capacity at constant pressure differ from that at constant volume in the region below the coexistence curve?

14.13   n-hexane-nitrobenzene system at 0.1 MPa pressure behaves like a regular binary mixture. Sketch the phase diagram of the system under this condition. Label all the phases and identify the coexistence curve and the critical temperature.

14.14   Sketch the phase diagram of liquid He^4 and explain it. How does it compare with the phase diagram of liquid He^3? Give a reasonable explanation for the superfluid property of liquid He(II).

15

Distribution Functions and Thermodynamics of Fermi–Dirac and Bose–Einstein Gases

15.1 INTRODUCTION

In Chapters 12 and 13 we have developed an understanding of the fundamental principles of statistical mechanics and their relationships to thermodynamics. This knowledge can be used to predict the physical properties of various systems, even the probability that the system is in a particular microstate. In principle, all we need to do is to obtain an expression for the energy of the system and evaluate the partition function. Once this is done, we can predict the distribution function for that system and all its thermodynamic properties. Therefore, it might appear that the applications of statistical mechanics should be a relatively simple matter, viz. the evaluation of the partition function for the system in question. However, in reality it is not simple. We are usually led to the problem of selecting a *model* for our physical system that has a sufficiently simple energy to make the evaluation of Z possible. Sometimes we may even have two possible models for one system. By examining which of these models are realistic, we can learn to identify which microscopic (molecular) properties are responsible for the various thermodynamic (macroscopic) properties of a system. Thus, one can acquire an insight into the molecular origin of the physical properties of various systems.

In this chapter we wish to discuss several distribution functions and apply these to consider behaviours of ideal gases. We shall also discuss several statistical models related to the heat capacity of solids and compare these with classical models. An elaborate discussion of the heat capacity of solids and gases is included in Chapter 16.

15.2 MAXWELLIAN DISTRIBUTION FUNCTIONS

In this section we shall consider the statistical descriptions of the translational motion of molecules in a perfect gas. The material discussed here will apply equally well to systems in which quantum statistic must be used as to those systems which can be adequately described by classical or prequantum mechanical statistics. The classical statistics are those which assume the canonical distribution of Boltzmann for the energy levels of the system. The material discussed below will be applicable to any system of molecules which is devoid of interaction energy between the molecules.

We assume that the energy of any molecule is composed of the sum of its kinetic energy of translation, vibration and rotation, and of its potential energy corresponding, for example, to its internal parameters as well as any potential energy arising from bodies external to the collection of molecules, such as from the effects of gravitational fields of the orientational potential energy due to a magnetic field. However, in all cases we exclude any mutual interaction energy. Under these conditions, the total internal energy of the system is the sum of the energies of the molecules forming the system without interaction of its constituents.

We assume that each molecule may be found in either a definite number (as in the quantum theory) or in an indefinite number (forming a continuous series) of states. We shall classify these states according to the energy associated with the states and they will be designated as energy levels. It will often happen that for a given energy there may correspond several different states, say m. In such cases, the energy level is said to be m-fold degenerate.

We also assume that we have a very large number N of these molecules, where N is of the order of Avogadro's number, and that there are many different ways of distributing these molecules among the given energy levels. Each of these possible distributions of the molecules corresponds to one macroscopic state. We assume that the most probable macroscopic state of the system is that one which is observed. The task of estimating the observed macroscopic state is then equivalent to counting the total number of corresponding distributions. To count the number of distributions accurately we must know, for example, whether the molecules are distinguishable or not, whether an arbitrary number of molecules can be placed into each state, or if the number is fixed. To each of the possible different rules there will correspond a different set of statistics, and we require experiments to show us which rules are valid and under what conditions.

Since we shall consider the statistical descriptions of the translational motion of molecules in a perfect gas, we can use classical mechanics to describe the translational motion of the molecules. This is justifiable, except in the cases of extremely low temperatures or very high densities. Perfect gases are usually not degenerate, except when the temperature is very low or the density is very large. Furthermore, we can ignore the structure of the molecules because we are only concerned at present with the translational motion of the molecules and not their rotational or vibrational motion. Thus, we can effectively depict the molecule as a monatomic molecule with a total mass m and a velocity of the centre of mass $\mathbf{v}$.

Consider an ideal gas inside a container of volume V and in equilibrium at the absolute temperature T. It is possible that this gas may contain several different types of molecule, and we assume that this gas is under such conditions (i.e. if the gas is sufficiently dilute so that $n = N/V$ is small, if T is sufficiently high and if the mass m of a molecule is not too small) that a classical treatment of its molecules is permissible. Let us consider one of the molecules of the gas, which is taken as a distinct small system in thermal contact with other molecules which form a heat reservoir at T. Under these conditions, the canonical distribution is applicable, and for simplicity we assume that the molecule under consideration is monatomic. We also assume that the gas is dilute enough to be ideal; therefore, any potential energy of interaction with other molecules can be taken to be negligible. If we neglect any external force (gravitational, etc.), the energy ε of this molecule is then simply its kinetic energy

$$\varepsilon = \frac{1}{2}m\mathbf{v}^2 = \frac{\mathbf{p}^2}{2m} \tag{15.1}$$

where $\mathbf{p} = m\mathbf{v}$ is the momentum of the molecule. Thus, the energy of the molecule at any place within the container is independent of the position vector ($\mathbf{r}$) of the molecule.

In terms of classical consideration, the position of the molecule can be described by its three position co-ordinates x, y, z, and the corresponding three momentum components p_x, p_y and p_z. Before we proceed any further we introduce two abbreviations:

$$d^3\mathbf{r} \equiv dxdydz$$
$$d^3\mathbf{p} \equiv dp_xdp_ydp_z \tag{15.2}$$

for an element of volume of real space and an element of volume of momentum space.

We wish to consider the probability that the position of the molecule lies in the range between r and $r + dr$ and that at the same time its momentum lies between p and $p + dp$. In fact, these ranges mean that its x co-ordinate lies between x and $x + dx$, the y co-ordinate lies between y and $y + dy$, and the z co-ordinate lies between z and $z + dz$; also, its x component of momentum lies between p_x and $p_x + dp_x$, the y component of momentum lies between p_y and $p_y + dp_y$, and the z component of momentum lies between p_z and $p_z + dp_z$. These ranges of position and momentum variables correspond to a "volume" of phase space of magnitude $dxdydzdp_xdp_ydp_z \equiv d^3\mathbf{r}d^3\mathbf{p}$.

The statistical description of a system in terms of classical mechanics is completely analogous to that in quantum mechanics. The difference is one of interpretation; whereas the microstate of a system refers to a particular quantum state of a system in the quantum theory, it refers to a particular cell of phase space in the classical theory. This makes it apparent that any general argument based upon the statistical postulates and the counting of states must remain equally valid in the classical description. In particular, it follows that the derivation of the canonical distribution remains applicable.

Let us assume that a complex system X (composed of many particles moving in all directions) can be described by some set of f co-ordinates $q_1, q_2, \ldots, q_f$ and f corresponding momenta $p_1, p_2, \ldots, p_f$, i.e. by a total of $2f$ independent co-ordinates ($2f$ degrees of freedom). To deal with these continuous variables in a manner where the possible states of the system can be counted, it is convenient to subdivide the possible values of the ith co-ordinate q_i into fixed small intervals of magnitude ∂q_i, and also the possible values of the ith momentum p_i into fixed small intervals of magnitude ∂p_i. Then, for each i the size of the subdivision interval can be chosen so that

$$\partial q_i \partial p_i = a \tag{15.3}$$

where a is some arbitrarily small constant of fixed magnitude independent of i.

We can now specify the state of the system by saying that its co-ordinates and momenta are such that the set of values ($q_1, q_2, \ldots, q_f; p_1, p_2, \ldots, p_f$) lies in a particular set of intervals. Geometrically, this set of values can be considered as a 'point' in a *phase space* of $2f$

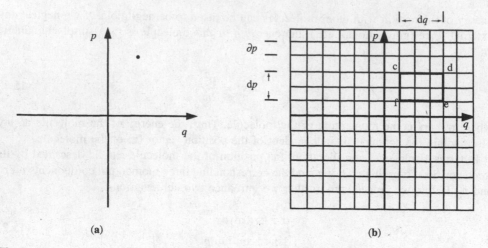

Figure 15.1   (a) Classical phase space for a single particle in one dimension. (b) A two-dimensional phase space subdivided into small cells of equal volume $\partial q \partial p = a$. The region *cdef* represents an element of volume with a size $dq dp$ and containing many cells of volume $\partial q \partial p$.

dimensions where each cartesian axis is labelled by one of the co-ordinates or momenta. This phase space is completely analoguous to a two-dimensional phase space, as shown in Figure 15.1*b*. The subdivision into intervals divides the space into equal small cells of volumes $(\partial q_1 \partial p_1), (\partial q_2 \partial p_2), \ldots, (\partial q_f \partial p_f)$, i.e.

$$(\partial q_1 \partial q_2 \cdots \partial q_f \ \partial p_1 \partial p_2 \cdots \partial p_f) = (a)^f \qquad (15.4)$$

The state of the system can then be described by specifying in which cell in phase space, i.e. in which particular set of intervals the co-ordinates $q_1, q_2, \ldots, q_f$ and momenta $p_1, p_2, \ldots, p_f$ of the system actually lie. We can label each cell in phase space or set of intervals by some index, say r, so that all these cells can be numbered as $r = 1, 2, 3, \ldots$ and can be listed.

If the complex system X, described classically, is in thermal equilibrium with a heat reservoir at the absolute temperature $T = 1/k\beta$ and it has an energy E_r in a particular state r, then the probability $\mathcal{P}_r$ of finding this system in the particular state r is given by

$$\mathcal{P}_r \propto \exp{(-\beta E_r)}. \qquad (15.5)$$

The state r refers to a particular cell of phase space where the co-ordinates and momenta of the system X have particular values $(q_1, q_2, \ldots, q_f; p_1, p_2, \ldots, p_f)$, and the energy E_r of the system is the energy E of X when its co-ordinates and momenta have these values, i.e.

$$E_r = E(q_1, q_2, \ldots, q_f; \ p_1, p_2, \ldots, p_f). \qquad (15.6)$$

This is because the energy of X is a function of its co-ordinates and momenta.

For convenience, let us express the canonical distribution, Equation 15.5, in terms of a probability density. The mathematical expression for the probability that the system X in contact with the heat reservoir is found to have its first co-ordinate in the range between q_1 and $q_1 + dq_1$, the second co-ordinate in the range between q_2 and $q_2 + dq_2, \ldots$, its fth co-ordinate in the range between q_f and $q_f + dq_f$, its first momentum in the range of p_1 and $p_1 + dp_1$, the second momentum in the range between p_2 and $p_2 + dp_2, \ldots$, and its fth momentum between the range p_f and $p_f + dp_f$, is

$$\mathscr{P}(q_1, q_2, \ldots, q_f; \ p_1, p_2, \ldots, p_f) dq_1 dq_2 \ldots dq_f dp_1 dp_2 \ldots dp_f. \tag{15.7}$$

In this expression there are two points to note: (i) the co-ordinate range dq_i and the momentum range dp_i are assumed to be small to the extent that the energy E of the system X does not change appreciably when q_i changes by an amount dq_i or p_i changes by an amount dp_i, and, (ii) both dq_i and dp_i are taken to be large compared to the size of the subdivision of phase space, i.e. $\partial q_i \ll dq_i$ and $\partial p_i \ll dp_i$. From Figure 15.1b it is apparent that the element of volume $(dq_1 dq_2 \cdots dq_f dp_1 dp_2 \cdots dp_f)$ of phase space contains many cells of volume $(\partial q_1 \partial q_2 \cdots \partial q_f \partial p_1 \partial p_2 \cdots \partial p_f) = (a)^f$ each. It is clear that the energy of the system X in each of these cells is nearly the same, so its probability given by Expression 15.5. Therefore, the probability of Expression 15.7 can be found by multiplying the probability of Expression 15.5 (which is the probability of finding X in a given cell of phase space) by the total number of cells, i.e. $(dq_1 dq_2 \cdots dq_f dp_1 dp_2 \cdots dp_f)/(a)^f$. Thus, we have

$$\mathscr{P}(q_1 q_2 \cdots q_f; \ p_1 p_2 \cdots p_f) dq_1 dq_2 \cdots dq_f dp_1 dp_2 \cdots dp_f$$
$$\propto \exp{(-\beta E_r)(dq_1 dq_2 \cdots dq_f dp_1 dp_2 \cdots dp_f)/(a)^f}. \tag{15.8}$$

By substituting the expression for E_r from Equation 15.6, we have

$$\mathscr{P}(q_1 q_2 \cdots q_f; \ p_1 p_2 \cdots p_f) dq_1 dq_2 \cdots dq_f dp_1 dp_2 \cdots dp_f$$
$$C_1 [\exp{\{-\beta E(q_1 q_2 \cdots q_f; \ p_1 p_2 \cdots p_f)\}}] dq_1 dq_2 \cdots dq_f dp_1 dp_2 \cdots dp_f \tag{15.9}$$

where C_1 is a constant of proportionality and it includes $(a)^f$ as well. Later, we shall see how to determine C_1.

We now go back to our monatomic gas in the container of volume V and in equilibrium at the absolute temperature T. Using the canonical distribution derived in Equation 15.9, we can now find the desired probability $\mathscr{P}(\mathbf{r}, \mathbf{p}) \, d^3r d^3p$ that the molecule under consideration has a position between $\mathbf{r}$ and $\mathbf{r} + d\mathbf{r}$ and a momentum between $\mathbf{p}$ and $\mathbf{p} + d\mathbf{p}$. Thus,

$$\mathscr{P}(\mathbf{r}, \mathbf{p}) \, d^3r d^3p \propto [\exp{\{-\beta(\mathbf{p}^2/2m)\}}] \, d^3r d^3p \tag{15.10}$$

Here, $\beta = 1/kT$ and we have used energy of the molecule (Equation 15.1). By using the velocity of the molecule ($\mathbf{v} = \mathbf{p}/m$), the probability $\mathscr{P}_1(\mathbf{r}, \mathbf{v}) d^3r d^3v$ that the molecule has a position between $\mathbf{r}$ and $\mathbf{r} + d\mathbf{r}$ and a velocity between $\mathbf{v}$ and $\mathbf{v} + d\mathbf{v}$ can be obtained. Thus,

$$\mathscr{P}_1(\mathbf{r}, \mathbf{v}) \, d^3r d^3v \propto \left[\exp\left\{ \frac{-\beta m \mathbf{v}^2}{2} \right\} \right] d^3r d^3v \tag{15.11}$$

where $d^3\mathbf{v}$ is $dv_x dv_y dv_z$.

The probabilities 15.10 and 15.11 are both very general results. The probability 15.10 provides information about the position and momentum of any molecule in the gas, while the probability 15.11 provides information about the position and velocity of any molecule in the gas. If the gas consists of a mixture of different types of molecule of different masses, Equation 15.11 can help to determine how many molecules of a given type have a velocity in any given velocity range.

Let us concentrate on the molecules of a particular type and try to find out the value of $\eta(\mathbf{v})\mathrm{d}^3\mathbf{v}$ which is the mean number of molecules of the particular type, per unit volume, which have a velocity between $\mathbf{v}$ and $\mathbf{v}+\mathrm{d}\mathbf{v}$. There are other results which we can obtain from Equation 15.11. For example, if we have a mixture of gases of different types of molecules with different masses, we may find out the number of molecules of a given type that have a velocity in any specified range. In addition, if we have a gas of only one kind of molecule, we may find how many molecules have a velocity in a given range. The gas constitutes a statistical ensemble of molecules because the N molecules of the ideal gas move independently of one another without appreciable mutual interaction. A fraction of these molecules given by the probability 15.11 have a position $\mathbf{r}$ and $\mathbf{r}+\mathrm{d}\mathbf{r}$ and a velocity $\mathbf{v}$ and $\mathbf{v}+\mathrm{d}\mathbf{v}$. Therefore, the mean number $\eta(\mathbf{v})\,\mathrm{d}^3\mathbf{v}$ can be found by multiplying the probability 15.11 by the total number N of molecules of this kind and then dividing by $\mathrm{d}^3\mathbf{r}$ (the volume element). Thus, we have

$$\eta(\mathbf{v})\,\mathrm{d}^3\mathbf{v} = \frac{N\mathscr{P}_1(\mathbf{r},\mathbf{v})\,\mathrm{d}^3\mathbf{r}\mathrm{d}^3\mathbf{v}}{\mathrm{d}^3\mathbf{r}}$$

$$= C\left[\exp\left(\frac{-\beta m\mathbf{v}^2}{2}\right)\right]\mathrm{d}^3\mathbf{v} \tag{15.12}$$

where C is a constant of proportionality. Equation 15.12 is known as the *Maxwell velocity distribution* function; it was derived by James Clerk Maxwell in 1857.

An interesting point to note is that the probability $\mathscr{P}_1$ of (15.11) and the mean number $\eta(\mathbf{v})\,\mathrm{d}^3\mathbf{v}$ of Equation 15.12 do not depend on the position $\mathbf{r}$ of the molecule. This is true by symmetry considerations; where there is no external force a molecule cannot have a preferred position in space. Another point to note is that both $\mathscr{P}_1$ and $\eta(\mathbf{v})$ depend only on the magnitude of the velocity and not on its direction, that is

$$\eta(\mathbf{v}) = \eta(\mathbf{v}) \tag{15.13}$$

where $|\mathbf{v}| = v$. This is also true by virtue of symmetry considerations, because there can be no preferred direction where the container of the gas, and hence the centre of mass of the entire gas, is taken to be at rest.

We mentioned the symmetry consideration when we said that $\mathscr{P}_1$ and η are independent of $\mathbf{r}$ but depend only on the magnitude of the velocity and not on its direction. This requires a little explanation. If there is no external force field acting on the gas, the molecules have an equal chance of being found in the region $\mathrm{d}^3\mathbf{r}$ of any point within the system. This makes sense, because we do not expect to find a density gradient in this system when it is in thermal equilibrium. Therefore, the spatial variable is really superfluous, and we can consider the probable number of molecules in the *whole system* that have velocities in the range d^3v of $\mathbf{v}$. This quantity is $\eta(\mathbf{v})\,\mathrm{d}^3v$.

We have introduced a constant of proportionality in Equation 15.12. Before we proceed further, let us determine the value of the constant C. After that, we shall discuss some of the features of the Maxwell velocity distribution functions and note what it means.

Determination of the constants C_1 and C: The value of the proportionality constant C_1 in Equation 15.9 is determined by the normalization requirement that the integral (i.e. the sum) of the probability 15.9 over all accessible co-ordinates and momenta of the system X should be 1, i.e.

$$\int \mathscr{P}(q_1 q_2 \cdots q_f; \, p_1 p_2 \cdots p_f) \mathrm{d}q_1 \mathrm{d}q_2 \cdots \mathrm{d}p_1 \mathrm{d}p_2 \cdots \mathrm{d}p_f = 1. \tag{15.14}$$

Here, the integral extends over the whole region of the phase space accessible to X. So, we can write

$$C_1 \int \exp \{-\beta E(q_1 q_2 \cdots q_f; \, p_1 p_2 \cdots p_f)\} \, \mathrm{d}q_1 \mathrm{d}q_2 \cdots \mathrm{d}q_f \mathrm{d}p_1 \mathrm{d}p_2 \cdots \mathrm{d}p_f = 1$$

or

$$\frac{1}{C_1} = \int \exp \{-\beta E(q_1 q_2 \cdots q_f; \, p_1 p_2 \cdots p_f)\} \, \mathrm{d}q_1 \mathrm{d}q_2 \cdots \mathrm{d}q_f \mathrm{d}p_1 \mathrm{d}p_2 \cdots \mathrm{d}p_f. \tag{15.15}$$

The constant C is determined by the requirement that the integral (i.e. the sum) of Equation 15.12 over all possible velocities must give the total mean number $n(=N/V)$ of molecules of the type considered per unit volume. Then,

$$C \int \exp \left(\frac{-\beta m v^2}{2} \right) \mathrm{d}^3 \mathbf{v} = n. \tag{15.16}$$

Considering all the components v_x, v_y and v_z of $\mathbf{v}$, we write

$$C \iiint \exp \left\{ \frac{-\beta m(v_x^2 + v_y^2 + v_z^2)}{2} \right\} \mathrm{d}v_x \mathrm{d}v_y \mathrm{d}v_z = n$$

or

$$C \int_{-\infty}^{\infty} \exp \left(\frac{-\beta m v_x^2}{2} \right) \mathrm{d}v_x \int_{-\infty}^{\infty} \exp \left(\frac{-\beta m v_y^2}{2} \right) \mathrm{d}v_y \int_{-\infty}^{\infty} \exp \left(\frac{-\beta m v_z^2}{2} \right) \mathrm{d}v_z = n.$$

Strictly speaking, the upper limit to the velocity of the monatomic gas molecule is the speed of light. However, due to the exponential term, no significant error is introduced by taking the upper limit as infinity. Each of the above integrals has the same value. [See Appendix 5 for an evaluation of integrals of the form

$$\int_{-\infty}^{\infty} \exp(-ax^2) x^n \, \mathrm{d}x]$$

$$\int_{-\infty}^{\infty} \exp \left(\frac{-\beta m v_x^2}{2} \right) \mathrm{d}v_x = \left(\frac{\pi}{\beta m/2} \right)^{1/2} = \left(\frac{2\pi}{\beta m} \right)^{1/2}.$$

Hence, we have

$$C = n\left[\left(\frac{\beta m}{2\pi}\right)^{1/2}\right]^3 = n\left(\frac{\beta m}{2\pi}\right)^{3/2}. \tag{15.17}$$

By substituting this value of C in Equation 15.12, we obtain

$$\eta(\mathbf{v})d^3\mathbf{v} = n\left(\frac{\beta m}{2\pi}\right)^{3/2} \exp\left(\frac{-\beta m v^2}{2}\right) d^3\mathbf{v} \tag{15.18}$$

where $\eta(\mathbf{v})\,d^3\mathbf{v}$ is equal to the probable number of molecules having velocities in the range $d^3\mathbf{v} \equiv dv_x dv_y dv_z$ of $\mathbf{v}$.

15.2.1   Features of the Maxwell velocity distribution function

Let us now study some of the features of the Maxwell velocity distribution function. Thereafter, we shall argue that the results 15.11 and 15.12 are very general results which are also valid in describing the motion of the centre-of-mass of a polyatomic molecule in an ideal gas.

We can represent the velocity of a molecule by a single point in a three-dimensional space defined by a set of co-ordinate axes (v_x, v_y, v_z) as shown in Figure 15.2. If we put a point in this space for the velocity of every molecule, then these points will form a cloud, which is densest at the origin of the axes and thins out as we move away from it. All of these points move around randomly and change their position each time the velocity of a molecule is changed due to a collision with another molecule or the walls of the container. The distribution function $\eta(\mathbf{v})$ gives the *probable* number of molecules that can be expected to be found in the region $dv_x\,dv_y\,dv_z$ about $\mathbf{v}$. Since the probable number of molecules depends only on v^2 (see Equation 15.18), on average, the points in Figure 15.2 will be symmetrically distributed about the origin $\mathbf{v} = 0$. Because the factor $\exp\left(-\beta m v^2/2\right)$ decreases as v^2 increases, we can expect the molecules to be predominantly situated near the origin (see Figure 15.2).

(i) Distribution of a velocity component: In some situation we may wish to know the probable number of molecules that have a certain value of some *component* of the velocity. For example, we want to know the probable number of molecules with an x component of the velocity in the range dv_x of v_x. In terms of our velocity space, we are wanting the probable number of points between v_x and $v_x + dv_x$. This number can be obtained by adding up the probable number of molecules between two finite planes perpendicular to the v_x axis. Then we are concerned with the quantity given in Equation 15.19, describing a given kind of molecule, i.e.

$$\eta(v_x)\,dv_x \equiv \text{the mean number of molecules,}$$
$$\text{per unit volume, that have an}$$
$$x \text{ component of velocity in the}$$
$$\text{range between } v_x \text{ and } v_x + dv_x \tag{15.19}$$

(this is true irrespective of the values of the other velocity components).

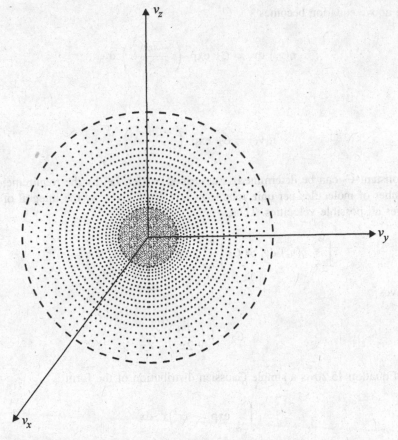

Figure 15.2 The velocity of a molecule in a (v_x, v_y, v_z) space

As said above, this number can be obtained by adding all the molecules which have an x component of velocity in the range v_x and $v_x + dv_x$. Therefore,

$$\eta(v_x)\ dv_x = \int_{v_y} \int_{v_z} \eta(\mathbf{v})\ d^3\mathbf{v}$$

Here, the integrations extend over all the possible y and z velocity components of the molecules. Using Equation 15.12, we obtain

$$\eta(v_x)\ dv_x = C \int_{v_y} \int_{v_z} \exp\left\{\frac{-\beta m}{2}\left(v_x^2 + v_y^2 + v_z^2\right)\right\} dv_x dv_y dv_z$$

$$= C\ \exp\left(\frac{-\beta m v_x^2}{2}\right) dv_x \int_{-\infty}^{\infty} \int_{-\infty}^{\infty} \exp\left\{\frac{-\beta m}{2}\left(v_y^2 + v_z^2\right)\right\} dv_y dv_z.$$

Quite clearly, the integration over all values of v_y and v_z will give some constant. This constant can then be combined with the constant C to give a new constant C_2 (say).

Then the above equation becomes

$$\eta(v_x) \, dv_x = C_2 \, \exp \left(\frac{-\beta m v_x^2}{2} \right) dv_x \tag{15.20}$$

or

$$\eta(v_x) = C_2 \, \exp \left(\frac{-\beta m v_x^2}{2} \right). \tag{15.21}$$

The constant C_2 can be determined as described earlier, i.e. by the requirement that the total number of molecules per unit volume $n(=N/V)$ is given by the integral of Equation 15.20 over all possible velocities. Thus,

$$\int_{-\infty}^{\infty} \eta(v_x) \, dv_x = C_2 \int_{-\infty}^{\infty} \exp \left(\frac{-\beta m v_x^2}{2} \right) dv_x = n \tag{15.22}$$

which gives

$$C_2 = n \left(\frac{\beta m}{2\pi} \right)^{1/2} \tag{15.23}$$

because Equation 15.20 is a simple Gaussian distribution of the form

$$\int_{-\infty}^{\infty} \exp \left(-ax^2 \right) x^n \, dx$$

(see Appendix 6). By substituting the value of C_2 into Equation 15.21, we have

$$\eta(v_x) = n \left(\frac{\beta m}{2\pi} \right)^{1/2} \exp \left(\frac{-\beta m v_x^2}{2} \right). \tag{15.24}$$

Obviously, a similar result holds for the y and z components of the velocity. These are (by replacing v_x in Equation 15.24 with v_y and v_z)

$$\eta(v_y) = n \left(\frac{\beta m}{2\pi} \right)^{1/2} \exp \left(\frac{-\beta m v_y^2}{2} \right) \tag{15.25}$$

$$\eta(v_z) = n \left(\frac{\beta m}{2\pi} \right)^{1/2} \exp \left(\frac{-\beta m v_z^2}{2} \right). \tag{15.26}$$

Remember that $\beta = 1/kT$ in all the above equations.

(ii) Symmetry of the velocity components: It is apparent from the results 15.24, 15.25 and 15.26 that any of the velocity components v_x, v_y and v_z is distributed symmetrically about the values $v_x = 0$, $v_y = 0$ or $v_z = 0$, respectively. Therefore, the *mean value* of any velocity component of a molecule must always be zero, i.e.

$$\bar{v}_x = \bar{v}_y = \bar{v}_z = 0. \tag{15.27}$$

This can be seen physically by symmetry, as the x component of velocity of a molecule is as likely to be positive as negative. It follows from the mathematical definition of the average as given by

$$\bar{v}_x \equiv \frac{1}{n} \int_{-\infty}^{\infty} \eta(v_x) v_x \, dv_x. \tag{15.28}$$

We already noted that $\eta(v_x)$ depends only on v_x^2. So, $\eta(v_x)$ remains unchanged when v_x changes its sign, i.e. $\eta(v_x)$ is an even function of v_x. On the other hand, the integrand changes its sign when v_x does so; therefore, it is an odd function of v_x. Hence, contributions to the integrand in Equation 15.28 from v_x and $-v_x$ will cancel each other. The same arguments apply to the other two velocity components, v_y and v_z.

We can plot $\eta(v_x)$ as a function of v_x, or for that matter, plot $\eta(v_y)$ as a function of v_y and $\eta(v_z)$ as a function of v_z, to physically represent the symmetry. Let us plot $\eta(v_x)$ against v_x; the result is shown in Figure 15.3. Two cases are shown, corresponding to two different values of β, i.e. two different temperatures, to show how different temperatures affect the distribution of velocities. $\eta(v_x)$ has its maximum value when $v_x = 0$ and decreases rapidly as $|v_x|$ increases. Actually, $\eta(v_x) \to 0$ when $|v_x| \gg (kT/m)^{1/2}$. Therefore, $\eta(v_x)$ becomes quite sharp at the peak when $v_x = 0$ if the absolute temperature T is decreased. This means that the mean kinetic energy of a molecule tends to decrease as $T \to 0$.

The area under both curves must be the same, because $\eta(v_x)$ is normalized to n

$$\int_{-\infty}^{\infty} \eta(v_x) \, dv_x = n.$$

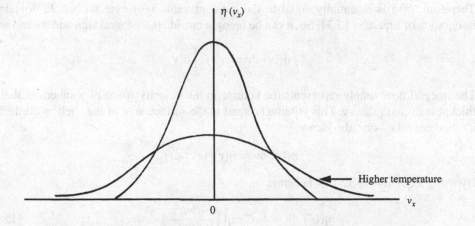

Figure 15.3 Schematic diagram of $\eta(v_x)$ vs. v_x at two different temperatures

This is an expression of the fact that all molecules have some velocity v_x. It is easy to see from Figure 15.3 that equal numbers of molecules are moving in the positive and negative x directions. As discussed above, this is a consequence of the fact that $\eta(v_x)$ depends only on v_x^2 (so also $\eta(v_y)$ and $\eta(v_z)$ depend on v_y^2 and v_z^2, respectively). We have noted the same symmetry about the origin in Figure 15.2. From the above discussion we see that similar results hold for the velocity components v_x, v_y and v_z. The simple reason for this is that all velocity components are completely equivalent by virtue of the symmetry of the situation.

(iii) The Maxwell–Boltzmann distribution of molecular speed: Besides $\eta(\mathbf{v})$ and $\eta(v_x)$, another distribution function is also of interest, which is the distribution of molecular speed. This distribution function tells us the probable number of molecules which are in the range v and $v + dv$ where $v = |\mathbf{v}|$. Let us consider a given kind of molecule and investigate the quantity

$$\eta_1(v)\ dv = \text{the mean number of molecules,}$$
$$\text{per unit volume, which have}$$
$$\text{a speed } v \text{ in the range } v \text{ and}$$
$$v + dv. \tag{15.29}$$

The speed v is related to the velocity $\mathbf{v}$ by the relation

$$v = \left(v_x^2 + v_y^2 + v_z^2\right)^{1/2}. \tag{15.30}$$

We can find the number $\eta_1(v)$ by adding all molecules with speeds in the given range, regardless of the directions of their velocity. From Figure 15.4 we see that the speed v is simply the distance from the origin in the velocity space within a spherical shell of inner radius v and outer radius $v + dv$. The number $\eta_1(v)\ dv$ can be determined by integrating $\eta(\mathbf{v})\ d^3\mathbf{v}$ over all velocities which satisfy the condition $v < |\mathbf{v}| < v + dv$, i.e. over all velocity vectors that end in velocity space within the spherical shell. Then,

$$\eta_1(v)\ dv = \int \eta(\mathbf{v})\ d^3\mathbf{v}. \tag{15.31}$$

Note that dv is infinitesimally small and $\eta(\mathbf{v})$ depends only on the magnitude of velocity $\mathbf{v}$. Therefore, $\eta(\mathbf{v})$ is essentially equal to the constant value $\eta(v)$ over the whole domain of integration of Equation 15.31. So, it can be brought outside the integral sign and we can write

$$\eta_1(v)\ dv = \eta(v) \int d^3\mathbf{v}.$$

The integral now simply represents the volume in the velocity space of a spherical shell of thickness dv and radius v. This volume is equal to the surface area of the shell multiplied by its thickness, i.e. $4\pi v^2\ dv$. Hence,

$$\eta_1(v)\ dv = \eta(v)[4\pi v^2\ dv].$$

By using Equation 15.12 this becomes

$$\eta_1(v)\ dv = 4\pi C \exp\left(\frac{-\beta m v^2}{2}\right) v^2\ dv. \tag{15.32}$$

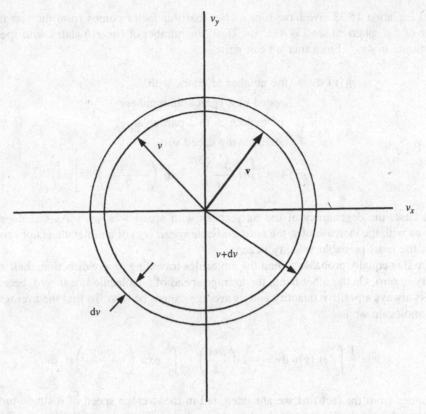

Figure 15.4 Velocity space in two dimensions (the v_z axis is pointing out of the paper; not shown). The spherical shell defined by v_x, v_y and v_z axes contains all molecules having velocity $\mathbf{v}$ so that $v < |\mathbf{v}| < v + dv$

By substituting the value of the constant of proportionality C from Equation 15.17, we have

$$\eta_1(v) \, dv = 4\pi n \left(\frac{\beta m}{2\pi} \right)^{3/2} \exp \left(\frac{-\beta m v^2}{2} \right) v^2 \, dv$$

$$= n \left(\frac{2}{\pi} \right)^{1/2} (\beta m)^{3/2} \exp \left(\frac{-\beta m v^2}{2} \right) v^2 \, dv$$

or

$$\eta_1(v) = n \left(\frac{2}{\pi} \right)^{1/2} (\beta m)^{3/2} \exp \left(\frac{-\beta m v^2}{2} \right) v^2. \tag{15.33}$$

Equation 15.33 is the *Maxwell–Boltzmann velocity distribution function*. Note that as v increases the exponential factor decreases, but the volume of the phase space available to the molecule is proportional to v^2 and this volume increases as v increases. Hence, the net result is a maximum. We find that the difference between $\eta_1(v)$ and $\eta(\mathbf{v})$ is the multiplicative factor

$4\pi v^2$ in Equation 15.33. We have seen earlier that this factor comes from the fact that the volume of the spherical shell is $4\pi v^2 \, dv$. Thus, the number of (micro) states with speed v is proportional to $4\pi v^2$. From this we can write

$$\eta_1(v) \, dv = \text{(the number of states with}$$
$$\text{speed } v) \times \text{(probable number}$$
$$\text{of molecules in a particular}$$
$$\text{state having speed } v)$$
$$= (4\pi v^2) \left[n \left(\frac{\beta m}{2\pi} \right)^{3/2} \exp \left(\frac{-\beta m v^2}{2} \right) dv \right]. \tag{15.34}$$

In this case, the degeneracy of the microstates with speed v is $4\pi v^2$. As this degeneracy increases with the increase of v, the *most probable speed* (v_m) of a molecule is not zero, even though the most probable velocity is zero.

Since it is equally probable to find the molecules travelling in any direction, their average velocity is zero. On the other hand, the average speed of a molecule is *not zero*, because the speed is always a positive quantity and its average cannot be zero. To find the average speed $\bar{v}$ of a molecule we use

$$\bar{v} = \frac{1}{N} \int_0^\infty \eta_1(v) v \, dv = \frac{4}{\sqrt{\pi}} \left(\frac{\beta m}{2} \right)^{3/2} \int_0^\infty \exp \left(\frac{-\beta m v^2}{2} \right) v^3 \, dv.$$

$1/N$ comes from the fact that we are interested in the average speed of a single molecule. The integral is of the Gaussian form (see Appendix 6). The value of the integral is

$$\frac{1}{2 \left(\dfrac{m}{2kT} \right)^2}.$$

Then,

$$\bar{v} = \frac{4}{\sqrt{\pi}} \left(\frac{\beta m}{2} \right)^{3/2} \left[\frac{2}{(\beta m)^2} \right] = \left(\frac{8}{\pi \beta m} \right)^{1/2}$$
$$= \left(\frac{8kT}{\pi m} \right)^{1/2}. \tag{15.35}$$

If $\eta_1(v) \, dv$ is summed over all possible speeds $v = |\mathbf{v}|$, the result will produce the total mean number n of molecules per unit volume ($n = N/V$), i.e.

$$\int_0^\infty \eta_1(v) \, dv = \frac{N}{V} = n. \tag{15.36}$$

Here, the lower limit of the integral conforms to the fact that the speed of a molecule cannot be negative. The schematic of a plot of $(1/n)\eta_1(v)$ against v is shown in Figure 15.5.

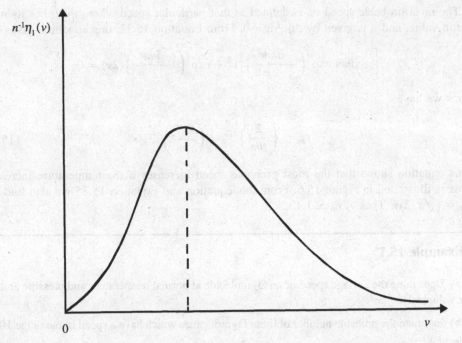

Figure 15.5 Maxwell distribution showing the mean number of molecules having a speed in the range v and $v + dv$

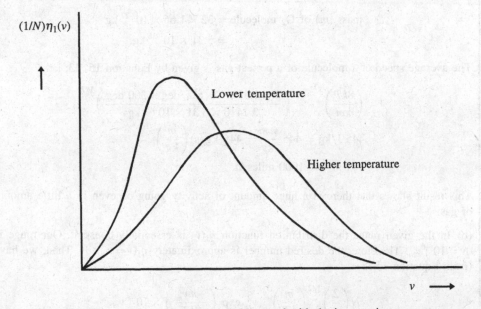

Figure 15.6   Increase in the most probable speed with the increase in temperature

The most probable speed v_m is defined as that particular speed where $\eta_1(v)$ has its maximum value, and it is given by $d\eta_1/dv = 0$. From Equation 15.32, this condition gives

$$\left[-\beta mv \exp\left(\frac{-\beta mv^2}{2}\right)\right]v^2 + \exp\left(\frac{-\beta mv^2}{2}\right)(2v) = 0.$$

Thus we have

$$v_m = \left(\frac{2}{\beta m}\right)^{1/2} = \left(\frac{2kT}{m}\right)^{1/2}. \tag{15.37}$$

This equation shows that the most probable speed increases if the temperature increases. This is illustrated in Figure 15.6. From this equation and Equation 15.35 we also find that $v_m = (\sqrt{\pi}/2)\bar{v}$. Thus $\bar{v}/v_m \simeq 1.13$.

Example 15.1

(a) Determine the average speed of an O_2 molecule at normal temperature and pressure and in a volume 10^{-4} m^3.

(b) Estimate the probable number of these O_2 molecules which have a speed in the range 10^{-6} v_m of $v = v_m$.

Solution

(a) We must first determine the mass of O_2 molecule. The molecular weight of oxygen is 32. The value of one atomic mass unit is 1.66×10^{-27} kg. Therefore,

$$\text{mass } (m) \text{ of } O_2 \text{ molecule} = 32 \times 1.66 \times 10^{-27} \text{ kg}$$
$$= 5.31 \times 10^{-26} \text{ kg}.$$

The average speed of a molecule of a perfect gas is given by Equation 15. 33, i.e.

$$\bar{v} = \left(\frac{8kT}{\pi m}\right)^{1/2} = \left(\frac{8 \times 1.38 \times 10^{-23} \text{ J/deg} \times 300 \text{ deg}}{3.1416 \times 5.31 \times 10^{-26} \text{ kg}}\right)^{1/2}$$
$$= 445 \text{ J/kg} = 445 \frac{nm}{\text{kg}} = 445(\text{kg/s})\left(\frac{m}{\text{kg}}\right)$$
$$= 445 \text{ m/s} \simeq 1000 \text{ miles/h}.$$

This result shows that there is a huge amount of activity going on even in a little amount of gas.

(b) In the given range the distribution function $\eta_1(v)$ is essentially constant. Our range is $dv = 10^{-6}v_m$. Therefore, the desired number is approximately $\eta_1(v = v_m)\,dv$. Then, we have (from Equation 15.33)

$$N\left(\frac{2}{\pi}\right)^{1/2}\left(\frac{m}{kT}\right)^{3/2} v_m^2 \exp\left(\frac{-mv_m^2}{2kT}\right) \times 10^{-6}v_m.$$

However, from Equation 15.37 we have $v_m = (2kT/m)^{1/2}$. Hence, the probable number of O_2 molecules in the speed range $10^{-6}v_m$ is

$$\left(\frac{4}{\sqrt{\pi}}\right)(N/e) \times 10^{-6} = 0.83 \times 10^{-6}N.$$

Now, we have to find the value of N. For a perfect gas we have $PV = NkT$. So, $N = PV/kT = (1.013 \times 10^5 n/m^2)\,(10^{-4}m^3) \times (1.38 \times 10^{-23}\,\text{J/deg} \times 300\,\text{deg})^{-1} = 2.45 \times 10^{21}$. Substituting this values of N, into the above expression, we have $0.83 \times 10^{-6} \times 2.45 \times 10^{21} \simeq 2.04 \times 10^{15}$. This result shows that even in this very small range of speeds $10^{-6}v_m \simeq 0.04\,\text{cm/s}$ the number of molecules is very large.

Note: We may come across situations where we want to know the probable number of molecules with a velocity component v_x which is larger than some value v_0, where v_0 is not necessarily zero. In such cases, we use

$$\int_{v_0}^{\infty} \eta(v_x)\,dv_x = N\left(\frac{\beta m}{2\pi}\right)^{1/2} \int_{v_0}^{\infty} \exp\left(\frac{-\beta m v_x^2}{2}\right)\,dv_x. \tag{15.38}$$

Such integrals are essentially incomplete Gaussian integrals because the lower limit v_0 is not necessarily equal to zero. The most convenient way to deal with such integrals is to use values for an incomplete Gaussian integral of this type. The standard integral of this type is known as the *error function* defined by

$$\text{erf}(x) = \frac{2}{\sqrt{\pi}} \int_0^x \exp\left(-u^2\right)\,du. \tag{15.39}$$

Due to the presence of $2/\sqrt{\pi}$, the error function $\text{erf}(x)$ goes from zero to 1 as x goes from 0 to ∞. All the intermediate values are given in Appendix 6.

In order to perform the integration in Equation 15.38 we use $\beta m v_x^2/2 = u^2$. Rearranging, we have

$$v_x = \left(\frac{2}{\beta m}\right)^{1/2} u$$

or

$$dv_x = \left(\frac{2}{\beta m}\right)^{1/2} du.$$

By substituting this into Equation 15.38, we have

$$\int_{v_0}^{\infty} \eta(v_x)\,dv_x = \frac{N}{\sqrt{\pi}}\left(\frac{\beta m}{2}\right)^{1/2} \int_{u_0}^{\infty} \exp\left(-u^2\right)\left(\frac{2}{\beta m}\right)^{1/2} du$$

$$= \frac{N}{\sqrt{\pi}} \int_{u_0}^{\infty} \exp\left(-u^2\right)\,du \tag{15.40}$$

where the lower limit $u_0 = (\beta m/2)^{1/2} v_0$. As can be seen, Equation 15.40 is still not in the desired form of an error function. So we rewrite it in the following form

$$\int_{v_0}^{\infty} \eta(v_x) \, dv_x = \frac{N}{\sqrt{\pi}} \int_0^{\infty} \exp(-u) \, du - \frac{N}{\sqrt{\pi}} \int_0^{u_0} \exp(-u^2) \, du$$

$$= \frac{N}{2} - \frac{N}{2} \, \text{erf}(u_0). \tag{15.41}$$

This can be determined using the values of erf(x) given in Appendix 6.

Example 15.2

Using the data for oxygen given in Example 15.1, calculate the probable number of molecules of oxygen with speed $v_x \geqslant 10^3 \, \text{m/s}$.

Solution

We have $T = 300 \, \text{K}$, mass of oxygen $m = 5.31 \times 10^{-26} \, \text{kg}$ (as calculated in Example 15.1) and $v_0 = 10^3 \, \text{m/s}$. Let us calculate the value of $(\beta m/2)^{1/2}$, i.e. $(m/2kT)^{1/2}$. By substituting the values for m, T and k, we have 2.53×10^{-3}. We know that $u_0 = (\beta m/2)^{1/2} v_0$, but $v_0 = 10^3 \, \text{m/s}$. Therefore, $u_0 = (2.53 \times 10^{-3})(10^3 \, \text{m/s}) = 2.53 \, \text{m/s}$. By substituting this value of u_0 into Equation 15.41 and using the value of erf(u_0) = erf(2.53) from Appendix 6, we have the most probable number of molecules as

$$\frac{N}{2}[1 - \text{erf}(2.53)] = \frac{N}{2}(1 - 0.99964)$$

$$= \left(\frac{2.45 \times 10^{21}}{2} \right)(36 \times 10^{-5})$$

$$\simeq 4.41 \times 10^{17}$$

15.2.2   Maxwellian distribution function for polyatomic molecules

The results derived in Section 15.2 were all applicable to monatomic molecules. Let us now find out whether these results are also valid for polyatomic molecules.

Let us assume that the gas under consideration consists of polyatomic molecules. Under the same conditions that we applied in Section 15.2 to consider a gas of monatomic molecules, the motion of the centre of mass of a polyatomic molecule can be treated by the classical approximation. However, the rotational and vibrational motions within a polyatomic molecule about its centre of mass must be discussed by the quantum mechanical concepts. Thus, the state of the polyatomic molecule can be described by the position **r** and momentum **p** of its centre of mass, and by specifying the particular quantum state that describes the intramolecular rotation and vibration.

If s is the specified particular quantum state, then the energy of the molecule is given by

$$\varepsilon = \mathbf{p}^2/2m + \varepsilon_s. \tag{15.42}$$

In this equation $\mathbf{p}^2/2m$ is the kinetic energy of the motion of the centre of mass of the molecule, and ε_s is simply the energy due to the rotation and vibration within the molecule in the state s. By virtue of canonical distribution we can write an expression for the probability $\mathscr{P}_s(\mathbf{r}, \mathbf{p})\, d^3\mathbf{r}d^3\mathbf{p}$ that the molecule can be found in that state where the position of its centre-of-mass is between $\mathbf{r}$ and $\mathbf{r} + d\mathbf{r}$ and the momentum of the centre-of-mass is between $\mathbf{p}$ and $\mathbf{p} + d\mathbf{p}$. Thus,

$$\mathscr{P}_s(\mathbf{r}, \mathbf{p})\, d^3\mathbf{r}d^3\mathbf{p} \propto \exp\left[-\beta(\mathbf{p}^2/2m + \varepsilon_s)\right]\, d^3\mathbf{r}d^3\mathbf{p}$$
$$\propto \exp\left(-\beta\varepsilon_s\right)\exp\left(-\beta\mathbf{p}^2/2m\right)\, d^3\mathbf{r}d^3\mathbf{p}. \tag{15.43}$$

In order to find the probability $\mathscr{P}_s(\mathbf{r}, \mathbf{p})\, d^3\mathbf{r}d^3\mathbf{p}$ it is only required to find the sum of Expression 15.43 over all possible intramolecular states s (irrespective of the state of intramolecular motion of the molecule). Note that the sum over all possible values of $\exp(-\beta\varepsilon_s)$ will give some constant value which will simply multiply the second term in Expression 15.43. Thus the result will be a result of the form of Expression 15.10 that describes the centre-of-mass of the molecule. We can conclude that Equation 15.11 and the Maxwell velocity distribution Equation 15.12 are pretty much general results which are also valid for the motion of the centre-of-mass of a polyatomic molecule in a gas.

15.3   THE FERMI–DIRAC DISTRIBUTION FUNCTION

The Maxwell–Boltzmann distribution law is a result of classical theory; it is valid for molecules of a gas under ordinary conditions. The Maxwell–Boltzmann law is, of course, a good approximation to the actual state of affairs only if the average spacing between particles is large in comparison with the de Broglie wavelength ($\lambda = h/p$, where h is Planck's constant and p is the momentum mv of the electron considered to be in free space) calculated using the velocity v of the particles. However, electrons are much lighter than molecules. Moreover, in metals the concentration of free electrons is about 10 000 times higher than that of molecules in a gas under normal temperature and pressure. Therefore, under these conditions the use of classical statistics is not a valid approximation to the correct quantum statistics. However, we shall show later that the Boltzmann distribution can be regarded as the limiting case of the Fermi–Dirac distribution.

We divide all particles into *macroparticles* and *microparticles*. The former are those whose motion is sufficiently well described by the laws of classical mechanics, while the latter obey the laws of quantum mechanics. Microparticles include electrons. Such particles cannot be distinguished in the quantum sense. This means that the macroproperties of a body are not affected by an interchange of particles between two microstates. Quantum statistics requires that we treat all electrons as *indistinguishable* and that each state of the system may be occupied by a maximum of one electron. A one-particle state of the free particle system is determined by the values of the quantum numbers n_x, n_y, n_z as well as the spin quantum number $m_z = \pm 1/2$ of the electron. If each state of a system is occupied by at most one electron, when we are dealing with large numbers of electrons, then even in the lowest state of the entire system many high quantum number states of the individual electrons will be occupied. Compared with this case, the Maxwell–Boltzmann case is quite different where any number of particles can have the same energy and momentum. Unlike the lowest state in a quantum system, in the lowest state of a classical system all particles can have zero energy and momentum.

Particles of half integral spin e.g. $\frac{1}{2}, \frac{3}{2}, \frac{5}{2}, \ldots$ obey Fermi–Dirac statistics. Such particles are called *fermions*. To explain the behaviour of a particle it is necessary to assign a mass and a charge to it. It is also necessary to assign a spin to a particle. The spin σ of a particle is related to an angular momentum which is always associated with the particle (referred to as intrinsic angular momentum). $\sigma = \frac{1}{2}$ for the fundamental particles proton, electron and neutron. However, for compound particles such as H_2, the spin is an integer if the compound particle is made of an even number of the elementary particles, whereas it is a half-integer if the compound particle is composed of an odd number of the elementary particles. An electron is said to be a fermion of spin $1/2$. Only one fermion from a group of several identical fermions can occupy an orbital (a single particle quantum state) with a given set of quantum numbers. For example, consider a group of electrons in an atom. Only one of these electrons can have a given set of values of the quantum numbers n, l, m and s. The electrons must differ in at least one of these quantum numbers.

We now find the distribution of particles inside a system. To do this, first of all we consider a cell defined by the quantum numbers n_x, n_y, n_z and m_z. Such a cell can have no particle, in which case the occupation number is zero, or can have one particle, in which case the occupation number is 1. Imagine that there is a set of g_s cells $c_1, c_2, c_3, \ldots$ with approximately the same amount of energy E given by

$$E_n = \frac{\hbar^2}{2m}\left(\frac{2\pi}{L}\right)^2 (n_x^2 + n_y^2 + n_z^2) = \frac{h^2 n^2}{2mV^{2/3}} \tag{15.44}$$

where n denotes the triplet (n_x, n_y, n_z), $n^2 = n_x^2 + n_y^2 + n_z^2$ and $V = L^3$ (L being a line of length within which an electron is being confined). Let us suppose that there are altogether n_s electrons in the set g_s such that, of these g_s cells, n_s are singly occupied (i.e. there is only one electron in each of these cells) and the remaining $g_s - n_s$ cells are empty. Since the distribution of electrons is rather without any particular preference for a cell, we can uniquely characterize the distribution by assigning to each cell its occupation number. Thus we can arrange, if we wish, in the order

cell	c_1	c_2	c_3	c_4	c_5	c_6	$\cdots$
occupation	0	1	0	0	1	1	

This allows us to give a complete characterization by identifying the cells which are occupied by one particle and which are not occupied (empty). Thus,

empty (0)	occupied (1)
$c_1 \, c_3 \, c_4 \, \cdots$	$c_2 \, c_5 \, c_6 \, \cdots$

We can choose the first cell in g_s ways, the second cell in $g_s - 1$ ways, and so on. Therefore, there are $g_s!$ ways in which we can write the names (c_i) of the g_s number of cells on a line. However, many of these sequences will be indistinguishable if the electrons are not distinguishable. Consider the sequence $c_5 c_6$. Here, interchanging the order $c_5 c_6 \rightarrow c_6 c_5$ is not an obvious distinguishable change.

The question is: how do we find the number of distinguishable sequences. We should not count those distributions as distinguishable if they differ from one another only by permutation of the n_s number of occupied cells or by $g_s - n_s$ of empty cells. The total number of sequences must be the number of distinguishable sequences multiplied by the number of indistinguishable sequences contained within each distinguishable sequence. Hence, if W_s represents the number of distinguishable sequences, we can write

$$W_s(g_s - n_s)!n_s! = g_s! \tag{15.45}$$

or

$$W_s = \frac{g_s!}{(g_s - n_s)!n_s!}. \tag{15.46}$$

So far we have considered only a selected set of g_s cells. Let us assume that there are g_p probable sets of cells in our system. Now, if we wish to cover the entire energy range by taking into consideration all the sets, we can write for the total number of distinguishable arrangements in the whole system

$$W = \prod_p (W_p) = \prod_p \left[\frac{g_p!}{(g_p - n_p)!n_p!} \right]. \tag{15.47}$$

In statistical mechanics, if a thermodynamic system is in equilibrium, its observable average properties are given sufficiently accurately by the properties of the most probable distribution. The most probable distribution can be obtained when W is at a maximum as a function of n_p, provided that the following conditions are fulfilled.

The total energy E of the system must be constant, i.e.

$$E = \sum n_p E_p = \text{constant} \tag{15.48}$$

where E_p is the energy of a particle in the set g_p. Secondly, the total number of particles N of the system must be constant, i.e.

$$N = \sum n_p = \text{constant}. \tag{15.49}$$

By taking the logarithm of both sides of Equation 15.47, we obtain

$$\ln W = \sum_p [\ln g_p! - \ln (g_p - n_p)! - \ln n_p!] \tag{15.50}$$

or

$$\ln W = \sum_p [g_p \ln g_p - (g_p - n_p) \ln (g_p - n_p) - n_p \ln n_p] \tag{15.51}$$

(expanding the logarithms by *Stirling's approximation* valid for large numbers, i.e. $\ln n \simeq n \ln n - n$. See Appendix 6).

As said earlier, the most probable distribution can be found when W is at a maximum as a function of n_p, provided that the conditions of Equations 15.48 and 15.49 are satisfied. To find the maximum of $\ln W$ we use a well-known mathematical technique, the method of Lagrangian multipliers, subject to the conditions of Equations 15.48 and 15.49. Thus,

$$\frac{\partial}{\partial n_p}\left[\ln W + \alpha\left(N - \sum n_p\right) + \beta\left(E - \sum n_p E_p\right)\right] = 0 \tag{15.52}$$

where α is the *Lagrangian multiplier* which is determined by Equation 15.49 and β is another constant which we shall determine later. From Equations 15.51, 15.49 and 15.48 we get

$$\frac{\partial}{\partial n_p}(\ln W) = \ln (g_p - n_p) - \ln n_p \tag{15.53}$$

$$\frac{\partial}{\partial n_p}\left[\alpha\left(N - \sum n_p\right)\right] = -\alpha \tag{15.54}$$

$$\frac{\partial}{\partial n_p}\left[\beta\left(E - \sum n_p E_p\right)\right] = -\beta E_p. \tag{15.55}$$

By combining Equations 15.52–15.55, we have

$$\ln (g_p - n_p) - \ln n_p - \alpha - \beta E_p = 0$$

or

$$\frac{g_p - n_p}{n_p} = \exp (\alpha + \beta E_p)$$

or

$$n_p = \frac{g_p}{\exp (\alpha + \beta E_p) + 1}. \tag{15.56}$$

Now, coming back to the constant β, it is actually equal to $1/kT$ (see Equation 13.28). At very high temperatures many states are accessible. Because of this, at very high temperatures the ratio n_p/g_p must be very much smaller than 1. Therefore, in the high-temperature range we have

$$n_p \simeq g_p \exp (-\beta E_p). \tag{15.57}$$

By comparing this equation with the Boltzmann distribution law valid in the high temperature range, we find that

$$\beta = 1/kT. \tag{15.58}$$

By rearranging Equation 15.56 we have the distribution function (probability) that a given state is occupied as

$$f = \frac{n}{g} = \frac{1}{\exp{(\alpha + \beta E)} + 1}. \tag{15.59}$$

This equation is called the *Fermi–Dirac distribution function*. It is further modified by defining an energy E_F (called the Fermi energy) such that

$$\alpha = -E_F/kT. \tag{15.60}$$

By substituting this in Equation 15.59, we obtain

$$f(E) = \frac{1}{\exp{\left(\frac{E - E_F}{kT}\right)} + 1}. \tag{15.61}$$

The Fermi–Dirac distribution function is sketched in Figure 15.7 for zero absolute temperature ($T = 0$) and for a low temperature.

The width of the region where the distribution is affected by temperature is about kT. Figure 15.7 is an attempt to represent the Fermi sphere in a one-dimensional diagram. One can readily see how the Fermi sphere begins to spread out over an energy range kT when $T \neq 0\,\mathrm{K}$.

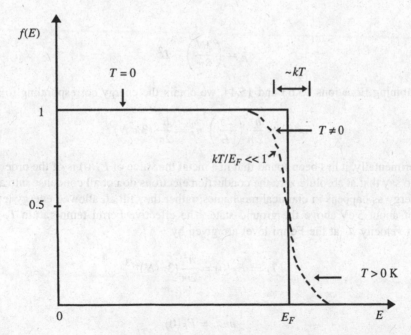

Figure 15.7 Schematic diagram of the Fermi–Dirac distribution functions for two different temperatures ($T = 0$ and for a low temperature)

From Equation 15.61 it can be seen that at $T = 0\,K$, $f(E) = 1$ when $E < E_F$ (at low values of E, much less than E_F, $f(E)$ is very close to unity) and $f(E) = 0$ when $E > E_F$. Thus, E_F is a sort of *cut-off* energy at absolute zero, i.e. all states with energy greater than E_F are empty and all states with energy less than E_F are completely filled. For $T = 0.1\,T_F$, the large changes in $f(E)$ are confined to an energy range of about kT on either side of E_F. As seen in Figure 15.7, with the increase in T the distribution rounds off. The significance of this is that states within about kT below E_F are partly depopulated and states within about kT above E_F are partly populated. In the energy range of about kT on either side of E_F, some of the fermions (which at $T = 0$ would have occupied single particle states with energy just below E_F) will, at a temperature $T \ll T_F$, be found in single particle states with energy just above E_F.

Later we shall show that the Fermi energy E_F is the electrochemical potential (μ) of the electrons or their partial molar free energy. The value of E_F can be obtained from Equation 15.60. It depends on the temperature; however, it can be shown that $E_F(T)$ is closely equal to its value at $0\,K$ when $kT/E_F \ll 1$. We call the distribution *degenerate* when $kT \ll E_F$ and *non-degenerate* when $kT \gg E_F$ (which is the classical limit). Actually, degeneracy of a system means *the number of sublevels of the same energy which can be occupied*.

There is considerable interest to know the value of the Fermi level E_F as a function of the electron concentration N. The number of states with n less than a certain value n_F is given by $2(4\pi/3)(n_F)^3$, because there are two independent states of different spin orientation per unit volume in n space, each of the three integral numbers n_x, n_y, n_z giving two states. To hold NL^3 electrons at absolute zero, all n values must be filled up to

$$2\left(\frac{4\pi}{3}\right)(n_F)^3 = NL^3 \tag{15.62}$$

or

$$n_F^2 = \left(\frac{3N}{8\pi}\right)^{2/3} L^2. \tag{15.63}$$

By combining Equations 15.63 and 15.44, we obtain the energy corresponding to n_F as

$$E_F(0) = \frac{\hbar^2}{2m}\left(\frac{2\pi}{L}\right)^2 n_F^2 = \frac{\hbar^2}{2m}(3\pi^2 N)^{2/3}. \tag{15.64}$$

Experimentally, it has been found that in a metal the value of $E_F(0)$ is of the order of $5\,eV$, that is to say that at absolute zero the conduction electrons do not all condense into a state of zero energy as happens in classical mechanics; rather they fill all allowed energy levels of a range of about $5\,eV$ above the ground state. The effective Fermi temperature T_F and the electron velocity u_F at the Fermi level are given by

$$kT_F = E_F(0) = \frac{\hbar^2}{2m}(3\pi^2 N)^{2/3} \tag{15.65}$$

$$\frac{1}{2}mu_F^2 = E_F(0) \tag{15.66}$$

where m is the mass of a free particle.

The density of state $g(E)$ i.e. the number of states per unit energy range per unit volume as a function of the energy E, is given as

$$N = \int g(E) \, dE = \frac{1}{3\pi^2} (2mE/\hbar^2)^{3/2} \tag{15.67}$$

(using Equation 15.64). This gives

$$g(E) = \frac{1}{2\pi^2} (2m/\hbar^2)^{3/2} E^{1/2}. \tag{15.68}$$

A sketch of $g(E)$ against the energy E is given in Figure 15.8. In this figure, $E_F(0)$ is the Fermi level at absolute zero and the hatched area indicates that the states up to $E_F(0)$ (at $T = 0\,\text{K}$) are filled. The broken line indicates the density of filled states at $kT \ll E_F$ (degenerate). At higher temperatures the filling curve rounds off. One can readily see how the Fermi sphere begins to spread out over an energy range kT when $T > 0$.

Note: In the non-relativistic limit, the number of single particle states (orbitals) for a free particle in a cubical box, which differ in at least one of the quantum numbers n_x, n_y, n_z, and spin numbers s, and which has energy in the energy range E to $(E + dE)$ is

$$g(E) \, dE = \frac{gV}{4\pi^2} (2m/\hbar^2)^{3/2} E^{1/2} \tag{15.69}$$

where $g = (2s + 1)$ is the spin degeneracy, i.e. if the particle has spin s then there are g states associated with each set of values of n_x, n_y and n_z, V is the volume of the cubical box.

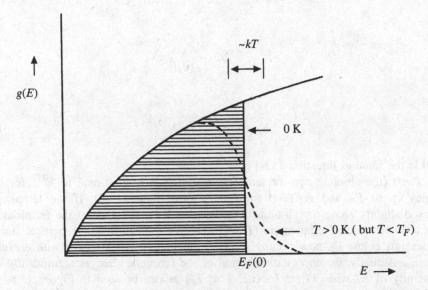

Figure 15.8 Schematic diagram of the density of state $g(E)$ versus the energy E

At thermal equilibrium, the number of electrons per unit volume with energy between E and $E + dE$ is given by

$$dn = f(E)g(E) \, dE. \tag{15.70}$$

By using the values of $f(E)$ and $g(E)$ from Equations 15.61 and 15.68 in Equation 15.70, we obtain

$$dn = \frac{1}{2\pi^2} \left(\frac{2m}{\hbar^2} \right)^{3/2} \frac{E^{1/2} \, dE}{\exp\left(\frac{E-E_F}{kT} \right) + 1}. \tag{15.71}$$

Since

$$\frac{1}{2\pi^2} \left(\frac{2m}{\hbar^2} \right)^{3/2}$$

is a constant, we may take this as C, so that Equation 15.71 can be written as

$$dn = \frac{CE^{1/2} \, dE}{\exp\left(\frac{E-E_F}{kT} \right) + 1}. \tag{15.72}$$

The value of E_F can be found by integrating this equation and taking the integral equal to N. We already noted that $f(E) = 1$ for $E < E_F$ (the values of E_F at 0 K) and $f(E) = 0$ for $E > E_F$. So all states are filled up to E_F and

$$\int dn = C \int_0^{E_F(0)} E^{1/2} \, dE$$

or

$$N = \frac{2C}{3} [E_F(0)]^{3/2} \tag{15.73}$$

or

$$E_F(0) = \frac{\hbar^2}{2m} (3\pi^2 N)^{2/3} \tag{15.74}$$

which is the same as Equation 15.64

At $T = 0$ (the absolute zero of temperature), $g(E)$ is proportional to $E^{1/2}$ for fermion energies up to E_F, and $g(E) = 0$ for $E > E_F$ (see Figure 15.8). If the temperature is increased slightly above the absolute zero, keeping $T \ll T_F$, some of the fermions, which at the absolute zero of temperature $(T = 0)$ would have been in single particle states with energies just below E_F, now occupy some of the single particle states with energies just above E_F. Actually, the energy distribution of the fermions changes significantly only in the vicinity of E_F when $kT \ll E_F$ (i.e. $T \ll T_F$) as can be seen in Figure 15.8. On the other hand, a large proportion of the fermions will be in single particle states with energies well above E_F at much higher temperatures when $kT \gg T_F$ (i.e. $T \gg T_F$). Later we shall see that, in the classical limit, the Fermi–Dirac energy distribution becomes the Maxwell–Boltzmann distribution when $T \gg T_F$ (nondegenerate) and for $T \ll T_F$, it is degenerate.

The Fermi temperature represents that temperature of the system below which the system is degenerate and above which the system is nondegenerate. The Fermi temperature is often referred to as the *degeneracy temperature* of the system.

Example 15.3

The Fermi energy of lithium calculated on the basis of the free electron model is 4.72 eV. What is the effective Fermi temperature of Li?

Solution

Fermi temperature

$$T_F = \frac{E_F}{k}.$$

Therefore,

$$T_F = \frac{4.72 \text{ eV}}{1.38 \times 10^{-23} \text{ J/K}} = \frac{4.72 \text{ eV K}}{1.38 \times 10^{-23} \text{ J}}$$

$$= \frac{4.72 \times 1.602 \times 10^{-19} \text{ J K}}{1.38 \times 10^{-23} \text{ J}} \quad (\because 1 \text{ eV} = 1.602 \times 10^{-19} \text{ J})$$

$$= 5.48 \times 10^4 \text{ K}.$$

15.3.1 To Show that the Fermi Level is the electrochemical potential of the electrons or their partial molar free energy

The entropy of a Fermi–Dirac system is expressed in terms of the number of different ways the particles (electrons) in the system can be distributed over all the available states of the system, i.e.

$$S = k \ln w \tag{15.75}$$

where k is the Boltzmann constant. For a Fermi–Dirac system we have (see Equation 15.46)

$$W = \prod_i \left[\frac{g_i!(g_i - n_i)}{n_i!} \right] \tag{15.76}$$

where $\prod$ represents the product and n_i is the number of particles in the ith state which has degeneracy g_i (i.e. the number of sublevels of the same energy which can be occupied).

The probability that a given state is occupied is given by $f = n/g$ (see Equation 15.59). Now, from Equation 15.76 we get

$$W = \prod_i \left[\left(\frac{g_i!}{n_i!} \right) g_i (1 - n_i/g_i) \right]$$

$$= \prod_i \left[g_i \frac{1}{f} (1 - f) \right].$$

However, f is given by Equation 15.61. By substituting the value of f into the above equation and then putting the value of W into Equation 15.75, we obtain

$$S = k \sum_i g_i \ln \left[\frac{1}{f} (1 - f) \right]$$

$$= k \sum_i g_i \ln \left(\frac{1}{f} \right) + k \sum_i \ln (1 - f)$$

$$= k \sum_i g_i \ln \left[1 + \exp \left(\frac{E_i - E_F}{kT} \right) \right] + k \sum_i \ln \left[1 - \frac{1}{1 + \exp \left(\frac{E_i - E_F}{kT} \right)} \right]$$

$$= k \sum_i g_i \ln \left[1 + \exp \left(\frac{E_i - E_F}{kT} \right) \right] + k \sum_i \ln \left[\exp \left(\frac{E_i - E_F}{kT} \right) \right]$$

$$= k \sum_i g_i \ln \left[1 + \exp \left(\frac{E_i - E_F}{kT} \right) \right] + \sum_i \left(\frac{E_i}{T} - \frac{E_F}{T} \right)$$

$$= k \sum_i g_i \ln \left[1 + \exp \left(\frac{E_i - E_F}{kT} \right) \right] + \frac{E}{T} - \frac{N}{T} E_F \qquad (15.77)$$

where the total energy E is given by Equation 15.48 and the total number of particles N is given by Equation 15.49. Differentiating with respect to N and keeping E constant, we obtain (since g_i, E_i and k are all constant)

$$-T \left(\frac{\partial S}{\partial N} \right)_E = E_F. \qquad (15.78)$$

By definition, the chemical potential is given by

$$\mu = \left(\frac{\partial G}{\partial N} \right)_{P, T} = -T \left(\frac{\partial S}{\partial N} \right)_{E, V} \qquad (15.79)$$

where G is the Gibbs free energy and V is the volume of the system.

By comparing Equations 15.78 and 15.79 we have

$$\mu = E_F. \qquad (15.80)$$

This means that *the electrochemical potential μ of the electrons or the partial molar free energy $(\partial G/\partial N)_{P,\ T}$ is equal to the Fermi energy E_F.* Then, Equation 15.61 can be written as

$$f(E) = \frac{1}{\exp\left(\frac{E-\mu}{kT}\right) + 1} \tag{15.81}$$

which is another form of the Fermi–Dirac distribution function.

Example 15.4

A mole of He^3 gas atoms has a volume of $0.0224\,m^3$ at 273 K. The mass of a He^3 gas atom is $5.11 \times 10^{-27}\,kg$ and it has the spin 1/2. Calculate the value of $\exp(-\mu/kT)$ and the mean occupancy of a single particle state.

Solution

In the high temperature classical limit, the chemical potential of an ideal monatomic gas is given by

$$\mu = -kT\left(\frac{\partial \ln Z}{\partial N}\right)_{V,\ T} = -kT \ln\left[\frac{gV}{N}\left(\frac{mkT}{2\pi\hbar^2}\right)^{3/2}\right]$$

or

$$\exp(-\mu/kT) = \left(\frac{gV}{N}\right)(mkT/2\pi\hbar^2)^{3/2} \gg 1 \tag{15.82}$$

where all the symbols have the same meanings as earlier. By substituting the values of $N = 6.02 \times 10^{23}$, $g = (2s+1) = 2 \times (1/2) + 1 = 2$, $m = 5.11 \times 10^{-27}\,kg$, $T = 273\,K$, $V = 0.0224\,m^3$, $k = 1.38 \times 10^{-23}\,J/K$ and $\hbar = 1.05 \times 10^{-34}\,Js$ in the above equation, we get

$$\exp(-\mu/kT) = 3.4 \times 10^5.$$

The mean kinetic energy of a monatomic gas atom in the classical limit is $3\,kT/2$. We choose this energy as equal to E to represent a typical fermion in the classical limit. Substituting this value of E and the value of $\exp(-\mu/kT) = 3.4 \times 10^5$ in Equation 15.81, we find that the mean occupancy of a single particle state is

$$f(E) = f(-\mu/kT) = 0.66 \times 10^{-6}.$$

This result shows that in the classical limit the mean occupancy of any particular single particle state is very small.

15.3.2   The chemical potential of the fermion

The average number of fermions $n(E)$ dE in single particle states which have energies in the range E and $E + dE$ can be found by multiplying the number of single particle states (orbitals), $g(E)$ dE, with energies in the range E to $E + dE$, by the mean number of fermions in each of these single particle states $f(E)$. Thus, from Equations 15.69 and 15.81 we obtain

$$n(E) \, dE = \frac{gV}{4\pi^2} \left(\frac{2m}{\hbar^2} \right)^{3/2} \left[\frac{E^{1/2} \, dE}{\exp\{(E - \mu)/kT\} + 1} \right]. \tag{15.83}$$

We have not yet determined the value of the chemical potential μ (N, V, T) in Equations 15.81 and 15.83. Equation 15.81 was derived for the case of ideal fermion gas on the assumption that there is no interchange of fermions between the ideal fermion gas and the container. Under these conditions, the single particle state is chosen as our small system for the use of Equation 15.81 and the fermions in the other single particle states act as the reservoir. The total number N of fermions in the ideal fermion gas can be found by integrating Equation 15.83 between the limits $E = 0$ to $E = \infty$. Hence,

$$N = \frac{gV}{4\pi^2} \left(\frac{2m}{\hbar^2} \right)^{3/2} \int_0^\infty \frac{E^{1/2} \, dE}{\exp\{(E - \mu)/kT\} + 1}. \tag{15.84}$$

This equation can be solved numerically to evaluate μ in terms of N, V and T. If any one of these variables changes, the value of μ also changes. Figure 15.9 shows schematically the variation of μ of an ideal fermion gas with the temperature.

By examining Equation 15.81, it can be seen that when $\mu = E$, $f(E) = 0.5$. Therefore, the chemical potential of an ideal fermion gas is equal to that value of energy at which the mean number of fermions in the single particle state, $f(E)$, is 0.5. At $T = 0$, $\mu(0)$ is equal to the Fermi energy E_F. When $T \ll T_F$, μ is only just below E_F. As the temperature increases, μ decreases significantly and becomes zero at a temperature just below T_F. When T increases above T_F, μ becomes more and more negative. When the temperature is just above the absolute zero, only those fermions which at $T = 0$ would have been in single particle states with energy just below E_F are in excited single particle states with energy just above E_F. The energy at which mean occupancy $f(E) = 0.5$ remains close to, but just below, E_F.

As the temperature is increased further, more and more fermions go into the excited single particle states above E_F. Then, the value of the energy at which $f(E) = 0.5$ shifts to lower and lower energies until finally $f(E)$ is equal to 0.5 for the single particle ground state. At the temperature where it happens, μ is equal to the energy of the single particle ground state. At even higher temperatures, more and more fermions occupy excited states above E_F and the mean occupancy of the single particle ground state is less than 0.5.

To see what happens to the chemical potential in the high temperature limit, let us put $E = 0$ and $f(E) < 0.5$ in Equation 15.81. This gives

$$\frac{1}{\exp(-\mu/kT) + 1} < 0.5$$

or

$$[\exp(-\mu/kT) + 1]^{-1} < 0.5$$

By expanding binomially and rearranging, we obtain

$$\exp(-\mu/kT) > 1$$

or

$$-\mu/kT > \ln 1 = 0$$

We conclude that μ is negative in the high temperature limit. In the high temperature limit, when $T \gg T_F$, the ideal fermion gas approximates to an ideal classical gas.

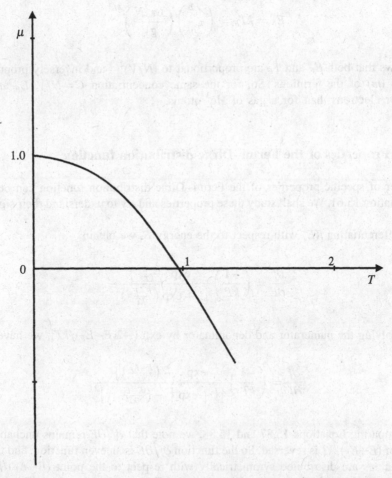

Figure 15.9   Schematic diagram of variation of the chemical potential μ of an ideal fermion gas with the absolute temperature T

When $T=0$, there are no fermions with energies higher than E_F. So the upper limit of Equation 15.84 can be taken as E_F and we can write

$$N = \frac{gV}{4\pi^2}\left(\frac{2m}{\hbar^2}\right)^{3/2}\int_0^{E_F} E^{1/2}\, dE = \frac{gV}{4\pi^2}\left(\frac{2m}{\hbar^2}\right)^{3/2}\left(\frac{2}{3}\right)E_F^{3/2}. \tag{15.85}$$

Hence,

$$E_F = \left(\frac{\hbar^2}{2m}\right)\left(\frac{6\pi^2 N}{gV}\right)^{2/3}.$$

Since $E_F = kT_F$, we have

$$E_F = kT_F = \left(\frac{\hbar^2}{2m}\right)\left(\frac{6\pi^2 N}{gV}\right)^{2/3}. \tag{15.86}$$

This shows that both E_F and T_F are proportional to $(N/V)^{2/3}$ and inversely proportional to the mass (m) of the fermions. So, for the same concentration $C = N/V$, E_F and T_F are bigger for electrons than for a gas of He^3 atoms.

15.3.3   Properties of the Fermi–Dirac distribution function

A number of specific properties of the Fermi–Dirac distribution function can be obtained from Equation 15.61. We shall study these properties and try to understand their significance.

(i) By differentiating $f(E)$ with respect to the energy E, we obtain

$$\frac{\partial f}{\partial E} = \left(\frac{-1}{kT}\right)\frac{\exp\left(\frac{E-E_F}{kT}\right)}{\left[1+\exp\left(\frac{E-E_F}{kT}\right)\right]^2}. \tag{15.87}$$

By multiplying the numerator and denominator by $\exp\left[-2(E-E_F)/kT\right]$, we have

$$\frac{\partial f}{\partial E} = \left(\frac{-1}{kT}\right)\frac{\exp\left\{-\left(\frac{E-E_F}{kT}\right)\right\}}{\left[1+\exp\left\{-\left(\frac{E-E_F}{kT}\right)\right\}\right]^2}. \tag{15.88}$$

By comparing Equations 15.87 and 15.88, we note that $\partial f/\partial E$ remains unchanged when the sign of $(E-E_F)/kT$ is reversed. So the function $\partial f/\partial E$ is an even function, and this means that its values are distributed symmetrically with respect to the point $(E-E_F)/kT = 0$, or which is equivalent with respect to the energy $E = E_F = \mu$. The symmetrical distribution of $\partial f/\partial E$ is shown in Figure 15.10.

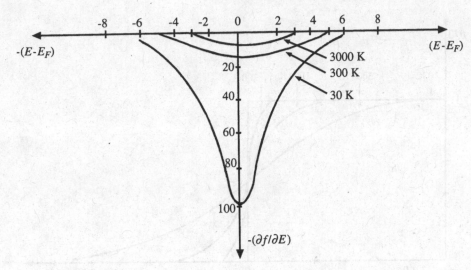

Figure 15.10 Schematic diagram of variation of $\partial f/\partial E$ with $E-E_F$ or $E-\mu$

(ii) It is easy to see that at $E=E_F$ (or if we use $E_F=\mu$, at $E=\mu$) Equation 15.88 reduces to

$$\frac{\partial f}{\partial E} = -\frac{1}{4kT}. \qquad (15.89)$$

As $\partial f/\partial E$ is equal to $\tan \alpha$, where α is the angle of the slope of the tangent to the curve $f(E)$ at the point $E=E_F(=\mu)$, it follows that at this point the function $f(E)$ decreases.

(iii) It follows from Equation 15.89 that the function $f(E)$ at the point $E=E_F(=\mu)$ becomes less steep when the temperature is lowered; at absolute zero, $\partial f/\partial E$ becomes $-\infty$. The function $f(E)$ acquires the step-like form shown in Figure 15.11, where $f(E)$ is shown as a function of E for $T=0$ K and higher temperatures.

(iv) If we assume that the energy E is the reference energy (i.e. it is the starting point with a value 0) in the Fermi–Dirac distribution, we can represent Equation 15.61 in the form

$$f(E) = \frac{1}{1 + \exp(-\mu^*)} \qquad (15.90)$$

where $\mu* = \mu/kT\ (=E_F/kT)$ is a dimensionless quantity and is known as the *reduced electrochemical potential* or the *reduced Fermi level*.

(v) We note from Equation 15.61 that when $E-E_F=0$ (or $E-\mu=0$) the value of $f(E)$ is equal to $1/2$, irrespective of the temperature. Consequently, we may identify the electrochemical potential with some energy E_F for which the probability of occupation is equal to $1/2$. A point to remember is that *we can identify E_F with the electrochemical potential only when the electron system is under equilibrium conditions. Under non-equilibrium conditions $E_F \neq \mu$.*

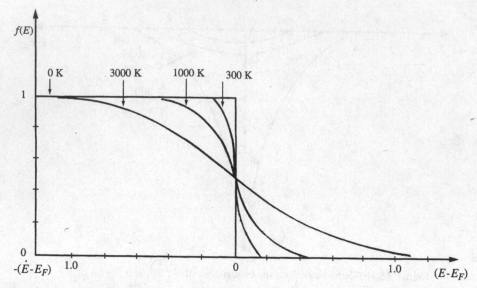

Figure 15.11 Schematic diagram of variation of $f(E)$ with temperature at the point $E = E_F (= \mu)$

(vi) At absolute zero, only the lowest-lying states are occupied; as the temperature rises the distribution rounds off as shown qualitatively in Figure 15.8. Quantitatively, the occupancy is described by the Fermi–Dirac distribution law.

Fermi–Dirac statistics are applicable to electrons as a system of indistinguishable particles with one particle allowed in each state of the system. On the other hand, Maxwell–Boltzmann statistics allow any number of particles to have exactly the same energy and momentum.

(vii) At $T = 0$ K, $f(E) = 1$ for all values of $E < E_F$, and $f(E) = 0$ for all values of $E > E_F$. Hence, according to Equation (15.61) or Equation (15.81), at absolute zero of temperature all the single particle states which have energy up to the value of the chemical potential $\mu(0)$ (or $E_F(0)$) at the absolute zero are occupied. All the single particle states with energies greater than $\mu(0)$ are empty at absolute zero. This result follows from the Pauli exclusion principle (which stipulates that there can only be a maximum of one fermion in each single particle state). For $T > 0$ K, $f(E) \simeq 1$ for all values of $E \ll E_F (= \mu)$, and $f(E) \simeq 0$ for all values of $E \gg E_F (= \mu)$. However, *for values of E not too far from E_F, $f(E)$ will have values intermediate between 0 and 1.*

If a system is made of N fermions, then all the N fermions cannot go into the single particle ground state at the absolute zero, but they occupy the N single particle states with the N lowest energy values. For fermions with spin 1/2 in a cubical box, there can be two fermions with spin quantum numbers $+ 1/2$ and $- 1/2$ associated with each set of the quantum numbers n_x, n_y and n_z. (The positive and negative spins are generally represented by $\uparrow$ and $\downarrow$, respectively).

15.4 EFFECT OF HEATING UP THE ELECTRONS FROM ABSOLUTE ZERO TO A HIGHER TEMPERATURE

Let us examine the effect of heating up the electrons from absolute zero to a higher temperature T, because warming up the electrons will affect their energy distribution. At room temperature the value of kT is approximately 0.02 eV. In classical mechanics, the general effect of warming the electrons is to increase the energy of each electron by an amount which is of the order of kT. As the temperature increases, kT increases and thus the energy of each electron increases; they are excited.

It has been observed that even at absolute zero the energy distribution in the conduction electrons extends over several eV. However, it is not likely that an electron in this distribution will be excited at room temperature if it is more than 0.1 eV below the Fermi level, because all the states within an energy range of kT of such an electron are completely filled. To excite such an electron an energy of at least 0.1 eV i.e. $5kT$, will be required. It is not likely that at room temperature the electron will gain this much energy to become excited to make a transition into an otherwise unoccupied state above E_F.

Unlike in the case of classical mechanics, in quantum mechanics upon warming from absolute zero not all electrons gain an energy approximately equal to kT. Here, the Pauli exclusion principle governs the thermal behaviour of the conduction electrons. Only those electrons which are already within an energy range of the order of kT of the Fermi level will become thermally excited. Thus, all the electrons within this range will gain an energy of the order of kT (see Figure 15.8).

This behaviour of the thermal properties of the conduction electrons in the case of quantum mechanics as compared with their classical behaviour is very useful to determine the heat capacity of these electrons. The electronic heat capacity is given by (el is used for 'electronic' to distinguish it from classical heat capacity)

$$C_V(\text{el}) = \partial U / \partial T. \tag{15.91}$$

If a system has N number of total electrons, then only a fraction of the order $T/[E_F(0)/k] = kT/E_F(0)$ can be excited thermally at T, since only these many electrons will lie within an energy range of the order of kT of the top of the energy distribution. Thus, a total of $NkT/E_F(0)$ individual electrons gain a thermal energy of the order of kT. If U is the total electronic thermal energy, then U is of the order of

$$U = \left[\frac{NkT}{E_F(0)} \right] kT = RT^2 / T_F \text{ per mole} \tag{15.92}$$

since $Nk/n = R$ (n being the number of moles) and $k/E_F(0) = T_F$ (see Equation 15.65). Hence,

$$C_V(\text{el}) = \frac{\partial U}{\partial T} \simeq \frac{RT}{T_F}. \tag{15.93}$$

The classical value of C_V at room temperature in $3R/2$, but the electronic heat capacity at room temperature is less than $3R/2$ by a factor of the order of 0.01 or less. We shall discuss the heat capacity of solids in Chapter 16.

Example 15.5

The Fermi temperature of Ag is found to be 64 000 degrees. Determine its electronic heat capacity at 330 K. Comment on your answer.

Solution

We are given $T = 330\,\mathrm{K}$, $T_F = 64\,000$ degrees and $R = 8.31\,\mathrm{J/mol \cdot deg}$. Then, by substituting in Equation 15.93, we have

$$C_V(\mathrm{el}) \simeq \frac{8.31 \times 330}{64\,000}\ \frac{\mathrm{J\ deg}}{\mathrm{mol\ deg\ deg}}$$

$$\simeq 0.043\,\mathrm{J/mol \cdot deg}$$

This value is lower than the actual electronic heat capacity of silver because of the approximation made in Equation 15.93.

15.5 THE BOLTZMANN DISTRIBUTION

Let us go back to Equation 15.61. When $(E - E_F) \gg kT$, in the *tail of the distribution of electron energies*, the distribution reduces approximately to

$$f(E) = \frac{1}{\exp\left(\frac{E - E_F}{kT}\right)} = \exp\left(\frac{E_F - E}{kT}\right) \tag{15.94}$$

because under this condition $\exp\left(E - E_F/kT\right)$ becomes very large compared with 1 so that we can ignore 1 in the denominator of Equation 15.61. Equation 15.94 is the *classical Boltzmann distribution* or *canonical distribution*. According to Equation 15.82

$$\exp\left(-\mu/kT\right) = (gV/N)(mkT/2\pi\hbar^2)^{3/2}$$

for an ideal monatomic gas in the classical limit, we have

$$\exp\left(-\mu/kT\right) \gg 1.$$

Since the energy E is always positive, $\exp\left(E/kT\right)$ is always greater than 1. Therefore, in the classical limit, when the above equation is satisfied, we have

$$\exp\left(E/kT\right)\ \exp\left(-\mu/kT\right) \gg 1$$

for all values of E. Thus, Equation 15.81 can be approximated to

$$f(E) = \exp\left(-E/kT\right)\ \exp\left(\mu/kT\right). \tag{15.95}$$

We now see that, in the classical limit, the Fermi–Dirac distribution function $f(E)$, which gives the probability that there is a fermion in a single particle state of energy E, is given by Equation 15.95. The quantity $\exp\left(\mu/kT\right)$ is called the *absolute activity*. Provided that $\exp\left[(E - \mu)/kT\right] \gg 1$, the classical Boltzmann distribution can also be used to compute $f(E)$ at high values of the energy E, even when $T_F > T$ (here again $f(E)$ is very small).

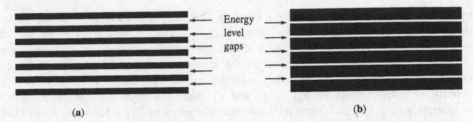

(a)   (b)

Figure 15.12 Distances between energy levels: (*a*) energy levels at low temperatures (these are widely separated), and (*b*) energy levels at high temperatures (the distances between levels reduce as they broaden)

This distribution applies to particles obeying the laws of classical mechanics. It may appear paradoxical that the quantum-indistinguishable particles (i.e. the electrons), having spin and obeying Pauli's exclusion principle, can be described by a *classical* distribution. This apparent paradox can be easily understood if we remember the most important difference between classical and quantum particles. The fundamental difference between classical and quantum particles is *the continuity of the energy spectrum of classical particles and the discrete nature of the energy spectrum of quantum particles*. In fact, *the condition* $(E-E_F) \gg kT$ *is equivalent to a transition from a discrete to a continuous spectrum*.

Actually, Equation 15.94 is better satisfied at higher temperatures. When the temperature is high, the energy levels broaden (see Figure 15.12) and the distances between them become so small that the energy spectrum is practically continuous. Conversely, as the temperature is lowered, the distribution of the particles in an electron gas differs more and more from the Boltzmann distribution. Thus, *the Boltzmann distribution can be regarded as the limiting case of the Fermi–Dirac distribution*.

An electron gas obeying the Fermi–Dirac distribution is known as *degenerate* and the value of the reduced Fermi level $\mu* = \mu/kT (= E_F/kT)$ is called the *degree of degeneracy*. On the other hand, an electron gas obeying the Boltzmann distribution is said to be *non-degenerate*. The division into degenerate and non-degenerate states is quite arbitrary, because it is difficult to draw a definite line between these states. An ideal non-degenerate gas corresponds to the limit $1 \gg \exp(\mu/kT)$. When $\exp(\mu/kT) < 1$, the gas is weakly degenerate and when $\exp(\mu/kT) > 1$, it is strongly degenerate.

The question is: what happens if $\exp(\mu/kT) \gg 1$? Obviously, this limit must correspond to a very degenerate gas, one that can be expected to be quite different from the classical gas. In fact, it is not so apparent that it is possible to have $\exp(\mu/kT) \gg 1$, because this requires that $\mu > 0$ and $\mu/kT \gg 1$. However, in a Fermi gas μ can be positive. In fact, as $T \rightarrow 0$, $\mu > 0$ (so $\mu/kT \gg 1$).

15.6 THE BOSE–EINSTEIN DISTRIBUTION FUNCTION

We have discussed the quantum statistics of non-interacting particles and noted that we must distinguish between two situations, when the particles are distinguishable and when they are not (indistinguishable). In the case of distinguishable particles there is not much problem in evaluating the partition function, except those due to the quantized energy levels. On the other hand, the situation is not that simple if the particles are indistinguishable. In this case, we have to introduce an idea referred to as *occupation numbers* (n_s) which is defined by the number of particles in the particle state s. Once this has been done, a microstate of a system is then specified by the values of all $n_s = (n_1, n_2, n_3, \ldots)$. If ε_s is the energy of the particle state s, then the total energy of the system of non-interacting particles is given by Equation 15.48 and the total number of particles in the system is given by Equation 15.49. For example, let a system have occupation numbers $n_1 = 2, n_2 = 1, n_3 = 1, n_4 = 0, n_5 = 1, \ldots$, and particles with energies ε_1, ε_2, ε_3, ε_4 and ε_5. Then the microstate of the system is represented by $(2, 1, 1, 0, 1, \ldots)$ and the total energy is given by $2\varepsilon_1 + 1\varepsilon_2 + 1\varepsilon_3 + 1\varepsilon_5$ (since $n_4 = 0$; there is no contribution from this).

For a system of indistinguishable particles, the partition function can be treated quite easily, provided that the system is non-degenerate, i.e. if the number of particle states with energy $\varepsilon_s < kT$ is much greater than N. In such a case, most of n_s are zero. However, at very low temperature or at high densities, a system may be degenerate, in which case quantum-mechanical effect can become very important. It has been found that nature apparently has only two types of particle. We have already seen that in the case of those particles which obey the Fermi–Dirac distribution law, only one particle from a group of several fermions can be in one particle state, i.e. a state can have one particle, in which case the occupation number is 1, or a state can have no particle, in which case the occupation number is zero. It is not possible for fermions that any number of them can be in the same particle state. On the other hand, there is another type of particle of integral spin $(0, 1, 2, \ldots)$ which obey Bose-Einstein statistics and are called bosons. For these particles there is no limit to the number of identical bosons that can be in any particular single particle state. In this case, the possible values of n_s are $0, 1, 2, \ldots, N$. Examples of bosons are H_2, He^4 or any particle with an integer spin, whereas H^+, He^3, electrons, protons, neutrons or any particle with a half-integer spin are examples of fermions. Thus, we can write

$$\left. \begin{array}{l} n_s = 0, 1, 2, 3, \ldots, N \quad \text{(bosons)} \\ n_s = 0, 1/2 \quad \text{(fermions)} \end{array} \right\} \tag{15.96}$$

We assign a mass and a charge to a particle to explain its behaviour. Likewise, we must assign a spin to particles. The spin σ of a particle is related to an angular momentum which is always associated with the particle. For electrons, protons and neutrons (which are the fundamental particles), the spin $\sigma = 1/2$. On the other hand, for compound particles, such as O_2, the spin could be integer or half-integer, depending on whether the compound particle is made of an even or odd number of the fundamental particles.

An important point to note is that in the nondegenerate case we do not have to distinguish between these two types of particle, since for most of the states the occupation number is zero, i.e. $n_s = 0$. Since according to Equation 15.96 for both bosons and fermions n_s could be zero, it is not necessary to distinguish between them if we are considering a nondegenerate case. However, in the case of a degenerate system (where the number of particle states with energy $\varepsilon_s < kT$ is very much less than N), the particles tend to occupy the lowest possible value of the energy ε_s. Since only one fermion can go into each particle state, so all of them cannot occupy the lowest energy state. On the other hand, since in the case of bosons any number of particles can be in the same particle state, they can all occupy the lowest energy state. As mentioned earlier, this distinction between bosons and fermions becomes very important at low temperatures (when a system is degenerate).

To derive the Bose–Einstein distribution law we shall use grand canonical distribution and hence the partition function for open systems. One of the reasons for this approach is that the partition function for closed systems (no exchange of particles, so N is fixed) is much more difficult to evaluate than the partition function for open systems (N variables). Besides, if the fluctuation in the number of particles in an open system is very small compared to the mean value $\bar{N}$, the thermodynamic properties of the system are the same as for a closed system with $N = \bar{N}$.

Let us consider a grand canonical system containing N identical bosons. As in the case of derivation of the Fermi–Dirac distribution, here also, we shall treat one of the orbitals (single particle states) as our small system for the application of the grand canonical distribution. Let ε be the single particle energy of the chosen single particle state. Under this condition, the bosons in the single particle states other than the chosen one will act as the particle reservoir for the chosen single particle state. Since all of the N bosons in our system are identical, the total number of bosons in our small system (the chosen single particle state) can have any of the values $0, 1, 2, 3, \ldots$. Then the corresponding total energies of the bosons in the small system are $0, \varepsilon, 2\varepsilon, 3\varepsilon, \ldots$. Thus, the possible states of the chosen single particle state are

$$
\begin{aligned}
N = 0, &\quad \text{energy } \varepsilon = 0 \\
N = 1, &\quad \text{energy } = \varepsilon \\
N = 2, &\quad \text{energy } = 2\varepsilon \\
N = 3, &\quad \text{energy } = 3\varepsilon \\
&\quad \cdots \qquad \cdots \\
&\quad \cdots \qquad \cdots
\end{aligned}
$$

Substituting these values in Equation 12.99, we have the grand partition function as

$$
\begin{aligned}
Z &= \exp(0) + \exp[\beta(\mu - \varepsilon)] + \exp[\beta(2\mu - 2\varepsilon)] \\
&\quad + \exp[\beta(3\mu - 3\varepsilon)] + \cdots \\
&= 1 + \sum_{N=1}^{\infty} x^N
\end{aligned}
\tag{15.97}
$$

where $x = \exp[\beta(\mu - \varepsilon)]$. However,

$$
1 + \sum_{N=1}^{\infty} x^N = (1 - x)^{-1}, \quad \text{if } x < 1.
\tag{15.98}
$$

Then, Equation 15.97 becomes

$$
Z = [1 - \exp\{\beta(\mu - \varepsilon)\}]^{-1}.
\tag{15.99}
$$

The average value or the mean value, (x) or $\bar{x}$, of a set of N numbers $x_1, x_2, x_3, \ldots, x_N$ is given by

$$
(x) = \bar{x} = \frac{x_1 + x_2 + \cdots + x_N}{N} = \frac{1}{N} \sum_{k=1}^{N} x_k
\tag{15.100}
$$

$$
= \sum x_k \mathscr{P}_k
\tag{15.101}
$$

where $\mathscr{P}_k$ represents the probability of getting the value x_k. The summation is over all possible values of x_k. If, when our small system is in a microstate in which it is composed of N particles and has total energy ε_k, the value of any variable x is $x_{N,k}$, then according to Equation 15.101 the ensemble average of the variable x is given by

$$\bar{x} = \sum_N \sum_k \mathscr{P}_{N,k} x_{N,k} = \left[\sum_N \sum_k x_{N,k} \exp\left(\mu\beta N - \beta\varepsilon_k\right) \right] \bigg/ Z \tag{15.102}$$

substituting the value of $\mathscr{P}(\varepsilon, N)$ from Equation 12.97. If we put $x_{N,k} = N$ in this equation, then we find that the mean (ensemble average) number of bosons, $b(\varepsilon)$, in our chosen single particle state (our small system) is

$$b(\varepsilon) = \bar{N} = \left(\frac{1}{Z}\right) \sum_N \sum_k N \exp\left(\mu\beta N - \beta\varepsilon_k\right)$$

$$= \left(\frac{1}{Z}\right) [0 \ \exp\ (0-0) + 1 \ \exp\ \{\beta(\mu - \varepsilon)\}$$

$$+ 2 \ \exp\ \{\beta(2\mu - 2\varepsilon)\} + \cdots] \tag{15.103}$$

(using $N = 0, 1, \ldots$ and the corresponding values of ε). This becomes

$$b(\varepsilon) = \left(\frac{1}{Z}\right) \sum Nx^N \tag{15.104}$$

where $x = \exp\ \{\beta(\mu - \varepsilon)\}$.

It can be shown for $x < 1$ and for all values of n from 0 to ∞ that

$$\sum x^n = (1-x)^{-1}; \quad x\frac{\mathrm{d}}{\mathrm{d}x}\sum x^n = x\sum nx^{n-1} = \sum nx^n.$$

In addition,

$$\sum nx^n = x\frac{\mathrm{d}}{\mathrm{d}x}(1-x)^{-1} = \frac{x}{(1-x)^2}. \tag{15.105}$$

By virtue of Equation 15.105, Equation 15.104 becomes

$$b(\varepsilon) = \frac{x}{1-x} = \frac{\exp\ \{\beta(\mu - \varepsilon)\}}{1 - \exp\ \{\beta(\mu - \varepsilon)\}} \tag{15.106}$$

which is the mean (ensemble average) number of bosons in the small system (the chosen single particle state) which has single particle energy ε. This is called the *Bose–Einstein distribution function*.

By multiplying the numerator and denominator on the right-hand side of Equation 15.106 by $\exp\{\beta(\varepsilon - \mu)\}$, we have

$$b(\varepsilon) = \frac{1}{[\exp\{\beta(\varepsilon - \mu)\}] - 1}$$

$$= \frac{1}{[\exp(\varepsilon - \mu)/kT] - 1} \quad \text{(since } \beta = 1/kT\text{)}. \tag{15.107}$$

When comparing this equation with Equation 15.81, we note that there is $(+1)$ in the denominator of the Fermi–Dirac distribution function, whereas there is (-1) in the denominator of the Bose–Einstein distribution function.

Whether a molecule or an atom obeys the Bose–Einstein or Fermi–Dirac statistics depends on the number of elementary particles (electrons, protons and neutrons) forming this molecule or atom. If the number of elementary particles is even, the molecule or the atom will obey the Bose–Einstein statistics; on the other hand, if the number is odd, the molecule or the atom will obey the Fermi–Dirac distribution law. It does not matter whether the elementary particles are in the nucleus or the electron shells.

It was said earlier that He^4 obeys the Bose–Einstein statistics, whereas He^3 obeys the Fermi–Dirac statistics. Let us now examine why this is so. He^4 has two protons and two neutrons in the nucleus, and two electrons outside the nucleus; the total number of elementary particles making up He^4 is 6 (even number). Hence, it obeys the Bose–Einstein statistics. On the other hand, there is only one neutron in the nucleus of He^3, so it obeys the Fermi–Dirac statistics.

This behaviour actually follows from the symmetry of the wave functions. Systems with an even number of elementary particles have symmetrical wave functions (in the sense that if the co-ordinates of any two bosons are interchanged, the wave function of the N-particle system is unchange), whereas systems with an odd number of elementary particles have antisymmetric wave functions. If two molecules are in the same energy level, exchange of these molecules must leave the wave function identical. In the anti-symmetrical case, if this operation changes the sign, the wave function must disappear. So we cannot have states with two molecules in the same energy level. If we have two molecules m_1 and m_2 with collections of co-ordinates X_1 and X_2, respectively, and if m_1 and m_2 are in different states, then the exchange of these molecules changes a function of X_1 and X_2 to a function of X_2 and X_1, i.e. $f(X_1, X_2) \rightarrow f(X_2, X_1)$. Since in this case the functional dependence on the first set of co-ordinates X_1 is different from that on the second set of co-ordinates X_2, therefore before exchange the function $f(X_1, X_2)$ will differ from the function $f(X_2, X_1)$.

Example 15.6

Given that for a mole of He^4 at 273 K, $\exp(-\mu/kT)$ is 2.55×10^5. Calculate the chemical potential of He^4 and the mean occupancies of the single particle states. Comment on your result. All the symbols have their usual meaning and assume that He^4 has an energy $3kT/2$.

Solution

He^4 is a boson gas. From $\exp(-\mu/kT) = 2.55 \times 10^5$ we have

$$\mu/kT = -12.45.$$

Therefore,

$$\mu = -12.4 \times (1.38 \times 10^{-23} \text{ JK}^{-1})(273 \text{ K})$$
$$= -4.67 \times 10^{-20} \text{ J}.$$

By taking $\varepsilon = \frac{3}{2}kT$ and $\exp(-\mu/kT) = 2.55 \times 10^5$ and using Equation 15.107, we get the mean occupancies of the single particle states $b(\varepsilon) = 8.7 \times 10^{-7}$.

This result shows that at STP the mean occupancies of the single particle states are very small.

15.6.1 Properties of the Bose–Einstein distribution function

Let us consider an ideal boson gas of N identical bosons, each of mass m contained in a cubical box of side L and volume $V(=L^3)$ kept at an absolute temperature T. We assume that the bosons do not interact with one another and that they occupy single particle states (orbitals). In the case of particles, the term single particle quantum state is used rather than a stationary wave solution or normal mode. The term 'orbital' is often used, particularly in chemistry, for a single particle quantum state. The orbitals can be determined by solving Schrödinger's time-independent equation in the way described for a particle in a cubical box (see Appendix 6).

(1) In Equation 15.107 T is the absolute temperature of the heat reservoir and μ is the chemical potential of the particle reservoir. In the present case, the particle reservoir is the bosons in all other single particle states. For bosons, there is no limit to the number that can go into any single particle state. Therefore, at the absolute zero of temperature, all the bosons in our system will be in the single particle ground state, i.e. at $T=0$, $b(\varepsilon)=N$ for the single

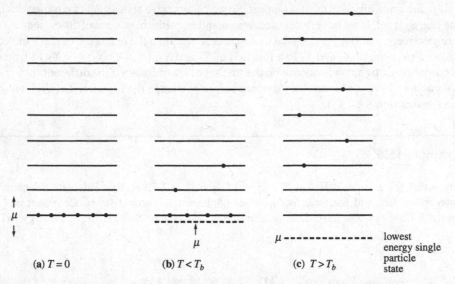

Figure 15.13 Distribution of particles in a system of 6 identical bosons with respect to temperature (the dotted line represents the level of μ)

particle ground state and $b(\varepsilon) = 0$ for all the excited single particle states. Figure 15.13 represents qualitatively the case of a system made up of six identical bosons of spin 0. Figure 15.13(a) shows that at $T = 0$ all the bosons are in the single particle ground state, (b) shows that at $T > 0$ but $< T_b$ (the Bose temperature), some of the bosons have escaped to the excited states but a significant number of them are still in the single particle ground state, and (c) shows that at $T > T_b$ there are very few or no bosons in the single particle ground state. An important point to remember is that at $T = 0$, μ is equal to the energy of the lowest energy single particle state, i.e. $\mu = 0$; at $T < T_b$, μ becomes marginally negative i.e. less than the energy of the lowest energy single particle state (but has a fairly constant numerical value just below zero), and as T becomes greater than T_b, μ becomes more negative. At high temperatures, the average number of bosons in any single particle state is much less than one. When $b(\varepsilon) \ll 1$ for all the single particle states, the situation then corresponds to the classical limit.

(2) To find the energy distribution of the ideal boson gas in a box, the number of single particle states (given by Equation 15.69) which have single particle energy in the range ε to $\varepsilon + d\varepsilon$ has to be multiplied by the mean number of bosons in each of these states, $b(\varepsilon)$. Thus, it follows that multiplying Equations 15.69 and 15.107 together the energy distribution $n(\varepsilon) \, d\varepsilon$ is

$$n(\varepsilon) \, d\varepsilon = \left(\frac{gV}{4\pi^2}\right) \left(\frac{2m}{\hbar^2}\right)^{3/2} \left[\frac{\varepsilon^{1/2} \, d\varepsilon}{\exp\left\{(\varepsilon - \mu)/kT\right\} - 1}\right] \tag{15.108}$$

where all the terms have the same meanings as given earlier. This equation gives the chemical potential μ in terms of N, V and T.

(3) We have already seen that in the high temperature classical limit, $\exp(-\mu/kT)$ becomes very large. Due to the large value of $\exp(-\mu/kT)$ in this limit, $\exp[(\varepsilon - \mu)/kT)] \gg 1$. Under this condition, both bosons and fermions are very small and the equation for $f(E)$ reduces to the classical Boltzmann distribution function (see Equation 15.95 for fermions), and for bosons we have

$$b(\varepsilon) = \exp(\mu/kT) \, \exp(-\varepsilon/kT) \tag{15.109}$$

where $\exp(\mu/kT)$ is the *absolute activity*. By putting $\varepsilon = \frac{1}{2}mv^2$ (where m is the mass and v is the velocity of a boson) into Equation 15.109 and with the help of Equation 15.82, i.e.

$$\exp(-\mu/kT) = \left(\frac{gV}{N}\right)(mkT/2\pi\hbar^2)^{3/2} \gg 1 \tag{15.110}$$

it can be shown that in the classical limit Equation 15.108 converts to the *Maxwell-Boltzmann velocity distribution* (see Equation 15.33) in the same way that the ideal fermion gas did.

(4) If an ideal boson gas consists of N bosons in a box, then N can be found by integrating Equation 15.108. Thus,

$$N = \left(\frac{gV}{4\pi^2}\right) \left(\frac{2m}{\hbar^2}\right)^{3/2} \int_0^\infty \frac{\varepsilon^{1/2} \, d\varepsilon}{\left\{\exp\left\{(\varepsilon - \mu)/kT\right\} - 1\right.} \tag{15.111}$$

If N, V and T are known, we can calculate μ from this equation.

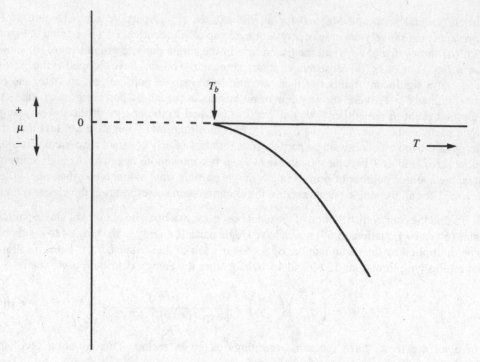

Figure 15.14 Schematic diagram of variation of the chemical potential μ of an ideal boson gas with temperature

(5) It can be seen that the denominator of Equation 15.107 cannot be negative, because the number of bosons in any single particle state cannot be negative. Therefore, for an ideal boson gas we always have $\exp{[(\varepsilon-\mu)/kT]} > 1$, which means $\varepsilon > \mu$. Thus, we can conclude that *the chemical potential μ of an ideal boson gas must always be less than the energy of the lowest energy single particle state (the single particle ground state) and it is always negative.* It might be convenient to choose the energy of the lowest energy single particle state as the zero of the energy scale in a discussion of the ideal boson gas. A point to remember is that if N and V are changed, keeping T constant, T_b will change, and so also μ. Variation of μ of an ideal boson gas with temperature is shown schematically in Figure 15.14. The value of the chemical potential μ of an ideal boson gas is practically zero below and up to T_b as shown by the broken line in Figure 15.14, and it becomes increasingly negative as T becomes greater than T_b.

(6) If U is the total energy of an ideal boson gas, then U can be found by multiplying $n(\varepsilon)\,d\varepsilon$ by the energy ε of each particle and then integrating the product within the limit 0 to ∞. Thus, using Equation 15.108, we can write

$$U = \int_0^\infty n(\varepsilon)\varepsilon\,d\varepsilon = \left(\frac{gV}{4\pi^2}\right)\left(\frac{2m}{\hbar^2}\right)^{3/2}\int_0^\infty \frac{\varepsilon^{3/2}\,d\varepsilon}{\exp{[(\varepsilon-\mu)/kT]}-1}. \tag{15.112}$$

When $T=0$, all the bosons in an ideal boson gas are in the lowest energy single particle state (the single particle ground state). Choosing this as the zero of our energy scale, variation of

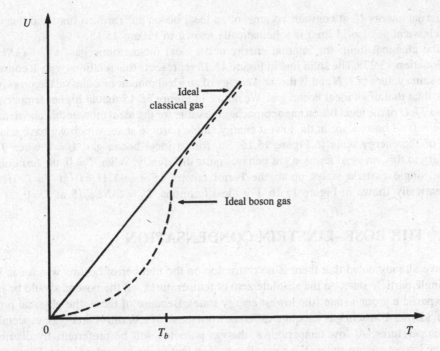

Figure 15.15 Schematic diagram of variation of the internal energy U of an ideal boson gas with temperature at constant volume

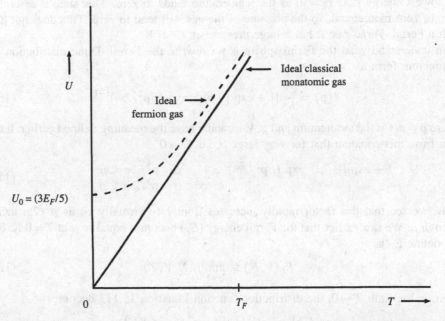

Figure 15.16 Schematic diagram of variation of the internal energy U of an ideal fermion gas with temperature at constant volume

the internal energy U, at constant volume, of an ideal boson gas (broken line) and that of an ideal classical gas (solid line) is schematically shown in Figure 15.15.

In the classical limit, the internal energy of an ideal monoatomic gas is $U = (3NkT)/2$ (see Equation 13.27). The solid line in Figure 15.15 represents this relation (with V constant). At the same values of T, N and V, the total energy of an ideal monatomic classical gas is always higher than that of an ideal boson gas. We notice in Figure 15.15 that at higher temperatures beyond T_b, U of the ideal boson gas approaches the value for the ideal monatomic classical gas.

At $T = 0$ all bosons are in the lowest energy single particle state, which we have taken as the 0 of the energy scale in Figure 15.15. So, for an ideal boson gas, $U = 0$ when $T = 0$. Contrary to this, an ideal fermion gas behaves quite differently. When $T = 0$, the fermions fill all the single particle states up to the Fermi energy (E_F) and $U \neq 0$ at $T = 0$. This is schematically shown in Figure 15.16. For ideal fermions, $U_0 = (3NE_F)/5$ at $T = 0$.

15.7   THE BOSE–EINSTEIN CONDENSATION

We have already noted that there is no restriction on the number of bosons which can be in any single particle state. At the absolute zero of temperatures, all the bosons should be in the single particle ground state (the lowest energy state). Because of these, the physical properties of a Bose–Einstein gas are quite different from those of a Fermi–Dirac gas, especially at low temperatures. At low temperatures, the gas particles will be preferentially distributed over the lowest energy states. We have already seen that in the case of a Fermi–Dirac gas all the particles are not distributed in the lowest energy state as the temperature tends to zero; *Pauli's exclusion principle* forces them to form the *Fermi sphere* distribution. However, as this principle is not applicable to Bose–Einstein particles, they all tend to be distributed in the lowest energy state ($\varepsilon = 0$) as the temperature tends to zero. This state is essentially a state of zero momentum, so the pressure of the gas will tend to zero. This does not happen with a Fermi–Dirac gas; it has a large pressure at $T = 0$ K.

To understand what the Fermi sphere is we rewrite the Fermi–Dirac distribution in the momentum form as

$$f(\mathbf{p}) = \frac{gV}{h^3} [1 + \exp(-\beta\mu) \exp(\beta\mathbf{p}^2/2m)]^{-1} \tag{15.113}$$

where $\mathbf{p} = m\mathbf{v}$ is the momentum and g, V, m, and $\mathbf{v}$ have the meaning defined earlier. It can be seen from this equation that for very large β, i.e. $T \to 0$

$$\left. \begin{array}{ll} 1 + \exp(-\beta\mu) \exp(\beta\mathbf{p}^2/2m) \simeq 1 & \text{if } \mathbf{p}^2/2m < \mu \\ \simeq \infty & \text{if } \mathbf{p}^2/2m > \mu \end{array} \right\}. \tag{15.114}$$

Thus, we see that this factor rapidly increases from 1 to virtually ∞ as $\mathbf{p}^2/2m$ increases through μ. We saw earlier that the Fermi energy (E_F) becomes equal to μ at $T = 0$ K. So, we can define E_F as

$$E_F(V, N) \equiv \lim_{T \to 0} \mu(N, V, T). \tag{15.115}$$

Then, in the limit $T \to 0$, the distribution function Equation 15.113 becomes

$$\left. \begin{array}{ll} f(\mathbf{p}) = gV/h^3 & \text{if } \mathbf{p}^2/2m < E_F \\ = 0 & \text{if } \mathbf{p}^2/2m > E_F \end{array} \right\}. \tag{15.116}$$

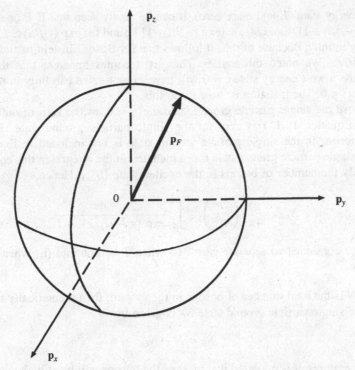

Figure 15.17   The Fermi sphere of uniform value equal to gV/h^3

This distribution is essentially different from the Maxwellian distribution $\eta(\mathbf{p})$ (put $\mathbf{p} = m\mathbf{v}$ in Equation 15.18 and compare). $f(\mathbf{p})$ is discontinuous at $\mathbf{p}^2/2m = E_F$. If $f(\mathbf{p})$ is represented in momentum space as in Figure 15.17 it will have a uniform value of gV/h^3 inside a sphere which has a radius equal to $(2mE_F)^{1/2}$ and a value of zero outside this sphere. The sphere so obtained is known as the *Fermi sphere* and its radius is called the *Fermi momentum*

$$p_F = (2mE_F)^{1/2}. \qquad (15.117)$$

We now know why and how the perfect Bose–Einstein particles distribute in the lowest energy state ($\varepsilon = 0$) as $T \to 0$, but the way in which the particles will begin to occupy the lowest energy state as $T \to 0$ is not so clear. We might think that the probable number of particles in the lowest energy state would increase uniformly as $T \to 0$. Although this is a reasonable expectation, at least for large systems this is not true. It has been found that in such systems the lowest energy level ($\varepsilon = 0$) is filled very abruptly at a certain critical temperature. This then leads to an abrupt change in the behaviour of all thermodynamic properties. This behaviour is unexpected and is the most interesting characteristic of the perfect Bose–Einstein gas.

We have seen that the formal difference between the Fermi–Dirac and the Bose–Einstein distributions is the term 1 (+ in the former and − in the latter). In the non-degenerate case $\exp(-\beta\mu) \gg 1$, this term is not important, and both distributions reduce to the classical distribution (Maxwellian). However, in the degenerate case $\exp(-\beta\mu) \gtrsim 1$ (i.e., $\beta\mu$ is a small negative number) the two distribution functions are completely different; this is more so for the lowest energy state (E and ε are zero). It can be easily seen that if E or ε is zero, then $f(E) = [\exp(-\beta\mu) + 1]^{-1}$ and $b(\varepsilon) = [\exp(-\beta\mu) - 1]^{-1}$, and for $\exp(-\beta\mu) \simeq 1$, $f(E) = \frac{1}{2}$ and $b(\varepsilon)$ is nearly infinity. Because of this, it follows that for Bose–Einstein particles, μ must be always negative. We noted this earlier. Thus, it becomes apparent that the occupation number for the lowest energy state $\varepsilon = 0$ will have to be treated carefully in the degenerate limit $\beta\mu \simeq$ or < 0. The question is: how to do this.

If we choose the single particle ground state, i.e. $\varepsilon = 0$, as the zero of our energy state, then $\sqrt{\varepsilon}$ in Equation 15.111 is zero for the single particle ground state. Therefore, the number of bosons in the single particle ground state is not included in Equation 15.111 because the single particle ground state is not included in the integral in this equation. It will then give only the number of bosons in the excited state (N_{ex}). Hence,

$$N_{ex} = \frac{gV}{4\pi^2}\left(\frac{2m}{\hbar^2}\right)^{3/2}\int_0^\infty \frac{\sqrt{\varepsilon}\,d\varepsilon}{\exp(\varepsilon - \mu)/kT - 1}. \tag{15.118}$$

We have to consider two separate cases: (i) when T is high and (ii) when T is near the absolute zero.

Case (i): If N is the total number of bosons in the system, then theoretically the number of bosons in the single particle ground state N_0 is given by

$$N_0 = N - N_{ex}. \tag{15.119}$$

At high temperatures, it is expected that most of the bosons will be at high energy levels, so the number of bosons in the single particle ground state will be very much smaller than N. Thus, $N_0 \ll N$. Under this condition, Equation 15.111 is a satisfactory approximation for the total number of bosons.

Case (ii): As mentioned earlier, near the absolute zero of temperature, all the N bosons tend to occupy the single particle ground state. Therefore, in the limit as $T \to 0$, the mean number of bosons in a single particle state having energy ε, i.e., the Bose–Einstein distribution function $b(\varepsilon)$, tends to N for the single particle gound state.

Now, if we choose the single particle ground state as our zero of energy and put $\varepsilon = 0$ in Equation 15.107, then in the limit as $T \to 0$, for the ground state we shall have

$$b(0) = [\exp(-\mu/kT) - 1]^{-1} \simeq N \tag{15.120}$$

or

$$-\mu/kT = \ln(1 + 1/N). \tag{15.120}$$

For large values of N, at temperatures very close to the absolute zero, we have

$$-\mu/kT \simeq 1/N \tag{15.121}$$

because $\ln(1 + 1/N) \simeq 1/N$. Therefore, Equation 15.121 shows that at $T = 0$, μ of an ideal boson gas is zero.

For a small macroscopic system (say, consisting of bosons of the order 10^{16}–10^{17}), the value of $-\mu/kT$ will be extremely small (as given by Equation 15.121). Therefore, at temperatures near the absolute zero, we can assume that for a small system $\exp(-\mu/kT) \simeq 1$ in Equation 15.118. We have already seen in Figure 15.14 that μ of an ideal boson gas is practically zero up to the Bose temperature. Therefore, this assumption is a good approximation up to the Bose temperature. Thus, we obtain

$$N_{ex} = \left(\frac{gV}{4\pi^2}\right)\left(\frac{2m}{\hbar^2}\right)^{3/2} \int_0^\infty \frac{\sqrt{\varepsilon}\, d\varepsilon}{\exp(\varepsilon/kT) - 1}. \tag{15.122}$$

Let $x = \varepsilon/kT$. Then, $dx = d\varepsilon/kT$, so $d\varepsilon = kT\, dx$. Also $\sqrt{x} \cdot \sqrt{kT} = \sqrt{\varepsilon}$. Hence, $\sqrt{\varepsilon}\, d\varepsilon = \sqrt{x}\, dx(kT)^{3/2}$. By substituting this in Equation 15.122, we obtain

$$N_{ex} = \left(\frac{gV}{4\pi^2}\right)\left(\frac{2mkT}{\hbar^2}\right)^{3/2} \int_0^\infty \frac{\sqrt{x}\, dx}{\exp(x) - 1}. \tag{15.123}$$

The integral in this equation is equal to the product of the gamma function $\Gamma(3/2) = \sqrt{\pi}/2$ and the Riemann zeta function $\zeta(3/2) = 2.612$. Hence, by substituting these values in Equation 15.123, we have

$$N_{ex} = \left(\frac{gV}{4\pi^2}\right)\left(\frac{2mkT}{\hbar^2}\right)^{3/2}\left(\frac{\sqrt{\pi}}{2}\right)(2.612)$$

$$= 2.612gV\left(\frac{mkT}{2\pi\hbar^2}\right)^{3/2}. \tag{15.124}$$

There is a temperature at which, according to Equation 15.124, all the bosons should be in excited states. This temperature is sometimes called the *Bose–Einstein temperature* and denoted by T_b. Thus, at $T = T_b$, $N_{ex} = N$, and Equation 15.124 becomes

$$N = 2.612gV\left(\frac{mkT_b}{2\pi\hbar^2}\right)^{3/2}. \tag{15.125}$$

Hence,

$$T_b = \frac{2\pi\hbar^2}{mk}\left(\frac{N}{2.612gV}\right)^{2/3} = \frac{3.31\hbar^2}{mk}\left(\frac{N}{gV}\right)^{2/3}. \tag{15.126}$$

Example 15.7

A He^4 gas is composed of 6.02×10^{23} atoms, each of mass 6.65×10^{-27} kg. It is confined in a container of volume 22.4×10^{-3} m^3. The atoms have zero spin, and we assume that there are no interactions between them. Calculate the temperature at which all the atoms will be in the excited state. Comment on your result if there were interactions between the particles. What would be the Bose–Einstein temperature of liquid helium if it were treated as an ideal boson gas?

Solution

He^4 is a boson gas. So the temperature at which all the atoms will be in excited state is the Bose–Einstein temperature T_b, which is given by Equation 15.126. Given that $N = 6.02 \times 10^{23}$, the mass of a boson $m = 6.65 \times 10^{-27}$ kg, the spin of a boson $s = 0$, so $g = 2s + 1 = 1$, the volume of the container $V = 22.4 \times 10^{-3}$ m^3, $\hbar^2 = 1.055 \times 10^{-34}$ J s and $k = 1.38 \times 10^{-23}$ JK^{-1}. We have to be careful about the units. By substituting these values in Equation 15.126, we have

$$T_b = \frac{(3.31)(1.055 \times 10^{-34} \text{Js})^2}{(6.65 \times 10^{-27} \text{kg})(1.38 \times 10^{-23} \text{J/K})} \left[\frac{6.02 \times 10^{23}}{(1)(22.4 \times 10^{-3} \text{ m}^3)} \right]^{2/3}$$

$$= 0.035 \frac{\text{J}^2 \text{ s}^2 \text{ K}}{\text{kg J m}^2}$$

$$= 0.035 \text{ K (since 1 J} = \text{kg m}^2 \text{ s}^{-2}).$$

It is an experimental fact that if there were no interactions between the bosons, the ideal boson gas would not liquefy. However, in actual practice, due to the interactions between the bosons, all real boson gases liquefy before they are cooled to T_b. Helium gas liquefies at 4.21 K at 0.1 MPa pressure, which is much higher than the result we obtained using Equation 15.126. The effect of this liquefaction is an increase in the number of bosons per unit volume (N/V).

If the liquefied helium were treated as an ideal boson gas, the value of T_b given by Equation 15.126 would have been 3.1 K.

We wish to find out the relative value of N_0 (the number of boson in the single particle ground state) with respect to the temperature. Dividing Equation 15.124 by Equation 15.125, to a good approximation we have (when $T_b > T$)

$$N_{ex} = N\left(\frac{T}{T_b}\right)^{3/2}. \tag{15.127}$$

By combining this equation with Equation 15.119, we obtain

$$N_0 = N[1 - (T/T_b)^{3/2}] \quad (T \leqslant T_b) \tag{15.128}$$

which, to a good approximation, gives the number of bosons in the single particle ground state for temperatures below T_b.

Equation 15.128 shows that at $T = 0$ K, all the particles are in the lowest energy state (i.e. $\mathbf{p}^2/2m = 0$, or $\mathbf{p} = 0$, where $\mathbf{p}$ is the momentum). As T increases, the number of particles in this state decreases, as shown in Figure 15.18, and the remainder are distributed over all the other particle states. These particles can be represented by a continuous distribution function in terms of momentum, as given by Equation 15.129.

$$b(\mathbf{p}) = \left(\frac{4\pi g V}{h^3}\right)(\mathbf{p}^2)[\exp{(\beta \mathbf{p}^2/2m)} \exp{(-\beta\mu)} - 1] \tag{15.129}$$

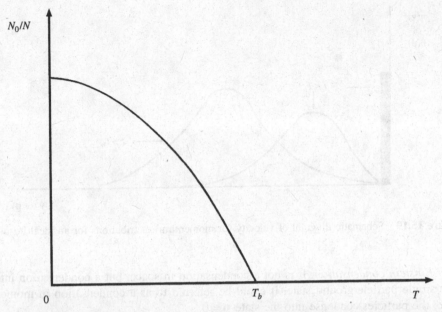

Figure 15.18 Schematic diagram of variation of the ratio N_0/N with the absolute temperature T for T T_b. N_0 is the number of bosons in the single particle ground state and N is the total number of bosons

where $\mathbf{p}$ is the momentum of the bosons and the other symbols have the meanings defined earlier. We can put $\mu = 0$ in this equation when $T < T_b$. A point to remember is that $b(\mathbf{p} = 0) = 0$ and $b(0)$ (as given by Equation 15.120) are not the same. $b(0)$ represents the occupation number of one state: the single particle ground state (the lowest energy state). On the other hand, $b(\mathbf{p})$ is an ordinary distribution function so that

$$b(\mathbf{p}) \, d\mathbf{p} = \text{average number of particles in}$$

$$\text{the range } \mathbf{p} \text{ and } \mathbf{p} + d\mathbf{p}.$$

For temperatures well below the Bose temperature T_b, there is a finite significant fraction of the particles in the single particle ground state (or single lowest-momentum state). This is illustrated in Figure 15.3b. At $T > T_b$ the number of bosons in the single particle ground state (if any at all) is very much less than N (the total number of bosons). This is shown in Figure 15.13c.

The velocity distribution or the distribution of an ideal boson gas over the momentum states is shown schematically for $T < T_b$ and $T > T_b$ in Figure 15.19. The bold line, which is marginally to the right of the ordinate, represents the bosons in the single particle ground state. It is to be noted that if the bosons are confined inside a container of finite dimensions, the velocity (hence the momentum) of the bosons in the single particle ground state cannot be zero because of the uncertainty principle. This is why the thick line is fractionally on the right of the ordinate. From the figure it can be seen that when $T > T_b$ the distribution tends to the Maxwell–Boltzmann distribution, but when $T < T_b$ there are significant differences.

For temperatures well below T_b there is a finite fraction of particles in the single particle ground state (single lowest-momentum state). This phenomenon is often called the

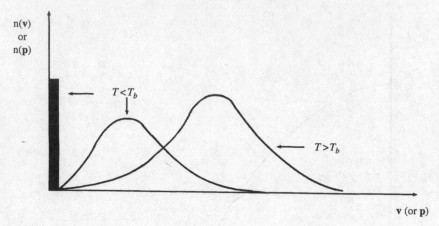

Figure 15.19   Schematic diagram of velocity (or momentum) distributions for an ideal boson gas

Bose–Einstein condensation. It is not a condensation in space, but a condensation into the same single particle ground state. It could be referred to as a condensation in momentum space; the particles condense into the state $\mathbf{p} = 0$.

This condensation of the particles into the state $\mathbf{p} = 0$ produces abrupt changes in the thermodynamic properties of the ideal Bose–Einstein gas when the temperature decreases through T_b. Since the particles that are in state $\mathbf{p} = 0$ have zero momentum, they do not contribute to the pressure or internal energy of the system. Therefore, the pressure and the internal energy rapidly decrease as the temperature tends to zero. In the next section we shall see that one of the most interesting results of this condensation is the abrupt change in the heat capacity at $T = T_b$.

15.8   THERMODYNAMICS OF IDEAL FERMION AND BOSON GASES

In the preceding sections we have studied in some detail the distributions of ideal Fermi–Dirac and Bose–Einstein gases. Although we now have a quantitative idea of the behaviour of these gases, we still do not have explicit expressions for the thermodynamic functions. In the subsequent sections, we wish to explore the thermodynamic properties of these gases and attempt to obtain suitable equations for the thermodynamic quantities such as internal energy, pressure, heat capacity, entropy, etc. The chemical potential of ideal fermion and boson gas has been discussed.

The properties of the perfect gases discussed in Section 2.5 were derived for a system of noninteracting gas molecules (no attraction or repulsion) which obeyed the second law of Newton, $F = m\mathbf{a}$. Statistical mechanics *per se* was not mentioned at all; the whole development was along the lines of what is traditionally thought of as thermodynamics and the kinetic theory of gases. We have seen that the equation of state, $PV = nRT$, of a gas without cohesion and covolume (i.e. the volume actually occupied by the individual molecules of an ideal gas under ordinary conditions), obeying any of the three statistics Maxwell–Boltzmann–Planck, Fermi–Dirac and Bose–Einstein, is obtained when the molecules of the gas are distinguishable. For distinguishable particles, it is possible to keep track of which particles we have assigned to particular energy states of the gas system.

It is now well established through experimental results that particles of a given type on the nuclear and atomic levels are indistinguishable, and so cannot be labeled for the purpose of identification. Although any electron, proton and photon can be distinguished from one another, they behave in exactly the same way; there is no way to number the various electrons, protons or photons. Not only are all atomic particles of a given class identical, but all classes of particles can also be classified according to the spin angular momentum (for short, called spin) of the particle. Spin is an inherent property of all particles found in nature, and it could be an integer or an odd half-integer multiple of $h/2\pi = \hbar$, where h is Planck's constant.

Particles with spin of odd half-integer multiples of $\hbar$ obey the *Fermi–Dirac statistics* and are known as *fermions*, whereas particles with spin of an integer multiple of $\hbar$ obey the Bose–Einstein statistics and are known as *bosons*. Some common fermions and bosons are listed in Table 15.1.

In the above table we observe that protons and neutrons are fermions, and they have spin $\hbar/2$; however, a composite particle, such as He^4, O^{16}, N^{14} and B^{10}, consisting of an even number of protons and neutrons behaves as a boson. On the other hand, a composite particle, such as He^3, O^{17}, N^{15} and B^{11}, consisting of an odd number of fermions behaves like a fermion. This may appear anomalous, but these observations are, in fact, consistent with the general rule that all particles, whether or not they are composites, which have integral spin obey the Bose–Einstein statistics and all particles which have odd half-integral spin obey the Fermi–Dirac statistics.

In the following sections we shall see that the internal energy, entropy and pressure of a boson gas tend to zero when the absolute temperature goes to zero. On the other hand, a

Table 15.1. Spin angular momenta of some fermions and bosons

Particle or element	Spin (units of $\hbar$)
Fermions	
Electron (e)	1/2
Proton (p)	1/2
Neutron (n)	1/2
Muon (μ)	1/2
He, 3 nucleus (He^3)	1/2
Oxygen, 17 nucleus (O^{17})	5/2
Nitrogen, 15 nucleus (N^{15})	1/2
Boron, 11 nucleus (B^{11})	3/2
Bosons	
Photon (γ)	1
π-meson (π)	0
He, 4 nucleus (He^4)	0
Oxygen, 16 nucleus (O^{16})	0
Nitrogen, 14 nucleus (N^{14})	1
Boron, 10 nucleus (B^{10})	3

fermion gas, at a finite volume, has zero entropy when the absolute temperature is zero, but the internal energy and pressure do not disappear at $T = 0$. For large values of $VT^{3/2}$, the equation of state of a weakly degenerate fermion gas may be written as

$$PV = RT \left[1 + \frac{Nh^3 (VT^{3/2})^{-1}}{16(m\pi k)^{3/2}} \right] \tag{15.130}$$

and that of a weakly degenerate boson gas as

$$PV = RT \left[1 - \frac{Nh^3 (VT^{3/2})^{-1}}{16(m\pi k)^{3/2}} \right] \tag{15.131}$$

where P, V and T are the pressure, volume and temperature, respectively, R is the universal gas constant, k and h are Boltzmann's and Planck's constants, respectively, and N is the number of particles of mass m. It can be seen that for sufficiently large values of $VT^{3/2}$ the second term in Equations 15.130 and 15.131 can be neglected and the equation of state in both cases reduces to $PV = RT$ which obeys the three classical gas laws (Boyle's, Charles's and Joule's laws). Then we can say that the gas is nondegenerate.

The statement that the fermion or boson gas becomes nondegenerate under the condition that $VT^{3/2}$ is sufficiently large means that they obey the three classical perfect gas laws. Therefore, at high temperatures or when the volume occupied by the molecules is large ideal gas laws are obeyed by the gas made up of real particles. Since these conditions are applicable to dilute gases, such gases are ideal. Quite clearly, at high temperatures the thermal energies of the gas molecules are high, so the number of allowed energy states for the system becomes large. Although the number of available energy states increases, the number of molecules available to fill these states does not change. Consequently, the probability that a particular quantum state will be occupied by a molecule of the gas becomes small. This is the condition under which the gas behaves classically, and it is then said to be nondegenerate.

15.8.1 Thermodynamics of perfect Fermi–Dirac gases

In this section we shall consider in some detail the thermodynamic properties of a system of noninteracting Fermi–Dirac particles (a perfect fermion gas). Perhaps the most frequent applications of these results is to *free electrons* in metals. It has long been thought that the electrons in metals (such as the highly metallic alkali metals) are free to move. These electrons may be considered approximately as an ideal gas of free electrons held in a container by the attractive forces of the positive particles. This model was suggested before the advent of the quantum mechanics, and was quite successful in explaining the high conductivity of heat and electricity in metals. However, it was very puzzling that there was no apparent contribution of the electrons to the specific heat. We shall discuss this in the next chapter.

Internal energy of perfect Fermi–Dirac gases

The average energy of the ideal fermion gas can be obtained using equation (15.183) as

$$U = \int_0^\infty E n(E)\, dE$$

$$= \frac{gV}{4\pi^2}\left(\frac{2m}{\hbar^2}\right)^{3/2} \int_0^\infty \frac{(E)^{3/2}\, dE}{\exp(E/kT)\exp(-\mu/kT)+1}. \tag{15.132}$$

Figure 15.16 shows the variation of U with the absolute temperature at constant volume V. The internal energy of an ideal monatomic classical gas is given by $3NkT/2$, where N is the total number of atoms. The solid line represents $U = 3NkT/2$. From this figure it can be seen that for same V, T and N, the energy of the perfect Fermi–Dirac gas is always higher than that of a perfect monatomic classical gas. We have already seen that when $T > T_F$, the behavior of the ideal fermion gas begins to follow the ideal monatomic classical gas laws. On the other hand, when $T < T_F$, the behaviour of the ideal fermion gas deviates quite markedly from the ideal classical gas laws. The ideal fermion gas is degenerate in this temperature range. However, when $T = 0$, there are no fermions with energies higher than the Fermi energy E_F, so we can replace the upper limit ∞ in Equation 15.132 with E_F. Then the mean energy $\bar{U}$ of fermions is given by

$$\bar{U} = \frac{1}{N}\int_0^{E_F} E n(E)\, dE = \frac{gV}{4\pi^2 N}\left(\frac{2m}{\hbar^2}\right)^{3/2}\int_0^{E_F} (E)^{3/2}\, dE \tag{15.133}$$

or

$$\bar{U} = \frac{gV}{10\pi^2 N}\left(\frac{2m}{\hbar^2}\right)^{3/2}(E_F)^{5/2}.$$

However,

$$E_F = (\hbar^2/2m)\left(\frac{6\pi^2 N}{gV}\right)^{2/3}$$

(see Equation 15.86). Hence,

$$\bar{U} = \left(\frac{3}{5}\right)E_F \quad \text{when} \quad T = 0. \tag{15.134}$$

If an ideal fermion gas consists of N noninteracting particles, then its total energy is given by the product of N and $\bar{U}$ of each fermion. Hence, at $T = 0$ the total energy of an ideal fermion gas is

$$U_0 = \left(\frac{3}{5}\right)NE_F \tag{15.135}$$

(see Figure 15.16).

Pressure of perfect Fermi–Dirac gases

Consider an ideal fermion gas inside a container of rigid adiabatic walls fitted with a movable frictionless piston of area A and made of a rigid adiabatic material. The piston can move forward and backward. Let us assume that the system is in its ith microstate with energy ε_i and that the pressure exerted on the piston by the system is P_i. If the piston moves a distance dx inward, the volume of the system will change by area $\times$ displacement $= A\,dx$, and the work done on the system will be $P_i A\,dx = -P_i\,dV$, where dV is the change in volume. If all the work is used to increase the, energy of the ith microstate from ε_i to $\varepsilon_i + d\varepsilon_i$, then

$$d\varepsilon_i = -P_i\,dV. \tag{15.136}$$

Hence, according to the *principle of virtual work*, the pressure of the system in its ith microstate with energy ε_i is

$$P_i = -\partial\varepsilon_i/\partial V. \tag{15.137}$$

Imagine that in a real change of volume the piston moves infinitely slowly, so that at every stage it is reversible. According to the time-dependent perturbation theory of quantum mechanics, during this extremely slow movement of the piston the probability that the system is in any particular microstate does not change. Therefore, the change takes place at constant entropy. The average energy of the system is given by

$$U = \sum_{\text{states}} \mathscr{P}_i \varepsilon_i \tag{15.138}$$

where $\mathscr{P}_i$ is the probability. By differentiating, we have

$$dU = \sum \varepsilon_i\,d\mathscr{P}_i + \sum \mathscr{P}_i\,d\varepsilon_i.$$

For any reversible adiabatic change, $\mathscr{P}_i$ is constant, so $d\mathscr{P}_i = 0$. Therefore, the above equation becomes

$$dU = \sum \mathscr{P}_i\,d\varepsilon_i.$$

By substituting the value of $d\varepsilon_i$ from Equation 15.136, we obtain

$$dU = -\left(\sum P_i\mathscr{P}_i\right)dV. \tag{15.139}$$

However, $\sum P_i\mathscr{P}_i$ is the average pressure of the system P. Therefore,

$$dU = -P\,dV \tag{15.140}$$

which is the same as Equation 5.7 developed from classical equilibrium thermodynamics. Taking

$$P = \sum_{\text{states}} P_i\mathscr{P}_i$$

and by using Equation 15.137 for P_i, the average pressure of the system is

$$P = \sum_{\text{states}} \mathscr{P}_i \left(\frac{-\partial \varepsilon_i}{\partial V} \right). \tag{15.141}$$

If a system consists of N particles and $E_i(N, V)$ is the energy of the ith N particle state, and if $\mathscr{P}_i$ is the probability that the N-particle system is in that N-particle state, then in view of Equation 15.141 the mean pressure of the perfect Fermi–Dirac gas can be written as

$$P = \sum \mathscr{P}_i \left[\frac{-\partial E_i(N, V)}{\partial V} \right]. \tag{15.142}$$

For an ideal fermion in a cubical box, the energies of the single particle states are given by Equation i in the solution of Exercise 12.25, that is

$$\varepsilon = (\hbar^2 \pi^2 / 2m V^{2/3})(n_x^2 + n_y^2 + n_z^2). \tag{15.143}$$

If there are no interactions between the fermions in an ideal Fermi–Dirac gas, the energy of the system consisting of N fermions is equal to the sum of the energies of all the N individual fermions. Therefore, the energy E_i of the ith microstate of the system is

$$E_i = (\hbar^2 \pi^2 / 2m V^{2/3}) \sum [(n_x)_s^2 + (n_y)_s^2 + (n_z)_s^2] \tag{15.144}$$

where the summation is over the N single particle states occupied by the N particles when the system is in its ith N particle state. By differentiating both sides of Equation 15.144 with respect to V and substituting the value of $\partial E / \partial V$ in Equation 15.141, we have

$$P = \frac{2}{3} \sum \mathscr{P}_i (E_i / V) = \left(\frac{2}{3} \right)(U/V) \tag{15.145}$$

(using Equation 15.138). Here U is the average energy of the N-fermion system. This equation is valid for an ideal fermion gas for all values of U provided there are no interactions between the particles. The same argument can be applied to derive Equation 15.145 for an ideal boson gas. This equation shows that the pressure of a perfect Fermi–Dirac gas is proportional to the internal energy U if the volume is constant. So, the variation of P with the temperature at constant volume should have the same characteristic as the variation of U with T (see Figure 15.16). We see from Figure 15.16 that the pressure of a perfect Fermi–Dirac gas is always higher than that of a perfect classical monatomic gas, provided that N, V and T are the same for both gases. Note that in the classical limit, the pressure of an ideal monatomic gas is equal to NkT/V.

If all the probabilities $\mathscr{P}_i$ are constant, the entropy S is constant. This can be easily seen from Boltzmann's definition of entropy which is

$$S = -k \sum \mathscr{P}_i \ln \mathscr{P}_i = \text{constant}. \tag{15.146}$$

Under this condition, $\partial \mathscr{P}_i / \partial V = 0$. Therefore, from Equation 15.141 we have, when S is constant,

$$P = -\frac{\partial}{\partial V}\left(\sum \mathscr{P}_i \varepsilon_i\right)_{S,N}$$

and by substituting the value of $\sum \mathscr{P}_i \varepsilon_i$ from Equation 15.138 we obtain

$$P = -\left(\frac{\partial U}{\partial V}\right)_{S,N}. \tag{15.147}$$

This is the equation as was developed from classical equilibrium thermodynamics (see Equation 13.29). It shows that there is a definite relationship between the pressure and the internal energy of an ideal Fermi–Dirac gas.

Entropy of perfect Fermi–Dirac gases

Since entropy is a function of the internal energy, the volume of the gas and the total number of particles, we can write

$$S = f(U, V, N).$$

Then,

$$dS = \left(\frac{\partial S}{\partial U}\right)_{V,N} dU + \left(\frac{\partial S}{\partial V}\right)_{U,N} dV + \left(\frac{\partial S}{\partial N}\right)_{U,V} dN.$$

Since N is constant, $dN = 0$ and we have

$$dS = \left(\frac{\partial S}{\partial U}\right)_{V,N} dU + \left(\frac{\partial S}{\partial V}\right)_{U,N} dV.$$

If the entropy is constant, $dS = 0$. Hence, the above equation becomes

$$0 = \left(\frac{\partial S}{\partial U}\right)_{V,N} dU + \left(\frac{\partial S}{\partial V}\right)_{U,N} dV.$$

By dividing both sides by dV, we have

$$0 = \left(\frac{\partial S}{\partial U}\right)_{V,N}\left(\frac{\partial U}{\partial V}\right)_{S,N} + \left(\frac{\partial S}{\partial V}\right)_{U,N}.$$

Using Equations 13.30 and 15.147, the above equation becomes

$$P = T\left(\frac{\partial S}{\partial V}\right)_{U,N}. \tag{15.148}$$

This is the same equation as Equation 7.29 of classical equilibrium thermodynamics.

Heat capacity of perfect Fermi–Dirac gases

There are two cases to be considered: the heat capacity (C_V) at high temperatures and that at low temperatures. For low-temperature cases, C_V is given by

$$C_V = \left(\frac{\partial U}{\partial T}\right)_V = \left(\frac{\pi^2 kT}{2E_F}\right)Nk \qquad (15.149)$$

when $E_F \gg kT$. As can be seen from Figure 15.20, in the low-temperature limit, C_V of the free electrons in a solid is proportional to the temperature and is equal to zero when $T = 0$. The reason for this is that when heat is added to a system only those electrons which are in the proximity of the Fermi surface are affected. The number of such electrons (which change their energy) is proportional to kT/E_F and each electron's contribution to C_V is approximately $3k/2$ to the heat capacity. Therefore, the heat capacity should be proportional to

$$\left(\frac{3k}{2}\right)\left(\frac{kT}{E_F}\right)$$

i.e. proportional to $k^2 T/E_F$.

One of the important observations that can be made from Equation 15.149 is that the electronic heat capacity should be much less than the classical value for all temperatures $E_F/k \gg T$. A little more detailed analysis might make this result apparent. According to the classical equiparition law the internal energy of the conduction electrons (which are near the

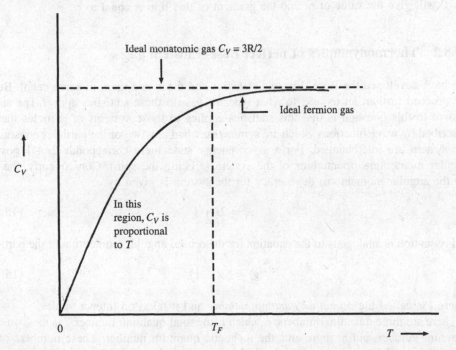

Figure 15.20 Schematic diagram of variation of C_V of an ideal fermion gas at high and low temperatures

Fermi surface) should be $3NkT/2$. Therefore, their contribution to heat capacity is $C_V = 3Nk/2$. On the other hand, the heat capacity of solids at high temperatures is $C_V \simeq 3kN_0$ (given by the Dulong–Petit law) where N_0 is the number of atoms, and this heat capacity is due to the motion of the atoms in the lattice. Thus, it follows that the classical heat capacity $(3Nk/2)$ must be much higher than the heat capacity of the free electrons in a solid. Equation 15.149 shows that the electronic heat capacity should indeed be much less than the classical heat capacity for all temperatures $T \ll E_F/k$. At very low temperatures, C_P and C_V are essentially equal.

At high temperatures beyond T_F, the heat capacity tends to the classical value of $3R/2$. A point to remember is that at very low temperatures the electronic heat capacity given by Equation 15.149 is very small; however, it may equal or exceed the heat capacity due to the lattice vibrations of the solid. According to the Debye T^3 law, the lattice heat capacity at low temperatures is proportional to T^3. So the total heat capacity of the solid at low temperatures becomes

$$C_V = \gamma T + \alpha T^3$$

or

$$C_V/T = \gamma + \alpha T^2 \tag{15.150}$$

where γ and α are the proportionality constants. This equation is known as the *Debye–Sommerfeld equation*. A plot of C_V/T against T^2 should be a straight line whose intercept at $T = 0$ will give the value of γ, and the gradient of this line is equal to α.

15.8.2   Thermodynamics of perfect Bose–Einstein gases

We have developed the Bose–Einstein statistics and discussed them in some detail. Before we proceed further, let us ask: to what types of gas do these statistics apply? The simple answer to this question is that this statistics applies to those systems of particles that are described by wave functions which are symmetric when any two of the particles constituting the system are interchanged. For a given energy state there corresponds $2s + 1$ possible angular momentum orientations of the system (s being the spin). Consequently, the spin (or the angular momentum) degeneracy for the system is given by

$$g_s = 2s + 1 \tag{15.151}$$

This equation is analogous to the equation for the orbital angular momentum of the particles:

$$g_l = (2l + 1) \tag{15.152}$$

where l is called the *azimuthal quantum number* and it takes on integer values.

There are three quantum numbers: n which is the total quantum number, l is the azimuthal quantum number and m represents the magnetic quantum number. These numbers of the Bohr–Sommerfeld prequantum mechanical treatment of the single-electron atom arise naturally from the postulates of quantum mechanics.

The total degeneracy for a system of particles which undergo translation is given by

$$g = (2s + 1)\left(\frac{4\pi V}{h^3}\right)(\mathbf{p}^2 \, d\mathbf{p}) \tag{15.153}$$

where $\mathbf{p}$ is the momentum and all other symbols have their usual meanings. The resultant angular momentum of the particles is given by $s\,\hbar$. Equation 15.153 gives the proper description of the density of states for systems of both fermions and bosons. The symmetry of the system which we mentioned above occurs whenever the resultant angular momentum quantum number (s) is zero or an integer. We have already noted that particles possessing integer spin angular momentum are by definition bosons.

In order to find whether a molecule or an atom obeys the Bose–Einstein statistics, we may have to find the value of s for the atom in the ground state (the internal state of least energy). Since a complete electron shell will have $s = 0$ value, for atoms with closed electron shells (e.g. alkaline ions, rare gases, etc.) the *Bose–Einstein statistics* will be applicable, and also the angular momentum quantum degeneracy will be 1.

We have seen in Table 15.1 that the helium atom is a boson ($s = 0$), that diatomic hydrogen at very low temperatures has $s = 0$, and that the Bose–Einstein statistics can be used to represent photon with an intrinsic spin angular momentum equal to $\hbar$. From the above discussion it is apparent that an investigation of the properties of the boson gas is not merely academic. In the following sections we shall find such properties as internal energy, pressure, entropy, specific heat, etc.

The internal energy of perfect Bose–Einstein gases

Let us consider the case of N non-interacting identical bosons inside a cubical box. The average energy of the ideal boson gas can be found using Equation 15.108 as

$$U = \int_0^\infty E n(E) \, dE$$

$$= (gV/4\pi^2)(2m/\hbar^2)^{3/2} \int_0^\infty \frac{E^{3/2} \, dE}{\exp(E/kT) \exp(-\mu/kT) - 1}.$$

From Equation 15.145 we get $PV = 2U/3$, and from Equation 12.113 we have $\Omega = -PV$. Hence, by combining all these equations we have

$$\Omega = -PV = -\frac{2U}{3} = -\left(\frac{gV}{6\pi^2}\right)\left(\frac{2m}{\hbar^2}\right)^{3/2} \int_0^\infty \frac{E^{3/2} \, dE}{\exp(E/kT) \exp(-\mu/kT) - 1}. \tag{15.154}$$

One of the requirements to use Equation 15.154 for the determination of the pressure P and the internal energy U is the value of the chemical potential μ. Once the value of μ of the boson gas has been calculated using Equation 15.111, Equation 15.154 can be used to calculate P and U. In Section 15.6.1, a method has been discussed (property 6) for the determination of U. However, a further discussion is given here.

Figure 15.14, shows that if we choose the energy of the single particle ground state as the zero of the energy scale, the chemical potential μ of a perfect Bose–Einstein gas is very close to zero for all temperatures less than T_b. So, in this temperature range and by taking $\mu = 0$, Equation 15.112 becomes

$$U = \left(\frac{gV}{4\pi^2}\right)\left(\frac{2m}{\hbar^2}\right)^{3/2} \int_0^\infty \frac{E^{3/2}\,d\varepsilon}{\exp(\varepsilon/kT) - 1}. \tag{15.155}$$

If we substitute $x = E/kT$ in this equation, we have

$$U = \left(\frac{gV}{4\pi^2}\right)\left(\frac{2m}{\hbar^2}\right)^{3/2} (kT)^{5/2} \int_0^\infty \frac{x^{3/2}\,dx}{e^x - 1}. \tag{15.156}$$

The integral on the right side of this equation is equal to the product of the gamma function $\Gamma(5/2) = \frac{3}{4}(\sqrt{\pi})$ and the Riemann zeta function $\zeta(5/2) = 1.34$. Then Equation 15.155 becomes

$$U = (1.329)(2)^{3/2}\left(\frac{1}{4\pi^2}\right)\left(\frac{gV}{N}\right)\left(\frac{mk}{\hbar^2}\right)^{3/2} (NkT)T^{3/2}$$

or

$$U = 0.094\left(\frac{gV}{N}\right)\left(\frac{mk}{\hbar^2}\right)^{3/2} (NkT)T^{3/2}.$$

By substituting the value of

$$\left(\frac{gV}{N}\right)\left(\frac{mk}{\hbar^2}\right)^{3/2}$$

from Equation 15.126 and simplifying, we obtain

$$U = (0.770)(NkT)(T/T_b)^{3/2} \tag{15.157}$$

in the temperature range $T < T_b$

Pressure of perfect Bose–Einstein gases

From Equations 15.154 and 15.157 we can write for the pressure of the ideal boson gas

$$-\Omega = PV = \frac{2U}{3} = (0.513)(NkT)(T/T_b)^{3/2} \tag{15.158}$$

in the range $T < T_b$. In this temperature range, if the volume remains constant, then both the pressure and the temperature are directly proportional to $T^{5/2}$.

Entropy of perfect Bose–Einstein gases

From Equation 15.158 the entropy of the ideal boson gas is (in the range $T < T_b$)

$$S = -\frac{\partial}{\partial T}(\Omega)_{V,\mu} = \frac{\partial}{\partial T}[(0.513)(NkT)(T/T_b)^{3/2}]$$

or

$$S = (0.513)(Nk)(T/T_b)^{3/2} + (0.513)(NkT)\left(\frac{3}{2}\right)(T/T_b)^{1/2}$$

or

$$S = \left(\frac{5}{2}\right)(0.513)(Nk)(T/T_b)^{3/2}$$

or

$$S = 1.283 Nk(T/T_b)^{3/2}. \tag{15.159}$$

This shows that when $T=0$, $S=0$. This is in agreement with the third law of thermodynamics.

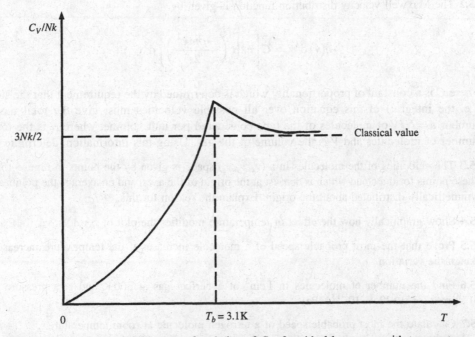

Figure 15.21   Schematic diagram of variation of C_V of an ideal boson gas with temperature

Heat capacity of perfect Bose–Einstein gases

The heat capacity of the ideal boson gases for the temperature range $T < T_b$ is given by

$$C_V = \left(\frac{\partial U}{\partial T} \right)_V = 1.925 Nk (T/T_b)^{3/2}. \qquad (15.160)$$

This shows that the heat capacity is directly proportional to $T^{3/2}$.

We have noted earlier that one of the most interesting results of the *Bose–Einstein condensation* is the abrupt change in the heat capacity at $T = T_b$. This is shown in Figure 15.21. At temperatures well above the T_b, C_V has the classical value $3Nk/2$ and it increases as $T \rightarrow T_b$. At the temperature T_b it abruptly changes its slope $(\partial C_V / \partial T)$ as can be seen in Figure 15.21. Actually, this change occurs due to the loss of particles to the lowest energy state. The temperature at which this change takes place is inversely dependent on the mass of the boson, which means that the temperature will be the highest if the mass of the boson is small. The Bose–Einstein particle with the smallest mass, which does not solidify at low temperatures, is He^4. The critical temperature for He^4 is 3.13 K.

EXERCISES 15

15.1 In the Maxwell velocity distribution function, $\eta(\mathbf{v})$ does not depend on the position of the molecule. However, it depends on the magnitude of the velocity of the molecule and not on its direction. Explain why.

15.2 The Maxwell velocity distribution function is given by

$$\eta(\mathbf{v}) \, d^3 \mathbf{v} = C \left[\exp \left(\frac{-\beta m \mathbf{v}^2}{2} \right) \right] d^3 \mathbf{v}$$

where C is a constant of proportionality which is determined by the requirement that the sum (i.e. the integral) of this equation over all possible velocities must give the total mean number $n = N/V$ of molecules of the type considered per unit volume, where N is the total number of molecules and V is the volume of the gas. Using this information, determine C.

15.3 The velocities of the molecules in a (v_x, v_y, v_z) space is given by the points in Figure 15.2. These points form a cloud which is densest at the origin of the axes, and on average the points are symmetrically distributed about the origin. Explain the reason for this.

15.4 Show graphically how the effect of temperature modifies the plot of $\eta(v_x)$ vs. v_x.

15.5 Prove that the most probable speed of a molecule increases if the temperature increases. Sketch the variation.

15.6 Find the number of molecules in $1 \, cm^3$ of a perfect gas at 300 K and at a pressure of 10^{-10} mm (or 0.132×10^{-13} MPa).

15.7 Calculate the most probable speed of a nitrogen molecule at room temperature.
    Find the root mean square velocity of a nitrogen molecule at 273 K.

15.8 Consider a system of helium gas at standard pressure and temperature in a container of volume 10^3 cm^3. Assuming that He is a perfect gas, calculate: (a) the number of molecules in the system, (b) the most probable speed of a He molecule, and (c) the average speed of a He molecule. Check whether the ratio of the average speed and the most probable speed is approximately 1.13.

15.9 Discuss whether Equations 15.10, 15.11 and 15.12 which are applicable to gases of monatomic molecules, are also valid for gases of polyatomic molecules.

15.10 Sketch the Fermi–Dirac distribution function for zero absolute temperature and for another low temperature. Explain the behaviour of the sketches with respect to T and indicate the conditions of the states within about kT below E_F and above E_F. Determine the value of $f(E)$ at $T = 0$ K for $E < E_F$ and $E > E_F$, and hence discuss the significance of E_F at this temperature.

15.11 (a) Show that the Fermi energy at absolute zero is given by

$$E_F(0) = \frac{\hbar}{2m}(3\pi^2 N)^{2/3}.$$

If the number of electrons per unit volume with energy between E and $E + dE$ is dn and the density of states $g(E)$ is given by Equation 15.68, show that the total number of electrons N is given by

$$N = \frac{2C}{3}[E_F(0)]^{3/2}$$

where C is a constant equal to

$$\frac{1}{2\pi^2}\left(\frac{2m}{\hbar^2}\right)^{3/2}.$$

Hence, find an equation for E_F and compare it with that given above.

(b) If the Fermi energy of copper is 7.04 eV, calculated on the basis of the free electron model, determine the Fermi temperature of Cu.

15.12 Show that the Fermi energy of a system is the electrochemical potential or the partial molar energy of the electrons in that system.

15.13 (a) Show that the electronic heat capacity of a metal with N electrons is approximately RT/T_F, where R is the universal gas constant, T is the temperature and T_F is the Fermi temperature.

(b) Calculate the electronic heat capacity of Au at 300 K, assuming that its Fermi temperature is 64 000 degrees. Comment on your result.

15.14 Show that the function $\partial f(E)/\partial E$ is an even function with respect to the point $(E - \mu)/kT = 0$.

15.15 Define the reduced Fermi level. Discuss the conditions under which E_F can be identified with the electrochemical potential. At $T = 0$ K, $f(E) = 1$ for all values of $E < E_F$, and $f(E) = 0$ for all values of $E > E_F$. What sort of values should $f(E)$ have (a) for $T > 0$ K, and (b) for values of E not too far from E_F?

15.16 What is the salient difference between classical and quantum particles? Under what condition it is possible to have a transition from a discrete to a continuous spectrum and hence from the Fermi–Dirace to the Boltzmann distribution. Is it correct to say that the Boltzmann distribution is in fact the limiting case of the Fermi–Dirac distribution? Why is it that the Boltzmann distribution is satisfied better at higher temperatures? How do the Fermi–Dirac and the Boltzmann distributions differ in respect of their applications?

15.17 Assume that in a system there are only three particle states with energies $\varepsilon(1)$, $\varepsilon(2)$ and $\varepsilon(3)$. Determine the maximum number of Fermions in this system.

15.18 Discuss the requirements that determine whether a molecule or an atom will obey the Bose–Einstein or Fermi–Dirac statistics. Explain whether He^4 and He^3 are bosons or fermions. What is the origin of their behaviour in relation to these statistics?

15.19 Show that the entropy of an ideal fermion gas at the absolute zero temperature is in agreement with the third law of thermodynamics.

15.20 A mole of He^3 gas atoms has a volume of $0.0244 \, m^3$ at 273 K. The mass of an He^3 atom is $5.11 \times 10^{-27} \, kg$ and it has the spin $1/2$. Calculate the chemical potential of He^3. Comment on your result.

15.21 Deduce an equation for the energy distribution of the ideal boson gas consisting of N bosons in a cubical box. Hence, find an appropriate equation for N.

15.22 A mole of He^4 gas is confined in a cubical box of volume $22.4 \times 10^{-3} \, m^3$ at 0.1 MPa pressure and 273 K. The atom of He^4 has a spin zero and each atom has a mass of $6.65 \times 10^{-27} \, kg$. Calculate the Bose–Einstein temperature. Comment on the condition of the gas at this temperature.

15.23 Show that the pressure of an ideal fermion gas confined in a cubical box of volume V is proportional to its internal energy if V remains constant and there are no interactions between the particles. Comment on your result and the variation of the pressure with the temperature at constant volume. How does the pressure of an ideal fermion gas differ from that of an ideal classical monatomic gas if N, V and T are the same for both?

15.24 An ideal Fermi–Dirac gas is contained in a cubical box of volume $22.4 \times 10^{-3} \, m^3$ at 0.1 MPa pressure and 273 K. Under this condition, the internal energy of the gas is 2.5 J. Assuming that there are no interactions between the particles, calculate the pressure of the gas.

15.25 The volume of a mole of liquid He^4 is $27 \times 10^{-6} \, m^3$ and the mass of a He^4 atom is $6.65 \times 10^{-27} \, kg$. Assuming that liquid He^4 is an ideal boson gas with spin zero, calculate the concentration of the bosons in this volume and the Bose temperature.

15.26 What reason do you have to justify that the λ-point in liquid He^4 is in some fashion related to a Bose–Einstein type condensation?

16
Heat Capacities of Solids and Gases

16.1 INTRODUCTION

In thermodynamics and statistical mechanics the quantities which are most accessible to experiment are the response functions. They give us information about how a specific state variable changes as other independent state variables are changed under controlled conditions. The response functions may be divided into thermal response functions such as the heat capacities, and mechanical response function such as compressibility and susceptibility.

The heat required to raise the temperature of 1 mol of a system by 1°C is called its *heat capacity* C. If the system is 1 mol of a pure substance, the heat capacity is defined as *molar heat capacity* of that substance. The analogous quantity for 1 g of substance is called the *specific heat capacity*, so that the molar heat capacity is the specific heat capacity multiplied by the molecular weight. The heat capacity, or in other words, the ratio of the energy input to the temperature increase ($C = dQ/dT$) depends strongly on the nature of the substance and its physical state. We talk about C as though it is a well-defined quantity, but of course the heat required to pass from one specified state of system to another will vary with the path taken. The heat capacity can then become a well-defined quantity only by virtue of a specification of path. In fact, we find useful two species of heat capacities, corresponding to two simply specified paths, namely that at constant volume (C_V) and that at constant pressure (C_P). For gases the difference between C_P and C_V is considerable, whereas it is much smaller for liquids or solids. However, at very low temperatures, for solids the difference between C_P and C_V is very small, so we take $C_P = C_V$. In Sections 2.2.2, and 5.1–5.3 we have discussed heat capacity quite qualitatively.

In this chapter, we shall discuss the thermal properties of solids and gases in the light of statistical mechanics. We consider the classical theory of heat capacity associated with the lattice vibrations of crystalline substances, that is we shall study the internal energy and the heat capacity that arise from the motion of the atoms in the lattice of solids. In our discussion we shall see that the classical theory of heat capacity cannot predict many of the observed features of the heat capacity. This is especially true at low temperatures. Not only do the observed heat capacities differ from the classical results (even at fairly high temperatures, say 200 K), but also the classical theory cannot explain the fact that the electrons fail to contribute their fair share to the heat capacity. Some models, such as Einstein's and Debye's models, were developed to explain these anomalies. We shall discuss these models and their successes.

16.2   THE CLASSICAL THEORY OF HEAT CAPACITY OF SOLIDS

We know that the average energy of a classical system is equal to $\frac{1}{2}kT$ per degree of freedom, where k is the Boltzmann constant and T is the temperature. For any system of free particles this result is true within the limits of classical theory. We can easily extend this result to particles interacting with harmonic forces. By harmonic forces we mean forces which vary directly as the relative displacement according to Hooke's law. It has been proved that the average kinetic energy per degree of freedom, $\frac{1}{2}kT$, does not change, and the average potential energy is just equal to the average kinetic energy. Also, it is known that for particles with harmonic interactions the average total energy is kT per degree of freedom. This result and also the appropriate extensions to more general interactions can be obtained from the result of classical statistical mechanics as discussed below.

Consider a system defined by the co-ordinates $\mathbf{q}$ and conjugate momenta $\mathbf{p}$. The theorem which is the central result of the classical statistical mechanics states that if the system is in thermal equilibrium, then the average value $\overline{M}$ of any quantity M (which is a function of $\mathbf{q}$ and $\mathbf{p}$) of the system is given by

$$\overline{M} = \frac{\int M(\mathbf{p}, \mathbf{q}) \exp \left[-E(\mathbf{p}, \mathbf{q})/kT\right] \, d\mathbf{p} d\mathbf{q}}{\int \exp \left[-E(\mathbf{p}, \mathbf{q})/kT\right] \, d\mathbf{p} d\mathbf{q}}. \tag{16.1}$$

$\mathbf{q}$ and $\mathbf{p}$ denote the set of all co-ordinates and momenta of the system under consideration, and $E(\mathbf{p}, \mathbf{q})$ is the energy of the system. If the co-ordinates are Cartesian and there are N particles in the system, all of mass m, then

$$\mathbf{q} \equiv (x_1, y_1, z_1; x_2, y_2, z_2; \cdots; x_N, y_N, z_N)$$

$$\mathbf{p} \equiv (p_{x1}, p_{y1}, p_{z1}; p_{x2}, p_{y2}, p_{z2}; \cdots; p_{xN}, p_{yN}, p_{zN}).$$

In addition, if m_i is the mass of the ith particle, then

$$p_{xi} = (m_i) \frac{dx_i}{dt}.$$

The integrals in Equation 16.1 are over the entire range of variation of each independent variable, and are usually taken from $-\infty$ to $+\infty$. The energy for N free particles, each one of mass m, is given by

$$E = \left(\frac{1}{2m}\right) \sum_{1}^{N} [(p_{xi})^2 + (p_{yi})^2 + (p_{zi})^2]. \tag{16.2}$$

If the system is in thermal equilibrium at temperature T, then E is the same as the internal energy U of the system. The energy is also independent of the set of all co-ordinates $\mathbf{q}$. So the integrals over $\mathbf{dq}$ in the numerator and denominator cancel. Thus, we have

$$U = \frac{\left(\frac{1}{2m}\right) \sum_1^{3N} \int_{-\infty}^{\infty} (p_j)^2 \prod_1^{3N} [\exp\left(-p_r^2/2mkT\right)] \, \mathbf{dp}}{\int_{-\infty}^{\infty} \prod_1^{3N} [\exp\left(-p_r^2/2mkT\right)] \, \mathbf{dp}} \tag{16.3}$$

where both j and r extend from 1 to $3N$ for three degrees of freedom. Further cancellations in the above equation reduce the expression for U to the sum of $3N$ terms. Each of these terms is actually equal to

$$\frac{\left(\frac{1}{2m}\right) \int_{-\infty}^{\infty} p_j^2 \exp\left(-p_j^2/2mkT\right) \, dp_j}{\int_{-\infty}^{\infty} \exp\left(-p_j^2/2mkT\right) \, dp_j} = D \text{ (say)}.$$

It can be easily seen that the numerator of the above expression is of the form

$$\int_{-\infty}^{\infty} y^2 \exp\left(-y^2\right) \, dy = \frac{\sqrt{\pi}}{2}$$

and the denominator is of the form

$$\int_{-\infty}^{\infty} \exp\left(-y^2\right) \, dy = \sqrt{\pi}.$$

Hence, we can write

$$D = (kT) \frac{\int_{-\infty}^{\infty} y^2 \exp\left(-y^2\right) \, dy}{\int_{-\infty}^{\infty} \exp\left(-y^2\right) \, dy} = (kT) \left(\frac{\sqrt{\pi}}{2} \bigg/ \sqrt{\pi}\right) = \frac{kT}{2}$$

where $y^2 = p_j^2/2mkT$. Thus, from Equation 16.3 we have for the sum of $3N$ terms

$$U = \left(\frac{3}{2}\right) NkT. \tag{16.4}$$

At the start of this section we stated that the average energy of a classical system is equal to $kT/2$ per degree of freedom. So, for $3N$ degrees of freedom of N atoms in three-dimensional space will be $\left(\frac{3}{2}\right)NkT$. This agrees with our result in Equation 16.4.

Let us now consider a harmonic oscillator with frequency ω and mass m. The energy of such an oscillator in one dimension is given by

$$E = \frac{p^2}{2m} + \frac{m\omega^2 q^2}{2}. \tag{16.5}$$

If the oscillator is in thermal equilibrium, then its average energy is given by (cancelling the common terms in numerator and denominator)

$$\overline{E} = \frac{\left(\frac{1}{2m}\right) \int_{-\infty}^{\infty} p^2 \exp\left(-p^2/2mkT\right) dp}{\int_{-\infty}^{\infty} \exp\left(-p^2/2mkT\right) dp} + \frac{\frac{m\omega^2}{2} \int_{-\infty}^{\infty} q^2 \exp\left(-m\omega^2 q^2/2kT\right) dq}{\int_{-\infty}^{\infty} \exp\left(-m\omega^2 q^2/2kT\right) dq}. \tag{16.6}$$

In this equation, according to our earlier argument the first term on the right is equal to $kT/2$. If we put $y^2 = m\omega^2 q^2/2kT$, then the second term can be written as

$$(kT) \frac{\int_{-\infty}^{\infty} y^2 \exp\left(-y^2\right) dy}{\int_{-\infty}^{\infty} \exp\left(-y^2\right) dy} = kT/2.$$

Thus, the average energy of a simple harmonic oscillator in thermal equilibrium is

$$\overline{E} = \left(\frac{1}{2}\right)kT + \left(\frac{1}{2}\right)kT = kT. \tag{16.7}$$

Hence, the internal energy of N harmonic oscillators in three dimensions is given by

$$U = 3N\overline{E} = 3NkT. \tag{16.8}$$

For one mole of a substance this becomes

$$U_m = 3RT \tag{16.9}$$

where R is the universal gas constant.

If we consider the atoms in a solid as harmonic oscillators which oscillate about their equilibrium positions, then according to the classical theory discussed above the molar heat capacity of a solid at constant volume is

$$C_V = \left(\frac{\partial U_m}{\partial T}\right)_V. \tag{16.10}$$

Then the lattice contribution to the molar heat capacity for one mole of atoms is

$$C_V = 3R \simeq 8.94 \, \text{J/deg}. \tag{16.11}$$

This value is known as *Dulong–Petit value*. It is in quite good agreement with the observed total heat capacity of many metals and non-metals from room temperature up to reasonably high temperatures. However, the agreement fails as the solid is cooled below room temperature. As an example, consider the heat capacity of metallic silver shown in Figure 16.1. The heat capacity drops rapidly as the temperature goes down below room temperature. It has been found that the heat capacity of metals tends to zero as T, as $T \to 0$, whereas the heat capacity of non-metallic solids approaches zero as T^3 with $T \to 0$. It has been proved experimentally that a T^3-approach is characteristic of lattice vibrations in cubic and near-cubic crystals; on the other hand, a T-approach is found to be characteristic of that part of the heat capacity which is associated with conduction electrons in the crystal.

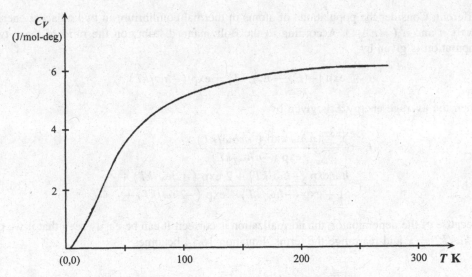

Figure 16.1   Schematic diagram of variation of C_V of metallic silver with temperature

16.3   THE EINSTEIN MODEL OF HEAT CAPACITY OF SOLIDS

The above discussion clearly brings out the difficulties for the classical theory to predict many of the observed features of the lattice heat capacity of solids, particularly those at low temperatures. The Einstein model was developed to give a fairly good representation of the drop in heat capacity at low temperatures below the value $3R$ per mole. Einstein treated the thermal properties of the vibration of a lattice of N atoms as a set of $3N$ independent harmonic oscillators in one dimension. He assumed that each oscillator has the identical frequency ν and then quantized the energies of the oscillators according to Planck's theory of black body radiation. Therefore, it would appear that any representation of the drop in heat capacity at low temperatures based on Einstein's model will depend on an appropriate choice of the oscillator frequency.

In classical theory an oscillator can have any arbitrary amplitude of oscillation, which means any arbitrary energy. According to Planck's theory the energy may have only the values given by

$$E = nh\nu \tag{16.12}$$

where n is any positive integer, $n = 0, 1, 2, \ldots$; h is Planck's constant and ν is the frequency of oscillation. This relation can be written as

$$E = n(h/2\pi)(2\pi\nu) = n\hbar\omega \tag{16.13}$$

where ω is the angular velocity and $\hbar = h/2\pi$. An energy level is labelled as level 1, level 2, ..., etc., by the value of $n = 1$, $n = 2$, ..., etc., and called the *quantum number*.

The expression for the average energy of an oscillator on the basis of classical theory is kT (see Equation 16.7). However, this expression on the basis of quantum theory is quite different. Consider the populations of atoms in thermal equilibrium in two adjacent energy levels n and n' ($=n+1$). According to the Boltzmann distribution the ratio of these two population is given by

$$\exp\left[-(E_{n'} - E_n)/kT\right] = \exp\left(-\hbar\omega/kT\right).$$

Then the average energy $\overline{E}$ is given by

$$\overline{E} = \frac{\sum_{n=0}^{\infty} n\hbar\omega \exp\left(-n\hbar\omega/kT\right)}{\sum_{n=0}^{\infty} \exp\left(-n\hbar\omega/kT\right)}$$
$$= \frac{\hbar\omega[\exp\left(-\hbar\omega/kT\right) + 2\exp\left(-2\hbar\omega/kT\right) + \cdots]}{1 + \exp\left(-\hbar\omega/kT\right) + \exp\left(-2\hbar\omega/kT\right) + \cdots}. \tag{16.14}$$

Because of the denominator, the normalization is correct. It can be easily seen that if we put $-\hbar\omega/kT = x$ and rearrange the terms, Equation 16.14 becomes

$$\overline{E} = \hbar\omega\frac{d}{dx} \ln\left(1 + e^x + e^{2x} + \cdots\right)$$
$$= \hbar\omega\frac{d}{dx} \ln\left(\frac{1}{1 - e^x}\right) = \frac{\hbar\omega}{\exp\left(-x\right) - 1}$$

or

$$\overline{E} = \frac{\hbar\omega}{\exp\left(\hbar\omega/kT\right) - 1}. \tag{16.15}$$

Equation 16.15 is known as the *Planck–Einstein distribution law* for the harmonic oscillator. Here we shall consider two cases: (i) when the temperature is high, and (ii) when the temperature is low.

(i) At high temperatures, $kT \gg \hbar\omega$. So the denominator of Equation 16.15 can be written as

$$\exp\left(\hbar\omega/kT\right) - 1 = 1 + (\hbar\omega/kT) + \cdots - 1$$
$$\simeq \hbar\omega/kT.$$

Then,

$$\overline{E} = \frac{\hbar\omega}{\exp\left(\hbar\omega/kT\right) - 1} = \frac{\hbar\omega}{\hbar\omega/kT} = kT. \tag{16.16}$$

This shows that at high temperatures the average energy of the harmonic oscillator is approximately equal to the classical average energy kT.

(ii) At low temperatures, $kT \ll \hbar\omega$. So, $\exp\left(\hbar\omega/kT\right) \gg 1$ and we have

$$\overline{E} \simeq \hbar\omega \exp\left(-\hbar\omega/kT\right). \tag{16.17}$$

Hence, since the heat capacity C_V is

$$C_V \simeq Nk(\hbar\omega/kT)^2 \exp(-\hbar\omega/kT)$$

it tends to zero as $T \to 0$.

In the limit, $\exp(-\hbar\omega/kT)$ is the dominant factor. Therefore, at low temperatures, variation of C_V on the basis of the Einstein model is as $\exp(-\hbar\omega/kT)$.

As mentioned earlier, experimental results show that the variation for the lattice contribution to heat capacity at low temperatures is not really exponential; rather, it varies as T^3. The T^3 variation is better explained by the *Debye model*. We define a temperature, known as the *Einstein temperature* (T_E), by

$$\hbar\omega = kT_E. \tag{16.18}$$

In general, for many solids T_E is in the range 100–300 K, but there are solids for which T_E is below or above this range. The heat capacity of a solid is given

$$C_V = \left(\frac{\partial U}{\partial T}\right)_V = \frac{\partial(N\overline{E})}{\partial T}$$

where U is the internal energy, $\overline{E}$ is the average energy of the system of N particles and T is the temperature. By substituting the value of $\overline{E}$ from Equation 16.17 and differentiating, we obtain

$$\begin{aligned}
C_V &= Nk(\hbar\omega/kT)^2 \left[\frac{\exp(\hbar\omega/kT)}{\{\exp(\hbar\omega/kT) - 1\}^2}\right] \\
&= Nk(T_E/T)^2 \left[\frac{\exp(T_E/T)}{\{\exp(T_E/T) - 1\}^2}\right]. \tag{16.19}
\end{aligned}$$

From this equation it can be seen that if we say T_E has a particular value, we mean that C_V gives a reasonably good fit for that value of T_E to the experimental values over a wide range of temperature (in which C_V is varying appreciably with T). If $T_E = 300$ K and ω is taken in radians/s, then by using Equation 16.18 it can be shown that the characteristic frequency $\nu = \omega/2\pi$ is approximately 5×10^{12} cycles/s, which is a reasonable magnitude for a characteristic frequency of atomic oscillation. Figure 16.2 shows the temperature variation of the molar heat capacity of a solid according to the Einstein and Debye models.

The Einstein model does not give an adequate representation of the dependence of heat capacity on temperature below room temperature. At low temperatures, the predicted values of the molar heat capacity of a solid tend to zero as $T \to 0$ faster than the experimental results. Experimentally, the variation for the lattice contribution at low temperatures follows a T^3 law. The mathematical form of Equation 16.19 is the same for all substances. If the heat capacity of a mole of atoms is plotted versus T/T_E, the values of C_V for all solids should lie on the same curve as plotted in Figure 16.2. *For solids below room temperature the difference between C_P and C_V is small and usually can be ignored.*

In deriving his model, Einstein treated each atom as an independent oscillator and assumed that the oscillator performs harmonic motion about a fixed point in space. However,

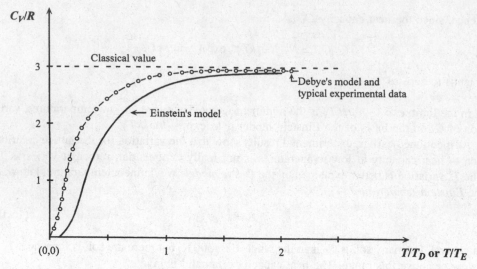

Figure 16.2 Schematic diagram of variation of molar heat capacity at constant volume of a solid according to the Einstein and Debye models

in reality the atoms oscillate relative to their neighbours in the lattice. If the wavelengths of atomic oscillations are long compared to the lattice spacing, the motions of neighbouring atoms are not really independent; rather large domains of the crystal lattice move together. However, such movements are coherent, but if the identical frequency is assigned to all $3N$ oscillations, it may lead to an over-simplification. This is quite apparent from the fact that the long wavelength motions can have quite low frequencies. At low temperatures the long wavelength motions become important. This is because in an infinite lattice there will always exist modes of vibrations for which $kT \gg \hbar\omega$ in spite of T being $\ll T_E$. Therefore, even at low temperatures some degrees of freedom of the crystal lattice will exhibit classical behaviour. Each one of these modes will contribute an energy approximately equal to kT. Thus, the total energy will not necessarily tend to zero exponentially (as predicted by the Einstein model).

16.4   DEBYE'S MODEL OF HEAT CAPACITY OF SOLIDS

In Einstein's theory of heat capacity, it was assumed that all the atoms in the lattice oscillate with simple harmonic motion in a steady potential well with the same angular frequency ω_E (known as the *Einstein frequency*). In reality, if one of the lattice atoms in a crystal is displaced from its position, this affects the restoring forces acting on the neighbouring atoms. Therefore, the collective oscillations of the atoms in a solid can be important. This is similar to sound waves in a solid. However, in Debye's theory, the solid is treated as a continuous elastic medium; this neglects the discrete structure of the crystal. Thus, the Debye theory assumes that the dispersion relation for sound waves $\omega = \kappa v$ holds, not just for the long-wavelength sound waves, but for all vibrations in the crystal. In this equation, v is the speed of sound, $\kappa = 2\pi/\lambda$ is the wave number and λ is the wavelength of the sound waves.

These waves form standing waves and have allowed wavelengths which are determined by the size of the crystal; $\lambda = 2L/n$, where $n = 1, 2, 3, \ldots, \infty$, and are the same as for a string which is pinned at both ends, and L is the length. Any given wave will have a dispersion relation given by

$$\omega_i^2 = v^2(\kappa_{xi}^2 + \kappa_{yi}^2 + \kappa_{zi}^2). \tag{16.20}$$

Using $\kappa = 2\pi/\lambda$ and $\lambda = 2L/n$, this equation becomes

$$\omega_i^2 = v^2\left[\left(\frac{\pi n_{xi}}{L_x}\right)^2 + \left(\frac{\pi n_{yi}}{L_y}\right)^2 + \left(\frac{\pi n_{zi}}{L_z}\right)^2\right] \tag{16.21}$$

where L_x, L_y and L_z are the lengths of the crystal in the x, y and z directions, respectively.

The allowed values of ω_i can be plotted as points in a three-dimensional frequency space, as shown in Figure 16.3. Consider the x, y and z directions separately. The distance between

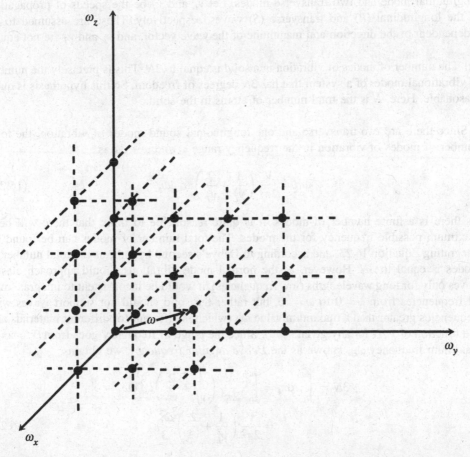

Figure 16.3 Schematic diagram of allowed values of frequency ω in a three-dimensional frequency space

points in the x-direction is $\pi v/L_x$, that in the y-direction is $\pi v/L_y$ and that in the z-direction is $\pi v/L_z$. The volume per point is

$$(\pi v/L_x)(\pi v/L_y)\pi v/L_z) = (\pi v)^3/L_x L_y L_z$$

and the number of points per unit volume is $(L_x L_y L_z)/(\pi v)3$ where $L_x L_y L_z = V$, the volume of the crystal. If we imagine a sphere of radius ω (the frequency is always positive), then, $\frac{1}{8}$ of this sphere is $\frac{1}{8}(4\pi\omega^3/3)$. Hence, the total number of allowed values of ω_i less than some value ω is given by $[\frac{1}{8}(4\pi\omega^3/3)[V/(\pi v)^3]$. If dn_i is the number of points in the range ω and $\omega + d\omega$, then

$$dn_i = \frac{V\omega^2 \, d\omega}{2\pi^2 v^3}. \tag{16.22}$$

The dispersion law can be used to calculate the specific heat of a solid if we make, along with Debye, two hypotheses, as follows.

(i) In a solid, there are three are three modes of vibration for each wave vector: one longitudinal mode and two transverse modes. Let v_L and v_T be the speeds of propagation of the longitudinal (P) and transverse (S) waves, respectively. These are assumed to be independent of the direction and magnitude of the wave vector, and v_L and v_T are not equal.

(ii) The number of modes of vibration in a solid is equal to $3N$. This is precisely the number of vibrational modes of a system that has $3N$ degrees of freedom. So this hypothesis is quite reasonable. Here, N is the total number of atoms in the solid.

Since there are two transverse and one longitudinal sound modes of vibration, the total number of modes of vibration in the frequency range ω and $\omega + d\omega$ is

$$dn_i = \frac{V}{2\pi^2}\left(\frac{1}{v_L^3} + \frac{2}{v_T^3}\right)\omega^2 \, d\omega. \tag{16.23}$$

As there is a finite number of-modes, it is quite reasonable to think that there will be a maximum possible frequency for the modes. The total number of modes can be found by integrating Equation 16.23, and according to Debye's second hypothesis the total number of modes is equal to $3N$. However, as the normal mode vibrations should approach elastic waves only for long wavelengths (low frequencies), it would be inconsistent to integrate over all frequencies from $\omega = 0$ to $\omega = \infty$, but rather the solid should not support waves with frequencies greater than a maximum value ω_D (which is different for different materials and is a function of V/N for any given solid). Since the range of frequency goes from 0 to some maximum frequency ω_D, known as the *Debye angular frequency*, we obtain

$$3N = \int_0^{\omega_D} dn_i = \int_0^{\omega_D} \frac{V}{2\pi^2}\left(\frac{1}{v_L^3} + \frac{2}{v_T^3}\right)\omega^2 \, d\omega$$
$$= \frac{V}{2\pi^2}\left(\frac{1}{v_L^3} + \frac{2}{v_T^3}\right)\frac{\omega_D^3}{3} \tag{16.24}$$

According to an atomic theory of solids there is a maximum frequency, i.e. an upper cut-off frequency when the wavelength of the standing wave is equal to double the separation distance of neighbouring atoms. By solving for ω_D^3, we have

$$\omega_D^3 = (18\pi^2)\left(\frac{N}{V}\right)\left(\frac{1}{v_L^3} + \frac{2}{v_T^3}\right)^{-1}. \tag{16.25}$$

This equation gives the Debye maximum angular frequency ω_D in terms of v_L and v_T of elastic (sound) waves in the solid. If we now substitute

$$V\left(\frac{1}{v_L^3} + \frac{2}{v_T^3}\right)$$

from Equation 16.25 into Equation 16.23, we have

$$dn_i = (9N\omega^2\,d\omega)/\omega_D^3. \tag{16.26}$$

According to this equation, dn_i is proportional to ω^2 up to ω_D. The variation of the number of normal modes (dn_i), per unit angular frequency range, with ω is shown in Figure 16.4.

Debye treated each normal mode (standing wave solution) as an independent linear harmonic oscillator in thermal equilibrium with the rest of the crystal, and the crystal acts as a heat reservoir of constant absolute temperature T for each normal mode. The total

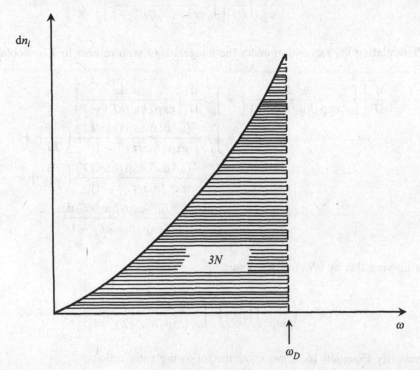

Figure 16.4 Schematic diagram of variation of the number of normal modes dn_i with frequency ω up to the maximum frequency ω_D

average energy of the crystal is given by

$$U = \int_0^{\omega_D} \bar{\varepsilon} \, dn_i \tag{16.27}$$

and the mean energy of a harmonic oscillator with natural angular frequency ω is given by

$$\bar{\varepsilon} = \hbar\omega/2 + \hbar\omega/[exp\,(\hbar\omega/kT) - 1]. \tag{16.28}$$

By combining Equations 16.27 and 16.28, substituting for dn_i from Equation 16.26 and then integrating, we have

$$U = \frac{9N\hbar\omega_D}{8} + \frac{9N\hbar}{\omega_D^3} \int_0^{\omega_D} \frac{\omega^3 \, d\omega}{exp\,(\hbar\omega/kT)^{-1}}. \tag{16.29}$$

It can easily be seen that the first term on the right of this equation is independent of temperature. By definition, the specific heat at constant volume is given by

$$C_V = \left(\frac{\partial U}{\partial T}\right)_V$$

$$= \left(\frac{9N\hbar}{\omega_D^3}\right) \frac{\partial}{\partial T} \left[\int_0^{\omega_D} \frac{\omega^3 \, d\omega}{exp\,(\hbar\omega/kT) - 1}\right]_V.$$

By differentiating the expression under the integral sign with respect to T, we obtain

$$\frac{d}{dT}\left[\int_0^{\omega_D} \frac{\omega^3 \, d\omega}{exp\,(\hbar\omega/kT) - 1}\right] = \int_0^{\omega_D} \frac{d}{dT}\left[\frac{\omega^3 \, d\omega}{exp\,(\hbar\omega/kT) - 1}\right]$$

$$= \int_0^{\omega_D} \frac{-(\omega^3 \, d\omega)\,exp\,(\hbar\omega/kT)}{[exp\,(\hbar\omega/kT) - 1]^2}\left(\frac{-\hbar\omega}{kT^2}\right)$$

$$= \int_0^{\omega_D} \frac{-(\omega^4 \, d\omega)\,exp\,(\hbar\omega/kT)}{[exp\,(\hbar\omega/kT) - 1]^2}\left(\frac{\hbar}{kT^2}\right)$$

$$= \frac{k\hbar}{(kT)^2}\int_0^{\omega_D} \frac{(\omega^4 \, exp\,(\hbar\omega/kT)\,d\omega}{[exp\,(\hbar\omega/kT) - 1]^2}.$$

By multiplying this by $9N\hbar/\omega_D^3$, we obtain

$$C_V = \left(\frac{9Nk}{\omega_D^3}\right)\left(\frac{\hbar}{kT}\right)^2 \int_0^{\omega_D} \frac{\omega^4 \, exp\,(\hbar\omega/kT)\,d\omega}{[exp\,(\hbar\omega/kT) - 1]^2}. \tag{16.30}$$

To simplify Equation 16.30 we make the following substitutions:

$$x = \hbar\omega/kT \tag{16.31}$$

which gives $\omega = (kT/\hbar)x$ and $d\omega = (kT/\hbar)\,dx$,

$$x_D = \hbar\omega_D/kT = T_D/T \tag{16.32}$$

where

$$T_D = \hbar\omega_D/k = (\hbar v_0/k)(6\pi^2 N/V)^{1/3} \tag{16.33}$$

which is the *Debye temperature*. Here v_0 is the velocity of sound and V is the volume of the solid. Substituting all these into Equation 16.30, we obtain

$$C_V = \frac{9Nk}{x_D^3} \int_0^{x_D} \frac{x^4 e^x\,dx}{(e^x - 1)^2}. \tag{16.34}$$

The function

$$F_D(T_D/T) = \int_0^{x_D} \frac{x^4 e^x\,dx}{(e^x - 1)^2} \tag{16.35}$$

is called the *Debye function*. This integral has to be evaluated numerically (see p. 316 of *Heat and Thermodynamics*, 5th Edition, M. W. Zemansky; Mc Graw-Hill, New York, 1957). Using Equation 16.35, Equation 16.34 can be written as

$$C_V = \left(\frac{9Nk}{x_D^3}\right) F_D(T_D/T)$$
$$= 9Nk(T/T_D)^3 F_D(T_D/T). \tag{16.36}$$

The variation of C_V with T/T_D, as predicted by Equation 16.36, is schematically shown in Figure 16.2.

The development of the Debye theory in this section has not been limited to crystalline solids; it is equally applicable to substances with an amorphous state and propagating transverse vibrations. The theory can also be extended to compounds such as FeS. In this compound, both the Fe and S atoms occupy sublattice points and there are $2 \times 3N$ degrees of freedom for the system, because the Fe and S atoms can be considered independent. In fact (according to the law of Woestyne) the Debye theory allows the atomic heat capacities of Fe and S to combine additively at high temperatures, whereas at low temperatures the T^3 dependence of C_V exists.

If we consider the absolute zero of temperature to correspond to the smallest energy realizable by the system, then this temperature will be attained by the system when all its oscillators are situated in the cell of minimum energy ($\varepsilon = nh\nu$ is zero in the Planck representation or $\varepsilon = (n+1/2)h\nu$ equal to $h\nu/2$ according to the quantum mechanics where ν is the frequency of radiation). There is only one way in which all the particles can be distributed so that they all are in the lowest energy configuration. Hence, from $S = k \ln \Omega$ we have at $T = 0$

$$S_{T=0} = k \ln 1 = 0 \quad \text{(since } \Omega = 1 \text{).}$$ (16.37)

Thus, the absolute entropy is zero at the absolute zero of temperature. This is in agreement with the somewhat imprecise statement of the Nernst theorem that at the absolute zero of temperature all perfect crystalline substances have zero absolute entropy.

Important parameters for the Debye model are the speed of sound and the maximum supposed frequency ω_D. We do not have the choice to select both of the parameters independently to get a fit between experiment and the Debye model. In the final expressions, ω_D is generally replaced by the *Debye characteristic temperature* $T_D = \hbar \omega / k$ to quote C_V. The general behaviour of the heat capacity of a solid as a function of temperature can be calculated from Equation 16.34.

16.4.1   The Debye temperature

The idea of the Debye temperature is useful in many topics in solid state physics other than specific heats. We might have noticed in the above discussion that the Debye theory of specific heat did not take into account the periodicity of a crystal lattice or the limitation imposed by the Brillouin zone on the physically realizable ranges of wave vector and frequency. This is not the whole scenario, because the ω_D is automatically comparable with the angular frequencies of phonons whose wave vectors are close to the Brillouin zone boundaries. In fact, these phonons are in the majority, and these are the ones which are excited at $T > T_D$. On the other hand, at temperatures much lower than T_D, the only phonons excited are those whose wave vectors are close to the centre of the Brillouin zone and well away from the zone boundaries. Therefore, thermal conduction (which is controlled by the anharmonic coupling of phonons to phonons) and electrical conduction (which is controlled by the scattering of electrons by phonons) behave differently above and below the Debye temperature T_D. At T above T_D, most phonons have wavelengths equal to only a few atomic spacings d, whereas at temperatures well below T_D the most probable phonon wavelength is of the order of $(T_D/T)d$. At sufficiently low temperatures, this wavelength could be hundreds, even thousands, of atomic spacings.

Debye followed Einstein in postulating that a solid with N atoms would have $3N$ vibrational modes, each with an energy given by Equation 16.28. However, he observed that the angular frequency ω of a mode must depend on its wave vector. There must be some maximum angular frequency ω_{max} such that the total number of distinguishable modes is given by Equation 16.24. Then, the same ω_{max} should be the upper limit for the integral describing the total vibrational energy given by Equation 16.29. This is not an easy task for the actual density of states in a real solid. Since the Debye model involves an over-simplification of the actual density of state, it is expected that the value of the Debye temperature T_D which fits C_V at a certain temperature will not necessarily be the optimum value for all temperatures. However, for any pair of measurements of C_V and T, it is possible to calculate the equivalent Debye temperature T_D, and a series of such measurements can allow us to plot T_D versus T. There is always some variation of T_D with temperature, although for most solids the variation is small. Let us consider how C_V behaves at temperatures above and below the Debye temperature T_D.

At high temperatures $x = T_D/T$ is small (very much less than unity). So, if we expand e^x and ignore the terms $x^2, x^3, \ldots$, etc., and also take $1 + x \simeq 1$, then we have

$$F_D(T_D/T) = \int_0^{x_D} \frac{x^4 e^x \, dx}{(e^x - 1)^2} \simeq \int_0^{x_D} x^2 \, dx = \frac{x_D^3}{3}.$$

Hence, at high temperatures, for one mole, Equation 16.36 becomes

$$C_V \simeq (9N_0k/x_D^3)(x_D^3/3) = 3N_0k = 3R \qquad (16.38)$$

where N_0 is the Avogadro number. This is shown in Figure 16.2.

We have seen earlier (see Equation 16.11) that Equation 16.38 is nothing but *Dulong and Petit's* law. Thus, at high temperatures the Debye approximation, like the Einstein approximation, also leads to Dulong and Petit's law. In fact, this result was expected, because according to Equation 16.24, for one mole of atoms there should be $3N_0$ independent harmonic oscillators. Each of these oscillators has an energy kT in the classical limit. Therefore, the total energy is $U = 3N_0kT$, and at high temperatures $C_V = (\partial U/\partial T)_V = 3N_0k = 3R$.

At *very low temperatures* the upper limit x_D ($= \hbar\omega/kT$) in Equation 16.34 becomes very large. We can say that as $T \to 0$, $x_D \to \infty$. Thus, no great error is introduced at very low temperatures if we replace the upper limit x_D in the integral in Equation 16.34 by infinity. Note that the value of $x = \hbar\omega/kT$ is very large in the proximity of x_D. At very low temperatures, when $x \gg 1$ and $e^x \gg 1$, we have

$$\frac{x^4 e^x}{(e^x - 1)^2} \simeq x^4 e^{-x}.$$

This shows that when $x \gg 1$, the value of $x^4 e^{-x}$ is dominated by the term e^{-x} and it becomes very small. Thus, at very low temperatures, the integrand in Equation 16.34 is very small in the vicinity of x_D but x_D itself is very large. This justifies the replacement of the upper limit x_D by infinity. With this modification, the definite integral in Equation 16.34 becomes a number which is equal to $4\pi^4/15$. Hence, at very low temperatures, C_V becomes

$$C_V = \frac{9R}{x_D^3}\left(\frac{4\pi^4}{15}\right) = \frac{12R\pi^4}{5}\left(T/T_D\right)^3$$

or

$$C_V = 1.95 \times 10^3 (T/T_D)^3 \, \text{J mol}^{-1} \, \text{K}^{-1}. \qquad (16.39)$$

This is the *Debye T^3 law*. At temperatures very much less than the Debye temperature ($T \ll T_D$), this law has been verified experimentally for insulators. T_D can be evaluated from experimental determinations of the variation of C_V with temperature. At very low temperatures, a plot of C_V against T^3 should give a straight line passing through the origin. The linear relationship confirms the Debye T^3 law at these temperatures. There should be an excellent agreement. However, in the case of a metal, the free electrons contribute significantly to the total heat capacity at low temperatures ($< 5\,\mathrm{K}$).

The Debye temperature T_D for a solid can be evaluated from the measured values of the elastic constants of that solid. From the definition of T_D, i.e. $T_D = \hbar \omega_D / k$, and Equation 16.25 we obtain

$$T_D^3 = \frac{18N\pi^2\hbar^3}{Vk^3[(2/v_T^3) + (1/v_L^3)]}. \tag{16.40}$$

However, according to the theory of elastic waves in solids v_T and v_L are given by

$$v_T = (G/\rho)^{1/2} \tag{16.41}$$

$$v_L = \left(\frac{K + 4G/3}{\rho}\right)^{1/2} \tag{16.42}$$

where G is the modulus of rigidity, ρ is the density and K is the bulk modulus. Although not exact, there is generally a reasonably good agreement between the measured and calculated values of T_D.

16.5 COMPARISON OF THE EINSTEIN AND THE DEBYE MODELS

By now we should have noticed that a term given by Equation 16.43 ought to occur in the expression for the internal energy of a diatomic molecule.

$$\begin{aligned}
\overline{E} &= \frac{h\nu}{\exp\left(h\nu/kT\right) - 1} \\
&= \frac{h\nu}{\exp\left(\beta h\nu\right) - 1}
\end{aligned} \tag{16.43}$$

where ν is the frequency of oscillation of each atom, $\overline{E}$ is the average energy and $\beta = 1/kT$. Three such terms with three main frequencies (ν_1, ν_2, ν_3) should appear in the internal energy of a simple crystal. The contribution to the specific heat capacity of the system of a simple harmonic oscillator should be the same (per degree of freedom) in both cases.

For simplicity, if we assume that in the case of a simple solid all three main frequencies are the same, then C_V of the system can be written as

$$C_V = 3N_0 \frac{\partial \overline{E}}{\partial T}. \tag{16.44}$$

Also,

$$\frac{\partial \overline{E}}{\partial T} = \left(\frac{\partial \overline{E}}{\partial \beta} \right) \left(\frac{\partial \beta}{\partial T} \right) = \left(\frac{-1}{kT^2} \right) \frac{\partial \overline{E}}{\partial \beta} \tag{16.45}$$

$$\frac{\partial \overline{E}}{\partial \beta} = -(\overline{E})^2 \exp(\beta h\nu). \tag{16.46}$$

By combining Equations 16.43, 16.44, 16.45 and 16.46, we have

$$
\begin{aligned}
C_V &= 3N_0 \left(\frac{1}{kT^2} \right) \left[\frac{h\nu}{\exp(\beta h\nu) - 1} \right]^2 \exp(\beta h\nu) \\
&= 3N_0 k \left(\frac{1}{k^2 T^2} \right) \left[\frac{h\nu}{\exp(h\nu/kT) - 1} \right]^2 \exp(h\nu/kT) \\
&= 3N_0 k \left[\frac{h\nu/kT}{\exp(h\nu/kT) - 1} \right]^2 \exp(h\nu/kT) \\
&= 3N_0 k \left[\frac{\theta/T}{\exp(\theta/T) - 1} \right]^2 \exp(\theta/T) \tag{16.47}
\end{aligned}
$$

where $\theta = h\nu/k$, and is known as the *characteristic temperature*. It is different for different solids and varies with V/N (where V is the volume and N is the total number of atoms in V) within the same crystal. It can be calculated by using the proper infrared frequency (the Reststrahl frequency) ν. If we write

$$f(\theta/T) = \left[\frac{\theta/T}{\exp(\theta/T) - 1} \right]^2 \exp(\theta/T) \tag{16.48}$$

then Equation 16.47 can be written as

$$C_V = 3Rf(\theta/T) \tag{16.49}$$

This equation can be interpreted to say that by measuring the temperature of different substances in terms of θ for that substance, all experimental curves for C_V (at least for all monatomic crystals) should coincide. This is an example of the *law of corresponding states*.

A familiar example of this law is the equation of state of a van der Waals gas, given in terms of the critical pressure, temperature and volume, so that the constants describing the different gases do not occur.

At high temperatures where $\theta \ll T$, $f(\theta/T) \to 1$, and we have the classical value of $C_V \to 3R$ as $T \to \infty$. However, when $\theta \gg T$ the exponential term in Equation 16.47 becomes dominant and the model predicts that

$$C_V \to 3R(\theta/T)^2 \exp(-\theta/T) \text{ as } T \to 0. \tag{16.50}$$

It can be seen from this equation that this *Einstein crystal* has a specific heat capacity which tends to zero as $T \to 0$, as required by experiment. However, C_V given by this equation tends to zero much faster (exponentially) than the values obtained experimentally. Therefore, although it seems that the behaviour of C_V is qualitatively correct, quantitatively there appears to exist a discrepancy between the experiment and the theory of the Einstein model of a crystal.

The Debye model is better verified by experiment than the simple Einstein model. In fact, the molar specific heat capacity at constant volume predicted by this theory is reminiscent of that given by the Einstein theory. We note that the Einstein temperature T_E ($= \hbar\omega/k$; see Equation 16.18) and the Debye temperature T_D ($= \hbar\omega_D/k$; see Equation 16.33) are of the same form, but there is a marked difference in the interpretations. The Debye temperature T_D refers to the cut-off frequency ω_D of the wave supported by the solid, whereas the Einstein temperature T_E refers to the only frequency propagated by the solid. It appears that the Debye theory, like Einstein's, leads to a law of corresponding states for C_V.

The Debye theory, while giving good agreement with experiment, is by no means perfect. In real solids the Debye temperature, obtained by comparing Equation 16.39 with heat capacity data, varies with temperature. This variation is a direct consequence of the limiting assumptions of the Debye theory.

As expected, the quantum-mechanical models differ from the classical results only at *low* temperatures. As we have seen, both the Einstein and Debye models predict that the heat capacity of a solid should tend to zero as the temperature tends to zero. In fact, they predict

$$\lim_{T \to 0} C_V = 3Nk\,(T_E/T)^2 \exp(-T_E/T) \text{ (Einstein model)}$$

$$\lim_{T \to 0} C_V = \left(\frac{12\,Nk\pi^4}{5}\right)(T/T_D)^3 \text{ (Debye model)} \tag{16.51}$$

These limiting forms for C_V are completely different functions of the temperature. According to the Einstein model the predicted values of C_V are much lower than those predicted by the Debye model as $T \to 0$. The Debye model predicts that C_V should go to zero as T^3, which is *Debye's T^3-law*. It is interesting to note that it has been found experimentally that this law does accurately describe the dependence of C_V on the temperature for many solids at low. temperatures. This is schematically shown in Figure 16.2.

In spite of the fact that the Debye model is very successful in many ways, it also has limitations. The fact that the large differences between the heat capacities of many solids can be represented by one parameter T_D (the Debye temperature) is a considerable triumph of the Debye model. Considering the variety of complex forces acting in the solids and the differences in their lattice structures, it is not at all surprising that the difference between the heat capacities of some solids cannot be accurately described by one parameter T_D. If

Equation 16.36 is used to define "the" Debye temperature, then it has been observed that for some solids T_D is not constant but varies with the temperature.

The Debye model also fails to explain the observed heat capacity at very low temperature (near 1 K). At these very low temperatures the free electrons in the solid contribute significantly to the heat capacity.

At low temperatures, the Debye model is better than the Einstein model. In the former, the concept of standing waves in a continuous solid is expected to be reasonably good at low temperatures, because at low temperatures only the low-frequency modes (i.e. the long-wavelength normal modes) are excited to any significant extent. When the wavelengths of the standing waves (normal modes) are considerably bigger than atomic spacings in the solid it is a reasonable approximation to treat the solid as a continuous body.

At high temperatures both theories agree reasonably well with the experimental results, and in both theories the molar heat capacity tends to the classical value of $3R$ when the energy of each of the $3N_0$ harmonic oscillators tends to kT.

In the intermediate temperature range, a marked difference is observed between the experiment and the Debye model. An atomic model must be used for the solid to obtain better agreement with the experimental results in this temperature range. This would then lead to a different expression for dn_i than that given in Equation 16.26.

16.6   THE HEAT CAPACITY OF CONDUCTION ELECTRONS IN METALS

Classical value of specific heat capacity given by the law of Dulong and Petit is generally verified by experiment at ordinary temperatures (with the exception of diamond). However, even here the classical theory is insufficient to account for all the experimental results. Consider the following cases.

(i) It has been observed that at low temperatures the specific heat capacity of all substances tends to zero as the absolute temperature tends to zero. This behaviour is not predicted by the classical theory.

(ii) Experimentally, it has been observed that the specific heat capacities of a metal and an insulator are approximately the same. However, there is plenty of experimental evidence to substantiate the fact that in metals there are free (conduction) electrons which form an electron gas; this electron gas should undergo translational motion. We know that in both the metal and the insulator there are six vibrational degrees of freedom associated with the positive ions (neutral atoms in the case of an insulator) about their lattice sites. However, in the metal, in addition to these six degrees of vibrational freedom, there should be three degrees of electronic translational freedom (which are not found in insulators). Hence, the classical theory predicts $C_V = \left(3 + \frac{3}{2}\right)R = 9R/2$ for metals and $3R$ for insulators. This difference is not confirmed experimentally. Fortunately, the hypothesis of Planck, introducing the constant h, reconciles all these difficulties.

When a metal is made up of atoms, the valence electrons of these atoms become the conduction electrons of the metal. These conduction electrons are free to move in the lattice of the metal; they can be accelerated by an applied electric field to generate a flow of electric current. If the conduction electrons behave as free classical particles, they would make a contribution C (electron) $= \left(\frac{3}{2}\right)Nk$ to the heat capacity of the metal (N is the number of free conduction electrons per unit volume of the metal). Then, as said above, it would be expected that at temperatures $T \gg T_D$ there should be a total C_V of $3R$ per mole for insulators and $9R/2$ per mole for metals. In fact, at high temperatures the C_V of metals is not very different to that of insulators. There is no experimental evidence to support the idea that the conduction electrons in metals contribute to C_V to anything like the extent predicted by the classical theory.

Pauli's exclusion principle prohibits any two electrons from occupying the same quantum state. This principle controls the actual thermal behaviour of the free conduction electrons in metals. For free electrons of mass m, according to the quantum theory the electronic heat capacity at low temperatures is given by

$$C_V \text{ (electronic)} = \gamma T.$$

Later, we shall derive an expression for C_V (electronic). At considerably low temperatures (generally below 3 K), the electronic contribution to the heat capacity of a metal becomes bigger than the lattice contribution. Actually, the lattice contribution decreases as T^3 while the electronic contribution decreases as T. Therefore, if the temperature is sufficiently low, the electronic specific heat in a metal will always be dominant. Hence, it is possible to determine γ in the above equation by measurements at low temperatures. To obtain sufficiently accurate values of specific heat capacity of metals, the contribution of the conduction (free) electrons to the heat capacity must be added to the contribution of the lattice given by the Debye T^3 law. Therefore, at low temperatures the experimentally determined variation of heat capacity of metals with temperature is given, to a good approximation, by an equation of the form

$$C_V = C_V(\text{lattice}) + C_V(\text{electronic})$$
$$= \alpha T^3 + \gamma T \tag{16.52}$$

where α is a constant given by

$$\alpha = \frac{12Nk\pi^4}{5T_D^3} \tag{16.53}$$

(see Equation 16.51).

There could be other contributions to the total heat capacity of a solid. For example, the paramagnetic ions in a solid give a large contribution to C_V of a solid in the temperature range where $kT/\mu B$ is approximately equal to 0.8. Significant anomalies in C_V have also been noted associated with other phase changes, such as for example, near the Curie temperature, when there is a transition from ferromagnetism to paramagnetism. However, we shall not discuss these contributions, as these are part of solid state physics; rather we wish to find out the electronic contribution to specific heat of a solid.

To find the electronic specific heat let us consider the energy content of an electron gas. At absolute zero of temperature the electron gas has a minimum total energy given by

$$U_0 = \int_0^{E_F(0)} g(E)E \, dE = \frac{[E_F(0)]^{5/2}}{5\pi^2} \left(\frac{2m}{\hbar^2}\right)^{3/2} \tag{16.54}$$

where $E_F(0)$ is the Fermi energy at absolute zero, m is the mass of the electron and $g(E)$ is the density of state given by Equation 15.68. The total density of electrons per unit volume is given by

$$n = \frac{N}{V} = \frac{1}{2\pi^2} \left(\frac{2m}{\hbar^2}\right)^{3/2} \int_0^{E_F(0)} \sqrt{E}\, dE = \frac{1}{3\pi^2} \left(\frac{2m}{\hbar^2}\right)^{3/2} [E_F(0)]^{3/2} \tag{16.55}$$

where N is the number of electrons in a volume V of a metal. By combining Equations 16.54 and 16.55, we obtain

$$U_0 = \left(\frac{3}{5}\right) n E_F(0). \tag{16.56}$$

This equation shows that there is an average energy of $3E_F(0)/5$ per electron without any thermal energy being supplied. Electrons move very rapidly. In equilibrium position, the motion of any electron at any instant is matched by the motion of another electron moving in the opposite direction somewhere in the same metallic crystal.

If the temperature of the metal is increased from zero, the electronic energy becomes slightly bigger than U_0. This is because at any higher temperature a few electrons are thermally excited from states just below the Fermi energy E_F to states higher than E_F. Thus, at any finite temperature

$$U = \int_0^{\infty} g(E) f(E) E\, dE$$

$$= \left(\frac{1}{2\pi^2}\right) \left(\frac{2m}{\hbar^2}\right)^{3/2} \int_0^{\infty} \frac{E^{3/2}\, dE}{\exp\left(\frac{E-E_F}{kT}\right) + 1}. \tag{16.57}$$

This equation involves the function $F_{3/2}(x_0)$ from the family of Fermi–Dirac integrals

$$F_i(x_0) = \int_0^{\infty} \frac{x^i\, dx}{\exp(x - x_0) + 1}. \tag{16.58}$$

Integrals of this type cannot be expressed in closed form for any arbitrary value of x_0. However, asymptotic forms do exist when x_0 is large and negative, or large and positive. Tables of numerical values of Fermi–Dirac integrals for integer and half-integer orders can be found in literature (J. McDougal and E. C. Stoner; *Phil. Trans.*, *Royal Soc.*, A237 (1968) 67).

For x_0 large and negative, the asymptotic form is given by

$$F_i(x_0) = \Gamma(i + 1) \exp(x_0), \quad -x_0 > 2 \tag{16.59}$$

For x_0 large and positive, the asymptotic form is given by

$$F_i(x_0) = \frac{(x_0)^{i+1}}{(i+1)} \left[1 + \frac{i(i+1)\pi^2}{6x_0^2} + \cdots\right], \quad x_0 \gg 1. \tag{16.60}$$

The third, fourth, etc., terms in the series are actually in ascending powers of $1/x_0^2$. The series ends when i is an integer, but $F_i(x_0)$ is a converging infinite series for i not being an integer. However, in any case the first two terms are sufficient when x_0 is a sufficiently large number.

For an electron density comparable with that in a metal, the asymptotic form of a Fermi–Dirac integral is that for x_0 large and positive. Hence, in our case we use Equation 16.60 to solve the integral in Equation 16.57.

$$I = \int_0^\infty \frac{E^{3/2}\,\mathrm{d}E}{\exp\left(\frac{E-E_F}{kT}\right)+1} \simeq \frac{2E_F^{5/2}}{5}\left[1+\frac{\left(3\frac{\pi^2}{2}\right)\left(\frac{5}{2}\right)}{6E_F^2}+\cdots\right]$$

$$\simeq \frac{2E_F^{5/2}}{5}\left[1+\frac{5\pi^2}{8E_F^2}\right], \quad E_F \gg kT. \tag{16.61}$$

The Fermi energy E_F at any finite temperature in a metal is related to that at absolute zero $E_F(0)$ by

$$E_F \simeq E_F(0)\left[1-\frac{(\pi kT)^2}{12\{E_F(0)\}^2}\right]. \tag{16.62}$$

Let us first substitute the value of E_F from Equation 16.62 in the term outside the parentheses in Equation 16.61. Thus,

$$I = \frac{2}{5}\left[E_F(0)-\frac{(\pi kT)^2}{12E_F(0)}\right]^{5/2}\left[1+\frac{5\pi^2}{8E_F^2}\right]$$

$$= \frac{2[E_F(0)]^{5/2}}{5}\left[1-\frac{(\pi kT)^2}{12\{E_F(0)\}^2}\right]^{5/2}\left[1+\frac{5\pi^2}{8E_F^2}\right].$$

By expanding and ignoring the higher power terms of $[\pi kT/E_F(0)]^2$, we obtain

$$I = \frac{2[E_F(0)]^{5/2}}{5}\left[1-\left(\frac{5}{24}\right)\left\{\frac{\pi kT}{E_F(0)}\right\}^2\right]\left[1+\frac{5\pi^2}{8E_F^2}\right].$$

Next we substitute the value of E_F from Equation 16.62 into the second set of parentheses of the above equation, expand it and ignore the higher powers. Thus,

$$I = \frac{2[E_F(0)]^{5/2}}{5}\left[1-\left(\frac{5}{24}\right)\left\{\frac{\pi kT}{E_F(0)}\right\}^2\right]\left[1+\left(\frac{5}{8}\right)\left\{\frac{\pi kT}{E_F(0)}\right\}^2\right]$$

$$= \left(\frac{2}{5}\right)\left[E_F(0)\right]^{5/2}\left[1+\left(\frac{5}{12}\right)\left\{\frac{\pi kT}{E_F(0)}\right\}^2\right]. \tag{16.63}$$

By substituting the value of I from Equation 16.63 into Equation 16.57 and simplifying, we have

$$U = \frac{[E_F(0)]^{5/2}}{5\pi^2} \left(\frac{2m}{\hbar^2}\right)^{3/2} \left[1 + \left(\frac{5}{12}\right)\left\{\frac{\pi kT}{E_F(0)}\right\}^2\right].$$

Using Equation 16.54, we obtain

$$U = \left[U_0 + \frac{k^2\{E_F(0)\}^{1/2}}{12}\left(\frac{2m}{\hbar^2}\right)^{3/2} T^2\right], \quad T \ll E_F(0)/k. \tag{16.64}$$

Therefore, to a first approximation the electronic specific heat is

$$C_V(\text{electronic}) = \left(\frac{\partial U}{\partial T}\right)_{\text{el}} = \frac{2k^2\{E_F(0)\}^{1/2}}{12}\left(\frac{2m}{\hbar^2}\right)^{3/2} T.$$

By substituting the value of $E_F(0)$ from Equation 15.74 and simplifying, we have

$$C_V(\text{electronic}) = \frac{mN^{1/3}k^2\pi^2 T}{\hbar^2(3\pi^2)^{2/3}} \tag{16.65}$$

or

$$C_V(\text{electronic}) = \gamma T \tag{16.66}$$

where

$$\gamma = \frac{mN^{1/3}k^2\pi^2}{\hbar^2(3\pi^2)^{2/3}}. \tag{16.67}$$

Equation 16.66 shows that the heat capacity of the free electrons in a metal goes to zero in proportion to T. This can be explained by considering the fact that only those electrons which are near the Fermi surface are affected when thermal energy is added to the system. The number of electrons that change their energy is proportional to kT/E_F and each electron should contribute approximately $3k/2$ of energy to the electronic heat capacity. Thus, the electronic C_V should be proportional to

$$\left(\frac{3k}{2}\right)\left(\frac{kT}{E_F}\right) \quad \text{i.e. } k^2T/E_F.$$

The numerical value of γ for metals is often of the order of $4 \times 10^{-4}\,\text{J}\,\text{deg}^{-2}\,\text{mol}^{-1}$. Therefore, at room temperature the electronic contribution is only of the order $300 \times 4 \times 10^{-4}\,\text{J}\,\text{deg}^{-1}\,\text{mol}^{-1}$, i.e. $12 \times 10^{-2}\,\text{J}\,\text{deg}^{-1}\,\text{mol}^{-1}$. This value is significantly small compared with the Dulong and Petit value for the lattice heat capacity of metals ($\simeq 24\,\text{J}\,\text{deg}^{-1}\,\text{mol}^{-1}$).

The result in Equation 16.65 is known as *Sommerfeld's equation* for the electronic specific heat. We should compare, this result with the value

$$C_V \,(\text{el}) = \frac{3Nk}{2} \tag{16.68}$$

expected for a classical electron gas. According to the quantum restrictions, the electronic part of the total specific heat is given by

$$C_V \,(\text{electronic}) = \left(\frac{\pi^2 kT}{3E_F} \right) C_V \,(\text{el}). \tag{16.69}$$

It is often stated that the electronic specific heat in a metal is *degenerated* by a factor of about $E_F/3kT$ from its classical value $3Nk/2$. An electron gas for which $kT \ll E_F$ is said to be a *degenerate gas*.

Note: Two states of a system (for example two excited states of an atom) are said to have degenerated when they have the same energy. Here, the word is used in the sense that the electronic specific heat has been degraded from the large value expected according to the classical theory. The electron gas in a semiconductor material can be described as 'degenerate' or 'non-degenerate' according to whether or not free electrons are numerous enough to make quantum restrictions on electrons important.

Equation 16.65 conforms in its magnitude and temperature dependence to the electronic component of specific heat actually observed in metals. As the electronic heat capacity, C_V (electronic), varies with T (the first power of the absolute temperature) and the lattice heat capacity varies more or less as T^3, the electronic portion of heat capacity can be measured most easily at very low temperatures.

Sommerfeld's result of Equation 16.65 explains one of the intriguing questions not answered by classical statistical theories. According to the classical theory the internal energy of conduction electrons should be $3NkT/2$. Therefore, they should contribute a heat capacity

$$C_V \,(\text{electronic}) = \left(\frac{\partial U}{\partial T} \right)_V = \frac{\partial}{\partial T} \left(\frac{3NkT}{2} \right) = \frac{3Nk}{2}$$

to the solid. It is known that at high temperatures the heat capacity of solids is approximately given by $C_V \simeq 3Nk$ (Dulong and Petit's law). Now, as this heat capacity is lattice heat capacity (i.e. due to the movement of the lattice atoms), it follows that the heat capacity of the free electrons must be much less than the classical value $3Nk/2$. It can be seen in Equation 16.65 that the electronic heat capacity is indeed much less than the classical value for temperatures $T \ll E_F/k$.

Even though the electronic heat capacity is very small at low temperatures (Equation 16.65), it can equal or become greater than the lattice heat capacity (due to the vibrations of the atoms in the solid). At sufficiently low temperatures (generally below 37 K), the electronic contribution to the heat capacity exceeds the lattice contribution, and the former decreases as the first power of absolute temperature, whereas the latter contribution decreases as T^3. Therefore, if we go low enough in temperature, the electronic contribution will always be dominant in metals. Hence, at low temperatures, it is possible to determine γ in Equation 16.66. Hence, it is clear that at low temperatures the electronic contribution cannot be ignored. If we add the electronic heat capacity to the lattice heat capacity, then the total heat capacity of the solid at low temperatures should be

$$C_V = \gamma T + \alpha T^3$$

(see Equation 16.52) or

$$C_V/T = \gamma + \alpha T^2 \tag{16.70}$$

where the value of the constants are given by Equations 16.53 and 16.67. Equation 16.70 is sometimes called the *Debye–Sommerfeld equation*.

If we plot C_V/T against T^2, we should obtain a straight line whose intercept at $T = 0$ gives the value of γ, while the slope of the line yields α and therefore the value of the Debye temperature T_D. It is found experimentally that the temperature dependence of the conduction electron heat capacity is actually linear in T for most metals; however, γ may have values quite different from those calculated using Equation 16.67. Figure 16.5 shows the total specific heat as a sum of the kind given by Equation 16.52.

As only those states which are within the energy range of a few kT of the Fermi level actually influence the measured properties of a degenerate electron gas, it is quite logical to express C_V (electronic) in those terms which relate specifically to that section of the energy

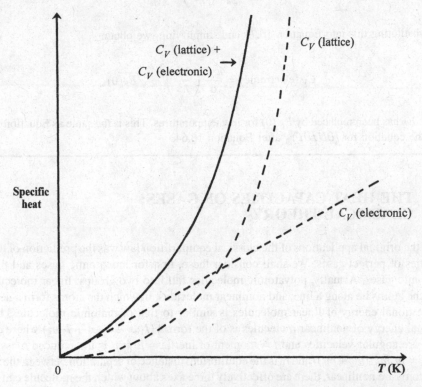

Figure 16.5 Schematic diagram of variation of C_V (electronic), C_V (lattice) and the sum of these with temperature

range. Thus, C_V (electronic) can be described in terms of the *density of states*, $g(E_F)$, at the Fermi energy as

$$C_V \text{ (electronic)} = \left(\frac{\pi^2 k^2 T}{3}\right) g(E_F). \tag{16.71}$$

This is equivalent to the classical specific heat $3k/2$ per electron for all the electrons lying in an energy range slightly more than $2kT$ wide, centred on the Fermi energy.

Example 16.1

Prove that Equation 16.71 is consistent with both Equation 16.65 and Equation 16.69.

Solution

To prove this we have to define density of state at the Fermi energy. For this we use Equation 15.68 and write

$$g(E_F) = \frac{1}{2\pi^2} \left(\frac{2m}{\hbar^2}\right)^{3/2} \sqrt{E_F}. \tag{16.72}$$

By substituting this into Equation 16.71 and simplifying, we obtain

$$C_V \text{ (electronic)} = \frac{k^2 T}{6} \left(\frac{2m}{\hbar^2}\right)^{3/2} \sqrt{E_F(0)}$$

where E_F has been replaced by $E_F(0)$ for low temperatures. This is the same as Equation 16.65 (see the equation for $(\partial U/\partial T)_{el}$ after Equation 16.64.

16.7 THE HEAT CAPACITIES OF GASES: CLASSICAL THEORY

One of the original applications of the classical equipartition law was the prediction of the heat capacities of perfect gases. We shall consider these, first for monatomic gases and then for polyatomic gases. Actually, polyatomic molecules fall into two groups: linear molecules, in which the atoms lie along a line, and nonlinear molecules, in which the atoms form a network. The rotational energy of linear molecules is similar to that of diatomic molecules, but the rotational energy of nonlinear molecules is of the form $\frac{1}{2}(I_1\omega_1^2 + I_2\omega_2^2 + I_3\omega_3^2)$ where ω_1, ω_2, and ω_3 are angular velocities and $I = $ moment of inertia $= \mu r_0^2$; μ is the reduced mass of the atoms ($= m_a m_b/(m_a + m_b)$) and r_0 is the equilibrium distance of separation between the atoms. If the atoms are nonlinear, there are effectively three axes about which the molecule can rotate. As the larger molecules have more ways in which to vibrate (some of which take very little energy), the contribution of the vibrational motion also tends to be larger for these molecules.

(i) *Ideal monatomic gases*: For such gases the classical energy of the system is given by (see Section 12.5: model for gases)

$$E = \sum_{i=1}^{N} \frac{1}{2}(m\mathbf{v}_i^2) \tag{16.73}$$

where m is the mass, $\mathbf{v}$ is the velocity of the atoms, and N is the number of atoms.

The translational energy (which is entirely due to translational motion of the molecules) is the sum of $3N$ quadratic terms. By the law of equipartition of energy, each of these quadratic terms should contribute $kT/2$ to the average energy of the system. Therefore, the internal energy will be $U = \frac{3}{2}NkT$ and the predicted heat capacity is given by

$$C_V = \left(\frac{\partial U}{\partial T}\right)_V = \frac{3Nk}{2} = \frac{3nR}{2} \tag{16.74}$$

where n is the number of moles. For an ideal gas we have $C_P = C_V + nR$. Therefore, using Equation 16.74, the predicted value of $C_P = 5nR/2$. Thus, according to the classical theory, the predicted value of the ratio of C_P and C_V is

$$\gamma = \frac{C_P}{C_V} = \left(\frac{5nR}{2}\right)\Big/\left(\frac{3nR}{2}\right) = \frac{5}{3} = 1.667.$$

Note that this γ is not the same as that in Equations 16.52 and 16.66. Typical experimental values of γ for some common and inert gases are given in Table 16.1. These values of γ for inert gases are in very good agreement with the predicted value 1.667. Therefore, clearly the classical theory is quite adequate for monatomic perfect gases. This is a consequence of the fact that the internal energy is due entirely to translational motion of molecules.

(ii) *Ideal polyatomic gases*: Here we shall consider perfect diatomic gases. The same principles are also applicable to gases with more than two atoms. The energy of a diatomic molecule is given by (see Equation 12.19)

$$\varepsilon = \varepsilon_t + \varepsilon_r + \varepsilon_v \tag{16.75}$$

where ε_t, ε_r and ε_v are translational, rotational and vibrational energies. According to the models for gases (see Section 12.5), we have the following equations for the energies:

$$\varepsilon = M\mathbf{V}^2/2 + I(\omega_1^2 + \omega_2^2)/2 \quad \text{(rigid dumbbell model)} \tag{16.76}$$

Table 16.1 Experimental values of γ for some monatomic and diatomic gases

Monatomic gases		Diatomic gases	
He	1.66	H_2	1.408
Ne	1.64	O_2	1.400
Ar	1.67	N_2	1.404
Kr	1.68	Cl_2	1.340
Xe	1.66	CO	1.404

$$\varepsilon = MV^2/2 + I(\omega_1^2 + \omega_2^2)/2 + \frac{\mu}{2}\left(\frac{dr_{ab}}{dt}\right)^2$$

$$+ \frac{\kappa}{2}(r_{ab} - r_0)^2 \quad \text{(harmonic model; see Equation 12.28)} \quad (16.77)$$

where κ is the force constant; I, μ, r_0 and ω are all defined above, and r_{ab} and dr_{ab}/dt are defined in Equation 12.31. In both models, $MV^2/2$ represents the translational energy of the entire molecule. In the dumbbell model it is assumed that there is no vibrational motion of the molecule, i.e. $\varepsilon_v = 0$, but in the harmonic model we have

$$\varepsilon_v = \frac{\mu}{2}\left(\frac{dr_{ab}}{dt}\right)^2 + \frac{\kappa}{2}(r_{ab} - r_0)^2. \quad (16.78)$$

It is important to determine which of these two models are appropriate in our case. As the forces between two atoms (diatomic) are finite, it is not possible that a perfectly rigid dumbbell can be formed in the classical theory. So, we can discard this model, and accept that on the basis of classical mechanics the only realistic model is the harmonic model given by Equation 16.77.

Let ε_i be the energy of all kinds of the ith molecule as given by Equation 16.75. Then the total energy of a perfect gas is

$$E = \sum_{i=1}^{N} \varepsilon_i. \quad (16.79)$$

In the classical theory, the law of equipartition of energy is applicable to Equation 16.79, and we can use Equation 16.77 for the molecular energy.

Note: $kT/2 (= 1/2\beta)$ represents the average value of the contribution of one (particular) degree of freedom of the kinetic energy. It can easily be seen that all degrees of freedom will contribute the same amount and in the very same way. Thus, the total contribution of one degree of freedom, from all N molecules, to the total ensemble energy will be $NkT/2 = nRT/2$ (n is the number of moles in the system), and the contribution to the specific heat capacity of one degree of freedom (from the kinetic energy) will be $R/2$. Thus, the energy is, on average, distributed equally over all types of motion given by the quadratic terms (kinetic energy terms) in E. This is known as the *law of equipartition of energy* (see Section 5.3.1). Alternatively, we can state this law as: *the contribution of the kinetic energy of the molecules in the system to the molar specific heat capacity of the system is given by*

$$C_V = (f)R/2 \quad (16.80)$$

where f is the number of degree of freedom of the molecules.

The molecular energy ε as given by Equation 16.77 is the sum of three translational, two rotational and two vibrational quadratic terms. In the translational term the variable is the velocity (v) of the centre of mass, in the rotational terms the variables are ω_1 and ω_2, and in the vibrational term the two variables are the velocity along the axis connecting the atoms (dr_{ab}/dt) and the quantity ($r_{ab} - r_0$). It is, therefore, obvious that the total energy given by Equation 16.79 consists of the sum of $7N$ quadratic terms. Hence, according to the law of equipartition of energy the internal energy of the system should be

$$U = 7N(kT/2) \quad (16.81)$$

$$C_V = \left(\frac{\partial U}{\partial T}\right)_V = 7Nk/2 \tag{16.82}$$

$$C_P = C_V + Nk = 7Nk/2 + Nk = 9Nk/2$$

so the harmonic model prediction is that

$$\gamma = C_P/C_V = 9/7 = 1.286. \tag{16.83}$$

From Table 16.1 we note that the γ-values for diatomic gases are about 9% higher. Therefore, $\gamma = 1.286$ predicted by the classical harmonic model is not accurate. Let us consider the result given by the rigid dumbbell model and investigate whether the choice of the model of the molecule makes any difference.

In the rigid dumbbell model, the energy of a molecule, as given by Equation 16.76, consists of the sum of five quadratic terms (there is no vibrational motion). Therefore, the total energy given by Equation 16.79 consists of $5N$ quadratic terms, and according to the law of equipartition of energy the internal energy of the system should be

$$U = 5N(kT/2) \tag{16.84}$$

and

$$C_V = \left(\frac{\partial U}{\partial T}\right)_V = 5Nk/2. \tag{16.85}$$

Hence, we see that the rigid dumbbell model predicts that

$$\gamma = 7/5 = 1.4. \tag{16.86}$$

This value closely matches with those given in Table 16.1. It would, therefore, appear that this model is more realistic than the harmonic model. However, we have already said that as the forces between the atoms are finite, there is no way to explain using classical mechanics why a molecule should behave like a rigid dumbbell. We must also note that a harmonic molecule which has a very large force constant corresponding to a very stiff spring between the atoms is not the same as an absolutely rigid dumbbell, because according to the law of equipartition of energy there should be an average energy of $kT/2$ for every quadratic variable, irrespective of the value of the force constant κ. Hence, on the basis of classical mechanics we cannot explain why in the dumbbell model the vibrational motion of diatomic molecules should be completely zero. There are diatomic molecules where the vibrational motion is not quite *frozen*, as the γ-values in such cases lie between 1.400 and 1.286 (e.g. Cl_2). These are mysterious from the point of view of classical mechanics.

16.8 THE HEAT CAPACITIES OF GASES: STATISTICAL THEORY

According to quantum mechanics, the average energy of a harmonic oscillator is

$$\bar{\varepsilon} = \frac{h\nu}{\exp(\beta h\nu) - 1} \tag{16.87}$$

where ν is frequency of vibration and $\beta = 1/kT$. We then ask: under what conditions does $\bar{\varepsilon}$ given by Equation 16.87 agree with the classical value $\bar{\varepsilon} = kT$? $kT = 1.38 \times 10^{-23}$ T(J), and $h\nu \simeq 10^{-20}$ to 10^{-21} J. If the temperature is high, then kT can be much larger than $h\nu$ and $\beta h\nu = h\nu/kT \ll 1$. In this case, Equation 16.87 becomes (expanding $\exp(\beta h\nu)$ and taking only the first two terms)

$$\bar{\varepsilon} \simeq \frac{h\nu}{(1 + \beta h\nu) - 1} = kT. \tag{16.88}$$

This result agrees with that predicted by the classical law of equipartition of energy. Here, the classical and quantum results are in agreement with each other if the difference between the energy levels of the oscillator ($h\nu$) is much smaller than kT. We can express this fact in terms of temperature by defining the *characteristic vibrational temperature* by

$$\theta_v \equiv \frac{h\nu}{k} = \frac{\hbar\omega}{k} \qquad \left(\text{since } \nu = \frac{\omega}{2\pi}\right). \tag{16.89}$$

Then the classical approximation is in agreement with the quantum result if $T \gg \theta_v$ (i.e. $\beta h\nu \ll 1$), as evident from Equation 16.88. It is useful to note that the magnitude of θ_v is of the order of several thousand degrees Kelvin.

At low temperatures, i.e. when $\theta_v \gg$ T or $\beta h\nu \gg 1$, Equation 16.87 becomes

$$\bar{\varepsilon} \simeq h\nu \exp(-\beta h\nu). \tag{16.90}$$

This average energy differs greatly from that given by the equipartition law, i.e. kT, when T is small. This shows that the classical theory is not accurate if $h\nu < kT$ or of the order of kT.

Therefore, we can say that classical statistical mechanics can produce accurate results if the energy difference between the quantum states of the oscillators is much less than kT.

Although this is a necessary condition, it is not sufficient for classical statistical mechanics to be accurate. There are other conditions which must be satisfied and these conditions become important at very low temperatures or very high density of particles. Suffice it to say that many of the discrepancies between the results predicted by classical statistical mechanics and the observed properties of systems can be ascribed to the fact that the condition $h\nu \ll kT$ is not satisfied.

From the above discussion we know that the average energy of a harmonic oscillator predicted by classical statistical mechanics is inaccurate if $h\nu > kT$ or of the order of kT, i.e. $\theta_v \equiv h\nu/k >$ or $\simeq T$. It has been said above that θ_v is of the order of several thousand degrees Kelvin. So, it is not surprising that the experimental values of γ for diatomic molecules at room temperature do not agree with the value given by the law of equipartition of energy. Excitation of the vibrational motion of most molecules requires an energy $h\nu \gg kT$, but in effect the available energy per molecule (kT) is too small to cause excitation. Thus, the vibrational motion is frozen and the molecules behave like rigid dumbbells.

16.9   VARIATION OF THE MOLAR HEAT CAPACITY OF AN IDEAL DIATOMIC GAS WITH TEMPERATURE

The energy of a diatomic molecule is given by Equation 16.75. There are three components: translational, rotational and vibrational. In this section we wish to examine how each of these components is related to the temperature and what is the extent of its contribution. Our discussion is of a very general nature without referring to any model.

Consider a mole of a perfect gas, consisting of N diatomic molecules. We define a temperature θ_r by

$$\theta_r = \hbar^2/2Ik \tag{16.91}$$

where I is the moment of inertia of the molecule about an axis through the centre of mass of the molecule perpendicular to the line joining the two atoms. According to Equation 16.91, θ_r is inversely proportional to I, so that θ_r is higher for lighter molecules; for example, for $HCl \rightarrow \theta_r$ is 15.2 K, whereas for the lighter $H_2 \rightarrow \theta_r$ is 85.5 K. Here, we shall consider three cases: $T \ll \theta_r$, $T \gg \theta_r$, and higher temperatures.

(i) *At temperatures $T \ll \theta_r$*: Only those translational states which are at higher energy are excited to any significant extent. We have already established that in this range the mean translational energy of each molecule is approximately $3kT/2$. So, the total energy of N diatomic molecules is approximately $3NkT/2$. Hence, the molar heat capacity at constant volume is

$$C_V = \left(\frac{\partial U}{\partial T}\right)_V \simeq \frac{\partial}{\partial T}\left(\frac{3NkT}{2}\right)_V \simeq \frac{3Nk}{2} = \frac{3R}{2}. \tag{16.92}$$

(ii) *At temperatures $T \gg \theta_r$*: As the temperature increases and becomes of the order of θ_r, the higher energy rotational states become excited to a significant extent. When T becomes very much greater than θ_r (i.e. $T \gg \theta_r$), the mean rotational energy per molecule is kT. Therefore, the total energy of N diatomic molecules is NkT and the molar heat capacity is (if $T \ll \theta_v$)

$$C_V \simeq \frac{\partial}{\partial T}\left[N\left(kT + \frac{3kT}{2}\right)\right]_V = \frac{5Nk}{2} = \frac{5R}{2}. \tag{16.93}$$

In the proximity where T equals θ_r, C_V increases from $3R/2$ to $5R/2$. The molar heat capacity of a diatomic gas at constant volume is generally about $5R/2$ at room temperature (which is generally between θ_r and θ_v).

(iii) *At higher temperatures*: As the temperature increases further, the higher energy vibrational states are excited at $T \simeq \theta_v$. When $T \gg \theta_v$, the mean vibrational energy per molecule is again kT. So the total energy of N diatomic molecules in this range is NkT and the molar heat capacity is given to a good approximation by

$$C_V = \frac{\partial}{\partial T}\left[N\left(\frac{3kT}{2} + kT + kT\right)\right]_V = \frac{7Nk}{2} = \frac{7R}{2}. \tag{16.94}$$

In the proximity where $T = \theta_v$, C_V increases from $5R/2$ to $7R/2$.

This discussion is only a simplified theory of the variation of C_V with T. To obtain more accurate results several corrections have to be applied. These were discussed in the previous section.

EXERCISES 16

16.1 Prove that the average energy of a simple harmonic oscillator in thermal equilibrium is given by $\overline{E} = kT$. Hence, find the internal energy of N harmonic oscillators in three dimensions and that of one mole of a substance.

16.2 If the atoms in a solid are considered as harmonic oscillators which oscillate about their equilibrium positions, then according to the classical theory what would be the molar heat capacity of a solid at constant volume? Hence, determine the Dulong–Petit value of the molar heat capacity of a solid at constant volume. Compare this value with the observed total heat capacity of metals and non-metals.

16.3 It has been found that the heat capacity of metals tends to zero as T, as $T \to 0$, whereas the heat capacity of non-metallic solids approaches zero as T^3 with $T \to 0$. Explain these behaviours in terms of the lattice and electrons in metals and non-metals. How does this affect the predicted classical theory of heat capacity of solids?

16.4 Explain how the Einstein model of heat capacity of solids attempted to give a fairly good representation of the drop in heat capacity at low temperatures below the value $3R$ per mole.

16.5 According to Planck's theory the energy E of an oscillator may have only the values $E = nh\nu$, where n is any positive integer $0, 1, 2, \ldots$, h is the Planck constant and ν is the frequency of oscillation. Starting with this equation, derive the Planck–Einstein distribution law.

Using this law, show that at high temperatures the average energy of a harmonic oscillator approaches the classical average energy kT and at low temperatures the lattice heat capacity at constant volume approaches zero as $T \to 0$ when the system is in thermal equilibrium.

16.6 Define the Einstein temperature T_E. Discuss the limits of the value of T_E. Derive an expression for heat capacity at constant volume of a system consisting of N harmonic oscillators in one dimension.

16.7 Discuss the drawbacks of the Einstein model of the lattice heat capacity at constant volume.

16.8 State the assumptions which Debye made to formulate his model for the lattice heat capacity at constant volume. Discuss the fundamental differences between the basis of his and Einstein's models. Define the Debye characteristic temperature.

16.9 Derive the Debye T^3 law. Discuss whether this law is applicable to both insulating and metallic solids. Suggest possible methods for evaluating T_D.

16.10 Discuss the successes and failures of the Einstein and the Debye models of heat capacity of solids.

16.11 Show that for isotropic crystals or quasi-isotropic solid materials the total number of acoustical (long wavelength) frequencies between ν and $\nu + d\nu$ follows as a special case of the relation

$$\frac{dn}{dV} = 4\pi\nu^2 \left[\frac{1}{c_x^3} + \frac{1}{c_y^3} + \frac{1}{c_z^3} \right] d\nu$$

where V is the volume of the (cubic) cavity of the crystal, ν is the frequency, and c_x, c_y and c_z are the components of the velocity of sound in the x, y and z directions in the medium.

16.12 Show that there is an average energy of $3E_F(0)/5$ per electron without any thermal energy being supplied, $E_F(0)$ being the Fermi energy at absolute zero of temperature.

16.13 For free electrons of mass m the quantum theory result is that the electronic heat capacity at low temperatures is linear in the temperature. Prove. Is it possible to determine the proportionality constant? How can you explain that the heat capacity of the free electrons in a metal goes to zero in proportion to T?

16.14 Explain why the electronic heat capacity in metals at low temperatures cannot be ignored.

16.15 The values of C_V ($J\,kg^{-1}\,K^{-1}$) of a metal at low temperatures are found to be

T(K)	1	2	3	4	6	8
$C_V \times 10^3$	12	28	53	91	230	470

Determine the values of γ and α in the Debye–Sommerfeld equation. If the Fermi energy of the metal is 8.28×10^{-19} J, calculate the value of γ and compare this value with that obtained above. Assume that the number of conduction electrons per unit volume for this metal is $85 \times 10^{21}\,cm^{-3}$.

16.16 Calculate the number of conduction electrons per unit volume for copper. Assume that the density of copper is $8.94\,g\,cm^{-3}$.

16.17 The specific heat at constant volume ($J\,mol^{-1}\,deg^{-2}$) of copper at different temperatures varies as follows:

T(K)	1.4	2	2.44	2.83	3.16	3.46	3.74
$\left(\dfrac{C_V}{T} \right)$	3.76	5.02	6.69	7.94	9.20	10.87	12.12

Calculate the value of γ in the equation $C_V = \gamma T + \alpha T^3$ and the Debye temperature T_D. Assume that the number of electrons per unit volume of copper is $85 \times 10^{21}\,cm^{-3}$.

16.18 Calculate the electronic specific heat per m^3 at 300 K for sodium and copper, given that the free electron concentrations in sodium and copper are $2.5 \times 10^{28}\,m^{-3}$ and $8.5 \times 10^{28}\,m^{-3}$, respectively.

16.19 Establish by verbal reasoning that the contribution of the kinetic energy of the molecules in a system to the molar specific heat capacity of that system is $C_V = (f)R/2$, where f is the number of degree of freedom of the molecules.

16.20 Derive an equation for the heat capacity at constant volume of an ideal Bose–Einstein gas in the temperature range $T < T_b$, where T_b is the Bose temperature.

Solutions to Exercises

EXERCISES 1

1.1 Properties such as pressure, volume, temperature, composition, etc., are the properties of matter in bulk and are called macroscopic properties. Microscopic properties of matter are those which belong to individual isolated molecules. Thermodynamics deals with the macroscopic properties of matter and provides a convenient and powerful method of relating, systematizing and discussing such properties.

1.2 Although thermodynamics is a powerful tool for solving many kinds of important problem and is exceedingly general in its applicability, it has its limitations. Some of the limitations are as follows.

(i) Thermodynamics can often tell us whether a process will occur but not how fast it will occur.

(ii) It can often provide a quantitative description of an overall change in state without giving any indication of the character of the process by which the change might take place.

(iii) It does not provide the deep insight into chemical and physical phenomena which is provided by microscopic models and theories.

The usefulness of thermodynamics is quite considerable. The thermodynamic laws and principles make it possible to predict whether a particular chemical process can take place under any given conditions. The amount of energy that must be put into the process and its maximum yield can also be determined, and the effect of changes in these conditions on the equilibrium can be predicted. With the help of thermodynamic laws and principles, equations may be written to correlate physical and chemical properties of substances, and formulae and laws discovered experimentally can be derived theoretically.

1.3 It is always important to express a measurement of some property in the form of a number and a unit because the unit indicates the defined reference against which the measurement is compared. A measurement without a unit does not always convey the intended meaning. By including the units in a measurement and any factor used in the calculations, we can make sure that the unwanted units cancel out and the proper unit remains.

The accepted systems of expressing units are

(i) the fps system (foot–pound–second system or the British system),

(ii) the cgs system (centimetre–gram–second system or the French system)

(iii) the MKS system (metre–kilogram–second system or the Système Internationale des Unités (SI units)). SI units are now most commonly used.

1.4 Physical quantities can be divided into two types, namely *base* or *fundamental quantities* and *derived quantities*. The corresponding units for these quantities are called *base units* and *derived units*. For the SI the base quantities and units are as follows.

Quantity	Unit	Unit symbol
Length	metre	m
Mass	kilogram	kg
Time	second	s
Temperature	kelvin	K
Amount of substance	mole	mol
Electric current	ampere	A
Luminous intensity	candela	cd

1.5 Usually the relationship between two different units in the measurements of the same property is defined or can be deduced. This is called a conversion factor. Conversion factors are very useful for solving numerical problems. Conversion factors that express a relationship between units can be used to convert a measurement expressed in terms of one unit to a measurement in terms of another unit.

1.6 $1 \, km = 0.621$ mile (1 mile $= 1.6 \, km$).
$1 \, km \, h^{-1} = 0.621$ mile h^{-1}.
$1 \, cal = 4.187 \, J$.

1.7 (a) In 1 year there are 3.15×10^7 s. Therefore, in 1 light year there are 3.15×10^7 s $\times$ $2.98 \times 10^8 \, m \, s^{-1} = 9.4 \times 10^{15} \, m$.

(b) 1 light year $= 9.4 \times 10^{15} \, m$. $1 \, AU = 1.50 \times 10^8 \, km = 1500 \times 10^8 \, m$. (AU $=$ astronomical unit.) Therefore the number of astronomical units in 1 light year $= (9.5 \times 10^{15})/(1500 \times 10^8) = 6.3 \times 10^4$.

1.8 (a) Incorrect. Thermodynamics is dependent on macroscopic properties of matter, such as pressure, volume, temperature and composition.

(b) Correct. Thermodynamics can often tell us whether a process will occur but not how fast it will occur, and it can often provide a quantitative description of an overall change in state without giving any indication of the character of the process by which the change might take place.

(c) Incorrect. Thermodynamics does not provide a deep insight into chemical and physical phenomena. A deep insight into such phenomena is provided by microscopic models and theories.

(d) Incorrect. By applying the laws and principles of thermodynamics it is possible to determine the maximum yield of a particular chemical process but thermodynamics does not provide information about the speed of any chemical process.

(e) Correct. Indeed, thermodynamics provides a convenient and powerful method of relating, systematizing and discussing macroscopic properties of matter.

(f) Correct. The subject of classical thermodynamics is based on the zeroth, first, second and third laws of thermodynamics.

(g) Incorrect. By applying the laws and principles of thermodynamics it is indeed possible to predict whether a particular chemical process can take place under any given conditions.

(h) Incorrect. Thermodynamics can often tell us whether a process will occur but not how fast it will occur.

1.9

Quantity	British	cgs	MKS (SI)
		Units	
Length	ft	cm	m
Mass	lb	g	kg
Time	s	s	s
Temperature	°F	°C	K
Force	pdl	dyn	N
Energy	ft lb	erg	J
Pressure	lb ft^{-2}	g cm^{-2}	Pa
Heat	BTU	cal	J
Acceleration	ft s^{-2}	cm s^{-2}	m s^{-2}
Density	lb ft^{-3}	g cm^{-3}	kg m^{-3}
Volume	ft^3	cm^3	m^3

1.10   It is very important to include the units in a measurement and any factor used in the calculations. The unit indicates the defined reference against which the measurement is compared. A measurement without a unit does not always convey the intended meaning. By including the units, we can make sure that the unwanted units cancel out and the proper units remain.

Conversion factors usually define the relationship between two different units in the measurements of the same property. Conversion factors are very useful for solving numerical problems and can also be used to convert a measurement expressed in one unit to a measurement in terms of another unit.

1.11   (a) In this case, the reactants $Zn + 2HCl$ and the products $ZnCl_2 + H_2$ constitute the system.

(b) The closed vessel together with the thermostat bath is the surroundings.

(c) No.

(d) No.

(e) No. Although there might be exchange of energy between the system and its surroundings, no exchange of matter occurs because the reaction takes place in a closed vessel.

(f) Yes. Exchange of both energy and matter takes place between our body and surroundings.

1.12   (a) An adiabatic process because there is no exchange of heat.

(b) An isothermal process because the temperature did not change.

(c) An isobaric (or constant pressure) process, since the pressure remains the same.

(d) An isobaric process.

(e) A cyclic process because the gas goes back to its initial state.

1.13   (a) Extensive variables are proportional to the amount of matter. Volume, heat capacity, work, heat, mole number and internal energy are extensive variables. Intensive variables, on the other hand, are independent of the amount of matter. Density, pressure and temperature are intensive variables.

(b) Volume, temperature, pressure and density are state variables; the others are not.

(c) The state of the ideal gas can be specified by the temperature, pressure and volume.

1.14   (a) C_V is the specific heat at constant volume which is defined as the measured heat capacity of a substance when it is heated with its volume constant. At constant volume, the heat absorbed goes to increase the energy of the system when the temperature is raised from T_1 to T_2. This gives $C_V(T_2 - T_1) = \Delta U$, the change in internal energy. For a small change in temperature, $C_V = dU/dT$.

If a gas is heated at a constant pressure, the measured heat capacity is called the heat capacity at constant pressure and is denoted by C_P. When heat is supplied to a system at constant pressure, expansion occurs and therefore work is done against the applied pressure. Consequently, more heat is required to produce a 1°C rise in temperature at constant pressure than at constant volume. The extra heat needed goes into work done; thus $C_P = C_V +$ work done in expansion. Also, $C_P = dH/dT$, where H is the enthalpy.

The enthalpy H is a thermodynamic quantity defined by the equation $H = U + PV$, where U, P and V are the internal energy, pressure and volume, respectively. It has the units of energy.

Q denotes the quantity of heat. It is a form of energy mainly due to the temperature. Heat can be transferred between a system and its surroundings solely because of a temperature difference. It is considered positive when it enters into a system and negative when it leaves a system.

(b) Most chemical experiments are carried out at constant pressure rather than at constant volume. Under such conditions the work done by the system is not zero but $dW = -P\Delta V$. Then $\Delta U = Q + W = Q_P - P\Delta V$. Substituting this value of ΔU into $\Delta H = \Delta U + P\Delta V$, we have $\Delta H = Q_P$ which is satisfactory. For an ideal gas, $\Delta(PV) = \Delta(nRT) = \Delta n\, RT$. Then $\Delta H = \Delta U + \Delta n\, RT$.

(c) (i) The measured enthalpy change is called the heat of reaction for the given reaction.

(ii) The reaction is endothermic because the negative sign indicates that heat is absorbed in the reaction.

(iii) Water in the gaseous form has a higher heat content than the liquid.

1.15 The system will perform work on its surroundings; this work is expansion work and is considered to be negative because it is done by the system on its surroundings. Such work does not impart kinetic energy to the movable parts of the boundary.

1.16 The heat of reaction is the difference between the heat contents of the products and the reactants at a constant pressure and at a definite temperature. This is denoted by ΔH and is measured in joules.

The heat of formation is the increase ΔH in heat content when 1 mol of a substance is formed from its elements. The heat of formation of any free element is always taken as zero.

The heat of formation of a substance has the same value as that of the heat of reaction but with the opposite sign.

1.17 Changes in state may be brought about along reversible or irreversible paths. A reversible path is one that may be followed in either direction; at any point in an expansion or a compression the direction may be reversed by an infinitesimal change in the variable such as the temperature or pressure.

Let us imagine that the gas is confined in a cylinder which is equipped with a piston capable of moving backwards and forwards and exerting pressure on the gas. In the problem, the pressure exerted by the gas on the piston is P_2. If P_2 is equal to the pressure P_{ext} exerted by the piston on the gas, then equilibrium exists. At the equilibrium state, by a slight decrease in P_{ext} the gas can be expanded. On the other hand, the gas can be contracted by slightly increasing P_{ext}. Now let us imagine that the gas is expanded by slowly decreasing P_{ext} in such a way that at all times it is very slightly smaller than P_2. Such an expansion would be reversible because at any stage of such an expansion the gas can be contracted by an infinitesimal increase in P_{ext}.

If the gas is expanded by rapidly decreasing P_{ext} so that the difference between P_2 and P_{ext} at any time is great, the expansion would be irreversible because in that case it would not be possible to change expansion into compression by an infinitesimal increase in P_{ext}.

1.18 The four variables of state by which the state of a system can be completely defined are the temperature, pressure, volume and composition (density). They are not independent of each other.

1.19 When the state of a system is specified by assigning values of all or a few of the state variables, the other properties of the system are restricted.

It is necessarily true that the change in any state function due to a change in the state of a system depends only on the initial and final state of the system and not on how the change is brought about.

1.20 There are conditions of a thermodynamic system which cannot be described in terms of state variables. These conditions include all those conditions in which the state variables change with time and space. These conditions are called *non-equilibrium states*. These are not dealt with in thermodynamics.

1.21 If the confined gas is suddenly compressed, its state cannot be described in terms of one pressure and temperature. When the piston is not moving within the cylinder, the state of the entire volume of the gas can be specified by assigning the values of its pressure and temperature. If the gas is suddenly compressed by moving the piston, the gas in the immediate proximity of the piston is compressed and heated but the gas at the far end of the cylinder is not. There is then no such thing as the pressure or temperature of the gas as a whole. The gas is said to be in a non-equilibrium state.

1.22 The following are some of the criteria for an equilibrium state.

(i) At equilibrium the chemical composition of a system must be uniform and there must be no net chemical reactions taking place. Any net chemical change would inevitably change the density, temperature, etc., of the system and make it impossible to specify its state.

(ii) The mechanical properties of the system must be uniform and constant. Any change in these properties would cause the volume to change continuously, making it difficult to specify the state of the system.

(iii) The temperature of the system must be uniform and the same as the temperature of its surroundings. In any system where heat flows owing to a temperature difference, the macroscopic properties are not uniform and may change with time. So such a system cannot be in equilibrium state.

1.23 Yes, these two bodies are at different temperatures. When there is no change in their thermal properties, they will be in thermal equilibrium.

1.24 From $P = P_0 + P_0 \alpha t$, where α is a constant, we have $t = (1/\alpha)[(P - P_0)/P_0]$. This shows that the temperature t is proportional to $(P - P_0)/P_0$.

When the freezing point of water is $0°$ and the boiling point is $100°$, $1/\alpha = 273.15$.

1.25 From $P = P_0 + P_0 \alpha t$ we have

$$\frac{P_1}{P_2} = \frac{P_0(1 + \alpha t_1)}{P_0(1 + \alpha t_2)} = \frac{1 + \alpha t_1}{1 + \alpha t_2}$$

If we write $T_1 = 1 + \alpha t_1$ and $T_2 = 1 + \alpha t_2$, we have

$$\frac{P_1}{P_2} = \frac{T_1}{T_2}.$$

1.26 On the Kelvin scale the difference between the freezing and boiling points of water is no longer exactly 100 K by definition, even though the assignment of 273.16 K to the triple point of water makes this difference close to 100 K. In other words, the temperature $t' = T - 273.15$ is not identical with the centigrade temperature, even though it is very close to it. The t' scale is called the *Celsius scale*.

The only aspect of the Kelvin scale which is not universal is that the size of the Kelvin is related to the properties of water.

The Kelvin temperature is very important in thermodynamics because it is independent of the detailed structure or properties of any material.

1.27 For a complete answer, refer to Section 1.9.

1.28 A reversible process is an idealized concept which can be carried out only in theory. Any actual expansion or contraction is irreversible.

Yes, a reversible process is an idealized process that represents the limit of a sequence of irreversible processes for which the parameter changes required to reverse the direction become smaller and smaller.

Irreversibility does not imply that it is impossible to force a process in the reverse direction; it only indicates that such a reversal cannot be achieved simply by changing parameters by infinitesimal amounts.

EXERCISES 2

2.1 When a system expands into a vacuum there is no external pressure acting on the system, i.e. $P = 0$. Therefore, from $W = -P\Delta V$ we have $W = 0$, i.e. the system does no work.

2.2 A given change in state of a gas may be achieved in many ways and the work W done may be different for any two of these ways. However, if the gas undergoes a change in state along a reversible path, all along this path $P_{exp} = P_{int}$ (except for an infinitesimal amount) and at no point could P_{ext} be increased (in the case of expansion), thus increasing the integral $\int_1^2 P_{ext}\, dV$ without changing the expansion into a compression, or at no point could P_{ext} be decreased (in the case of compression) thus decreasing $\int_1^2 P_{ext}\, dV$ without changing the compression into an expansion. The outside pressure may, however, be made smaller or greater than the internal pressure P_{int} for any part of the path. For this changed portion of the path, the contribution to $\int_1^2 P_{ext}\, dV$ would be either decreased or increased, and in addition the expansion or the compression would not be reversible because it could not be changed into a compression or expansion by an infinitesimal increase or decrease in P_{ext}. Therefore, it is apparent that the reversible path is the only path for which the integral $\int_1^2 P_{ext}\, dV$ has its maximum value.

2.3 Work $W = $ pressure $\times$ change in volume or $W = 1\, atm \times 1\, cm^3 = (1.0 \times 10^5\, N\, m^{-2})$ $(1 \times 10^{-6}\, m^3)$ (because $1\, atm = 1.0 \times 10^5\, N\, m^{-2}) = 0.1\, N\, m = (0.1\, kg\, m\, s^{-2})\, m$ (because $1\, N = 1\, kg\, m\, s^{-2}) = 0.1\, kg\, m^2\, s^{-2} = 0.1\, J$ (because $1\, kg\, m^2\, s^{-2} = 1\, J$).

2.4 From $H = U + PV$ we have

$$\Delta H = \Delta U + \Delta(PV)$$

or

$$\Delta H - \Delta U = \Delta(PV) = \Delta(nRT).$$

Assuming that the water vapour at 100°C behaves as an ideal gas, $n = 1$ mol, $R = 8.314$ J K^{-1} mol^{-1} and $T = 373$ K, then $\Delta(nRT) = (1 \text{ mol})(8.314 \text{ J K}^{-1} \text{ mol}^{-1}) (373 \text{ K}) = 3.101$ kJ. Hence $\Delta H - \Delta U = 3.101$ kJ.

2.5 Heat lost by the block of metal must be equal to the heat gained by the water as its temperature rises from 294 to 300 K. If C_M is the heat capacity of the metal,

$$(1 \text{ kg})(400 \text{ K})C_M = (0.5 \text{ kg})(300 - 294 \text{ K})(4200 \text{ J K}^{-1} \text{ kg}^{-1})$$

or

$$C_M = \frac{(0.5)(6)(4200)}{400} \text{ J K}^{-1} \text{ kg}^{-1} = 31.5 \text{ J K}^{-1} \text{ kg}^{-1}.$$

2.6 1 kg of water at 100°C occupies a volume of 1000 cm^3 and the volume of 1.0 kg of steam at 100°C is 1.67 m^3. Therefore, the work done is given by

$$W = \text{pressure} \times \text{change in volume}$$

or

$$W = P(V_2 - V_1) = (1.01 \times 10^5 \text{ N m}^{-2})(1.67 \text{ m}^3 - 1.0 \times 10^{-3} \text{ m}^3)$$
$$= 1.686 \text{ N m} = 1.686 \text{ kg m s}^{-2} \text{ m} = 1.686 \text{ J}.$$

2.7 Just as the thermodynamic work associated with a given change in state depends on the means by which the change is brought about, so also the heat attending a given change in state depends on the path through which it is transferred.

The energy of a system is what is stored in it, but heat and work are two ways of transferring energy across the boundaries of a system and its surroundings. Once energy has been imparted to a system, it is not possible to tell whether the energy was transferred as heat or work. Therefore, the term *heat of a substance* is as meaningless as the term *work of a substance*.

2.8 The heat capacity C_V at constant volume is the heat required to increase the temperature of 1 mol of a substance by 1°C when the volume remains constant.

The heat capacity C_P at constant pressure is the heat required at constant pressure to increase the temperature of 1 mol of a substance by 1°C.

The heat capacity C is given by the ratio of energy input to temperature increase, i.e. $C = dQ/dT$. It strongly depends on the nature of the substance and its physical state.

From $C = dQ/dT$, it is possible to compute the heat required to increase the temperature of a substance over any temperature interval within the range of validity of the data. Let us assume that C is given by the equation $C = a + bT + cT^2$ and we wish to increase the temperature from T_1 to T_2. Then,

$$dQ = C \, dT$$

or

$$Q = \int_{T_1}^{T_2} C \, dT = \int_{T_1}^{T_2} (a + bT + cT^2) \, dT$$

which gives

$$Q = a(T_2 - T_1) + \frac{b}{2}(T_2^2 - T_1^2) + \frac{c}{3}(T_2^3 - T_1^3).$$

2.9 The enthalpy H is defined by the equation $H = U + PV$, where U, P and V are the internal energy, pressure and volume, respectively. The units of enthalpy are those of U. Since U, P and V are all state functions, consequently H is also a function only of the state of a system.

2.10 In Section 2.1 we have shown that $dW = -P_{ext} \, dV$. For a finite change in volume the external pressure P_{ext} (or simply P) need not remain constant during the change in volume. The external pressure P is supposed to be known for each point of the path, along which the system expands. P may be given as a function of the temperature T and the volume V of the system, i.e. $P = f(T, V)$. Therefore, for a finite change in volume, we may write

$$dW = -P(T, V) \, dV \text{ or } W = -\int_{V_1}^{V_2} P \, dV.$$

In the important special case, in which the outside pressure P remains constant during a finite expansion of the system, P may be brought outside the integral sign to write

$$W = -P \int_{V_1}^{V_2} dV = -P(V_2 - V_1) = -P\Delta V.$$

This equation shows that, when $P = 0$, i.e. when the external pressure is zero, $W = 0$. A zero external pressure can be achieved in a vacuum. Thus a system which expands into a vacuum performs no work.

2.11 (a) Heat flow in and out of the system can be stopped by insulating the whole system. Such an insulated system is called a *thermally insulated system*.

(b) When $P = P'$, the system is in pressure equilibrium. The gas inside the cylinder in this state can be either compressed or expanded by a small change in a variable such as the pressure or temperature provided that the process is carried out very very slowly. Such a process is called a *reversible process*. At any stage of a reversible compression the process can be reversed to cause expansion and vice versa.

(c) As the piston moves towards the left, the gas inside is compressed; there will be an increase in the temperature (unless the process is isothermal) and a decrease in the volume. This will cause an increase in the molecular motion, thus increasing the energies of the molecules (since the kinetic energy of the molecules is a function of the temperature). The change in the temperature will be the observable variable that can be used to detect the overall change in the average energy.

2.12 (a) Correct.

(b) Incorrect.

(c) Incorrect.

(d) Incorrect.

(e) Correct.

(f) Incorrect.

(g) Incorrect.

(h) Incorrect because the reversible path is the only path for which the integral $\int_1^2 P \, dV$ has its maximum value.

(i) The molar heat capacity of a substance is its specific heat multiplied by its molecular weight.

(j) Correct.

(k) Correct.

(l) Incorrect.

(m) Incorrect, which is evident from the first law of thermodynamics.

2.13 This is an incorrect statement because heat is not measured in degrees; only the temperature is measured in degrees. Heat is a form of energy; so it should be expressed in the units appropriate for energy.

At the absolute zero temperature all matter has zero energy but in practice it is not possible to attain the absolute zero temperature.

2.14 As the substance is receiving heat from a source, the transferred heat should increase the temperature of the substance. Since the temperature of the substance is constant, this means that the substance is undergoing an isothermal heat transfer process. This happens when the substance gives out heat to its surroundings.

2.15 When two bodies at different temperatures are allowed to interact, heat flows from the body at a higher temperature to the body at a lower temperature and as a result the temperature of the latter rises. It is an irreversible process; it is reversible only at temperature inequality. When the two bodies attain thermal equilibrium, the energy transfer between them will be the same.

2.16 Just as the work associated with a given change in state depends upon the means by which the change was brought about, so the heat attending a given change in state also depends on the path.

The heat capacity C of a body is defined as the ratio of energy input to temperature increase, i.e. $C = dQ/dT$. If this equation is integrated over a finite interval of temperature, we have

$$Q = \int_{T_1}^{T_2} C \, dT.$$

If the experimental values of C are available over the temperature interval of interest, the value of Q in the above equation may be obtained graphically. When C is known as a function of temperature in the form of an empirical equation such as $C = a + bT + cT^2 + \cdots$, the above integral becomes

$$Q = \int_{T_1}^{T_2} (a + bT + cT^2 + \cdots) \, dT.$$

2.17 The initial volume $V_1 = nRT/P_1$. $P_1 = 0.50 \, \text{MPa}$, $n = 1 \, \text{mol}$, $T = 273.15 + 25 = 298.15 \, \text{K}$ and $R = (0.082)(0.1013) \, 1 \, \text{MPa} \, \text{mol}^{-1} \text{K}^{-1}$ (because $0.1 \, \text{MPa} = 1 \, \text{atm}$ and $R = 0.082 \, \ell \, \text{atm} \, \text{mol}^{-1} \, \text{K}^{-1}$). Therefore,

$$V_1 = \frac{(1 \, \text{mol})(0.082)(0.1013)1 \, \text{MPa} \, \text{mol}^{-1} \, \text{K}^{-1}(298.15 \, \text{K})}{0.50 \, \text{MPa}} = 4.95 \, \ell.$$

Since, at constant temperature, $P \propto 1/V$, so the final volume V_2 will be five times larger, i.e. $V_2 = 5 \times 4.95 \, \ell = 24.77 \, \ell$. The work done by the gas is $W = -P(V_2 - V_1)$ or

$$W = -(0.10 \, \text{MPa})(24.77 \, \ell - 4.95 \, \ell) = 1.98 \, \ell \, \text{MPa} = -2.007 \, \text{kJ}$$

2.18 In this case the initial and the final volumes will be the same as in the above problem, i.e.

$$V_1 = 4.95 \, \ell \quad \text{and} \quad V_2 = 24.77 \, \ell$$

The work done in a reversible isothermal process is given by

$$W = -\int_{V_1}^{V_2} P \, dV = -nRT \int_{V_1}^{V_2} \frac{dV}{V} = -nRT \ln\left(\frac{V_2}{V_1}\right)$$

where P is the external pressure. Thus,

$$W = -(1 \, \text{mol})(8.314 \, \text{J} \, \text{mol}^{-1}\text{K}^{-1})(298.15 \, \text{K}) \, 2.303 \, \log(5) = -3.989 \, \text{kJ}.$$

2.19 See Sections 2.5, 2.6 and 2.7, respectively.

2.20 See Section 2.5.2.

2.21 (a) See Section 2.5.1 for the derivation of Equation 2.34. For the next part of the exercise see Section 2.5.2.

2.22 See Section 2.5.2. Equation 2.42 indicates that at the same temperature the molecules of all ideal gases have the same kinetic energy.

2.23 See Section 2.5.

2.24 The kinetic theory of gases accounts for the properties of ideal gases, but some modification is required in order to apply it to real gases, because real gases do not in actual practice obey the gas laws. The necessary modifications are

(i) that the volume occupied by the molecules of a real gas under ordinary conditions may not be negligible.

(ii) the force exerted by the molecules of a real gas on one another may not be negligible.

(For a more detailed discussion see Section 2.6).

2.25   See Section 2.7.

2.26   From Equation 2.48 we have $P = \frac{1}{3}\rho u^2$. Rearranging this equation gives

$$u^2 = \frac{3P}{\rho} = \frac{3(1.013 \times 10^5\,\text{N m}^{-2})}{0.0899\,\text{kg m}^{-3}} = 3.38 \times 10^6\,\text{N m kg}^{-1}.$$

Therefore, the root mean square speed is given by

$$\sqrt{u^2} = (3.38 \times 10^6\,\text{N m kg}^{-1})^{1/2} = 1.84 \times 10^3\,\text{m s}^{-1}$$

(because $1\,\text{N} = 1\,\text{kg m s}^{-2}$).

2.27   Consider a 1 kg mol of nitrogen. At standard conditions, i.e. at a pressure of $1.013 \times 10^5\,\text{n m}^{-2}$, the volume of 1 mol of an ideal gas (nitrogen is an ideal gas) is $22.4\,\text{m}^3\,\text{mol}^{-1}$. From Equation 2.34 we have

$$u^2 = \frac{3PV}{Nm}$$

where N is the number of moles and m is the molecular weight. Then by substituting the given values, we have

$$u^2 = \frac{3(1.013 \times 10^5\,\text{N m}^{-2})(22.4\,\text{m}^3\,\text{mol}^{-1})}{(1\,\text{mol})(28\,\text{kg})} = 2.43 \times 10^5\,\text{m}^2\,\text{s}^{-2}.$$

Therefore, the root mean square speed $\sqrt{u^2} = (2.43 \times 10^5\,\text{m}^2\,\text{s}^{-2})^{1/2} = 492\,\text{m s}^{-1}$.

2.28   According to the kinetic theory of gases we must assume that individual molecules of a particular ideal gas travel at different speeds. However, owing to the numerous collisions, the speed of the molecules varies from virtually zero to nearly the speed of light. Because of molecular collisions and consequential exchanges of energy, the speed of a given molecule changes with each collision. However, since a very large number of molecules is involved, the distribution of speeds of the total number of molecules remains constant. The Maxwell–Boltzmann distribution law describes the nature of the distribution of molecular speeds. According to this law, at a given temperature, very few molecules move at very low or very high speeds; the number of molecules with intermediate speeds increases rapidly up to a maximum and drops off rapidly. At a high temperature, more molecules have low speeds. This is expected from the kinetic theory of gases.

2.29 We rewrite the van der Waals equation in the form

$$Z = \frac{1}{1 - b/V} - \frac{a}{RTV}.$$

Now, expanding $(1 - b/V)^{-1}$ binomially, we obtain

$$\left(1 - \frac{b}{V}\right)^{-1} = 1 + \frac{b}{V} + \frac{b^2}{V^2} + \frac{b^3}{V^3} + \cdots.$$

This series is a convergent series for $b < V$, and b is always less than V. Then the above van der Waals equation becomes

$$Z = \left(1 + \frac{b}{V} + \frac{b^2}{V^2} + \frac{b^3}{V^3} + \cdots\right) - \frac{a}{RTV} = 1 + \frac{b - a/RT}{V} + \frac{b^2}{V^2} + \frac{b^3}{V^3} + \cdots$$

2.30 For 1 mol of a gas the van der Waals equation is

$$\left(P + \frac{a}{V^2}\right)(V - b) = RT$$

or

$$V = \frac{RT}{P - a/V^2} + b.$$

The simple way to solve this equation for V is to use the method of successive approximations. This method involves substitution of any numerical value for V on the right of this equation to find a value for V on the left. We call this value V_1. Then again substitute V_1 for V on the right and obtain another value V_2 for V, and continue this process until we get a value for V which remains constant for the subsequent substitution. To perform this method we shall first try with the value of V obtained from $PV = RT$. (Remember that any value of V will do for the first trial.) Using the given data, $V_{ideal} = 0.25\ \ell$. Substituting this value of V on the right

$$V_{first} = \frac{24.6}{10.1 + 0.137/(0.25)^2} + 0.032 = 0.233\ \ell$$

$$V_{second} = \frac{24.6}{10.1 + 0.137/(0.233)^2} + 0.032 = 0.229\ \ell$$

$$V_{third} = \frac{24.6}{10.1 + 0.137/(0.229)^2} + 0.032 = 0.227\ \ell$$

It seems that further trials with the value $0.227\ \ell$ will give $0.227\ \ell$. So, $0.227\ \ell$ is the desired answer.

2.31 The van der Waals equation may be written in the form

$$P = \frac{RT}{V - b} - \frac{a}{V^2} = \frac{RT}{V(1 - b/V)} - \frac{a}{V^2}.$$

Now, by expanding $(1 - b/V)^{-1}$ binomially, we obtain

$$P = \frac{RT}{V}\left(1 + \frac{b}{V} + \frac{b^2}{V^2} + \frac{b^3}{V^3} + \cdots\right) - \frac{a}{V^2}$$

$$= \frac{RT}{V}\left[1 + \left(b - \frac{a}{RT}\right)\frac{1}{V} + \cdots\right].$$

Comparing this equation with the given virial equation we find that

$$B' = b - \frac{a}{RT}$$

EXERCISES 3

3.1 $98.6°F = \left(\frac{5}{9}\right)(98.6 - 32)°C = 37.0°C$

3.2 $t°C = \left(\frac{5}{9}\right)(t°F - 32)$

or

$$1.8t°C + 32 = t°F$$

or

$$(-79)(1.8) + 32 = -110.2°F.$$

3.3

$$-252.8°C = [(-252.8)(1.8) + 32]°F = -423.04°F.$$
$$t°F = T°R - 459.67°.$$

Therefore, $-423.04°F = (-423.04 + 459.67)°R = 36.63°R.$

3.4 At $-40°$, a Celsius and a Fahrenheit thermometer would be the same.

3.5 (a) $t°C = T - 273.15$. Therefore, $0°C = (T - 273.15)$ K or $T = 273.15$ K.

(b) $100\,K = 100 \times \frac{9}{5} = 180$ units of the other units.

(c) The size of the degree Celsius is of the same magnitude as that of the Kelvin but the zero point of the Celsius scale is shifted to 273.15. This makes $0\,K$ correspond to $-273.15°C$.

3.6 The zeroth law is an important principle of thermodynamics. The importance of this law to the temperature concept was not fully realized until after the first, second and third laws of thermodynamics had reached an advanced state of development. Hence the unusual name zeroth law.

The zeroth law is actually the law of thermal equilibrium. It gives us an operational definition of temperature which does not depend on the physiological sensation of 'hotness' or 'coldness'. This law is based on the experience that systems in thermal contact are not in complete equilibrium with one another until they have the same degree of hotness, i.e. the same temperature.

3.7 See Section 3.3.

3.8 It is convenient to distinguish between levels of temperature and temperature intervals. Sometimes actual temperatures are specified in terms of degrees Kelvin (or Centigrade) and temperature intervals or differences are given in terms of Kelvin (or Centigrade) degrees. 80 degrees centigrade specify the actual temperature, whereas 80 centigrade degrees give the temperature interval or the temperature difference. The ice points in degrees Celsius, degrees Rankine and degrees Fahrenheit corresponding to 273.15 K are 0°C, 491.67°R and 32°F, respectively. The normal boiling point of water in degrees Celsius, degrees Rankine and degrees Fahrenheit corresponding to 373.15 K are 100°C, 671.67°R and 212.0°F, respectively.

EXERCISES 4

4.1 See Section 4.5.

4.2 Consider an isolated system. For such a system there is no exchange of heat between the system and its surroundings. This means that $Q = 0 = W$. Therefore, from the first law of thermodynamics we have $\Delta U = 0$. Hence $\Delta U = Q = W = 0$. (An isothermal, isochoric, i.e. constant-volume, expansion in vacuum will give the same results.)

4.3 See Section 4.6.

4.4 See Section 4.7.1.

4.5 See Section 4.7.2.

4.6 See Section 4.7.2.

4.7 (a) Firstly, $n = 1$, initial volume is V_1, final volume $V_2 = 2V_1$ and $T = 298.15$ K. Then the work is given by

$$W = -\int_{V_1}^{V_2} P \, dV = -nRT \ln\left(\frac{V_2}{V_1}\right)$$
$$= (-1)(8.314 \text{ J mol}^{-1} \text{ K}^{-1})(298.15 \text{ K})2.303 \log 2 = -1.71 \text{ kJ}$$

Since this is an isothermal expansion, there is no change in the internal energy, i.e. $\Delta U = 0$. Hence, from the first law, $Q = -W = 1.71$ kJ.

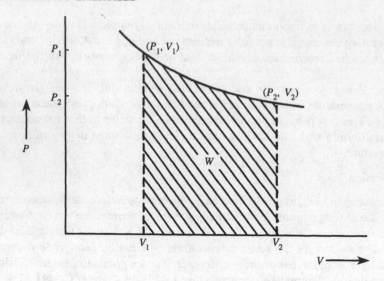

(b) The initial volume before the expansion was

$$V_1 = \frac{nRT}{P_1} = \frac{(1\,\text{mol})(0.082 \times 0.101\,\ell\,\text{MPa mol}^{-1}\,\text{K}^{-1})(298.15\,\text{K})}{0.202\,\text{MPa}} = 12.25\,\ell$$

Since the pressure has dropped from 0.202 to 0.101 MPa, the volume must have doubled, i.e. the final volume is $2(12.25\,\ell) = 24.50\,\ell$. The work done is given by

$$W = -P_{\text{ext}}(V_2 - V_1) = -(0.101\,\text{MPa})(24.50\,\ell - 12.25\,\ell) = 12.25\,\ell\,\text{atm} = -1.24\,\text{kJ}$$

For an isothermal expansion, $\Delta U = 0$ and $W = -Q$. Therefore, in this case $Q = 1.24\,\text{kJ}$.

4.8 From Equation 4.15 we have $W = -\int_{V_1}^{V_2} P\,dV$. For an isothermal expansion $P = nRT/V$. Therefore,

$$W = -\int_{V_1}^{V_2} \frac{nRT}{V}\,dV = -nRT\,\ln\left(\frac{V_2}{V_1}\right).$$

Since, at constant, T, $V \propto 1/P$, we have

$$W = -nRT\,\ln\left(\frac{P_1}{P_2}\right).$$

4.9 (a) The work done by the gas in the isothermal expansion is

$$W = -nRT\,\ln\left(\frac{V_2}{V_1}\right) = -(2\,\text{mol})(8.314\,\text{J mol}^{-1}\,\text{K}^{-1})(300\,\text{K})\,\ln 2$$

$$= -3.46\,\text{kJ}$$

To calculate the heat added to the gas, we have to calculate the ΔU first. Since this is an isothermal process, there is no change in the internal energy of the gas, i.e. $\Delta U = 0$. (It must be remembered that it is not necessarily true that for an isothermal process there is no change in the internal energy of a real gas.) Hence from $\Delta U = Q + W$ we have $Q = -W = 3.46\,\text{kJ}$.

(b) From the figure given in the problem we can see that this process consists of two parts, i.e. the path AB at constant volume and the path BC at constant pressure. The work done along AB is zero, i.e. $W_{AB} = 0$ because $\int P\,dV = 0$ as $dV = 0$. The work done along BC is

$$W_{BC} = -P_2(V_2 - V_1) = -nRT_2\left(1 - \frac{V_1}{V_2}\right)$$

$$= -(2.0\,\text{mol})(8.314\,\text{J mol}^{-1}\,\text{K}^{-1})(300\ \text{K}) \times \left(1 - \frac{3.5\,\text{m}^3}{7.0\,\text{m}^3}\right) = -2.49\,\text{kJ}$$

Therefore, the total work done along the paths AB and BC is $W_{AB} + W_{BC} = 0 - 2.49\,\text{kJ} = -2.49\,\text{kJ}$.

Since the initial and the final temperatures are the same (300 K in each case), $\Delta U = 0$. Therefore, $\Delta U = Q + W$ gives $Q = -W = 2.49\,\text{kJ}$.

We can solve this problem by another method if we know the specific heat C_P at constant pressure and the specific heat C_V at constant volume of an ideal gas. We have already defined these terms in Chapter 1. For an ideal gas, $C_P = 20.79\,\text{J mol}^{-1}\,\text{K}^{-1}$ and $C_V = 12.47\,\text{J mol}^{-1}\,\text{K}^{-1}$. Here we shall make use of the equation $Q = mC\Delta T$, where Q is the amount of heat, m is the mass, C is the specific heat and ΔT is the change in temperature.

Using the ideal-gas law we find that the volume at the point B is $V_B = 3.50\,\text{m}^3$ and the corresponding temperature is $T_B = T_2(V_B/V_2)$, or $T_B = 300(3.50/7.0) = 150\,\text{K}$. Hence, for the process AB we have $Q_{AB} = nC_V\,\Delta T$ (since it is a constant-volume process). Then,

$$Q_{AB} = (2\,\text{mol})(12.47\,\text{J mol}^{-1}\,\text{K}^{-1})(150\,\text{K} - 300\,\text{K}) = -3.740\,\text{kJ}$$

The minus sign indicates that the heat was lost along the path AB.

For the process BC we have $Q_{BC} = nC_P\,\Delta T$ (since it is a constant-pressure process). Then,

$$Q_{BC} = (2\,\text{mol})(20.79\,\text{J mol}^{-1}\,\text{K}^{-1})(300\,\text{K} - 150\,\text{K}) = 6.234\,\text{kJ}$$

Now the total heat $Q = Q_{AB} + Q_{BC} = -3.740\,\text{kJ} + 6.234\,\text{kJ} = 2.494\,\text{kJ}$. This is what we found in the previous method. Since we now know Q and W, we can calculate ΔU, which is $Q + W = 2.494\,\text{kJ} - 2.494\,\text{kJ} = 0$. This is in line with our results.

4.10   The volume of 1.0 kg of water at 100°C is $1.0 \times 10^{-3}\,\text{m}^3$ and the volume of 1.0 kg of steam at 100°C is $1.67\,\text{m}^3$. The work done in the process is

$$W = -P(\text{volume of steam} - \text{volume of water})$$

$$= -0.101\,\text{MPa}(1.67\,\text{m}^3 - 1.0 \times 10^{-3}\,\text{m}^3) = 170.0\,\text{kJ}$$

The heat Q of vaporization of water is given as 2260.0 kJ. Then $\Delta U = Q + W$ gives $\Delta U = 2090.0\,\text{kJ}$. This shows that only 7.5% of the heat added is used to perform work and the remaining 92.5% is used to increase the internal energy of the water.

4.11   (a) $P = $ constant, then the equation of state $P(V - b) = RT$ can be written

$$V - b = \frac{RT}{P} = \frac{RT}{\text{constant}}$$

or

$$V = \frac{RT}{\text{constant}} + b.$$

This shows that, when $T = 0$, $V = b$. Then the plot of V versus T is a straight line with gradient R/P and intercept b. This is shown in (a).

(b) If T is constant (isotherm), the equation of state can be written

$$P = \frac{\text{constant}}{V - b}$$

(since R and T are both constants) or

$$P \propto \frac{1}{V - b}.$$

It can be easily seen that, when $V \to \infty$, $P \to 0$ and, when $V \to b$, $P \to \infty$. Thus the curve of P versus V is as shown in (b).

(c) If V is constant, $P = RT/\text{constant}$. Here again the curve is a straight line with gradient $R/(V - b)$ and zero intercept. This is shown in (c).

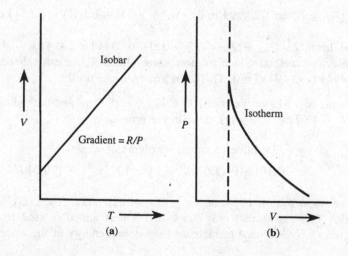

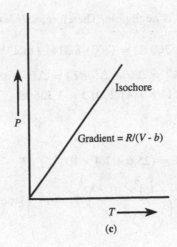

Isochore

P

Gradient $= R/(V - b)$

$T \longrightarrow$

(c)

EXERCISES 5

5.1 (a) For an adiabatic process, $PV^\gamma = $ constant, so that

$$\frac{P_1}{P_2} = \left(\frac{V_2}{V_1}\right)^\gamma$$

or

$$\frac{V_2}{V_1} = \left(\frac{P_1}{P_2}\right)^{1/\gamma}$$

where V_1 is the initial volume, V_2 is the final volume, P_1 is the initial pressure and P_2 is the final pressure which is equal to $P_1/2$. Therefore, $V_2/V_1 = 2^{3/5} \doteq 1.52$, i.e. the final volume is 1.5 times the initial volume.

(b) For an isothermal process, $P_1V_1 = P_2V_2$ because $T_1 = T_2$. Hence $V_2/V_1 = P_1/P_2 = 2.0$. Therefore, in this case the final volume is twice the initial volume.

5.2 (a) This molecule may exhibit translational motion in three directions, rotational motion in two directions, or vibrational motion in one direction.

(b) $U(\text{total}) = U(\text{translational}) + U(\text{rotational}) + U(\text{vibrational}) + U(\text{electrons}) + U(\text{nuclear})$.

(c) $C_P = C_V + R = \frac{5}{2}R + R = \frac{7}{2}R$. If $R = 8.314\,\text{J mol}^{-1}\,\text{K}^{-1}$

$$C_P = \frac{7}{2}(8.314\,\text{J mol}^{-1}\,\text{K}^{-1}) = 29.099\,\text{J mol}^{-1}\,\text{K}^{-1}$$

$$\gamma = \frac{C_P}{C_V} = \frac{7}{5} = 1.4$$

(d) $\Delta U = \int_{T_1}^{T_2} C_V \, dT$. To solve this we have to assume firstly that C_V is constant over the temperature range $T_1 = 27°C$ to $T_2 = 227°C$ and secondly that oxygen behaves as an ideal gas and its contribution to C_V is negligible. Therefore, $\Delta U = C_V(T_2 - T_1)$, i.e.

$$\Delta U = \frac{5}{2}R(500\,\text{K} - 300\,\text{K}) = (200)(8.314)\,\text{J mol}^{-1} = 1.662\,\text{kJ mol}^{-1}$$

$$\Delta H = \Delta U + \Delta(PV) = \Delta U + \Delta(nRT) = \Delta U + nR\Delta T$$
$$= 1.662\,\text{kJ} + (1)(8.314)(200)\,\text{J} = 3.326\,\text{kJ}$$

(e) (i) At constant pressure,

$$Q_P = \Delta H = \int_{T_1}^{T_2} C_P\,\text{d}T = (25.6 + 1.4 \times 10^{-3}T)\,\text{d}T$$

$$= 25.6(T_2 - T_1) + \left[\left(\frac{1.4 \times 10^{-3}}{2}\right)(T_2^2 - T_1^2)\right]\text{J mol}^{-1} = 5.188\,\text{kJ mol}^{-1}.$$

(ii) At constant volume,

$$\Delta U = \Delta H - \Delta(PV) = \Delta H - nR\Delta T$$

$$= 5.188\,\text{kJ mol}^{-1} - \frac{(1)(8.314)(200)}{1000}\,\text{kJ mol}^{-1} = 3.526\,\text{kJ mol}^{-1}.$$

5.3   (a) The heat effect Q in a process for a change in temperature from T_1 to T_2 is given by

$$Q = C\,\Delta T = C(T_2 - T_1)$$

where C is the heat capacity. At constant volume, over a temperature range in which C is constant, the above equation becomes $Q_V = C_V\Delta T$. However, at constant volume, by definition $Q_V = \Delta U$. Hence, $\Delta U = C_V\Delta T$.

(b) At constant pressure, over a temperature range in which C is constant, the equation $Q = C\Delta T$ becomes $Q_P = C_P\Delta T$. However, at constant pressure, by definition $Q_P = \Delta H$. Hence, $\Delta H = C_P\Delta T$.

5.4   See Section 5.3.

5.5   (a) With a monatomic gas all the energy input goes into increasing the energy of translation motion. On the other hand, in the case of polyatomic gases, part of the energy input is used to increase the energies of rotational and vibrational motions. Therefore, the total heat input required to produce the standard increase in kinetic energy corresponding to 1°C increase in temperature is greater for a polyatomic gas than for a monatomic gas. The greater these 'rotational and vibrational motions', i.e. 'the degree of freedom' active in absorbing energy, the greater will be C_V and the smaller the ratio $\gamma = C_P/C_V$.

It is interesting to note that, since all except the monatomic gases have such extra degrees of freedom which are fully operative at and above room temperature, $\gamma = 1.67$ becomes a highly distinctive mark of a monatomic gas.

(b)   See the note in Section 5.3.

5.6   See the second part of Section 5.4. (b).

5.7 The steady radial flow Q of heat between two concentric spheres of radii r_1 and r_2 is given by

$$Q = 4\pi K(T_1 - T_2)\left(\frac{1}{r_1} - \frac{1}{r_2}\right)^{-1}$$

(For a mathematical derivation of this equation, see Section 5.4. (c).

5.8 In order to solve this problem we shall have to make use of the relation $(\partial U/\partial V)_T = a/V^2$, which we shall deduce at a later stage when we have discussed the second law of thermodynamics.

$$W = -\int_{V_1}^{V_2} P \, dV.$$

Rearranging the given equation of state (for $n = 1$), we have

$$P = \frac{RT}{V - b} - \frac{a}{V^2}.$$

Substituting this in the equation for W gives

$$W = -\int_{V_1}^{V_2} \left(\frac{RT}{V - b} - \frac{a}{V^2}\right) dV = -RT \ln\left(\frac{V_2 - b}{V_1 - b}\right) + a\left(\frac{1}{V_1} - \frac{1}{V_2}\right).$$

Using the relation $(\partial U/\partial V)_T = a/V^2$, we have

$$\Delta U = \int_{V_1}^{V_2} \frac{a}{V^2} \, dV = a\left(\frac{1}{V_1} - \frac{1}{V_2}\right)$$

for an isothermal reversible process. From the first law, $Q = \Delta U - W$. Therefore,

$$Q = a\left(\frac{1}{V_1} - \frac{1}{V_2}\right) + RT \ln\left(\frac{V_2 - b}{V_1 - b}\right) - a\left(\frac{1}{V_1} - \frac{1}{V_2}\right) = RT \ln\left(\frac{V_2 - b}{V_1 - b}\right).$$

To find ΔH, we use $\Delta H = \Delta U + \Delta(PV)$. From the given equation of state we have

$$P = \frac{RT}{V - b} - \frac{a}{V^2}$$

or

$$PV = \frac{RTV}{V - b} - \frac{a}{V}$$

or

$$\Delta(PV) = RT\left(\frac{V_2}{V_2 - b} - \frac{V_1}{V_1 - b}\right) - \frac{a}{V_2} + \frac{a}{V_1}.$$

To simplify $V_2/(V_2 - b) - V_1/(V_1 - b)$, we proceed as follows:

$$\frac{V_2}{V_2 - b} - \frac{V_1}{V_1 - b} = \frac{V_2(V_1 - b) - V_1(V_2 - b)}{(V_2 - b)(V_1 - b)} = \frac{-b(V_2 - V_1)}{(V_2 - b)(V_1 - b)}$$

$$= \frac{-b[(V_2 - b) - (V_1 - b)]}{(V_2 - b)(V_1 - b)} = \frac{b}{V_2 - b} - \frac{b}{V_1 - b}.$$

Substituting this in the above equation for $\Delta(PV)$ gives

$$\Delta(PV) = RTb\left(\frac{1}{V_2 - b} - \frac{1}{V_1 - b}\right) - \frac{a}{V_2} + \frac{a}{V_1}$$

Then,

$$\Delta H = RTb\left(\frac{1}{V_2 - b} - \frac{1}{V_1 - b}\right) + 2a\left(\frac{1}{V_1} - \frac{1}{V_2}\right).$$

5.9   Here we shall make use of an equation given by $(\partial U/\partial V)_T = -P + T(\partial P/\partial T)_V$. This relation can be deduced from $dU = T\,dS - P\,dV$ which we shall deal with under the Maxwell relationships at a later stage. .

$$W = -\int_{V_1}^{V_2} P\,dV = -\int_{V_1}^{V_2} \frac{RT}{V - B}\,dV = -RT\ln\left(\frac{V_2 - B}{V_1 - B}\right) = RT\ln\left(\frac{P_2}{P_1}\right).$$

For ΔU we use $(\partial U/\partial V)_T = -P + T(\partial P/\partial T)_V$. Then,

$$\Delta U = \int_{V_1}^{V_2}\left(\frac{\partial U}{\partial V}\right)_T dV = \int_{V_1}^{V_2}\left[-P + T\left(\frac{\partial P}{\partial T}\right)_V\right] dV.$$

To find the integral we have to find $(\partial P/\partial T)_V$. The equation of state is $P = RT/(V - B)$. Then,

$$\left(\frac{\partial P}{\partial T}\right)_V = \frac{R}{V - B} + \frac{RT}{(V - B)^2}\frac{dB}{dT}.$$

(This is achieved by differentiating $RT/(V - B)$ with respect to T first and then differentiating $1/(V - B)$ with respect to $V - B$, followed by differentiation of $V - B$ with respect to T). Therefore,

$$T\left(\frac{\partial P}{\partial T}\right)_V = \frac{RT}{(V - B)} + \frac{RT^2}{(V - B)^2}\frac{dB}{dT}.$$

Hence,

$$\Delta U = RT^2\frac{dB}{dT}\int_{V_1}^{V_2}\frac{dV}{(V - B)^2} = -RT^2\frac{dB}{dT}\left(\frac{1}{V_2 - B} - \frac{1}{V_1 - B}\right)$$

$$= -T\frac{dB}{dT}\left(\frac{RT}{V_2 - B} - \frac{RT}{V_1 - B}\right) = T\frac{dB}{dT}(P_1 - P_2).$$

From this equation it can be seen that the isothermal change in the internal energy depends on dB/dT and generally does not vanish when B is temperature dependent. It is worthwhile to note that with the equation $P(V-b)=RT$, where b is not temperature dependent, the isothermal change in the internal energy is always zero. However, if $dB/dT=0$, both the equations of state give $\Delta U=0$. To calculate Q we use the first law; so

$$Q = \Delta U - W = T\frac{dB}{dT}(P_1 - P_2) + RT \ln\left(\frac{P_1}{P_2}\right)$$

For ΔH,

$$\Delta H = \Delta U + \Delta(PV)$$
$$\Delta(PV) = \Delta(RT + BP) = (RT + BP_2) - (RT + BP_1) = B(P_2 - P_1).$$

Hence,

$$\Delta H = T\frac{dB}{dT}(P_1 - P_2) + B(P_2 - P_1) = \left(B - T\frac{dB}{dT}\right)(P_2 - P_1).$$

5.10 (a) $W = RT \ln(P_2/P_1) = -(8.314)(298) \ln(1.01/0.101) = -5.732\,\text{kJ}$. For ideal behaviour, $\Delta U = 0$ for an isothermal expansion; $Q = \Delta U - W = 0 + 5.732\,\text{kJ} = 5.732\,\text{kJ}$, and $\Delta H = \Delta U + \Delta(PV) = 0 + 0 = 0$.

(b) $B(25°C) \simeq (-21.0 - 11.0)/2 = -16.0\,\text{m}\ell\,\text{mol}^{-1}$. Therefore, $P(V + 16.0) = RT$. In this case the work W is equal to that in the ideal case, i.e. $W = W_{ideal} = 5.732\,\text{kJ}$. Similarly, $\Delta U = (\Delta U)_{ideal} = 0, Q = \Delta U - W = 0 - (-5.732) = 5.732\,\text{kJ}$ and $\Delta H = \Delta U + \Delta(PV) = B\,(P_2 - P_1) = (-16.0)(0.101 - 1.01) = 14.4\,\text{m}\ell\,\text{MPa} = (14.4)(8.314)/(82.05)(0.101)\,\text{J} = 14.446\,\text{J}$.

5.11 See Equation 5.34.

5.12 (a) $dW = -P\,dV$ or $W = -\int_{V_1}^{V_2} P\,dV$. The given equation of state is $P = RT/(V-b)$. Therefore,

$$W = -\int_{V_1}^{V_2} \frac{RT}{V-b}\,dV = -RT \ln\left(\frac{V_2 - b}{V_1 - b}\right).$$

The equation $P(V-b) = RT$ is identical with the van der Waals equation $(P + a/V^2)(V - b) = RT$ if $a = 0$. Thus, the gas model for the given equation is that of hard spheres with no attractive forces between the molecules, b being related to the actual space occupied by the gas molecules. Therefore, for an isothermal process involving such a gas, $\Delta U = 0$ just as in the case of an ideal gas. An alternative method to prove that in this case $\Delta U = 0$ is to use the identity

$$\left(\frac{\partial U}{\partial V}\right)_T = T\left(\frac{\partial P}{\partial T}\right)_V - P = T\left(\frac{\partial[RT/(V-b)]}{\partial T}\right)_V - P$$

on substituting for P. Finally this equals $RT/(V-b) - P = P - P = 0$. Therefore, $\Delta U = 0$. It follows from the first law that $Q = -W = RT \ln[(V_2 - b)/(V_1 - b)]$.

(b) For an adiabatic reversible expansion we have $Q = 0$. Hence $\Delta U = W$ or $dU = dW$, but $dU = C_V\,dT$ and $dW = -P\,dV$. Therefore,

$$C_V\,dT = -P\,dV = -\frac{RT}{V-b}\,dV$$

or

$$\int_{T_1}^{T_2} \frac{C_V\,dT}{T} = -\int_{V_1}^{V_2} \frac{R}{V-b}\,dV$$

or

$$C_V \ln\left(\frac{T_2}{T_1}\right) = -R\ln\left(\frac{V_2-b}{V_1-b}\right)$$

or

$$\frac{T_2}{T_1} = \left(\frac{V_2-b}{V_1-b}\right)^{-R/C_V} = \left(\frac{V_2-b}{V_1-b}\right)^{-(\gamma-1)}.$$

Therefore,

$$W = \Delta U = C_V(T_2 - T_1) = C_V T_1\left[\left(\frac{V_1-b}{V_2-b}\right)^{\gamma-1} - 1\right].$$

(c) (i) $T_1 = T_2 = 300\,\text{K}$; $V_1 = 0.220\,\ell$, $V_2 = 20.020\,\ell$ and $b = 0.020\,\ell\,\text{mol}^{-1}$. For an isothermal reversible expansion,

$$W = -Q = -RT_1 \ln\left(\frac{V_2-b}{V_1-b}\right) = -(8.314)(300)\ln\left(\frac{20.0}{0.20}\right)$$

$$= -11.548\,\text{kJ}\,\text{mol}^{-1}.$$

$$\Delta U = 0.$$

(ii) For an adiabatic reversible expansion, $Q = 0$. Therefore,

$$\Delta U = W = C_V T_1\left[\left(\frac{V_1-b}{V_2-b}\right)^{\gamma-1} - 1\right].$$

Given that $C_V = \frac{3}{2}R$, $\gamma = \frac{5}{3}$ and $T_1 = 300\,\text{K}$, then

$$\Delta U = W = (3R/2)(300)[(200/20\,000)^{2/3} - 1] = -3.554\,\text{kJ}.$$

5.13   (a) For isothermal expansion of an ideal gas, $\Delta U = \Delta H = 0$ and $Q = W$. The expansion is irreversible. Using the ideal gas law, $PV = RT$, we get

$$V_1 = \frac{(273)(0.082\,\ell)(0.101\,\text{MPa})}{1.01\,\text{MPa}} = 2.24\,\ell$$

$$V_2 = \frac{(273)(0.082\,\ell)(0.101\,\text{MPa})}{0.041\,\text{MPa}} = 55.15\,\ell.$$

$R = 0.082 \ell$ atm mol^{-1} K^{-1} $= 0.082 \times 0.101 \ell$ MPa mol^{-1} K^{-1}. Therefore,

$$Q = W = -P\,dV = -P_{ext}(V_2 - V_1) = -0.041(55.15 - 2.24)$$
$$= -2.16 \ell \,MPa = -2.168 \,kJ.$$

(b) Here the isothermal expansion is reversible. Again $\Delta U = \Delta H = 0$.

$$Q = W = -P\,dV = -RT\frac{dV}{V} = -RT\ln\left(\frac{V_2}{V_1}\right) = -RT\ln\left(\frac{P_1}{P_2}\right)$$

$$= (-8.314)(273)\ln\left(\frac{10}{0.4}\right) = -7.351 \,kJ.$$

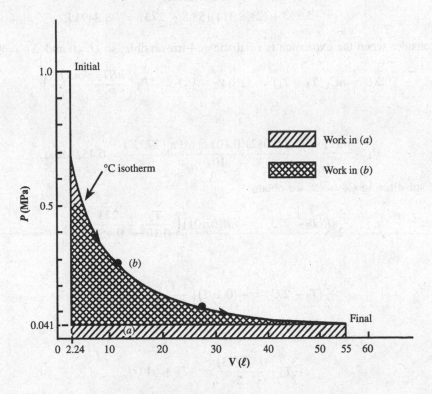

5.14   (a) (i) For a reversible adiabatic process

$$P_1V_1^{\gamma} = P_2V_2^{\gamma} \text{ or } T_1V_1^{\gamma-1} = T_2V_2^{\gamma-1} \text{ or } \frac{T_1}{T_2} = \left(\frac{V_2}{V_1}\right)^{\gamma-1}.$$

Because the gas is monatomic and ideal, $C_V = \frac{3}{2}R$, $C_P = C_V + R = \frac{5}{2}R$ and $\gamma = \frac{5}{3}$.
Substituting the values $V_1 = 10 \ell$, $V_2 = 100 \ell$ and $T_1 = 273$ K into the above equation we obtain

$$T_2 = \frac{T_1}{\left(\frac{V_2}{V_1}\right)^{\gamma-1}} = \frac{273}{10^{2/3}} = 58.8 \,K.$$

If P_2 is the pressure after the expansion, $P_2 V_2 = nRT_2$, i.e.

$$P_2 = \frac{nRT_2}{V_2} = \frac{(2)(0.082)(58.8)}{100} = 0.096 \, \text{atm} = 0.0097 \, \text{MPa}.$$

(ii) Since the expansion is adiabatic, $Q = 0$. Hence, from the first law

$$\Delta U = W = \int_{T_1}^{T_2} nC_V \, dT = nC_V(T_2 - T_1)$$

$$= (2)\left(\frac{3}{2}\right)(8.314)(58.8 - 273) = -5.343 \, \text{kJ}$$

$$\Delta H = \Delta U + \Delta(PV) = \Delta U + nR\Delta T$$

$$= -5.343 + (2)(8.314)(58.8 - 273) = -8.849 \, \text{kJ}.$$

(b) Consider when the expansion is adiabatic and irreversible; so $Q = 0$ and $\Delta U = W$.

$$\Delta U = nC_V(T_2 - T_1) = -P_2(V_2 - V_1) = -P_2\left(\frac{nRT_2}{P_2} - \frac{nRT_1}{P_1}\right)$$

but

$$P_1 = \frac{nRT_1}{V_1} = \frac{(2)[(0.082)(0.101 \, \ell) \, \text{MPa}](273 \, \text{K})}{10\ell} = 0.452 \, \text{MPa}.$$

Then, substituting $C_V = \frac{3}{2}R$, we obtain

$$\frac{3}{2}nR(T_2 - 273) = -nR(0.101)\left(\frac{T_2}{0.101} - \frac{273}{0.452}\right)$$

or

$$\frac{3}{2}(T_2 - 273) = -(0.101)\left(\frac{T_2}{0.101} - \frac{273}{0.452}\right)$$

or

$$\frac{3}{2}T_2 - \frac{(3)(273)}{2} = -T_2 + 61.00$$

or

$$T_2 = 188.2 \, \text{K}.$$

Consider when $Q = 0$ and $\Delta U = W = nC_V(T_2 - T_1)$. Therefore,

$$\Delta U = (2)\left[\frac{3}{2}(8.314)(188.2 - 273)\right] = -2.115 \, \text{kJ}$$

$$\Delta H = \Delta U + \Delta(PV) = \Delta U + nR\Delta T = -2.115 + (2)(8.314)(188.2 - 273)$$

$$= -3.525 \, \text{kJ}.$$

5.15   We shall use Equation 5.2, i.e. $Q = -KA(T_1 - T_2)/d$. Here, $Q = 0.048\,\mathrm{kJ\,s^{-1}}$, the area $A = \pi r^2 = \pi(0.076)^2 = 0.018\,\mathrm{m}^2$, the thickness $d = 0.0016\,\mathrm{m}$ and $T_1 - T_2 = 100 - 20 = 80°C$. Then,

$$K = \frac{-Qd}{A(T_1 - T_2)} = \frac{(0.048\,\mathrm{kJ\,s^{-1}})(0.0016\,\mathrm{m})}{(0.018\,\mathrm{m}^2)(80°C)} = 5.3 \times 10^{-5}\,\mathrm{kJ\,s^{-1}\,m^{-1}\,°C^{-1}}$$

$$= 5.3 \times 10^{-5}\,\mathrm{W\,m^{-1}\,°C^{-1}}.$$

5.16   Proceed as in Problem 5.15 above.

5.17   Refer to Equation 5.24. In that equation Q/t is the rate of heat flow which is given by

$$\frac{\mathrm{d}Q}{\mathrm{d}t} = \frac{\text{area } A}{\text{thickness } d}\,\mathrm{d}T \times \text{constant.}$$

The constant in this equation is the thermal conductivity of the material of the wall, which is given by $a + bT$. Then

$$\frac{Q}{t} = \frac{A}{d}\int (a + bT)\,\mathrm{d}T.$$

Integrating between the limits T_1 and T_2

$$\frac{Q}{t} = \frac{A}{d}\left(a(T_1 - T_2) + \frac{b}{2}(T_1^2 - T_2^2)\right) = \frac{A(T_1 - T_2)}{d}\left(\frac{a + b(T_1 - T_2)}{2}\right).$$

5.18   Refer to Equation 5.33, which gives $Q = 4\pi K(T_1 - T_2)/(1/r_1 - 1/r_2)$. If $T_1 = 300°C$, $T_2 = 100°C$, $r_1 = 60\,\mathrm{cm}$, $r_2 = 75\,\mathrm{cm}$ and $K = 1.22 \times 10^{-3}\,\mathrm{J\,s^{-1}\,cm^{-1}\,°C^{-1}}$, then

$$Q = \frac{(4)(3.14)(1.22 \times 10^{-3})(200)}{1/60 - 1/75}\,\mathrm{J\,s^{-1}} = 928.68\,\mathrm{J\,s^{-1}} \text{ outward.}$$

5.19   From Equation 5.30, $Q/l = 2\pi K(T_1 - T_2)/\ln(r_2/r_1)$. Substituting the values for the parameters,

$$\frac{Q}{l} = \frac{(2)(3.14)(1.5 \times 10^{-3})(50)}{\ln 6}\,\mathrm{J\,s^{-1}\,cm^{-1}} = 0.263\,\mathrm{J\,s^{-1}\,cm^{-1}}.$$

5.20   For the first three expressions see Equations 5.45, 5.46 and 5.42, respectively. For the next three expressions refer to Equations 5.48, 5.49 and 5.47, respectively.

5.21   For the first three expressions see Equations 5.44, 5.50 and 5.51, respectively, and for the next three expressions refer to Equations 5.52, 5.53 and 5.54.

5.22   For a reversible isobaric (constant-pressure) process

$$\Delta U = U_2 - U_1 = Q_p - P(V_2 - V_1)$$

or

$$(U_2 + PV_2) - (U_1 + PV_1) = Q_P$$

where Q_P is the heat absorbed at constant pressure. Hence, $H_2 - H_1 = \Delta H = Q_P$.

5.23   Describe Joule's experiment (see Section 5.6.2).

5.24   The Joule–Thomson coefficient is defined as the change in temperature per unit change in pressure when the enthalpy is constant. In terms of partial derivatives, $\mu = (\partial T / \partial P)_H$. This implies that, if a gas cools in the process of effusing through the plug, the Joule–Thomson coefficient μ is positive because the pressure always decreases in the Joule–Thomson experiment. On the other hand, a negative Joule–Thomson coefficient implies an increase in temperature.

The Joule–Thomson effect provides a sensitive method of studying gas imperfection by a pressure drop. The Joule–Thomson coefficient should be zero for an ideal gas; it is non-zero for real gases.

5.25   In the Joule–Thomson experiment if we consider enthalpy H as a function of temperature and pressure, the total differential of H is given by $H = f(T, P)$, or $dH = (\partial H / \partial T)_P dT + (\partial H / \partial P)_T dP$. By definition $(\partial H / \partial T)_P = C_P$. Therefore, $dH = C_P \, dT + (\partial H / \partial P)_T \, dP$. Since there is no change in enthalpy in the Joule–Thomson experiment, $dH = 0$. Then,

$$0 = C_P \, dT + \left(\frac{\partial H}{\partial P} \right)_T dP = C_P \left(\frac{\partial T}{\partial P} \right)_H + \left(\frac{\partial H}{\partial P} \right)_T$$

or

$$\left(\frac{\partial H}{\partial P} \right)_T = -\mu C_P.$$

5.26   Most gases at room temperature have a positive Joule–Thomson coefficient. As the temperature increases, μ decreases to zero and then changes sign. The temperature at which this change of sign occurs is called the *Joule–Thomson inversion temperature* T_i. At this temperature $\mu = 0$, and below it μ is negative.

Below certain limiting pressures, there are two inversion temperatures characteristic of each gas and dependent on pressure. Between the upper and lower inversion temperatures, μ is negative; below the lower and above the upper inversion temperatures, μ is positive; at the inversion temperatures, $\mu = 0$.

5.27   See Section 5.7 for derivation of the Joule–Thomson coefficient for a van der Waals gas. From Equation 5.97 we get $\mu = \partial T / \partial P = -[(b - 2a/RT)/C_P]$. Since, at the inversion temperature T_i, $\mu = 0$, we get $\mu = -[(b - 2a/RT_i)/C_P] = 0$ or $b - (2a/RT_i) = 0$ or $T_i = 2a/Rb$.

5.28   From Equation 5.94, $H = U_0 + (C_V + R)T + P(b - 2a/RT)$. Partially differentiating both sides with respect to T and keeping P constant, $(\partial H / \partial T)_P = (C_V + R) + 2aP/RT^2$, but $(\partial H / \partial T)_P = C_P$. Therefore, $C_P = C_V + R + 2aP/RT^2$ or $C_P - C_V = R + 2aP/RT^2$. This shows that $C_P - C_V$ in the case of van der Waals gases is greater than that in the case of ideal gases by the factor $2aP/RT^2$.

5.29 See the derivations of Equations 5.56 and 5.58.

5.30 The enthalpy change for a van der Waals gas is given by $\Delta H = \Delta U + \Delta(PV)$. From Equation 5.73, $\Delta U = -a/V_2 + a/V_1$. From the van der Waals equation, $P = RT/(V-b) - a/V^2$ or $PV = RTV/(V-b) - a/V$. Therefore,

$$\Delta(PV) = RT\left(\frac{V_2}{V_2 - b} - \frac{V_1}{V_1 - b}\right) - \frac{a}{V_2} + \frac{a}{V_1}$$

$$= bRT\frac{V_1 - V_2}{(V_2 - b)(V_1 - b)} - \frac{a}{V_2} + \frac{a}{V_1}$$

$$= bRT\frac{(V_1 - b) - (V_2 - b)}{(V_2 - b)(V_1 - b)} - \frac{a}{V_2} + \frac{a}{V_1}$$

$$= bRT\left(\frac{1}{V_2 - b} - \frac{1}{V_1 - b}\right) - \frac{a}{V_2} + \frac{a}{V_1}.$$

Therefore,

$$\Delta H = bRT\left(\frac{1}{V_2 - b} - \frac{1}{V_1 - b}\right) - \frac{2a}{V_2} + \frac{2a}{V_1}.$$

5.31

State	P (MPa)	V (ℓ)	T (K)
1	0.101	22.4	273
2	0.202	22.4	546
3	0.101	44.8	546

Step	Name of process	W (kJ)	Q (kJ)	ΔU (kJ)
A	Isochoric	0	$Q = C_V\Delta T$ 3.431	$\Delta U = Q + W$ 3.431
B	Isothermal	$W = -nRT\ln(V_2/V_1)$ −3.155	$Q = W$ 3.155	0
C	Isobaric	$W = -P\Delta V$ −2.242	$Q_P = C_P\Delta T$ 5.293	3.051
	Cycle	−5.397	11.879	6.482

5.32 (a) The weight of water equals 18.02 g. The number of moles of water equals $18.02/18.02 = 1$ mol, the heat Q_{vap} of vaporization equals 40.710 kJ, the enthalpy change $(\Delta H)_{vap}$ due to vaporization is $Q_{vap}/n = 40.710/1 = 40.710$ kJ mol^{-1} and $\Delta V = V_{vap} - V_\ell = (1.677 - 0.001)(18.02) = 30.201\ \ell$. Therefore, work $W = -P\Delta V = -(0.101)(30.201)\ \ell$ MPa = $-3.050\ \ell$ MPa, $(\Delta H)_{vap} = (\Delta U)_{vap} + \Delta(PV)$ or $(\Delta U)_{vap} = (\Delta H)_{vap} - \Delta(PV) = (\Delta H)_{vap} - P\Delta V$, or $(\Delta U)_{vap} = 40.710\ \text{kJ} - (30.201)(24.22)(4.184 \times 10^{-3})\ \text{kJ} = 37.65\ \text{kJ}$. If the volume of liquid water is neglected, $\Delta V \simeq V_{vap}$. In this case, work $W = -P\Delta V = -(0.101)$ $(1.677)(18.02)\,\ell\,\text{MPa}$ or $W = -(0.101)(1.677)(18.02)(1003.86)\,\text{J} = -3.063\,\text{kJ}$. $(1\,\ell\,\text{MPa} = 1003.86\,\text{J}.)$

(b) Since the process is an isothermal expansion in an evacuated space, there is no absorption or evolution of heat, i.e. $\Delta U = 0$. Therefore, from the first law, $Q = -W = -nRT$ $\ln(V_2/V_1)$, $n = 1$ (calculated above), $T = 100^\circ\text{C} = 373\,\text{K}$, $R = 8.314\,\text{J mol}^{-1}\,\text{K}^{-1}$, V_1 (volume of water) $= (0.0010)(18.02) = 0.018\,\ell$ and V_2 (volume of vapour) $= (1.677)(18.02) = 30.219\,\ell$. Therefore, $Q = -W = -(1)(8.314)(373)\ln(30.219/0.018) = -22.949\,\text{kJ}$, and $\Delta H = \Delta U + P\Delta V = 0 + 22.949\,\text{kJ} = 22.949\,\text{kJ}$.

5.33 $P_{initial} = P_{final} = P_{system} = P_{surroundings}$, $W = -P_{ext}\Delta V$, $\Delta V = V_2 - V_1 = (33.10 - 24.79) = 8.31\,\ell$. Therefore, $W = -(0.101)(8.31) = -0.839\,\ell\,\text{MPa} = -(0.839)(1003.86)\,\text{J} = -842.55\,\text{J}$. Since the process is a constant-pressure process, $Q = Q_P = \Delta H = \int_{300\,\text{K}}^{400\,\text{K}} C_P\,dT$ (assuming that C_P is constant). Therefore, $Q = Q_P = \Delta H = 100 C_P$. For H_2, $C_V = 3R/2 + R = 5R/2$. Therefore,

$$C_P = C_V + R = 7R/2.$$

Hence,

$$Q_P = \Delta H = 100\frac{7R}{2} = 2.908\,\text{kJ}$$

but

$$\Delta H = \Delta U + P\Delta V$$
$$\Delta U = \Delta H - P\Delta V = 2.908 - 0.842\,\text{kJ} = 2.077\,\text{kJ}.$$

5.34 $2C_6H_6(\ell) = 2C_6H_6(v)$; $(\Delta H)_{vap} = nQ_{vap} = (2)(30.8)\,\text{kJ} = 61.6\,\text{kJ}$. At the boiling point of C_6H_6, $Q_P = \Delta H = 61.6\,\text{kJ}$, but $Q_P = Q_V + \Delta(PV) = Q_V + \Delta(nRT)$ because benzene vapour behaves ideally. Therefore,

$$Q_V = Q_P - \Delta(nRT) = 61.6 - (2)(8.314)(353.2 \times 10^{-3})\,\text{kJ} = 55.727\,\text{kJ}.$$

5.35 For this constant pressure process, $\Delta H = Q_P = \int_{T_1}^{T_2} C_P\,dT$ or

$$\Delta H = \int_{298\text{K}}^{348\text{K}} (26.96 + 5.85 \times 10^{-3}T - 3.38 \times 10^{-7}T^2)\,dT$$

$$= 26.96(348 - 298) + \frac{5.85 \times 10^{-3}}{2}(348^2 - 298^2) - \frac{3.38 \times 10^{-7}}{3}(348^3 - 298^3) = 1.442\,\text{kJ}.$$

5.36 Since heat is liberated, the reaction is exothermic, i.e. $Q = -86.9\,\text{kJ}$, $\Delta U = Q + W$ and $W + \Delta(PV)$. $\Delta(PV)_{reaction} = \Delta(PV)_\ell + \Delta(PV)_s + \Delta(PV)_g$. Since $\Delta(PV)_g \gg \Delta(PV)_s$ or $\Delta(PV)_\ell$, therefore, $\Delta(PV)_{reaction} = \Delta(PV)_g = \Delta(nRT)$ (because the liberated hydrogen behaves as an ideal gas). Hence, $\Delta(PV)_{reaction} = W = (1)(8.314)(291) = 2.419\,\text{kJ}$, and $\Delta U = -86.9 + 2.419 = -84.481\,\text{kJ}$.

5.37 The molecular weight of C_6H_{12} is 84.2. So the number of moles involved in the process is 10/84.2.

$$(\Delta H)_{vapn} = Q_{vapn} \, J \, mol^{-1}.$$

Therefore, $(\Delta H)_{vapn} = (84.2/10)(3.573) kJ \, mol^{-1} = 30.08 \, kJ \, mol^{-1}$. Since the pressure is constant, we have

$$(\Delta U)_{vapn} = (\Delta H)_{vapn} - PV_{vapn} = (\Delta H)_{vapn} - P\frac{RT}{P}$$

assuming ideal behaviour. Thus,

$$(\Delta U)_{vapn} = (\Delta H)_{vapn} - RT = 30.08 - (8.314)(334 \times 10^{-3}) \, kJ \, mol^{-1}$$
$$= 27.304 \, kJ \, mol^{-1}.$$

It is apparent that in converting C_6H_{12} from liquid to vapour we have increased its energy by only 27.304 kJ mol^{-1}. Of the 30.08 kJ mol^{-1} heat energy put into the system, 2.773 kJ were used to perform the PV work of expansion of the vapour phase against the applied pressure.

5.38 The terms which are difficult to determine are

$$U(\text{electrons}) = \text{potential energy interactions} + \text{kinetic energies of all}$$
$$\text{the electrons involved in bonding } U(\text{nuclear})$$
$$= \text{stored energy in nucleus.}$$

An absolute value of the internal energy is very difficult to measure experimentally or to calculate.

Because the evaluation of some of the energy terms which contribute to the total internal energy is virtually impossible, an arbitrary reference point of zero internal energy has been set for all elements. This reference point is called the *standard state*. The specific values of temperature and pressure which fix the standard state are 298.15 K and 0.1 MPa ($= 101.3 \, kN \, m^{-2} = 1 \, atm$), respectively. The internal energy of all elemental substances is taken as zero when the element is in its most stable state under the standard conditions. In the case of hydrogen the most stable state is a gas. So, for the given conditions, the internal energy of hydrogen gas is zero.

5.39 Equation 5.49 provides the relation between P and V for an adiabatic reversible process. Such a process is also called an isentropic process. Now we have $PV^\gamma = $ constant $= \beta$ (say), or $P = BV^{-\gamma}$. Differentiating both sides with respect to V gives

$$\frac{dP}{dV} = \beta(-\gamma)V^{(-\gamma-1)} = -\gamma\frac{P}{V}. \tag{i}$$

As all the terms on the right of this equation are positive, dP/dV is a negative quantity because of the negative sign on the right. Again, for an isotherm we have $PV = nRT$ and for 1 mol of an ideal gas this becomes $PV = RT$ or $P = RT/V$.

Differentiating with respect to V gives

$$\frac{dP}{dV} = -RTV^{-2} = \frac{RT}{V} V^{-1} = -P/V. \tag{ii}$$

Here again, dP/dV is negative. Comparing (i) and (ii), we see that dP/dV in the case of the reversible adiabatic curve is γ times greater than that in the isothermal curve at the same values of P and V. As γ is greater than 1, dP/dV in (i) is greater than that in (ii). A schematic diagram is shown below.

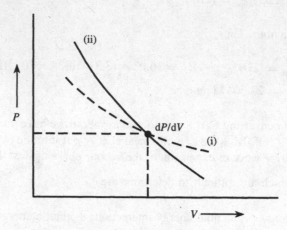

5.40   Fe, He and $Cl_2(g)$ have $U = H = 0$ at 298.15 K and 0.1 MPa.

EXERCISES 6

6.1   Equation 6.4 gives $C_V = (\partial U/\partial T)_V$. This shows that measurement of C_V gives the rate of change in internal energy with respect to temperature when the volume is constant, and $(\partial U/\partial T)_V$ may be replaced by C_V in any equation in which it appears, even if the equation refers to a process in which the volume is not constant.

6.2   Equation 6.2 gives $dQ = (\partial U/\partial T)_V\, dT + [P + (\partial U/\partial V)_T]\, dV$. Replacing $(\partial U/\partial T)_V$ by C_V we get $dQ = C_V\, dT + [P + (\partial U/\partial V)_T]\, dV$.

6.3   See the derivation of Equation 6.16.

6.4   See the derivation of Equation 6.19.

6.5   See the derivation of Equation 6.26.

6.6   See the derivation of Equation 6.30.

6.7   See the derivation of Equation 6.31.

6.8 $PV = RT$, so for constant temperature we have

$$V\left(\frac{\partial P}{\partial V}\right)_T + P\left(\frac{\partial V}{\partial V}\right)_T = \frac{\partial(RT)}{\partial V} = 0$$

or

$$V\left(\frac{\partial P}{\partial V}\right)_T = -P$$

or

$$\left(\frac{\partial P}{\partial V}\right)_T = -\frac{P}{V}.$$

For constant volume we have

$$V\left(\frac{\partial P}{\partial T}\right)_V + P\left(\frac{\partial V}{\partial T}\right)_V = R\left(\frac{\partial T}{\partial T}\right)_V$$

or

$$V\left(\frac{\partial P}{\partial T}\right)_V = R$$

or

$$\left(\frac{\partial P}{\partial T}\right)_V = \frac{R}{V}.$$

For constant pressure,

$$P\left(\frac{\partial V}{\partial T}\right)_P + V\left(\frac{\partial P}{\partial T}\right)_P = R\left(\frac{\partial T}{\partial T}\right)_P$$

or

$$P\left(\frac{\partial V}{\partial T}\right)_P = R$$

or

$$\left(\frac{\partial V}{\partial T}\right)_P = \frac{R}{P}.$$

6.9 Equation 6.36 gives the internal energy of a van der Waals gas. This equation shows that the internal energy of such gases depend on their volume and temperature.

6.10

$$\left(P + \frac{a}{V^2}\right)(V - b) = RT$$

or

$$PV - bP + \frac{a}{V} - \frac{ab}{V^2} - RT = 0$$

or

$$P\left(\frac{\partial V}{\partial T}\right)_P - V\left(\frac{\partial P}{\partial T}\right)_P - b\left(\frac{\partial P}{\partial T}\right)_P - \frac{a}{V^2}\left(\frac{\partial V}{\partial T}\right)_P + \frac{ab}{V^3}\left(\frac{\partial V}{\partial T}\right)_P - R = 0$$

or

$$P\left(\frac{\partial V}{\partial T}\right)_P - \frac{a}{V^2}\left(\frac{\partial V}{\partial T}\right)_P + \frac{ab}{V^3}\left(\frac{\partial V}{\partial T}\right)_P - R = 0$$

or

$$\left(\frac{\partial V}{\partial T}\right)_P = \frac{R}{P - a/V^2 + ab/V^3}$$

or

$$\frac{1}{V}\left(\frac{\partial V}{\partial T}\right)_P = \frac{R}{P/V - a/V^3 + ab/V^4}$$

or

$$\beta = \frac{RV^4}{PV^3 - aV + ab}.$$

6.11 See the derivation of Equations 6.43 and 6.44 for the total work done and the total heat absorbed by the system. The efficiency is given by Equation 6.48.

6.12 Here $\theta_1 = 30°C = 303$ K, $\theta_2 = 200°C = 473$ K. The efficiency $\eta = 1 - \theta_1/\theta_2 = 1 - 303/473 = 36\%$.

6.13 Decreasing the temperature θ_1 and keeping θ_2 constant is more convenient because all the engines run more or less at the maximum temperature.

6.14 The coefficient of performance of a refrigerator is given by

$$\eta = \frac{Q}{W} = \frac{\theta_1}{\theta_2 - \theta_1}$$

where θ_2 is the higher temperature equal to 303 K, θ_1 is the lower temperature equal to 263 K, Q is the heat withdrawn equal to 4.184 kJ and W is the amount of work required to withdraw Q amount of heat. Therefore,

$$W = Q\frac{\theta_2 - \theta_1}{\theta_1} = 4.184\frac{303 - 263}{263}\text{kJ} = 636\,\text{J}.$$

6.15 The efficiency is $\eta = $ work output/heat input $= (\theta_2 - \theta_1)/2$. When $\theta_2 = 100 + 273 = 373$ K, $\theta_1 = 0 + 273 = 273$ K and heat input $Q = 1000$ J, therefore, $\eta = (373 - 273)/373 = 0.268$, i.e. 26.8%. Work output $= \eta$ (heat input) $= 0.268 \times 1000\,\text{J} = 268\,\text{J}$. Heat rejected $=$ heat input $-$ work output $= 1000 - 268 = 732\,\text{J}$.

6.16   The coefficient of performance of the Carnot refrigerator is given by

$$\eta = \frac{\theta_2 - \theta_1}{\theta_2} = \frac{400 - 200}{400} = 1/2.$$

If the coefficient of performance of the refrigerator in question is one-half that of the Carnot refrigerator, then the value of its coefficient of performance is 0.25. Then $(Q_2 - Q_1)/Q_2 = 0.25$ or $(Q_2 - 600)/Q_2 = 0.25$ or $Q_2 = 800$ J, which is the amount of heat rejected to the high-temperature reservoir.

6.17   Equation 6.50 gives $Q_2/\theta_2 - Q_1/\theta_1 = 0$. Rewriting, this equation gives $Q_2/Q_1 = \theta_2/\theta_1$ which shows that the absolute magnitude of the quantities of heat absorbed and rejected are proportional to the temperature of the heat reservoir in either circle.

6.18   See Section 6.9.

6.19   Here $\theta_2 = 100 + 273 = 373$ K, $\theta_1 = 0 + 273 = 273$ K and $Q_1 = 1000$ J. The coefficient of performance is $\eta = 100/373 = 0.268$. Then $(Q_2 - Q_1)/Q_2 = 0.268$ or $Q_2 - 1000 = 0.268Q_2$, i.e. $Q_2 = 1000/0.732 = 1366$ J.

6.20   The maximum coefficient of performance is given by

$$\eta_{max} = \frac{T_{cool}}{T_{hot} - T_{cool}} = \frac{240}{300 - 240} = 4.$$

It is noteworthy that, unlike the thermal efficiency of a heat engine, the coefficient of performance of a refrigerator could be many times greater than unity.

6.21   Work W/heat absorbed $Q = (T_h - T_c)/T_h$, where W is the work done by the engine when a quantity of heat Q is transferred to it from the reservoir at a high temperature T_h. In our case, $Q = 250$ kJ, $T_h = 680$ K and $T_c = 273$ K. Then $W = [(T_h - T_c)/T_h]\, Q = [(680 - 273)/680] \times 250 = 149.63$ kJ.

6.22   Refer to Figure 6.1. At the point A, the volume, temperature and pressure are $V_1 = 20\,\ell$, $\theta_2 = 27°C$ and pressure P_1. From A $\rightarrow$ B is an isothermal reversible process and for this step we have $\Delta\theta = 0$, $\Delta U = 0 = \Delta H$, $\Delta V = V_2 - V_1 = 20\,\ell$ and $\Delta P = R\theta_2(1/V_2 - 1/V_1) = (0.082)(300)\,(1/40 - 1/20) = -0.61$ atm $= -0.061$ MPa.

Step B $\rightarrow$ C. This is an adiabatic reversible expansion. For this, $\Delta\theta = \theta_3 - \theta_2 = \theta_3 - \theta_1 = 200 - 300 = -100$ K.

$$\Delta V = V_3 - V_2 = V_2 \left(\frac{\theta_1}{\theta_3}\right)^{1/(\gamma - 1)} - V_2 = V_2\left[\left(\frac{\theta_1}{\theta_3}\right)^{C_V/R} - 1\right]$$

$$= 40\left[\left(\frac{300}{200}\right)^{3/2} - 1\right] = 34\,\ell$$

$$\Delta P = P_3 - P_2 = P_2\left(\frac{V_2}{V_3}\right)^\gamma - P_2 = \frac{R\theta_2}{V_2}\left[\left(\frac{V_2}{V_3}\right)^\gamma - 1\right]$$

$$= \frac{R\theta_1}{V_2}\left[\left(\frac{\theta_3}{\theta_1}\right)^{C_P/R} - 1\right] = (0.082)(300)\left(\frac{1}{40}\right)\left[\left(\frac{200}{300}\right)^{5/2} - 1\right] = -(0.61)(0.64)$$

$$= -0.39\text{ atm} = 0.039\text{ MPa}$$

$$\Delta U = C_V(\theta_3 - \theta_2) = C_V(\theta_3 - \theta_1) = \frac{3}{2}R(-100) = -1255.2 \text{ J}$$

$$\Delta H = \Delta U + \Delta(PV) = \Delta U + R\Delta\theta = (C_V + R)\Delta\theta = C_P\Delta\theta$$

$$= \frac{5}{2}(8.314)(-100) = -2078.5 \text{ J}$$

Step $C \to D$. This is an isothermal reversible compression. For this, $\Delta\theta = 0$ and $\Delta U = \Delta H = 0$. Since steps $B \to C$ and $D \to A$ are reversible adiabatic, $S_3 = S_2$ and $S_4 = S_1$, where S represents the entropy. We shall see in a later chapter that, for an adiabatic reversible process, $\Delta S = S_4 - S_3 = S_1 - S_2 = R \ln(V_1/V_2) = R \ln(V_4/V_3)$. This gives $V_1/V_2 = V_4/V_3$, but here $V_3 = 34 + 40 = 74 \ell$ (from step $B \to C$). Hence, $V_4 = (V_1/V_2)V_3 = (20/40)(74) = 37 \ell$.

$$\Delta V = V_4 - V_3 = 37 - 74 = -37 \ell$$

and $\theta_3 = \theta_4 = -73°C = 200 \text{ K}$. $P_1V_1 = R\theta_1$ or $P_1 = R\theta_1/V_1 = (0.082)(300)(1/20) = 1.23 \text{ atm}$. Therefore, $P_2 = 1.23 - 0.61 = 0.62 \text{ atm}$ (from step $A \to B$), $P_3 = 0.62 - 0.39 = 0.23 \text{ atm}$ (from step $B \to C$) and $P_4 = R\theta_4/V_4 = 0.44 \text{ atm}$. Hence,

$$\Delta P = P_4 - P_3 = 0.44 - 0.23 = 0.21 \text{ atm.}$$

Step $D \to A$. This is an adiabatic reversible compression. For this, $\Delta\theta = \theta_1 - \theta_4 = 300 - 200 = 100 \text{ K}$.

$$\Delta V = V_1 - V_4 = 20 - 37 = -17 \ell$$

$$\Delta P = P_1 - P_4 = 1.23 - 0.44 = 0.79 \text{ atm}$$

$$\Delta U = C_V\Delta\theta = C_V(\theta_1 - \theta_4) = \frac{3}{2}R(100) = 1255.2 \text{ J}$$

$$\Delta H = C_P\Delta\theta = C_P(\theta_1 - \theta_4) = \frac{5}{2}R(100) = 2092.0 \text{ J}$$

6.23   A reversible Carnot cycle consists of four steps, namely step $1 \to 2$ (path a) is an isothermal expansion, step $2 \to 3$ (path b) is an adiabatic expansion, step $3 \to 4$ (path c) is an isothermal compression and step $4 \to 1$ (path d) is an adiabatic compression.

(a) The pressure versus volume diagram is as follows.

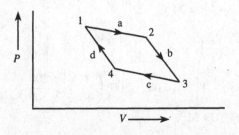

(b) The temperature versus pressure diagram is as follows.

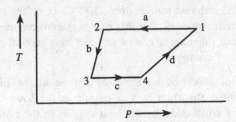

(c) The diagram in (b) also represents the internal energy U versus pressure P diagram.

(d) The temperature versus pressure diagram also represents enthalpy H versus pressure P diagram.

(e) The volume versus temperature diagram is as follows.

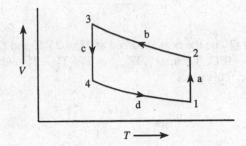

(f) The volume versus temperature diagram also represents the volume versus internal energy U diagram.

(g) The internal energy versus enthalpy diagram is as follows.

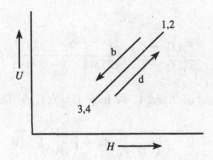

(h) The internal energy versus enthalpy diagram also represents the internal energy versus temperature diagram.

(i) The diagram for internal energy versus enthalpy also represents the diagram for enthalpy versus temperature.

6.24   Here, the lower temperature $T_\ell = 300 + 273 = 573$ K and the higher temperature $T_h = 600 + 273 = 873$ K. The maximum efficiency is given by $\eta = (T_h - T_\ell)/T_h = 0.34$. This engine is 34% efficient. The exhaust temperature, 300°C, is rather high. High exhaust temperatures are typical of steam engines. For this reason, steam engines are often connected in series so that the exhaust of one engine can be utilized by a second or third engine as intake, thus minimizing the loss of energy.

6.25   Let Q_2 be the heat absorbed by the heat engine at the higher temperature $T_2 = 1000°C$, and Q_1 be the heat absorbed by this engine at the lower temperature $T_1 = 25°C$. If W_{cycle} is the amount of work done by the heat engine in the complete cycle we have

$$\eta = \frac{W_{cycle}}{Q_2} = \frac{(T_2 - T_1)}{T_2}$$

or

$$W_{cycle} = \frac{Q_2(T_2 - T_1)}{T_2}.$$

For this refrigerator, let Q_2' be the heat absorbed at $T_2' = 25°C$ and Q_1' be the heat absorbed at the temperature $T_1' = 0°C$. Then, $Q_1'/W_{cycle}' = T_1'/(T_2' - T_1')$ where W_{cycle}' is the work done by the refrigerator. This gives

$$W_{cycle}' = \frac{Q_1'(T_2' - T_1')}{T_1'}.$$

Since $W_{cycle} = W_{cycle}'$, we get

$$\frac{Q_2(T_2 - T_1)}{T_2} = \frac{Q_1'(T_2' - T_1')}{T_1'}$$

or

$$\frac{Q_2}{Q_1'} = \frac{T_2(T_2' - T_1')}{T_1'(T_2 - T_1)} = \frac{1273(298 - 273)}{273(1273 - 298)} = 0.12.$$

6.26   (a) If U is a function of T and V, we have $U = f(T, V)$. Then,

$$dU = \left(\frac{\partial U}{\partial T}\right)_V dT + \left(\frac{\partial U}{\partial V}\right)_T dV. \tag{i}$$

Dividing both sides by dT and holding U constant give

$$0 = \left(\frac{\partial U}{\partial T}\right)_V + \left(\frac{\partial U}{\partial V}\right)_T \left(\frac{\partial V}{\partial T}\right)_U$$

but $C_V = (\partial U/\partial T)_V$. Then by rearranging, we have

$$-\left(\frac{\partial U}{\partial T}\right)_V = -C_V = \left(\frac{\partial U}{\partial V}\right)_T \left(\frac{\partial V}{\partial T}\right)_U.$$

(b) For the second identity

$$C_P = \left(\frac{\partial H}{\partial T}\right)_P = \left(\frac{\partial(U+PV)}{\partial T}\right)_P = \left(\frac{\partial U}{\partial T}\right)_P + P\left(\frac{\partial V}{\partial T}\right)_P. \tag{ii}$$

Dividing both sides of (i) by dT and holding P constant, we obtain

$$\left(\frac{\partial U}{\partial T}\right)_P = \left(\frac{\partial U}{\partial T}\right)_V + \left(\frac{\partial U}{\partial V}\right)_T \left(\frac{\partial V}{\partial T}\right)_P. \tag{iii}$$

Substituting (ii) in (i) and taking $C_V = (\partial U/\partial T)_V$, we find that

$$C_P = C_V + \left[P + \left(\frac{\partial U}{\partial V}\right)_T\right]\left(\frac{\partial V}{\partial T}\right)_P. \tag{iv}$$

Again, dividing (i) by dV and holding P constant, we have

$$\left(\frac{\partial U}{\partial V}\right)_P = \left(\frac{\partial U}{\partial T}\right)_V \left(\frac{\partial T}{\partial V}\right)_P + \left(\frac{\partial U}{\partial V}\right)_T. \tag{v}$$

(This step is required to convert $(\partial U/\partial V)_T$ in (iv) to $(\partial U/\partial V)_P$.) Rearranging (v) gives

$$\left(\frac{\partial U}{\partial V}\right)_T = \left(\frac{\partial U}{\partial V}\right)_P - \left(\frac{\partial U}{\partial T}\right)_V \left(\frac{\partial T}{\partial V}\right)_P. \tag{vi}$$

Now by substituting (vi) into (iv), we obtain

$$C_P = C_V + \left[P + \left(\frac{\partial U}{\partial V}\right)_P - \left(\frac{\partial U}{\partial T}\right)_V \left(\frac{\partial T}{\partial V}\right)_P\right]\left(\frac{\partial V}{\partial T}\right)_P$$

$$= P\left(\frac{\partial V}{\partial T}\right)_P + \left(\frac{\partial V}{\partial T}\right)_P \left(\frac{\partial U}{\partial V}\right)_P$$

or

$$C_P = \left[P + \left(\frac{\partial U}{\partial V}\right)_P\right]\left(\frac{\partial V}{\partial T}\right)_P$$

or

$$\left(\frac{\partial U}{\partial V}\right)_P = C_P\left(\frac{\partial T}{\partial V}\right)_P - P.$$

For 1 mol of an ideal gas, $PV = RT$. In addition, $C_P - C_V = R$. From this

$$P = R\left(\frac{\partial T}{\partial V}\right)_P$$

(by differentiating $PV = RT$ with respect to V) and

$$\frac{P}{R} = \left(\frac{\partial T}{\partial V}\right)_P.$$

Then,

$$\left(\frac{\partial U}{\partial V}\right)_P = C_P \frac{P}{R} - P = \frac{P}{R}(C_P - R) = \frac{PC_V}{R}.$$

(c) For the third identity

$$C_P = \left(\frac{\partial H}{\partial T}\right)_P = \left(\frac{\partial (U + PV)}{\partial T}\right)_P = \left(\frac{\partial U}{\partial T}\right)_P + P\left(\frac{\partial V}{\partial T}\right)_P.$$

Therefore,

$$\left(\frac{\partial U}{\partial T}\right)_P = C_P - P\left(\frac{\partial V}{\partial T}\right)_P.$$

6.27   See Equation 6.29 for the derivation of

$$\left(\frac{\partial U}{\partial V}\right)_T = T\left(\frac{\partial P}{\partial T}\right)_V - P. \tag{i}$$

For 1 mol of a van der Waals gas,

$$\left(P + \frac{a}{V^2}\right)(V - b) = RT$$

or

$$P = \frac{RT}{V - b} - \frac{a}{V^2}. \tag{ii}$$

Differentiating both sides with respect to T and keeping V constant, we have

$$\left(\frac{\partial P}{\partial T}\right)_V = \frac{R}{V - b}.$$

Substituting this into (i), and using (ii)

$$\left(\frac{\partial U}{\partial V}\right)_T = \frac{RT}{V - b} - P = \frac{a}{V^2}.$$

The quantity $(\partial U/\partial V)_T$ has the dimensions of energy per volume, which is equivalent to force per area, and hence to pressure. Physically, it is regarded as a measure of the pressure in the interior of a van der Waals gas. Therefore, it is frequently referred to as the internal pressure, i.e. the internal pressure that occurs from the attractions and repulsions of the molecules. This internal pressure has now been found to be equal to a/V^2. Thus, it is possible to conclude that, since the energy content of a real gas, unlike that of an ideal gas, is not independent of volume, it is necessary to introduce the a/V^2 term into the van der Waals equation of state for a real gas.

EXERCISES 7

7.1 (a) Incorrect.

(b) Correct.

(c) Incorrect.

(d) Correct.

(e) Correct.

(f) Correct.

(g) Incorrect.

(h) Incorrect.

(i) Correct.

(j) Correct.

7.2 (a) Generally, a large number of possible microstates (i.e. a state in which all the details that can be and are known about atoms and molecules) of a system correspond to a given macrostate of that system. These microstates are indistinguishable at the macroscopic level and are called *realizations* of the macrostate. In any system, the number of microscopic realizations of a macroscopic state of *fixed energy* is denoted by Ω.

(b) Entropy is given by the equation $S = k \ln \Omega$ because it is more convenient to work with the logarithms of the Ω-values, as Ω possesses an additive property $\ln \Omega = \ln \Omega_1 + \ln \Omega_2$.

(c) We can define an instantaneous property of an assembly by an arbitrary multiple of $\ln \Omega$ and, therefore, introduce the instantaneous entropy as $\bar{S} \equiv K_s \ln \Omega$. The bar on S signifies that the entropy is fluctuating and the constant k_s can be selected to express the entropy in whatever units we choose. Then the time-averaged entropy is given by $S \equiv k_s \ln \Omega$. Careful quantum-statistical calculations show that selection of the entropy constant k_s as the Boltzmann constant (the universal gas constant per molecule) makes the thermodynamic-temperature and the empirical perfect-gas-temperature scales identical. Hence the choice of the Boltzmann constant as the entropy constant.

(d) Just as the first law of thermodynamics deals with the energy, similarly the second law of thermodynamics deals with entropy which is a measure of randomness of atoms and molecules in a substance.

(e) The second law of thermodynamics states: 'The entropy of an isolated assembly must increase or, in the limit, remain constant.' This statement does not necessarily mean that the entropy of a non-isolated assembly can never decrease. Transfer of energy between a non-isolated system and its surroundings can definitely result in a lowering of the entropy of the system. However, the probability that the entropy of a system will decrease is extremely low.

(f) Let us consider an isolated assembly not in equilibrium state. If we could determine the instantaneous macrostate and compute the number of microscopic realizations (Ω-values), we would expect the result to be as shown below.

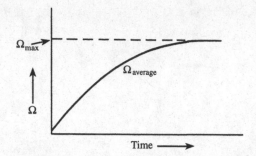

As equilibrium is approached, the system would be in the more probable macrostates most of the time, and Ω would fluctuate below its maximum value Ω_{max}. The average value of Ω would increase gradually, eventually levelling off at some value near to but below the value of Ω for the most probable macrostate. The least probable macrostates are highly ordered, and there are very few ways to form these ordered states. As time progresses, we would expect an isolated system to drift towards more probable, more disordered macrostates. Thus, the value of Ω seems to be a measure of the disorder of a system. Since the entropy of a system is a function of Ω, the latter is considered as a measure of the disorder of a system.

(g) The conceptual basis of the thermodynamic temperature is the concept of entropy of a system. The thermodynamic temperature of a simple compressible substance is defined by Equation 7.39. Although this equation is restricted to simple compressible substances, this definition may be extended to other classes of substances.

(h) The temperature is a measure of the sensitivity of the entropy of a system to changes in its internal energy produced solely by energy transfer as heat. At low temperatures, a slight increase in internal energy results in large increases in the microscopic disorder. In fact, at the absolute zero of temperature, an infinitesimal increase in internal energy produces a significant increase in the disorder. As the temperature increases, the sensitivity of the disorder to the changes in internal energy decreases. So, a state of infinite temperature would be one for which a change in the internal energy would produce no change in the disorder of the system. A state of negative thermodynamic temperature would be one for which an addition of energy as heat would tend to make the microscopic structure more ordered.

If the temperature of a simple compressible substance can be expressed functionally as $T = f(U, V)$, and as it can be shown that this function must be such that $[\partial(1/T)/\partial U]_V < 0$, the thermodynamic temperature could only occur at higher internal energies than positive temperatures, i.e. in states more energetic than a state of infinite temperature. However, this does not seem to be a practical reality; consequently, we are interested only in positive thermodynamic temperatures.

(i) The thermodynamic pressure of a simple compressible substance is defined by Equation 7.29. The necessary conditions for thermal and mechanical equilibrium between two simple compressible bodies, say x and y, are $T_x = T_y$ and $P_x = P_y$ where T_x and T_y are the temperatures and P_x and P_y are the pressures of x and y, respectively.

(j) The right-hand side of Equation 7.62 is considered as entropy production.

(k) Let us consider a substance enclosed in a cylinder fitted with a frictionless piston attached to a spring as shown below.

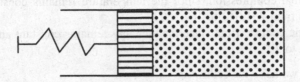

This system may be considered as an oscillator, i.e. an isolated system whose state changes in an infinitely repetitive cyclic manner. (The entropy of an isolated reversible oscillator remains unchanged during oscillation). Let us assume that the piston is adiabatic and the substance undergoes slow reversible adiabatic compression and expansion. Since the entropy of a reversible oscillator remains constant during its oscillation and as neither the entropy of the piston mass nor that of the spring changes, the entropy of the substance must not change either. Hence, we can say that a reversible adiabatic process is one of constant entropy. In fact, the same result can be obtained for every reversible adiabatic process that we may examine. Therefore, we conclude that reversible adiabatic process is isentropic, i.e. the entropy of the matter undergoing an adiabatic process does not change if the process is reversible.

It is important to note that reversible transfer of energy as macroscopic work does not alter the entropy of a system but its entropy is changed by reversible energy transfer as heat.

(l) Since the thermodynamic temperature scale is equivalent to the Kelvin scale, $-10°C$ is a negative thermodynamic temperature.

7.3   See the derivation of Equation 7.8.

7.4   See the derivation of the equation $S = S_1 + S_2 = k \ln \Omega$.

7.5   See the derivation of Equation 7.9.

7.6   (a) The steps of the Carnot cycle and the direction along each step are shown in the following diagram.

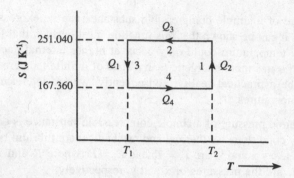

Step 1. Isothermal expansion as the temperature remains constant at T_2 and the entropy increases from 167.360 to 251.040 J.

Step 2. Adiabatic expansion as the entropy remains constant at 251.040 J and the temperature decreases from T_2 to T_1 (273 K).

Step 3. Isothermal compression, since the temperature remains constant at T_1 and the entropy decreases from 251.040 to 167.360 J.

Step 4. Adiabatic compression, since the enthalpy remains constant and the temperature increases from T_1 to T_2.

(b) In the second we have the following.

Step 1. Since this is a reversible isothermal expansion, $(\Delta S)_1 = Q_2/T_2$ where Q_2 is the heat absorbed, but $(\Delta S)_1 = 251.040 - 167.360 = 83.680$ J. Therefore, $Q_2 = 83.680 T_2$.

Step 2. Since this is a reversible adiabatic expansion, $(\Delta S)_2 = Q_3 = 0$.

Step 3. As this is a reversible isothermal compression, $(\Delta S)_3 = Q_1/T_1$ where Q_1 is the heat evolved (negative), but $(\Delta S)_3 = 167.360 - 251.040 = -83.680$ J. Therefore, $Q_1 = -83.680 T_1 = -(83.680)(273)$ J $= -22.845$ kJ.

Step 4. This step is a reversible adiabatic compression. Therefore, $(\Delta S)_4 = Q_4 = 0$.

For a complete cycle, $(\Delta U)_{cycle} = 0$ and $W_{cycle} = Q_{cycle} = Q_1 + Q_2 + Q_3 + Q_4$, or $418.400 = -22.845$ kJ $+ Q_2$ or $Q_2 = 23.263$ kJ. However, $Q_2 = 23.263$ kJ $= 83.680 T_2$. Therefore, $T_2 = 278$ K.

7.7 (a) $H_2O(\ell)$ ($-10°C$) $\rightarrow H_2O(s)$ ($-10°C$). This process is irreversible because it is the freezing of supercooled water at $-10°C$. In order to calculate the entropy of this process we need to devise an equivalent reversible path because $dS = dQ_{rev}/T$. Such a path is given below.

Step 1. $H_2O(\ell)$ ($-10°C$) $\rightarrow H_2O(\ell)(0°C)$.

Step 2. $H_2O(\ell)$ ($0°C$) $\rightarrow H_2O(s)(0°C)$.

Step 3. $H_2O(s)(0°C) \rightarrow H_2O(s)(-10°C)$.

A combination of these three steps gives the given irreversible process.

Step 1.
$$(\Delta S)_1 = \int_{T_1}^{T_2} (C_P)_{H_2O(\ell)} \frac{dT}{T} = C_P \ln\left(\frac{T_2}{T_1}\right)$$

because C_P is constant over the given temperature range.

Given that the specific heat of water is $4.184 \, \mathrm{J \, K^{-1} \, g^{-1}}$, therefore $C_P = 18 \times 4.184 = 75.312 \, \mathrm{J \, K^{-1} \, mol^{-1}}$. $T_2 = 0°\mathrm{C} = 273 \, \mathrm{K}$ and $T_1 = -10°\mathrm{C} = 263 \, \mathrm{K}$. Therefore, $(\Delta S)_1 = 75.312 \ln(273/263) = 2.803 \, \mathrm{J \, K^{-1} \, mol^{-1}}$.

Step 2. $(\Delta S)_2 = Q_{\mathrm{rev}}/T = (\Delta H)_f/T_f$ where $(\Delta H)_f$ is the enthalpy of fusion and T_f is the fusion temperature. $(\Delta H)_f = -334.720 \, \mathrm{J \, g^{-1}} = -334.720 \times 18 \, \mathrm{J \, mol^{-1}} = -6024.96 \, \mathrm{J \, mol^{-1}}$. $T_f = 273 \, \mathrm{K}$, then $(\Delta S)_2 = (-6024.96)/273 = -22.069 \, \mathrm{J \, K^{-1} \, mol^{-1}}$.

Step 3.
$$(\Delta S)_3 = \int_{T_1}^{T_2} (C_P)_{\mathrm{H_2O(s)}} \frac{dT}{T} = (C_P)_s \ln\left(\frac{T_2}{T_1}\right)$$

because (C_P) is given to be constant over the temperature range. The specific heat of ice is $2.092 \, \mathrm{J \, K^{-1} \, g^{-1}}$. So, C_P for ice is $(18)(2.092) = 37.656 \, \mathrm{J \, K^{-1} \, mol^{-1}}$. $T_2 = -10°\mathrm{C} = 263 \, \mathrm{K}$ and $T_1 = 0°\mathrm{C} = 273 \, \mathrm{K}$. Therefore,

$$(\Delta S)_3 = 37.656 \ln\left(\frac{263}{273}\right) = -1.423 \, \mathrm{J \, K^{-1} \, mol^{-1}}.$$

Hence, the total entropy change for the given process is

$$(\Delta S)_{\mathrm{process}} = (\Delta S)_1 + (\Delta S)_2 + (\Delta S)_3 = -20.689 \, \mathrm{J \, K^{-1} \, mol^{-1}}.$$

(b) A change in the entropy of the thermostat, i.e. the surroundings, would occur because heat is evolved on freezing. If $(\Delta S)_{\mathrm{surr}}$ is the change in entropy of the surroundings, then $(\Delta S)_{\mathrm{surr}} = (\Delta H)_f(-10°\mathrm{C})/T_f(-10°\mathrm{C})$. However, $(\Delta H)_f(0°\mathrm{C}) = (\Delta H)_{T_1} = -334.720 \, \mathrm{J \, g^{-1} \, K^{-1}}$ and $(\Delta H)_f$ at $-10°\mathrm{C}$ is equal to $(\Delta H)_{T_2}$ but not given. To find $(\Delta H)_{T_2}$ we use the Kirchhoff equation, i.e.

$$(\Delta H)_{T_2} = (\Delta H)_{T_1} + \int_{T_1}^{T_2} \Delta C_P \, dT = (\Delta H)_{T_1} + \Delta C_P(T_2 - T_1)$$

assuming that C_P is constant over the given temperature range. Since the process is $\mathrm{H_2O(\ell)}$ $(-10°\mathrm{C}) \rightarrow \mathrm{H_2O(s)}(-10°\mathrm{C})$, we have $\Delta C_P = C_P(\mathrm{H_2O(s)}) - C_P(\mathrm{H_2O})(\ell)) = 2.092 - 4.184 = -2.092 \, \mathrm{J \, K^{-1} \, g^{-1}}$. Therefore, $(\Delta H)_{T_2} = (\Delta H)_{T_1} - 2.092(263 - 273) = -355.64 \, \mathrm{J \, K^{-1} \, g^{-1}}$. The surroundings absorb the heat evolved on fusion. Hence, the heat absorbed by the surroundings at $-10°\mathrm{C} = (355.64)(18) = 6401.52 \, \mathrm{J \, mol^{-1}}$. Therefore $(\Delta S)_{\mathrm{surr}} = 6401/263 = 24.340 \, \mathrm{J \, K^{-1} \, mol^{-1}}$.

(c) The total entropy change for the process and the surroundings, i.e. the entropy change of the isolated system, is $(\Delta S)_{\mathrm{process}} + (\Delta S)_{\mathrm{surr}} = -20.689 + 24.340 = 3.651 \, \mathrm{J \, K^{-1} \, mol^{-1}}$. So, there is an increase in the entropy of the isolated system. As the second law of thermodynamics says that, if a process is spontaneous or irreversible, its entropy increases; so the given process is spontaneous.

7.8 (a) At constant pressure, the entropy change per mole of an ideal gas is given by

$$\Delta S = \int_{T_1}^{T_2} C_P \frac{dT}{T}.$$

For 2 mol of nitrogen heated from 300 to 600 K

$$\Delta S = 2 \int_{300\,K}^{600\,K} (27.0 + 0.0060T) \frac{dT}{T}$$

$$= 2\left[27.0\ln\left(\frac{600}{300}\right) + 0.0060(600 - 300)\right] = 41.000\,J\,K^{-1}.$$

At constant volume, the ideal gas also undergoes a change in pressure, i.e. $(P_1, V, T_1) \rightarrow (P_2, V, T_2)$. The entropy change per mole for such a process is given by

$$\Delta S = \int_{T_1}^{T_2} C_P \frac{dT}{T} - R \int_{P_1}^{P_2} \frac{dP}{P}.$$

At constant volume, $P_1/P_2 = T_1/T_2 = 300/600 = 1/2$. Therefore,

$$\Delta S = 2 \int_{300\,K}^{600\,K} \left(\frac{27.0}{T} + 0.0060\right) dT + 2R\ln\left(\frac{P_1}{P_2}\right)$$

$$= 41.000 + (2)(8.314)\ln\left(\frac{1}{2}\right) = 29.5\,J\,K^{-1}$$

7.9   See the derivation of Equation 7.23.

7.10   See the derivation of Equation 7.29.

7.11   See the derivation of Equation 7.39.

7.12   See the derivation of Equation 7.46.

7.13   See the derivation of Equation 7.52.

7.14   See the derivation of Equation 7.58, which shows that the entropy of a body undergoing an adiabatic process must increase or, in the limit of a reversible process, remain constant.

7.15   (a) Entropy change of an ideal gas for a reversible process is given by one of the following equations:

$$\Delta S = \int_{T_1}^{T_2} C_V \frac{dT}{T} + \int_{V_1}^{V_2} \frac{R}{V} \, dV$$

or

$$\Delta S = \int_{T_1}^{T_2} C_P \frac{dT}{T} - \int_{P_1}^{P_2} \frac{R}{P} \, dP.$$

We use the second equation and, as T is constant, we have

$$\Delta S = -R\ln\left(\frac{P_2}{P_1}\right) = -8.314\ln\left(\frac{1}{5}\right) = 13.400\,J\,K^{-1}\,mol^{-1}.$$

(b) Since S is a state function, ΔS depends only on the initial and final states of the gas. T_1, P_1 and P_2 are the same for both the reversible and the irreversible isothermal expansion. Therefore, ΔS must also be the same as in the case above i.e. $\Delta S = 13.400 \, \text{J K}^{-1} \, \text{mol}^{-1}$.

7.16 The total heat absorbed in the process is given by

$$Q_{total} = \text{molecular weight} \times \text{heat of vaporization} = (120)(209.200)$$
$$= 25.104 \, \text{kJ mol}^{-1}.$$

As the vapour behaves as an ideal gas, we have $W = -PV = -nRT$. Therefore, $W = -RT$ (since $n = 1$) or $W = -8.314(273 + 60) = -2.769 \, \text{kJ mol}^{-1}$.

$$\Delta U = Q + W = 25.104 - 2.769 = 22.335 \, \text{kJ mol}^{-1}$$
$$\Delta H = \Delta U + P\Delta V$$

(since the pressure is constant). Therefore,

$$\Delta H = \Delta U + RT = 22.335 + 8.314(273 + 60) = 25.104 \, \text{kJ mol}^{-1}$$

This shows that $\Delta H = Q_{total}$. Since the process is reversible at a constant temperature, $\Delta S = Q_{total}/T = 25.104/333 = 75.387 \, \text{J K}^{-1} \, \text{mol}^{-1}$.

7.17 See the derivation of Equation 7.67.

7.18 See Section 7.15.

7.19 The rule of Pictet and Trouton states that for many liquids the entropy of vaporization at the normal boiling point is approximately the same, i.e.

$$(\Delta S)_{vaporization} \simeq 87.864 \, \text{J K}^{-1} \, \text{mol}^{-1}.$$

This rule is convenient for finding an approximate value of the heat of vaporization of a liquid, if its boiling point is known. Since according to this rule the entropy of vaporization of many liquids is approximately equal to $87.864 \, \text{J K}^{-1} \, \text{mol}^{-1}$, it signifies that the increase in molar disorder on vaporization is almost the same for many liquids. In contrast, there is no general rule for entropies of fusion of solids at the melting temperature.

For the limitations of the rule of Pictet and Trouton, see Section 7.16.

7.20 Hildebrand's rule is a modification of Trouton's rule. He argued that for comparison it would be more appropriate to evaluate $(\Delta S)_{vaporization}$ for a change in state chosen so that the molar volumes of different vapours are always the same, rather than comparing at 1 atm (0.1 MPa) pressure. Hildebrand found that, if T' is the boiling temperature of a liquid at which its vapour occupies $22.4 \, \ell$, then the change in entropy for its vaporization may be expressed empirically by

$$[(\Delta S)_{vaporization}]_{T'} = \frac{(\Delta H)_{vaporization}}{T'} = 84.935 \, \text{J K}^{-1}.$$

7.21 According to the kinetic molecular model of matter, in a solid a near-perfect molecular or atomic order exists. In contrast, a liquid is a less ordered structure and there is nearly complete disorder in a gas. Since the entropy is a measure of the degree of randomness or disorder in a given state, the molar entropy of fusion, i.e. the entropy change associated with the process solid → liquid for 1 mol of the substance, is less than the molar entropy of vaporization, i.e. the entropy change per mole of substance associated with the process liquid → vapour.

7.22 For convenience, the unit of entropy (eU) is sometimes used instead of joules per kelvin per mole ($J K^{-1} mol^{-1}$). In fact, $1 eU = 1 J K^{-1} mol^{-1}$.

Chloroform is a non-polar substance. Trouton's rule can be used for this substance. The boiling temperature T' of chloroform is 61°C, i.e. 334 K. From Trouton's rule

$$(\Delta S)_{\text{vaporization}} = \frac{(\Delta H)_{\text{vaporization}}}{T'} \simeq 84.935 \text{ J K}^{-1}.$$

Therefore, $(\Delta H)_{\text{vaporization}} = (84.935)(334) = 28.368 \text{ kJ mol}^{-1}$.

7.23

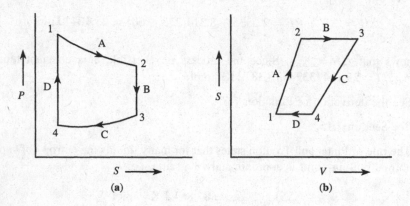

(a)   (b)

For (c) and (d) Enthalpy versus entropy is represented by the temperature versus entropy diagram, taking C_V on the energy axis and C_P on the enthalpy axis. This diagram also represents internal energy versus entropy.

7.24 As the block of ice melts very slowly at 0°C, the process is carried out at a constant temperature of 273 K and is also reversible. Therefore,

$$(\Delta S)_{\text{ice}} = \frac{1}{T} \int dQ = \frac{Q}{T}.$$

The heat required to melt the ice is $Q = $ mass of ice × heat of fusion $= (2.0 \text{ kg})$ $(333.0 \text{ kJ kg}^{-1})$. Therefore, $(\Delta S)_{\text{ice}} = (2.0 \text{ kg})(333.0 \text{ kJ kg}^{-1})/273 = 2.439 \text{ kJ K}^{-1}$.

An amount of heat Q is removed from the container at the constant temperature of 273 K to supply to the ice for melting. Therefore,

$$(\Delta S)_{\text{container}} = -\frac{Q}{T} = -2.439 \text{ kJ K}^{-1}.$$

The total entropy change for the system is

$$(\Delta S)_{\text{ice}} + (\Delta S)_{\text{container}} = 2.439 - 2.439 = 0.$$

7.25 Obviously the process is irreversible but we may assume that for this process the entropy change will be the same as that for a reversible process. The quantity of heat dQ that will be transferred from the ball to the lake equals the mass of the ball × the specific heat of iron × the difference in temperature. Then,

$$(\Delta S)_{\text{iron}} = \int_{T_1}^{T_2} \frac{dQ}{T} = -\int_{280\,\text{K}}^{800\,\text{K}} (4\,\text{kg})(460.240\,\text{J\,K}^{-1}\,\text{kg}^{-1}) \frac{dT}{T}$$

$$= -1840.96 \ln\left(\frac{880}{280}\right) \text{J\,K}^{-1} = -2.091\,\text{kJ\,K}^{-1}.$$

Since the temperature change of the lake is insignificant, its initial and final temperatures will be the same; however, the lake heats up in the region where the ball drops before going back to the equilibrium temperature at 280 K. The heat dQ received by the lake from the ball is the same as that given up by the ball, i.e. $dQ = (4\,\text{kg})(460.240\,\text{J\,K}^{-1}\,\text{kg}^{-1})\,(880{-}280\,\text{K}) = 1104.576\,\text{kJ}$. Although this is an irreversible process, we may take it as equivalent to a reversible isothermal heat transfer. Therefore,

$$(\Delta S)_{\text{lake}} = \frac{1104.576\,\text{kJ}}{280\,\text{K}} = 3.945\,\text{kJ\,K}^{-1}.$$

The total change in entropy of the iron ball and the lake is $-2.09 + 3.945 = 1.854\,\text{kJ\,K}^{-1}$. This shows that, although the entropy of the ball decreases (negative), the total change in entropy is positive.

7.26 (a) To calculate ΔS we need to know Q. Since the process is an isothermal reversible expansion, $\Delta U = 0$. From the first law of thermodynamics $\Delta U = Q + W$, we have $Q = -W$. For an isothermal reversible expansion the work is given by

$$W = -\int_{V_1}^{V_2} P\,dV = -RT \int_{V_1}^{V_2} dV = -(8.314)(298) \ln\left(\frac{40}{20}\right) = -1.715\,\text{kJ\,mol}^{-1}.$$

Therefore, $Q = -W = 1.715\,\text{kJ\,mol}^{-1}$. Hence, $\Delta S = Q/T = 1.715/298 = 5.757\,\text{J\,K}^{-1}\,\text{mol}^{-1}$.

(b) In the second case, the process is an isothermal expansion; therefore, $\Delta U = 0$. As there is no opposing pressure, no work is done, i.e. $W = 0$. Although the process is irreversible, we may assume that for this process the entropy change is the same as that for a reversible change. Then, taking Q_{rev} from the first case

$$\Delta S = \frac{Q_{\text{rev}}}{T} = \frac{1715}{298} = 5.757\,\text{J\,K}^{-1}\,\text{mol}^{-1}.$$

7.27 (a) The spring absorbs 4.184 J at a constant temperature of 300 K. The entropy change of the spring for this isothermal reversible process is given by

$$(\Delta S)_{\text{spring}} = \frac{Q_{\text{rev}}}{T} = \frac{4.184\,\text{J}}{300\,\text{K}} = 13.947 \times 10^{-3}\,\text{J\,K}^{-1}.$$

(b) The reaction of the spring is a spontaneous process. The change in entropy in this case is the same in magnitude as above but is opposite in sign, i.e. $(\Delta S)_{\text{spring}}$ for the reaction is $-13.947 \times 10^{-3}\,\text{J}\,\text{K}^{-1}$.

(c) The entropy change of the thermostat in the first case is equal in magnitude but is opposite in sign to the entropy change of the spring, i.e. $-13.947 \times 10^{-3}\,\text{J}\,\text{K}^{-1}$. So in this case the total entropy change for the system (i.e. the spring) and the surroundings (i.e. the thermostat) is zero.

In the second case, the entropy change for the surroundings is given by $(\Delta S)_{\text{thermostat}} = 14.644\,\text{J}/300\,\text{K} = 48.813 \times 10^{-3}\,\text{J}\,\text{K}^{-1}$. Therefore, the total change in entropy for the system and the surroundings is $48.813 \times 10^{-3} - 13.947 \times 10^{-3} = 34.866 \times 10^{-3}\,\text{J}\,\text{K}^{-1}$. This is also the total change in entropy for the entire process.

7.28   The processes involved are as follows.

Stage 1. $H_2O(s)(100\,\text{K}, 0.1\,\text{MPa}) \rightarrow H_2O(s)(273\,\text{K}, 0.1\,\text{MPa})$
Stage 2. $H_2O(s)(273\,\text{K}) \rightarrow H_2O(\ell)(273\,\text{K})$
Stage 3. $H_2O(\ell)(273\,\text{K}) \rightarrow H_2O(\ell)(373\,\text{K})$
Stage 4. $H_2O(\ell)(373\,\text{K}) \rightarrow H_2O(\text{vapour})(373\,\text{K})$
Stage 5. $H_2O(\text{vapour})(373\,\text{k}) \rightarrow H_2O(\text{vapour})(500\,\text{K})$

$$(\Delta S)_1 = \int_{100\,\text{K}}^{273\,\text{K}} C_P \frac{dT}{T} = \int_{100\,\text{K}}^{273\,\text{K}} (0.50 + 0.03T) \frac{dT}{T} = 23.815\,\text{J}\,\text{K}^{-1}\,\text{mol}^{-1}$$

$$(\Delta S)_2 = \frac{6.004\,\text{kJ}\,\text{mol}^{-1}}{273\,\text{K}} = 21.993\,\text{J}\,\text{K}^{-1}\,\text{mol}^{-1}$$

$$(\Delta S)_3 = \int_{273\,\text{K}}^{373\,\text{K}} C_P \frac{dT}{T} = \int_{273\,\text{K}}^{373\,\text{K}} \frac{75.312}{T}\,dT = 23.506\,\text{J}\,\text{K}^{-1}\,\text{mol}^{-1}$$

$$(\Delta S)_4 = \frac{40.292\,\text{kJ}\,\text{mol}^{-1}}{373\,\text{K}} = 108.02\,\text{J}\,\text{K}^{-1}\,\text{mol}^{-1}$$

$$(\Delta S)_5 = \int_{373\,\text{K}}^{500\,\text{K}} \frac{1}{T}(7.256 + 2.30 \times 10^{-3}T + 2.83 \times 10^{-7}T^2)\,dT$$

$$= 10.184\,\text{J}\,\text{K}^{-1}\,\text{mol}^{-1}.$$

Therefore, the total change in entropy for the entire process is

$$(\Delta S)_{\text{total}} = (\Delta S)_1 + (\Delta S)_2 + (\Delta S)_3 + (\Delta S)_4 + (\Delta S)_5 = 187.519\,\text{J}\,\text{K}^{-1}\,\text{mol}^{-1}.$$

7.29   (a) For a reversible isothermal expansion of an ideal gas, $\Delta U = 0$ and the work done is given by

$$W = -RT \int_{P_1}^{P_2} \frac{dP}{P} = -RT \ln\left(\frac{P_2}{P_1}\right) = -(8.314)(273.15)\ln\left(\frac{1}{10}\right)$$

$$= -5.230\,\text{kJ}\,\text{mol}^{-1}.$$

From the first law of thermodynamics, $\Delta U = Q + W$, or $Q = \Delta U - W = 5.230\,\text{kJ}\,\text{mol}^{-1}$. Therefore, $\Delta S = Q/T = 5.230/273.15 = 19.147\,\text{J}\,\text{K}^{-1}\,\text{mol}^{-1}$.

If the gas expands freely, it does no work. Therefore, $\Delta U = W = Q = 0$; no change in entropy.

(b) The entropy change for the surroundings when the expansion of the gas is isothermal and reversible is equal in magnitude but opposite in sign to that of the gas, i.e. $-19.147 \, \mathrm{J\,K^{-1}}$ $\mathrm{mol^{-1}}$. Then the total entropy change for the entire isolated system, i.e. the gas and its surroundings is zero. Similarly, for a free expansion of the gas, no work is done and there is no change in entropy. So, the total change in entropy for the isolated system is zero.

7.30 The entropy is a function of T and P. Therefore, to calculate the change in entropy we have to use the following equation:

$$\Delta S = \int_{T_1}^{T_2} C_P \frac{dT}{T} - R \int_{P_1}^{P_2} \frac{dP}{P}.$$

Since cadmium vapour behaves as an ideal monatomic gas, $C_P = C_V + R$ and $C_V = \frac{3}{2}R$. Hence, $C_P = \frac{5}{2}R = 20.785 \, \mathrm{J\,K^{-1}\,mol^{-1}}$. Given $T_1 = 1040 \, \mathrm{K}$, $T_2 = 1300 \, \mathrm{K}$, $P_1 = 0.1 \, \mathrm{MPa}$ and $P_2 = 0.6 \, \mathrm{MPa}$. Substituting these values into the above equation, we obtain

$$\Delta S = \int_{1040\,\mathrm{K}}^{1300\,\mathrm{K}} \frac{20.785}{T} \, dT + 8.314 \ln\left(\frac{1}{6}\right) = -10.330 \, \mathrm{J\,K^{-1}\,mol^{-1}}.$$

7.31 Heating of sulphur from 300 to 410 K involves two isothermal and three non-isothermal stages. We shall calculate the entropy change for each stage separately and add up to obtain the total entropy change for the entire process.

Stage 1. S(rhombic)(300 K) → S(rhombic)(368.6 K)

$$(\Delta S)_1 = \int_{T_1}^{T_2} C_P \frac{dT}{T} = \int_{300\,\mathrm{K}}^{368.6\,\mathrm{K}} (3.58 + 6.24 \times 10^{-3}T)\frac{dT}{T}$$

$$= 3.58 \ln\left(\frac{368.6}{300}\right) + (6.24 \times 10^{-3})(368.6 - 300)$$

$$= 1.165 \, \mathrm{cal\,K^{-1}\,mol^{-1}} = 4.874 \, \mathrm{J\,K^{-1}\,mol^{-1}}.$$

Stage 2. S(rhombic)(368.6 K) → S(monoclinic)(368.6 K)

$$(\Delta S)_2 = \frac{(\Delta H)_{\text{transition}}}{368.6 \, \mathrm{K}} = \frac{86 \, \mathrm{cal\,mol^{-1}}}{368.6 \, \mathrm{K}} = 0.233 \, \mathrm{cal\,K^{-1}\,mol^{-1}} = 0.975 \, \mathrm{J\,K^{-1}\,mol^{-1}}.$$

Stage 3. S(monoclinic)(368.6 K) → S(monoclinic)(392 K)

$$(\Delta S)_3 = \int_{368.6\,\mathrm{K}}^{392\,\mathrm{K}} (3.56 + 6.96 \times 10^{-3}T)\frac{dT}{T}$$

$$= 3.56 \ln\left(\frac{392}{368.6}\right) + (6.96 \times 10^{-3})(392 - 368.6)$$

$$= 0.388 \, \mathrm{cal\,K^{-1}\,mol^{-1}} = 1.602 \, \mathrm{J\,K^{-1}\,mol^{-1}}.$$

Stage 4. S(s)(monoclinic)(392 K) → S(ℓ)(392 K)

$$\Delta S_4 = \frac{(\Delta H)_f}{T_{mp}} = \frac{300\,cal\,mol^{-1}}{392\,K} = 0.765\,cal\,K^{-1}\,mol^{-1} = 3.20\,J\,K^{-1}\,mol^{-1}.$$

Stage 5. S(ℓ)(392 K) → S(ℓ)(410 K)

$$\Delta S_5 = \int_{392\,K}^{410\,K} (5.40 + 5.0 \times 10^{-3}T)\frac{dT}{T} = 0.333\,cal\,K^{-1}\,mol^{-1} = 1.393\,J\,K^{-1}\,mol^{-1}.$$

Therefore, the total change in entropy when sulphur is heated from 300 to 410 K is given by

$$(\Delta S)_{total} = (\Delta S)_1 + (\Delta S)_2 + (\Delta S)_3 + (\Delta S)_4 + (\Delta S)_5 = 2.879\,cal\,K^{-1}\,mol^{-1}$$
$$= 12.046\,J\,K^{-1}\,mol^{-1}.$$

7.32 (a) Because of hydrogen bonding and consequently greater ordering of H_2O molecules in the liquid state, a larger increase in disorder occurs when the water is vaporized, compared with that of benzene on vaporization, as hydrogen bonding does not occur in benzene.

(b) Nitrogen is a gas at 298 K but water is not. So the entropy of nitrogen at 298 K is greater than that of water at 298 K.

(c) At 298 K, mercury is a liquid but copper is not. Therefore, as mercury is a liquid it is more disordered than copper.

(d) Hardness and low entropy are manifestations of the presence of strong directional bonds between atoms. Diamond has strong tetrahedrally directed covalent bonds and therefore has a rigid ordered structure. Metals such as copper have metallic bonds which are non-directional and hence are softer and have higher entropies.

7.33 (a) Ba(s) + $\frac{1}{2}O_2$(g) → BaO(s). The reactants represent a more disordered state of the reaction because the oxygen molecules are not located in fixed sites, whereas in the product the oxygen is part of the BaO solid structure and is located in fixed sites. So, considering the degree of disorder of the reactants and the product, it seems that the reaction will result in a negative entropy.

(b) $BaCO_3$(s) → BaO(s) + CO_2(g). In $BaCO_3$(s), CO_2 is part of the $BaCO_3$ solid structure and is located in fixed sites, whereas in the products the CO_2 molecules are not located in fixed sites. Therefore, the products will have a greater entropy than the reactant, i.e. the reaction will result in a positive entropy.

(c) H_2(g) + Br_2(ℓ) → 2HBr(g). In the product HBr, bromine molecules are in a gaseous state and thus are more disordered than bromine in the reactants where it is in a liquid state. The entropy change associated with hydrogen in the reaction is not significant because hydrogen is in gaseous form in both the product and the reactants.

7.34 The relationship between Ω and the entropy of a system is given by $S = k \ln \Omega$. Given that $S = 41.840 \, \mathrm{J\,K^{-1}}$ and the Boltzmann constant $k = 1.38 \times 10^{-23} \, \mathrm{J\,K^{-1}}$. Substituting these values into the above equation, we get

$$41.840 = 1.38 \times 10^{-23} \ln \Omega = (1.38 \times 10^{-23})(2.303)\log \Omega$$

or

$$\log \Omega = \frac{41.840 \times 10^{23}}{(1.38)(2.303)} = 1.32 \times 10^{24}$$

or

$$\Omega = 10^{(1.32 \times 10^{24})}.$$

EXERCISES 8

8.1 See the derivation of Equation 8.32.

8.2 Equation 8.12 gives $C_P - C_V = T(\partial P/\partial T)_V(\partial V/\partial T)_P$. This equation clearly shows that the difference between C_P and C_V can be easily found for any substance for which $(\partial P/\partial T)_V$ and $(\partial V/\partial T)_P$ are known. Of course, $C_P - C_V$ can be computed when both C_P- and C_V-values are available.

8.3 See the derivation of Equation 8.53.

8.4 β for an ideal gas is defined by $\beta = (1/V)(\partial V/\partial T)_P$. The equation of state is $PV = RT$. Then, $(\partial V/\partial T)_P = R/P$ and $R = PV/T$. Substituting these in $\beta = (1/V)(\partial V/\partial T)_P$ we get $\beta = (1/V)(1/P)(PV/T) = 1/T$.

For an ideal gas, $Z = -(1/V)(\partial V/\partial P)_T$. From $PV = RT$, $(\partial V/\partial P)_T = -RT/P^2$. Then, $Z = -(1/V)(-RT/P^2) = PV/P^2V = 1/P$.

8.5 (a) $P(V - b) = RT$ or $V = RT/P + b$. Differentiating, $(\partial V/\partial T)_P = R/P$. Then,

$$\beta = \frac{P}{RT + Pb} \frac{R}{P} = \frac{R}{RT + Pb}$$

or

$$\beta = \frac{1/T}{1 + bP/RT}.$$

(b) From $P(V - b) = RT$ we have $V = RT/P + b$. Differentiating, $(\partial V/\partial P)_T = -RT/P^2$. Substituting in $Z = -(1/V)(\partial V/\partial P)_T$ we get $Z = (-1/V)(-RT/P^2)$. This gives $Z = [P/(RT + Pb)] (RT/P^2) = RT/P(RT + Pb) = (1/P)/(1 + bP/RT)$.

8.6 $PV = RT(1 + B/V) = RT + RTB/V$, or $PV^2 - RTV = RTB$. Differentiating with respect to T and keeping P constant, we obtain

$$2PV\left(\frac{\partial V}{\partial T}\right)_P - RT\left(\frac{\partial V}{\partial T}\right)_P - RV = RB + RT\,\frac{dB}{dT}$$

or

$$\left(\frac{\partial V}{\partial T}\right)_P (2PV - RT) = RV + RB + RT \frac{\mathrm{d}B}{\mathrm{d}T}$$

or

$$\left(\frac{\partial V}{\partial T}\right)_P = \frac{R[V + B + T(\mathrm{d}B/\mathrm{d}T)]}{2PV - RT}.$$

Therefore,

$$\beta = \frac{1}{V}\left(\frac{\partial V}{\partial T}\right)_P = \frac{R[V + B + T(\mathrm{d}B/\mathrm{d}T)]}{V(2PV - RT)}.$$

or

$$\beta = \frac{R[V + B + T(\mathrm{d}B/\mathrm{d}T)]}{V[2PV - PV/(1 + B/V)]}.$$

(on substituting for RT). Thus,

$$\beta = \frac{R(V + B + T(\mathrm{d}B/\mathrm{d}T))(1 + B/V)}{V(2PV + 2PB - PV)} = \frac{R[V + B + T(\mathrm{d}B/\mathrm{d}T)](1 + B/V)}{PV(V + 2B)}$$

$$= \frac{R[V + B + T(\mathrm{d}B/\mathrm{d}T)](1 + B/V)}{RT(1 + B/V)(V + 2B)}$$

(on substituting for PV). Therefore,

$$\beta = \frac{1}{T}\frac{V + B + T(\mathrm{d}B/\mathrm{d}T)}{V + 2B}.$$

8.7 S is a function of T and P, i.e. $S = f(T, P)$. Isobarically this means that P is constant. We know that $\mathrm{d}H = T\,\mathrm{d}S + V\,\mathrm{d}P$ and $C_P = (\partial H/\partial T)_P$. Therefore, $C_P = T(\partial S/\partial T)_P$, or $\mathrm{d}S = C_P\frac{\mathrm{d}T}{T}$, or $\mathrm{d}S = C_P\,\mathrm{d}(\ln T)$. Integrating both sides, we get $S = C_P \ln T + B$ where B is an unspecified integration constant. This equation shows that a plot of S versus $\ln T$ is a straight line having a slope C_P, if C_P does not change with temperature.

8.8 S is a function T and V, $S = f(T, V)$. Isochorically this means that V is constant. We know that $\mathrm{d}U = T\,\mathrm{d}S - P\,\mathrm{d}V$ and $C_V = (\partial U/\partial T)_V$. Therefore, $C_V = T(\partial S/\partial T)_v$, or $C_V\,\mathrm{d}(\ln T) = \mathrm{d}S$. Integrating both sides of this equation, we obtain $S = C_V \ln T + B'$ where B' is an unspecified integration constant. This equation shows that a plot of S versus $\ln T$ is a straight line with a slope C_V in any isochoric curve in which C_V does not change with temperature.

8.9   (a) Entropy change at constant pressure is given by $\Delta S = nC_p \ln(T_2/T_1)$. Given that $T_1 = 30$ K, and $T_2 = 300$ K and $C_P = \frac{5}{2}R\,\mathrm{J\,K^{-1}\,mol^{-1}} = \frac{5}{2}(8.314)\,\mathrm{J\,K^{-1}\,mol^{-1}} = 13.857\,\mathrm{J\,K^{-1}}$ $\mathrm{mol^{-1}}$. Substituting these values in the above equation, we have $\Delta S = (3)(13.857)$ $\ln(300/30) = 95.738\,\mathrm{J\,K^{-1}}$.

(b) Entropy change at constant pressure is given by $\Delta S = nC_V \ln(T_2/T_1)$. $C_P - C_V = R$ and this gives $C_V = 5.543 \, \text{J K}^{-1} \, \text{mol}^{-1}$. Hence,

$$\Delta S = (3)(5.543) \ln(300/30) = 38.294 \, \text{J K}^{-1}$$

8.10 See the derivation of Equation 8.64.

8.11 See the derivation of Equation 8.76.

8.12 The reaction is $2C(\text{graphite}) + 2H_2(g) = C_2H_4(g)$. Then,

$$(\Delta S^\circ)_{\text{reaction}} = \sum_{\text{product}} S^\circ - \sum_{\text{reactants}} S^\circ = (S^\circ)_{C_2H_4} - 2S^\circ_C - 2S^\circ_{H_2}$$

$$= 219.451 - 2(5.690 + 130.587) = -53.103 \, \text{J K}^{-1}.$$

It is not possible to conclude from this answer that the reaction is not spontaneous because in order to determine whether the reaction is spontaneous or not we need to know the total entropy change for the system and its surroundings, i.e. $(\Delta S)_{\text{sys}} + (\Delta S)_{\text{surr}}$.

8.13 (a) For the first case, the reaction is $H_2 + Cl_2 = 2HCl$. Therefore,

$$(\Delta S^\circ)_{\text{reaction}} = \sum_{\text{product}} S^\circ - \sum_{\text{reactants}} S^\circ$$

$$= 2 \times 44.617 - 31.211 - 53.286 = 4.737 \, \text{J K}^{-1}.$$

$4.737 \, \text{J K}^{-1}$ is the entropy change for the formation of 2 mol of HCl. So, the entropy change for the formation of 1 mol of HCl is half of this, i.e. $2.368 \, \text{J K}^{-1} \, \text{mol}^{-1}$.

(b) For the second case, the reaction is $CaCO_3 = CaO + CO_2$. Therefore,

$$(\Delta S^\circ)_{\text{reaction}} = (S^\circ)_{CaO} + (S^\circ)_{CO_2} - (S^\circ)_{CaCO_3}$$

$$= 39.748 + 213.639 - 92.885 = 160.502 \, \text{J K}^{-1} \, \text{mol}^{-1}.$$

8.14 Let us denote 0°C and 100°C by $T_{273 \, \text{K}}$ and $T_{373 \, \text{K}}$, respectively. The two gases in effect exchange heat under constant-pressure conditions until they reach the same final temperature, say T_f. If C_P is constant, then we can write

$$n_1 C_P (T_f - T_{273 \, \text{K}}) = -n_2 C_P (T_f - T_{373 \, \text{K}})$$

where n_1 is the number of moles of the gas at 0°C and n_2 is the number of moles of the gas at 100°C. This gives $T_f = (n_1 T_{273 \, \text{K}} + n_2 T_{373 \, \text{K}})/(n_1 + n_2) = 323 \, \text{K}$.

A point to note is that, since the process is adiabatic, there is no entropy change of the surroundings. We know that, for a constant-pressure process, the change in entropy is given by

$$dS = nC_P \frac{dT}{T}$$

or

$$\Delta S = nC_P \ln \left(\frac{T_{\text{final}}}{T_{\text{initial}}} \right).$$

Then,

$$(\Delta S)_{273\,\text{K}} = 1 \times \frac{5}{2} R \ln\left(\frac{323}{273}\right) = 3.489\,\text{J K}^{-1}\text{mol}^{-1}$$

$$(\Delta S)_{373\,\text{K}} = 1 \times \frac{5}{2} R \ln\left(\frac{323}{373}\right) = -2.992\,\text{J K}^{-1}\text{mol}^{-1}.$$

Since there is no entropy change of the surroundings, the total entropy change due to mixing is the sum of the entropy changes of the gas at 0°C, i.e. $(\Delta S)_{273\,\text{K}}$, and the gas at 100°C, i.e. $(\Delta S)_{373\,\text{K}}$, which is $3.489 - 2.992 = 0.497\,\text{J K}^{-1}\text{mol}^{-1}$.

8.15   The increase in entropy which results when two or more ideal gases are mixed at constant temperature and pressure is given by $(\Delta S)_m = -R\sum_i (n_i \ln x_i)$ where n_i is the number of moles of component i and x_i is the mole fraction of component i. We have $n_{N_2} = 4$ and $n_{O_2} = 1$, $x_{N_2} = 4/(1+4) = 0.8$ and $x_{O_2} = 1/(1+4) = 0.2$. Therefore, $(\Delta S)_m = -8.314[4\ln(0.8) + 1\ln(0.2)] = 20.8\,\text{J K}^{-1}$.

8.16   See the derivation of Equations 8.88 and 8.89. Since the mole fraction x_1 and x_2 are less than 1, Equation 8.89 shows that $(\Delta S)_m$ is positive, i.e. mixing of two or more ideal gases results in an increase in entropy. Equation 8.89 also shows that $(\Delta S)_m$ does not depend on temperature.

8.17   For the complete solution of this problem see the derivation of Equations 8.33, 8.34 and 8.35.

8.18   From Equation 5.83 we obtain

$$\left(\frac{\partial T}{\partial P}\right)_H = \frac{[T(\partial V/\partial T)_P - V]}{C_P}. \tag{i}$$

Now

$$\left(\frac{\partial (V/T)}{\partial T}\right)_P = -\frac{V}{T^2} + \frac{1}{T}\left(\frac{\partial V}{\partial T}\right)_P$$

or

$$T^2\left(\frac{\partial (V/T)}{\partial T}\right)_P = -\left[V - T\left(\frac{\partial V}{\partial T}\right)_P\right].$$

Then (i) can be written as

$$\left(\frac{\partial T}{\partial P}\right)_H = \frac{T^2}{C_P}\left(\frac{\partial (V/T)}{\partial T}\right)_P = \frac{V}{C_P}(\beta T - 1).$$

For an ideal gas, $PV = nRT$. Then $(\partial V/\partial T)_P = nR/P$ or $(1/V)(\partial V/\partial T)_P = nR/PV$. Substituting this in $(\partial T/\partial P)_H = (V/C_P)(\beta T - 1)$ we obtain

$$\left(\frac{\partial T}{\partial P}\right)_H = \frac{1}{C_P}\left(\frac{nRT}{P} - V\right) = 0.$$

8.19 We have $H = U + PV$, or $U = H - PV$. Differentiating both sides with respect to P, but keeping T constant we get $(\partial U/\partial P)_T = (\partial H/\partial P)_T - V - P(\partial V/\partial P)_T$. However, $(\partial H/\partial P)_T = V - T(\partial V/\partial T)_P$. Therefore,

$$\left(\frac{\partial U}{\partial P}\right)_T = V - T\left(\frac{\partial V}{\partial T}\right)_P - V - P\left(\frac{\partial V}{\partial P}\right)_T$$

$$= -TV\frac{1}{V}\left(\frac{\partial V}{\partial T}\right)_P + PV\left(-\frac{1}{V}\right)\left(\frac{\partial V}{\partial P}\right)_P$$

$$= -\beta TV + ZPV.$$

8.20 To prove this result we need to know an identity. Assume that we have three variables x, y and z, but only two of these are independent, i.e. the third is a function of the other two. Let us take x as a function of y and z, i.e. $x = f(y, z)$. Then, $\mathrm{d}x = (\partial x/\partial y)_z \, \mathrm{d}y + (\partial x/\partial z)_y \, \mathrm{d}z$. Dividing both sides by $\mathrm{d}y$ and keeping $\mathrm{d}x$ constant, we get $(\mathrm{d}x/\mathrm{d}y)_x = (\partial x/\partial y)_z + (\partial x/\partial z)_y (\partial z/\partial y)_x$. Since $\mathrm{d}x$ is constant, the left-hand side is zero. Therefore, $-(\partial x/\partial y)_z = (\partial x/\partial z)_y (\partial z/\partial y)_x$ or

$$\left(\frac{\partial x}{\partial z}\right)_y \left(\frac{\partial z}{\partial y}\right)_x \left(\frac{\partial y}{\partial x}\right)_z = -1. \tag{i}$$

Similarly, if we have three thermodynamic variables P, V and T but only two of these are independent, we can write

$$\left(\frac{\partial P}{\partial T}\right)_V \left(\frac{\partial T}{\partial V}\right)_P \left(\frac{\partial V}{\partial P}\right)_T = -1. \tag{ii}$$

Equations (i) and (ii) are perfectly symmetrical and quite easy to remember. Now given that

$$C_P - C_V = T\left(\frac{\partial P}{\partial T}\right)_V \left(\frac{\partial V}{\partial T}\right)_P = TV\frac{1}{V}\left(\frac{\partial V}{\partial T}\right)_P \left(\frac{\partial P}{\partial T}\right)_V = TV\beta\left(\frac{\partial P}{\partial T}\right)_V.$$

From (ii) we obtain

$$\left(\frac{\partial P}{\partial T}\right)_V = -1/(\partial T/\partial V)_P (\partial V/\partial P)_T = -(\partial V/\partial T)_P/(\partial V/\partial P)_T$$

because $(\partial T/\partial V)_P = -1/(\partial V/\partial T)_P$. Therefore,

$$\left(\frac{\partial P}{\partial T}\right)_V = -\frac{(1/V)(\partial V/\partial T)_P}{(1/V)(\partial V/\partial P)_T} = \frac{\beta}{Z}.$$

Hence,

$$C_P - C_V = \frac{TV\beta^2}{Z}.$$

8.21 Two points should be remembered here: firstly, since the process is adiabatic, there is no change in the entropy of the surroundings, and secondly the two quantities of water exchange heat on mixing under constant-pressure conditions until they attain the final equilibrium temperature. Let us assume that the final temperature reached is T_f, and $Q_{400\,K}$ and $Q_{300\,K}$ are the quantities of heat associated with the masses of water 3 kg and 4 kg at 400 K and 300 K, respectively. Then,

$$Q_{300\,K} = 4C_P(T_f - 300) \qquad Q_{400\,K} = 3C_P(T_f - 400).$$

Since the water at the higher temperature loses heat and the water at the lower temperature gains heat, we can write

$$Q_{300\,K} = -Q_{400\,K}$$

i.e.

$$4C_P(T_f - 300) = -3C_P(T_f - 400).$$

This gives

$$T_f = \frac{(4)(300) + (3)(400)}{3 + 4} = 343\ \text{K}.$$

The change in entropy of a system for a constant pressure process is given by $\Delta S = C_P \ln (T_{\text{final}}/T_{\text{initial}})$. Then the entropy changes of the two quantities of water are

$$(\Delta S)_{\text{water at 300 K}} = 4180 \ln\left(\frac{343}{300}\right) = 559.903\ \text{J K}^{-1}\text{kg}^{-1}$$

$$(\Delta S)_{\text{water at 400 K}} = 4180 \ln\left(\frac{343}{400}\right) = -642.609\ \text{J K}^{-1}\text{kg}^{-1}.$$

As there is no entropy change of the surroundings, the total entropy change is $-642.609 + 559.903 = -82.706\ \text{J K}^{-1}\text{kg}^{-1}$.

8.22 (a) We know that $dU = C_V\, dT$ for 1 mol of an ideal gas. Substituting this into Equation 8.1, we obtain $C_V\, dT = T\, dS - P\, dV$ or $dS = C_V\, dT/T + (P/T)\, dV = C_V\, dT/T + R\, dV/V$. This shows that the entropy of an ideal gas is a function of T and V. Integrating this equation, we have $S = C_V \ln T + R \ln V + S_1$, where S_1 is a constant of integration. For an ideal gas, $PV = RT$ (when $n = 1$), i.e. $T = PV/R$. Substituting this in the above equation, we obtain

$$S = C_V \ln P + C_V \ln V - C_V \ln R + R \ln V + S_1$$
$$= (C_V \ln V + R \ln V) + C_V \ln P + (S_1 - C_V \ln R)$$
$$= (C_V + R) \ln V + C_V \ln P + S = C_P \ln V + C_V \ln P + S$$

where $S = S_1 - C_V \ln R$ is constant. In $S = C_P \ln V + C_V \ln P + S'$ the volume is a molar volume.

(b) To find the value of S' we have to find the value of V first. For an ideal gas for $n = 1$, $V = RT/P$. Then,

$$V = \frac{(8.314\ \text{J K}^{-1}\text{mol}^{-1})(273.15\ \text{K})}{10 \times 10^6\ \text{Pa}} = 2271 \times 10^{-7}\text{m}^3\text{mol}^{-1}.$$

Since the molar entropy of the gas at the given temperature and pressure is zero, we have

$$0 = 13.0 \ln(10^7) + 21.0 \ln(2271 \times 10^{-7}) + S'$$

or

$$S' = 926.477.$$

The proper units are

$$J K^{-1} mol^{-1} = (J K^{-1} mol^{-1}) \ln(Pa) + [J K^{-1} mol^{-1} \ln(m^3 mol^{-1})] + 926.477$$

8.23 (a) The second law of thermodynamics stipulates that for a reversible adiabatic process the entropy change of an ideal gas is zero. Since this expansion is reversible and adiabatic, there is no change in entropy.

(b) For a free expansion against no opposing pressure, i.e. in a vacuum, the entropy change of an ideal gas is given by $\Delta S = R \ln (V_2/V_1)$. As the gas expands to double its volume, $V_2/V_1 = 2$. Then, $\Delta S = 8.314 \ln 2 = 5.763 \, J \, K^{-1}$.

EXERCISES 9

9.1 The essential criterion for application of the third law is the maximum order in the solid.

This law applies to perfect crystalline substances only. Unfortunately, perfect crystalline substances are difficult to identify. As the entropy of a substance is a measure of randomness or disorder, $S_0 = 0$ implies a highly ordered state of matter. The most highly ordered state of matter that we can think of is the crystalline state at the absolute zero of temperature, because at this temperature even the rotations and vibrations of the molecules are at their minimum state. A glassy or amorphous substance is not completely ordered even at the absolute zero of temperature. Therefore, any disorder remaining at the absolute zero of temperature gives a finite value of S_0.

If one or more phase changes occur between 0 K and the temperature of interest, appropriate entropy change for each phase change must be included to calculate the third-law entropy.

9.2 For the statements of the third law, see Section 9.2.

The fundamental difference between the statements of Planck and Nernst is that the former emphasizes that $S_0 = 0$ only for pure solids and liquids, whereas the latter assumes that it is applicable to all condensed phases, including solutions.

9.3 See Section 9.2.1 and Section 9.2.2 for the consequences of the third law and its corollaries.

There are many uses of the third law of thermodynamics. These include determination of free energies in reactions by calorimetry, computation of partition functions by the statistical spectroscopic method, and from these data the calculation of entropy and other thermodynamic parameters, determination of heat of reactions, equilibrium constants, electromotive forces, heat capacities of substances at very low temperatures, etc. The most useful application of the third law is the calculation of absolute entropies of pure substances at temperatures other than 0 K from their heat capacities and heat of transition.

9.4 See Section 9.4 for an experimental procedure for verification of the third law. Such experiments can only show that the difference in entropies of reactants and products at 0 K is almost always zero within the limits of accuracy of experiment.

9.5 See Section 9.5 for the principle of unattainability of the absolute zero of temperature.
   Consider Equation 9.10. The left-hand side of this equation is positive because $T_i > 0$ and $C_i > 0$ for any non-zero temperature T_i. Consequently, the right-hand side of this equation is also positive. This is possible only when $T_f \neq 0$, which implies that the final temperature T_f cannot be equal to 0 K. Therefore, we find that for a reversible cooling process the absolute zero of temperature cannot be achieved. Similar analysis of Equation 9.12 shows that for an irreversible cooling process the absolute zero of temperature cannot be attained. Hence, we can conclude that the absolute zero of temperature is unattainable.

9.6 See Section 9.7.

9.7 The absolute entropy of any substance is denoted by S, unless it refers to the standard state and 1 mol of the substance concerned, whereas standard entropy is simply denoted by S° and this refers to the entropy at 298.15 K. S° is the entropy of formation of a compound from its constituent elements at the standard state.

9.8 The entropy change is given by

$$S = \int_{273\,\mathrm{K}}^{T_f} C_P \frac{\mathrm{d}T}{T} + \frac{(\Delta H)_f}{T_f}$$

where the initial temperature is $0°C = 273$ K, the final temperature is $T_f = 25 + 273 = 298$ K, $(\Delta H)_f$ is the molar enthalpy of fusion of ice at $0°C$ and is equal to $6.025\,\mathrm{kJ\,mol^{-1}}$, and $C_P = 75.312\,\mathrm{J\,K^{-1}mol^{-1}}$. Substituting these values into the above equation

$$S = \int_{273\,\mathrm{K}}^{298\,\mathrm{K}} \frac{75.312}{T}\,\mathrm{d}T + \frac{6.025}{298} = 75.312 \ln\left(\frac{298}{273}\right) + 20.218 = 26.816\,\mathrm{J\,K^{-1}}.$$

9.9 The masses of ice and water in effect exchange heat under constant pressure conditions until they reach the same final temperature T. Thus, for constant C_P we have

$$m_{\mathrm{ice}}(C_P)_{\mathrm{ice}}(T - T_{\mathrm{ice}}) = m_{\mathrm{water}}(C_P)_{\mathrm{water}}(T_{\mathrm{water}} - T).$$

Now $m_{\mathrm{ice}} = 10$ g, $m_{\mathrm{water}} = 100$ g, $T_{\mathrm{ice}} = 263$ K, $T_{\mathrm{water}} = 298.15$ K, $(C_P)_{\mathrm{ice}} = 2.092\,\mathrm{J\,K^{-1}g^{-1}}$ and $(C_P)_{\mathrm{water}} = 4.184\,\mathrm{J\,K^{-1}g^{-1}}$. Substituting these values into the above equation, we have the final temperature $T = 294.82$ K. The entropy change for mixing is given by

$$(\Delta S)_{\mathrm{m}} = \int_{T_{\mathrm{ice}}}^{273.15\,\mathrm{K}} (C_P)_{\mathrm{ice}} \frac{\mathrm{d}T}{T} + \int_{273.15\,\mathrm{K}}^{T} (C_P)_{\mathrm{water}} \frac{\mathrm{d}T}{T} + \frac{(\Delta H)_f}{T}$$

where $(\Delta H)_f$ is the enthalpy of fusion of ice and is equal to $333.465\,\mathrm{J\,g^{-1}}$. Then,

$$(\Delta S)_{\mathrm{m}} = 2.092 \ln\left(\frac{273.15}{263}\right) + 4.184 \ln\left(\frac{294.82}{273.15}\right) + \frac{333.465}{294.82} = 1.530\,\mathrm{J\,K^{-1}}.$$

9.10 Equation 9.5 gives the heat capacity of solids for temperatures approaching the absolute zero. We know that the difference between C_P and C_V of solids is negligible, especially at low temperatures. So, at low temperatures we can replace C_P by C_V and vice versa. Then, for low T,

$$S = \int_0^T C_P \frac{dT}{T} = \int_0^T C_V \frac{dT}{T} = \int_0^T aT^3 \frac{dT}{T} = \left[\frac{aT^3}{3} \right]_0^T = \frac{aT^3}{3} = \frac{1}{3}C_V.$$

9.11 The entropy change in the transformation $AgI(\beta) \rightarrow AgI(\alpha)$ is simply the difference between the entropy of the α-form from 0 to 146.5°C and that of the β-form from 0 to 146.5°C, and this difference is equal to the heat of transition divided by the transition temperature. Then we may write

$$(S_{0\rightarrow 146.5°C})_\alpha - (S_{0\rightarrow 146.5°C})_\beta = \frac{-6401.52}{146.5 + 273.15} = -15.259 \, \text{J K}^{-1}\text{mol}^{-1}.$$

This is the entropy change involved in the transformation of 1 mol of the β form to the α form. Then, for 2 mol of the β form, the entropy change is $(2)(-15.259) = -30.519 \, \text{J K}^{-1} \text{mol}^{-1}$.

9.12 Let $(C_P)_s$ and $(C_P)_\ell$ be the heat capacities of the solid and liquid propane. Then,

$$(\Delta S)_s = \int_0^{85.45 \, \text{K}} (C_P)_s \frac{dT}{T}$$

and

$$(\Delta S)_\ell = \int_{85.45 \, \text{K}}^{231.04 \, \text{K}} (C_P)_\ell \frac{dT}{T}.$$

These give

$$41.505 = (C_P)_s \times 4.448 \qquad 88.115 = 0.995(C_P)_\ell$$

Hence, $(C_P)_s = 9.331 \, \text{J K}^{-1}$ and $(C_P)_\ell = 88.558 \, \text{J K}^{-1}$. The entropy of gaseous propane at 231.04 K is given by

$$S_{231.04 \, \text{K}} = \int_0^{84.45 \, \text{K}} (C_P)_s \frac{dT}{T} + \frac{(\Delta H)_f}{85.45} + \int_{85.45 \, \text{K}}^{231.04 \, \text{K}} (C_P)_\ell \frac{dT}{T} + \frac{(\Delta H)_\text{vaporization}}{231.04}$$

from Equation 9.1. Thus,

$$S_{231.04 \, \text{K}} = 9.331 \ln(85.45) + \frac{3523.765}{85.45} + 88.558 \ln\left(\frac{231.04}{85.45} \right) + \frac{18\,773}{231.04}$$

$$= 252.104 \, \text{J K}^{-1}.$$

9.13   The absolute standard entropy of Cl_2 at 298.15 K is equal to the entropy change of the process $Cl_2(s)(0\,K) \rightarrow Cl_2(g)(298.15\,K)$. The entropy change of this process is given by the sum of the entropies of the following steps.

Step 1. $Cl_2(s)(0\,K) \rightarrow Cl_2(s)(15\,K)$

$$(\Delta S)_1 = \int_0^{15} C_P \frac{dT}{T}.$$

For solid substances, the difference between C_P and C_V is negligible, especially at low temperatures. Between 0 and 15 K, C_P values are given by the Debye cube law, i.e. $C_V = aT^3$ where a is a constant. Hence, we can write $C_P \simeq C_V = aT^3$. Therefore,

$$(\Delta S)_1 = \int_0^{15} aT^3 \frac{dT}{T} = \left[\frac{aT^3}{3} \right]_0^{15} = \frac{3.724}{3} = 1.241 \, J\,K^{-1}mol^{-1}$$

(since $C_P = 0$ at 0 K). Also, by the third law, $S_0 = 0$. Therefore,

$$(\Delta S)_1 = S_{15} - S_0 = S_{15} = 1.241 \, J\,K^{-1}mol^{-1}.$$

Step 2. $Cl_2(s)(15\,K) \rightarrow Cl_2(s)(172.12\,K)$

$$(\Delta S)_2 = \int_{15\,K}^{172.12\,K} C_P \frac{dT}{T} = \int_{15\,K}^{172.12\,K} C_P \, d(\ln T).$$

By graphical integration, using any one of the plots of C_P/T versus T, C_P versus $\ln T$ given below, we get $(\Delta S)_2 = 70.166 \, J\,K^{-1}mol^{-1}$.

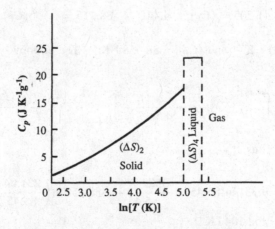

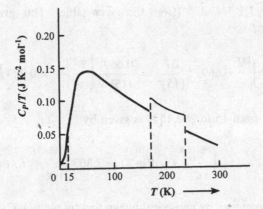

Step 3. $Cl_2(s) \xrightarrow{172.12 \, K} Cl_2(\ell)$

$$(\Delta S)_3 = \frac{(\Delta H)_f}{T_{mp}} = \frac{6.406}{172.12} = 37.218 \, J \, K^{-1} mol^{-1}$$

Step 4. $Cl_2(\ell)(172.12 \, K) \rightarrow Cl_2(\ell)(239.05 \, K)$

$$(\Delta S)_4 = \int_{172.12 \, K}^{239.05 \, K} C_P \frac{dT}{T} = \int_{172.12 \, K}^{239.05 \, K} C_P \, d(\ln T) = 22.259 \, J \, K^{-1} mol^{-1}$$

(by graphical integration).

Step 5. $Cl_2(\ell) \xrightarrow{239.05 \, K} Cl_2(vap)$

$$(\Delta S)_5 = \frac{(\Delta H)_{vaporization}}{T_{bp}} = \frac{20409}{239.05} = 85.375 \, J \, K^{-1} mol^{-1}.$$

Step 6. $Cl_2(g)(239.05 \, K) \rightarrow Cl_2(g)(298.15 \, K)$

$$(\Delta S)_6 = \int_{239.05 \, K}^{298.15 \, K} C_P \frac{dT}{T} = C_P \ln\left(\frac{298.15}{239.05}\right) = 7.561 \, J \, K^{-1} mol^{-1}.$$

Therefore,

$$\Delta S^\circ = (\Delta S)_1 + (\Delta S)_2 + (\Delta S)_3 + (\Delta S)_4 + (\Delta S)_5 + (\Delta S)_6 = 223.820 \, J \, K^{-1} mol^{-1}.$$

9.14 Below 15 K, the entropy of solid silver obeys the Deybe cube law, i.e.

$$S_{0 \, K \rightarrow 15 \, K} = \int_0^{15} aT^3 \frac{dT}{T}$$

where a is a proportionality constant. To evaluate a we make use of the fact that, for solids, $C_P \simeq C_V$, especially at low temperatures. From the Debye cube law, $C_V = aT^3$. Then,

$C_P = C_V = aT^3 = 0.669\,\text{J K}^{-1}\text{mol}^{-1}$ (from the given table). This gives $a(15)^3 = 0.669$ or $a = 0.669/(15)^3$. Then,

$$S_{0\,\text{K}\to 15\,\text{K}} = \int_{0\,\text{K}}^{15\,\text{K}} 0.669 T^2 \frac{\mathrm{d}T}{(15)^3} = \frac{(0.669)}{(15)^3}\left[\frac{T^3}{3}\right]_0^{15} = 0.223\,\text{J K}^{-1}\text{mol}^{-1}.$$

The absolute entropy from 15 to 298.15 K is given by

$$S_{15\,\text{K}\to 298.15\,\text{K}} = \int_{15\,\text{K}}^{298.15\,\text{K}} C_P\,\mathrm{d}(\ln T) = 2.303 \int_{15\,\text{K}}^{298.15\,\text{K}} C_P\,\mathrm{d}(\log T).$$

This integral can be computed by numerical integration by plotting C_P versus $\log T$, or C_P/T versus T, and calculating the area under the curve between the limits 15 to 298.15 K.

Numerical integration gives $S_{15\,\text{K}\to 298.15\,\text{K}} \simeq 42.258\,\text{J K}^{-1}\text{mol}^{-1}$. Hence, $S_{0\,\text{K}\to 298.15\,\text{K}} = 42.258 + 0.223 = 42.481\,\text{J K}^{-1}\text{mol}^{-1}$.

$T(\text{K})$	$\log[T(\text{K})]$	$C_P\,(\text{J K}^{-1}\text{mol}^{-1})$
15	1.176	0.669
20	1.301	1.715
30	1.477	4.774
40	1.602	8.383
50	1.699	11.648
70	1.845	16.334
90	1.945	19.133
110	2.041	20.961
130	2.114	22.129
150	2.176	22.970
170	2.230	23.615
190	2.279	24.087
210	2.322	24.422
230	2.362	24.732
270	2.431	25.313
290	2.462	25.439
300	2.477	25.501

9.15 Since, for solids, $C_V \simeq C_P$ (especially at low temperatures) and the heat capacity of benzothiophene obeys the Debye cube law below 12 K, the entropy change from 0 to 12 K is given by

$$(S_{0\,\text{K}\to 12\,\text{K}})_{\text{form 1}} = \int_{0\,\text{K}}^{12\,\text{K}} aT^3\frac{\mathrm{d}T}{T} = \frac{(4.469)(12)^3}{(12)^3(3)} = 1.489\,\text{J K}^{-1}\text{mol}^{-1}$$

$$(S_{0\,\text{K}\to 12\,\text{K}})_{\text{form 2}} = \int_{0\,\text{K}}^{12\,\text{K}} aT^3\frac{\mathrm{d}T}{T} = \frac{(6.573)(12)^3}{(12)^3(3)} = 2.191\,\text{J K}^{-1}\text{mol}^{-1}.$$

Therefore,

$$(S_{0\,K \to 261.6\,K})_{form\,1} = \int_{0\,K}^{12\,K} aT^3 \frac{dT}{T} + \int_{12\,K}^{261.6\,K} C_P \, d(\ln T) + (S_{0\,K})_{form\,1}$$
$$= 149.594 + (S_{0\,K})_{form\,1}$$

$$(S_{0\,K \to 261.6\,K})_{form\,2} = \int_{0\,K}^{12\,K} aT^3 \frac{dT}{T} + \int_{12\,K}^{261.6\,K} C_P \, d(\ln T) + (S_{0\,K})_{form\,2}$$
$$= 2.191 + 152.733 + (S_{0\,K})_{form\,2}$$

where $S_{0\,K}$ is the entropy at $0\,K$. Now

$$(S_{0\,K \to 261.6\,K})_{form\,2} - (S_{0\,K \to 261.6\,K})_{form\,1} = (S_{0\,K})_{form\,2} - (S_{0\,K})_{form\,1} + 5.330.$$

However, $(S_{0\,K \to 261.6\,K})_{form\,2} - (S_{0\,K \to 261.6\,K})_{form\,1} = 3012.480/261.6$. Therefore, $(S_{0\,K})_{form\,2} - (S_{0\,K})_{form\,1} = 11.516 - 5.330 = 6.186\,J\,K^{-1}mol^{-1}$. This shows that at $0\,K$ the high-temperature form 2 has a residual entropy 6.186 eU. According to the third law of thermodynamics, a perfect crystal has no entropy at the absolute zero of temperature. Therefore, the high-temperature crystal form is not a perfect crystal.

9.16 We consider the reaction as one which proceeds as shown below.

$$H_2(g)_{(298.15\,K)} + \frac{1}{2}\,O_2(g)_{(298.15\,K)} \xrightarrow{\Delta S^\circ} H_2O(g)_{(298.15\,K)}$$

$$1 \downarrow (\Delta S^\circ)_1 \qquad 2 \downarrow (\Delta S^\circ)_2 \qquad 4 \uparrow (\Delta S^\circ)_4$$

$$H_2(s)_{(0\,K)} \qquad + \frac{1}{2}\,O_2(s)_{(0\,K)} \xrightarrow[(\Delta S^\circ)_3]{3} H_2O(s)_{(0\,K)}$$

This reaction means that either gases H_2 and O_2 react directly to give gaseous H_2O, or gaseous H_2 and O_2 change first to their solids at $0\,K$, react to form solid H_2O at $0\,K$, which then goes to gaseous H_2O at $298.15\,K$. As the standard entropy is a state function, we may write

$$\Delta S^\circ = (\Delta S^\circ)_1 + (\Delta S^\circ)_2 + (\Delta S^\circ)_3 + (\Delta S^\circ)_4.$$

The absolute entropies of $H_2(s)$, $O_2(s)$ and $H_2O(s)$ are zero at the absolute zero of temperature, because these are all pure substances. This gives $(\Delta S^\circ)_3 = 0$.

$$(\Delta S^\circ)_1 = (S^\circ_{H_2(s)})_{0\,K} - (S^\circ_{H_2(g)})_{298\,K} = 0 - (S^\circ_{H_2(g)})_{298\,K}.$$

Therefore,

$$(\Delta S^\circ)_1 = -(S^\circ_{H_2(g)})_{298\,K}$$
$$(\Delta S^\circ)_2 = \frac{1}{2}(S^\circ_{O_2(s)})_{0\,K} - \frac{1}{2}(S^\circ_{O_2(s)})_{298\,K} = 0 - \frac{1}{2}(S^\circ_{O_2(g)})_{298\,K}.$$

Hence,

$$(\Delta S^\circ)_2 = -\frac{1}{2}(S^\circ_{O_2(g)})_{298\,K}$$

$$(\Delta S^\circ)_4 = (S^\circ_{H_2O(g)})_{298\,K} - (S^\circ_{H_2O(s)})_{0\,K} = (S^\circ_{H_2O(g)})_{298\,K} - 0.$$

Hence,

$$(\Delta S^\circ)_4 = (S^\circ_{H_2O(g)})_{298\,K}.$$

This gives

$$(\Delta S^\circ) = -(S^\circ_{H_2(g)})_{298\,K} - \frac{1}{2}(S^\circ_{O_2(g)})_{298\,K} + O + (S^\circ_{H_2O(g)})_{298\,K}.$$

Substituting the values of the respective standard entropy from Table 9.1, we obtain

$$\Delta S^\circ = -130.58 - \frac{1}{2}(205.03) + 188.71 = -44.37\,J\,K^{-1}$$

9.17 The entropy change for the reaction is given by

$$\Delta S^\circ = (S^\circ_{H_2O(g)})_{298\,K} - (S^\circ_{H_2O(\ell)})_{298\,K}.$$

Now the absolute entropy of $H_2O(\ell)$ at 298 K is $(S^\circ_{H_2O(\ell)})_{298\,K}$, and equals

$$(S^\circ_{H_2O(\ell)})_{298\,K} - (S^\circ_{H_2O(s)})_{298\,K} = 69.91 - 0 = 69.91\,J\,K^{-1}\,mol^{-1}.$$

Now the absolute entropy of $H_2O(g)$ at 298 K is $(S^\circ_{H_2O(g)})_{298\,K}$, and equals

$$(S^\circ_{H_2O(g)})_{298\,K} - (S^\circ_{H_2O(s)})_{298\,K} = 188.70 - 0 = 188.70\,J\,K^{-1}\,mol^{-1}.$$

Therefore, $\Delta S^\circ = 188.70 - 69.91 = 118.79\,J\,K^{-1}$.

The positive value of ΔS° indicates that the molecules in the gaseous H_2O are in much more rapid and random motion than in the liquid H_2O.

EXERCISES 10

10.1 See Section 10.1 for a discussion of *permitted* and *forbidden processes*. Equations 10.1 and 10.2 provide the criteria of such processes.

10.2 See Section 10.2 for a complete discussion.

10.3 If an isothermal process is carried out at constant volume, $dV = 0$ and therefore $W = 0$. Then from Equation 10.4, for a constant-volume isothermal process we have $\Delta A \leqslant 0$. This shows that for such a process the work content ΔA tends to decrease spontaneously, as the process proceeds, until it reaches the minimum value. Once the minimum value is attained, the system attains the equilibrium state.

10.4 The Helmholtz free energy of any system is denoted by A and given by the equation $A = U - TS$ where U, T and S are the internal energy, temperature and entropy of the system concerned.

The Gibbs free energy of a system is denoted by G and given by the equation $G = A + PV = H - TS$ where H is the enthalpy of the system concerned.

It is not possible to calculate the absolute value of either work content or the Gibbs free energy in any case. However, in principle, it is possible to calculate ΔA and ΔG for any specified change.

See the derivation of Equation 10.12. This equation shows that at constant temperature and pressure the Gibbs free energy tends to decrease until it reaches the minimum value and, once the minimum value is attained, the system goes into equilibrium.

10.5 For an isothermal process, $\Delta A = \Delta U - Q_{rev} = W_{max}$ (see the derivation of Equation 10.15). This shows that in an isothermal process the maximum work W_{max} done by a system can be obtained by a change in the Helmholtz free energy ΔA of that system.

10.6 See the derivation of Equations 10.18 and 10.19. As the entropy of any substance is positive, $(\partial A / \partial T)_V = -S$ shows that the Helmholtz free energy A of any substance decreases as the temperature increases. The rate of decrease in A with T increases as the entropy of the substance increases. For gases, which have high entropies, the rate of decrease in A with T is larger than that for liquids and solids which have comparatively small entropies.

10.7 See the derivation of Equations 10.28 and 10.29, which give the variations in the Gibbs free energy with temperature and pressure. As the entropy of any substance is positive, $(\partial G / \partial T)_P = -S$ shows that the Gibbs free energy G of any substance decreases as the temperature increases, and the rate of this decrease increases as the entropy of the substance increases. For gases, which have high entropies, the rate of decrease in G with T is larger than that for liquids and solids.

10.8 For a discussion of the fundamental properties of the Helmholtz and Gibbs free energies, see Section 10.4.2.

The standard free-energy change $\Delta G°$ is the difference between the free energies of the products and the reactants in their standard state.

If the standard enthalpies and entropies of the reactants and the products are known, it is possible to calculate the standard free-energy change for that reaction from $\Delta G° = \Delta H° - T\Delta S°$.

10.9 The process is $H_2O(\ell) \rightarrow H_2O(g)$. Given that $V_2 = 31.0 \times 10^{-3} \, m^3$ and $V_1 = 18.8 \times 10^{-6} \, m^3$. Therefore, $\Delta V = V_2 - V_1 = 30.98 \times 10^{-3} \, m^3$ and $P\Delta V = (101.3 \times 10^3 \, N\,m^{-2})(30.98 \times 10^{-3}) = 3138 \, J$.

For the phase change in question, $\Delta G = 0$, i.e. no mechanical work can be done, if the process occurs at 0.1 MPa pressure and the normal boiling temperature. Hence, $\Delta G = \Delta A + \Delta(PV)$, or $\Delta A = \Delta G - P\Delta V$ (since P is constant). Therefore, $\Delta A = 0 - 3138 = -3138 \, J$.

10.10   The actual reaction is $CH_3COCH_3(\ell) + H_2(g) = CH_3CHOHCH_3(\ell)$

$$\Delta G^\circ = G^\circ_{products} - G^\circ_{reactants} = \sum_{products} \Delta G^\circ_f - \sum_{reactants} \Delta G^\circ_f$$

$$= (-184.096) - (-155.728) - (\Delta G^\circ_f)_{H_2(g)}$$

$$= -28.368 + 0 = -28.368 \text{ kJ mol}^{-1}.$$

10.11   The process is $2C_6H_6(\ell) \rightarrow 2C_6H_6(g)$. The molecular weight of benzene is 78.06 g mol^{-1}. The boiling point is 353.35 K and the latent heat of vaporization is 422.584 J.

$$(\Delta S)_{353.35 \text{ K}} = (\Delta S)_{vap} = \frac{(\Delta H)_{vap}}{T_b}.$$

Therefore,

$$T(\Delta S)_{vap} = (\Delta H)_{vap}$$

$$(\Delta G)_{vap} = (\Delta G)_{353.35 \text{ K}} = (\Delta H)_{vap} - T(\Delta S)_{vap} = 0$$

$$(\Delta A)_{vap} = (\Delta A)_{353.35 \text{ K}} = (\Delta U)_{vap} - T(\Delta S)_{vap} = (\Delta H)_{vap} - \Delta(PV) - T(\Delta S)_{vap}.$$

Since the benzene vapour behaves ideally, we have $\Delta(PV) = \Delta(nRT)$. Hence,

$$(\Delta A)_{vap} = -\Delta(nRT)$$

(because $(\Delta H)_{vap} = T(\Delta S)_{vap}$). Thus,

$$(\Delta A)_{vap} = -(2)(8.314)(353.35) = -5875.50 \text{ J}.$$

Since the final pressure is 0.05 MPa, the process may be considered to consist of an evaporation followed by an isothermal expansion, i.e.

$$2C_6H_6(\ell) \xrightarrow{80.2^\circ C} 2C_6H_6(g)(0.1 \text{ MPa}) \xrightarrow{80.2^\circ C} 2C_6H_6(g)(0.05 \text{ MPa}).$$

So,

$$\Delta G = (\Delta G)_{vaporization} + (\Delta G)_{expansion} = 0 + nRT \ln\left(\frac{P_2}{P_1}\right)$$

(because $(\partial G/\partial P)_T = V$). Thus,

$$\Delta G = (2)(8.314)(353.35)(2.303) \log(0.5) = -4072.917 \text{ J}$$

$$\Delta A = (\Delta A)_{vaporization} + (\Delta A)_{expansion}$$

but

$$(\Delta A)_{vaporization} = -5875.503 \text{ J}$$

$$(\Delta A)_{expansion} = (\Delta U)_{expansion} - T(\Delta S)_{expansion}$$

For isothermal expansion of an ideal gas, $\Delta U = 0$. Therefore, $(\Delta A)_{\text{expansion}} = -T(\Delta S)_{\text{expansion}}$, but

$$(\Delta S)_{\text{expansion}} = \frac{Q}{T} = nR \ln\left(\frac{P_1}{P_2}\right)$$

$$= (2)(8.314)(2.303) \log\left(\frac{0.1}{0.05}\right) = 11.506 \, \text{J K}^{-1}.$$

Then,

$$(\Delta A)_{\text{expansion}} = -(353.35)(11.506) = -4065.645 \, \text{J}.$$

Hence,

$$\Delta A = (\Delta A)_{\text{vaporization}} + (\Delta A)_{\text{expansion}} = -5875.50 - 4065.645 = -9941.145 \, \text{J}.$$

10.12 The process is $H_2O(\ell) \rightarrow H_2O(g)(0.1 \, \text{MPa})$. For a process at constant temperature, $\Delta G = \Delta H - T\Delta S$.

(a) For the process to be at equilibrium, $\Delta G = 0$. Therefore, under this condition, $\Delta H = T \Delta S$ or $T = \Delta H/\Delta S = 40125/107.529 = 373.16 \, \text{K}$.

(b) For the process to be spontaneous, ΔG is negative. Therefore, $T \, \Delta S > \Delta H$, i.e. $T > \Delta H/\Delta S$ or $T > 373.16 \, \text{K}$.

(c) For the process to be non-spontaneous, ΔG is positive, i.e. $T < 373.16 \, \text{K}$.

10.13 See Section 10.6 for the definitions and differences between spontaneous and non-spontaneous reversible processes.
    See Section 10.6(c) for a discussion of the statement.

10.14 See the derivations of Equations 10.42 and 10.43.

10.15 Equation 10.44 implies that, for a transformation in an isolated system, dS must be positive, i.e. the entropy must increase. The entropy of an isolated system continues to increase so long as changes occur within the system. When there is no more change, the system has attained the equilibrium state, and the entropy has reached the maximum value. Therefore, it is necessarily true that the condition of equilibrium in an isolated system is that the system has the maximum entropy.

10.16 $\Delta G = 13\,580 + 16.1T \log T - 72.59T$, but $\Delta G = \Delta H - T\Delta S$ and $d(\Delta G)/dT = -\Delta S$. Then, $d(\Delta G)/dT = 16.1 \log T + 16.1/2.303 - 72.59 = -25.768$. Therefore, $-\Delta S = -25.768$ or $\Delta S = 25.768 \, \text{eU}$. Then, from $\Delta G = \Delta H - T\Delta S$ we have $13.580 + 16.1T \log T - 72.59T = \Delta H - (298.15)(25.768)$ or $\Delta H = 11.495 \, \text{kJ}$.

10.17 The general criterion of equilibrium in a reversible process is given by $T \, dS - dU + dW = 0$ (see Equation 10.42).

(a) When S and P are constant, we have $dS = 0$. Then the required criterion of equilibrium is $dW = dU$.

(b) Constant H and P means that $dH = 0 = dP$, but $H = U + PV$. This gives $dH = dU + P\,dV + V\,dP$ or $dU = -P\,dV$. Substituting this in the general equation for equilibrium, we get $T\,dS + P\,dV + dW = 0$, but $dW = -P\,dV$. Then the criterion for equilibrium becomes $T\,dS = 0$.

(c) Constant S and H means that $dS = 0$ and $dH = dU + P\,dV + V\,dP = 0$, or $dU = -P\,dV - V\,dP$. Then the equation $T\,dS - dU + dW = 0$ becomes $T\,dS - dU - P\,dV = 0$ or $0 - (-P\,dV - V\,dP) - P\,dV = 0$ or $V\,dP = 0$ which is the required criterion.

(d) Constant S and V means that $dS = 0$ and $dV = 0$. Then the required criterion for equilibrium is $dU = 0$.

(e) Constant S and U gives $dS = 0$ and $dU = 0$, which reduces the general equation $T\,dS - dU - P\,dV = 0$ to $P\,dV = 0$.

10.18 Calculation of $\Delta G° = \sum_{products} \Delta G_f° - \sum_{reactants} \Delta G_f°$ shows that $\Delta G°$ has a negative value for the reaction $C_2H_5OH(\ell) + O_2(g) = CH_3COOH(\ell) + H_2O(\ell)$ and a positive value for the other reaction. Since at constant T and P the condition for spontaneity is $\Delta G < 0$, therefore the first reaction is spontaneous.

10.19 The Clausius–Clapeyron equation shows the change in temperature necessarily accompanying a change in pressure occurring in a system containing two phases of a pure substance in equilibrium (see the derivation of Equation 10.49).

The significance of this equation is that, if one solid form of a pure substance transforms to another solid form, the variation in the transition temperature T with pressure P is given by this equation.

10.20 The atomic weight of sodium is 23, and the melting temperature T is 370.95 K. Then the latent heat of fusion is $\Delta H = 2635.920/23 = 114.605\,J\,g^{-1}$. This conversion is necessary in order to be consistent with the units. Substituting the known values of the parameters in the Clausius–Clapeyron equation, i.e. in $dP/dT = \Delta H/(T\,\Delta V)$ or $dT/dP = T\,\Delta V/\Delta H$, we have

$$\frac{dT}{dP} = \frac{(370.95\,K)(27.90 \times 10^{-6}\,\ell\,g^{-1})}{114.605\,J\,g^{-1}} = 90.306 \times 10^{-6}\,K\,\ell\,J^{-1}.$$

These units $K\,\ell\,J^{-1}$ must now be converted to more conventional units. To do this, we have to multiply by a quantity in units of $J\,\ell^{-1}\,atm^{-1}$ to obtain dT/dP in units of $K\,atm^{-1}$, i.e $(K\,\ell\,J^{-1})(J\,\ell^{-1}\,atm^{-1}) = K\,atm^{-1}$. Since $1\,J = 9.869 \times 10^{-3}\,\ell\,atm$, we multiply dT/dP by this factor. So $dT/dP = (90.306 \times 10^{-6}\,K\,\ell\,J^{-1})(1/9.869 \times 10^{-3}\,J\,\ell^{-1}\,atm^{-1}) = 0.009\,15\,K\,atm^{-1}$.

This shows that, for an increase of pressure to 10 atm ($= 1\,MPa$), an increase of 0.09 K results in the melting temperature of metallic sodium. This is typical of the small effect of pressure on the melting temperature of pure metals.

10.21 For the reaction $\Delta G° = 18\,660 - 14.42T\,\log T - 6.07T + 8.24 \times 10^{-3}T^2$ we have $-\Delta S° = d(\Delta G°)/dT = (-14.42/2.303) - 14.42\,\log T - 6.07 + (2)(8.24 \times 10^{-3})T$. When $T = 25°C = 298\,K$, $-\Delta S° = d(\Delta G°)/dT = -43.098$. Therefore, $\Delta S° = 43.098\,J\,mol^{-1}$. Substituting this value of $\Delta S°$ into $\Delta G° = \Delta H° - T\Delta S°$ and taking $T = 298\,K$, we obtain $\Delta H° = 19.794\,kJ\,mol^{-1}$.

10.22 Since the volume of a liquid is very much smaller than the volume of its vapour, the volume change on vaporization of the liquid will be approximately equal to the volume of its vapour. Then, ΔV in the case of vaporization of liquid CCl_4 is approximately equal to

$V_{vapour} = RT/P$. Using the Clapeyron equation $dP/dT = \Delta H/(T \Delta V)$, we have

$$\frac{dP}{dT} = \frac{\Delta H}{TV_{vap}} = \frac{P \Delta H}{RT^2}$$

or

$$\frac{d(\ln P)}{dT} = \frac{\Delta H}{RT^2}$$

or

$$\frac{d(\log P)}{dT} = \frac{\Delta H}{(2.303)(8.314T^2)} = \frac{\Delta H}{19.147T^2}.$$

Since $\log P = -2400/T - 2.303 \log T + 23.6$, we have

$$\frac{d(\log P)}{dT} = \frac{2400}{T^2} - \frac{2.30}{2.303T} = \frac{2400}{T^2} - 1/T.$$

Therefore, $\Delta H = 19.147T^2 (2400T^{-2} - 1/T)$. Then at $77°C = 350.15 \, K$, $\Delta H = 39\ 248.145 \, J$ mol^{-1}.

It should be noted that, in the problem, P is not given any units. This is because the units are immaterial when $\log P$ is differentiated in order to substitute in the Clapeyron equation.

10.23 (a) For the first case $\Delta A = \Delta U - T \Delta S$ at constant T, and $\Delta S = -R \ln(P_2/P_1)$ where $P_1 = 0.5 \, MPa$ is the initial pressure and $P_2 = 0.1 \, MPa$ is the final pressure. Then $\Delta S = -8.314 \ln(\frac{1}{5}) = 13.4 \, J \, K^{-1} \, mol^{-1}$. Substituting this value of ΔS in $\Delta A = \Delta U - T \Delta S$, we get $\Delta A = 0 - (298)(13.4) = -4.0 \, kJ \, mol^{-1}$ (this is also $-W_{max}$ for reversible expansion).

(b) For the second case, both S and A are state functions. Therefore, ΔS and ΔA depend only on the initial and final states of the gas. T_1, P_1 and P_2 are the same for both the reversible and the irreversible isothermal expansion. So, ΔS and ΔA must also be the same, i.e. $\Delta S = 13.4 \, J$ $K^{-1} mol^{-1}$ and $\Delta A = -4.0 \, kJ \, mol^{-1}$ ($> W_{irreversible}$).

10.24 (a) All reactions for which ΔG is negative. Changes in the Gibbs free energy can be made to do work.

(b) All reactions for which ΔG is positive. However, in such cases the reverse reactions would be spontaneous.

(c) (i) Since the reaction is carried out under standard conditions, i.e. 0.1 MPa pressure and 298.15 K temperature, $\Delta G = W_{rev} = W$ (useful), i.e. the maximum useful work is $-262.0 \, kJ$.

(ii) If the reaction is carried out irreversibly, W_{irrev} can have any value in the range from 0 to $-262 \, kJ$.

(iii) $\Delta G = -260 \, kJ$ and $W_{irrev} = -182 \, kJ$. Therefore, the difference is 80 kJ. This 80 kJ has turned into heat which we may call Q_{irrev}.

10.25 (a) For an isothermal reversible expansion, $\Delta U = 0$ and $\Delta H = 0$. In addition $\Delta S = Q_{rev}/T$. To find Q, we need to know W so that we can use the first law. Now

$$W = -\int P\,dV = -nRT \ln\left(\frac{V_2}{V_1}\right)$$

but $V_2/V_1 = P_1/P_2$. Then $W = -RT \ln(P_1/P_2) = -(8.314)(273.15) \ln(10/1)\,\text{J mol}^{-1} = -5230.042\,\text{J mol}^{-1}$. Then, from $\Delta U = Q + W$, $Q = 5230.042\,\text{J mol}^{-1}$ and $\Delta S = Q/T = 5230.042/273.15 = 19.147\,\text{J K}^{-1}\text{mol}^{-1}$, $\Delta G = \Delta H - T\Delta S = 0 - 5230.042 = -5230.042\,\text{J mol}^{-1}$ and $\Delta A = \Delta U - T\Delta S = 0 - 5230.042 = -5230.042\,\text{J mol}^{-1}$.

(b) The values of ΔG and ΔA obtained above are for the system. For the surroundings, these quantities are $\Delta G = 5230.042\,\text{J mol}^{-1}$ and $\Delta A = 520.042\,\text{J mol}^{-1}$. Then, for the entire isolated system, $\Delta G = \Delta A = 0$.

(c) If the gas is allowed to expand freely so that no work is done by it, we have $\Delta U = \Delta H = W = Q = 0$. Hence, $\Delta S = \Delta G = \Delta A = 0$.

(d) For the entire isolated system, $\Delta G = 0$ and $\Delta A = 0$.

10.26 For the reaction $H_2O(g) + CO(g) = H_2(g) + CO_2(g)$, $(\Delta G^\circ)_{298\,K}$ is given by

$$(\Delta G^\circ)_{298\,K} = \sum_{products} G_f^\circ - \sum_{reactants} G_f^\circ = (G_f^\circ)_{H_2(g)}$$
$$+ (G_f^\circ)_{CO_2(g)} - (G_f^\circ)_{H_2O(g)} - (G_f^\circ)_{CO(g)}.$$

Now using the values of G_f° of the products and of the reactants from Table 10.1, we have

$$(\Delta G^\circ)_{298\,K} = 0 - 394.384 + 228.614 + 137.277 = -28.493\,\text{kJ}.$$

10.27

$$-\Delta S^\circ = \frac{d(\Delta G^\circ)}{dT} = \frac{d}{dT}(7700 + 4.2T \log T - 27.80T)$$
$$= 4.2\log T + \frac{4.2}{2.303} - 27.80$$

or

$$\Delta S^\circ = 25.976 - 4.2 \log T \text{ eU}.$$

10.28 (a) The total energy available from the given reaction to do mechanical work is the change ΔG in the Gibbs free energy.

(b) $\Delta U = 0$ for any isochoric adiabatic process, $\Delta S = 0$ for any reversible adiabatic process and $\Delta G = 0$ for any reversible constant-T and constant-P process.

(c) The reduction of CdS(g) with hydrogen is given by the reaction

$$CdS(g) + H_2(g) = Cd(g) + H_2S(g).$$

Now

$$H_2(g) + \frac{1}{2}S_2(g) = H_2S(g)$$

$$\Delta G^\circ = (\Delta G^\circ_f)_{H_2S(g)} = -48.953 \text{ kJ mol}^{-1}$$

$$Cd(g) + \frac{1}{2}S_2(g) = CdS(g)$$

$$\Delta G^\circ = (\Delta G^\circ_f)_{CdS(g)} = -127.194 \text{ kJ mol}^{-1}.$$

For the reaction

$$\Delta G = G^\circ_{products} - G^\circ_{reactants} = \sum_{products} \Delta G^\circ_f - \sum_{reactants} \Delta G^\circ_f$$

$$= [(\Delta G^\circ_f)_{Cd} + (\Delta G^\circ_f)_{H_2S}] - [(\Delta G^\circ_f)_{CdS} - (\Delta G^\circ_f)_{H_2}]$$

$$= (0 - 48.953) - (-127.194 + 0) = 78.241 \text{ kJ mol}^{-1}.$$

This gives a positive ΔG at 1100°C for the reaction. Therefore, the reaction is not thermodynamically feasible.

10.29 For the conversion reaction,

$$\Delta S^\circ = S^\circ_{H_2} + S^\circ_{1,3-butadiene} - S^\circ_{1-butene} = 130.583 + 278.738 - 307.106$$

$$= 102.215 \text{ J mol}^{-1}$$

$$\Delta H^\circ = (\Delta H^\circ)_{H_2} + (\Delta H^\circ)_{1,3-butadiene} - (\Delta H^\circ)_{1-butene}$$

$$= 0 + 111\,922.0 - 1171.520 = 110.750 \text{ kJ}$$

$$\Delta G^\circ = \Delta H^\circ - T\Delta S^\circ = 110\,750.0 - (298)(102.215) = 80.289 \text{ kJ}.$$

10.30 A positive value of ΔS° of a reaction favours products. A positive value of ΔH° means that reactants are energetically more stable, and so positive ΔH° favours reactants. If ΔG° of a reaction is positive, the reaction is not favoured.

10.31 To calculate the free-energy change ΔG for the irreversible process, $H_2O(1100°C, 700 \text{ Torr}) = H_2O(100°C, 700 \text{ Torr})$, we have to break this process into a sequence of equivalent reversible processes:
the isothermal compression of liquid water

$$H_2O(\ell)(1100°C, 700 \text{ Torr}) = H_2O(1100°C, 760 \text{ Torr}) \qquad \text{(i)}$$

the isobaric vaporization of water

$$H_2O(\ell)(1100°C, 760 \text{ Torr}) = H_2O(g)(100°C, 760 \text{ Torr}) \qquad \text{(ii)}$$

and the isothermal expansion of water vapour

$$H_2O(g)(100°C, 760 \text{ Torr}) = H_2O(g)(100°C, 700 \text{ Torr}) \qquad \text{(iii)}$$

The sum of (i)–(iii) gives the given irreversible process

$$H_2O(1100°C, 700 \text{ Torr}) = H_2O(100°C, 700 \text{ Torr}).$$

Therefore, $\Delta G = (\Delta G)_{(i)} + (\Delta G)_{(ii)} + (\Delta G)_{(iii)}$ because ΔG is a state function. As $(\partial G/\partial P)_T = V$, we have

$$(\Delta G)_{(i)} = n \int_{P_1}^{P_2} \bar{V} \, dP$$

where $\bar{V}$ is the molar volume of water. Assuming that the molar volume is independent of P,

$$(\Delta G)_{(i)} = n\bar{V}(P_2 - P_1).$$

The molar volume of water is $0.018 \, \ell$ and $n = 1$. Therefore,

$$(\Delta G)_{(i)} = (0.018)(\ell)\left(1 - \frac{700}{760}\right) = 1.422 \times 10^{-3} \, \ell \, \text{atm}$$

$$\Delta G_{(ii)} = 0$$

as the process (ii) is a free expansion.

$$(\Delta G)_{(iii)} = \int_{P_1}^{P_2} V \, dP = nRT \int_{P_1}^{P_2} \frac{dP}{P}$$

$$= nRT \ln\left(\frac{P_2}{P_1}\right) = (0.082)(373)(2.303) \log\left(\frac{700}{760}\right)$$

$$= -2.516 \, \ell \, \text{atm}$$

Then, $\Delta G = 1.422 \times 10^{-3} + 0 - 2.516 = -2.515 \, \ell \, \text{atm} = -254.429 \, \text{J}$.

10.32 The reaction is $CO(g) + 2H_2(g) = CH_3OH(g)$. The standard enthalpy $(\Delta H°)_{298 \, K}$ for this reaction is

$$(\Delta H°)_{298 \, K} = H°_{\text{products}} - H°_{\text{reactants}} = \sum_{\text{products}} \Delta H°_f - \sum_{\text{reactants}} \Delta H°_f$$

$$= [(\Delta H°_f)_{CH_3OH(\ell)} - (\Delta H°)_{\text{vap}}] - [(\Delta H°_f)_{CO(g)} + 2(\Delta H°_f)_{H_2(g)}].$$

Taking the $\Delta H°_f$ values for the respective substances from Appendix 5 we have

$$(\Delta H°)_{298 \, K} = (-238.572 + 37.405) - (-110.520 + 0) = -90.647 \, \text{kJ} \, \text{mol}^{-1}.$$

From the Kirchhoff equation,

$$\frac{\partial(\Delta H)}{\partial T} = \Delta C_P$$

or

$$(\Delta H°)_T = \Delta H° + \int_0^T \Delta C_P \, dT$$

where

$$\Delta C_P = \sum_{\text{products}} C_P - \sum_{\text{reactants}} C_P = (C_p)_{\text{CH}_3\text{OH}} - [(C_P)_{\text{CO}} + 2(C_P)_{\text{H}_2}]$$

$$= (18.401 + 101.546 \times 10^{-3}T - 286.813 \times 10^{-7}T^2)$$

$$- [26.861 + 6.966 \times 10^{-3}T - 8.201 \times 10^{-7}T^2$$

$$+ 2(29.041 - 0.837 \times 10^{-3}T + 20.117 \times 10^{-7}T^2)]$$

$$= -66.547 + 92.927 \times 10^{-3}T - 318.846 \times 10^{-7}T^2.$$

Therefore,

$$(\Delta H^\circ)_T = (\Delta H^\circ)_{0\,\text{K}} + \int_0^T (-66.547 + 92.927 \times 10^{-3}T - 318.846 \times 10^{-7}T^2)\mathrm{d}T$$

$$= (\Delta H^\circ)_{0\,\text{K}} - 66.547T + 46.464 \times 10^{-3}T^2 - 106.282 \times 10^{-7}T^3.$$

However, we have calculated that $(\Delta H^\circ)_{298\,\text{K}} = -90.647\,\text{kJ mol}^{-1}$. Therefore,

$$(\Delta H^\circ)_{0\,\text{K}} = -90\,647 + (66.547)(298) - (46.464 \times 10^{-3})(298)^2$$

$$+ (106.282 \times 10^{-7})(298)^3 = -74.789\,\text{kJ mol}^{-1}.$$

Hence,

$$(\Delta H^\circ)_T = -74\,789 - 66.547T + 46.464 \times 10^{-3}T^2 - 106.282 \times 10^{-7}T^3.$$

At constant T, $\Delta G^\circ = \Delta H^\circ - T\,\Delta S^\circ$ or $\Delta G^\circ/T = \Delta H^\circ/T - \Delta S^\circ$. Then the Gibbs–Helmholtz equation becomes

$$\frac{\mathrm{d}(\Delta G^\circ/T)}{\mathrm{d}T} = -\frac{\Delta H^\circ}{T^2}$$

or

$$\frac{\Delta G^\circ}{T} = -\int \frac{\Delta H^\circ}{T^2}\,\mathrm{d}T + C$$

where C is an integration constant. Hence,

$$\frac{\Delta G^\circ}{T} = -\int(-74\,789 - 66.547T + 46.464 \times 10^{-3}T^2 - 106.282 \times 10^{-7}T^3)\frac{\mathrm{d}T}{T^2} + C$$

$$= -\frac{74\,789}{T} + (66.547)(2.303)\log T - 46.464 \times 10^{-3}T + 53.141 \times 10^{-7}T^2 + C$$

or

$$(\Delta G^\circ)_T = -74\,789 + 153.258\log T - 46.464 \times 10^{-3}T^2 + 53.141 \times 10^{-7}T^3 + CT.$$

Now

$$(\Delta G^\circ)_{298\,\text{K}} = G^\circ_{\text{products}} - G^\circ_{\text{reactants}} = \sum_{\text{products}} \Delta G^\circ_{\text{f}} - \sum_{\text{reactants}} \Delta G^\circ_{\text{f}}$$

$$= (\Delta G^\circ_{\text{f}})_{\text{CH}_3\text{OH}} - (\Delta G^\circ_{\text{f}})_{\text{CO}} - 2(\Delta G^\circ_{\text{f}})_{\text{H}_2}.$$

Taking the values of ΔG_f° for the respective substances from Table 10.1, we have $(\Delta G^\circ)_{298\,K} = -166.230-(-137.277)-0 = -28.953\,kJ\,mol^{-1}$. Therefore,

$$-28\,953 = -74\,789 + 153.258 \log T - 46.464 \times 10^{-3}T^2$$
$$+ 53.141 \times 10^{-7}T^3 + CT$$
$$= -74\,789 + 153.258 \log (298) - (46.464 \times 10^{-3})(298)^2$$
$$+ (53.141 \times 10^{-7})(298)^3 + 298C$$

or

$$C = 165.914.$$

Hence,

$$(\Delta G^\circ)_T = -74\,789 + 153.258T - 46.464 \times 10^{-3}T^2 + 53.141 \times 10^{-7}T^3 - 165.914.$$

10.33 Follow the same procedure as in problem 10.31 above. $\Delta G = 4560\,J$.

10.34 See the derivation of Equation 10.36 (or Equation 10.60) for the Gibbs–Helmholtz equation. See the derivation of Equation 10.74 to prove that

$$\Delta G^\circ = \Delta H^\circ - aT \ln T - bT^2/2 - cT^3/6 - \cdots + AT.$$

10.35 See the derivation of Equation 10.78.

10.36 $G = G^\circ + RT \ln P$ is strictly applicable to ideal gases, since the ideal-gas equation $PV = nRT$ is used to derive this equation. However, if the details of non-ideal behaviour are ignored, it can be used and assumed to apply approximately for all gases.

10.37 Refer to Section 10.14.

10.38 $PV = RT + BP + CP^2$. Rearranging, we obtain $V - RT/P = B + CP$. Substituting this value of $V - RT/P$ into Equation 10.94, i.e. $RT \ln \gamma = \int_0^P (V - RT/P)dP$ and integrating, we have $RT \ln \gamma = BP + CP^2/2$ which is the required equation of $\ln \gamma$ as a function of P.

If $T = 223.2\,K$, $B = -3.69 \times 10^{-2}$, $C = 1.79 \times 10^{-4}$ and $P = 100\,atm$, then from the above equation, i.e. $RT \ln \gamma = BP + CP^2/2$, we get

$$(0.082)(223.2) \ln \gamma = (-3.69 \times 10^{-2})(100) + \frac{1}{2}(1.79 \times 10^{-4})(100)^2$$

or

$$18.302 \ln \gamma = -2.795$$

or

$$\gamma = 0.858.$$

10.39 At normal temperature and pressure an ideal gas has a volume of $22.4\,\ell$. The P–V relationship of oxygen is given by $PV = 298.15R - 0.0211P$ or $P(V + 0.0211) = (298.15)(0.082)$, or $P(22.4 + 0.0211) = 24.448$ or $P = 1.090\,atm$. Since for ideal behaviour $f/P = 1$, so we have $f = 1.090\,atm$.

10.40 See the derivations of Equation 10.91 for the first part of the problem and of Equation 10.94 for the second part.

10.41 See the derivation of Equation 10.96. It should be noted that, although Q in this equation has the same form as the general mass action law equilibrium constant, the two are not equivalent. However, they are closely related.

10.42 The reaction is $2Ag_2S(s) + 2H_2O(\ell) \rightarrow 4Ag(s) + 2H_2S(g) + O_2(g)$. Then,

$$\Delta G° = \sum_{products} G_f° - \sum_{reactants} G_f° = 4(G_f°)_{Ag(s)} + 2(G_f°)_{H_2S(g)} + (G_f°)_{O_2(g)} - 2(G_f°)_{Ag_2S(s)} - 2(G_f°)_{H_2O(\ell)}$$

$$= (4)(0) + (2)(-33.59) + (2)(0) - (2)(-40.7) - (2)(-237.19) = 488.60 kJ.$$

This gives a positive value of $\Delta G°$, which indicates that at 25°C and 0.1 MPa the forward reaction should not occur spontaneously but the reverse reaction, for which $\Delta G° = -488.60 kJ$, should be spontaneous.

10.43 The free-energy change for the reaction is given by $\Delta G° = \Delta H° - T\Delta S°$.

$$\Delta H° = [(\Delta H)_f]_{CaSO_4(s)} - [(\Delta H)_f]_{CaO(s)} - [(\Delta H)_f]_{SO_3(g)}.$$

Using the ΔH_f-values for the respective compounds from Appendix 5 we have

$$\Delta H° = -1432.7 - (-635.131) - (-395.179) = -402.39 kJ$$
$$\Delta S° = S_{CaSO_4(s)}° - S_{CaO(s)}° - S_{SO_3(g)}°.$$

Again, using the $S°$-values for the respective compounds from Table 8.1, we get $\Delta S° = 106.0 - 39.748 - 256.228 = -189.976 J K^{-1}$. Then, $\Delta G° = \Delta H° - T\Delta S° = -402\ 390 J - (298 K)(-189.976 J K^{-1})$ or $\Delta G° = -345.777 kJ$.

10.44 To calculate ΔG under non-standard conditions at 298 K, we use the equation $\Delta G = \Delta G° + RT \ln Q$ where $Q = [CaSO_4]/[CaO][SO_3]$. As CaO and $CaSO_4$ are solids, their concentrations (or activities) are taken as 1. On the other hand, as SO_3 is a gas, its concentration (or activity) is taken as its pressure. For easy calculation, we use the pressure in atmospheres. Then,

$$Q = \frac{1}{(1)(0.2)} = 5.$$

In problem 10.43, we found that $\Delta G° = -345.777 kJ$. Using this value, we get $\Delta G = \Delta G° + RT \ln Q = -345.777 + (8.314)(298) \ln(5) \times 10^{-3} = 341.790 kJ$.

10.45 The reaction is $H_2O(\ell) \rightarrow H_2O(g)$. As the liquid and the vapour states are in equilibrium, the free-energy change $\Delta G°$ in going from the liquid to the vapour is zero. Also, we assume that $\Delta H°$ and $\Delta S°$ are independent of temperature. Now, $\Delta G° = \Delta H° - T\Delta S°$ or $0 = \Delta H° - T\Delta S°$ or $T = \Delta H°/\Delta S°$. First of all we shall have to determine $\Delta H°$ and $\Delta S°$. $\Delta H°$ is given by

$$\Delta H° = [(\Delta H)_f]_{H_2O(g)} - [(\Delta H)_f]_{H_2O(\ell)} = -241.8 - (-285.8) = 44.0 kJ.$$

In problem 9.17 we have calculated $\Delta S°$ for this problem, which is $118.79 J K^{-1}$. Then, $T = \Delta H°/\Delta S° = (44\ 000 J)/118.79 J K^{-1} = 370 K$.

One can easily see a discrepancy in the final answer. The correct answer should have been 373 K (or 100°C), which is the boiling point of water at 0.1 MPa pressure. The discrepancy is due to the assumption that $\Delta H°$ and $\Delta S°$ are independent of temperature. For exact calculations, $\Delta H°$ and $\Delta S°$ values at the temperature in question must be used.

10.46   (a) Firstly, we use equation $\Delta G° = \Delta H° - T\Delta S°$, or simply $\Delta G° = \sum_{products} G_f° - \sum_{reactants} G_f° = (G_f°)_{Cu(s)} + (G_f°)_{H_2S(g)} - (G_f°)_{CuS(s)} - (G_f°)_{H_2(g)}$. Using the $G_f°$-values from Table 10.1. $\Delta G° = 0 + (-33.012) - (-53.6) - 0 = 20.588$ kJ. (The positive $\Delta G°$ means that the reaction is not spontaneous.)

(b) To apply $\Delta G° = \Delta H° - T\Delta S°$ we need to know $\Delta H°$ and $\Delta S°$. Now,

$$\Delta H° = [(\Delta H)_f]_{Cu(s)} + [(\Delta H)_f]_{H_2S(g)} - [(\Delta H)_f]_{CuS(s)} - [(\Delta H)_f]_{H_2(g)}.$$

Using the $(\Delta H)_f$ values from Appendix 5

$$\Delta H° = 0 + (-20.146) - (-53.1) = 32.954 \text{ kJ}.$$

(The positive $\Delta H°$ means the reaction is endothermic.) To calculate $\Delta S°$ at 298.15 K we use $\Delta G° = \Delta H° - T\Delta S°$ or $\Delta S° = (\Delta H° - \Delta G°)/T$ or $\Delta S° = (32\,954 - 20\,588)/298.15 = 41.476$ J K^{-1}.

Now, knowing $\Delta H°$ and $\Delta S°$ at 298.15 K, we can calculate $\Delta G°$ at 800 K using $\Delta G° = \Delta H° - T\Delta S°$, i.e. $\Delta G° = 32\,954 - (800\,\text{K})(41.476\,\text{J K}^{-1}) = 226.80$ J.

(c) Finally, to calculate the temperature at which $\Delta G° = 0$, we again use $\Delta G° = \Delta H° - T\Delta S°$. This gives $T = \Delta H°/\Delta S°$. Using the values of $\Delta S°$ and $\Delta H°$ from (b) above, $T = (32\,954\,\text{J})/(41.476\,\text{J K}^{-1}) = 794.53$ K.

EXERCISES 11

11.1   For properties of equilibrium constant, see Section 11.2. See the derivation of Equations 11.7 and 11.8 for the mathematical relationships between K_C and K_P.

In a reaction, if the number of molecules of the reactants is equal to that of the products, $\sum n$ or Δn is necessarily equal to zero. Then from Equations 11.7 and 11.8 we get $K_C = K_P$. This implies that there is no volume change on reaction. If there is a volume change on reaction, $K_C \neq K_P$. A volume change could be either an increase or decrease in volume. For an increase in volume on reaction, $\sum n$ or Δn is positive, and from Equations 11.7 and 11.8 we get $K_P > K_C$. On the other hand, if there is a decrease in volume, $\sum n$ or Δn is negative and numerically $K_P < K_C$.

11.2   For the given reaction, the equilibrium constant K is given by $\Delta G° = -RT \ln K$. From Table 10.1 the value of $(\Delta G_f°)_{298.15\,K}$ for $CO_2(g)$ is -394.384 kJ mol^{-1}. Therefore, substituting this value in the above equation, we obtain

$$-394\,384 \text{ J} = (-8.314\,\text{J K}^{-1})(298.15\,\text{K}) \ln K$$

or

$$\log K = \frac{-394\,384\,\text{J}}{(-2.303)(8.314\,\text{J K}^{-1})(298.15\,\text{K})} = 69.085$$

or

$$K = 1.216 \times 10^{69}.$$

11.3   To solve this problem we use Equation 11.19c. To use this equation we need to know $K_{298\,K}$, $\Delta G°$ for 298 K and $(\Delta H)_f$ for 298 K for the formation of benzene. We know

$$\Delta G° = (G_f°)_{C_6H_6(\ell)} - 3(G_f°)_{C_2H_2(g)}.$$

Using the respective values of $G_f°$ given in Table 10.1, we find that

$$\Delta G° = 124\ 516 - (3)(209\ 200) = -503.084\ J$$

which is the value of $\Delta G°$ for 298 K. Substituting this value of $\Delta G°$ in $\Delta G° = -RT \ln K_{298\,K}$, where $K_{298\,K}$ is the equilibrium constant for $T = 298$ K we get

$$-503\ 084\ J = -(8.314)(298)\ln K_{298\ K}$$

or

$$\ln K_{298\ K} = \frac{503\ 084\ J}{(8.314\ J\ K^{-1})(298\ K)} = 203.055.$$

We now have to determine $\Delta H°$ for the reaction at 298 K:

$$\Delta H° = (\Delta H_f°)_{C_6H_6(\ell)} - 3(\Delta H_f°)_{C_2H_2(g)}.$$

Using the respective $\Delta H_f°$ values given in Appendix 5,

$$\Delta H° = 49\ 036\ J - (3)(226\ 731\ J) = -631\ 157.0\ J.$$

Now, substituting the values of $\Delta H°$ and $\ln K_{298}$ into Equation 11.19c, we get

$$\ln\left(\frac{K_{773\ K}}{K_{298\ K}}\right) = -\frac{(631\ 157.0)(773 - 298)}{(8.314)(298)(773)}$$

or

$$\ln K_{773\ K} = -156.560 + 203.055 = 46.495.$$

Therefore, $K_{773\,K} = 1.542 \times 10^{20}$ which is the required equilibrium constant at 500°C for the reaction. In this problem it has been assumed that $\Delta H°$ and $\Delta S°$ are independent of temperature.

11.4   See the derivation of Equation 11.9 for the relationship between the standard free-energy change and the activities of the substances in a reaction. See the derivation of Equation 11.11 for $\Delta G° = -RT \ln K$.

11.5   See the derivation of Equations 11.13–11.19. Each of the equations 11.14–11.19 uses its appropriate units and gives the same value for ΔG. Thus, ΔG is determined by the ratio of Q to K, but $\Delta G°$ is related only to the equilibrium reaction quotient.

11.6   The sign of ΔG at a particular temperature depends on the relative magnitudes of Q and K. Three cases may arise, as seen from Equation 11.13. If $Q = K$, $\Delta G = 0$, i.e. no reaction. If $Q < K$, ΔG is negative, i.e. the process will occur spontaneously from left to right and, if $Q > K$, ΔG is positive, i.e. the reverse reaction is thermodynamically spontaneous, with the consequence that there will be conversion of some products with an associated reduction in the value of Q.

The physical significance of negative ΔG is that the relative concentrations of product species present are smaller than those at equilibrium. As the reaction proceeds, large amounts of products are formed and the value of Q increases until $K = Q$ and equilibrium is established.

Although a negative value of ΔG makes a reaction theoretically possible, it has no relationship to the speed of the reaction. A reaction may only become spontaneous and proceed at an observable speed when the pressure is favourable, or perhaps in the presence of a specific catalyst.

11.7   From Equation 11.11, $\Delta G° = -RT \ln K$, or $\Delta G° = -2.303\ RT \log K$. $(\log K)_T = 277.4/T + 0.3811\ J$. Rearranging this equation, $T(\log K)_T = 277.4 + 0.3811T$. Substituting this in $\Delta G° = -2.303RT \log K$ and taking $T = 298.15\ K$, we get

$$\Delta G° = -(2.303)(8.314)[277.4 + (0.3811)(298.15)] = -7487.010\ J.$$

11.8 $K = 8.28$ for the reaction $CO_2(g) + H_2(g) = CO(g) + H_2O(g)$.
Substituting this value of K in $\Delta G° = -RT \ln K$ we get

$$\Delta G° = -(2.303)(8.314)(3500)\log(8.28) = -61.520\ kJ.$$

Consider

$$\Delta G = -\Delta G° + RT\ln\left(\frac{P_{CO}P_{H_2O}}{P_{CO_2}P_{H_2}}\right)$$

where P_{CO}, P_{H_2O}, P_{CO_2} and P_{H_2} are the pressures of CO, H_2O, CO_2 and H_2, respectively. Now, substituting $T = 3227 + 273 = 3500\ K$, $R = 8.314\ J\ mol^{-1}$ and $\Delta G° = -61\ 520\ J$ in this equation, we obtain

$$\Delta G = -61\ 520 + (8.314)(3500)\ \ln\left(\frac{(2)(2)}{(0.1)(0.1)}\right) = 112.853\ kJ.$$

11.9   (a) Firstly, at 1000 K, the percentage decomposition is 2.0×10^{-5} for CO. So, the partial pressure is $(2.0 \times 10^{-5})/100 = 2.0 \times 10^{-7}$. Since the partial pressure is proportional to the molar concentration and as 3 mol of the product are involved, the partial pressure of O_2 is 1.0×10^{-7}. At 1000 K, we take the partial pressure of CO_2 as 1 because only a very small amount decomposes.

At 1400 K, the partial pressure of CO is $(1.27 \times 10^{-2})/100 = 1.27 \times 10^{-4}$. So, the partial pressure of O_2 is half of that of CO, i.e. 0.635×10^{-4}. As in the case of 1000 K, here also, only a very small amount of CO_2 decomposes. So, we take 1 as the partial pressure of CO_2.
Then,

$$K_{1000 \text{ K}} = \frac{[CO]^2[O_2]}{[CO_2]^2} = \frac{(4.0 \times 10^{-14})(1.0 \times 10^{-7})}{(1)} = 4.0 \times 10^{-21}$$

$$K_{1400 \text{ K}} = \frac{(1.27 \times 10^{-4})^2(0.635 \times 10^{-4})}{(1)} = 1.024 \times 10^{-12}.$$

(b) Secondly,

$$(\Delta G°)_{1000 \text{ K}} = -RT \ln K_{1000 \text{ K}}$$

or

$$(\Delta G°)_{1000 \text{ K}} = -(8.314)(1000)(2.303)\log(4.0 \times 10^{-21}) = 390.563 \text{ kJ mol}^{-1}.$$

(c) Thirdly, to calculate $\Delta S°$ at 727°C, we need to know $\Delta H°$ at this temperature. If we assume that $\Delta H°$ is independent of temperature, we can use Equation 11.19c, which gives

$$\ln \left(\frac{K_{1400 \text{ K}}}{K_{1000 \text{ K}}} \right) = \frac{\Delta H°}{R} \left(\frac{1}{T_1} - \frac{1}{T_2} \right)$$

or

$$\log K_{1400 \text{ K}} - \log K_{1000 \text{ K}} = \frac{\Delta H°}{(8.314)(2.303)} \frac{400}{(1000)(1400)}$$

or

$$-11.99 + 20.40 = \frac{4\Delta H°}{268.060 \times 10^3}$$

or

$$\Delta H° = 563.596 \times 10^3 \text{ J}.$$

Hence,

$$(\Delta G°)_{1000 \text{ K}} = (\Delta H°)_{1000 \text{ K}} - T(\Delta S°)_{1000 \text{ K}}$$

or

$$(\Delta S°)_{1000 \text{ K}} = \frac{(\Delta H°)_{1000 \text{ K}} - (\Delta G°)_{1000 \text{ K}}}{1000}$$
$$= \frac{(563\ 596 - 390\ 563)}{1000} = 173.033 \text{ JK}^{-1} \text{ mol}^{-1}.$$

11.10   Substituting $K = 0.0198$ and $T = 444 + 273 = 717\,\text{K}$ into the equation $\Delta G^\circ = -RT \ln K$, we find that $\Delta G^\circ = -(8.314)(717)(2.303)\log(0.0198) = 23.380\,\text{kJ}$. This is the free energy of formation of 2 mol of HI at 444°C. Therefore, ΔG° for 1 mol is $\frac{1}{2}(23.380\,\text{kJ}) = 11.900\,\text{kJ}$.

11.11   See the derivation of Equation 11.20. The van't Hoff equation can be expressed as

$$\frac{d(\log K_P)}{d(1/T)} = -\frac{\Delta H^\circ}{2.303\ R}.$$

Assuming that ΔH° is essentially independent of temperature, this equation may be integrated to given

$$\log K_P = -\frac{\Delta H^\circ}{2.303\ R}\frac{1}{T} + C$$

where C is an integration constant. This equation shows that a plot of $\log K_P$ versus $1/T$ should be a straight line, i.e. K_P is directly proportional to $1/T$.

It must be noted that for exact integration of Equation 11.20, ΔH° must be known as a function of temperature.

11.12   Combining $\Delta G^\circ = -RT \ln K$ and $\Delta G^\circ = \Delta H^\circ - T\Delta S^\circ$, we get

$$\ln K = -\frac{\Delta H^\circ}{RT} + \frac{\Delta S^\circ}{R}.$$

This equation shows that, if ΔH° and ΔS° are constants independent of temperature, then $\ln K$ is a simple linear function of $1/T$.

The entropy change of a reaction at any temperature is given by the equation

$$(\Delta S^\circ)_T = (\Delta S^\circ)_{298\ \text{K}} + \int_{298\ \text{K}}^{T} \Delta C_P \frac{dT}{T}$$

where ΔC_P is the difference in the heat capacities of reactants and products. This equation shows that if, $\Delta C_P = 0$, ΔS° is independent of temperature, and also $\ln K$ is a linear function of $1/T$.

11.13   See the derivation of Equation 11.28 for the relationship between K_x and K_P. This equation shows that, unless $\Delta n = 0$, K_x is a function of pressure because, for ideal gases, K_P is independent of pressure. Taking logarithms of both sides of this equation, we get

$$\ln K_x = \ln K_P + \Delta n \ln P.$$

Differentiating both sides with respect to P we have

$$\frac{d}{dP}(\ln K_x) = \frac{\Delta n}{P} = \frac{\Delta V}{RT}.$$

Because K_P is independent of P, the differentiation of $\ln K_P$ with respect to P is zero. At constant pressure $P\Delta V = \Delta n \, RT$.

It should be noted that, if $\Delta n = 0$, i.e. if there is no change in the number of moles in a reaction, the equilibrium constant K_P is the same as K_x or K_C, and for ideal gases the position of equilibrium does not depend on the total pressure.

11.14 According to Equation 11.8 we have $K_C = K_P(RT)^{\Delta n}$ where R is the universal gas constant, T is the temperature of reaction and Δn is the difference between the sum of the number of molecules of the products and that of the number of molecules of the reactants. For the given reaction, $\Delta n = 2 - 1 = 1$, $T = 300\,\text{K}$, and by taking $R = 8.314\,\text{J}\,\text{K}^{-1}\,\text{mol}^{-1}$ we get $K_C/K_P = [(8.314)(300)]^1 = 2494.2$.

11.15 At equilibrium, we have $\Delta G^\circ = -RT \ln K$, where K is the equilibrium constant. Then, $27.0\,\text{kJ} = -(8.314\,\text{JK}^{-1})(298.15\,\text{K}) \ln \text{K}$. This gives $K = 1.8 \times 10^{-5}$.

Since the given value of ΔG° is positive, we assume that the reaction is not spontaneous.

11.16 The value of $(\Delta G_f^\circ)_{298.15\,\text{K}}$ for $H_2O(g)$ is $-228\,614\,\text{J}\,\text{mol}^{-1}$ (see Table 10.1). Then from $\Delta G^\circ = -RT \ln K$ we get

$$-228\,614 \text{ J} = -(8.314 \text{ JK}^{-1})(298.15 \text{ K})\ln K$$

or

$$K = 1.17 \times 10^{40}.$$

11.17 (a) In problem 10.46 we have obtained $\Delta G^\circ = 20.588\,\text{kJ}$ for the given reaction. Substituting this value into $\Delta G^\circ = -RT \ln K$, we get

$$\ln K = \frac{\Delta G^\circ}{-RT} = \frac{-20\,588\,\text{J}}{(8.314\,\text{J}\,\text{K}^{-1})(298.15\,\text{K})}$$

or

$$K = 3.13 \times 10^{-4} \text{ at } 298.15 \text{ K}.$$

This value of the equilibrium constant is less than unity, which indicates that the quantities of the reactants in the reaction are more than those of the products. This, in turn, means that the equilibrium lies to the left.

(b) If the equilibrium constant K_{T_1} and $(\Delta H^\circ)_{T_1}$ are known at a temperature T_1, the equilibrium constant K_{T_2} at another temperature T_2 can be obtained using the equation

$$\ln\left(\frac{K_{T_2}}{K_{T_1}}\right) = \frac{(\Delta H^\circ)_{T_1}(T_2 - T_1)}{RT_1T_2}.$$

For the given reaction $K_{298\,\text{K}} = 3.13 \times 10^{-4}$ (see (a)) and $(\Delta H^\circ)_{298\,\text{K}} = 32.954\,\text{kJ}$ (see problem 10.46(a), $T_1 = 298\,\text{K}$, $T_2 = 798\,\text{K}$ and $R = 8.314\,\text{J}\,\text{K}^{-1}\,\text{mol}^{-1}$. Then,

$$\ln\left(\frac{K_{798\,\text{K}}}{K_{298\,\text{K}}}\right) = \frac{(32\,954\,\text{J})(798 - 298\,\text{K})}{(8.314\,\text{J}\,\text{K}^{-1})(798\,\text{K})(298\,\text{K})}$$

or

$$K_{798\,\mathrm{K}} = 4.163 \times 10^3.$$

11.18   For a complete discussion of the various types of partial molar quantity, see Sections 11.8.1–11.8.3.

Equations 11.35 and 11.36 show that partial molar volumes of the constituents of a mixture can be obtained from the average molar volume.

11.19   The free energy of mixing of two ideal gases is given by

$$(\Delta G)_{\mathrm{mix}} = RT(Nx_1 \ln x_1 + Nx_2 \ln x_2)$$

(see Equation 11.63) where N is the total number of moles of all the gases in the mixture. This equation can be applied to a mixture of any number of ideal gases. Therefore,

$$(\Delta G)_{\mathrm{mix}} = NRT \sum_i (x_i \ln x_i).$$

Since each of the terms $x_i \ln x_i$ on the right-hand side is negative (because x_i is a fraction and therefore $\ln x_i$ is negative), the sum of all such terms is negative. This makes $(\Delta G)_{\mathrm{mix}}$ negative.

The entropy of mixing of ideal gases is given by $(\Delta S)_{\mathrm{mix}} = -NR \sum_i (x_i \ln x_i)$. Then we can write

$$(\Delta G)_{\mathrm{mix}} = (\Delta H)_{\mathrm{mix}} - T(\Delta S)_{\mathrm{mix}}$$

or

$$(\Delta H)_{\mathrm{mix}} = (\Delta G)_{\mathrm{mix}} + T(\Delta S)_{\mathrm{mix}} = NRT \sum_i (x_i \ln x_i) - NRT \sum_i (x_i \ln x_i) = 0.$$

This shows that there is no change in the enthalpy associated with the formation of an ideal mixture.

11.20   The volume change for an ideal mixing is given by

$$(\Delta V)_{\mathrm{mix}} = \frac{\partial}{\partial P} [(\Delta G)_{\mathrm{mix}}]_{T,n_i}$$

but $(\Delta G)_{\mathrm{mix}}$ is independent of pressure. Therefore,

$$\frac{\partial}{\partial P} [(\Delta G)_{\mathrm{mix}}]_{T,n_i} = 0$$

i.e.

$$(\Delta V)_{\mathrm{mix}} = 0.$$

We conclude that there is no change in volume for an ideal mixing.

See the derivation of Equations 11.70 and 11.71 for the variation in the chemical potentials of the various species present in a phase in internal equilibrium when the composition of the phase is varied at constant temperature and pressure.

11.21 For a mixture of two components, the Gibbs–Duhem equation is written as (see Equation 11.70)

$$n_1 \, \mathrm{d}\mu_1 + n_2 \, \mathrm{d}\mu_2 = 0 \quad (T \text{ and } P \text{ constant})$$

or

$$\mathrm{d}\mu_2 = -\frac{n_1}{n_2} \mathrm{d}\mu_1.$$

This equation shows that, if $\mathrm{d}\mu_2$ is positive, i.e. μ_2 increases, $\mathrm{d}\mu_1$ must be negative and μ_1 must decrease at the same time, and vice versa.

11.22 See the derivations of Equations 11.76 and 11.77. In order to derive these equations in the case of gaseous mixtures, the assumption involved is that the gases behave ideally. However, no assumption as to the ideality is required to apply these equations to a liquid mixture.

11.23 See the derivation of Equation 11.81. Indeed, it is possible to derive the same equation for a reaction at constant temperature and volume by using the condition of equilibrium, $\mathrm{d}A = 0$ (where A is the Helmholtz free energy), for such cases.

11.24 For the definitions of the excess functions of mixing, see Section 11.15.

Equation 11.94 enables the determination of the excess chemical potential of one component from knowledge of the excess chemical potential of the other component as a function of composition.

11.25 We shall use the equation

$$\frac{\mathrm{d}(\ln K_P)}{\mathrm{d}T} = \frac{\Delta H^\circ}{RT^2}$$

to express K_P as a function of temperature. First of all, we have to calculate ΔH° for the reaction $CO(g) + H_2O(g) \rightarrow H_2(g) + CO_2(g)$. Now,

$$(\Delta H^\circ)_{298 \text{ K}} = \Delta H_f^\circ(H_2) + \Delta H_f^\circ(CO_2) - \Delta H_f^\circ(CO) - \Delta H_f^\circ(H_2O).$$

Using the standard heat of formation values given in Appendix 5,

$$(\Delta H^\circ)_{298 \text{ K}} = 0 - 393.514 + 110.520 + 241.826 = -41.168 \, \text{kJ mol}^{-1}$$
$$(\Delta G^\circ)_{298 \text{ K}} = \Delta G_f^\circ(H_2) + \Delta G_f^\circ(CO_2) - \Delta G_f^\circ(CO) - \Delta G_f^\circ(H_2O).$$

Using the standard free energy of formation values given in Table 10.1,

$$\Delta G^\circ = 0 - 394.384 + 137.277 + 228.614 = -28.493 \, \text{kJ mol}^{-1}$$
$$\Delta C_P = C_P(CO_2) + C_P(H_2) - C_P(CO) - C_P(H_2O)$$
$$= -0.515 + 6.233 \times 10^{-3}T - 31.512 \times 10^{-7}T^2.$$

We use the Kirchhoff equation $(\partial/\partial T)[(\Delta H)_T] = \Delta C_P$ to calculate $(\Delta H^\circ)_T$ at any temperature T. Then,

$$(\Delta H^\circ)_T = A - 0.515T + \frac{1}{2}(6.233 \times 10^{-3}T^2) - \frac{1}{3}(31.512 \times 10^{-7}T^3)$$

where A is an integration constant. To calculate this constant, we use the condition that, at $T = 298\,\mathrm{K}$, $(\Delta H^\circ)_{298\,\mathrm{K}} = -41\,168\,\mathrm{J\,mol^{-1}}$. Then,

$$A = -41\,168 + 153.5 - 276.8 + 27.8 = -41\,263.5.$$

Hence,

$$(\Delta H^\circ)_T = -41\,263.5 - 0.515T + 3.117 \times 10^{-3}T^2 - 10.504 \times 10^{-7}T^3$$

Now from

$$\frac{\mathrm{d}}{\mathrm{d}T}(\ln K_P) = \frac{(\Delta H_T^\circ)}{RT^2}$$

we get

$$\ln K_P = \frac{1}{R}\int\left(\frac{-41\,263.5}{T^2} - \frac{0.515}{T} + 3.117 \times 10^{-3} - 10.504 \times 10^{-7}T\right)\mathrm{d}T.$$

Integrating,

$$\ln K_P = \frac{1}{R}\left(\frac{41\,263.5}{T} - 0.515\ln T + 3.117 \times 10^{-3}T - 5.252 \times 10^{-7}T^2\right) + A'$$

where A' is an integration constant. To evaluate A', we use $(\Delta G^\circ)_{298\,\mathrm{K}} = -RT\ln[(K_P)_{298\,\mathrm{K}}]$. Hence, $\ln[(K_P)_{298\,\mathrm{K}}] = 28\,493/(8.314)(298) = 11.50$. Then,

$$A' = 11.50 - 16.655 + 0.353 - 0.056 + 0.003 = -4.855.$$

Therefore,

$$\ln K_P = -4.855 + \frac{4963.134}{T} - 0.062\ln T + 0.374 \times 10^{-3}T - 0.632 \times 10^{-7}T^2$$

which is the required expression for the equilibrium constant as a function of temperature.

11.26   We shall use $\Delta G^\circ = -RT\ln K$ to calculate the equilibrium constant at 298.15 K. We know ΔH°; we need the value of ΔS° to evaluate ΔG° from $\Delta G^\circ = \Delta H^\circ - T\Delta S^\circ$.

$$(\Delta S^\circ)_{298.15\,\mathrm{K}} = S^\circ(CH_3NH_3Cl) - S^\circ(HCl) - S^\circ(CH_3NH_2)$$
$$= 138.616 - 186.690 - 243.592 = -291.666\,\mathrm{J}.$$

Then,

$$(\Delta G^\circ)_{298.15\,\mathrm{K}} = -182\,506 + (298.15)(291.666) = -95\,545.79\,\mathrm{J}.$$

Substituting this value in $\Delta G° = -RT \ln K$ we get

$$\ln K = \frac{-\Delta G°}{RT} = \frac{95\,545.79}{(8.314)(298.15)} = 38.55.$$

Therefore, $K = 5.5 \times 10^{16}$.

11.27   Given that $\Delta G° = 7700 + 4.2T \log T - 27.80T$. Then, using $\Delta G° = -RT \ln K_P$,

$$2.303 \log K_P = \frac{-\Delta G°}{RT} = -\frac{7700 + 4.2 \log T - 27.80T}{(2.303)(8.314)T}$$

$$= -\frac{402}{T} - 0.219 \log T + 1.452.$$

11.28   The reaction is $CO(g) + H_2(g) \rightleftharpoons CH_2O(g)$.

$$\Delta C_P = C_P(CH_2O) - C_P(CO) - C_P(H_2) = -8.79 + 12.31 \times 10^{-3}T - 5.73 \times 10^{-7}T^2.$$

(a) At 298.15 K, $\ln K_P(-\Delta G°)/RT = (-27\,196)/(8.314)(298.15) = -10.97$.

$$\frac{d}{dT}(\ln K_P) = \frac{\Delta H°}{RT^2}.$$

If $\Delta H°$ is independent of temperature, on integration we get

$$\ln[(K_P)_{1000\,K}] - \ln[(K_P)_{298.15\,K}] = (\Delta H°)_{298.15\,K} \int_{298.15\,K}^{1000\,K} \frac{dT}{RT^2}$$

or

$$\ln[(K_P)_{1000\,K}] = -10.97 + \frac{(\Delta H°)_{298.15\,K}}{R}\left[-\frac{1}{T}\right]_{298.15\,K}^{1000\,K}$$

$$= -10.97 + \frac{5439}{8.314}\left(\frac{1}{1000} - \frac{1}{298.15}\right) = -12.509$$

or

$$K_P \text{ at } 1000\,K \text{ is } 3.66 \times 10^{-6}.$$

(b) If $\Delta H°$ is dependent on temperature, we use the Kirchoff equation $(\partial/\partial T)(\Delta H) = \Delta C_P$. On integration,

$$(\Delta H°)_T = A - 8.79T + 6.108 \times 10^{-3}T^2 - 1.91 \times 10^{-7}T^3$$

where A is an integration constant. To obtain the value of this constant we use the condition that, when $T = 298.15$ K, $(\Delta H°)_{298.15\,K} = -5439$ J. Then,

$$A = -5439 + 2620.739 - 542.961 + 5.062 = -3356.159.$$

Hence,

$$\frac{d}{dT}(\ln K_P) = \frac{-3356.159 - 8.79T + 6.108 \times 10^{-3}T^2 - 1.91 \times 10^{-7}T^3}{8.314T^2}.$$

Integrating between 298.15 and 1000 K gives

$$(\ln K_P)_{1000K} = (\ln K_P)_{298.15K} \int_{298.15K}^{1000K} \left(\frac{403.676}{T^2} + \frac{1.057}{T} - 0.735 \times 10^{-3} + 0.229 \times 10^{-7}T\right)dT$$

$$= -10.97 \text{(from}(a)) - \left[\frac{-403.676}{T} + 1.057\ln T\right.$$

$$\left. -0.735 \times 10^{-3}T + 0.114 \times 10^{-7}T^2\right]_{298.15K}^{1000K} = -12.693.$$

Therefore, K_P at 1000 K is 3.10×10^{-6}.

11.29   The reaction is $CO(g) + H_2O(g) = CO_2(g) + H_2(g)$, The van't Hoff reaction isotherm is

$$\Delta G = \Delta G^\circ + RT \ln\left(\frac{P_{CO_2}P_{H_2}}{P_{CO}P_{H_2O}}\right) = -RT \ln K_P + RT\ln\left(\frac{P_{CO_2}P_{H_2}}{P_{CO}P_{H_2O}}\right).$$

Substituting $K_P = 0.71$, $P_{CO_2} = P_{H_2} = 1.5$ atm, $P_{CO} = 10$ atm, $P_{H_2O} = 5$ atm and $T = 973$ K into the above equation, we obtain

$$\Delta G = -(8.314)(973)\ln(0.71) + (8.314)(973)\ln\left(\frac{(1.5)(1.5)}{(10)(5)}\right) = -22.343 \text{ kJ}.$$

Since ΔG is negative, it is theoretically possible to perform the reaction under the given conditions.

11.30   The reaction is $\frac{1}{2}N_2 + \frac{1}{2}O_2 \to NO$.

$$\Delta H^\circ = H^\circ_{\text{products}} - H^\circ_{\text{reactants}} = \sum_{\text{products}} \Delta H^\circ_f - \sum_{\text{reactants}} \Delta H^\circ_f$$

$$= \Delta H^\circ_f(NO) - \frac{1}{2}\Delta H^\circ_f(N_2) - \frac{1}{2}\Delta H^\circ_f(O_2)$$

because N_2 and O_2 are elements; this gives

$$\Delta H^\circ = \Delta H^\circ_f(NO) = 90.374 \text{ kJ mol}^{-1}$$

$$\Delta S^\circ = S^\circ_{\text{products}} - S^\circ_{\text{reactants}} = S^\circ(NO) - \frac{1}{2}S^\circ(O_2) - \frac{1}{2}S^\circ(N_2)$$

$$= 210.623 - \frac{1}{2}(205.058 + 191.502) = 12.343 \text{ JK}^{-1} \text{ mol}^{-1}.$$

Then,

$$\Delta G^\circ = \Delta H^\circ - T\Delta S^\circ = 90\,374 - (298)(12.343) = 86.696 \text{ kJ}.$$

Now,

$$\Delta G^\circ = -nRT \ln K$$

or

$$\ln K = -\frac{-\Delta G^\circ}{nRT} = \frac{(-86\,696)}{(8.314)(298)} = -34.992$$

Therefore, $K = 1.57 \times 10^{15}$. If ΔG is negative, $K > 1$.

11.31 (a) The reaction is

$$CH_2{=}CH{-}CH_2{-}CH_3 \rightarrow CH_2{-}CH{-}CH{=}CH_2 + H_2.$$

For this reaction,

$$\Delta G^\circ = 80.289\,kJ \text{ (see problem 10.29), but } \Delta G^\circ = -nRT \ln K$$

i.e.   ln $K = (-\Delta G^\circ)/nRT = (-80289)/(8.314)(298) = -32.406$.   Therefore, $K = 8.438 \times 10^{-15}$.

(b) $\Delta G^\circ = \Delta H^\circ - T\Delta S^\circ$. If ΔH° is independent of temperature, we have

$$(\Delta G^\circ)_{500\,K} = 110\,750 - (500)(102.215) = 59\,642.50\,J$$

(see problem 10.29 for the values of ΔH° and ΔS°).

(c) $(d/dT)(\ln K_P) = \Delta H^\circ/RT^2$. If ΔH° is independent of temperature, then by integrating the above equation between T_1 and T_2 we have

$$\ln\left(\frac{K_2}{K_1}\right) = -\frac{\Delta H}{R}\left(\frac{1}{T_2} - \frac{1}{T_1}\right)$$

where K_1 and K_2 are equilibrium constants at T_1 and T_2, respectively. Then,

$$\ln K_2 - \ln K_1 = -\frac{110\,750}{8.314}\left(\frac{1}{1000} - \frac{1}{298}\right)$$

or

$$\ln K_2 + 32.406 = 31.380$$

or

$$\ln K_2 = -1.026$$

or

$$K_2 = 0.358$$

11.32 The reaction is $2NH_3 {\rightleftharpoons} N_2(g) + 3H_2(g)$ with the concentrations $(1-0.98)$ mol for NH_3, $(1)(0.98/2)$ mol for N_2 and $(3)(0.98/2)$ mol for H_2. Thus, the total number of moles is $(1-0.98)+(0.98/2)+(3)(0.98/2)=1.98$. Hence, the partial pressure P_{NH_3} of ammonia is $[(1-0.98)/(1+0.98)](10)=0.10$ atm. The partial pressure P_{N_2} of nitrogen is $(0.98)(10)/2(1+0.98)=2.47$ atm. The partial pressure P_{H_2} of hydrogen is $(3)(0.98)(10)/2(1+0.98)=7.41$ atm. Then,

$$K_P = \frac{(P_{N_2})(P_{H_2})^3}{(P_{NH_3})^2} = \frac{(2.47)(7.41)^3}{(0.10)^2} = 1 \times 10^5$$

$(\Delta G°)_{673\,K} = -RT \ln K_P = -(8.314)(673)(2.303)\log(10^5) = -64\,430.13 \text{ J mol}^{-1}$.

11.33 The process at 1850 K is $Al(\ell) \rightleftharpoons Al(g)$.

$$\Delta G° = -RT \ln K = -RT \ln \left(\frac{P}{P_0}\right)$$

where $P_0 = 1$ atm and $\Delta G° = 50.836$ kJ mol^{-1}. Therefore,

$$50\,836 = -(8.314)(1850) \ln P$$

or

$$\ln P = -\frac{50\,836}{15\,380.9} = -3.305$$

or the vapour pressure P of molten aluminium equals 3.67×10^{-2} atm $= 3.67 \times 10^{-3}$ MPa.

11.34 The van't Hoff equation is $(\partial/\partial T)(\ln K_P) = \Delta H°/RT^2$. On the assumption that $\Delta H°$ is independent of temperature, integration of this equation gives

$$\ln K_P = -\frac{\Delta H°}{RT} + C$$

where C is an integration constant. This equation represents a straight line. Therefore, a plot of $\ln K_P$ versus $1/T$ will have a slope $-\Delta H°/R$. If $\log K_P$ is plotted against $1/T$, the slope will be $-\Delta H°/2.303\,R$.

Temperature (K)	$(1/T) \times 10^4$	$K_P \times 10^4$	$\log K_P$
1900	5.26	2.31	-3.63
2000	5.00	4.08	-3.39
2100	4.76	6.86	-3.17
2200	4.54	11.0	-2.96
2300	4.34	16.9	-2.78
2400	4.16	25.1	-2.61
2500	4.00	36.0	-2.45
2600	3.84	50.3	-2.29

A plot of $\log K_P$ against $1/T$ with the above data gives a slope of -9510. Then,

$$\Delta H° = -(-9510)(2.303)(8.314) = 182\,089.32 \text{ J}.$$

According to Le Chatelier's principle, an increase in temperature at constant pressure shifts an equilibrium in the direction in which it absorbs heat. We found that ΔH° for the reaction is positive, i.e. the reaction is endothermic. Therefore, by Le Chatelier's principle an increase in temperature increases the product concentration of the reaction, i.e. K_P increases. This is in line with the given data.

11.35 In problem 10.32 we found that

$$(\Delta G^\circ)_T = -74789 + 153.258T - 46.464 \times 10^{-3}T^2$$
$$+ 53.141 \times 10^{-7}T^3 - 165.914.$$

Now $(\Delta G^\circ)_T = -RT \ln K_P$; K_P here is the equilibrium constant at the temperature T. Then,

$$\ln K_P = -\frac{(\Delta G^\circ)_T}{RT} = \frac{1}{RT}(74789 - 153.258T + 46.464 \times 10^{-3}T^2 - 53.141 \times 10^{-7}T^3 + 165.914)$$

which is the required expression for K_P.

EXERCISES 12

12.1 There are two special cases: all the 4 molecules can be in the left half of the box, and all the 4 molecules could be in the right half. Then there are 16 possible ways in which $N=4$ molecules 1, 2, 3 and 4 can be distributed between the two halves as shown below (L = left and R = right). $C(N_L)$ represents the number of possible configurations of the molecules when N_L of them are in the left half of the box.

1	2	3	4	N_L	N_R	$C(N_L)$
L	L	L	L	4	0	1
L	L	L	R	3	1	1 ⎫
L	L	R	L	3	1	1 ⎪
L	R	L	L	3	1	1 ⎬ 4
R	L	L	L	3	1	1 ⎭
L	L	R	R	2	2	2 ⎫
L	R	L	R	2	2	2 ⎪
L	R	R	L	2	2	2 ⎪
R	L	L	R	2	2	2 ⎬ 6
R	L	R	L	2	2	2 ⎪
R	R	L	L	2	2	2 ⎭
L	R	R	R	1	3	3 ⎫
R	L	R	R	1	3	3 ⎪
R	R	L	R	1	3	3 ⎬ 4
R	R	R	L	1	3	3 ⎭
R	R	R	R	0	4	1

12.2 The total number of accessible microstates for a macroscopic system is given by 2^N, where N is the total number of distinguishable particles in the system. In this case, $N = 6.02 \times 10^{23}$. Hence, the total number of microstates $= (2)^{6.02 \times 10^{23}} = (10)^{1.8 \times 10^{23}}$.

12.3 According to the Clausius statement of the second law of thermodynamics, heat will only flow from the system 1 to the system 2, never the other way round. This flow is an irreversible process. After 1 J of heat has been transferred from system 1 to system 2 through the diathermic partition and this partition has been replaced with an adiabatic partition, there will be no more transfer of heat because of the adiabatic partition and each system will reach a new state of internal thermodynamic equilibrium. We may assume that the changes in the temperatures of the two systems are small enough to be ignored. The entropy change is given by the equation $\Delta S = \Delta Q / T$. Therefore, the change in the entropy, ΔS_1, of the system 1 is $\Delta S_1 = -1/400 = -0.0025\,\text{JK}^{-1}$, and that for the system 2 is $\Delta S_2 = 1/300 = 0.0033\,\text{JK}^{-1}$. The total change in entropy is

$$\Delta S_1 + \Delta S_2 = -0.0025 + 0.0033 = 0.0008\,\text{JK}^{-1}.$$

The total entropy increases. However, if 1 J of heat flowed from system 2 to system 1, the total entropy would go down by $0.0008\,\text{JK}^{-1}$. This process is never observed in practice. Therefore, the direction of heat flow is given by

$$(\Delta S_1 + \Delta S_2) > 0.$$

If the difference between T_1 and T_2 is very small, according to the second law of thermodynamics heat would still flow from the left to the right. When $T_1 = T_2$, the change would be reversible and $(\Delta S_1 + \Delta S_2) = 0$.

If the diathermic partition was not replaced by an adiabatic one, heat would continue to flow from system 1 to system 2 until $T_1 = T_2$ and a thermal equilibrium is reached, when $S_1 + S_2$ is a maximum.

12.4 See Section 12.4.3 for the definitions of phase space, μ-space and γ-space.

The general form of the kinetic energy of the form given in the question can be derived using Equations 12.8, 12.9, 12.10 and $\partial \varepsilon_k / \partial \dot{r}_i = p_i$.

12.5 In classical mechanics, a system with constant energy cannot make a transition from one state to another of same energy if the system has to pass through a situation where the total energy is less than the potential energy. If it does, the kinetic energy will be negative. In such cases, the spacing of the energy levels then may seem superficially to be irregular, and energy differences between successive levels of varying magnitudes will exist, corresponding to the various characteristic times. These situations are much more complicated than those considered by Planck's equation.

12.6 If U units of energy are distributed between N distinguishable particles of zero spin, then the total number of accessible microstates Ω in the system is given by

$$\Omega = (N + U - 1)! / (N - 1)! U!$$

For $N = 3$ and $U = 10^2$, 10^3, 10^4, 10^5 and 10^6, $\Omega = 5.15 \times 10^3$, 5.0×10^5, 5.00×10^7, 5.00×10^9 and 5.00×10^{11}, respectively.

12.7 Moment of inertia $I = \mu r_0^2$ where μ is the reduced mass and r_0 is the distance of separation.

$$\mu = \left[\frac{(1)(35.5)}{(35.5 + 1)} \right] \times 1.66 \times 10^{-24}\,\text{g} = 1.62 \times 10^{-24}\,\text{g}.$$

Then,

$$r_0^2 = I/\mu$$

or

$$r_0 = \sqrt{I/\mu} = \left(\frac{2.65 \times 10^{-40}\,\text{g cm}^2}{1.62 \times 10^{-24}\,\text{g}} \right)^{1/2}$$

$$= 1.28 \times 10^{-8}\,\text{cm} = 1.28\text{Å}.$$

Note that this value is not the same as that given in Table 12.1, because we said that ε_0 and r_0 values for HCl are dependent on temperature.

12.8

Dipole moment = (electronic charge) × (distance of separation between an electron and a proton)

Electronic charge $= 4.80 \times 10^{-10}\,\text{esu} = 1.60 \times 10^{-19}$ coulomb.

The distance of separation between an electron and a proton is given as 1 angstrom. So,

$$\text{diapole moment} = 4.80 \times 10^{-10} \times 10^{-8} = 4.80 \times 10^{-18}\,\text{esu cm}$$

$$= 1.60 \times 10^{-19}\,\text{C} \times 10^{-10}\,\text{m} = 1.60 \times 10^{-29}\,\text{coulomb m}$$

The esu-cm unit is called a *debye*, and the coulomb metre unit is the mks unit.

12.9 See the derivation of Equation 12.40.

12.10 Constant energy occurs on a surface in a space formed by the quantum numbers n_x, n_y and n_z marked off along the three co-ordinate axes. Equation 12.40 is a mathematical representation of this surface. Actually, it is the equation of an ellipsoid which has volume in n_x, n_y and n_z space equal to

$$\frac{4\pi}{3}(8M\varepsilon_t)^{3/2}V/h^3$$

where V is the volume of the box containing the molecule. The volume in the n_x, n_y and n_z space is equal to the number of combinations of integral values included in the volume, since this will be the number of unit cubes inside that volume. Therefore, the number of energy levels below ε_t is the volume of the part of the ellipsoid, where n_x, n_y and n_z are all positive or

$$\left(\frac{1}{8} \right)\left(\frac{4\pi}{3} \right)(8M\varepsilon_t)^{3/2}V/h^3 = \frac{4\pi}{3}(2M\varepsilon_t)^{3/2}V/h^3.$$

12.11 Given that

Momentum $= 1.66 \times 10^{-19}\,\text{g cm/s}$

Velocity $= 10^5\,\text{cm/s}$

$n_x = 5 \times 10^7$

Translational energy $\varepsilon_t = 0.83 \times 10^{-21}$ J

Mass of hydrogen atom $= 1.66 \times 10^{-24}$ g

Since the box has a volume 1 cm^3, $L_x = L_y = L_z = 1$

Momentum in the x direction is $n_x h/2L_x \simeq 1.66 \times 10^{-19}$ g cm/s (1 J $= 1$ kg m^2/s^2).

If n_x is changed by 1, then the total momentum is $(n_x + 1)h/2L_x$ so the change in momentum is 3.30×10^{-27} g cm/s and the corresponding change in the velocity is 2×10^{-3} cm/s. The change in the translational energy is given approximately by $(h^2/8m)(2n/L_x^2) = (2/n)\varepsilon_t = (2/5 \times 10^7)\varepsilon_t = (4 \times 10^{-8})(0.83 \times 10^{-21}) = 3.32 \times 10^{-29}$ J.

12.12 For $J = 0$, the rotational energy is zero. Therefore, the difference in the rotational energy between the quantum states $J = 0$ and $J = 1$ is

$$
\begin{aligned}
\Delta \varepsilon_r &= \left[\left(\frac{h^2}{8\pi^2 I} \right) J(J+1) \right]_{J=1} - 0 \\
&= \frac{2h^2}{8\pi^2 I} = \frac{2(6.626 \times 10^{-27} \text{ erg s})^2}{8(3.142)^2 \times (1.34 \times 10^{-40} \text{ g cm}^2)} \\
&= 5.292 \times 10^{-15} \text{ erg}^2 \text{ s}^2/\text{g cm}^2 \\
&= 5.292 \times 10^{-15} \left(\frac{\text{g cm}^2}{\text{s}^2} \right)^2 \frac{\text{s}^2}{\text{g cm}^2} \\
&= 5.292 \times 10^{-15} \text{ g cm}^2/\text{s}^2 \\
&= 5.292 \times 10^{-15} \text{ ergs} = 5.292 \times 10^{-22} \text{ J}.
\end{aligned}
$$

Comparing this result with that of HCl molecule calculated in Example 1.4, we see that the order of magnitude for HCl and HF is the same.

12.13 According to Equation 12.45, the minimum value of the energy of a harmonic oscillator is given by $n = 0$, which is $h\nu/2$. This energy is known as the *zero-point energy*. Since both h and ν are constants, the zero-point energy is also a constant. Such a constant term can be neglected in the energy if we define what we mean by zero energy. Equation 12.45 incorporated the state of zero energy as that state in which the oscillator is at rest. The significance of the zero-point energy is that it indicates that even when the oscillator has its lowest values of the energy it cannot be at rest. (The state in which an oscillator is at rest corresponds, in the classical case, to a particle at its equilibrium position where $\partial \phi/\partial x = 0$, ϕ being the potential energy.)

The zero-point energy has no effect on the difference in vibrational energy between two quantum states.

12.14 The difference in energy between the quantum states $n = 0$ and $n = 1$ is

$$
\begin{aligned}
\Delta \varepsilon_v &= \left[\left(n + \frac{1}{2} \right) h\nu \right]_{n=1} - \left[\left(n + \frac{1}{2} \right) h\nu \right]_{n=0} \\
&= h\nu = (6.45 \times 10^{13} \text{ s}^{-1})(6.626 \times 10^{-27} \text{erg s}) \\
&= 42.74 \times 10^{-14} \text{erg} = 4.27 \times 10^{-20} \text{ J}.
\end{aligned}
$$

Frequency $\nu = (1/2\pi)(\kappa/\mu)^{1/2}$, where κ is the harmonic force constant and μ is the reduced mass of the molecule. This gives

$$\kappa = 39.49\, \mu\nu^2$$

$$\mu = \frac{(\text{atomic mass of carbon})(\text{atomic mass of oxygen})}{(\text{atomic mass of carbon}) + (\text{atomic mass of oxygen})}$$

$$= \frac{(12.011)(15.999)}{12.011 + 15.999} = 6.86.$$

Therefore, $\kappa(39.49)(6.86)(6.45 \times 10^{13})^2 = 1.127 \times 10^{30}$
The units of κ are dynes cm^{-1}.

12.15 For $n = 0$ and $n = 1$,

$$\Delta\varepsilon_v = h\nu = (6.626 \times 10^{-27}\ \text{erg s})(1.68 \times 10^{13}\ \text{s}^{-1})$$
$$= 11.13 \times 10^{-14}\ \text{erg} = 1.113 \times 10^{-20}\ \text{J}.$$

This shows that $\Delta\varepsilon_v$ of a CO molecule is almost 4 times bigger than that of a Cl$_2$ molecule. This is because a CO molecule is much lighter than a Cl$_2$ molecule.

Force constant $\kappa = (39.49)(17.73)(1.68 \times 10^{13}) = 1.97 \times 10^{30}$

Angular velocity $\omega = (\kappa/m)^{1/2} = \left(\dfrac{1.97 \times 10^{30}}{70.91}\right)^{1/2} = 52.70 \times 10^5\ \text{rad/s}$

Period $= \dfrac{2\pi}{\omega} = 1.19 \times 10^{-6}$.

12.16 Two physically identical macroscopic systems means both have the same number and type of atoms. We label the systems as A and B. The total energy of a macroscopic system is the sum of the total kinetic energy (E_k), plus the total potential energy due to external forces (ϕ_{ext}) plus the total potential energy due to the interactions within the system. If we assume that A and B are made of same type of N atoms, then the total kinetic energy of A or B is simply the sum of the individual kinetic energies of N particles, and is given by

$$E_k = \frac{1}{2}m_1v_1^2 + \frac{1}{2}m_2v_2^2 + \cdots + \frac{1}{2}m_Nv_N^2$$
$$= \sum_i \frac{1}{2}m_iv_i^2 \tag{1}$$

where m_i and v_i are the mass and velocity of particle i and $i = 1, 2, 3, \ldots, N$.

The total potential energy of the N particles due to external forces, such as electric field or gravity, is the sum of the potential energy of N individual particles, and is given by

$$\phi_{\text{ext}} = \sum_i \phi(r_i) \tag{2}$$

where $i = 1, 2, 3, \ldots, N$ and $\mathbf{r}_i$ represents the position of each particle.

The potential energy due to the interactions within the system is given by

$$\frac{1}{2}\sum_{i \neq j}^{N} \Phi(r_{ij}) \tag{3}$$

where r_{ij} is the distance between the particles and i does not equal j (a particle does not interact with itself). Note that the factor $\frac{1}{2}$ is due to the fact that the sum contains identical terms, such as $\Phi(r_{12})$ and $\Phi(r_{21})$, which should only be considered once. Using Equations 1, 2 and 3, we obtain the total energy of a macroscopic system:

$$E = \sum_i \frac{1}{2} m_i v_i^2 + \sum_i \phi(\mathbf{r}_i) + \frac{1}{2} \sum_{i \neq j} \Phi(r_{ij}) \tag{4}$$

where $i = 1, 2, 3, \ldots, N$.

The total energy E_{A+B} of the composite system $A + B$ is given by Equation 4, where the sums are over all particles in both A and B. Equation 4 can be written as

$$E_{A+B} = E_A + E_B + E_{int}. \tag{5}$$

In this equation E_A or E_B is given by Equation 4, except that the summations are only over the particles in A or B. The interaction energy E_{int} between the systems is

$$E_{int} = \sum_i \sum_j \Phi(r_{ij}). \tag{6}$$

In these sums, i is to be summed only over the particles in A, and j is to be summed only over the particles in B. Since A and B are macroscopic systems, E_{int} is small compared with E_A and E_B. This is because most of the terms in Equation 6 are zero unless the distances between particles (r_{ij}) is very small ($\simeq 3$ to 4Å). This implies that particles i and j must be very close to the partition between A and B, i.e. the interaction region. Then, for macroscopic systems Equation 5 can be approximated to

$$E_{A+B} \simeq E_A + E_B.$$

12.17 A schematic of microstate of a hypothetical one-dimensional system in its phase space is shown here.

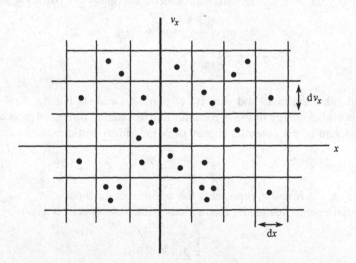

The microstate of a system in a space is specified by the co-ordinates (x, y, z, v_x, v_y, v_z), which is known as phase space. Each point in this space gives the position and velocity of a particle. It is rather difficult to draw a space with six co-ordinates, However, the concept is illustrated in the figure. The grid in this space marks off areas of size $dx dv_x$. The points move around as the position and velocity of the particles change with time. Therefore, the above figure actually represents the system at some particular instant of time.

From the point of view of classical mechanics, the microscopic state or configuration changes, i.e. we have a new microstate when any point moves out of any area defined by $dx dv_x$. The probability of these microstates is given by 12.57.

12.18 If all the balls are to be in one or other of the holes, we first have to choose a hole. Since there are N holes, we have N choices. Once a hole has been chosen, $N-1$ holes and X balls will be left. We can then choose another hole, in which case the first hole will be left empty, or we can pick a ball to put into the first hole. If the balls are distinguishable, this choice can be made in $N+X-1$ ways. The next choice can be made in $N+X-2$ ways, and so on until a situation arises where we have a hole which is empty or we have a ball to put into the last hole. Thus, the total number of ways we can put X distinguishable balls into N distinguishable holes is $N(N+X-1)!$, and for each choice of the order of the boxes we have $X!$ ways of selecting the order in which the X distinguishable balls are but into the boxes.

If the balls are indistinguishable, it does not matter in what order they are put into the holes; we must divide $N(N+X-1)!$ by $X!$ because there are $N!$ ways of selecting N distinguishable holes. However, if we are only interested in how many indistinguishable balls are in each hole but not interested in the order in which we choose the holes, then also we must divide by $N!$. Therefore, the number of ways in which X indistinguishable balls can be put into N distinguishable holes is given by

$$n = \frac{N(N+X-1)!}{N!X!} = \frac{(N+X-1)!}{(N-1)!X!}.$$

12.19 Molecular mass $= 6.64 \times 10^{-24}$ g; $T = 300$ K. Following the procedure in Example 12.5, we find that $N/V \ll 7.3 \times 10^{23}$ cm^{-3}. $V = 2 \times 2 \times 2$ cm$^3 = 8$ cm^3 is the volume of the box, so it is also the total volume of the gas. Therefore, the number density N/V is $N/8$ cm$^3 \ll 7.3 \times 10^{23}$ cm^{-3}, i.e. $N \ll 8 \times 7.3 \times 10^{23}$. Thus, the total number of particles in the total volume of the gas is less than 10^{25}.

Since this number is not particularly high, the gas is not likely to be degenerate. At room temperature, a density of this size would correspond to a pressure approximately 3000 MPa. So, at ordinary pressures, perfect gases are non-degenerate.

12.20 See Section 12.12.

12.21 The microstates of an open system are defined in the same way as for closed systems, except that now a system is also considered to be in a different microstate if the total number of particles N has changed. Therefore, we can apply the same reasoning as used for closed systems to extend the basic assumption to open systems to write that all (micro)-states with the same number of particles N and the same amount of energy E are assumed to be equally probable.

12.22 See the derivation of Equation 12.97.

12.23 The probability derived in Exercise 12.22 is equally valid for the classical case where the energy varies continuously, except that the distribution function 12.100 must be used.

12.24 See the derivations of Equations 12.114, 12.115, 12.116 and 12.117.

12.25 The energies for a particle of mass m, in a cubical box of side L and volume V, are given by (see Equation 12.3)

$$\varepsilon = \left(\frac{\hbar^2 \pi^2}{2mV^{2/3}} \right) (n_x^2 + n_y^2 + n_z^2) \tag{i}$$

where n_x, n_y and n_z are the quantum numbers. These can have the positive integer values $1, 2, 3, \ldots$. According to quantum mechanics, Equation (i) above will give one of the energies of the particle. The lowest energy, which is the ground state, has the quantum numbers $n_x = n_y = n_z = 1$. The energy of this state is then given by

$$\varepsilon_{1,1,1} = \left(\frac{\hbar^2 \pi^2}{2mV^{2/3}} \right) (1^2 + 1^2 + 1^2) = 3 \left(\frac{\hbar^2 \pi^2}{2mV^{2/3}} \right)$$

(a) By substituting the values of $\hbar^2$, π, m and V into the above equation, we obtain

$$\varepsilon_{1,1,1} = 3(1.039 \times 10^{-40}) = 3.12 \times 10^{-40} \text{ J} \tag{ii}$$
$$= 1.95 \times 10^{-21} \text{ eV} \quad (1 \text{ eV} = 1.602 \times 10^{-19} \text{ J}). \tag{iii}$$

(b) The lowest energy level given by the quantum numbers $n_x = n_y = n_z = 1$ is not degenerate. The next set of quantum states have quantum numbers 2, 1, 1; 1, 2, 1; and 1, 1, 2 for n_x, n_y and n_z, respectively, corresponding to three independent solutions of Schrödinger's equation having the same energies. This energy level is threefold degenerate and the energy is

$$\varepsilon = (2^2 + 1^2 + 1^2)(1.039 \times 10^{-40}) = 6(1.039 \times 10^{-40} \text{ J})$$

which is double the value of $\varepsilon_{1,1,1}$.

The next higher energy level having quantum numbers 2, 2, 1 is also threefold degenerate (2, 2, 1; 2, 1, 2; and 1, 2, 2). The energy is

$$\varepsilon = (2^2 + 2^2 + 1^2)(1.039 \times 10^{-40}) = 9(1.039 \times 10^{-40} \text{ J})$$

which is three times the value of $\varepsilon_{1,1,1}$.

(c) According to the kinetic theory of gases, the mean energy $(\bar{\varepsilon})$ of a helium atom in a helium gas at 273 K is $\frac{3}{2}(kT)$. Therefore,

$$\bar{\varepsilon} = \frac{3 \times 1.38 \times 10^{-23} \text{ J K}^{-1} \times 273 \text{ K}}{2} = 5.65 \times 10^{-21} \text{ J}.$$

This value of $\bar{\varepsilon}$ is about 1.81×10^{19} times bigger than the ground state energy found in (a) above.

(d) Taking the energy given by Equation (i) to be equal to $3kT/2$, we have

$$\frac{3kT}{2} = \left(\frac{\hbar^2 \pi^2}{2mV^{2/3}}\right)(n_x^2 + n_y^2 + n_z^2)$$

or

$$n_x^2 + n_y^2 + n_z^2 = \frac{5.65 \times 10^{-21} \, \text{J}}{1.039 \times 10^{-40} \, \text{J}} = 5.43 \times 10^{19}.$$

For a rough estimate of the order of magnitude of n_x, n_y and n_z for a helium gas atom of mean energy $3kT/2$ at 273 K, we use the special case $n_x = n_y = n_z$. Then, from the above value of $n_x^2 + n_y^2 + n_z^2$ we have

$$n_x^2 = n_y^2 = n_z^2 = 1.81 \times 10^{19}$$

or

$$n_x = n_y = n_z \simeq 4 \times 10^9.$$

This shows that a helium gas atom at 273 K in a cubical box of the size given in the problem is likely to be in a single particle quantum state which has quantum numbers in the range 10^9–10^{10}.

(e) It is quite apparent from Equation (i) that the values of the energies of the single particle states for a particle in a cubical box of side L and volume V are inversely proportional to $(V)^{2/3}$. This dependence actually arises through the boundary conditions applied to solve Schrödinger's equation. If V is increased, keeping n_x, n_y and n_z fixed, then according to Equation (i) the values of all the energies decrease. On the other hand, if the volume of the box is decreased, keeping n_x, n_y and n_z fixed, the values of all the energies increase.

EXERCISES 13

13.1 (a) When the composite system is in thermal equilibrium with the reservoir, both A and B are free to exchange energy with the reservoir and they interact with it independently of each other.

(b) In view of the answer above, at any instant of time the state of system A is independent of the state of system B is in.

(c) The total energy of the composite system is the sum of the energies of the systems A and B, i.e. $1.60 + 1.95 = 3.55 \, \text{kJ}$.

13.2 The total energy of the composite system is equal to $E_X + E_Y$. So, the probability of finding the composite system at any instant of time in a particular state with energy $E_X + E_Y$ is $\mathcal{P}_{(X+Y)}(E_{X+Y}) = \mathcal{P}_{(X+Y)}(E_X + E_Y)$. This is the same as the intersection of two separate events: (i) the system X is in a particular state with energy E_X, and (ii) the system Y is in a particular state with energy E_Y. This may be written as $\mathscr{P}[X(E_X)Y(E_Y)]$. Although X and Y are now parts of the composite system, they are free to exchange energy with any external body independently of each other. This means that at any instant of time the state that the system X is in is independent of the state of the system Y is in. So, the two events (i) and (ii) are independent. For two independent events the probability of their intersection is equal to the product of the probabilities of the individual events, i.e. $\mathscr{P}(XY) = \mathscr{P}(X)\mathscr{P}(Y)$. Hence,

$$\mathscr{P}_{(X+Y)}(E_{X+Y}) = \mathscr{P}_{(X+Y)}(E_X + E_Y) = \mathscr{P}[X(E_X)Y(E_Y)] = \mathscr{P}_X(E_X)\mathscr{P}_Y(E_Y).$$

13.3 Given that $E(A) = -1$ and $E(B) = 1$; $\mathscr{P}(A) = 0.70$ and $\mathscr{P}(B) = 0.30$. By substituting these values into the given equation $\mathscr{P}(E) = C \exp(-\beta E)$, we have

$$0.70 = C \exp[-\beta(-1)] = C \exp \beta$$
$$0.30 = C \exp[-\beta(1)] = C \exp(-\beta).$$

By dividing the first equation by the second, we have $2.33 = \exp(2\beta)$. Taking natural logarithm of both sides and solving for β, we obtain $\beta = 0.424$. Substituting this into the first equation, we have $C = 0.460$.

13.4 The total energy of the system is the sum of the energies of the three molecules i.e. $E = (2.4 + 4.8 + 7.2) \times 10^{-21} \, \text{J} = 14.4 \times 10^{-21} \, \text{J}$. $T = 290 \, \text{K}$ and the Boltzmann constant $k = 1.38 \times 10^{-23} \, \text{JK}^{-1}$. Then,

$$\beta = \frac{1}{kT} = \frac{1}{1.38 \times 10^{-23} \, \text{JK}^{-1})(290 \, \text{K})} = 0.25 \times 10^{21} \, \text{J}^{-1}$$

$$\exp(-\beta E) = \exp[(-0.25 \times 10^{21} \, \text{J})(14.4 \times 10^{-21} \, \text{J})]$$

$$= \exp(-3.6) = 0.027.$$

13.5 See the derivations of Equations 13.26 and 13.28. Although $\beta = 1/kT$ has been derived for a perfect gas, it has nothing to do with a perfect gas; it holds for all systems. In the derivation of Equation 13.10 we showed that β does not depend on the atomic composition of the substance, forces, etc., in the system. Thus, β can be determined as equal to $1/kT$, using any system.

13.6 Since the partition is a diathermic partition, it keeps the volumes V_A and V_B of the systems A and B constant, i.e. the energy eigenvalues of the single particle quantum states of systems A and B are unchanged. There is no exchange of particles between A and B, but energy, in the form of heat, can flow across the diathermic partition, provided that the total energy of the composite system remains equal to the sum of the initial energies of A and B. Under these conditions, our composite system may be considered a closed composite system. There are seven ways in which the total energy of the composite is distributed: $U_A = 6$, $U_B = 0$; $U_A = 5$, $U_B = 1$; $U_A = 4$, $U_B = 2$; $U_A = 3$, $U_B = 3$; $U_A = 2$, $U_B = 4$; $U_A = 1$, $U_B = 5$; $U_A = 0$, $U_B = 6$. For each of these sets of values the microstates accessible to the closed composite system are as follows. There are four distinguishable particles.

$U_A = 6$ and $U_B = 0$				Total number of accessible microstates
6, 0, 0, 0	0, 6, 0, 0			
5, 1, 0, 0	1, 5, 0, 0			$\Omega_1 = 7$
4, 2, 0, 0	2, 4, 0, 0			
3, 3, 0, 0				

$U_A = 5$ and $U_B = 1$				
5, 0, 1, 0	4, 1, 1, 0	3, 2, 1, 0		
5, 0, 0, 1	4, 1, 0, 1	3, 2, 0, 1		
0, 5, 1, 0	1, 4, 1, 0	2, 3, 1, 0		$\Omega_2 = 12$
0, 5, 0, 1	1, 4, 0, 1	2, 3, 0, 1		

$U_A = 4$ and $U_B = 2$				
4, 0, 2, 0	0, 4, 2, 0	3, 1, 2, 0	1, 3, 2, 0	2, 2, 2, 0
4, 0, 0, 2	0, 4, 0, 2	3, 1, 0, 2	1, 3, 0, 2	2, 2, 0, 2	$\Omega_3 = 15$
4, 0, 1, 1	0, 4, 1, 1	3, 1, 1, 1	1, 3, 1, 1	2, 2, 1, 1

$U_A = 3$ and $U_B = 3$				
3, 0, 3, 0	0, 3, 3, 0	2, 1, 3, 0	1, 2, 3, 0	
3, 0, 0, 3	0, 3, 0, 3	2, 1, 0, 3	1, 2, 0, 3	$\Omega_4 = 16$
3, 0, 2, 1	0, 3, 2, 1	2, 1, 2, 1	1, 2, 2, 1	
3, 0, 1, 2	0, 3, 1, 2	2, 1, 1, 2	1, 2, 1, 2	

$U_A = 2$ and $U_B = 4$				
2, 0, 4, 0	0, 2, 4, 0	1, 1, 4, 0		
2, 0, 0, 4	0, 2, 0, 4	1, 1, 0, 4		
2, 0, 3, 1	0, 2, 3, 1	1, 1, 3, 1		$\Omega_5 = 15$
2, 0, 1, 3	0, 2, 1, 3	1, 1, 1, 3		
2, 0, 2, 2	0, 2, 2, 2	1, 1, 2, 2		

$U_A = 1$ and $U_B = 5$				
1, 0, 5, 0	0, 1, 5, 0	1, 0, 0, 5	0, 1, 0, 5	
1, 0, 1, 4	0, 1, 4, 1	1, 0, 1, 4	0, 1, 1, 4	$\Omega_6 = 12$
1, 0, 3, 2	0, 1, 3, 2	1, 0, 2, 3	0, 1, 2, 3	

$U_A = 0$ and $U_B = 6$							
0, 0, 6, 0	0, 0, 0, 6	0, 0, 5, 1	0, 0, 1, 5	0, 0, 4, 2	0, 0, 2, 4	0, 0, 3, 3	$\Omega_7 = 7$

In this table, in any set of microstates the first two numbers are for the two distinguishable particles in system A and the second two numbers are for the two distinguishable particles in system B.

When the two systems are in thermal contact there are altogether $\Omega_1 + \Omega_2 + \Omega_3 + \Omega_4 + \Omega_5 + \Omega_6 + \Omega_7 = 84$ microstates accessible to the composite system. It is evident from this table that in seven cases system A has gained energy from 5 to 6, and system B has lost energy from 1 to 0. Further analysis shows that in 65 cases system B gains energy, and in 12 cases the energies of systems A and B remain unchanged.

According to classical thermodynamics, the temperature of a system increases when the internal energy of the system increases and *vice versa*, provided the heat capacity of the system is positive. In our case, the single particle quantum states of systems A and B have identical energy values; the systems have the same number of distinguishable particles, and initially system A has more energy than system B. So it is reasonable to assume that at the beginning system A is hotter than system B. From the above table it appears that in 65 cases system B has gain energy, i.e. heat has gone from the hot system A to the cold system B. However, we find that in seven cases heat has gone from the cold system B to the hot system A. This simple calculation shows that, according to statistical mechanics, it is possible for heat to flow from a cold body to a hot body.

From the above table we find that out of the total 84 microstates accessible to the composite system, U_B has 0, 1, 2, 3, 4, 5 and 6 units of energy in $\Omega_1(=7)$, $\Omega_2(=12)$, $\Omega_3(=15)$, $\Omega_4(=16)$, $\Omega_5(=15)$, $\Omega_6(=12)$ and $\Omega_7(=7)$ cases, respectively. If we average these values over the 84 equally probable microstates accessible to the closed composite system, we get the average energy of system B. Thus,

$$\overline{U}_B = \frac{(7 \times 0) + (12 \times 1) + (15 \times 2) + (16 \times 3) + (15 \times 4) + (12 \times 5) + (7 \times 6)}{84} = 3.$$

In a similar way, the average energy of system A is $\overline{U}_A = 3$. Note that the mean energies are equal, $\overline{U}_A = \overline{U}_B = 3$. Generally, the mean energies are not equal, but in the present case they are equal because the energy eigenvalues of the single particle states and the total number of distinguishable particles are the same in the two systems.

Note that since $\overline{U}_B = 3$, U_B has gone up from its initial value of 1, whereas U_A has decreased from an initial value of 5 to an average value of $\overline{U}_A = 3$. Therefore, we can say that on average heat has flown from the hot system to the cold system.

Consider any particular distribution of energy from the above table. For example, we choose the case $U_A = 2$ and $U_B = 4$. At thermal equilibrium, the probability of finding the closed composite system in any one of the 84 microstates accessible to the closed composite system is $\frac{1}{84}$. The probability that at thermal equilibrium the closed composite system can be in any one of the 15 microstates accessible to it, where $U_A = 2$ and $U_B = 4$ is $\frac{15}{84}$. Therefore, the probability that at thermal equilibrium system B has energy $U_B = 4$ is $15/84$ (see Equation 13.42).

13.7 There are seven possible distributions of the total energy between systems A and B. Remembering that there are only two distinguishable particles in each system, the microstates accessible to systems A and B over all the seven possible distributions of energy are given in the table below. From Exercise 13.6 we know that at thermal equilibrium the composite system is equally likely to be found in any one of the total 84 microstates accessible to it.

U_A	Microstates accessible to system A		$\Omega_A(U_A)$	U_B	Microstates accessible to system B		$\Omega_B(U_B)$
0	0, 0		1	6	6, 0 5, 1 4, 2 3, 3	0, 6 1, 5 2, 4	7
1	1, 0	0, 1	2	5	5, 0 4, 1 3, 2	0, 5 1, 4 2, 3	6
2	2, 0 1, 1	0, 2	3	4	4, 0 3, 1 2, 2	0, 4 1, 3	5
3	3, 0 2, 1	0, 3 1, 2	4	3	3, 0 2, 1	0, 3 1, 2	4
4	4, 0 3, 1 2, 2	0, 4 1, 3	5	2	2, 0 1, 1	0, 2	3
5	5, 0 4, 1 3, 2	0, 5 1, 4 2, 3	6	1	1, 0	0, 1	2
6	6, 0 5, 1 4, 2 3, 3	0, 6 1, 5 2, 4	7	0	0, 0		1

From this table we find that for:

$$U_B = 6: \quad \Omega_B = 7 \quad \text{and} \quad \Omega_A = 1; \quad \Omega_A\Omega_B = 7$$
$$U_B = 5: \quad \Omega_B = 6 \quad \text{and} \quad \Omega_A = 2; \quad \Omega_A\Omega_B = 12$$
$$U_B = 4: \quad \Omega_B = 5 \quad \text{and} \quad \Omega_A = 3; \quad \Omega_A\Omega_B = 15$$
$$U_B = 3: \quad \Omega_B = 4 \quad \text{and} \quad \Omega_A = 4; \quad \Omega_A\Omega_B = 16$$
$$U_B = 2: \quad \Omega_B = 3 \quad \text{and} \quad \Omega_A = 5; \quad \Omega_A\Omega_B = 15$$
$$U_B = 1: \quad \Omega_B = 2 \quad \text{and} \quad \Omega_A = 6; \quad \Omega_A\Omega_B = 12$$
$$U_B = 0: \quad \Omega_B = 1 \quad \text{and} \quad \Omega_A = 7; \quad \Omega_A\Omega_B = 7$$

Using Equation 13.42, we find that $\mathscr{P}_B(U_B) = (\Omega_A\Omega_B)/\Omega_T$ where Ω_T is the total microstates accessible to the composite system. Hence,

$$U_B = 6, \quad \mathscr{P}_B(U_B) = 7/84; \qquad U_B = 5, \quad \mathscr{P}_B(U_B) = 12/84;$$
$$U_B = 4, \quad \mathscr{P}_B(U_B) = 15/84; \qquad U_B = 3, \quad \mathscr{P}_B(U_B) = 16/84;$$
$$U_B = 2, \quad \mathscr{P}_B(U_B) = 15/84; \qquad U_B = 1, \quad \mathscr{P}_B(U_B) = 12/84;$$
$$U_B = 0, \quad \mathscr{P}_B(U_B) = 7/84$$

13.8   See Section 13.4.5.

13.9   See the derivation of Equation 13.63.

13.10 (a) Generally, a large number of possible microstates of a system correspond to a given macrostate of that system. These microstates are indistinguishable at the macroscopic level. In any system, the number of microstates of *fixed energy* is denoted by the symbol Ω.

(b) Entropy is given by the equation $S = k \ln \Omega$ as it is more convenient to work with the logarithms of the Ω-values, because Ω has an additive property, e.g. $\Omega = \Omega_1 \Omega_2$ and $\ln \Omega = \ln \Omega_1 + \ln \Omega_2$.

(c) Careful quantum-statistical calculations show that selection of the entropy constant as the Boltzmann constant makes the thermodynamic temperature and the empirical perfect-gas temperature scales identical. For this reason the Boltzmann constant is chosen as the entropy constant.

(d) At the absolute zero of temperature, an infinitesimal increase in internal energy of a system produces a significant increase in the disorder of that system. As the temperature increases, the sensitivity of the disorder to the changes in internal energy decreases. Therefore, a state of infinite temperature would be one for which a change in the internal energy would produce no change in the disorder of the system. A state of negative thermodynamic temperature would be one for which an addition of energy as heat would tend to make the microscopic structure more disordered, i.e. an increase in the entropy.

13.11 See Section 13.7.1.

13.12 See Section 9.5.

13.13 Consider Equation 9.10. The left-hand side of this equation is positive because $T_i > 0$ and $C_i > 0$ for any non-zero temperature T_i. Consequently, the right-hand side of this equation is also positive. This is possible only when $T_f \neq 0$, which implies that the final temperature T_f cannot be equal to $0\,\mathrm{K}$. Therefore, we find that for a reversible cooling process the absolute zero of temperature cannot be achieved. Similarly, analysis of Equation 9.12 shows that for an irreversible cooling process the absolute zero of temperature cannot be attained. Hence, we can conclude that the absolute zero of temperature is unattainable.

13.14 In the microscopic approach, the entropy S of an N-particle system of energy U and volume V is given by $S = k \ln \Omega$. Since at room temperatures the entropy given by this equation is of the order of Nk, for the entropy to be negligible at the absolute zero we must have $\ln \Omega_0 \ll N$, where Ω_0 is the degeneracy of the ground state (the lowest energy level). Hence, for the third law to be valid for a macroscopic system at $T = 0$, we must have $\Omega_0 \ll e^N$.

13.15 We have $(\partial U / \partial T)_{V,N} = C_V$ (see Equation 6.4) and we are given that $C_V = \alpha nR$. By combining these two equations, we have

$$\left(\frac{\partial U}{\partial T} \right)_{V,N} = \alpha nR. \tag{1}$$

From Equation 8.10

$$\left(\frac{\partial U}{\partial V} \right)_{T,N} = T \left(\frac{\partial P}{\partial T} \right)_{V,N} - P. \tag{2}$$

We have $PV = nRT$. By differentiating both sides with respect to T, and keeping V and n constant, we obtain

$$\left(\frac{\partial P}{\partial T}\right)_{V,N} V = nR$$

or

$$\left(\frac{\partial P}{\partial T}\right)_{V,N} = \frac{nR}{V}. \tag{3}$$

By substituting this into (2), we have

$$\left(\frac{\partial U}{\partial V}\right)_{T,N} = \frac{nRT}{V} - P = P - P = 0. \tag{4}$$

By integrating (1) and (4) and combining, we have

$$U = U_0 + \alpha nR(T - T_0) \tag{5}$$

where U_0 and T_0 are integration constants.

13.16 (a) From Equation 8.8 we have

$$\left(\frac{\partial S}{\partial T}\right)_{V,N} = \frac{C_V}{T} = \frac{\alpha nR}{T} \quad (\because C_V = \alpha nR)$$

$$\left(\frac{\partial S}{\partial V}\right)_{T,N} = \left(\frac{\partial P}{\partial T}\right)_{V,N} = \frac{nR}{V} \quad \text{(see (3) above)} \tag{6}$$

By integrating this and rearranging, we obtain

$$S = S_0 + nR \ln [(T/T_0)^{\alpha}(V/V_0)]. \tag{7}$$

(b) To show that the internal energy depends on S and V, we combine Equations (5) and (7) to have

$$U = U_0 + \alpha nRT_0 \left[\left(\frac{V_0}{V}\right)^{1/\alpha} \exp\left(\frac{S - S_0}{\alpha nR}\right) - 1\right]. \tag{8}$$

This is the fundamental equation for the internal energy of an ideal gas.

(c) To find the enthalpy in terms of S and P, we use $PV = nRT$, i.e. $V = nRT/P$. From Equation (7),

$$S - S_0 = nR \ln [(T/T_0)^{\alpha}(V/V_0)]$$

$$= nR \ln \left[(T/T_0)^{\alpha}\left(\frac{P_0}{P}\right)(T/T_0)\right] \quad \text{(using } V = nRT/P)$$

$$= nR \ln [(T/T_0)^{1+\alpha}(P_0/P)]$$

or

$$(1+\alpha)\ \ln\ (T/T_0) = \frac{S-S_0}{nR} - \ln\ (P_0/P)$$

or

$$T = T_0\left[\exp\ \left\{\frac{S-S_0}{nR(1+\alpha)}\right\} + (P/P_0)^{1/(1+\alpha)}\right] \tag{9}$$

Now, $H=U+PV$. Therefore, $H_0=U_0+P_0V_0$, i.e. $U_0=H_0-P_0V_0$.
From Equation (5)

$$U = U_0 + \alpha nR(T - T_0) = H_0 - P_0V_0 + \alpha nR(T - T_0)$$
$$= H_0 + \alpha nRT - \alpha nRT_0 - P_0V_0$$
$$= H_0 + \alpha nRT - \alpha nRT_0 - nRT_0$$
$$= H_0 + \alpha nRT - (1 + \alpha).$$

By substituting the value of T from Equation (9) above, we obtain

$$H = H_0 - (1+\alpha)nRT_0\left[1 - (P/P_0)^{1/(1+\alpha)} - \exp\ \left\{\frac{S-S_0}{nR(1+\alpha)}\right\}\right]. \tag{10}$$

This is the fundamental equation for the enthalpy of an ideal gas.

(d) To find the Helmholtz free energy $A=U-TS$, we use $A_0=U_0-T_0S_0$. Using Equations (5) and (7) and simplifying, we obtain

$$A = A_0 + (\alpha nR - S_0)(T - T_0) - nRT\ \ln\ [(T/T_0)^{\alpha}(V/V_0)] \tag{11}$$

This is the fundamental equation for the Helmholtz free energy of an ideal gas.

(e) The Gibbs free energy can be obtained from $G=A+PV$. This gives $G_0=A_0+P_0V_0$. Also, using $PV=nRT$ and Equation (11) above and simplifying, we obtain

$$G = G_0 + [(1+\alpha)nR - S_0](T - T_0) - nRT\ \ln\ \left[\left(\frac{T}{T_0}\right)^{1+\alpha}\left(\frac{P}{P_0}\right)\right]. \tag{12}$$

From Equations (7), (8), (10), (11) and (12) we find that in the fundamental equations of U, S, H, A and G, these thermodynamic quantities depend on $(S, V), (T, V), (S, P), (T, V)$ and (T, P), respectively.

13.17

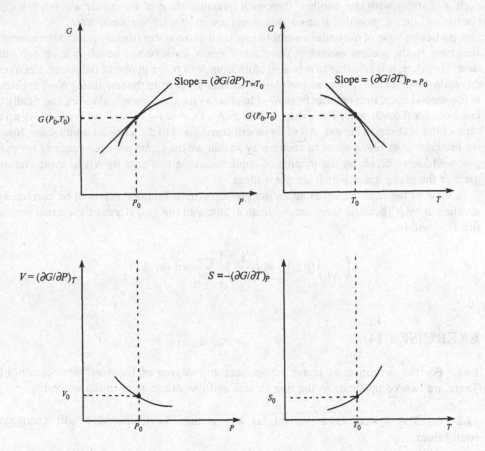

The slope of the curve of G vs. P is $(\partial G/\partial P)_{T=T_0}$ which is the volume V and that of the curve G vs. T is $(\partial G/\partial T)_{P=P_0}$ which is the negative of S. The point $G(P_0, T_0)$ is shown on the graph.

13.18   The given partition function is not correct for the given system. Here, we are dealing with a perfect gas of N identical molecules which can move freely and be exchanged in positions (these positions being included). In the given system, if two gas atoms exchange their state, i.e. the first one goes into the state of the second and *vice versa*, the final state of the whole gas becomes indistinguishable from the initial state. Then the quantum state of the entire gas must be considered identical in the final and initial states; in fact, there is only one quantum state.

Exchange of state of a pair of atoms or molecules actually corresponds to a shift of the point that represents the assembly in the phase space, because each atom has a set of co-ordinates and momenta (generally including spin co-ordinates and momenta) and each of these provides a co-ordinate in the phase space for the assembly of atoms. If a pair of atoms in the given system is exchanged, a phase volume of h^f will shift to another phase volume of the same magnitude, and each quantum state will be counted more than once if we add over the entire phase space. If we have an assembly of N atoms, as in this question, which can exchange states with one another, then each quantum state of the whole assembly will be counted $N!$ times, provided that no two atoms are in exactly the same state.

A particular state of the entire assembly implies that the states of each atom in the assembly are given. In the present assembly, there are N atoms, each one of which is in its particular state. However, it is possible to select an atom from N to place in one of the states. Let us call this state the first state. Then, we can choose another atom from the remaining $N-1$ to place it in the second state, then another from $N-2$ to place in the third state, and so on, and finally the last atom for the Nth state. Thus, there will be $N(N-1)(N-2)\cdots 1 = N!$ ways of setting up the state of the N atoms in the gas. All of these will correspond to the same quantum state. Now, if we integrate over phase space in such a way so that we duplicate possible states of the entire gas, we have to divide by the number of duplications (in this case by $N!$) to avoid counting parts of the phase space which are not distinct.

In view of the above consideration, the given partition function needs to be corrected by dividing it by $N!$ because there are N identical atoms in the gas. Hence, the correct partition function will be

$$\left(\frac{1}{N!}\right)(\text{p.f.})^N = \left(\frac{1}{N!}\right)\left[\sum_{\ell} \exp\left(\varepsilon_\ell/kT\right)\right]^N$$

EXERCISES 14

14.1 For the definition of phase, component and degree of freedom, see Section 14.2. There are two components in the first system and one component in the second.

14.2 (a) The system is univariant, as any change in temperature will change the equilibrium.

(b) The system is bivariant, as any change in temperature or pressure will change the equilibrium.

14.3 There are four one-phase, five two-phase, three three-phase and one four-phase equilibria possible.

14.4 There is one degree of freedom.

14.5 See the derivation of Equation 14.2

14.6 Plot $\log P$ against $1/T$. This would give a straight line and the slope of this straight line would be $(-\Delta H)_{vap}/2.303R$, where $(\Delta H)_{vap}$ is the heat of vaporization (see Equation 14.6).

14.7 The temperature is 386 K and the pressure is 0.117 atm.

14.8 The increase in vapour pressure is 0.0017 Torr.

14.9 From Equation 14.12, $\Delta T/\Delta P = T_m(V_\ell - V_s)/(\Delta H)_f$. In this case, $\Delta T = 0.47°C$, $\Delta P = 174$ atm and $(\Delta H)_f = 6276$ J mol^{-1}. It must be remembered that the value of $(\Delta H)_f$

must be converted from joules per mole to litre-atmosphere per mole to obtain a value for $\Delta T/\Delta P$ in atmospheres per degree Celsius. Therefore, we write $(\Delta H)_f = 6276\,\text{J mol}^{-1} = 61.898\,\ell\,\text{atm}$. Hence, the change in volume on melting is

$$V_\ell - V_s = \frac{(0.47°\text{C})(61.898\,\ell\,\text{atm})}{(125°\text{C})(174\,\text{atm})} = 0.00134\,\ell$$

This volume change is mainly due to the change in volume of the liquid because a liquid has larger volume.

14.10 At a phase transition, the chemical potentials of all phases involved must be equal and the phases can coexist. The Gibbs free energy is related to the chemical potential by $G = \sum n_i \mu_i$, where μ_i is the chemical potential per mole and n_i is the number of moles. Hence, at a phase transition, the Gibbs free energy of each phase must have the same value, and $(\partial G/\partial n_i)_{P,T} = \mu_i$ must be equal.

14.11 The amounts of liquid and vapour which coexist at the point W is given by the lever rule. The total molar volume v_W is given in terms of the molar volume of the liquid at the point Z and the molar volume of vapour at the point X. If v_ℓ is the molar volume of the liquid at Z an v_g is that of the vapour at X, then $v_W = x_\ell v_\ell + x_g v_g$ where x_ℓ is the mole fraction of liquid at W and, x_g is the mole fraction of vapour at W, and $x_\ell + x_g = 1$. By multiplying the left side of $v_W = x_\ell v_\ell + x_g v_g$ by $x_\ell + x_g$ (which is 1) and rearranging, we have $x_\ell/x_g = (v_g - v_W)/(v_W - v_\ell)$. Hence the idea is proved.

14.12 Let n_v, n_ℓ be the number of moles of vapour and liquid at point W, u_v be the internal energy per mole at point X and u_ℓ be the internal energy per mole at point Z. Let v_v be the volume per mole of vapour at point X and v_ℓ be that of liquid at point Z. If x_ℓ and x_v are the mole-fractions of liquid and vapour, respectively, then

$$x_\ell + x_v = 1 \tag{1}$$

$$V_m = x_\ell v_\ell + x_v v_v \tag{2}$$

where V_m is the total volume per mole at point W. The total internal energy at point W is

$$U_T = n_\ell u_\ell + n_v u_v \tag{3}$$

where u_ℓ is a function of T and v_ℓ, and u_v is a function of T and v_v. Then, the total energy per mole u_T at point W is

$$u_T = x_\ell u_\ell + x_v u_v \tag{4}$$

where u_v is a function of v_v and T, and u_ℓ is a function of v_ℓ and T.

The heat capacity per mole at point W can be found by differentiating Equation (4) with respect to T. Thus

$$c_V = \left(\frac{\partial u_T}{\partial T}\right)_{V_m} = x_\ell \left(\frac{\partial u_1}{\partial T}\right)_{V_m} + u_\ell \left(\frac{\partial x_\ell}{\partial T}\right)_{V_m} + x_v \left(\frac{\partial u_v}{\partial T}\right)_{V_m} + u_v \left(\frac{\partial x_v}{\partial T}\right)_{V_m}$$

$$= x_\ell \left(\frac{\partial u_\ell}{\partial T}\right)_{V_m} + x_v \left(\frac{\partial u_v}{\partial T}\right)_{V_m} + \left[u_v \left(\frac{\partial x_v}{\partial T}\right)_{V_m} + u_\ell \left(\frac{\partial x_\ell}{\partial T}\right)_{V_m}\right]$$

$$= x_\ell \left(\frac{\partial u_\ell}{\partial T}\right)_{V_m} + x_v \left(\frac{\partial u_v}{\partial T}\right)_{V_m} + \left[-u_v \left(\frac{\partial x_\ell}{\partial T}\right)_{V_m} + u_\ell \left(\frac{\partial x_\ell}{\partial T}\right)_{V_m}\right]$$

because $x_v = 1 - x_\ell$ and $\partial x_v / \partial T = -\partial x_\ell / \partial T$. Hence,

$$c_V = x_v \left(\frac{\partial u_v}{\partial T}\right)_{V_m} + x_\ell \left(\frac{\partial u_\ell}{\partial T}\right)_{V_m} + (u_\ell - u_v)\left(\frac{\partial x_\ell}{\partial T}\right)_{V_m}. \tag{5}$$

From the Clausius–Clapeyron equation (see Equation 10.49) we have $(dP/dT)_{coex} = \Delta H / T \Delta V$ and also $\Delta H = \Delta U + \Delta(PV)$. By combining these two equations, we have

$$\left(\frac{dP}{dT}\right)_{coex} = \frac{\Delta u}{T \Delta V_m} + P/T$$

or

$$T\left(\frac{dP}{dT}\right)_{coex} - P = \frac{\Delta u}{\Delta V_m}$$

or

$$\Delta u = \left[\left\{T\left(\frac{dP}{dT}\right) - P\right\}(\Delta V_m)\right]_{coex}.$$

However, $\Delta V_m = v_v - v_\ell$ and $\Delta u = u_v - u_\ell$. Then,

$$u_v - u_\ell = \left[\left\{T\left(\frac{dP}{dT}\right) - P\right\}(v_v - v_\ell)\right]_{coex}. \tag{6}$$

Next we want to find an expression for $(\partial x_\ell / \partial T)_{V_m}$. Let us differentiate Equation (2) with respect to T, keeping V_m constant. Then,

$$0 = x_v \left(\frac{\partial v_v}{\partial T}\right)_{V_m} + v_v \left(\frac{\partial x_v}{\partial T}\right)_{V_m} + x_\ell \left(\frac{\partial v_\ell}{\partial T}\right)_{V_m} + v_\ell \left(\frac{\partial x_\ell}{\partial T}\right)_{V_m}.$$

However, $x_\ell + x_v = 1$. Therefore $(\partial x_v / \partial T)_{V_m} = -(\partial x_\ell / \partial T)_{V_m}$. By substituting this in the second term of the above equation, we obtain

$$(v_v - v_\ell)\left(\frac{\partial x_\ell}{\partial T}\right)_{V_m} = x_\ell \left(\frac{\partial v_\ell}{\partial T}\right)_{V_m} + x_v \left(\frac{\partial v_v}{\partial T}\right)_{V_m} \tag{7}$$

or

$$\left(\frac{\partial x_\ell}{\partial T}\right)_{V_m} = \left[x_\ell \left(\frac{\partial v_\ell}{\partial T}\right)_{V_m} + x_v \left(\frac{\partial v_v}{\partial T}\right)_{V_m}\right] / (v_v - v_\ell) \tag{8}$$

However,

$$\left(\frac{\partial v_\ell}{\partial T}\right)_{V_m} = \left(\frac{\partial v_\ell}{\partial T}\right)_{coex}. \tag{9}$$

By definition V_m lies on the coexistence curve. Hence, Equation (8) becomes

$$\left(\frac{\partial x_\ell}{\partial T}\right)_{V_m} = \left[x_\ell\left(\frac{\partial v_\ell}{\partial T}\right) + x_v\left(\frac{\partial v_v}{\partial T}\right)\right]_{coex}\Big/(v_v - v_\ell). \tag{10}$$

By virtue of the mathematical relation 14.27, we can write

$$\left(\frac{\partial u_\ell}{\partial T}\right)_{V_m} = \left(\frac{\partial u_\ell}{\partial T}\right)_{v_\ell} + \left(\frac{\partial u_\ell}{\partial v_\ell}\right)_T\left(\frac{\partial v_\ell}{\partial T}\right)_{coex}. \tag{11}$$

Similarly, we can write

$$\left(\frac{\partial u_v}{\partial T}\right)_{V_m} = \left(\frac{\partial u_v}{\partial T}\right)_{v_v} + \left(\frac{\partial u_v}{\partial v_v}\right)_T\left(\frac{\partial v_v}{\partial T}\right)_{coex}. \tag{12}$$

However, $(\partial u_\ell/\partial T)_{v_\ell} = c_{v_\ell}$, the heat capacity per mole of the liquid phase and $(\partial u_v/\partial T)_{v_v} = c_{v_v}$, the heat capacity per mole of the vapour phase. Now by combining Equations (5), (6), (10), (11) and (12), we have

$$c_v = x_v c_{v_v} + x_\ell c_{v_\ell} + x_v\left(\frac{\partial u_v}{\partial v_v}\right)_T\left(\frac{\partial v_v}{\partial T}\right)_{coex} + x_\ell\left(\frac{\partial u_\ell}{\partial v_\ell}\right)_T\left(\frac{\partial v_\ell}{\partial T}\right)_{coex}$$
$$- \left[T\left(\frac{dP}{dT}\right) - P\right]_{coex}\left[x_\ell\left(\frac{\partial v_\ell}{\partial T}\right) + x_v\left(\frac{\partial v_v}{\partial T}\right)\right]_{coex}. \tag{13}$$

However,

$$\left(\frac{\partial u_v}{\partial v_v}\right)_T = T\left(\frac{\partial P_v}{\partial T}\right)_{v_v} - P_v. \tag{14}$$

Again, using the mathematical relation 14.27 we can write for the vapour and liquid phases

$$\left(\frac{\partial P_v}{\partial T}\right)_{v_v} = \left(\frac{dP}{dT}\right)_{coex} - \left(\frac{\partial P_v}{\partial v_v}\right)_T\left(\frac{\partial v_v}{\partial T}\right)_{coex} \tag{15}$$

$$\left(\frac{\partial P_\ell}{\partial T}\right)_{v_\ell} = \left(\frac{dP}{dT}\right)_{coex} - \left(\frac{\partial P_\ell}{\partial v_\ell}\right)_T\left(\frac{\partial v_\ell}{\partial T}\right)_{coex}. \tag{16}$$

Now by combining Equations (13), (14), (15) and (16) we obtain

$$c_v = x_v\left[c_{v_v} - T\left(\frac{\partial P_v}{\partial v_v}\right)_T\left(\frac{\partial v_v}{\partial T}\right)^2_{coex}\right] + x_\ell\left[c_{v_\ell} - T\left(\frac{\partial P_\ell}{\partial v_\ell}\right)_T\left(\frac{\partial v_\ell}{\partial T}\right)^2_{coex}\right]$$

which is the required heat capacity per mole inside the coexistence curve in Figure 14.9. The heat capacity at constant volume can remain finite below the coexistence curve, but the heat capacity at constant pressure is always infinite below the coexistence curve. If we add heat at constant pressure, liquid will become vapour but there will be no change in the temperature. Hence, $C_P = (dQ/dT)_P = \infty$ in the region where two phases exist.

14.13

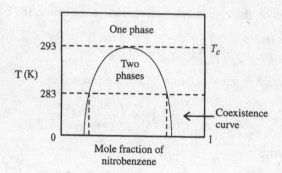

The diagram shows:
- One phase (top region)
- T_c (at 293)
- Two phases (middle region)
- 283
- T (K) (y-axis label)
- Coexistence curve
- 0 and 1 on x-axis
- Mole fraction of nitrobenzene

14.14 For the phase diagram of liquid He^4 and its explanation, see Section 14.10 (B) and Figure 14.18.

The phase diagram of liquid He^3 does not have a λ-point and liquid He^3 does not have the same superfluid properties of liquid He(II). However, liquid He^3 does show superfluid properties below 0.003 K, which is attributed to the formation of Cooper-type pairs of He^3 atoms that behave as bosons.

According to the ideal boson gas model, a significant proportion of the He^4 atoms in liquid helium should correspond to the single particle ground state below the Bose temperature $T_b = 3.1$ K. These particles may constitute the superfluid component of liquid He(II), while the He^4 atoms in excited single particle states could make up the normal component of liquid He(II). Thus, it is reasonable to think that the superfluid properties of liquid He(II) could be due, at least partly, to a Bose–Einstein type condensation at the λ-point.

EXERCISES 15

15.1 This is true by symmetry considerations. If there is no external force field acting on the gas, the molecules have an equal chance of being found in the region d^3r of any point within the system. This makes sense, because we do not expect to find a density gradient in this system when it is in thermal equilibrium. Therefore, the spatial variable is really superfluous, and we can consider the probable number of molecules in the whole system that have velocities in the range d^3v of v. This quantity is $\eta(v)\, d^3v$.

The point that $\eta(v)$ depends on the magnitude of the velocity and not on its direction is true also by virtue of symmetry considerations, because there can have no preferred direction where the container of the gas, and hence the centre-of-mass of the entire gas, is taken to be at rest.

15.2 See the derivation of Equation 15.17.

15.3 The distribution function $\eta(v)$ gives the probable number of molecules that can be expected to be found in the region $dv_x dv_y dv_z$ about v. Since the probable number of molecules depends only on v^2, on average the points in Figure 15.2 will be symmetrically distributed about the origin $\mathbf{v} = 0$. Because the factor $\exp(-\beta m \mathbf{v}^2/2)$ decreases as $\mathbf{v}^2$ increases, we can expect the molecules to be predominantly situated near the origin.

15.4 See Figure 15.3. The lower the temperature, the narrower the distribution curve. However, the area under any curve at a lower temperature is the same as that under another curve at a higher temperature.

15.5 See the derivation of Equation 15.37. The variation of v_m is shown in Figure 15.6.

15.6 For a perfect gas $PV = NkT$, where P is the pressure, V is the volume, N is the number of molecules, T is the temperature and k is the Boltzmann constant, equal to 1.38×10^{-23} J deg^{-1}.

We are given that $T = 300$ K and $P = 10^{-10}$ mm $= 10^{-10}/760$ atmosphere.

$$1 \text{ atmosphere} = 1.013 \times 10^5 \text{ newtons m}^{-2}$$

$$\frac{10^{-10}}{760} \text{ atmosphere} = 0.0132 \times 10^{-11} \times 1.013 \times 10^5 \text{ n m}^{-2}$$

$$1 \text{ cm}^3 = 10^{-6} \text{ m}^3$$

Therefore,

$$N = \frac{(0.0132 \times 1.013 \times 10^{-6} \text{ n/m}^2)(10^{-6} \text{ m}^3)}{(1.38 \times 10^{-23} \text{ J deg}^{-1})(300 \text{ deg})}$$

$$= \frac{0.01337 \times 10^{-12} \text{ n m}}{4.14 \times 10^{-23} \text{ J}} \quad [1 \text{ J} = 1 \text{ n m}]$$

$$= 0.00322 \times 10^9 = 3.22 \times 10^6.$$

15.7 The most probable speed of a molecule of a perfect gas is given by Equation 15.37, i.e. $v_m = (2kT/m)^{1/2}$. We are given that $T = 300$ K and $k =$ Boltzmann constant $= 1.38 \times 10^{-23}$ J K^{-1}. The molecular weight of N_2 is 28 and Avogadro's number is 6.02×10^{23} mol^{-1}. So the mass of an N_2 molecule is $m \simeq 28/6.02 \times 10^{23} \simeq 4.65 \times 10^{-23}$ g. Using this value, we have

$$v_m \simeq \left(\frac{2 \times 1.38 \times 10^{-23} \times 300}{4.65 \times 10^{-26}} \right)^{1/2} \frac{(\text{J/K})(\text{K})}{\text{kg}}$$

$$\simeq 4.24 \text{ m s}^{-1}.$$

The root mean square velocity v_{rms} of a diatomic molecule such as nitrogen and oxygen is given by

$$v_{rms} = \left(\frac{3kT}{m} \right)^{1/2}$$

where k is the Boltzmann's constant, T is the temperature and m is the mass of the molecule.

The mass of a nitrogen molecule is 4.65×10^{-26} kg, $k = 1.38 \times 10^{-23}$ J K^{-1} and $T = 273$ K. By substituting these values in the above equation, we have $v_{rms} = 493$ m s^{-1}.

15.8   For a perfect gas, $PV = NkT$. Given $P = 1.013 \times 10^5 \, \text{n m}^{-2}$, $V = 10^3 \, \text{cm}^3$, $T = 300 \, \text{K}$ and $k = 1.38 \times 10^{-23} \, \text{J K}^{-1}$. Then the number of molecules

$$N = \frac{PV}{kT} = \frac{1.013 \times 10^5 \times 10^{-3}}{1.38 \times 10^{-23} \times 300} = 2.45 \times 10^{22}.$$

The most probable speed is given by $v_m = (2kT/m)^{1/2}$. The mass m of He is $m = 4.002 \times 1.66 \times 10^{-27} \, \text{kg} = 6.64 \times 10^{-27} \, \text{kg}$. Then,

$$v_m = \left(\frac{2 \times 1.38 \times 10^{-23} \times 300}{6.64 \times 10^{-27}} \right)^{1/2} = 1.12 \times 10^3 \, \text{m s}^{-1}.$$

The average speed is given by $\bar{v} = (8kT/\pi m)^{1/2}$. Then,

$$\bar{v} = \left(\frac{8 \times 1.38 \times 10^{-23} \times 300}{3.14 \times 6.64 \times 10^{-27}} \right)^{1/2} = 1.26 \times 10^3 \, \text{m s}^{-1}.$$

The ratio $\bar{v}/v_m = \dfrac{1.26 \times 10^3}{1.12 \times 10^3} \simeq 1.13$

15.9   See Section 15.2.2.

15.10   For sketches, see Figure 15.7. As T increases, the distribution rounds off; states within about kT below E_F are partly depopulated and states within about kT above E_F are partly populated. At $T = 0 \, \text{K}$, we have (from Equation 15.61) $f(E) = 1$ for $E < E_F$, and $f(E) = 0$ for $E > E_F$. Hence, at absolute zero temperature E_F acts as a cut-off energy: all states with energy less than E_F are completely filled, whereas all states with energy greater than E_F are empty.

15.11   (a) To find the Fermi energy $E_F(0)$ at absolute zero, see the derivation of Equation 15.64. dn is given by

$$\mathrm{d}n = f(E)g(E) \, \mathrm{d}E.$$

By substituting the values of $f(E)$ from Equation 15.61 and $g(E)$ as given, we obtain

$$\mathrm{d}n = \frac{1}{2\pi^2} \left(\frac{2m}{\hbar^2} \right)^{3/2} \frac{E^{1/2} \, \mathrm{d}E}{\exp\left(\frac{E - E_F}{kT} \right) + 1}$$

$$= \frac{CE^{1/2} \, \mathrm{d}E}{\exp\left(\frac{E - E_F}{kT} \right) + 1}.$$

$f(E) = 0$ for $E > E_F$ and $f(E) = 1$ for $E < E_F$. Thus, all states are filled up to E_F and E_F is determined in terms of the number of electrons per unit volume N by

$$\int \mathrm{d}n = \int \frac{CE^{1/2} \, \mathrm{d}E}{\exp\left(\frac{E - E_F}{kT} \right) + 1}$$

or

$$N = \frac{2C}{3}[E_F(0)]^{3/2}.$$

By rearranging this equation and substituting the value of C as given, we obtain

$$E_F(0) = \frac{\hbar}{2m}(3\pi^2 N)^{3/2}$$

which is the same as derived above. This shows that $E_F(0)$ can be determined in two independent ways which give the same result.

(b) The Fermi temperature is given by

$$T_F = \frac{E_F}{k}$$

where k is the Boltzmann constant. So, T_F of copper is

$$T_F = \frac{7.04 \times 1.602 \times 10^{-19} \text{ J K}}{1.38 \times 10^{-23} \text{ J}}$$

$$\simeq 8 \times 10^4 \text{ K}.$$

15.12   See the derivation of Equation 15.80.

15.13   (a) See the derivation of Equation 15.93.

(b) Given that $T = 300$ K and $T_F = 64\,000$ degrees. Taking $R = 8.31 \text{ J mol}^{-1} \text{deg}^{-1}$ and substituting these values in Equation 15.93 we obtain

$$C_V(\text{el}) = \frac{8.31 \times 300}{64,000} \frac{\text{J deg}}{(\text{mol deg})\text{deg}}$$

$$= 0.038 \text{ J mol}^{-1} \text{ deg}^{-1}.$$

This value is lower than the actual value at 300 K because of the approximation in Equation (15.93).

15.14   See the property of $f(E)$ in Section 15.3.2.

15.15   The reduced Fermi level μ^* (also called the reduced electrochemical potential) is defined by the ratio μ/kT or E_F/kT. It is a dimensionless quantity.

The Fermi level can be identified with the electrochemical potential only when the electron system is under equilibrium conditions. Under non-equilibrium conditions, $E_F \neq \mu$.

(a) For $T > 0$ K, $f(E) \simeq 1$ for all values of $E \ll E_F$ and $f(E) \simeq 0$ for all values of $E \gg E_F$.

(b) For values of E not too far from E_F, $f(E)$ will have values between 0 and 1.

15.16   The salient difference between classical and quantum particles is the continuity of the energy spectrum of classical particles and the discrete nature of the energy spectrum of quantum particles.

The distribution of discrete energy levels is given by Equation 15.61. When $(E - E_F)$ is very much greater than kT, this equation makes a transition to Equation 15.94 which is classical Boltzmann distribution applicable to continuous energy spectrum. Hence, under the condition $(E - E_F) \gg kT$, it is possible to make a transition from discrete to continuous spectrum, and hence from the Fermi–Dirac to the Boltzmann distribution.

The Boltzmann distribution is satisfied better at higher temperatures, because when the temperature is high, the energy levels broaden and the distances between them become so small that the energy spectrum can be regarded as practically continuous. If the temperature is lowered, the distribution of the particles in an electron gas differs more and more from the Boltzmann distribution. Hence, the Boltzmann distribution can be considered as the limiting case of the Fermi–Dirac distribution.

The Fermi–Dirac distribution law applies to electrons as a system of indistinguishable particles with one particle allowed in each state of the system. On the other hand, the Boltzmann distribution allows any number of particles to have exactly the same energy and momentum.

15.17   In the case of fermions which obey the Fermi–Dirac statistics the particle states are $n_s = 0$ or $n_s = 1$. The sum on N (the total number of particles) must be restricted to $N = 0, 1, 2$ and 3, because if N were larger than 3, then two particles will have to be in the same state, and this is not allowed in the Fermi–Dirac case.

15.18   Whether a molecule or an atom obeys the Bose–Einstein or Fermi–Dirac statistics depends on the number of elementary particles (electron, proton and neutron) making up this molecule or atom. If the number of elementary particles is even, the molecule or the atom will obey the Bose–Einstein statistics. On the other hand, if the number is odd, the molecule or the atom will obey the Fermi–Dirac statistics. It does not matter whether the elementary particles are in the nucleus or the electron shells.

He^4 has two protons and two neutrons in the nucleus, and two electrons outside the nucleus. So the total number of elementary particles is 6 (even). Hence, it obeys the Bose–Einstein statistics. He^3 has only one neutron in the nucleus and obeys the Fermi–Dirac statistics.

This behaviour follows from the symmetry of the wave functions. Wave functions for systems which have an even number of elementary particles are symmetrical, whereas wave functions for systems with an odd number of elementary particles are antisymmetric. If two molecules are in the same energy level, exchange of these molecules must leave the wave function identical. In the anti-symmetrical case, it changes the sign and such a wave function must disappear. Therefore, we cannot have states with two molecules in the same energy level. If the two molecules are in the different states, the exchange of two particles changes a function of the collection of co-ordinates of the two molecules, e.g. x_1 (for the first), x_2 (for the second) to a function of x_2, x_1. Since in this case the functional dependence on the first set of co-ordinates (x_1) is different from the dependence on the second set of co-ordinates (x_2), before exchange, the function $f(x_1, x_2)$ will differ from the function $f(x_2, x_1)$.

15.19   At $T = 0$ all the single particle states up to the Fermi energy E_F are occupied. Since the fermions are identical, this corresponds to only one state of the N-particle system. This means that at $T = 0$ the number of states $W = 1$. Hence, $S = k \ln W = k \ln 1 = 0$. This is in agreement with the third law of thermodynamics which states that at absolute zero of temperature the entropy of a substance is zero.

15.20 In Example 15.4 it was shown that exp $(-\mu/kT) = 3.4 \times 10^5$ and $\mu/kT = -12.74$. Therefore, at 273 K we have $\mu = -4.79 \times 10^{-20}$ J. This result shows that, in the classical limit, the chemical potential of a typical monatomic gas is negative and exp $(-\mu/kT)$ is very much greater than unity.

15.21 For an ideal boson gas the number of single particle states which have single particle energy between the range ε and $\varepsilon + d\varepsilon$ is given by Equation 15.69, i.e.

$$g(\varepsilon)\, d\varepsilon = \frac{gV}{4\pi^2}\left(\frac{2m}{\hbar^2}\right)^{3/2}\sqrt{\varepsilon}.$$

To find the energy distribution we need to multiply this by the mean number of bosons in each of these states, $b(\varepsilon)$, given by Equation 15.107. Thus, the energy distribution is given by

$$n(\varepsilon)\, d\varepsilon = \frac{gV}{4\pi^2}\left(\frac{2m}{\hbar^2}\right)^{3/2}\left[\frac{\sqrt{\varepsilon}\, d\varepsilon}{\exp\{(\varepsilon - \mu)/kT\} - 1}\right].$$

The total number of bosons is then found by integrating this equation between the limits 0 and ∞, i.e.

$$N = \int_0^\infty n(\varepsilon)\, d\varepsilon = \frac{gV}{4\pi^2}\left(\frac{2m}{\hbar^2}\right)^{3/2}\int_0^\infty \frac{\sqrt{\varepsilon}\, d\varepsilon}{\exp\{(\varepsilon - \mu)/kT\} - 1}.$$

15.22 A mole of He^4 under the given conditions contains 6.02×10^{23} atoms, so the total number of particles in the gas $N = 6.02 \times 10^{23}$. Given that mass of a He^4 atom $m = 6.65 \times 10^{-27}$ kg and $g = (2s+1) = 1$ (since spin $s = 0$). Substituting all these and $\hbar^2 = 1.055 \times 10^{-34}$ J s into Equation 15.126, we have

$$T_b = \frac{(3.31)(1.055 \times 10^{-34}\text{J s})^2}{(6.65 \times 10^{-27}\text{kg})(1.38 \times 10^{-23}\text{J K}^{-1})}\left[\frac{6.02 \times 10^{23}}{(1)(22.4 \times 10^{-3}\ \text{m}^3)}\right]^{2/3}.$$

$$= (0.401 \times 10^{-18})(8710 \times 10^{13})\frac{\text{J}^2\ \text{s}^2\ \text{K}}{\text{kg J m}^2}$$

$$= 0.035\ \text{K} \quad (1\ \text{J} = \text{kg m}^2\ \text{s}^{-2})$$

He^4 is a Boson gas and it liquefies at 4.21 K at 0.1 MPa pressure, which is much higher than 0.035 K. So, at our calculated T_b He^4 is a liquid and not a perfect gas.

15.23 Follow the derivation of Equation 15.143. This is the required equation which shows that the pressure is proportional to the internal energy if V is constant and there are no interactions between the particles.

This equation is valid for an ideal fermion gas for all values of U, provided that there are no interactions between the particles. According to this equation, the variation of P with the temperature at constant volume should have the same characteristics as that of U with T (see Figure 15.16).

From Figure 15.16 it is apparent that the pressure of a perfect Fermi–Dirac gas is always higher than that of a perfect classical monatomic gas, provided that N, V and T are the same for both.

15.24   Under the given conditions, Equation 15.143 is valid for all values of the internal energy. So, we can use this equation to calculate P. Given that $U = 2.5\,\mathrm{J}$ and $V = 22.4 \times 10^{-3}\,\mathrm{m}^3$. Hence,

$$P = \frac{(2)(2.5\,\mathrm{J})}{(3)(22.4 \times 10^{-3}\,\mathrm{m}^3)} = 0.22\,\mathrm{J\,m}^{-3}$$
$$= 0.22\ (\mathrm{kg\,m^2\,s^{-2}})(\mathrm{m}^{-3}) = 0.22\,\mathrm{kg\,m^{-1}\,s^{-2}}$$
$$= 0.22\,\mathrm{N\,m}^{-2} = 0.22\ \mathrm{atm}.$$

15.25   One mole of an ideal boson gas contains 6.02×10^{23} atoms. Therefore, the concentration is

$$N/V = \frac{6.02 \times 10^{23}}{27 \times 10^{-6}\,\mathrm{m}^3} = 2.2 \times 10^{28}\,\mathrm{m}^{-3}.$$

Since the spin is zero, $g = 2s + 1 = 1$. By substituting these values and $m = 6.65 \times 10^{-27}\,\mathrm{kg}$ in Equation 15.126, we obtain the Bose temperature $T_b = 3.1\,\mathrm{K}$.

15.26   Liquid He^3 does not show a λ-transition as we find in the case of liquid He^4. The interatomic forces within liquid He^3 and He^4 should be quite similar. So the only significant difference between these two systems is that He^3 consists of Fermi–Dirac particles and He^4 consists of Bose–Einstein particles. This difference is one of the reasons to believe that the λ-point in liquid He^4 is related to a Bose–Einstein type condensation.

EXERCISES 16

16.1   See the derivation of Equation (16.7). The internal energy of N harmonic oscillators in three dimensions is given by $U = 3N\overline{E} = 3NkT$ and that of one mole of a substance is $U_m = 3RT$.

16.2   According to the classical theory the molar heat capacity of a solid at constant volume is $C_V = (\partial U_m/\partial T)_V$. Using the value of U_m from Question 16.1, we obtain the lattice contribution to the molar heat capacity for one mole of atoms as $C_V = 3R \simeq 8.94\,\mathrm{J\,deg}^{-1}$ which is the Dulong–Petit value.

At reasonably high temperatures, and often up to room temperature, this value is in quite good agreement with the observed total heat capacity of many metals and non-metals. However, the agreement fails as the solid is cooled below room temperature. As an example, below room temperature the heat capacity of metallic silver drops rapidly (see Figure 16.1).

16.3   It has been proved experimentally that a T-approach of heat capacity of metals as $T \to 0$ is characteristic of that part of the heat capacity which is associated with free (conduction) electrons in the crystal, whereas a T^3-approach is characteristic of lattice vibrations in cubic and near-cubic crystals.

These behaviours of heat capacity bring out the difficulties for the classical theory to predict many of the observed features of the lattice heat capacity of solids, particularly those at low temperatures.

16.4 Einstein treated the thermal properties of the vibration of a lattice of N atoms in a solid as a set of $3N$ independent harmonic oscillators in one dimension. He assumed that each oscillator has the identical frequency ν and then quantized the energies of the oscillators according to Planck's theory of black body radiation. Therefore, it would appear that any representation of the drop in heat capacity at low temperatures based on Einstein's model will depend on an appropriate choice of the oscillatory frequency.

16.5 See the derivation of Equation 16.15 for the Planck–Einstein distribution law.

The average energy of a harmonic oscillator at high temperatures can be derived from Equation 16.15, and is given by $\overline{E} = kT$ (see Equation 16.16). This shows that at high temperatures the average energy of the harmonic oscillator is approximately equal to the classical average energy kT.

At low temperatures, the average energy of the oscillator is given by Equation 16.17. This gives the heat capacity at constant volume of a system in thermal equilibrium as

$$C_V \simeq Nk(\hbar\omega/kT)^2 \, \exp\,(-\hbar\omega/kT)$$

which tends to zero as $T \to 0$. In this limit, the dominant factor is $\exp\,(-\hbar\omega/kT)$. Therefore, at low temperatures, the variation of C_V on the basis of Einstein's model is as $\exp\,(-\hbar\omega/kT)$.

16.6 The Einstein temperature is defined by Equation 16.18. In general, T_E is in the range 100–300 K, but these are solids for which T_E can be below or above this range.

For an expression of C_V of a system consisting of N harmonic oscillators in one dimension, see the derivation of Equation 16.19.

16.7 See the text following Equation 16.19.

16.8 Debye assumed that: (i) in a solid there are three modes of vibration for each wave vector, one longitudinal mode and two transverse modes, and (ii) the number of modes of vibration in a solid is equal to $3N$, where N is the total number of atoms in the solid. Precisely, $3N$ is the number of vibrational modes of a system that has $3N$ degrees of freedom. So Debye's second hypothesis is quite reasonable.

In Einstein's theory, it was assumed that all the atoms in the lattice oscillate with simple harmonic motion in a steady potential well with the same angular frequency ω_E (known as the Einstein frequency). The collective oscillations of the atoms in a solid is similar to sound waves in a solid. However, in Debye's theory the solid is treated as a continuous elastic medium; it neglects the discrete structure of the crystal. Thus, the Debye theory assumes that the dispersion relation for sound waves $\omega = \kappa v$ (where v is the speed of sound and $\kappa = 2\pi/\lambda$ is the wave number and λ is the wavelength of the sound waves) holds, not just for the long-wavelength of the sound waves, but for all vibrations in the crystal.

The Debye characteristic temperature is given by Equation 16.33.

16.9 See the derivation of Equation 16.39. At temperatures very much less than the Debye temperature, this law has been verified experimentally for insulators. The development of the Debye theory is not limited to crystalline solids; it is equally applicable to substances with an amorphous state and propagating transverse vibrations. The theory can also be extended to compounds such as AB, where A is a cation and B is an anion. According to the law of

Woestyne, the Debye theory allows the atomic heat capacities of A and B to combine additively at high temperatures. However, at low temperatures the T^3-dependence of C_V exists.

The Debye temperature T_D for a solid can be evaluated from the measured values of elastic constants of that solid (see Equations 16.40, 16.41 and 16.42). Although not exact, there is generally a reasonably good agreement between the measured and calculated values of T_D.

16.10   For a complete discussion see Section 16.5.

16.11   According to the Debye hypothesis, in a solid there are three modes of vibration for each wave vector: one longitudinal and two transverse modes. If we write v_L and v_T as the velocities of the longitudinal (P−) and transverse (S−) waves, respectively, then for a solid the given equation becomes

$$\frac{dn}{dV} = 4\pi\nu^2 \left[\frac{1}{v_L^3} + \frac{2}{v_T^3} \right] d\nu. \tag{i}$$

The number of harmonics within the interval $d\lambda$ about λ (wavelength) for a longitudinal wave (such as a sound wave in air) inside the volume of the cubic cavity of a solid is given by

$$dn_L = \left(\frac{4\pi V}{\lambda^4} \right) d\lambda.$$

On the other hand, for transverse vibrations, to each wavelength there corresponds an elliptical wave with a semimajor and a semiminor axis, so the number of vibrations given by the above equation for longitudinal waves must be doubled for transverse waves. (This brings out the fact that longitudinal waves are unpolarized and transverse waves have two orthogonal polarization directions.) Thus, total number of harmonics for transverse waves is

$$dn_T = \left(\frac{8\pi V}{\lambda^4} \right) d\lambda.$$

Combining these two equations and with the help of $\nu\lambda = v$, the total number of acoustical (long wavelength) frequencies between ν and $\nu + d\nu$ is given by

$$dn = 4\pi V\nu^2 \left[\frac{1}{v_L^3} + \frac{2}{v_T^3} \right] d\nu$$

or

$$\frac{dn}{dV} = 4\pi\nu^2 \left[\frac{1}{v_L^3} + \frac{2}{v_T^3} \right] d\nu.$$

which is the same as (i) above obtained from the given equation.

16.12   See the derivation of Equation 16.56.

16.13   See the derivation of Equation 16.66, i.e. C_V (electronic) $= \gamma T$, where $\gamma = (mN^{1/3}k^2\pi^2)/[\hbar^2(3\pi^2)^{2/3}]$; m is the mass of the electron and N is the number of electrons and N is the number of electrons.

If we plot C_V/T against T^2, we should get a straight line whose intercept at $T=0$ gives the value of the proportionality constant γ, while the slope of the line yields α (see Equation 16.53) of $C_V/T = \gamma + \alpha T^2$.

The result that the heat capacity of the free electrons in a metal goes to zero in proportion to T can be explained by considering the fact that only those electrons which are near the Fermi surface are affected when thermal energy is added to the system. The number of electrons that change their energy is proportional to kT/E_F and each electron should contribute approximately $3k/2$ amount of energy to the electronic heat capacity. Thus, electronic C_V should be proportional to

$$\left(\frac{3k}{2}\right)\left(\frac{kT}{E_F}\right) \simeq \frac{k^2 T}{E_F}.$$

This means that electronic C_V goes to 0 as $T \to 0$.

16.14 Even though the electronic heat capacity at low temperatures is small (see Equation 16.65), it can equal or become greater than the lattice heat capacity (due to the vibrations of the atoms in the solid). At sufficiently low temperatures (generally below 4 K), the electronic contribution to the heat capacity exceeds the lattice contribution. The electronic contribution decreases as the first power of absolute temperature, whereas at low temperatures the lattice contribution decreases as T^3. Therefore, if we go low enough in temperature, the electronic contribution will always be dominant in metals. Hence, it is clear that at low temperatures the electronic contribution to heat capacity cannot be ignored. In this temperature range, the total heat capacity of a metal is actually the sum of the lattice heat capacity and electronic heat capacity, i.e. $C_V = \gamma T + \alpha T^3$.

16.15 Plot C_V against T^2. This should give a straight line. The slope of this line is α and the intercept at $T^2 = 0$ is γ. You may find $\alpha \simeq 7.45 \times 10^{-10}\,\mathrm{J\,kg^{-1}\,K^4}$. γ is related to the Fermi energy E_F by the equation

$$\gamma = \frac{\pi^2 k^2 N}{2E_F} \quad \text{(see Equations 16.68 and 16.69).}$$

Substitute the values of π, k, N and E_F into this equation and obtain the value of γ. This value is less than that obtained from the graph.

16.16 The configuration of the ground state of the free electron in copper is $1s^2$, $2s^2$, $2p^6$, $3s^2$, $3p^6$, $3d^{10}$, 4s. The valence electron of copper is the 4s electron; the other electrons in the closed shell form the ion core. It is reasonable to assume that each atom of copper in the metal contributes one valence electron to the conduction band. On this assumption, the concentration N of conduction electrons will be equal to the number of copper atoms per unit volume. Then,

$$N = \frac{\text{Avogadro's number } (N_0)}{\text{Molar volume of Cu}}$$

$$\text{Molar volume} = \frac{\text{Molecular weight}}{\text{Density}} = \frac{63.5\,\mathrm{g}}{8.94\,\mathrm{g\,cm^{-3}}}$$

$$= 7.1\,\mathrm{cm^3}.$$

Therefore,

$$N = \frac{6.025 \times 10^{23}}{7.1 \, \text{cm}^3} = 85 \times 10^{21} \, \text{cm}^{-3} = 85 \times 10^{27} \, \text{m}^{-3}$$

16.17 Plot C_V/T against T^2. The intercept at $T^2 = 0$ is γ (about $2.5 \, \text{J} \, \text{mol}^{-1} \, \text{deg}^{-2}$). The gradient of the line is $\propto = 234Nk/T_D^3$ (see Equation 16.53), where N is the number of electrons, k is the Boltzmann constant and T_D is the Debye temperature. From the graph the approximate value of α is 0.75. This gives $T_D \simeq 7.30 \, \text{K}$.

16.18 The electronic specific heat of a metal is given by Equation 16.65

$$C_V \, (\text{electronic}) = \frac{mN^{1/3}k^2\pi^2 T}{\hbar^2(3\pi^2)^{2/3}}$$

where k is the Boltzmann constant, $\hbar = h/2\pi$ (h is the Planck constant), T is the temperature, N is the number of electrons and m is the mass (for copper and sodium these are 63.5 and 22.99 g, respectively). Substitute the values of all the parameters into the above equation to work out C_V (electronic) for Na and Cu.

16.19 See the derivation of Equation 16.80.

16.20 The internal energy of an ideal Bose–Einstein gas is given by Equation 15.57. Therefore, the heat capacity at constant volume in the temperature range $T < T_b$ is

$$C_V = \left(\frac{\partial U}{\partial T}\right)_V = (0.770)(Nk)\frac{\partial}{\partial T}[T(T/T_b)^{3/2}]$$
$$= 1.925Nk(T/T_b)^{3/2}$$

which is the required equation.

Appendix 1

FUNDAMENTAL CONSTANTS

Physical constant	Symbol	Value
Absolute zero of temperature		$-273.15°\,C = 0\,K$
Acceleration due to gravity	g	$9.81\,m\,s^{-2}$
Angström	Å	$1\,Å = 10^{-1}\,nm = 10^{-4}\,\mu m = 10^{-8}\,cm$ $= 10^{-10}\,m$
Atmospheric pressure	P	$1\,atm = 0.1\,MPa = 1.0135 \times 10^{5}\,Nm^{-2}$
Atomic mass unit (amu) $(=\frac{1}{2}\,mass\,C^{12} = 1/N_0)$	M_0	$1.6603 \times 10^{-24}\,g = 1.6603 \times 10^{-27}\,kg$ $= 931.478\,MeVc^{-2}$
Avogadro constant	N_0	$6.022\ 17 \times 10^{23}\,mol^{-1}$ $= 6.022\ 17 \times 10^{26}\,kmol^{-1}$
Bohr radius	a_0	$0.529\ 17\,Å = 5.2917 \times 10^{-11}\,m$
Boltzmann constant	$k = R/N_0$	$1.380\ 66 \times 10^{-23}\,J\,K^{-1}$
Charge of an electron	e	$1.602\ 189 \times 10^{-20}\,emu$ $= 4.80 \times 10^{-10}\,esu$ $= 1.602\ 189 \times 10^{-19}\,C$
Charge-to-mass ratio	e/m_e	$1.758\ 796 \times 10^{11}\,C\,kg^{-1}$ $= 1.758\ 796 \times 10^{7}\,emu\,g^{-1}$
Compton wavelength of an electron	$\lambda_c = h/m_e c$	$2.426 \times 10^{-12}\,m$
Electron rest mass	m_0	$9.1095 \times 10^{-31}\,kg = 5.485\ 803 \times 10^{-4}\,amu$ $= 0.511\,MeVc^{-2}$
Electronvolt	eV	$1.602\ 18 \times 10^{-19}\,J = 23.053\,kcal\,mol^{-1}$
Energy equivalent of 1 amu	$M_0 c^2$	$931\,MeV$
Electron rest mass	m_0	$9.1095 \times 10^{-31}\,kg = 5.485\ 803 \times 10^{-4}\,amu$ $= 0.511\,MeVc^{-2}$
Electronvolt	eV	$1.602\ 18 \times 10^{-19}\,J = 23.053\,kcal\,mol^{-1}$
Energy equivalent of 1 amu	$M_0 c^2$	$931\,MeV$
Faraday constant	F	$9.648\ 46 \times 10^{4}\,C\,mol^{-1}$
First radiation constant	$c_1 = 8\pi hc$	$4.992 \times 10^{-15}\,erg\,cm = 4.992 \times 10^{-24}\,J\,m$
Gas constant	R	$0.082\,\ell\,atm\,K^{-1}\,mol^{-1} = 0.082 \times 0.101\,\ell\,MPa\,K^{-1}\,mol^{-1}$ $= 8.314\ 41\,J\,K^{-1}\,mol^{-1} = 8.314\ 41 \times 10^{7}\,erg\,K^{-1}\,mol^{-1}$ $= 1.99\,cal\,K^{-1}\,mol^{-1}$
Gravitational constant	G	$6.672\ 04 \times 10^{-11}\,N\,m^2\,kg^{-2}$
Hydrogen atom rest mass	H	$1.6733 \times 10^{-24}\,g = 1.6733 \times 10^{-27}\,kg$

Physical constant	Symbol	Value
Mechanical equivalent of heat		$4.1840\,\mathrm{J\,cal^{-1}}$
Neutron rest mass	m_n	$1.674\,954 \times 10^{-27}\,\mathrm{kg} = 1.008\,665\,\mathrm{amu}$ $= 939.6\,\mathrm{MeV}\,c^{-2}$
Permeability of free space (in vacuum)	μ_0	$1.256\,63 \times 10^{-8}\,\mathrm{H\,cm^{-1}} = 4\pi \times 10^{-7}\,\mathrm{H\,m^{-1}}$ $= 4\pi \times 10^{-7}\,\mathrm{T\,m\,\mathring{A}^{-1}}$
Permittivity of free space (in vacuum)	ε_0	$8.854\,188 \times 10^{-12}\,\mathrm{C^2N^{-1}m^{-2}}$ $= 8.854\,188 \times 10^{-14}\,\mathrm{F\,cm^{-1}c^{-2}}\mu_0^{-1}$
Planck constant	h	$6.626\,176 \times 10^{-34}\,\mathrm{J\,s}$
Reduced Planck constant	$\hbar = h/2\pi$	$1.055 \times 10^{-34}\,\mathrm{J\,s} = 6.582 \times 10^{-16}\,\mathrm{eVs}$
Proton rest mass	m_p	$1.672\,648 \times 10^{-27}\,\mathrm{kg} = 938.3\,\mathrm{MeV}c^{-2}$
Proton-to-electron mass ratio	m_p/m_e	1.836×10^3
Rydberg constant (for infinite mass)	$R_\infty = \mu_0^2 m_e e^4 c^3 (8h^3)^{-1}$	$1.907 \times 10^5\,\mathrm{cm^{-1}} = 1.907 \times 10^7\,\mathrm{m^{-1}}$
Second radiation constant	$c_2 = hc/k$	$1.439\,\mathrm{cm\,K} = 1.439 \times 10^{-2}\,\mathrm{m\,K}$
Speed of light in vacuum	c	$2.997\,925 \times 10^8\,\mathrm{m\,s^{-1}} = 2.997\,925 \times 10^{10}$ $\mathrm{cm\,s^{-1}}$
Standard volume of ideal gas	V_0	$2.241\,36 \times 10^{-2}\,\mathrm{m^3\,mol^{-1}}$ $= 22.4136 \times 10^3\,\mathrm{(kg\,molecule)^{-1}}$
Stefan–Boltzmann constant	$\sigma = 2\pi^5 k^4 (15h^3 c^2)^{-1}$	$5.670 \times 10^{-5}\,\mathrm{erg\,cm^{-2}\,K^{-4}\,s^{-1}}$ $= 5.670 \times 10^{-8}\,\mathrm{J\,m^{-2}\,K^{-4}\,s^{-1}}$ $= 5.670 \times 10^{-8}\,\mathrm{W\,m^{-2}\,K^{-4}}$
Thermal voltage at 300 K	kT/e	$0.0259\,\mathrm{V}$
Wavelength of 1 eV photon	λ	$1.239\,77\,\mu\mathrm{m} = 1.239\,77 \times 10^{-6}\,\mathrm{m}$

Appendix 2

CRITICAL AND VAN DER WAALS CONSTANTS FOR SIMPLE GASES

| Gas | Critical constants | | | van der Waals constants | |
	Temperature (K)	Pressure (atm)	Density (g cm^{-3})	a (atm ℓ^2)	b (ℓ)
He	5.2	2.26	0.069	0.004	0.0236
Ne	44	25.9	0.484	0.242	0.0174
Ar	151	48	0.531	1.35	0.0323
H_2	33	12.8	0.031	0.242	0.0265
N_2	126	33.5	0.331	1.29	0.0377
O_2	154	49.7	0.430	1.30	0.0311
Cl_2	417	76	0.573	6.21	0.0550
CO	134	35	0.311	1.39	0.0384
CO_2	304	73	0.460	3.49	0.0418
H_2O	647	218	0.40	5.22	0.0298
NH_3	405	112	0.235	3.98	0.0363
SO_2	430	77.7	0.52	6.47	0.0555

Appendix 3

STANDARD MOLAR ENTHALPIES OF FORMATION, STANDARD MOLAR FREE ENERGIES OF FORMATION AND ABSOLUTE STANDARD ENTROPIES AT 298.15 K AND 0.1 MPa

Substance	$(\Delta H_f^\circ)_{298.15K}$ (kJ mol^{-1})	$S_{298.15K}^\circ$ (J K^{-1} mol^{-1})	$(\Delta G_f^\circ)_{298.15K}$ (kJ mol^{-1})
Ag(g)	284.66	172.91	245.69
Ag(s)	0	42.70	0
AgBr(s)	−99.58	–	–
AgCl(s)	−127.02	96.10	−109.70
AgI(s)	−62.34	–	–
Ag$_2$O(s)	−30.56	121.71	−11.21
AgNO$_3$(s)	−123.13	–	–
Ag$_2$S(s)	−32.61	145.60	−40.69
Al(g)	0	28.32	0
Al(s)	325.85	164.38	286.12
Al$_2$O$_3$(s)	−1676.09	51.03	−1581.98
AlF$_3$(s)	−1504.01	66.49	−1425.03
AlCl$_3$(s)	704.29	110.68	−628.95
AlCl$_3$·6H$_2$O(s)	2691.99	–	–
Al$_2$S$_3$(s)	−723.95	–	−492.38
Al$_2$(SO$_4$)$_3$(s)	−3440.83	238.97	−3100.15
Ar(g)	–	154.74	–
As(g)	302.98	174.12	261.05
As$_4$(g)	144.07	313.98	92.57
AsCl$_3$(g)	−258.58	327.14	−246.03
AsH$_3$(g)	66.48	222.74	68.96
As(s)	0	34.97	0
As$_4$O$_6$(s)	−1314.04	214.03	−1152.54
As$_2$O$_5$(s)	−924.91	105.06	−782.48
As$_2$S$_3$(s)	−169.03	164.05	−169.04
AsH$_3$O$_4$(s)	−906.37	–	–

Substance	$(\Delta H_f^\circ)_{298.15\,K}$ $(kJ\,mol^{-1})$	$S_{298.15\,K}^\circ$ $(J\,K^{-1}\,mol^{-1})$	$(\Delta G_f^\circ)_{298.15\,K}$ $(kJ\,mol^{-1})$
$B(g)$	562.81	153.43	518.82
$B_2H_6(g)$	35.97	231.98	86.71
$BF_3(g)$	−404.01	290.02	−388.82
$B_3N_3H_6(\ell)$	−540.99	200.03	−393.01
$B(s)$	0	5.90	0
$B_2O_3(s)$	−1273.02	54.03	−1193.82
$B(OH)_3(s)$	−1094.29	89.01	−968.98
$HBO_2(s)$	−794.33	39.97	−723.47
$Ba(g)$	175.58	170.27	144.79
$Ba(s)$	0	66.94	0
$BaO(s)$	558.14	70.29	528.39
$BaCO_3(s)$	−1216.70	112.13	−
$BaCl_2(s)$	−860.10	125.98	−810.91
$BaSO_4(s)$	−1465.35	131.98	−1353.04
$Be(g)$	320.58	136.19	282.77
$Be(s)$	0	9.58	0
$BeO(s)$	−611.02	14.09	−581.58
$Bi(g)$	206.99	186.93	167.98
$Bi(s)$	0	56.81	0
$BiCl_3(s)$	−379.02	176.98	−315.02
$Bi_2O_3(s)$	−573.91	150.97	−493.82
$Bi_2S_3(s)$	−143.03	199.98	−140.98
$Br(g)$	112.04	175.01	82.38
$Br_2(g)$	31.03	245.35	3.14
$BrF_3(g)$	−255.70	292.37	−230.01
$Br_2(\ell)$	0	152.29	0
$C(g)$	716.70	158.01	672.96
$CO(g)$	−110.52	197.91	−137.27
$CO_2(g)$	−393.51	213.64	−394.38
$CH_4(g)$	−74.85	186.19	−50.79
$CH_3OH(g)$	−200.67	239.68	−161.02
$CH_3COOH(g)$	−432.29	282.02	−373.98
$C_2H_2(g)$	226.73	200.82	209.20
$C_2H_4(g)$	52.30	219.45	68.12
$C_2H_6(g)$	−84.68	229.49	−32.89
$C_2H_5OH(g)$	−235.13	282.59	−168.63
$C_2H_5Cl(g)$	−112.19	276.03	−60.49
$C_2N_2(g)$	309.01	241.89	297.38
$C_3H_8(g)$	−103.85	270.02	−23.47
$C_4H_8(g)$	−13.98	−	−
$C_4H_{10}(g)$	−124.73	−	−
$C_5H_{12}(g)$	−165.98	−	−
$C_6H_6(g)$	82.93	269.25	129.66
$C_6H_{14}(g)$	−167.19	−	−
$CH_2Cl_2(g)$	−92.53	270.13	−66.02
$CH_3Cl(g)$	−81.49	234.57	−57.43

Substance	$(\Delta H_f^\circ)_{298.15\,K}$ (kJ mol^{-1})	$S_{298.15\,K}^\circ$ (J K^{-1} mol^{-1})	$(\Delta G_f^\circ)_{298.15\,K}$ (kJ mol^{-1})
CH$_3$CHO(g)	−166.36	–	–
CHCl$_3$(g)	−103.18	295.72	−70.39
HCN(g)	135.02	201.68	124.69
CCl$_4$(g)	−103.03	309.68	−61.02
CS$_2$(g)	117.47	237.76	67.21
CH$_3$OH(ℓ)	−238.67	126.78	−166.23
CH$_3$COOH(ℓ)	−487.02	159.83	−392.46
C$_2$H$_5$OH(ℓ)	−277.61	160.66	−174.77
CH$_2$Cl$_2$(ℓ)	−121.57	177.98	−67.37
C$_2$H$_5$Cl(ℓ)	−136.49	191.02	−59.46
C$_6$H$_6$(ℓ)	49.04	172.79	124.52
C$_6$H$_{12}$(cyclo)(ℓ)	–	298.24	26.65
CHCl$_3$(ℓ)	−134.49	201.97	−73.55
CCl$_4$(ℓ)	−136.33	216.43	−66.62
CS$_2$(ℓ)	89.86	151.28	65.30
HCOOH(ℓ)	−409.19	128.95	−346.02
HCN(ℓ)	109.02	113.03	125.01
C(diamond)(s)	1.89	2.44	2.85
C(graphite)(s)	0	5.69	0
Ca(g)	192.59	154.88	159.01
Ca(s)	0	41.63	0
CaO(s)	−635.49	39.75	−604.17
CaC$_2$(s)	–	70.29	–
CaCO$_3$(s)	−1206.67	92.89	−1128.76
Ca(OH)$_2$(s)	−985.75	76.12	−896.76
CaSO$_4$(s)	−1432.70	106.90	−1320.32
CaSO$_4\cdot$2H$_2$O(s)	−2021.09	193.98	−1795.80
CaSO$_3\cdot$2H$_2$O(s)	−1762.02	183.99	−1564.98
Cd(g)	111.98	167.73	77.51
Cd(s)	0	52.03	0
CdCl$_2$(s)	−391.67	115.32	−343.98
CdO(s)	−257.98	54.87	−227.97
CdSO$_4$(s)	−933.33	123.12	−822.81
CdS(s)	−161.97	65.03	−156.04
Cl(g)	121.70	165.11	105.68
Cl$_2$(g)	0	222.95	0
ClF(g)	−54.50	217.86	−56.02
ClF$_3$(g)	−162.98	281.48	−122.98
Cl$_2$O(g)	80.29	266.13	98.03
Cl$_2$O$_7$(g)	271.96	–	–
Cl$_2$O$_7$(ℓ)	238.02	–	–
HClO$_4$(ℓ)	−40.63	–	–
Co(s)	0	26.36	30.03
CoO(s)	−238.01	53.03	−214.19
Co$_3$O$_4$(s)	−891.09	103.04	−773.98
Co(NO$_3$)$_2$(s)	−420.48	–	–

Substance	$(\Delta H_f^\circ)_{298.15\,K}$ (kJ mol^{-1})	$S_{298.15\,K}^\circ$ (J K^{-1} mol^{-1})	$(\Delta G_f^\circ)_{298.15\,K}$ (kJ mol^{-1})
Cr(g)	397.03	174.43	352.06
Cr(s)	0	23.92	0
CrO$_3$(s)	−589.52	–	–
Cr$_2$O$_3$(s)	−1140.10	81.25	−1057.98
(NH$_4$)$_2$Cr$_2$O$_7$(s)	−1806.97	–	–
Cu(g)	338.34	166.29	298.48
Cu(s)	0	33.30	0
CuO(s)	−157.38	43.02	−130.02
Cu$_2$O(s)	−169.03	93.84	−145.98
CuCl(s)	–	–	−118.83
CuS(s)	−53.10	66.52	−53.60
Cu$_2$S(s)	−79.49	120.92	−86.22
CuSO$_4$(s)	−771.42	109.03	662.03
Cu(NO$_3$)$_2$(s)	−302.98	–	–
F(g)	79.00	158.70	62.01
F$_2$(g)	0	203.34	0
F$_2$O(g)	−21.98	247.28	−4.62
Fe(g)	415.98	180.42	371.02
Fe(CO)$_5$(g)	−734.00	445.25	−697.39
Fe(CO)$_5$(ℓ)	−773.98	337.96	705.39
Fe(s)	0	27.16	0
FeAsS(s)	−42.01	119.98	−50.02
FeO(s)	−271.97	53.97	–
Fe(OH)$_2$(s)	−569.05	87.98	−486.70
Fe(OH)$_3$(s)	−823.04	106.97	−696.57
FeS(α)(s)	–	67.36	–
FeS(s)	24.97	105.03	20.40
FeS$_2$(s)	53.14	–	–
FeSeO$_3$(s)	−1200.03	–	–
Fe$_2$O$_3$(s)	−824.16	89.60	740.99
Fe$_3$C(s)	25.02	105.01	19.97
Fe$_3$O$_4$(s)	−1117.93	146.44	−1014.20
H(g)	218.03	114.59	203.26
H$_2$(g)	0	130.59	0
HBr(g)	−36.23	198.48	−53.22
HCl(g)	−92.31	186.68	−95.27
HF(g)	−270.61	173.61	−273.03
HI(g)	25.94	206.33	1.69
H$_2$O(g)	−241.83	188.73	−228.61
H$_2$O$_2$(g)	−136.32	232.99	−105.59
H$_2$S(g)	−20.45	205.64	−33.61
H$_2$Se(g)	29.93	219.01	16.03
HCHO(g)	–	218.66	−109.62
H$_2$O(ℓ)	−285.84	69.87	−237.19
H$_2$O$_2$(ℓ)	−190.13	109.98	−119.97

Substance	$(\Delta H_f^\circ)_{298.15\,K}$ (kJ mol^{-1})	$S_{298.15\,K}^\circ$ (J K^{-1} mol^{-1})	$(\Delta G_f^\circ)_{298.15\,K}$ (kJ mol^{-1})
Hg(g)	61.36	175.02	31.76
Hg(ℓ)	0	77.40	0
HgCl$_2$(s)	−223.95	145.97	−179.01
Hg$_2$Cl$_2$(s)	−265.17	192.05	−210.66
HgO(red)(s)	−90.71	71.97	−58.59
HgO(yellow)(s)	−90.51	71.15	−57.30
HgS(red)(s)	−58.24	82.39	−51.00
HgS(black)(s)	−54.01	88.28	−48.02
HgSO$_4$(s)	−708.00	−	−
I(g)	106.87	181.02	70.17
I$_2$(g)	62.45	260.58	19.37
IBr(g)	41.02	258.72	3.69
ICl(g)	17.90	247.48	−5.46
IF(g)	95.70	236.20	−118.53
IF$_7$(g)	−944.01	345.98	−818.39
I$_2$(s)	0	116.73	0
K(g)	90.01	160.33	61.20
K(s)	0	63.59	0
KBr(s)	−	96.44	−
KCl(s)	−435.89	82.67	−408.31
KClO$_3$(s)	−	142.97	−
KClO$_4$(s)	−	151.04	−
KF(s)	−562.61	66.57	−533.19
KI(s)	−	104.35	−
KNO$_3$(s)	−	−	−393.12
KOH(s)	−425.85	−	−
Li(g)	155.00	139.00	122.00
Li(s)	0	27.98	0
Li$_2$CO$_3$(s)	−1216.01	90.49	−1131.98
LiF(s)	−612.00	36.01	−583.97
LiH(s)	−90.50	24.98	−70.01
Li(OH)(s)	−487.40	50.30	−440.00
Mg(s)	−	32.59	−
MgO(s)	−601.83	26.78	−
MgCl$_2$(s)	−	−	−592.33
MgCO$_3$(s)	−	65.69	−
Mn(g)	280.96	174.00	237.98
Mn(s)	0	31.75	0
MnO(s)	−385.19	60.25	−363.00
MnO$_2$(s)	−520.05	53.14	−464.84
Mn$_2$O$_3$(s)	−960.00	110.00	−881.27
Mn$_3$O$_4$(s)	−1387.96	155.98	−1282.97
N(g)	472.79	153.20	455.55
NO(g)	90.37	210.62	86.69

Substance	$(\Delta H_f^\circ)_{298.15\,K}$ (kJ mol^{-1})	$S_{298.15\,K}^\circ$ $(\text{J K}^{-1}\,\text{mol}^{-1})$	$(\Delta G_f^\circ)_{298.15\,K}$ (kJ mol^{-1})
$NO_2(g)$	33.85	240.45	51.84
$N_2(g)$	–	191.49	–
$N_2H_4(g)$	95.39	238.35	159.36
$N_2O(g)$	81.55	219.99	103.59
$N_2O_3(g)$	83.75	312.19	139.37
$N_2O_4(g)$	9.66	304.30	98.28
$N_2O_5(g)$	10.98	355.97	115.01
$NH_3(g)$	−46.19	192.51	−16.65
$HNO_3(g)$	−135.05	266.19	−74.83
$N_2H_4(\ell)$	50.59	121.36	149.18
$HNO_3(\ell)$	−173.23	155.60	−80.82
$NH_4Br(s)$	271.00	112.98	−175.00
$NH_4Cl(s)$	−314.39	94.58	−203.87
$NH_4I(s)$	−201.37	117.02	−112.98
$NH_4NO_2(s)$	−256.01	–	–
$NH_4NO_3(s)$	−365.58	151.20	−183.96
$Na(g)$	108.63	72.82	−376.98
$Na(s)$	0	51.04	0
$NaCl(s)$	−411.00	72.38	−384.00
$NaF(s)$	–	58.57	–
$NaHCO_3(s)$	–	–	−851.86
$NaNO_3(s)$	–	116.31	–
$NaOH(s)$	−426.72	–	–
$Na_2CO_3(s)$	–	–	−1047.67
$Na_2O(s)$	−416.01	72.80	−376.98
$Na_2SO_4(s)$	–	149.49	–
$O(g)$	249.21	161.00	230.12
$O_2(g)$	0	205.03	0
$O_3(g)$	142.98	238.78	163.43
$P(g)$	59.01	280.00	24.49
$P_4(g)$	314.62	163.10	279.91
$PCl_3(g)$	−306.35	312.92	−268.02
$PCl_5(g)$	−398.94	352.71	−304.98
$PF_3(g)$	–	268.29	–
$PH_3(g)$	5.39	210.15	12.97
$POCl_3(g)$	−558.52	325.31	−513.01
$POCl_3(\ell)$	−591.13	222.58	−521.01
$H_3PO_4(\ell)$	−1267.03	–	–
$P(s)$	0	41.11	0
$P_4(s)$	–	177.40	–
$P_4O_6(s)$	−1639.98	–	–
$P_4O_{10}(s)$	−3012.48	229.02	−2697.97
$HPO_3(s)$	−948.65	–	–
$H_3PO_2(s)$	−604.65	–	–
$H_3PO_3(s)$	−964.42	–	–
$H_3PO_4(s)$	−1281.14	110.49	−1119.02
$H_4P_2O_7(s)$	−2240.98	–	–

Substance	$(\Delta H_f^\circ)_{298.15\,K}$ (kJ mol^{-1})	$S_{298.15\,K}^\circ$ (J K^{-1} mol^{-1})	$(\Delta G_f^\circ)_{298.15\,K}$ (kJ mol^{-1})
Pb(g)	195.01	175.33	161.97
Pb(s)	0	64.89	0
PbCl$_2$(s)	359.38	136.06	−313.97
Pb(NO$_3$)$_2$(s)	−452.00	–	
PbO(yellow)(s)	−217.29	68.80	−188.00
PbO(red)(s)	−219.10	67.81	−189.00
Pb(OH)$_2$(s)	−516.01	–	–
PbO$_2$(s)	−276.97	68.58	−217.39
PbS(s)	−99.99	91.22	−99.01
S(g)	278.82	167.80	238.29
SCl$_2$(g)	−19.98	–	–
SF$_4$(g)	−775.01	292.00	−731.38
SF$_6$(g)	−1210.00	291.68	−1104.98
SOCl$_2$(g)	−212.99	309.68	−198.02
SO$_2$(g)	−296.89	248.52	−300.36
SO$_3$(g)	−395.17	256.22	−370.36
S$_2$Cl$_2$(g)	−18.01	331.38	−32.03
SO$_2$Cl$_2$(g)	−364.03	311.83	−319.98
SCl$_2$(ℓ)	−49.97	–	–
S$_2$Cl$_2$(ℓ)	−59.43	–	–
SOCl$_2$(ℓ)	−245.98	–	–
SO$_2$Cl$_2$(ℓ)	−393.98	–	–
H$_2$SO$_4$(ℓ)	−814.00	157.01	690.10
S(rhombic)(s)	0	31.82	0
H$_2$S$_2$O$_7$(s)	−1273.98	–	–
Sb(g)	262.01	180.24	221.98
SbCl$_3$(g)	−314.02	−337.68	−301.06
SbCl$_5$(g)	−394.28	402.02	−334.32
Sb(s)	0	45.72	0
Sb$_4$O$_6$(s)	−1441.03	220.98	1268.07
Sb$_2$S$_3$(s)	−174.97	181.99	−174.02
SbCl$_3$(s)	−382.26	183.97	−323.73
SbOCl(s)	−373.97	–	–
Si(g)	455.58	168.03	411.02
SiCl$_4$(g)	−657.00	330.63	−617.02
SiF$_4$(g)	−1615.03	284.51	−1506.24
SiH$_4$(g)	33.98	203.76	57.02
SiCl$_4$(ℓ)	−687.00	240.03	−620.00
Si(s)	0	18.72	0
SiC(s)	−65.29	16.63	−63.01
SiO$_2$(s)	−859.39	41.84	−856.70
H$_2$SiO$_3$(s)	−1188.98	130.01	−1092.01
H$_4$SiO$_4$(s)	−1481.01	190.04	−1332.98
Sn(g)	302.02	168.40	266.97
SnCl$_4$(g)	−471.49	365.98	−432.19
SnCl$_4$(ℓ)	−511.52	259.03	−440.23

Substance	$(\Delta H_f^\circ)_{298.15\,K}$ (kJ mol^{-1})	$S_{298.15\,K}^\circ$ (J K^{-1} mol^{-1})	$(\Delta G_f^\circ)_{298.15\,K}$ (kJ mol^{-1})
Sn(s)	0	51.46	0
SnO(s)	−286.02	56.48	−257.02
SnO$_2$(s)	−580.68	52.29	−519.73
Ti(g)	470.00	180.21	425.11
TiCl$_4$(g)	−763.19	354.78	−726.81
TiCl$_4$(ℓ)	−804.23	252.71	−737.19
Ti(s)	0	30.29	0
TiO$_2$(s)	−944.69	50.24	−889.52
W(g)	849.39	173.86	807.12
W(s)	0	32.63	0
WO$_3$(s)	−842.89	76.00	−764.10
Zn(g)	130.76	160.90	95.20
Zn(s)	0	41.63	0
ZnCl$_2$(s)	−415.12	111.54	−369.27
ZnCO$_3$(s)	−812.80	82.39	−731.59
ZnO(s)	−347.98	43.93	−318.19
ZnS(s)	−202.92	57.76	−201.29
ZnSO$_4$(s)	−982.83	119.98	−874.50

Appendix 4

INTERNATIONAL SYSTEM OF UNITS AND CONVERSION FACTORS

Quantity	Unit	Conversion factor
Length	1 in	$25.334 \text{ mm} = 2.533 \text{ cm} = 2.533 \times 10^{-5} \text{ km}$
	1 ft	$0.3048 \text{ m} = 0.3048 \times 10^{-3} \text{ km}$
	1 cm	$0.3937 \text{ in} = 0.0328 \text{ ft}$
	1 m	$39.370 \text{ in} = 3.281 \text{ ft} = 10^{10} \text{ Å}$
	1 km	0.6214 mile
	1 mile	$5280 \text{ ft} = 1.61 \text{ km}$
	1 nautical mile	$6080 \text{ ft} = 1.85 \text{ km}$
	1 fermi	$1 \text{ fm} = 10^{-15} \text{ m}$
	1 Å	$10^{-10} \text{ m} = 0.1 \text{ μm}$
	1 light year	$9.46 \times 10^{15} \text{ m}$
Area	1 in^2	$6.4516 \text{ cm}^2 = 6.4516 \times 10^{-4} \text{ m}^2$
	1 ft^2	$0.0929 \text{ m}^2 = 0.0929 \times 10^{-6} \text{ km}^2$
	1 m^2	10.76 ft^2
Volume	1 in^3	$16.387 \text{ cm}^3 = 16.387 \times 10^{-6} \text{ m}^3$
	1 ft^3	$0.0283 \text{ m}^3 = 0.0283 \times 10^{-9} \text{ km}^3$
	1 m^3	$35.31 \text{ ft}^3 = 10^6 \text{ cm}^3 = 264.172 \text{ US gal}$
	1 litre	$1000 \text{ ml} = 1000.028 \text{ cm}^3 = 10^{-3} \text{ m}^3$
Mass	1 lb	$453.4 \text{ g} = 0.4534 \text{ kg}$
	1 slug	14.5939 kg
	1 g	0.0022 lb
	1 kg	$2.2046 \text{ lb} = 1000 \text{ g} = 0.0685 \text{ slug}$
	1 amu	$1.6606 \times 10^{-27} \text{ kg}$
Mass per unit length	1 oz in^{-1}	$11.16 \text{ g cm}^{-1} = 1.116 \text{ kg m}^{-1}$
	1 lb ft^{-1}	1.488 kg m^{-1}
	1 lb in^{-1}	17.858 kg m^{-1}

Quantity	Unit	Conversion factor
Mass per unit area	$1\,oz\,ft^{-2}$	$0.305\,kg\,m^{-2}$
	$1\,lb\,in^{-2}$	$703.07\,kg\,m^{-2}$
	$1\,lb\,ft^{-2}$	$4.88\,kg\,m^{-2}$
Time	$1\,s$	$10^3\,ms$
		$10^6\,\mu s$
		$10^9\,ns$
	$1\,day$	$8.64 \times 10^4\,s = 1.44 \times 10^3\,min$
	$1\,year$	$3.156 \times 10^7\,s = 5.26 \times 10^5\,min$
Pressure	$1\,atm$	$760\,mmHg = 14.7\,lb\,in^{-2}$
		$1.013\,bar = 760\,Torr$
		$1.013 \times 10^6\,dyn\,cm^{-2} = 1.013 \times 10^5\,N\,m^{-2}$
		$1.013 \times 10^5\,Pa = 0.1013\,MPa$
	$1\,Pa$	$1\,N\,m^{-2} = 10\,dyn\,cm^{-2} = kg\,m^{-1}\,s^{-2}$
		$1.45 \times 10^{-4}\,lb\,in^{-2}$
		$10^{-5}\,bar = 7.5 \times 10^{-3}\,Torr = 9.869 \times 10^{-6}\,atm$
	$1\,Torr$	$1000\,\mu mHg = 1\,mmHg$
Temperature	$1\,K$	$273.15 + t\,°C$
Density	$1\,lb\,in^{-3}$	$27.679\,g\,cm^{-3}$
	$1\,lb\,ft^{-3}$	$16.018\,kg\,m^{-3}$
	$1\,slug\,ft^{-3}$	$515.379\,kg\,m^{-3}$
Specific volume	$1\,in^3\,lb^{-1}$	$36.127\,cm^3\,kg^{-1}$
	$1\,ft^3\,lb^{-1}$	$0.062\,m^3\,kg^{-1}$
Velocity	$1\,ft\,s^{-1}$	$0.3048\,m\,s^{-1} = 1.09\,km\,h^{-1} = 0.682\,mile\,h^{-1}$
	$1\,m\,s^{-1}$	$3.28\,ft\,s^{-1} = 3.60\,km\,h^{-1}$
	$1\,km\,h^{-1}$	$0.278\,m\,s^{-1} = 0.621\,mile\,h^{-1}$
	$1\,mile\,h^{-1}$	$1.47\,ft\,s^{-1} = 1.61\,km\,h^{-1} = 0.447\,m\,s^{-1}$
Acceleration	$1\,ft\,s^{-2}$	$0.3048\,m\,s^{-2}$
Mass flow rate	$1\,lb\,h^{-1}$	$0.1259\,g\,s^{-1} = 1.259\,kg\,s^{-1}$
Force or weight	$1\,dyne$	$1\,g \times 1\,cm \times 1\,s^{-1} = 10^{-5}\,N$
	$1\,pdl$	$1\,lb \times 1\,ft \times 1\,s^{-1} = 0.138\,N$
	$1\,lb$	$4.45\,N$
	$1\,ft\,lb$	$1\,lb \times 1\,ft \times 1\,s^{-1} \times 32.6 = 4.448\,N$
	$1\,kg\,ft$	$9.806\,N$
	$1\,N$	$0.225\,lb = 10^5\,dyn = 10^5\,kg\,m\,s^{-2}$
Energy	$1\,erg$	$1\,g \times 1\,cm^2 \times 1\,s^{-2} = 10^{-7}\,J$
	$1\,J$	$1\,kg\,m^2\,s^{-2}$
		$9.96 \times 10^{-4}\,\ell\,MPa = 9.96\,\ell\,atm$
	$1\,ft\,lb$	$1.36\,J = 1.29 \times 10^{-3}\,Btu = 3.25 \times 10^{-4}\,kcal$
	$1\,hp$	$2.684\,MJ = 550\,ft\,lb\,s^{-1} = 0.745\,kW$
	$1\,ev$	$1.602 \times 10^{-19}\,C\ (or\ J)$
	$1\,kWh$	$3.80 \times 10^6\,J = 860\,kcal$

Quantity	Unit	Conversion factor
Power	1 W	$1\,J\,s^{-1} = 0.738\,ft\,lb\,s^{-1} = 1\,kg\,m^2\,s^{-1}$
	1 hp (US)	$550\,ft\,lb\,s^{-1} = 746\,W$
	1 hp (metric)	$750\,W$
Heat	1 cal	$4.1868\,J = 3.97 \times 10^{-3}\,Btu$
	1 Btu	$1.055\,kJ = 252\,cal = 778\,ft\,lb$
Specific heat	$1\,Btu\,lb^{-1}\,°F^{-1}$	$4.1868\,kJ\,kg^{-1}\,°C^{-1}$
	$1\,cal\,g^{-1}\,°C^{-1}$	$4.1868\,kJ\,kg^{-1}\,K^{-1}$
Heat flow rate	$1\,Btu\,h^{-1}$	$0.293\,W$
	$1\,cal\,s^{-1}$	$4.2\,W$
Specific enthalpy	$1\,Btu\,ft^{-3}$	$0.037\,J\,cm^{-3}$
	$1\,Btu\,lb^{-1}$	$2.326\,kJ\,kg^{-1}$
	$1\,cal\,g^{-1}$	$4.186\,J\,g^{-1}$
	$1\,kcal\,m^{-3}$	$4.186\,kJ\,m^{-3}$
Specific entropy	$1\,Btu\,lb^{-1}\,°F^{-1}$	$4.186\,kJ\,kg^{-1}\,K^{-1}$
Thermal conductivity	$1\,cal\,cm^{-1}\,s^{-1}\,°C^{-1}$	$4.186\,W\,m^{-1}\,K^{-1}$
	$1\,Btu\,ft^{-1}\,h^{-1}\,°F^{-1}$	$1.730\,W\,m^{-1}\,K^{-1}$
Angle	1 rad	$57.30° = 57.18'$
	1°	$0.01745\,rad$
	$1\,rev\,min^{-1}$	$0.1047\,rad\,s^{-1}$
Magnetic field strength	1 T	$1\,kg\,Å^{-1}\,s^{-2} = 1\,Wb\,m^{-2}$
Magnetic flux	1 Wb	$1\,kg\,m^2\,Å\,s^{-2}$
Concentration	$1\,mol\,m^{-3}$	$10^{-3}\,mol\,\ell^{-1} = 10^{-3}\,mol\,dm^{-3}$
	$1\,mg\,\ell^{-1}$	$1\,\mu g\,cm^{-3} = 1\,ppm = 10^{-3}\,g\,dm^{-3}$
	$1\,\mu g\,g^{-1}$	$1\,ppm = 10^{-6}\,g\,g^{-1}$
	$1\,ng\,cm^{-3}$	$10^{-6}\,g\,dm^{-3}$
	$1\,ng\,dm^{-3}$	$1\,pg\,cm^{-3}$
	$1\,pg\,g^{-1}$	$1\,ppb = 10^{-12}\,g\,g^{-1}$
Viscosity	$1\,kg\,m^{-1}\,s^{-1}$	
	1 cP	$100\,P$

Appendix 5

HEATS OF FORMATION OF VARIOUS SUBSTANCES AT 298 K

Substance	ΔH_f (kJ mol^{-1})	Substance	ΔH_f (kJ mol^{-1})	Substance	ΔH_f (kJ mol^{-1})
AgCl(s)	−127.026	CH$_3$OH(ℓ)	−238.672	CH$_4$(g)	−74.852
AgBr(s)	−99.579	C$_2$H$_5$OH(ℓ)	−277.608	C$_2$H$_6$(g)	−84.684
AgI(s)	−62.342	CH$_3$COOH(ℓ)	−487.017	C$_3$H$_8$(g) (propane)	−103.847
Al$_2$O$_3$(s)	−1676.09	C$_6$H$_6$(ℓ)	49.036	C$_4$H$_{10}$(g) (n-butane)	−124.725
Ag$_2$O(s)	−30.568	CHCl$_3$(ℓ)	−134.496	C$_4$H$_{10}$(g) (isobutane)	−131.587
AgNO$_3$(s)	−123.135	CCl$_4$(ℓ)	−136.327	C$_5$H$_{12}$(g) (neopentane)	−165.979
BaO(s)	−558.146	CS$_2$(ℓ)	89.864	C$_5$H$_{12}$(g) (isopentane)	−154.473
BaCO$_3$(s)	−1216.707	HCOOH(ℓ)	−409.195	C$_5$H$_{12}$(g) (n-pentane)	−146.440
BaSO$_4$(s)	−1465.352	H$_2$O(ℓ)	−285.838	CH$_3$Cl(g)	−82.006
CaO(s)	−635.491	H$_2$O$_2\infty$(aq)	−190.125	C$_2$H$_4$(g)	52.300
Ca(OH)$_2$(s)	−985.750	H$_2$SO$_4$(ℓ)	−811.319	C$_2$H$_2$(g)	226.731
CaCO$_3$(s)	−1206.666	HNO$_3$(ℓ)	−173.234	C$_4$H$_8$(g) (cyclobutane)	−13.975
CaSO$_4$(s)	−1432.70	N$_2$H$_4$(ℓ)	50.172	C$_6$H$_{14}$(g) (n-hexane)	−167.193
CuO(s)	−157.318			C$_6$H$_6$(g)	82.927
Cu$_2$O(s)	−169.034			CH$_3$CHO(g)	−166.356
Cu(s)	−53.1			CO(g)	−110.520
C(diamond)(s)	1.896			CO$_2$(g)	−393.514
C(graphite)(s)	0.00			H$_2$O(g)	−241.826
Fe$_2$O$_3$(s)	−824.156			HF(g)	−270.613
Fe$_3$O$_4$(s)	−1117.928			HCl(g)	−92.312
HgO(s)	−90.709			HBr(g)	−36.233
H$_3$PO$_4$(s)	−1281.141			HI(g)	25.941
KOH(s)	−425.848			H$_2$S(g)	−20.446
KCl(s)	−435.868			NO(g)	90.374
MgO(s)	−601.827			NO$_2$(g)	33.853
NaCl(s)	−411.003			N$_2$O(g)	81.546
NaOH(s)	−426.726			N$_2$O$_4$(g)	9.660

Substance	ΔH_f (kJ mol^{-1})	Substance	ΔH_f (kJ mol^{-1})	Substance	ΔH_f (kJ mol^{-1})
$P_4O_{10}(s)$	−3012.480			$NH_3(g)$	−46.191
$SiO_2(s)$	−859.394			$PCl_3(g)$	−306.352
$SrO(s)$	−590.362			$PCl_5(g)$	−398.944
$ZnO(s)$	−347.983			$SO_2(g)$	−296.897
$ZnS(s)$	−202.924			$SO_3(g)$	−395.179

Appendix 6

USEFUL MATHEMATICAL FORMULAE

Combinations of distinguishable and indistinguishable particles

Let us first consider the case of distinguishable particles. Say we have three particles coloured red (R), blue (B) and green (G), and we wish to arrange these particles. There are six ways to arrange these, viz. R, G, B; R, B, G; G, R, B; G, B, R; B, R, G; and B, G, R. The first particle could be chosen as red, blue or green, i.e. in three ways. Once the first particle has been chosen, we have two choices for the second particle for each of the three first choices. Thus, the first two particles chosen can be arranged in $3 \times 2 = 6$ ways, which are R, G; R, B; G, R; G, B; B, R; and B, G. After this exercise we have only one choice left for the third particle in each of these cases. Therefore, the total number of ways of arranging our three red, blue and green particles is $3 \times 2 \times 1 = 6$. This can be written as 3! (factorial 3).

We can now generalize this procedure for N number of distinguishable particles. We can choose the first in N ways. Once the first particle has been chosen, there are $(N-1)$ particles left, so we can choose the second particle in $(N-1)$ ways for each of the N possibilities for the first choice. Thus, the first two particles can be chosen in $N(N-1)$ ways. Then, the third particle can be chosen in $(N-2)$ ways, and so on. Therefore, the total number of ways of arranging the N number of distinguishable particles is $N(N-1)(N-2)\ldots = N!$

Let us now consider the case where we have N distinguishable particles and we want to place these into two distinguishable boxes, say 1 and 2. It does not matter in which order the particles are put into the boxes 1 and 2; what matters is which particles are in which box. If we assume that W is the number of ways we can choose n_1 distinguishable particles for box 1 and $n_2 = (N-n_1)$ distinguishable particles for box 2, then

$$W n_1! n_2! = N!$$

or

$$W = \frac{N!}{n_1! n_2!} = \frac{N!}{n_1!(N-n_1)!}. \tag{A6.1}$$

The number of ways in which n particles can be chosen from N different particles is called the number of combinations of N particles taken n at a time. This is denoted by N_{C_n} or

$$\binom{N}{n}$$

Hence, from Equation A6.1 we have

$$N_{C_n} = \frac{N!}{n!(N-n)!}. \tag{A6.2}$$

If we have X distinguishable boxes, N distinguishable particles, and there are n_1 distinguishable particles in box 1, n_2 distinguishable particles in box 2, etc., then from Equation A6.1 we can generalize

$$W = \frac{N!}{n_1!n_2!\ldots n_X!}. \tag{A6.3}$$

Remember that $0! = 1$.

Let us now consider the case where we want to calculate the number of ways we can put N indistinguishable particles into X distinguishable boxes without any restriction on how many of the N particles are in each box. If we want to put all the particles in one or other of the X boxes, we have to choose the box first. Here, it can be done in X ways, leaving $(X-1)$ boxes and N particles. Then we can either choose another box (which means that the first box will be empty), or take a particle to put it into the first box. If the particles were all distinguishable, this choice could be made in $(X+N-1)$ ways. The next choice could be made in $(X+N-2)$ ways and so on. If we continue the process, we will end up with either a particle to go into the last box or with a box which is empty. Thus, the total number of ways in which we can put N distinguishable particles into X distinguishable boxes is $X(X+N-1)!$. For each choice of the order of the boxes, there are $N!$ ways of choosing the order in which N distinguishable particles can be put into the boxes.

However, in our case all the particles are indistinguishable. So, the order in which they are placed in the X boxes does not matter, but we must divide $X(X+N-1)!$ by $N!$. As there are $X!$ ways of choosing X distinguishable boxes (if we are not concerned about the order in which we choose the boxes, but interested only in how many indistinguishable particles there are in each box), we have to divide $X(X+N-1)!$ by $X!$ as well. Thus, the number of ways of putting N indistinguishable particles into X distinguishable boxes is

$$W = \frac{X(X+N-1)!}{N!X!} = \frac{(X+N-1)!}{N!(X-1)!}. \tag{A6.4}$$

Stirling's approximation

The calculation of $N!$ becomes very laborious when N is large. We would like to find a simple approximation useful for calculating $N!$ in such cases. By definition

$$N! = 1 \times 2 \times 3 \times \cdots \times N.$$

Hence,

$$\ln N! = \ln 1 + \ln 2 + \ln 3 + \cdots + \ln N$$

$$= \sum_{m=1}^{N} \ln m. \tag{A6.5}$$

The first few terms in Equation A6.5 are the smallest. If N is large, all terms in this sum, except the first few, correspond to values of m large enough so that $\ln m$ varies only slightly when m is increased by 1. The sum is equal to the area under the rectangles in Figure A6.1 up to $x = N$. This figure also shows $\ln x$ vs. x. It can be seen that for large N, the area under the curve of $\ln x$ vs. x is approximately equal to the sum of the areas of the rectangles. Hence, for large N, the sum in Equation A6.5 can be approximated with little error by an integral which gives the area under the continuous curve of Figure A6.1. Thus, Equation A6.5 becomes

$$\ln N! \simeq \int_{1}^{N} \ln x \, dx = [x \ln x - x]_{1}^{N}$$

$$= N \ln N - N + 1. \tag{A6.6}$$

Hence, if $N \gg 1$,

$$\ln N! \simeq N \ln N - N \tag{A6.7}$$

because the contribution of the lower limit in Equation A6.6 is then negligible.

A better approximation is given by Stirling's formula:

$$\ln N! = N \ln N - N + \frac{1}{2} \ln (2\pi N). \tag{A6.8}$$

This is good within less than 1% error in $N!$ even when N is as small as 10. When N is quite big, $N \gg \ln N$ so that Equation A6.8 reduces to the simpler form of Equation A6.7.

The reader can check the accuracy of Equations A6.7 and A6.8 by taking any arbitrary value of N.

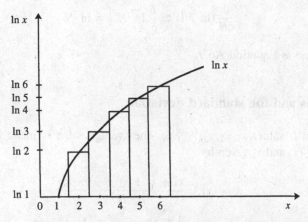

Figure A6.1   Plot of $\ln x$ versus x where $x = 0,1,2,\ldots$. $\ln N!$ is equal to the sum of the areas under the rectangles

Derivative of ln N! for large N

Consider ln N! when N is large. In such cases ln N! changes only by a small fraction of itself if N is changed by a small integer, so it can be considered as an almost continuous function of N. So by increasing N by 1 we obtain

$$\frac{d}{dN}(\ln N!) \simeq \frac{\ln (N+1)! - \ln N!}{1}$$

$$= \ln \left[\frac{(N+1)!}{N!} \right] = \ln (N+1).$$

Now, as $N \gg 1$, $N+1 \simeq N$. Hence, the above equation becomes

$$\frac{d \ln N!}{dN} \simeq \ln N \tag{A6.9}$$

when $N \gg 1$. This result can be obtained from Equation A6.7.

Alternatively, the derivative of ln N! can be obtained in terms of any small *integral* increment δ by the relation

$$\frac{d}{dN}(\ln N!) = \frac{\ln (N+\delta)! - \ln N!}{\delta}. \tag{A6.10}$$

Hence,

$$\frac{d}{dN}(\ln N!) = \frac{1}{\delta} \ln \left[\frac{(N+\delta)!}{N!} \right]$$

$$= \frac{1}{\delta} \ln [(N+\delta)(N+\delta-1)\cdots(N+1)].$$

Since $N \gg \delta$, we have

$$\frac{d}{dN}(\ln N!) \simeq \frac{1}{\delta} \ln [N^\delta] = \ln N$$

which is the same as Equation A6.9.

Average values and the standard deviation

Consider a set of N values $x_1, x_2, x_3, \ldots, x_N$. The average value or mean value of this set is denoted by $\bar{x}$ or $\langle x \rangle$ and is given by

$$\bar{x} = \langle x \rangle = \frac{x_1 + x_2 + x_3 + \cdots + x_N}{N}$$

$$= \frac{1}{N} \sum_{i=1}^{N} x_i. \tag{A6.11}$$

The summation is over all the N values of x_i. For example, if the values of x_i are 5, 6, 5, 6, 6, 7, 8, 6, 4, 7, the average value of x is the sum of all the 10 numbers divided by the total number 10, i.e.

$$\bar{x} = \langle x \rangle = (5 + 6 + 5 + 6 + 6 + 7 + 8 + 6 + 4 + 7)/10 = 6.$$

Note that in the values of x_i there are two 5, four 6, two 7, one 8 and one 4. Therefore, we can write the expression for $\bar{x}$ in the form

$$\bar{x} = \langle x \rangle = (1 \times 4 + 2 \times 5 + 4 \times 6 + 2 \times 7 + 1 \times 8)/10 = 6.$$

Alternatively,

$$\bar{x} = \langle x \rangle = 4\left(\frac{1}{10}\right) + 5\left(\frac{2}{10}\right) + 6\left(\frac{4}{10}\right) + 7\left(\frac{2}{10}\right) + 8\left(\frac{1}{10}\right).$$

From this expression it can be easily seen that the probability of getting a 4 is 1/10, that of getting a 5 is 2/10, etc. Hence, we can generalize Equation A6.11 in the form

$$\bar{x} = \langle x \rangle = \sum x_i \mathscr{P}_i \tag{A6.12}$$

where $\mathscr{P}_i$ is the probability of getting the value x_i.

In the above example the values of $\mathscr{P}_i$ are 0.1, 0.2, 0.4, 0.2 and 0.1 for 4, 5, 6, 7 and 8, and all other $\mathscr{P}_i$ values are zero.

The standard deviation σ and the *variance* σ^2 are defined by the equation

$$\sigma^2 = \langle (x_i - \bar{x})^2 \rangle = \frac{1}{N} \sum_1^N (x_i - \bar{x})^2. \tag{A6.13}$$

Alternatively,

$$\sigma^2 = \sum (x_i - \bar{x})^2 \mathscr{P}_i \tag{A6.14}$$

where $\mathscr{P}_i$ is the probability of getting the value of x_i, and the summation is over all possible values of x_i. Using $\bar{x} = 6$, from our above numerical example and Equation A6.14 we have

$$\sigma^2 = [0.1(4 - 6)^2 + 0.2(5 - 6)^2 + 0.4(6 - 6)^2 + 0.2(7 - 6)^2 + 0.1(8 - 6)^2]$$
$$= 1.2.$$

Therefore, $\sigma = \sqrt{1.2} = 1.10$. The standard deviation is a good measure of the width of the probability distribution.

The Maclaurin and the Taylor series

Consider the function $f(x)$ of the variable x expanded in the power series

$$f(x) = a_0 + a_1 x + a_2 x^2 + a_3 x^3 + \cdots. \tag{A6.15}$$

By differentiating this series term by term, we obtain

$$\frac{df}{dx} = f'(x) = a_1 + 2a_2 x + 3a_3 x^2 + \cdots$$

$$\frac{d^2 f}{dx^2} = f''(x) = 2a_2 + 6a_3 x + \cdots.$$

$$\vdots$$

By substituting $x = 0$ in $f(x), f'(x), f''(x), \ldots$ etc., we have

$$f(0) = a_0; \; f'(0) = a_1; \; f''(0) = 2! a_2; \cdots.$$

By putting these values into Equation A6.15 in the range where the series is convergent, we have

$$f(x) = f(0) + x f'(0) + \left(\frac{x^2}{2!}\right) f''(0) + \cdots + \left(\frac{x^n}{n!}\right) f^n(0) + \cdots. \tag{A6.16}$$

This is *Maclaurin's theorem*.

This is an important theorem in many ways. This theorem can be used to derive the *binomial theorem* and *Taylor series*. Consider the function $f(x) = (1 + x)^n$. Then $f(0) = 1$. By differentiating $f(x) = (1 + x)^n$ term by term and putting $x = 0$, we obtain $f'(0) = n$, $f''(0) = n(n-1), \ldots$, etc. Substituting these values into Equation A6.16, we have

$$(1 + x)^n = 1 + nx + \frac{n(n-1)}{2!} x^2 + \cdots \tag{A6.17}$$

which is the *binomial expansion*.

To derive the Taylor series let us consider a function $y = f(x)$. A plot of y against x is shown in Figure A6.2. We consider three points A, B and C. A and B have abscissae $x = 0$, $x = a$, respectively. Let the point C have an abscissa x. We want to express the value of $f(x)$ at the point C in terms of the values of the differential coefficients $f', f'', \ldots$, etc., evaluated at the point B. We draw a perpendicular from B to $0x$ and choose $0'$ as our new origin, where $00' = a$. According to Figure A6.2 the variable h is clearly equal to $x - a$. So the abscissa of the point C in our new co-ordinate system is $h = x - a$, and the new abscissa of B is $h = 0$. Now, by using Equation A6.16 and h as the variable instead of x, we can write the value of the function $f(x)$ at the point C as

$$f_c(x) = f(h = 0) + h\left(\frac{df}{dh}\right) + \left(\frac{h^2}{2!}\right)\left(\frac{d^2 f}{dh^2}\right) + \cdots. \tag{A6.18}$$

The differential coefficients have to be evaluated at $h = 0$. Now, since both df/dh evaluated at $h = 0$ and df/dx evaluated at $x = a$ are equal to the slope of the curve at the point B, we have $df/dh = df/dx$. Thus, for df/dh in Equation A6.18 we can write $f'(a)$, for $d^2 f/dh^2$ we

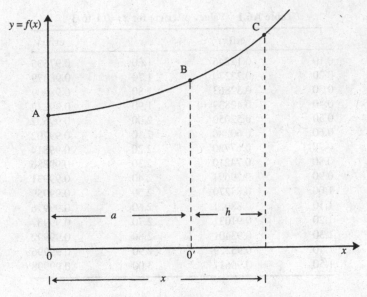

Figure A6.2   Plot of function $y = f(x)$ vs. x

write $f''(a)$, and so on. Since $h = x - a$ and $f(h = 0) = f(a)$, then we can write Equation A6.18 in the form

$$f_c(x) = f(a) + (x - a)f'(a) + \left[\frac{(x - a)^2}{2!}\right]f''(a) + \cdots. \tag{A6.19}$$

The differential coefficients in this equation have been evaluated at the point B which has the abscissa $x = a$ in the co-ordinate system with 0 as the origin. Since $h = x - a$, we have $x = a + h$. By substituting this into Equation A6.19, we obtain

$$f_c(x) = f_c(a + h) = f(a) + hf'(a) + \left(\frac{h^2}{2!}\right)f''(a) + \cdots$$

$$= \sum_{n=0}^{\infty} \left(\frac{h^n}{n!}\right)f^n(a) \tag{A6.20}$$

where the differential coefficients $f^n(a)$ have been evaluated at the point $h = 0$. Equations A6.19 and A6.20 are two different forms of the *Taylor series*.

The error function

An integral of the type

$$\frac{2}{\sqrt{\pi}} \int_0^x \exp\left(-x^2\right) \, \mathrm{d}x$$

Table A6.1   Values of erf(x) for $x = 0.1$ to 3

x	erf(x)	x	erf(x)
0.10	0.11246	1.60	0.97635
0.20	0.22270	1.70	0.98379
0.30	0.32863	1.80	0.98909
0.40	0.42839	1.90	0.99279
0.50	0.52050	2.00	0.99532
0.60	0.60386	2.10	0.99702
0.70	0.67780	2.20	0.99814
0.80	0.74210	2.30	0.99886
0.90	0.79691	2.40	0.99931
1.00	0.84270	2.50	0.99959
1.10	0.88021	2.60	0.99976
1.20	0.91031	2.70	0.99987
1.30	0.93401	2.80	0.99992
1.40	0.95229	2.90	0.99996
1.50	0.96611	3.00	0.99998

is known as the *error function* and written as

$$\text{erf}(x) = \frac{2}{\sqrt{\pi}} \int_0^x \exp\left(-x^2\right) \, dx. \tag{A6.21}$$

There are two cases: for all values of x we write

$$\text{erf}(x) = \frac{2}{\sqrt{\pi}} \left(x - \frac{x^3}{3 \times 1!} + \frac{x^5}{5 \times 2!} - \frac{x^7}{7 \times 3!} + \cdots \right) \tag{A6.22}$$

and for large positive values of x we can use

$$\text{erf}(x) \simeq 1 - \frac{\exp(-x^2)}{x\sqrt{\pi}} \left[1 - \frac{1}{2x^2} + \frac{1 \times 3}{(2x^2)^2} - \cdots \right]. \tag{A6.23}$$

The values of erf(x) corresponding to the values $x = 0.1$ to $x = 3$ are listed in Table A6.1

Evaluation of the Gaussian integrals

Integrals involving the Gaussian function $\exp\left(-x^2\right)$ often occur in statistical mechanics. The indefinite integral $\int \exp\left(-x^2\right) \, dx$ cannot be evaluated in terms of elementary functions. Let I denote the desired definite integral

$$I \equiv \int_{-\infty}^{\infty} \exp\left(-x^2\right) \, dx. \tag{A6.24}$$

This integral can be determined by using the properties of the exponential function. We can write Equation A6.24 in terms of a different variable of integration as

$$I \equiv \int_{-\infty}^{\infty} \exp\left(-y^2\right) dy. \tag{A6.25}$$

By multiplying Equations A6.24 and A6.25, we obtain

$$\begin{aligned}
I^2 &= \int_{-\infty}^{\infty} \exp\left(-x^2\right) dx \int_{-\infty}^{\infty} \exp\left(-y^2\right) dy \\
&= \int_{-\infty}^{\infty} \int_{-\infty}^{\infty} \exp\left(-x^2\right) \exp\left(-y^2\right) dxdy \\
&= \int_{-\infty}^{\infty} \int_{-\infty}^{\infty} \exp\left[-(x^2+y^2)\right] dxdy. \tag{A6.26}
\end{aligned}$$

The double integral extends over the whole of the xy-plane. Let us express the integration over this plane in terms of the polar co-ordinates r and ϕ, where $r^2 = x^2 + y^2$. The element of area in these co-ordinates is $rdrd\phi$. To cover the whole plane, r and ϕ must range over the values $0 < r < \infty$ and $0 < \phi < 2\pi$. Hence, Equation A6.26 can be rewritten in the form

$$\begin{aligned}
I^2 &= \int_{0}^{\infty} \int_{0}^{2\pi} \exp\left(-r^2\right) r \, drd\phi \\
&= 2\pi \int_{0}^{\infty} \exp\left(-r^2\right) r \, dr. \tag{A6.27}
\end{aligned}$$

Since the integration over ϕ is apparent. The factor r in this integrand makes the evaluation of this last integral trivial. Hence,

$$\begin{aligned}
I^2 &= 2\pi \int_{0}^{\infty} \left(-\frac{1}{2}\right) d[\exp\left(-r^2\right)] = -\pi[\exp\left(-r^2\right)]_{0}^{\infty} \\
&= -\pi(0-1) = \pi
\end{aligned}$$

or

$$I = \sqrt{\pi}.$$

Alternatively, put $r^2 = u$. Then, $du = 2r \, dr$ and

$$I^2 = \pi \int_{0}^{\infty} \exp\left(-u\right) du = \pi[-\exp\left(-u\right)]_{0}^{\infty} = \pi.$$

Thus,

$$\int_{-\infty}^{\infty} \exp\left(-x^2\right) dx = \sqrt{\pi}. \tag{A6.28}$$

Since the function $\exp(-x^2)$ is symmetrical about $x = 0$, i.e. it assumes the same value for x and $-x$, it also follows that

$$\int_{-\infty}^{\infty} \exp(-x^2)\,dx = 2\int_0^{\infty} \exp(-x^2)\,dx.$$

Hence,

$$\int_0^{\infty} \exp(-x^2)\,dx = \sqrt{\pi}/2. \tag{A6.29}$$

Evaluation of integrals of the form $\int_0^{\infty} \exp(-ax^2)\,x^n\,dx$

Before we integrate this expression we want to integrate two other integrals

$$I_0 = \int_0^{\infty} \exp(-ax^2)\,dx$$

$$I_1 = \int_0^{\infty} a\exp(-ax^2)\,dx.$$

Let $u = a^{1/2}x$. Then, $du = a^{1/2}\,dx$. By using Equation A6.29, we have

$$I_0 = \frac{1}{\sqrt{a}}\int_0^{\infty} \exp(-u^2)\,du = \frac{1}{2}\left(\frac{\pi}{a}\right)^{1/2}. \tag{A6.30}$$

Again, let $v = ax^2$. Then, $dv = 2ax\,dx$ and

$$I_1 = \left(\frac{1}{2a}\right)\int_0^{\infty} \exp(-v)\,dv$$

$$= \left(\frac{1}{2a}\right)[-\exp(-v)]_0^{\infty} = \frac{1}{2a}. \tag{A6.31}$$

Now let

$$\int_0^{\infty} \exp(-ax^2)\,x^n\,dx = I_n.$$

From Equations A6.30 and A6.31 it is clear that all other integrals, where n is any integer such that $n \geqslant 2$ can be calculated in terms of I_0 or I_1 by successive integration by parts. Now,

$$I_n \stackrel{.}{=} \int_0^{\infty} \exp(-ax^2)\,x^n\,dx = -\left(\frac{1}{2a}\right)\int_0^{\infty} (x^{n-1})\frac{d}{dx}[\exp(-ax^2)]\,dx$$

$$= -\left(\frac{1}{2a}\right)[\exp(-ax^2)\,x^{n-1}]_0^{\infty} + \left(\frac{n-1}{2a}\right)\int_0^{\infty} \exp(-ax^2)x^{n-2}\,dx.$$

Since the first term on the right-hand side vanishes for both $x = 0$ and $x = \infty$, we have

$$I_n = \left(\frac{n-1}{2a}\right) \int_0^\infty \exp\left(-ax^2\right) x^{n-2} \, dx$$

$$= \left(\frac{n-1}{2a}\right) I_{n-2}. \qquad (A6.32)$$

We can summarise the results as follows:

$$I_0 = \int_0^\infty \exp\left(-ax^2\right) dx = \frac{1}{2}\left(\frac{\pi}{a}\right)^{1/2} \qquad \text{(Equation A6.30)}$$

$$I_1 = \int_0^\infty \exp\left(-ax^2\right) x \, dx = \frac{1}{2a} \qquad \text{(Equation A6.31)}$$

$$I_2 = \int_0^\infty \exp\left(-ax^2\right) x^2 \, dx = \frac{I_0}{2a} = \frac{\sqrt{\pi}}{4a^{3/2}} \qquad (A6.33)$$

$$I_3 = \int_0^\infty \exp\left(-ax^2\right) x^3 \, dx = \frac{I_1}{a} = \frac{1}{2a^2} \qquad (A6.34)$$

$$I_4 = \int_0^\infty \exp\left(-ax^2\right) x^4 \, dx = \left(\frac{3}{2a}\right) I_2 = \frac{3\sqrt{\pi}}{8(a)^{5/2}} \qquad (A6.35)$$

$$\vdots \qquad \vdots \qquad \vdots \qquad \vdots \qquad \vdots$$

The simple rule to evaluate any integral of this form is to take the actual value of n, such as $3, 4, \ldots$, etc., and substitute this value into Equation A6.32. This will give I_n in terms of either I_0 or I_1. Then, use Equations A6.30 or A6.31 as required to find the value of I_n.

The Gaussian distribution

Before we actually derive an expression for the Gaussian distribution it is better to know another distribution called the *binomial distribution*. We shall require this distribution to discuss the Gaussian distribution.

Consider an ideal system of N spins $\frac{1}{2}$, each one of which has an associated magnetic moment μ_0. We suppose that the spin system is situated in an external magnetic field $\mathbf{B}$, so that each magnetic moment can point either *upward* (parallel to field $\mathbf{B}$), or *downward* (antiparallel to $\mathbf{B}$) as shown in Figure A6.3.

Let us assume that the spin system is in equilibrium. Thus, a statistical ensemble of N such spin systems is independent of time. Considering any one spin, we denote by p the

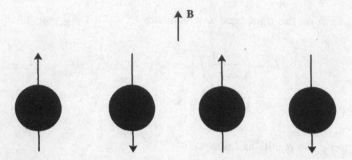

Figure A6.3 A system consisting of N number of spins $1/2$ (where, $N=4$). **B** is the external magnetic field

probability that its magnetic moment points upward and by q the probability that its magnetic moment points downward. As these two orientations complete all the possibilities, by the normalization requirement we have

$$p + q = 1 \qquad (A6.36)$$

Suppose that in an experiment a particular outcome is labelled by r and that there are (among the N systems of ensemble) N_r systems which exhibit this outcome. Then the fraction $\mathscr{P} \equiv N_r/N$ (where $N \to \infty$) is called the *probability of occurrence of the outcome r*. Suppose that experiments on some system X can lead to any of m mutually exclusive outcomes. The index r can then refer to any of the m numbers $r = 1, 2, 3, \ldots$, or m. In an ensemble of similar systems, N_1 of them will exhibit event 1, N_2 of them will exhibit event 2, ..., and N_m of them event m. As the m events are mutually exclusive and exhaust all possibilities, we have

$$N_1 + N_2 + N_3 + \cdots + N_m = N.$$

By dividing both sides by N, we obtain

$$\frac{N_1}{N} + \frac{N_2}{N} + \cdots + \frac{N_m}{N} = 1$$

or

$$\mathscr{P}_1 + \mathscr{P}_2 + \cdots + \mathscr{P}_m = 1$$

or

$$\sum_{r=1}^{m} \mathscr{P}_r = 1 \quad \text{where} \quad \mathscr{P}_r \equiv N_r/N$$

This relation, which states that the sum of all possible probabilities adds up to 1, is called the *normalization condition* for probabilities.

From Equation A6.36 it follows that $q = 1-p$. When there is no external field, i.e. **B** $= 0$, there is no preferred direction in space so that $p = q = 1/2$. However, in the presence of a magnetic field, it is more likely that a magnetic moment will point along the field than opposite to it, so that $p > q$. As the spin system is ideal, there is almost negligible interaction between the spins. Therefore, their orientations can be considered statistically independent, and the probability that any magnetic moment points upward is not affected by whether any other magnetic moment in the system points upward or downward.

Of the N magnetic moments of the spin system, let us specify how many of them point upward and how many downward. Let n be the number of magnetic moments that point upward and let n_1 be the number of magnetic moments which point downward. Thus, we have

$$n + n_1 = N. \tag{A6.37}$$

It is apparent that the number n of magnetic moments pointing upward is not the same in each system in the statistical ensemble we are considering here. However, n can assume any of the possible values $0, 1, 2, \ldots, N$. Now the question is: for each of these possible values of n, what is the probability $\mathscr{P}(n)$ that n of N magnetic moments point upward? It can be shown that this probability is given by

$$\mathscr{P}(n) = \frac{(N!)p^n(q)^{N-n}}{n!(N-n)!} = \frac{(N!)p^n(q)^{n_1}}{n!n_1!}. \tag{A6.38}$$

When there is no external magnetic field, i.e. $\mathbf{B} = 0$, so that $p = q = 1/2$, Equation A6.38 becomes

$$\mathscr{P}(n) = \frac{N!}{n!(N-n)!}\left(\frac{1}{2}\right)^N. \tag{A6.39}$$

The probability $\mathscr{P}(n)$ is called the *binomial distribution*. For a given number N, it is a function of n.

In this discussion we have dealt with a specific problem of a system of N independent spins. Here the occurrence of an event is merely represented by a spin pointing upward, while the nonoccurrence of an event is represented by a spin pointing downward. Actually, we have obtained the probability for a general problem. To establish this claim, consider N events which are statistically independent. Assume that each of these N events occurs with a probability p. Then, the probability that it does not occur is given by $q = 1 - p$ (see Equation A6.36). We can ask the question: what is the probability $\mathscr{P}(n)$ that any n out of N events do occur, while the remaining $N-n$ events do not occur? This question is answered by the binomial distribution.

For small N, the computation of $\mathscr{P}(n)$ poses very little problem. However, when N is large its calculation becomes difficult, as it then involves the calculation of the factorials of large numbers. Therefore, it is possible to use approximations which allow us to transform Equation A6.38 into a simple form.

When N is large, the probability $\mathscr{P}(n)$ tends to show a pronounced maximum. Consequently, $\mathscr{P}(n)$ tends to become negligibly small whenever the difference between n and the particular value of n where $\mathscr{P}(n)$ is at a maximum (we take this value as n') becomes significant. Thus, the region where the probability $\mathscr{P}(n)$ is not negligible consists only of those values of n which do not differ appreciably from n'. Although this region is relatively small, it is possible to find an approximate expression for $\mathscr{P}(n)$. This expression can then be used for the entire domain where a knowledge of $\mathscr{P}(n)$ is of interest (i.e. for all values of n where $\mathscr{P}(n)$ is not negligibly small.) Therefore, it is sufficient to examine the behaviour of $\mathscr{P}(n)$ in the close vicinity of n' of its maximum.

However, before we proceed any further let us remember two useful observations. The value of n' is neither very close to 0 nor to N if p or q are not very close to 0. Thus, n' itself is a large number if N is large, so the numbers n in the region of interest close to n' are also large. However, the changes of $\mathscr{P}(n)$ are relatively small if n changes by 1, i.e. $|\mathscr{P}(n+1)-\mathscr{P}(n)| \ll \mathscr{P}(n)$. This means that $\mathscr{P}(n)$ is a function which varies slowly. Thus, $\mathscr{P}(n)$ can be considered, to a good approximation, as a smooth function of a continuous variable n, but only integral values of n will be of physical relevance. Second, we noted above that $\mathscr{P}(n)$ is a slowly varying function of n. However, $\ln \mathscr{P}(n)$ is an even more slowly varying function of n than $\mathscr{P}(n)$ itself. Because of this, it is easier to examine the behaviour of logarithm of $\mathscr{P}(n)$ and to find a good approximation for it which is valid in wide region of n.

After discussing the binomial distribution, we now develop the Gaussian distribution on the basis of this for the special case when both N and n are very large. By taking the natural logarithm of Equation A6.38, we have

$$\ln \mathscr{P}(n) = \ln N! - \ln n! - \ln (N-n)! + n \ln p + (N-n) \ln q. \tag{A6.40}$$

The particular value of n where $\mathscr{P}(n)$ has its maximum, i.e. $n = n'$, is then determined by the condition

$$\frac{d\mathscr{P}(n)}{dn} = 0 \tag{A6.41}$$

or by the condition that $\ln \mathscr{P}(n)$ is maximum

$$\frac{d\ln \mathscr{P}(n)}{dn} = \frac{1}{\mathscr{P}(n)} \frac{d\mathscr{P}(n)}{dn} = 0. \tag{A6.42}$$

To differentiate Equation A6.40, we note that all the numbers occurring as factorials are large compared with 1. We can overcome the associated problem of differentiation in this case in two different ways: we can apply to each of these the approximation given by Equation A6.9 which asserts that for any number $M \gg 1$, $(d \ln M!)/dM \simeq \ln M$, or we can use Stirling's approximation (given by Equation A6.7) to Equation A6.40 before differentiating it. Let us use the second method. Thus, we obtain

$$\ln \mathscr{P}(n) = N \ln N - n \ln n - (N-n) \ln (N-n) + n \ln p + (N-n) \ln q.$$

Now, by differentiating this with respect to n and keeping N, p and q constant, we obtain

$$\frac{d \ln \mathscr{P}(n)}{dn} = -\ln n + \ln (N-n) + \ln p - \ln q$$

$$= \ln \left[\frac{(N-n)p}{nq} \right]. \tag{A6.43}$$

To find the maximum of $\mathscr{P}(n)$ we equate Equation A6.43 to zero according to Equation A6.42. Thus,

$$\ln \left[\frac{(N-n)p}{nq} \right] = 0$$

or

$$\frac{(N-n)p}{nq} = 1$$

or

$$Np = n(p+q).$$

Since $p+q=1$, the value $n=n'$ where $\mathscr{P}(n)$ has its maximum is then given by

$$n' = Np. \qquad (A6.44)$$

To examine the behaviour of $\ln \mathscr{P}(n)$ near n', i.e. near its maximum, we need only expand it in a Taylor series about the value n'. Let $n = n' + x$. Then we can write

$$\ln \mathscr{P}(n) = \ln \mathscr{P}(n') + \left[\frac{d \ln \mathscr{P}(n)}{dn} \right] x + \frac{1}{2!} \left[\frac{d^2 \ln \mathscr{P}(n)}{dn^2} \right] x^2$$

$$+ \frac{1}{3!} \left[\frac{d^3 \ln \mathscr{P}(n)}{dn^3} \right] x^3 + \cdots . \qquad (A6.45)$$

The derivatives in the square brackets are to be evaluated at $n = n'$, i.e. where $x = 0$. Since the expansion is about a maximum where Equation A6.42 is satisfied, the first derivative vanishes. To find the other derivatives we have to differentiate Equation A6.43 successively. Thus,

$$\frac{d^2 \ln \mathscr{P}(n)}{dn^2} = -\frac{1}{n} - \frac{1}{N-n} = -\frac{N}{n(N-n)}.$$

Where $n = n'$, i.e. where $n = Np$ (see Equation A6.44) and $(N-n) = N(1-p) = Nq$, the value of the above derivative becomes

$$\left[\frac{d^2 \ln \mathscr{P}(n)}{dn^2} \right] = -\frac{1}{Npq}.$$

By substituting this into Equation A6.45 and ignoring the terms of order x^3 and higher, we obtain

$$\ln \mathscr{P}(n) = \ln \mathscr{P}(n') - \frac{x^2}{2Npq}$$

or

$$\mathcal{P}(n) = \mathcal{P}(n') \exp\left(\frac{-x^2}{2Npq}\right) \qquad \text{(A6.46)}$$

$$= \mathcal{P}(n') \exp\left[\frac{-(n-n')^2}{2Npq}\right]. \qquad \text{(A6.47)}$$

The probability given by the right side of this equation is known as the *Gaussian distribution*.

It is interesting to note that the probability $\mathcal{P}(n)$ in Equations A6.46 and A6.47 becomes negligible compared with its maximum value $\mathcal{P}(n')$ when x is so large that $x^2/(Npq) \gg 1$, i.e. $|x| \gg (Npq)^{1/2}$, because the exponential factor is then very much smaller than 1. Hence, the probability $\mathcal{P}(n)$ is appreciable only in the region where $|x| \leqslant (Npq)^{1/2}$. However, in this domain x is generally small enough for the terms in Equation A6.45 involving x^3 and higher powers of x to be ignored (they are negligibly small, compared with the leading term involving x^2, which we retain). This is true to the extent that $(Npq)^{1/2} \gg 1$. The above argument allows us to conclude that Equation A6.46 is a good approximation to the probability $\mathcal{P}(n)$ in the whole domain where this probability has appreciable magnitude.

Since the point n' where $\mathcal{P}(n)$ has the maximum is a particular value, $\mathcal{P}(n')$ is a constant. Its value can be expressed directly in terms of p and q by using the normalization condition

$$\sum_n \mathcal{P}(n) = 1. \qquad \text{(A6.48)}$$

Here, the summation is over all possible values of n. It is possible to replace this sum by an integration, because the probability $\mathcal{P}(n)$ changes slightly between any two successive integral values of n. If we divide n into a range of values of magnitude dn ($\gg 1$), then this range will contain dn possible values of $\mathcal{P}(n)$. Therefore, the normalization condition given by Equation A6.48 can be written as

$$\int \mathcal{P}(n)\, dn = \int \mathcal{P}(n') \exp\left(-x^2/2Npq\right)\, dx$$

$$= \mathcal{P}(n') \int \exp\left(-x^2/2Npq\right)\, dx = 1.$$

Since $\mathcal{P}(n)$ is negligibly small anywhere whenever $|n-n'|$ becomes sufficiently large, we can extend the range of integration from $-\infty$ to $+\infty$. Hence,

$$\int \mathcal{P}(n)\, dn = \mathcal{P}(n') \int_{-\infty}^{\infty} \exp\left(-x^2/2Npq\right)\, dx = 1. \qquad \text{(A6.49)}$$

In view of Equation A6.30 this integral becomes

$$\mathcal{P}(n')(2\pi Npq)^{1/2} = 1$$

or

$$\mathscr{P}(n') = \frac{1}{(2\pi Npq)^{1/2}}.$$ (A6.50)

By using Equations A6.44 and A6.50 in Equation A6.47, we have

$$\mathscr{P}(n) = \left[\frac{1}{(2\pi Npq)^{1/2}}\right] \exp\left[\frac{-(n - Np)^2}{2Npq}\right].$$ (A6.51)

A probability of the form given by the right side of Equation A6.51 is also known as a *Gaussian distribution*. This is much easier to evaluate than that given by the right side of Equation A6.38 because it does not involve the calculation of any factorials. Gaussian distributions are frequently used in statistical arguments whenever the numbers being considered are large. A schematic diagram of a typical Gaussian distribution is shown in Figure A6.4.

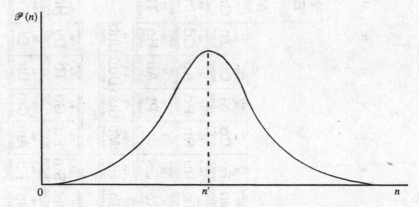

Figure A6.4   Schematic diagram of the Gaussian distribution of the probability $\mathscr{P}(n)$ as a function of n

Appendix 7

Key:

element name
atomic number
symbol
1997 atomic weight (mean relative mass)

1	2	3	4	5	6	7	8	9	10	11	12	13	14	15	16	17	18
hydrogen 1 **H** 1.00794(7)																	helium 2 **He** 4.002602(2)
lithium 3 **Li** 6.941(2)	beryllium 4 **Be** 9.012182(3)											boron 5 **B** 10.811(7)	carbon 6 **C** 12.0107(8)	nitrogen 7 **N** 14.00674(7)	oxygen 8 **O** 15.9994(3)	fluorine 9 **F** 18.9984032(5)	neon 10 **Ne** 20.1797(6)
sodium 11 **Na** 22.98976928(2)	magnesium 12 **Mg** 24.3050(6)											aluminium 13 **Al** 26.981538(2)	silicon 14 **Si** 28.0855(3)	phosphorus 15 **P** 30.973761(2)	sulfur 16 **S** 32.066(6)	chlorine 17 **Cl** 35.4527(9)	argon 18 **Ar** 39.948(1)
potassium 19 **K** 39.0983(1)	calcium 20 **Ca** 40.078(4)	scandium 21 **Sc** 44.955910(8)	titanium 22 **Ti** 47.867(1)	vanadium 23 **V** 50.9415(1)	chromium 24 **Cr** 51.9961(6)	manganese 25 **Mn** 54.938049(9)	iron 26 **Fe** 55.845(2)	cobalt 27 **Co** 58.933200(9)	nickel 28 **Ni** 58.6934(2)	copper 29 **Cu** 63.546(3)	zinc 30 **Zn** 65.39(2)	gallium 31 **Ga** 69.723(1)	germanium 32 **Ge** 72.61(2)	arsenic 33 **As** 74.92160(2)	selenium 34 **Se** 78.96(3)	bromine 35 **Br** 79.904(1)	krypton 36 **Kr** 83.80(1)
rubidium 37 **Rb** 85.4678(3)	strontium 38 **Sr** 87.62(1)	yttrium 39 **Y** 88.90585(2)	zirconium 40 **Zr** 91.224(2)	niobium 41 **Nb** 92.90638(2)	molybdenum 42 **Mo** 95.94(1)	technetium 43 **Tc** [97.9072]	ruthenium 44 **Ru** 101.07(2)	rhodium 45 **Rh** 102.90550(2)	palladium 46 **Pd** 106.42(1)	silver 47 **Ag** 107.8682(2)	cadmium 48 **Cd** 112.411(8)	indium 49 **In** 114.818(3)	tin 50 **Sn** 118.710(7)	antimony 51 **Sb** 121.760(1)	tellurium 52 **Te** 127.60(3)	iodine 53 **I** 126.90447(3)	xenon 54 **Xe** 131.29(2)
caesium 55 **Cs** 132.90545(2)	barium 56 **Ba** 137.327(7)	lutetium 71 **Lu** 174.967(1)	hafnium 72 **Hf** 178.49(2)	tantalum 73 **Ta** 180.9479(1)	tungsten 74 **W** 183.84(1)	rhenium 75 **Re** 186.207(1)	osmium 76 **Os** 190.23(3)	iridium 77 **Ir** 192.217(3)	platinum 78 **Pt** 195.078(2)	gold 79 **Au** 196.96655(2)	mercury 80 **Hg** 200.59(2)	thallium 81 **Tl** 204.3833(2)	lead 82 **Pb** 207.2(1)	bismuth 83 **Bi** 208.98038(2)	polonium 84 **Po** [208.9824]	astatine 85 **At** [209.9871]	radon 86 **Rn** [222.0176]
francium 87 **Fr** [223.0197]	radium 88 **Ra** [226.0254]	lawrencium 103 **Lr** [262.110]	rutherfordium 104 **Rf** [263.1125]	dubnium 105 **Db** [262.1144]	seaborgium 106 **Sg** [266.1219]	bohrium 107 **Bh** [264.1247]	hassium 108 **Hs** [269.1341]	meitnerium 109 **Mt** [268.1388]	ununnilium 110 **Uun** [272.1463]	unununium 111 **Uuu** [272.1535]	ununbium 112 **Uub** [277]		ununquadium 114 **Uuq** [289]		116 **Uuh** [289]		ununoctium 118 **Uuo** [293]

57–70 *
89–102 **

*lanthanoids

lanthanum 57 **La** 138.9055(2)	cerium 58 **Ce** 140.116(1)	praseodymium 59 **Pr** 140.90765(2)	neodymium 60 **Nd** 144.24(3)	promethium 61 **Pm** [144.9127]	samarium 62 **Sm** 150.36(3)	europium 63 **Eu** 151.964(1)	gadolinium 64 **Gd** 157.25(3)	terbium 65 **Tb** 158.92534(2)	dysprosium 66 **Dy** 162.50(3)	holmium 67 **Ho** 164.93032(2)	erbium 68 **Er** 167.26(3)	thulium 69 **Tm** 168.93421(2)	ytterbium 70 **Yb** 173.04(3)

**actinoids

actinium 89 **Ac** [227.0277]	thorium 90 **Th** 232.0381(1)	protactinium 91 **Pa** 231.03588(2)	uranium 92 **U** 238.0289(1)	neptunium 93 **Np** [237.0482]	plutonium 94 **Pu** [244.0642]	americium 95 **Am** [243.0614]	curium 96 **Cm** [247.0703]	berkelium 97 **Bk** [247.0703]	californium 98 **Cf** [251.0796]	einsteinium 99 **Es** [252.0830]	fermium 100 **Fm** [257.0951]	mendelevium 101 **Md** [258.0984]	nobelium 102 **No** [259.1011]

Index